Bailey's Industrial
Oil and Fat Products

EDITORIAL ADVISORY BOARD

BAILEY'S INDUSTRIAL OIL AND FAT PRODUCTS

Fifth Edition

Volume 1

Edible Oil and Fat Products: General Applications

Edited by

Y. H. HUI

Technology and Commerce, International

A Wiley-Interscience Publication

JOHN WILEY & SONS, INC.

New York · Chichester · Brisbane · Toronto · Singapore

Cover drawing courtesy of De Smet Process & Technology, Inc. The De Smet Multistock deodorizer incorporates a commercially proven single-vessel design to deodorize different feedstocks at capacities of over 720 TPD.

This text is printed on acid-free paper.

Library of Congress Cataloging in Publication Data

Bailey, Alton Edward, 1907–1953.
[Industrial oil and fat products]
Bailey's industrial oil and fat products. —5th ed. / edited by Y. H. Hui.
p. cm.
ISBN 0-471-59424-5 (alk. paper)
1. Oils and fats. I. Hui, Y. H. (Yiu H.) II. Title.
TP670.B28 1996
665—dc20 95-9528

Printed in the United States of America

10 9 8 7 6 5 4 3 2

Contributors

ROBERT ACKMAN: Technical University of Nova Scotia, Halifax, Canada, Fatty acids in newer fats and oils.

QINYUN CHEN: Rutgers University, New Brunswick, New Jersey, Flavor compounds in fats and oils.

JAMES GUZINSKI: Kalsec, Inc., Kalamazoo, Michigan, Oleoresins and essential oils.

ELIZABETH S. HANDS: ESHA Research, Inc., Salem, Oregon, Lipid composition of selected foods.

BERNIE HENNIG: University of Kentucky, Lexington, Kentucky, Dietary fat and health.

CHI TANG HO: Rutgers University, New Brunswick, New Jersey, Flavor compounds in fats and oils.

DAVID KITTS: University of British Columbia, Vancouver, Canada, Toxicity and safety of fats and oils.

JANE A. LOVE: Iowa State University, Ames, Iowa, Animal fats.

NANCY R. MORGAN: United States Department of Agriculture, Washington, D.C., Oilseeds and product trading.

W.W. NAWAR: University of Massachusetts, Amherst, Massachusetts, Chemistry.

FRANK T. ORTHOEFER: Riceland Foods, Inc., Stuttgart, Arizona, Vegetable oils.

SCOTT SANFORD: United States Department of Agriculture, Washington, D.C., Oilseeds and product trading.

ENDRE F. SIPOS: E.F. Sipos and Associates, Lake Geneva, Wisconsin, Lecithins.

JAMES STANTON: Consultant, St. Paul, Minnesota, Fat substitutes.

BERNARD F. SZUHAJ: Central Soya Company, Fort Wayne, Indiana, Lecithins.

MICHAL TOBOREK: University of Kentucky, Lexington, Kentucky, Dietary fat and health.

KATHLEEN WARNER: United States Department of Agriculture, Peoria, Illinois, Flavors and sensory evaluation.

BRUCE ALAN WATKINS: Purdue University, West Lafayette, Indiana, Dietary fat and health.

RUIBAO ZHOU: Rutgers University, New Brunswick, New Jersey, Flavor compounds in fats and oils.

Reviewers

To assure accuracy of the information, each chapter in this five-volume text has been reviewed by experts in industry, government, and academia. Although most chapters have one or two reviewers, some have as many as five or six. A list of the reviewers is presented. However, in studying this list, please note:

1. The authors of the chapters have also served as reviewers for chapters other than their own. They are not included in this list.
2. This list is incomplete for a variety of reasons. The review was spread over three years and some names are misplaced, some reviewers wish to remain anonymous, and so on. A note of appreciation and/or apology is extended to those reviewers not included in the list.

INDIVIDUALS

Verlin Allbritton: Pure Flo Product Group, Chicago, Illinois
Robert Allen: Consultant, Scroggins, Texas
David J. Anneken: Henkel Corporation/Emery Group, Cincinnati, Ohio
Rahaman Ansari: Quest International Fragrances, Mount Olins, New Jersey
Ralph Astarita: Consultant, Chester, New York
Anthony Athanassiadis: Consultant, Belgium
Imre Balazs: Central Soya Company, Fort Wayne, Indiana
Thomas J. Bartus: Van Leer Chocolate Corporation, Jersey City, New Jersey
Christopher Beharry: Procter & Gamble, Cincinnati, Ohio
Elliot Berlin: U.S. Department of Agriculture, Beltsville, Maryland
Karen Bett: U.S. Department of Agriculture, New Orleans, Louisiana
Bruce H. Booth: Zapata Protein USA, Reedville, Virginia
Charles Brah: Patrick Cudahy Company, Cudahy, Wisconsin

Thomas L. Cain: Van Der Bergh Foods, Joliet, Illinois
George Cavanagh: Consultant, Fresno, California
Marolyn R. Chambers: Riceland Foods Inc., Little Rock, Arkansas
Chuck Chittick: Geka Thermal Systems Inc., Atlanta, Georgia
William E. Cochran: Liberty Vegetable Oil Company, Santa Fe Spring, California
Richard Copeland: Central Soya Company, Fort Wayne, Indiana
Ian Cottrell: Rhône-Poulenc Food Ingredients, Cranbury, New Jersey
C. Dayton: Central Soya Company, Fort Wayne, Indiana
Etienne Deffense: Tirtiaux, Belgium
Robert Delashunt: Consultant, Pickwick Dam, Tennessee
Giorgio Dell'acqua: Fratelli Gianazza, Italy
Joseph F. Dempsey: Karlshamns USA Inc., Harrison, New Jersey
Albert J. Dijkstra: n.v. Vandemoortele, Belgium
Mark Dreher: Nabisco Foods, Parsippany, New Jersey
Walter Farr: Owensboro Grain Company, Owensboro, Kentucky
Fielden Fraley: Pure Flo Product Group, Chicago, Illinois
Herbert Gehring: Koerting Hannover AG, Germany
James Geyer: Wisconsin Dairies, Baraboo, Wisconsin
Lewrene Glaser: U.S. Department of Agriculture, Washington, D.C.
Paul Halberstadt: Swift Eckrich Inc., Downers Grove, Illinois
George W. Halek: Rutgers University, New Brunswick, New Jersey
Earl G. Hammond: Iowa State University, Ames, Iowa
Anthony J. Harper: De Smet Process & Technology, Belgium
Richard Heinze: Griffith Laboratories, Alsip, Illinois
Virginia H. Holsinger: U.S. Department of Agriculture, Philadelphia, Pennsylvania
Arthur House: Lou Ana Foods, Opelousas, Louisiana
Tom Hurley: Lou Ana Foods, Opelousas, Louisiana
Lewis G. Jacobs: Distillation Product Industry, Rochester, New York
Lawrence Johnson: Iowa State University, Ames, Iowa
Donald V. Kinsman: Henkel Corporation/Emery Group, Cincinnati, Ohio
Linda Kitson: Archer Daniels Midland Company, Decatur, Illinois
D. Lampert: Cargill Inc., Minneapolis, Minnesota
David Lawrence: Safeway, Deninson, Texas
Zalman Leibovitz: H.L.S., Israel
Stacey Levine: Bunge Foods, Bradley, Illinois
Mary Locniskar: University of Texas, Austin, Texas
Robert Loewe: Keebler Inc., Elmhurst, Illinois
Earnie Louis: Perdue Farms Inc., Salisbury, Maryland

TED MAG: Consultant, Canada
EUGENE MATERN: Ed Miniat Inc., Chicago, Illinois
WILLIAM MCPHERSON: EMI Corporation, Des Plaines, Illinois
KEN MCVAY: Henkel-Emery Group, Cincinnati, Ohio
JEANNE COCHRANE MILEWSKI: RTM Winners, Atlanta, Georgia
RON MOELLER: Cargill Inc., Minneapolis, Minnesota
DON MORTON: Premier Edible Oils Corporation, Portland, Oregon
JERRY MURPHY: Rangen Inc., Buhl, Idaho
BRIAN F. OSBORNE: Hutrel Engineering, Vancouver Canada
BOB PIERCE: Fats and Oils Consultant, Tucson, Arizona
ANTHONY SACCONE: Karlshamns, USA, Harrison, New Jersey
FOUAD Z. SALEEB: General Foods, White Plains, New York
TIMOTHY SAUNDERS: North Carolina State University, Raleigh, North Carolina
MARY SCHMIDL: Sandox Nutrition, Minneapolis, Minnesota
RICHARD W. SCHOENFELD: SVO Enterprises, Eastlake, Ohio
JAMES SCHWARTZ: Procter & Gamble, Cincinnati, Ohio
JACQUES SEGERS: OPAU (Unilever), Holland
PETER SJOBERG: Deodorization, Tetra-Laval, Sweden
MARTIN SILGE: Mariani Packing Inc., San Jose, California
PETER SLEGGS: Campro Agra Ltd., Canada
BARRY SMITH: Crown Iron Works, Minneapolis, Minnesota
FRANK E. SULLIVAN: Consultant, San Diego, California
BEVERLY TEPPER: Rutgers University, New Brunswick, New Jersey
DAVID TRECKER: Pfizer Food Science Group, Groton, Connecticut
STEPHEN W. TURNER: Henkel Corporation, Cincinnati, Ohio
PHILLIPPE VAN DOOSSELEARE: De Smet Process & Technology, Belgium
PETER WAN: U.S. Department of Agriculture, New Orleans, Louisiana
KLAUS WEBER: Krupp Machinentechnik, Germany
ANTHONY E. WINSTON: Church & Dwight Company, Inc., Princeton, New Jersey
VERNON C. WITTE: Kraft General Foods, Glenview, Illinois.
JOHN S. WYATT: Grindsted Products Inc., Industrial Airport, Kansas
JIM YEATES: Ag Processing Inc., Omaha, Nebraska

COMPANIES

De Smet Process & Technology, Belgium
Hixson, Cincinnati, Ohio

A Tribute to Alton E. Bailey

ALTON E. BAILEY

Alton Edward Bailey was born in Midland, Texas, in 1907 and died 46 years later in Memphis, Tennessee. During his relatively short professional career, A.E. Bailey made an imprint on the science and technology of fats and oils unequaled by any other person either before or since. Of his accomplishments, it is agreed that the most important was the legacy of his book, *Bailey's Industrial Oil and Fat Products*, first published in 1945. It immediately became the Bible of the fats and oils industry and continues to be so regarded, even today. While updated and expanded several times since his death in order to include more recent scientific findings and engineering developments, the 1945 first edition can still be perused with the reader hardly being aware that A.E. Bailey wrote the words 50 years previously.

In our present age, when having a Ph.D. from a prestigious university is almost mandatory in order to be taken seriously as a research investigator or authoritative author, it is interesting to note that "Ed" Bailey's university education ended upon receiving a B.S. in Chemical Engineering from the University of New Mexico in 1927.

Following graduation, A.E. Bailey was employed in the laboratories of the Cudahy Packing Company, first in Omaha, Nebraska, and later in Memphis, Tennessee. He left Cudahy in 1941 to accept a position as head of the oil processing research section of the United States Department of Agriculture at their Southern Regional Research Laboratory in New Orleans. It was in the pilot plant there that Ed Bailey tested out and quantified many of his ideas. His published laboratory work relied heavily on the use of the then new tool of dilatometry to explain and understand the functional characteristics of fats.

Five years later, Alton Bailey resigned from the USDA to join the Votator Division of the Girdler Corporation in Louisville, Kentucky, as Chief Process Engineer of their Oil and Fat Section. In that position, he was instrumental in the development of the semicontinuous deodorizer which, in addition to producing products of very high quality, permitted rapid changeover from one product to another. In 1950, Mr. Bailey returned to Memphis as Vice President and Director of Research for The Humko Company, a position he occupied until his death.

Alton E. Bailey joined the American Oil Chemists' Society in 1935 and subsequently made many contributions to the technical publications, education programs, and organization management of AOCS. He was elected to the Governing Board in 1949 and served the Society as its president in 1951–1952.

According to his contemporaries, Ed Bailey possessed an exceptional intellect, intense curiosity, and extremely high motivation. Those attributes were coupled with an almost photographic memory, along with significant ability to organize material and put it into writing in a straightforward and understandable manner.

When Fred Astaire died, Mikhail Baryshnikov commented, "Mr. Astaire was an *artiste*, the rest of us are dancers." We, who are authors of individual chapters of this fifth edition, feel the same relationship with Mr. Alton E. Bailey. We hope you will find our efforts worthy of being published in a book bearing his name.

ROBERT C. HASTERT

October 1995

BAILEY'S LEGACY IS DECADES OF FATS AND OILS RESEARCH

Following in the innovative footsteps of Alton E. Bailey, agricultural research scientists continue to have a significant impact on the fats and oils industry of today and tomorrow. Working in the scientific environment that cultured Bailey's creative genius, researchers at the regional research centers of the USDA's Agricultural Research Service (ARS) have contributed to decades of advancements in the processes and applications of fats and oils, both for food and industrial uses.

An extended program of research enabled scientists to help transform the soybean into the major source of high-protein feeds and food oil products of today. In the early 1940s the flavor of soybean oil was variously described by consumers as "grassy" or "beany" or "fishy," and it tasted even worse after it had been stored for awhile. In consultation with colleagues in the fats and oils industry, scientists decided that establishment of some uniformity of judgment about how various soybean oils tasted was an essential first step in the research. The technique of flavor evaluation, which they developed involving selected, trained taste panels and numerical rating of the flavors, was the first significant milestone in improving soybean oil

Guided by judgments of taste panels, researchers identified the source of many of the off-flavors in soybean oil as trace metals, particularly iron and copper. Even extremely small amounts of these contaminants sped oxidation of the oil, shortening its storage life and promoting undesirable flavors. Responding to these findings, industry removed brass valves in refineries and substituted stainless steel for the cold, rolled steel in equipment that came in contact with soybean oil. These actions alone improved the flavor of the oil. It was further discovered that the addition of citric acid to the deodorized oil would deactivate the trace metals in soybean oil. Today, practically all soybean oil is protected by adding citric acid during processing.

But many questions remained unanswered. Metal contaminants could speed the development of off-flavors, but chemists wondered what caused the flavors to develop in the first place. One of the principal causes turned out to be linolenic acid, a fatty acid that makes up from 7 to 9% of soybean oil. Of all the major edible vegetable oils on the market only soybean and rapeseed oils contained linolenic acid at these levels. It is this constituent of soybean oil that was a major contributor to deterioration on the shelf or when the oil was heated repeatedly in deep fryers. The industry turned to the nickel-catalyzed, partial hydrogenation process to lower the linolenic acid content of soybean oil to about 3.0%. In the 1960s, it was this lightly hydrogenated product that enabled soybean oil to displace cottonseed oil as the major edible oil in the world.

At the same time, the discovery that linolenic acid was a major reason soybean oil went bad spurred plant breeders in research to find soybean and rapeseed lines with lower linolenic acid content, research that continues to this day.

Today, while hydrogenation of soybean oil is still required for use as a high temperature cooking oil, in shortenings and margarines, and for long-term storage, evaluations by taste panels have made it clear that the industry's implementation of improved processing techniques and protection of the oil during processing produces an oil that is stable at ambient temperatures and has good shelf-life stability.

Other researchers modified cottonseed oil, giving it properties similar to that of cocoa butter. The confectionery fats derived from cottonseed and soybean oil are now being used for many applications in the industry. Further, the fatty acids from vegetable oils were combined with sucrose or table sugar to form sucrose esters, used today as emulsifiers, stabilizers, and texturizers in baked goods, baking mixes, biscuit mixes, frozen dairy desserts, and whipped milk products.

Over the years, ARS has devoted much of its resources to discovering new industrial uses for fats and oils. A new market for fats and oils was created by the discovery of a process called epoxidation, in which hydrogen peroxide was used to insert an atom of oxygen into the hydrocarbon chain of fatty acids. Epoxidized oils, when used as plasticizers, blend well with commonly used resins. They also eliminate the need for poisonous salts of lead, barium, or cadmium in vinyl plastics, which turn the plastics cloudy or opaque. Seventy-five percent of the 50,000 tons of epoxidized oils now used are based on soybean oil. The discovery also helped to create a billion-dollar plastics industry. Today, about 75% of the plasticizers for flexible vinyl plastics are made from soybean oil.

Modification of vegetable oil fatty acids has been a significant alternative to petrochemicals for chemical feedstocks. Oleic acid is converted to emulsifiers, cosmetic ingredients, and other specialty chemicals and is used in textile mills in lubricants and antistatic agents. Acetoglycerides, derived from fatty acids, can be formed into thin, stretchable films for a variety of uses in the food and cosmetics industries. Another development was a group of multipurpose chemicals called isopropenyl esters from fatty acids. These can be used to make paper and cotton repel water, to coat glass to reduce breakage in bottling lines, and in other applications where they have proven superiority to chemicals now in use.

A significant market for soybean oil was created in 1988 when scientists formulated a 100% soybean-oil-based printing ink that not only has a lower cost than petroleum-based inks, but also gives superior penetration of pigment into newsprint. These inks are adjustable to a wide range of viscosity and tackiness for news offset printing and have rub-off characteristics equal to those formulated and marketed as low-rub inks.

In 1959, the North Central Section, American Oil Chemists' Society, established the Alton E. Bailey award to annually recognize research and/or service in the field of oils, fats, and related disciplines. The list of 35 scientist recipients from academia, industry, and government is testament to the lasting impact of Bailey's legacy of research in the fats and oils field.

TIMOTHY L. MOUNTS

October 1995

Preface

This fifth edition of *Bailey's Industrial Oil and Fat Products* differs from the fourth edition in many ways:

1. There are five volumes instead of three.
2. All five volumes are published at one time instead of over several years.
3. In the fourth edition, chapters on edible and nonedible products were distributed over the three volumes. However, the fifth edition is clearly divided into the two groups: edible (Volumes 1–4) and nonedible (Volume 5).
4. Volume 1 serves as an introduction covering several subjects, some of which are basic to the discipline while others, because of their unique themes, cannot be placed in other volumes. As a result, Volume 1 covers such topics as chemistry, nutrition, toxicology, vegetable oil, animal fats, flavors, analysis, and sensory evaluations. Each of Volumes 2 to 4 covers one specific topic on edible oils or fats. Volume 2 covers individual oil seeds; Volume 3 discusses the application of oils and fats; and Volume 4 concentrates on the technology and engineering of processing vegetable oils.
5. Volume 5 covers the application of oils and fats in nonedible products including consumer goods such as soap, paints, leather, textiles, pharmaceuticals, and cosmetics. Other topics include rendering, fatty acids, and glycerine.

As observed, this new edition does not exactly update the fourth edition. Rather, it is more comprehensive and covers more applied information. Many professionals have expressed to me the following:

1. Much of the information in the older editions will always be useful.
2. Each new edition does not always update information in the last edition.

3. Each new edition always covers more information than the last edition.
4. Each new edition provides more applied information than the last edition.

In any book with multiple authors, the editor faces the same difficulties: length, content, format, delay, and updates. In spite of the difficulties encountered in an undertaking of this magnitude, I feel that through the excellent work of the authors I have achieved the major goal of this edition. This five-volume text will provide professionals in the oil and fat industry an excellent reference source on the subject matter with a special emphasis on edible products. Most of the authors are from the industry, with a limited number from academia and government. They have worked hard to make this edition a success and I will forever be grateful for their participation. But, of course, you are the final judge of the usefulness of this work.

Y. H. HUI

Acknowledgment

Most of us are aware that to prepare and publish a professional book of this magnitude is invariably the result of teamwork.

To overcome the technical difficulties, I have been advised by two groups of experts: members of the editorial advisory board and my professional colleagues. It is not an exaggeration to state that this work could not have been completed without their counsel during the last four years. Additionally, through the years of professional consultation for this project, several individuals have actually become my indispensable advisors. They have helped me above and beyond the normal call of professional courtesy. They are well-known individuals in the oils and fats industry. As a matter of fact, two of them were past presidents of the American Oil Chemists' Society. I will always be grateful to Ken Carlson, Bob Hastert, Tom Smouse, and Peter Wan; I value their friendship.

Every time we read a book we evaluate two things in our minds: the information and the quality of the book. I have worked with the production staff at Wiley in 1978, 1986, 1987, and 1991. They have consistently produced excellent quality books that I have authored. This one is no exception. They are wonderful people to work with and they are professionals to the core. No, this project could not have been completed without this group of dedicated individuals. I am sure you will agree with me.

Let me express my most sincere appreciation to everyone who has participated in the completion of this work. And, lastly, I am grateful to the support of my family.

Contents

Bailey's Industrial Oil and Fat Products

1

Animal Fats

The major edible animal fats (animal fats also are termed meat fats) are lard and tallow. Lard and rendered pork fat are produced from the fat of pigs (*Sus scrofa*). Edible tallow is produced mainly from the fat of cattle (*Bos taurus*), but also can be produced from sheep (*Ovis aries*) fat. Tallow from domestic cattle is known as beef tallow and tallow from sheep is termed mutton tallow.

PRODUCTION OF ANIMAL FATS

Worldwide, the proportion of fat and oil production that is due to animal fat has declined. Mielke (1) reported that in 1960, tallow and lard made up 15.6 and 14.2%, respectively, of the world production of 12 major fats and oils. By 1985, comparable shares for these fats were 11.3 and 9.1%. He forecasts that tallow and lard will make up only 8.4 and 7.3% of the production of the 12 major fats and oils in 1995. Preliminary USDA agricultural statistics for 1991–1992 showed that world production of tallow and grease was 6,995,000 metric tons out of a total production of fats and oils of 72,598,000 metric tons (2).

Dugan (3) has summarized trends in the production of animal fats in the United States from 1951 to 1983. The changes in lard production (an increase, then a decrease) during this period have been attributed both to the increased number of animals slaughtered and the development of pigs with less adipose tissue. He reported that lard production per 100 pounds of pig live weight was 13.9 pounds in 1959, 10.8 pounds in 1965 and 4.6 pounds in 1983. This trend has been a result of two factors: consumer preferences for lean meat and also economic pressure to produce animals more efficiently, together, have resulted in a reduction in the fat content of domestic meat animals. In addition, efforts to achieve further decreases have continued.

The increased production of inedible tallows and greases during the 1951–1983 period has been said to reflect increased sources of animals for slaughter and improved yields of fat from the available sources. Dugan (3) states that from 1951 to 1983 technological advances resulting in the increased use of edible tallow, made it possible to use tallow in shortenings, margarines and frying fats. Previously, flavor and performance problems had limited the use of tallow in these products. In 1990, Brady (4) estimated current potential production of edible tallow in the U.S. to be around a million tons, although domestic demand was only 300,000 to 400,000 tons.

Direct use of lard in the United States has decreased. Sonntag (5) reported that direct edible use of lard in the United States in 1976 was 568 million pounds, down from 1,683 million pounds in 1966. Per capita disappearance of lard decreased from 6.3 pounds in 1965 to 1.8 pounds in 1985 (6). Consumption of lard in 1989 in the United States was reported to be 442 million pounds, or 1.8 pounds per capita (2).

Statistics on the use of U.S. produced lard and tallow in margarines and shortenings are presented in Table 1.1. Of particular note is the decrease in the use of edible tallow in shortenings: from 1015 million pounds in 1985 to 462 million pounds in 1991.

Recent trends in edible fat and oil usage indicate a move away from animal fats to partially hydrogenated vegetable oil-based products and liquid oils. Reasons for these changes in animal fat usage may relate to inconsistency of supplies, variation in composition and properties, technological advances in processing vegetable oils and consumer preferences for bland flavor for most fat uses (5). Decreased direct use of fats such as butter and lard also may reflect a shift from the preparation of foods in the home to the use of convenience products and foods eaten away from the home. A major recent factor has been health-related concerns about the cholesterol and saturated fatty acids in animal fats. Consumer attitudes toward animal fat have affected the

Table 1.1 Animal fats used in the manufacture of margarine and shortenings in the United States[a]

	Shortening		Margarine		
	Lard	Tallow	Lard	Tallow	Animal Fats
Year	(Millions of Pounds)				
1947	115	64	4	8	
1957	345	226	19	9	
1963	594	413	84	11	
1968	601	487	153	15	
1975	165	165	45	7	
1980	378	673			104
1985	289	1015			65
1990	264	637			35
1991	274	462			43

[a] Data from 1947–1975 are from a publication by Dugan (3). Data from 1980–1991 are from USDA Agricultural Statistics (2). Data for 1991 are preliminary figures.

type of fat used by manufacturers in processed foods and by the foodservice industry. Trends in the baking industry have been to replace lard, tallow and hydrogenated vegetable shortening with liquid vegetable oils or a mixture of oils and surfactants (7). Haumann (8) and Orthoefer (9) discuss the impact of consumer attitudes on the choice of shortenings and frying fats by the fast food and food service industries.

Preliminary figures for 1991 U.S. exports of lard were 54,705 metric tons. The comparable figure for edible and inedible tallow, greases and oils was 1,160,515 metric tons (2). A considerable portion of the animal fat that is produced appears on the market as nonedible fats.

SOURCES AND IDENTIFYING CHARACTERISTICS OF ANIMAL FATS

In pigs, cattle and sheep, fatty tissues accumulate in some parts of the body when the animals are fed at above maintenance levels to a commercial slaughter weight. Major fat depots include the subcutaneous fat (located under the skin and overlying superficial muscles) and the intermuscular fat (located between muscles). Appreciable amounts of fat also are deposited in the abdominal cavity and other internal sites. The distribution of fat between the different sites varies somewhat with animal species, breed and degree of finish.

Some of the fat deposits, for example those located around the kidney, heart, caul and intestines, are stripped from the animal when it is slaughtered. These fats are called the killing-floor or killing fats. Additional fat is separated from the carcass when it is cut apart to give wholesale or retail cuts or for processed meat products. This fat is called cutting-floor or cutting fat. In general, the fatty deposits increase in hardness from surface subcutaneous locations through the inter- and intramuscular fat to deep abdominal and kidney fats in cattle, sheep and pigs. Thus the internal fats from these species, especially those surrounding the kidney, tend to be harder than the cutting and trimming fats and products made from kidney fat will be firmer and have better flavor stability.

The *Code of Federal Regulations* of the United States defines lard as the fat rendered from clean and sound edible swine tissues. Tissues to be used for lard are to be reasonably free from blood and shall not include stomachs, livers, spleens, kidneys and brains, or settlings and skimmings. "Leaf Lard" is prepared from fresh leaf (abdominal) fat. Lard (when properly labelled) may be hardened by the use of lard stearin or hydrogenated lard or both and may contain refined lard and deodorized lard, if so labelled (10). Dugan (3) provides a detailed compilation of the killing and cutting fats to be used in producing lard and rendered pork fat.

The *Codex Alimentarius* (11) contains international standards for four products from animal sources: Codex Stan 28-1981 (Rev. 1-1989) for lard, Codex Stan 29-1981 (Rev. 1-1989) for rendered pork fat, Codex Stan 30-1981 (Rev.

1-1989) for premier jus, and Codex Stan 31-1981 (Rev. 1-1989) for edible tallow. Premier jus (or oleo stock) is the product obtained by low-temperature rendering of the fresh fat of heart, kidney and mesentery of bovine animals. The raw material for this product does not include cutting fats. This fat has a creamy white to light yellow color and a characteristic mild flavor. Cutting and trimming fats can be utilized for tallow (this product also can be labelled dripping). Any product designated edible tallow must be produced exclusively from bovine fat. Any product designated mutton tallow must be exclusively from sheep fat. Lard, rendered pork fat and edible tallow may contain certain further processed forms of the rendered fat such as refined or hydrogenated product, or stearines, as long as labelling regulations are followed.

The Codex descriptions specify that all edible animal fats must come from animals determined to be in good health at the time of slaughter and fit for human consumption as judged by a competent authority recognized in national legislation. The main Codex analytical identity standards for lard, rendered pork fat, edible tallow and oleo stock are given in Table 1.2. The ranges for fatty acid composition specified in the Codex standards for lard, rendered pork fat, edible tallow and oleo stock are given in Table 1.3.

A number of tests have been used to attempt to detect the presence of beef fat in lard. The best known of these tests is the Bömer test. The Bömer test is based on the difference between melting points of glycerides and the fatty acids prepared from them. The difference is large for unhydrogenated pork fat and small for tallow. This test is invalidated by the presence of hydrogenated fat in the lard. It has been suggested that more than 0.01%,

Table 1.2 Codex standards[a] for lard, rendered pork fat, premier jus and edible tallow (11)

Characteristic	Lard	Rendered Pork Fat	Premier Jus	Edible Tallow
Relative density (40°C/water at 20°C)	0.896–0.904	0.894–0.906	0.893–0.898	0.893–0.904
Refractive index (n_D^{40})	1.448–1.460	1.448–1.461	1.448–1.460	1.448–1.460
Titre (°C)	32–45	32–45	42.5–47	40–49
Saponification value (mgKOH/g fat)	192–203	192–203	190–200	190–202
Iodine values (Wijs)	45–70	45–70	32–47	32–50
Unsaponfiable matter	Not more than 10g/kg	Not more than 12g/kg	Not more than 10g/kg	Not more than 12g/kg

[a] Elaborated by the Codex Committee on Fats and Oils. Adopted by the Codex Alimentarius Commission of the Food and Agriculture Organization of the United Nations.

Table 1.3 GLC ranges of fatty acid composition (%) specified by the Codex standards[a] for lard, rendered pork fat, premier jus, and edible tallow (11)

Fatty acid	Lard and Rendered Pork Fat	Premier Jus and Edible Tallow
C<14	<0.5	<2.5
C14:0	0.5–2.5	1.4–7.8
C14:ISO		<0.3
C14:1	<0.2	0.5–1.5
C15:0	<0.1	0.5–1.0
C15:ISO	<0.1	<1.5
C15:ANTI ISO		
C16:0	20–32	17–37
C16:1	1.7–5.0	0.7–8.8
C16:2		<1.0
C16:ISO	<0.1	<0.5
C17:0	<0.5	0.5–2.0
C17:1	<0.5	<1.0
C17:ISO		<1.5
C17:ANTI ISO		
C18:0	5.0–24	6.0–40
C18:1	35–62	26–50
C18:2	3.0–16	0.5–5.0
C18:3	<1.5	<2.5
C20:0	<1.0	<0.5
C20:1	<1.0	<0.5
C20:2	<1.0	
C20:4	<1.0	<0.5
C22:0	<0.1	

[a] Elaborated by the Codex Commission on Fats and Oils. Adopted by the Codex Alimentarius Commission of the Food and Agriculture Organization of the United Nations.

0.05% and 0.05% of branched chain C14, C15 and C16 fatty acids, respectively, in lard indicates the presence of tallow (12). However, when pigs are fed tallow, they incorporate some of the branched fatty acids into their depot fat (13).

According to Kirk and Sawyer (12) triacylglycerol profiles by HPLC may indicate adulteration of pork by beef fat, because the fatty acids in the 2-position of pork fat are significantly different from those in beef fat. They also state that the presence of 5% or more of pork fat in beef or mutton tallow can be detected and quantified by analysis of fatty acids in the 2-position of the triacylglycerols. The ratios of C16:0/C18:1ω9 for lard is about 5 whereas for edible tallow it is about 0.4.

QUALITY INDICATORS FOR EDIBLE FATS

The United States *Code of Federal Regulations* (10) specifies that lard shall have a maximum free fatty acid value of 0.5% (as oleic) or an acid value of 1.0, as milligrams of KOH per gram of sample. The maximum peroxide value should be 5.0 (as milliequivalents of peroxide per kilogram of fat). A maximum of 0.2% moisture and volatile matter is specified, and for insoluble impurities a maximum of 0.05%. The color should be white, and a maximum of 3.0 red units in a 5 ¼ inch cell on the Lovibond scale is specified.

In the Codex standards (11), maximum acid and peroxide values for lard are 1.3 mg KOH/g fat and 10 milliequivalents peroxide oxygen/kg fat, respectively. Comparable values for rendered pork fat and edible tallow are 2.5 and 16. Standards for premier jus are acid value not more than 2 mg KOH/g fat and not more than 10 milliequivalents peroxide oxygen/kg fat. The Codex standards also specify levels for antioxidants and antioxidant synergists, and the maximum amounts of impurities, soaps and certain metals in the edible fats.

PROCESSING OF ANIMAL FATS

4.1 Rendering of Meat Fats

The fatty tissues separated from meat animals at slaughter and during cutting consist of fat deposited in a connective tissue matrix containing protein and water. Through the years a variety of processes have been developed to remove the fat from these tissues as efficiently as possible. These processes are referred to as rendering. The aim of rendering is to obtain as complete a separation of the fatty tissue components as is possible. Most rendering systems rely on heat to release fat from the cells of the fatty tissues. In other methods, however, the heat is kept low and fat is released by mechanical rupture of the cells.

Some products are produced by "dry-rendering" which means that no water or steam is added to the fat during the process. In dry rendering, the chopped fatty tissues are heated, usually in a horizontal steam-jacketed vessel, to disintegrate the fat cells, release the melted fat and drive off moisture. Most batch cookers for dry rendering are equipped with rotating agitators which may be steam-heated. Agitation aids in heat transfer. When the fat has been released and sufficient moisture removed, the material is strained or filtered to remove the cooked proteinaceous residue (the cracklings). Dry rendering can be carried out at atmospheric pressure, under vacuum or at elevated pressure.

In wet rendering processes the fatty tissues are heated in the presence of water, usually at lower temperatures than in dry rendering. Fats rendered at lower temperatures typically have less color and a milder flavor than those that are dry-rendered. Lard and edible tallow are made primarily by a steam rendering process. In this process, live steam is injected into a closed vessel containing the fatty tissues, and the rendering takes place under pressure to shorten the cooking time. Lard produced by this process is called prime steam lard.

Continuous processes have been developed for both wet and dry rendering. High-quality edible products can be produced by continuous low-temperature rendering systems. In these systems, the fat is separated from the protein fraction at low temperatures in a relatively short time. The resulting products are light in color, mild in flavor and quite low in free fatty acids. The features and operation of a low temperature system used for edible fat processing have been described in detail (14).

If fatty tissues are handled and rendered properly, the resulting lard is suitable for use as a food fat without further treatment. Sometimes meat fats are chemically refined. Sonntag (15) reported that tallows may be alkali refined if the free fatty acid content is greater than about 0.3% or if collagen or proteinaceous material is present. Edible animal fats also may be subjected to bleaching, hydrogenation, deodorization, interesterification or fractional crystallization to improve their characteristics or produce fats for specialized use.

4.2 Bleaching

Bleaching is carried out to remove components that give fats an undesirable color. Most lards do not require bleaching, but tallows may be bleached to remove colored materials. Bleaching is usually accomplished by adding natural or acid-activated clays, and to a lesser extent, activated carbons. These materials absorb the color bodies and certain degradation products in the fats. Patterson (16) gives specific recommendations for bleaching lard and tallow. For lard, a contact time of 15 minutes at 95–100°C is suggested, with a maximum of 0.5% mildly activated clay or 0.25% moderately activated earth. For top grades of tallow, up to 1% mildly activated clay or about 0.3% well activated clay is suggested, and a contact time of 20 minutes at 95–100°C is normal.

Color and peroxide values were compared for tallows bleached at atmospheric pressure and under vacuum at several temperatures (17). At 90°C, atmospheric bleaching gave better color than bleaching under vacuum, but peroxide values were lower for the vacuum bleached product (2 versus 16). At higher temperatures, color was better under vacuum, and peroxide values also were lower under vacuum. A bleaching temperature of 90 to 110°C was recommended for beef tallow.

Treatment of lard with bleaching earth decomposes the peroxides and increases the content of conjugated trienes, which absorb at 268 nm. This characteristic has been the basis of a quality control procedure for determining whether lard has been bleached (16).

4.3 Hydrogenation

Tallows are very firm fats and are seldom hydrogenated for edible purposes except to a very slight degree. Mild hydrogenation of tallows delays or prevents the development of undesirable flavors thought to be due to the peroxidation of trienoic and tetraenoic fatty acids (3).

Unhydrogenated lard does not necessarily have the desired melting point or plastic range. The melting range characteristics of lard from different animals can vary greatly. A consistent product can be obtained by blending lards from different sources. The firmness of unhydrogenated lards can be increased and their working characteristics improved by blending them with lards that have been hydrogenated to almost complete saturation (an iodine value of five or less) (3).

4.4 Deodorization

Animal fats are subjected to deodorization processes when a very bland flavor or essentially flavorless fat is desired. Heated fats are stripped with dry steam under vacuum to efficiently achieve the removal of the materials that impart flavors. In addition to flavor components, free fatty acids and other minor constituents are stripped out of the fat. During deodorization, the content of free fatty acids and other minor constituents may be reduced.

4.5 Interesterification

Interesterification involves the exchange of acyl residues among the triacylglycerols making up a fat. Interesterification will lead to a random distribution of fatty acids if the process is carried out in a single phase with the usual types of chemical catalysts. Interesterification can be directed away from randomness if it is carried out at a sufficiently low temperature to cause the trisaturated triacylglycerols to crystallize out. These solid triacylglycerols no longer participate in the reaction, which proceeds in the liquid phase. The selective crystallization upsets the equilibrium and promotes the formation of more trisaturated acylglycerols. Rozenaal (18) discusses catalysts, kinetics and mechanisms of interesterification, as well as conditions for batch and continuous interesterification processes. Haumann (19) gives an overview of interesterification of fats and oils.

Interesterification was developed as a way to modify lard so that it would function more effectively as a shortening for products like cakes. Shortenings for these products must cream readily to facilitate incorporation of air into the batter. Unmodified lard has a narrow plastic range, a grainy texture and does not cream well. The grainy texture is due to the coarse β-type crystals. This crystal structure is due to the occurrence in lard of a relatively high proportion of a certain type of triacylglycerol. Lard contains approximately 27% disaturated glycerides, which is mostly oleopalmitostearin (OPS).

Randomization decreases the proportion of palmitic acid in the 2-position of the lard triacylglycerols from about 64% to 24% (20), thus producing a mixture of disaturated acylglycerols which has a substantially lower melting point than the original OPS. Randomized lard crystallizes in the β' form, which is characteristic of hydrogenated vegetable shortenings.

Interesterification of liquid oils with hard fats could be used as an alternative to partial hydrogenation to produce plastic fats for margarine. Chobanov and Chobanova (21) interesterified sunflower oil with lard and tallow. Lo and Handel (22) interesterified soybean oil with beef tallow to produce a plastic fat suitable for use in making tub-type margarine. The interesterified blend of 60% soybean oil and 40% tallow contained 3–3.4% trans-fatty acids from the tallow.

Directed interesterification can be employed to produce lard with an increased solids content at high temperatures. This product would be plastic over a greater range of temperature. Lard produced by randomization requires the addition of stearin for high-temperature stability.

Directed interesterification also could be carried out by replacing chemical catalysts with enzymes. Forssell and co-workers (23) interesterified tallow by using a commercial immobilized 1,3-specific *Mucor miehi* preparation as a catalyst, and also interesterified tallow and rapeseed oil by using this lipase. The altered composition of triacylglycerols in the interesterified rapeseed oil-tallow mixtures was reflected in significant changes in the solid fat content in the temperature range of 0 to 45°C. The melting point reduction achieved depended on the mass fraction of the substrates: the lower the mass fraction of tallow, the larger the reduction of the melting point.

4.6 Fractionation

The characteristics of the liquid portions of meat fats have been considered especially desirable for some uses. To obtain these materials, the fats were melted, then held at temperatures that resulted in the crystallization of the more highly saturated triacylglycerols. The chilled material was then pressed to separate it into the liquid (olein) and solid fractions (stearines). This process is termed dry fractionation, and is the oldest method for separating tallow fractions.

Solvent fractionation ("wet" fractionation) is another type of fractionation whereby the fat is dissolved in organic solvents, then the solution is brought to a temperature suitable to allow partial crystallization of the higher melting glycerides. The crystals are separated by decanting, filtering or centrifuging. Crystals of fat produced by this process retain less oil than with dry processes, but the solvent fractionation process is more costly.

A third method of fractionating fats is detergent (aqueous) fractionation. In this process, separation is achieved by adding an aqueous solution containing a surfactant and an electrolyte to a partially crystallized oil which preferentially wets the crystals into the aqueous solution (24). The dispersed crystalline stearin is separated from the olefin by using a centrifugal separator. The stearin can be recovered by heating the aqueous mixture to melt the stearin and centrifuging to yield the liquid stearin.

Tallow can be fractionated to yield several fractions with quite different melting ranges. Tirtiaux (25) has described a process for fractionating beef tallow to give products ranging from a very hard stearin (m.p. 56°C) to an oil (m.p. less than 20°C). Luddy and co-workers (26) partitioned beef tallow that had been dissolved in acetone into five fractions by a multistep crystallization process. Two of the five fractions were crystalline, one was a plastic solid and two were liquid. The properties of the plastic solid were similar to those of cocoa butter, and this fraction was reported to have excellent compatibility with cocoa butter. Use of specific tallow fractions for confectionery purposes and for hardening shortening or margarine base stocks was described by Luddy and co-workers (27, 28).

Grompone (29) described the relationship among composition, melting point, titer and solid fat index of beef tallow and its liquid and solid fractions obtained by dry fractional crystallization. This study was conducted with Urguayan tallow, which has been reported to have a higher titer (43.2–47.8°C) and melting point (45.0–48.8°C) than is average for beef tallow (30).

Taylor and co-workers (31) investigated the use of tallow fractions as cost-effective ingredients to substitute for or extend cocoa butter. Haumann (8) reported that a pourable tallow oil shortening for deep-fat frying could be prepared by fractionation. C. DeFouw, M. Zabik and J. Gray (32) reported that tallow fractions performed well in producing french-fried potatoes.

FATTY ACID COMPOSITION AND GLYCERIDE STRUCTURE OF ANIMAL FATS

The fatty acid composition of meat fats is influenced by several factors, notably diet. Fatty acids values from a few recent publications (33,34) dealing with the impact of diet on fat composition, are given in Table 1.4. Nonruminants readily incorporate the unsaturated fatty acids of the diet into depot fat. Data in this table demonstrate the effectiveness of feeding unsaturated oils in increasing the unsaturation of the pork fat. Typically, lard contains considerable palmitic acid, stearic acid, oleic acid, and linoleic acid. There are lesser quantities of myristic acid, palmitoleic acid, linolenic acid and arachidonic acid. Other fatty acids also have been reported as components of pork depot fat. For example, Morgan and co-workers (35) have reported 20:3, 22:5 and 22:6 as fatty acid components of the back fat of pigs fed barley and soybean meal-based diets containing 50 g/kg either of tallow or soy oil. These research-

Table 1.4 Fatty acid composition (% of total) of pork fat and beef adipose tissue

	Pork Fat (34)		Pork Fat (33)		Beef Fat (33)	
Fatty Acid	Tallow Diet[a]	Soy Oil[a]	Control[b]	Canola Oil[b]	Control[c]	Rapeseed[c]
C14:0	1.0	0.8	1.3	0.5	3.9	3.6
C16:0	26.6	22.1	25.6	10.8	26.7	24.3
C16:1	4.1	2.5	0.9	0.4	2.0	2.0
C18:0	12.1	11.3	12.9	4.3	20.2	20.5
C18:1	40.5	33.2	46.9	56.1	41.2	43.0
C18:2	11.2	24.4	11.5	21.5	4.8	5.5
C18:3	4.0	4.9	0.9	6.5	—	—
C20:4	0.3	0.5	—	—	—	—
C22:6	0.2	0.2	—	—	—	—
Total saturates	39.6	34.2	39.8	15.6	50.8	48.4

[a] Diets contained 3% tallow or soy oil.
[b] The control diet was based on corn–soybean meal and the canola oil diet contained 20% canola oil.
[c] The control was a high-energy corn–cottonseed meal diet and the rapeseed diet had 20% rapeseed at the expense of 20% corn.

ers have reported that the diet with soy oil gave high levels of linoleic acid in the back fat, but the altered fat did not pose any problems due to carcass appearance or softness during processing. Also, Larick and co-workers (36) have fed pigs a corn–soybean meal diet with added tallow, safflower oil, or a combination of tallow and safflower oil, with the result that the increased levels of safflower oil resulted in less C16:0 and C18:1 and more C18:2, C20:2 and C20:3 in the subcutaneous fat.

The depot fats of ruminants are very complex. Tallow is known to contain hundreds of fatty acids, although most are present only in small or trace amounts. The saturated moieties of ruminant glycerides contain normal and methyl-branched components; the latter may be of the iso and anteiso species as well as those with the branches on the even numbered carbons (relative to the carboxyl group). Both geometrical and positional isomers of unsaturated fatty acid moieties usually are present. The presence of trans and positional isomers of oleic and linoleic acids is characteristic of ruminants and other animals with ruminant-like digestive systems. In tallow, as in pig fat, the C16 and C18 cis-fatty acids are the major fatty acids and have the biggest effect on the melting point of the fat. The data in Table 1.5 (37) show that sheep fat typically has a higher percentage of saturated fatty acids than does beef fat, with the major differences in the unsaturated fatty acids being in the C16:1 and C18:1. In general, the concentration of saturated fatty acids is higher and that of linoleic acids is lower in ruminant than in pig fat, and fat from cattle and sheep is, in general, firmer than that from pigs.

Table 1.5 Fatty acid composition (% of total) of beef and lamb fat (37)

Fatty Acid	Bovine Fat-Subcutaneous	Ovine Fat-Subcutaneous	Fatty Acid	Bovine Fat-Subcutaneous	Ovine Fat-Subcutaneous
C12:0	—	0.8[a]	C20:0	—	0.8
C14:0	3.2	5.2	C14:1	0.9	—
C15:0	0.5	0.6	C16:1	4.4	1.6
C16:0	24.8	24.6	C18:1	46.9	38.7[b]
C17:0	1.8	1.0	C18:2	1.9	1.2
C18:0	13.7	22.9	C18:3	0.2	0.6
			C20:1	—	0.2

[a] Also includes C10:0, 0.4%; 14-iso, traces; 15-branched, 0.6%; 16-iso, 0.2%; 17-branched, 0.5% and 18-iso, 0.5% were present.
[b] Includes 0.8% *trans*-16-octadecenoic acid.

For ruminant fat to become directly responsive to dietary unsaturated fats, it is necessary to protect the lipids against saturation by the action of the rumen microorganisms. The data in Table 1.4 from Ref. 33 illustrate the lesser effect of added dietary vegetable oils on beef fat, compared to pork fat. McDonald and Scott (38) describe alterations in lipid content in mutton that have been achieved by feeding protected oil supplements. Rule and Beitz (39) have demonstrated that a diet of extruded soybeans increased the linoleic acid and linolenic acid contents of steer adipose tissue.

A number of factors besides diet have influenced the fatty acid composition of the fatty tissue in meat animals of a given species. These factors have included genetics and sex effects (37,40). J. Marchello, D. Cramer and L. Miller (41) demonstrated that for sheep, colder temperatures resulted in softer body fats with lower melting points and higher iodine numbers. In general diet has had a more marked effect on fat quality than breed or sex, especially in nonruminant animals which are susceptible to alteration of tissue fatty acid by dietary modification.

As noted in a previous section, the texture of lard is due to the pattern in which the fatty acids are distributed in the triacylglycerols. The positional distribution of fatty acids in several animal fats is shown in Table 1.6 (42). In pig fat, the 2-position contains 72 mol% of palmitic acid, 10 mol% is in the 1-position and trace amounts of palmitic acid are found in the 3-position. In tallow, palmitic acid is located at all three positions of the triacylglycerol molecule. Of the fatty acids located in the 2-position, 17 mol% is palmitic, in the 1-position, palmitic acid makes up 41% and in the 3-position, 22%.

Beef tallow contains 17–18% mol trisaturated triacylglycerols and approximately 40 mol% disaturates. Lard triacylglycerols contain about 7 mol% trisaturates and 30 mol% disaturates. Lard has a greater proportion of triacylglycerols with two double bonds than does beef tallow and about twice the mol% of triacylglycerols with three double bonds.

A number of minor constituents, some of them other lipids, are present in lard and tallow. Animal fats contain only a low amount of phospholipid. A

Table 1.6 Positional distribution of fatty acids in triacylglycerols of animal depot fats (42)

Fat	Position	Fatty acid, mol%						
		14:0	16:0	16:1	18:0	18:1	18:2	18:3
Pig	1	1	10	2	30	51	6	
(outer back fat)	2	4	72	5	2	13	3	
	3	—	tr	2	7	73	18	
Cattle	1	4	41	6	17	20	4	1
(subcutaneous fat)	2	9	17	6	9	41	5	1
	3	1	22	6	24	37	5	1
Sheep	1	1	35	2	47	4[a]	—	
(perineal fat)	2	4	14	2	15	52[a]	5	
	3	3	16	1	42	26[a]	2	

[a] Results for 18:1 cis-isomers only. 18:1 trans was present in positions *sn*-1, *sn*-2 and *sn*-3 as 5, 2 and 6% respectively.

minor lipid constituent of considerable interest from a nutritional standpoint is cholesterol. Punwar and Dearse (43) report cholesterol contents of 75, 101 and 145 mg/100g for lard, leaf lard and edible tallow.

Trace metals such as iron, copper, cobalt and nickel can be introduced during refining or processing and have an undesirable effect on oxidative stability. Patterson (16) has suggested that lower trace metal contents than those specified by the Codex are desirable for lard; specifically 1.0 ppm maximum for iron and 0.05 ppm for copper.

ANTIOXIDANTS IN ANIMAL FATS

The tocopherol contents in animal fats have generally been much lower than in vegetable oils and fats, even after refining and hydrogenation of the latter. Parker (44) concluded that animal fats are generally 2–3 orders of magnitude less concentrated in tocopherols than are vegetable oils. Data from Kanematsu and co-workers (45), as reported by Parker (44), indicated that α-tocopherol represented 90.7, 94.6 and 69.8% of the total tocopherols in the body fat of beef, lamb and pork, respectively. USDA Handbook 8-4 (46) has given the total tocopherol content of beef tallow as 2.7 mg/kg, and pork fat total tocopherols as 1.3 mg/kg. Increasing the level of tocopherols fed to meat animals has increased the tissue levels of tocopherol, and has provided some protection against fatty tissue lipid oxidation, even in pig fat with elevated levels of unsaturated fatty acids (34).

Lard and tallow respond well to the addition of antioxidants, and numerous studies have been conducted to evaluate the protection afforded to lard by various antioxidants and metal sequestering agents. In lard, Griewahn and

Daubert (47) have found that tocopherols increased in effectiveness in the following order: α, β, γ and δ. Cort (48) has found that a concentration of 0.02%, γ-tocopherol was more effective than the same concentration of BHA or BHT in chicken, pork and beef fat. Patterson (16) has stated that lard stability is improved by the addition of up to 250 ppm tocopherol, while for BHA and BHT the optimum effective dose is 200 ppm. Tocopherols also have had carry-through properties in baked and fried products prepared from lard (49). Sherwin and Thompson (50) have demonstrated the protective effects of major synthetic antioxidants on lard and their carry-through effects in products made with lard. Dugan (3) has given the best combinations of antioxidants and chelating agents to use in animal fats for particular applications. Natural antioxidants besides the tocopherols also have afforded protection to animal fats (51,52).

COMPOSITION, STRUCTURE AND TEXTURE OF ANIMAL FAT SHORTENINGS

Traditionally, the solids content, crystal structure and working characteristics of lard have made it the shortening of choice for pie crust. Vegetable shortenings can, however, be formulated to have characteristics similar to those of lard, and these shortenings have more favorable nutritional characteristics (less saturated fat, no cholesterol). Deodorized and stabilized lard and tallow are examples of the lowest cost shortening type suitable for cookies (53).

L. deMan, J. deMan and B. Blackman (53) studied a number of North American vegetable and animal fat shortenings for their physical and textural characteristics. Selected data from this study have been presented in Table 1.7. There was not much difference in the vegetable and animal fat shortenings as far as melting and crystallization temperatures, but the polymorphic forms were different. The animal fat shortenings were mostly in the β form, except for a tallow and a tallow-vegetable frying fat. The vegetable-based shortenings were mostly in the β′ form. The texture of several of the meat–vegetable shortenings was comparable to that of the vegetable–palm shortenings, although the meat–vegetable samples contained higher solids. The lards had high values for % deformation at breaking force, while the tallows were not as pliable. Also, shortenings containing high levels of palm oil were able to withstand large deformations without breakage. The tallows and tallow-lard blends were very hard.

The major fatty acid components of some United States and Canadian animal fat or animal fat-vegetable shortenings are presented in Table 1.8 (54), along with the trans fatty acid content. The trans fatty acid content of the vegetable shortenings (data not shown) was, with one exception, much higher (13.4–37.3%) than that of the animal fats or meat-vegetable combinations (0.9–6.0%). There was not much difference in the content of saturated plus trans fatty acids in the animal and vegetable fat products. The solids of the vegetable shortenings in the β form contained about 20% of C16:0, those in

Table 1.7 Selected characteristics of meat–fat or meat–vegetable fat shortenings (53); reprinted by permission of AOCS

Sample and Source	Dropping Point (°C)	Softening Point (°C)	Polymorphic Form	Hardness Index (g/mm)	SFC (%) at 20°C (Tempered at 30°C)	Air Content
Shortening						
Lard, U.S.	45.2	44.4	β	4.8	26.9	2.3
Lard, Canada	40.6	41.9	$\beta>>>\beta'$	5.1	26.2	6.2
Lard, Canada	38.2	37.7	β	4.5	25.3	10.0
Tallow–lard, Canada	42.7	42.3	$\beta>>>>\beta'$	11.8	27.0	8.8
Tallow–lard, Canada	42.3	43.7	β	14.6	28.7	9.0
Meat–vegetable, U.S.	44.6	45.8	$\beta'>>>\beta$	5.4	25.0	19.5
Meat–vegetable, U.S.	45.1	46.0	$\beta'=\beta$	5.9	26.9	21.0
Vegetable–tallow, Canada	50.6	50.7	β	8.1	26.6	4.0
Frying Fat						
Tallow, Canada	45.8	46.4	β'	8.9	—	—
Tallow–vegetable, Canada	44.8	44.7	β'	6.9		

the β' form, 30% or more. The animal fat shortenings were mainly in the β form. Their solids, however, contained 30% or more of C16:0.

CHARACTERISTICS OF ANIMAL FAT-BASED FRYING FATS

In the United States, animal fats have traditionally been used as fats for deep frying many types of foods. Meat fats provide good stability and have generally been economical to use. The flavors imparted by meat fats have been consid-

Table 1.8 Composition of the major fatty acids (% of total) of animal fat shortenings (54); reprinted by permission of AOCS

Sample	16:0	18:0	18:1	18:2	Trans
Lard, U.S.	24.6	15.8	42.9	9.9	0.9
Lard, Canada	24.0	14.8	41.9	11.6	2.0
Lard, Canada	24.8	14.8	42.3	9.6	1.7
Tallow–lard, Canada	25.5	19.3	39.8	5.3	3.4
Tallow–lard, Canada	25.2	20.2	39.2	5.7	3.0
Meat–vegetable, U.S.	20.9	16.4	44.5	10.0	6.0
Meat–vegetable, U.S.	22.6	17.1	42.5	9.8	4.3
Vegetable–tallow, Canada	17.7	19.7	43.2	9.9	1.6

ered desirable for some foods. For example, the flavors imparted by tallow have been considered desirable for french-fried potatoes. The content of saturated fatty acids and cholesterol in meat fats has, however, prompted a shift away from the use of all-meat fats for frying. Blends of beef fat and vegetable oil provide some beef-like flavor notes to fried foods; however, foods fried in vegetable oils lack the characteristic flavors imparted by beef tallow. Ha and Lindsay (55) showed that many of the volatile fatty acids present in raw tallow and in a fractionated pourable beef tallow shortening were found in potatoes that had been fried in a beef tallow–hydrogenated vegetable oil blend.

Attempts have been made to extract and concentrate beef fat volatiles using supercritical carbon dioxide (56). Total volatiles were concentrated over controls by 10–100 fold, with the lowest pressure extraction conditions yielding the highest concentration of volatiles. Um and co-workers (57) found that the flavor volatiles of heated pork fat could be fractionated with supercritical carbon dioxide.

The cholesterol present in tallow used for frying undergoes oxidative changes. Parks and Addis (58,59) have characterized the products of cholesterol oxidation in heated tallow. Ryan and Gray (60) have demonstrated the presence of cholesterol oxides in french-fried potatoes at concentrations approximately four times as high as those which existed in the heated tallow used for frying. Park and Addis (61) have found oxidation products of cholesterol in commercial french fries. Some of the values cited for cholesterol oxidation products in french fries may have included contributions from oxidized plant sterols (62). Nourooz-Zadeh and Appelqvist (63) have found certain cholesterol oxides in samples of Swedish lard.

In conclusion, supercritical fluid extraction may be a technique that can be used to reduce the cholesterol content of animal fats. R. Chao, S. Mulvaney and H. Huang (64) have manipulated extraction and separation conditions to disunite small quantities of lipid fractions with markedly enriched cholesterol contents from beef tallow. Fractions containing high cholesterol generally have contained high levels of myristic and palmitic acids and have been low in stearic and oleic acids.

REFERENCES

1. S. Mielke, *J. Am. Oil Chem. Soc.* **64,** 294–298 (1987).
2. U.S. Department of Agriculture, *Agricultural Statistics*, U.S. Government Printing Office, Washington, D.C., 1992, pp. 135–136, 293.
3. L.R. Dugan Jr., "Meat Animal By-Products and Their Utilization, Part I, Meat Fats," in J.F. Price and B.S. Schweigert, eds., *The Science of Meat and Meat Products*, 3rd ed., Food and Nutrition Press, Inc., Westport, CT, 1987, pp.507–529.
4. K.M. Brady, *INFORM* **1,** 547–549 (1990).
5. N.O.V. Sonntag, "Sources, Utilization, and Classification of Oils and Fats," in D. Swern, ed., *Bailey's Industrial Oil and Fat Products*, vol. 2, 4th ed., John Wiley & Sons, Inc., New York, 1979, pp. 271–289.

6. U.S. Department of Agriculture, *Statistical Bulletin 749*, Economic Research Service, U.S. Government Printing Office, Washington, D.C., 1985, p.18.
7. B.S. Kamel, *J. Am. Oil Chem Soc.* **69,** 794–796 (1992).
8. B.F. Haumann, *J. Am. Oil Chem. Soc.* **64,** 789–795 (1987).
9. F.T. Orthoefer, *J. Am. Oil Chem. Soc.* **64,** 795–799 (1987).
10. *Code of Federal Regulations*, title 9, Sec. 319.702., Washington, D.C., 1994.
11. *Codex Alimentarius*, vol. 8, "Fats, Oils and Related Products," part 3. Food and Agriculture Organization of the United Nations World Health Organization, Rome, 1993.
12. R.S. Kirk and R. Sawyer, *Pearson's Composition and Analysis of Foods*, 9th ed., Langman Scientific and Technical, Essex, U.K., 1991, pp. 611–612.
13. M.L. Meara, "Problems of Fats in the Food Industry," in R.J. Hamilton and A. Bhati, eds., *Fats and Oils: Chemistry and Technology*, Applied Science, London, 1980, p.196.
14. G. Pietroszek, *The National Provisioner*, 8–19 (June 27, 1987).
15. N.O.V. Sonntag, "Composition and Characteristics of Individual Fats and Oils," in D. Swern, ed., *Bailey's Industrial Oil and Fat Products*, vol. 2, 4th ed., John Wiley & Sons, Inc., New York, 1979, pp. 289–477.
16. H.B.W. Patterson, *Bleaching and Purifying Fats and Oils: Theory and Practice*, AOCS Press, Champaign, Ill., 1992.
17. *INFORM* **1,** 638–644 (1990).
18. A. Rozenaal, *INFORM* **3,** 1232–1237 (1992).
19. B.F. Haumann, *INFORM* **5,** 668–678 (1994).
20. B. Sreenivasan, *J. Am. Oil Chem. Soc.* **55,** 796–805 (1978).
21. D. Chobanov and R. Chobanova, *J. Am. Oil Chem. Soc.* **54,** 47–50 (1977).
22. Y.C. Lo and A.P. Handel, *J. Am. Oil Chem. Soc.* **60,** 815–818 (1983).
23. P. Forssell and co-workers, *J. Am. Oil Chem. Soc.* **69,** 126–129 (1992).
24. D.A. Glassner, E.A. Grulke, and J.I. Gray, *J. Am. Oil Chem. Soc.* **61,** 1919–1924 (1984).
25. A. Tirtiaux, *J. Am. Oil Chem. Soc.* **60,** 473 (1983).
26. U.S. Pat. 3,944,585 (1976), F.E. Luddy, J.W. Hampson, S.F. Herb and H.L. Rothbart (to the U.S. Secretary of Agriculture).
27. U.S. Pat. 4,072,766 (1978), F.E. Luddy, J.W. Hampson, S.F. Herb and H.L. Rothbart (to the U.S. Secretary of Agriculture).
28. U.S. Pat. 4,049,839 (1977), F.E. Luddy, J.W. Hampson, and S.F. Herb (to the U.S. Secretary of Agriculture).
29. M.A. Grompone, *J. Am. Oil Chem. Soc.* **66,** 253–255 (1989).
30. M.A. Grompone, *J. Am. Oil Chem. Soc.* **67,** 980 (1990).
31. H.H. Taylor, F.E. Luddy, J.W. Hampson, and H.L. Rothbart, *J. Am. Oil Chem. Soc.*, **53,** 491 (1976).
32. C.L. DeFouw, M.E. Zabik, and J.I. Gray, *J. Food Sci.* **46,** 452–456 (1981).
33. L.D. St. John, and co-workers, *J. Anim. Sci.* **64,** 1441–1447 (1987).
34. F.J. Monahan, D.J. Buckley, P.A. Morrissey, P.B. Lynch and J.I. Gray, *Meat Sci.* **31,** 229–241 (1992).
35. C.A. Morgan, R.C. Noble, M. Cocchi, and R. McCartnery, *J. Sci. Food Agric.* **58,** 357–368 (1992).
36. D.K. Larick, B.E. Turner, W.P. Schoenherr, M.T. Coffey and D.H. Pilkington, *J. Anim. Sci.* **70,** 1397–1403 (1992).
37. A.M. Pearson, J.D. Love and F.B. Shoreland, "Warmed-Over Flavor in Meat, Poulty, and Fish," in C.O. Chichester, E.M. Mrak, and G.F. Stewart, eds., *Advances in Food Research*, vol. 23, Academic Press, Inc., New York, 1977, pp. 2–74.

38. I.W. McDonald and T.W. Scott, *World Rev. Nutr. Diet.* **26,** 144–207 (1977).

39. D.C. Rule and D.C. Beitz, *J. Am. Oil Chem. Soc.* **63,** 1429–1436 (1986).

40. J.D. Wood, "Fat Deposition and the Quality of Fat Tissue in Meat Animals," in J. Wiseman, ed., *Fats in Animal Nutrition*, Butterworth & Co. Ltd., London, 1984, pp. 407–435.

41. J.A. Marchello, D.A. Cramer and L.J. Miller, *J. Anim. Sci.* **26,** 294–297 (1967).

42. W.W. Christie, "The Positional Distribution of Fatty Acids in Triglycerides," in R.J. Hamilton and J.B. Russell, eds., *Analysis of Fats and Oils*, Elsevier Science Publishing Co., Inc., New York, 1986.

43. J.K. Punwar and P.H. Dearse, *J. Assoc. Off. Anal. Chem.* **61,** 727–730 (1978).

44. R.S. Parker, "Dietary and Biochemical Aspects of Vitamin E," in J.E. Kinsella, ed., *Advances in Food Research*, vol. 33, Academic Press, Inc., New York, 1989, pp. 157–232.

45. H. Kanematsu and co-workers, *J. Jpn. Oil Chem. Soc.* (*Yukaguba*) **32,** 51–53 (1983).

46. Consumer and Food Economics Institute, *Composition of Foods, Fats and Oils: Raw, Processed, Prepared,* Agriculture Handbook 8-4, U.S. Department of Agriculture, 1979.

47. J. Griewahn and B.F. Daubert, *J. Am. Oil Chem. Soc.* **25,** 26–27 (1948).

48. W.M. Cort, *J. Am. Oil Chem. Soc.* **51,** 321–325 (1974).

49. L.R. Dugan, Jr. and H.R. Kraybill, *J. Am. Oil Chem. Soc.* **33,** 527–528 (1956).

50. E.R. Sherwin and J.W. Thompson, *Food Technol.* **21,** 912–916 (1967).

51. C.M. Houlihan and C-T. Ho, "Natural Antioxidants," in D.B. Min and T.H. Smouse, eds., *Flavor Chemistry of Fats and Oils*, AOCS Press, Champaign, IL, 1985, pp. 117–143.

52. C. Banias, V. Oreopoulou, and C.D. Thanopoulos, *J. Am. Oil Chem. Soc.* **69,** 520–524 (1992).

53. L. deMan, J.M. deMan, and B. Blackman, *J. Am. Oil Chem. Soc.* **68,** 63–69 (1991).

54. L. deMan, V. D'Souza, J. M. deMan, and B. Blackman, *J. Am. Oil Chem. Soc.* **69,** 246–250 (1992).

55. J.K. Ha and R.C. Lindsay, *J. Am. Oil Chem. Soc.* **68,** 294–298 (1991).

56. J.A. Merkle and D.K. Larick, *J. Food Sci.* 59, 478–483 (1994).

57. K.W. Um, M.E. Bailey, A.D. Clark, and R.R. Chao, Paper No. 185, presented at the *1992 Annual Meeting of the Inst. of Food Technologists*, New Orleans, LA, 1992.

58. S.W. Park and P.B. Addis, *J. Agric. Food Chem.* **34,** 653–659 (1986).

59. S.W. Park and P.B. Addis, *J. Food Sci.* **51,** 1380–1381 (1986).

60. T.C. Ryan and J.I. Gray, *J. Food Sci.* **49,** 1390–1391, 1393 (1984).

61. S.W. Park and P.B. Addis, *J. Food Sci.* **50,** 1437–1441, 1444 (1985).

62. K. Lee, A.M. Herion, and N.A. Higley, *J. Food Protect.* **48,** 158–161 (1985).

63. J. Nourooz-Zadeh and L.A. Appelqvist, *J. Am. Oil Chem. Soc.* **66,** 586–592 (1985).

64. R.R. Chao, S.J. Mulvaney, and H. Huang, *J. Am. Oil Chem. Soc.* **70,** 139–143 (1993).

2

Vegetable Oils

INTRODUCTION

Vegetable oils are neither animal or marine oils. The largest source of vegetable oils are annual plants such as soybeans, corn, cottonseed or peanuts. However, other sources are oil-bearing perennials such as the palm, olive, or coconut.

Some 250,000 plant species are known with 4,500 species being examined for oil. Few annual plants are cultivated only for the oil, the castor plant being the exception. Soybean or canola are generally utilized for high protein feed production with the oil as a coproduct. The annual world production of vegetable oils for consumption (Table 2.1) is in excess of 59 million metric tons compared to the total world production for all fats and oils of 72.6 million metric tons for 1991–1992 (1).

In the United States in 1993, the total domestic disappearance of edible vegetable oils was 20,867 billion pounds. The per capita use of all fats and oils including animal fats was 81.4 pounds per person with soybean being the dominant oil at 51.3 pounds per person (1). Most oil consumption is used in salad and cooking oil applications, accounting for about 44% of the total soybean oil consumed. Other uses include baking, frying, and margarine applications.

THE CHANGING CONSUMER

Vegetable oil consumption has been increasing and currently surpasses animal fat consumption. This trend represents a change in the fatty acid composition of the diet. Animal fats were less than 10% of the total domestic disappearance

Table 2.1 World production of fats and oils for 1991–1992 (2)

Fat/Oil	Metric Tons
Soybean	16.80
Cottonseed	4.17
Peanut	3.41
Sunflowerseed	7.61
Rapeseed	9.11
Olive	2.14
Coconut	2.91
Palm kernel	1.49
Palm	11.49
Total	59.14

in 1993 and about 20% of the world production of fats and oils (2). With the change in consumption that has occurred over the past 50 years, dietary linoleic acid utilization has increased the most and saturated fatty acids consumption has decreased.

The current recommendations by the American Heart Association include decreasing fat consumption to 30% of daily calories and decreasing saturated fatty acids to less than 10% of calories (3). Food manufacturers have responded. The composition of margarines and shortenings has been modified by formulating with lipid components that are less hydrogenated than previously and meet the proposed guidelines.

In formulating food products with vegetable oils, there is often a need to partially hydrogenate the oil to impart functional characteristics and/or to improve the oxidative stability. Hydrogenation, however, results in the formation of fatty acid isomers referred to as trans fatty acids. Clinical studies with trans fatty acids have shown an adverse effect on serum lipids; however, the biological significance is unknown (4–7).

THE INDUSTRIAL CHALLENGE

The changing consumer has brought new challenges to the oils industry, from the producer to the refiner and to the formulator. Major business reorganizations, structural issues, and new competitors are evolving (8). The food industry is challenged by demands for profitability, revenue, new product successes, and international expansion. Food fats and oil businesses have similar challenges (9).

This re-definition has been described as a by-product of: the changing taste of the consumer; the aging population; the change in value consciousness of more astute, demanding customers; and the expanded nutritional awareness and consolidation of forces into fewer, more powerful trade channels.

Successful strategies by food ingredient producers have involved leveraging of brand identity by new product introductions, cost reductions, acquisitions, and consolidations. Also, aggressive international expansions are being pursued.

In addition, biotechnology is being applied to edible fats and oils (10,11). Traditional breeding has resulted in the development of high oleic safflower. It is, perhaps, the first commercial success in oil modification using biotechnology. High oleic sunflower with nearly 90% oleic acid in its triglyceride structure represents new technology used in both specialty food and industrial applications. Successful modification of rapeseed oil to the low erucic acid canola has also been a commercial success. More recent attempts with rapeseed include elevated erucic acid levels, lauric acid production by gene transfer, and even lower (< 6%) saturated fatty acid content rapeseed or canola oil (12,13).

Genetic and mutation modifications of soybeans have produced low linolenic acid oil (about 3.0%), high oleic acid oils, and high saturated fatty acid (> 30%) oils (14). Success in these modified oils requires trait stability and identity preservation through production, storage, extraction, and refining. Costs involved in identity preservation have been a major factor in bringing the modified oils to market. Most breeders, seed companies, and oil producers are reluctant to bring products to the marketplace. In some instances, there is an inability to demonstrate a significantly functional difference from the traditional product. In this case, there is little or no market interest.

OIL SYNTHESIS IN PLANTS

The interest in fatty acid synthesis and subsequent triglyceride synthesis in plants has increased because of the promise of biotechnology to modify oils. It may become possible to manipulate precisely fatty acid synthesis and position of the acyl groups on the glycerol backbone producing more valuable, desirable, and functional commodities (15). Several examples have been presented of oil modification accomplished by selective breeding, mutation breeding, and recombinant DNA technology.

4.1 Oils in Seed Tissues

Oils accumulate in seed tissues in discreet, about 1.0 μm, oil bodies (16). The deposition follows defined kinetics depending upon the plant. Safflower cotyledons, for example, after a lag of 12 to 14 days after pollination, actively accumulate triacylglycerides for 8 to 10 more days. Almost 75% of the oil is laid down from 16 to 20 days after pollination. Each species varies in the time period for oil deposition and is influenced by plant hormones, factors affecting photosynthesis, cell division and expansion, and genetics of the tissue during development.

Microsomal membranes during the early phase of oil deposition are the most active at synthesizing triacylglycerols (17). Preparations from the developing cotyledons contain all the necessary enzymes for the acylation of glycerol phosphate and assembly of the triacylglycerols.

4.2 Fatty Acid Distribution

In many species, the fatty acids are arranged in a non-random distribution, as shown in Figure 2.1, on the triacylglycerols (16). The more saturated fatty acids dominate position one. Monounsaturated and polyunsaturated fatty acids are dominant at position two as shown for Borage oil (Table 2.2). The fatty acids on position three have a more general mix although in some plants a particular fatty acid may dominate. Assymetry in fatty acid distribution, at least in phosphatidylcholine, exists with the saturated and unsaturated fatty acids species at position *sn*-1 with only the 18 carbon unsaturated fatty acids occurring at position two. Most oil seed microsomal preparations have proved active in acylation of *sn*-glycerol 3-phosphate yielding phosphatidic acid. It is the enzymes involved in acylation of *sn*-glycerol 3-phosphate that control the symmetrical distribution of saturated and unsaturated acyl species. A phosphatidase catalyzes the cleavage of phosphatidic acids yielding a diacylglycerol. This phosphatidase likely plays a regulatory role in triacylglycerol synthesis.

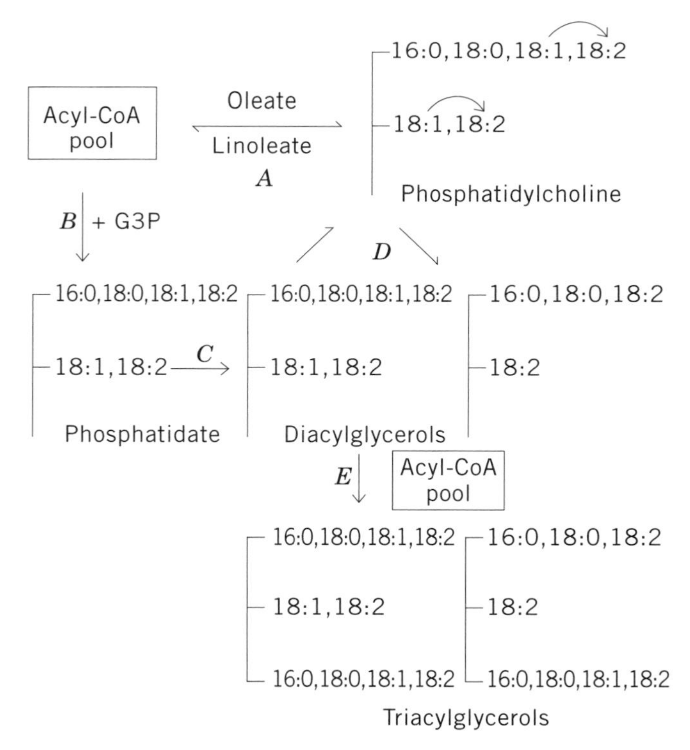

Figure 2.1 Synthesis of linoleate-rich triacylglycerols. *A*, acyl-CoA: lysophosphatidylcholine acyltransferase; *B*, glycerophosphate acyltransferases; *C*, phosphatidate phosphohydrolase; *D*, cholinephosphotransferase; *E*, diacylglycerol acyltransferase (16).

Table 2.2 Sterospecific distribution of fatty acids in triacylglycerols of* Borago officinalis *(16)

	Fatty acid (mol %)					
sn-carbon	16:0	18:0	18:1	18:2	γ18:3	20:1
sn-1	23	6	16	51	0	4
sn-2	0	0	17	39	44	0
sn-3	5	0	7	31	53	4

The final step in triacylglycerol synthesis is the acylation of the diacylglycerol at *sn*-3. This process is catalyzed by diacylglycerol acyl transferase with some likely specificity of diacylglycerol acyl transferase.

4.3 Fatty Acid Synthesis

Acetyl CoA is the starting point for fatty acid biosynthesis (18). Acetyl CoA is the product of photosynthetic fixation of CO_2 although carbohydrates are utilized as the carbon source in developing seeds (Figure 2.2) Pyruvate may

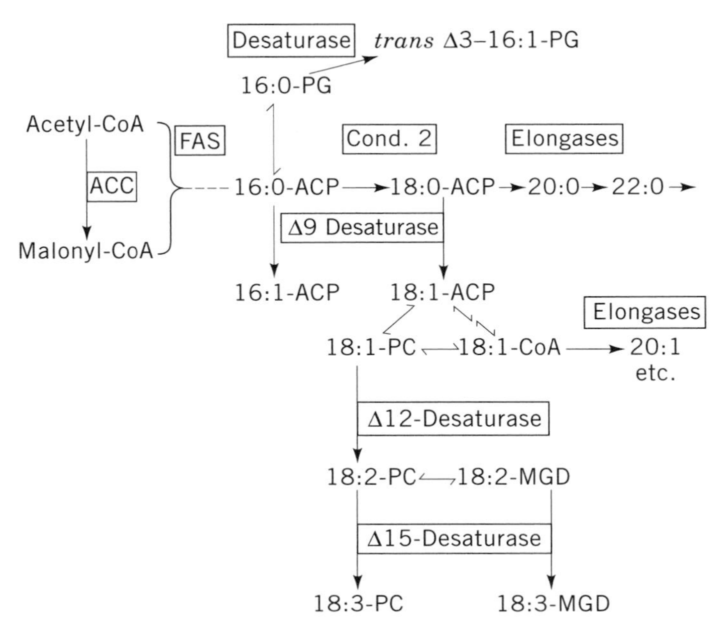

Figure 2.2 Major pathways of fatty acid synthesis in plants. ACC, Acetyl-CoA Carboxylase; FAS, fatty acid synthetase; Cond. 2, β-ketoacyl-ACP II. The elongases for saturated fatty acid synthesis work mainly with acyl-CoA substrates. Δ15-Desaturase activity uses mainly phosphatidylcholine (PC) as substrate in developing seeds and mainly monogalactosyldiacylglycerol (MGD) in leaf tissues (18).

also be the source of acetyl CoA through the pyruvate dehydrogenase/decarboxylase complex. Acetyl CoA and bicarbonate through ATP dependent acetyl CoA carboxylase results in the formation of malonyl CoA.

4.4 Desaturation and Synthesis

Oleate esterified to phosphatidylcholine is the base for desaturation to linoleate. The linoleate is enzymatically returned to the Acetyl CoA pool by acyl transferase. A 12 desaturase also operates at position one of *sn* phosphatidylcholine. For linoleic acid found in *Borago officinales,* a 6 desaturase specific for linoleate at *sn*-2 of phosphatidylcholine occurs.

4.5 Triacylglycerol Formation

The basic fatty acids (oleate, palmitate, stearate) are synthesized in the plastid. Saturated fatty acids are utilized in the acylation of *sn*-1 of *sn* glycerol-3-

Table 2.3 Typical compositions and chemical constants of common edible fats and oils (30)

	Butyric	Caproic	Caprylic	Capric	Undecanoic	Lauric	Tridecanoic	Myristic	Myristoleic	Pentadecanoic	Pentadecenoic	Palmitic	Palmitoleic	Margaric
Carbon atom: double bonds	4:0	6:0	8:0	10:0	11:0	12:0	13:0	14:0	14:1	15:0	15:1	16:0	16:1	17:0
Babassu		0.4	5.3	5.9		44.2		15.8				8.6		
Butterfat	3.8	2.3	1.1	2.0	0.1	3.1	0.1	11.7	0.8	1.6		26.2	1.9	0.7
Canola oil												3.9	0.2	
Chicken fat						0.2		1.3	0.2			23.2	6.5	0.3
Citrus seed oil						0.1		0.5				28.4	0.2	
Cocoa butter								0.1				25.8	0.3	
Coconut oil		0.5	8.0	6.4		48.5		17.6				8.4		
Cohune oil		0.3	8.7	7.2	0.1	47.3		16.2				7.7		
Corn oil												12.2	0.1	
Cottonseed oil								0.9				24.7	0.7	0.1
Lard				0.1		0.1		1.5		0.2		24.8	3.1	0.5
Linseed oil												4.8		
Murumuru tallow		0.1	1.3	1.5		46.2		32.4				5.6	0.1	
Oat oil								0.2				17.1	0.5	
Olive oil												13.7	1.2	
Palm oil						0.3		1.1				45.1	0.1	
Palm kernel oil		0.3	3.9	4.0		49.6		16.0				8.0		
Peanut oil								0.1				11.6	0.2	0.1
Rapeseed oil								0.1				2.8	0.2	
Rice bran oil			0.1	0.1		0.4		0.5				16.4	0.3	
Safflower oil								0.1				6.5		
Safflower oil (high oleic)								0.1				5.5	0.1	
Sesame oil												9.9	0.3	
Soybean oil								0.1				11.0	0.1	
Sunflower oil						0.5		0.2				6.8	0.1	
Tallow (beef)				0.1		0.1		3.3	0.2	1.3	0.2	25.5	3.4	1.5
Tallow (mutton)				0.2		0.3		5.2	0.3	0.8	0.3	23.6	2.5	2.0
Tucum oil		0.2	2.9	2.3		51.8		22.0				6.8		

phosphate. The acylation of *sn*-2 has a strong preference for linoleate. Oleate is transferred to position two of phosphatidylcholine for desaturation and the polyunsaturates are returned to the CoA pool. Phosphatidate is hydrolyzed to diacylglycerol and can interconvert with phosphatidylcholine during triacylglycerol formation. Diacylglycerol is acylated at *sn*-3 to yield the triacylglycerol.

CLASSIFICATION OF FATS AND OILS

Traditionally, fats and oils have been classified as to their animal or vegetable origin. Vegetable oils are further classified by iodine value or number into drying, semidrying, and nondrying oils. The conventional classification results in 10 groups that are based on six vegetable oils for edible products, a conjugated acid group, and a hydroxy acid group. The fatty acid composition of the major vegetable oils is shown in Table 2.3.

Table 2.3 *(Continued)*

Margaroleic	Stearic	Oleic	Linoleic	Linolenic	Nonadecanoic	Arachidic	Gadoleic	Eicosadienoic	Arachidonic	Behenic	Erucic	Docosadienoic	Lignoceric	Iodine Value	Saponification Value
17:1	18:0	18:1	18:2	18:3	19:0	20:0	20:1	20:2	20:4	22:0	22:1	22:2	24:0		
	2.9	15.1	1.7			0.1								13–18	247–254
0.2	12.5	28.2	2.9	0.5			0.2		0.1					25–42	210–240
	1.9	64.1	18.7	9.2		0.6	1.0			0.2			0.2	110–115	
0.1	6.4	41.6	18.9	1.3										76–80	194–204
	3.5	23.0	37.8	5.7		0.8								99–106	192–197
	34.5	35.3	2.9			1.1								32–40	190–200
	2.5	6.5	1.5			0.1								7–13	248–264
	3.2	8.3	1.0											8–14	250–260
	2.2	27.5	57.0	0.9		0.1								110–128	186–196
	2.3	17.6	53.3	0.3		0.1								99–121	189–199
0.3	12.3	45.1	9.9	0.1		0.2	1.3	0.1	0.4					53–68	192–203
	4.7	19.9	15.9	52.7											
	2.2	8.9	1.5			0.2								8–13	237–247
	1.4	33.4	44.8			0.2	2.4							105–110	180–198
	2.5	71.1	10.0	0.6		0.9								76–90	188–196
	4.7	38.8	9.4	0.3		0.2								45–56	195–205
	2.4	13.7	2.0			0.1								14–24	243–255
	3.1	46.5	31.4			1.5	1.4	0.1		3.0			1.0	84–102	188–196
	1.3	23.8	14.6	7.3		0.7	12.1	0.6		0.4	34.8	0.3	1.0	97–110	168–183
	2.1	43.8	34.0	1.1		0.5	0.4			0.2			0.1	92–109	181–195
	2.4	13.1	77.7			0.2								138–151	186–198
	2.2	79.7	12.0	0.2		0.2								85–93	185–195
	5.2	41.2	43.3	0.2										104–118	187–196
	4.0	23.4	53.2	7.8		0.3				0.1				125–138	188–195
	4.7	18.6	68.2	0.5		0.4								122–139	186–196
0.7	21.6	38.7	2.2	0.6	0.1	0.1			0.4					33–50	190–202
0.5	24.5	33.3	4.0	1.3	0.8				0.4					35–46	192–198
	2.3	9.3	2.4											10–14	240–250

The main vegetable oil categories are (19):

Principal Fatty Acid	Oil Source
Lauric	Coconut oil
	Palm kernel oil
	Babassu oil
Palmitic	Palm
Oleic	Olive
	Canola
	Peanut
	High oleic sunflower, safflower
Linoleic (medium)	Corn
	Cottonseed
	Sesame
	Soybean
Linoleic (high)	Sunflower
	Safflower
Erucic	Rapeseed

A nonconventional classification of vegetable oils by characteristic fatty acid composition is more useful in today's market. This classification consists of:

Characteristic Fatty Acid

- Lauric
- Palmitic
- Stearic
- Oleic/linoleic
  - Delta linoleic
  - Gamma linoleic
- Linolenic
- Erucic
- Confectionery
- Specialty oils
  - Conjugated acid oils
  - Hydroxy acid oils

Classification by the characteristic fatty acids that are present permits inclusion of the genetically modified oils where the characteristic fatty acid may be altered. An example would be regular sunflower oil versus high oleic sunflower oil.

The largest source for oils are the annual plants, e.g., corn, soybean, peanut, rapeseed, and sunflower (Table 2.4). These oils are produced in temperate climatic regions (20). The perennial oil bearing trees (coconut, palm, olive, tung) are grown in the tropical regions. The production of the major oils by

Table 2.4 World oil production (33) (in 1,000 metric tons)

Oil	1991–1992
Soybean	16,281.5
Palm	11,879.5
Rapeseed	9,420.0
Sunflower	7,732.2
Cottonseed	4,221.9
Peanut	3,986.3
Coconut	2,848.9
Olive	2,156.0
Corn	1,547.7
Palm kernel	1,528.1

region is shown in Table 2.5. In the case of palm and olive, the oil is obtained from both the seed and the pulp. Both coconut and oil palm produce more oil per unit of land than the annual oilseed crops.

Major differences in the flexibility of production and labor required occurs between tree oils and annual plants. With annual plants, oil production is

Table 2.5 Top 1991–1992 oil producing nations (33) (in 1,000 metric tons; 1989–1990 total in parentheses)

Country	Units		Country	Units		Country	Units	
Soybean oil			Palm oil			Rapeseed/canola oil		
United States	6,325	(5,898)	Malaysia	6,560	(6,095)	China	2,235	(1,831)
Brazil	2,620	(2,974)	Indonesia	3,000	(2,413)	India	1,889	(1,331)
Argentina	1,288	(1,080)	Nigeria	630	(580)	Germany	1,180	(1,042)
Japan	653	(680)	Colombia	292	(226)	Japan	803	(763)
Netherlands	617	(609)	Ivory Coast	280	(270)	Canada	647	(488)
Germany	482	(477)	Thailand	270	(226)	United Kingdom	486	(420)
China	458	(497)	Papua/			France	443	(412)
Taiwan	425	(407)	New Guinea	206	(138)	Belgium	311	(263)
Spain	382	(418)	Ecuador	147	(120)	Poland	258	(316)
India	344	(246)	Cameroon	107	(108)	Netherlands	152	(140)
			Zaire	105	(101)			
Sunflowerseed oil			Cottonseed oil			Peanut oil		
Argentina	1,320	(1,458)	China	1,112	(801)	India	1,580	(1,850)
CIS	1,952	(2,199)	CIS	662	(708)	China	1,331	(1,199)
France	557	(422)	United States	623	(438)	Nigeria	167	(123)
Spain	408	(398)	Pakistan	362	(300)	Senegal	152	(199)
United States	364	(217)	India	314	(371)	United States	120	(75)
Hungary	298	(287)	Brazil	186	(154)	Myanmar		
China	283	(219)	Turkey	132	(127)	(Burma)	114	(101)
India	283	(182)	Argentina	75	(62)	Argentina	91	(27)
Turkey	276	(407)	Egypt	66	(67)	Mali	72	(24)
Romania	206	(223)	Australia	58	(42)	Sudan	72	(77)
			Paraguay	58	(54)			

flexible. With the tree-derived oils, production is only semi-flexible with limited manipulation of the annual production. The coconut palm begins to bear fruit in 5–6 years and continues production for 60 years. The oil palm starts to bear fruit in 4–5 years developing a full yield in 15 years with continued production for 25 years. Olive trees, on the other hand, may produce for 200 years. The consumption of tree-derived oils for the United States was less than 300 million pounds in 1992–1993. Worldwide, tree oils represent a much larger share of oil consumption.

The labor required for annual plant versus tree oil production differs. Annual plants allow for mechanical harvesting and processing, particularly in mechanized societies. Tree crops, on the other hand, are not as easily mechanized for harvesting and handling resulting in labor intensive industries.

Soybean oil production, a by-product of soy protein meal, was previously projected to increase rapidly with the increased demand for soybean protein. The projections for the large increase in vegetable protein consumption of the 1970s did not occur. Palm oil production, mostly in Malaysia, has increased dramatically. Significant increases in canola or low erucic rapeseed oil has occurred in Canada but is, worldwide, less of a significant supply.

COMMERCIAL UTILIZATION OF FATS AND OILS

Previously, shortenings were the most consumed oil product in the U.S.; however, now salad and cooking oils are the main products consumed (Table 2.6). The U.S. consumer has become more health conscious with the promotion of polyunsaturated and monounsaturated fatty acid oils. In Europe, margarine is the major product made out of oil.

The type of oils consumed also differs by region of the world. Olive oil is preferred in the Mediterranean. Soybean, corn, and cottonseed are dominant in the U.S. Peanut or ground nut is the major oil for parts of Europe. In Eastern Europe and the Orient, rapeseed, sesame and sunflower oils are the major oils consumed. In Western Europe, rapeseed and low erucic acid rapeseed are the most important oilseed crops. Both high and low erucic rapeseed oils are produced for food and nonfood applications. *Crambe abyssinica*, as a source of erucic and behenic acid, may be used as a substitute for rapeseed in North America.

Table 2.6 Fats and oils used in edible products, by use, United States, 1986/87–1992/93 (1)[a]

	1986/87	1987/88	1988/89	1989/90	1990/91	1991/92	1992/93
Baking or frying fats	5,275	5,430	5,319	5,724	5,793	5,702	5,960
Margarine	1,973	1,909	1,865	2,117	2,168	2,173	2,186
Salad or cooking oil	6,351	6,527	6,205	6,262	6,222	6,536	6,374
Other edible	348	310	330	290	368	351	439
Total edible	13,947	14,175	13,719	14,383	14,491	14,765	14,959

[a] Million pounds.

Soybean and cottonseed oils are the dominant oils in China. In South America, it is sunflower oil. Also sunflower oil had been dominant in the former Soviet Union.

Much of the vegetable oil for Europe originates in Africa, India, China, Manchuria, and East India. Palm oil is a major oil because of price considerations. The Philippines are the source of coconut oil. South America also exports oils. Brazil is a major exporter of castor bean, babassu, oiticica oil and soybean oil. Argentina exports flaxseed and sunflower.

The U.S. is self-sufficient in vegetable oils except for coconut and tung oils. Oils such as palm and canola are imported depending upon market conditions.

CONSUMPTION

The consumption patterns of fats and oils continue to change. In the U.S., butter in the diet has decreased in favor of margarine. Animal fats have decreased in favor of vegetable oils. Consumption of vegetable oils in margarines and salad dressings and in cooking and frying have increased. Further increases have occurred in convenience and snack food consumption and in fast food preparation.

In nonfood products, vegetable oil derivatives continue to be utilized at approximately six billion pounds per year (Table 2.7). Substitution of fatty acid soaps in bar soaps to detergents has not occurred. Drying oil utilization in protective coatings has continued to decrease.

CHARACTERISTICS OF OIL CLASSES

8.1. Lauric Acid Group

The lauric acid group mainly is made up of oils from the seeds of palms; specifically, coconut, babassu, and palm kernel oils. The lauric acid content ranges from about 40 to 50%. Oils from the lauric acid group are higher in short chain fatty acids having 8, 10, and 14 carbon chain lengths fatty acids (Table 2.8). Unsaturated fatty acids present are mainly oleic and linoleic acid. The long chain saturated fatty acids are palmitic and stearic acids. These are present in relatively minor quantities. The high content of short chain fatty acids combined with the high degree of saturation imparts both high solids and a low melting point. The lauric acid oils are very stable due to the high content of saturated fatty acids. Combined with the desirable melting points, they are very adaptable to soap manufacture, giving preferred solubilities and lathering characteristics.

The lauric acid oils are the primary source of short chain fatty acids for production of fatty acid derivatives. These include the C-6, C-8, C-10, and C-12 acids produced by fractional distillation of hydrogenated coconut oil.

Table 2.7 Total fats and oils consumption, with inedible by category, United States, 1985/86–1992 (1)[a]

Year 1/	Total Consumption	Total Edible	Total Inedible	Soap	Paint or Varnish	Feed	Resins and Plastics	Lubricants 2/	Fatty Acids	Other Products
1985/86	19,249.6	13,972.6	5,277.0	754.9	228.0	1,705.7	173.4	103.4	1,968.9	342.6
1986/87	d[b]	d	5,990.6	d	d	d	d	d	d	d
1987/88	20,241.2	14,175.5	6,065.7	868.6	179.1	1,967.6	196.3	107.8	2,203.8	542.8
1988/89	19,426.7	13,542.0	5,884.7	744.5	180.3	2,079.3	202.3	115.8	2,074.1	488.4
1989/90	20,036.0	14,382.7	5,653.3	792.0	89.5	2,143.5	222.4	157.1	1,944.7	304.1
1991	20,332.1	14,613.0	5,719.1	832.9	106.8	1,974.0	182.6	101.7	2,234.7	286.4
1992	20,751.7	14,847.3	5,904.4	738.8	123.8	2,176.5	165.5	109.4	2,041.2	549.3

[a] Million pounds.
[b] d = Data withheld to avoid disclosing figures for individual companies.

Table 2.8 Coconut's (CO) and palm kernel oils' (PKO) typical fatty acid composition (34)

	Percent by weight	
Fatty acid	CO	PKO
Saturated		
Caproic	0.5	0.2
Caprylic	8.0	4.0
Capric	7.0	3.9
Lauric	48.0	50.4
Myristic	17.0	17.3
Palmitic	9.0	7.9
Stearic	2.0	2.3
Unsaturated		
Palmitoleic	0.2	
Oleic	6.0	11.8
Linoleic	2.3	2.1

Nonfood uses, particularly cleaning products and chemically derived emulsifiers, account for the major uses of this imported oil.

The food uses of the lauric oils captures (1) the resistance to oxidation and (2) desirable melting characteristics. Fully hydrogenated coconut oil with a melting point of 33° C (92° F) compared to the native 24° C (76° F) melting point has the required stability for nut roasting or applications to cereal products requiring long shelf lives. The melting characteristics of fully hydrogenated coconut oil with added hardfats produced from palm oil or cottonseed oil results in the desired melting suitable for use in confectionery butters and butterfat. Rearranged products, interesterified products, and fractionated lauric acid oils are commercially produced for compound coatings.

Coconut oil is the major oil in this group. It is obtained from Copra or dried coconut meat. Copra may contain from 65 to 68% oil. Coconut has one of the highest yields per acre of all oil producing plants and is higher than the annual oilseed crops. Most is produced in the Philippines (Table 2.9). The oil is expressed from the Copra using expellers.

Coconut oil uses are divided between edible and inedible products, mostly inedible in the U.S. (Table 2.10). Edible products rely upon high stability, high solids, and sharp melting characteristics. In recent years, the edible uses in the United States have declined. Nonfood uses rely upon the lower melting temperature that is characteristic of lauric acid.

Palm kernel oil, the second most consumed lauric acid group oil in the U.S., is derived from the dried kernels of the oil palm, *Elaeis guineensis.* Malaysia dominates production of palm kernel oil (Table 2.11). It is similar

Table 2.9 Coconut oil production (33) (in 1,000 metric tons)

Country/Region	1990–1991[a]	1991–1992[b]
Philippines	1,313.2	1,062.9
Indonesia	765.4	710.5
India	260.1	258.8
Mexico	108.7	106.6
Vietnam	98.0	103.3
Sri Lanka	61.5	62.8
Papua New Guinea	41.4	41.8
Malaysia	36.7	40.2
Mozambique	39.5	40.1
Ivory Coast	38.1	39.8
Thailand	38.4	37.4
Other countries	346.2	344.7
Total	3,147.2	2,848.9

[a] Estimated. October–September marketing year.
[b] Forecast. October–September marketing year.

to coconut oil but is somewhat less saturated and lower in medium chain length fatty acids (capric and caprylic acids).

The American palm kernel oils are Babassu, Tucum, Murumuru, Ouricuri, and Cohune (Table 2.12). These are generally similar to the African and Malaysian palm kernel oils.

8.2 Palmitic Acid Oils

Palm oil, the principal palmitic acid oil, is obtained from the palm, *Elaesis guineensis*. Two distinctly different oils are obtained from the palm fruit: palm oil from the pulp and palm kernel oil from the kernel (21). The world consumption of palm oil had been 11.49 million metric tons in 1991–1992, second behind soybean oil. The U.S. consumption of palm oil has decreased because of nutritional labeling and popular resistance to consumption of this oil (22). Palm oil contains 32 to 47% palmitic acid and 40 to 52% oleic acid.

Palm oil is produced primarily in Malaysia with nearly 60% of the total production. Other production is in West and Central Africa, Indonesia, and limited quantities in Brazil, Central and South America. Much of the palm oil is exported to Europe, England, and Japan. Palm oil is utilized mainly in shortenings, margarines, and in soap formulations. Overall, palm oil enters the food supply as a component of processed foods.

The palm fruit has an active lipase that hydrolyzes the oil during harvesting and handling resulting in a very high free fatty acid content in the crude oil. Palm oil has a distinctive orange–red color due to the presence of carotene. The major carotenoids in crude palm oil are alpha carotene (186 μg/g) and

Table 2.10 Coconut oil consumption, with inedible by category, United States, 1985/86–1992 (1)[a]

Year 1/	Total Consumption	Total Edible	Total Inedible	Soap	Paint or Varnish	Feed	Resins and Plastics	Lubricants 2/	Fatty Acids	Other Products
1985/86	634.7	332.8	301.9	123.2	d[b]	0.0	d	d	59.7	d
1986/87	858.2	319.4	538.8	216.1	d	0.0	d	d	95.7	d
1987/88	788.6	233.4	555.4	213.8	d	0.0	7.2	d	131.4	d
1988/89	688.8	211.2	477.6	130.6	1.4	d	14.6	d	121.9	206.6
1989/90	525.2	160.5	364.6	156.9	2.1	0.0	9.7	4.0	134.6	57.3
1991	815.6	153.0	662.6	158.0	d	d	2.4	d	426.7	72.8
1992	875.4	176.3	699.1	121.7	d	0.0	3.2	d	d	d

[a] Million pounds.
[b] d = Data withheld to avoid disclosing figures for individual companies.

Table 2.11 Palm kernel oil production (33) (in 1,000 metric tons)

Country/Region	1990–1992[a]	1991–1992[b]
Malaysia	783.1	801.6
Indonesia	258.3	290.7
Nigeria	155.8	172.4
Colombia	24.6	28.7
Ivory Coast	24.3	25.6
Thailand	21.5	24.1
Cameroon	23.1	23.0
Other countries	155.8	162.0
Total	1,446.5	1,528.1

[a] Estimated, October–September marketing year.
[b] Forecast, October–September marketing year.

beta carotene (356 μg/g) (23). Gamma, beta, and delta tocotrienols are also present. The oil is bleached by hydrogenation to the normal yellow color of vegetable oils.

Palm oil has nearly equal concentrations of saturated and unsaturated fatty acid. 85% of the triglycerides of palm contain an unsaturated fatty acid at the 2 position of the glycerol backbone. Palm oil is often available fractionated into olein and stearine fractions.

Table 2.12 Fatty acid composition of several American palm kernel oils (19)

Analysis	Babassu	Tucum	Murumuru	Ouricuri	Cohune
Characteristics					
Iodine number	16.1	15.8	11	14.7	0.8
Saponification number	249	241	242	256.9	251.0
Fatty acids, wt. %					
Caproic	0.2			1.8	
Caprylic	4.8	1.3	1.1	9.8	7.5
Capric	6.6	4.4	1.6	8.2	6.6
Lauric	44.1	48.9	42.5	45.8	46.4
Myristic	15.4	21.6	36.9	9.0	16.1
Palmitic	8.5	6.4	4.6	7.7	9.3
Stearic	2.7	1.7	2.1	2.3	3.3
Arachidic	0.2			0.1	
Total saturated	82.5	84.3	88.8	84.7	89.2
Oleic	16.1	13.2	10.8	13.1	9.9
Linoleic	1.4	2.5	0.4	2.2	0.9
Total unsaturated	17.5	15.7	11.2	15.3	10.8

8.3 Oleic/Linoleic Acid Group

The commercially important oils in the oleic/linoleic acid group are corn, cottonseed, peanut, olive, sunflower, sesame, safflower, sunflower, and rice bran oil. Newer members to this group of oils are high oleic safflower oil and high oleic sunflower oil. Others that are less commercially important are tobaccoseed, poppyseed, teaseed, kapok, sorghum, almond, apricot, pecan, grapeseed, and tomatoseed oils. Selected fatty acid compositions are shown in Table 2.13.

The oleic/linoleic group consists primarily of oils containing mainly unsaturated fatty acids. The saturated fatty acid content is generally less than 20%, with the highly unsaturated fatty acids and trisaturated triglycerides being almost entirely absent from this group. Both annual plants and tree oils are included.

This oleic/linoleic group is the most widely used and adaptable of all fats and oils. They are considered the premium oils since they have desirable antioxidant properties and do not undergo flavor reversion. They may be hydrogenated to plastic fats with varying degrees of hardness.

Industrially, the oleic/linoleic oils may be used for soap manufacture. The native oils are generally too unsaturated for the preferred degree of hardness and stability. The hydrogenated forms will form suitable hard fats for soaps. Blends may also be used to obtain the desired degree of hardness.

Corn oil is a by-product of either dry milling or wet milling of corn, *Zea Mays*. Historically, corn oil has been in short supply. The current supply of corn oil has increased due to the increased demand for corn sweeteners. The oil is recovered from the corn germ. Annual production of the oil in the United States is more than 1 billion pounds. Its image is that of being more healthful than alternative oils, particularly animal fats or saturated fats. Corn oil is a premium cooking and frying oil.

Some 80% of the fatty acids of corn oil consist of oleic and linoleic acid. Less than 1% is linolenic acid although its fatty acid composition depends on variety and locations. The oil has a high tocopherol content which, along with its fatty acid content, contributes to the oxidative stability. Ferulic acid has been detected in corn oil as the dihydro-beta-sitosterol ester. This same phenolic ester may also account for the oxidative stability of rice bran oil.

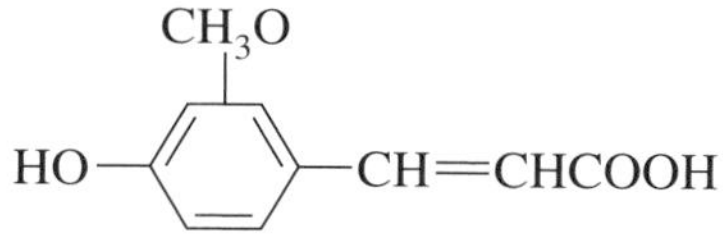

CH_3O

HO—C_6H_3—CH=CHCOOH

Ferulic acid

Cottonseed oil, a by-product of cotton fiber production, was the first edible oil produced in the United States (24). It is the third most consumed vegetable oil in the U.S. behind soybean and corn oils. Crude cottonseed oil contains

Table 2.13 Major fatty acids of selected oils

Fatty acid	Tobaccoseed[a]	Poppyseed[a]	Teaseed[a]	Tomatoseed[b]	Grapeseed[b]	Almond[b]	Apricot[b]	Pecan[b]	Sorghum[b]
18:1	12.0–15.9	28.3	74–87	34.1	15.6	44.5	62.0	66.9	24
18:2	73.1–77.3	58.5	7–14	27.1	72.2	40.5	28.1	22.1	59
Total saturated	9.6–12.2	7.2	7–11						

[a] From Swern, 1979 (19).
[b] From Chow, 1992 (32).

almost 2% of nonglyceride materials consisting of gossypol, phosphatides, sterols, resins, and pigments. Gossypol is most notable, consisting of complex phenolic substances that are toxic. Cottonseed oil production has lost ground due to the increased value of the whole seed in formulated feeds. Cottonseed oil is used almost entirely for edible applications in baking, shortening, salad oils, frying, and margarine. The oil contains small quantities, up to 0.5%, of cyclopropenoid fatty acids, giving rise to the characteristic Halpen test. These fatty acids are sterculic and malvalic acids. Their structures consist of:

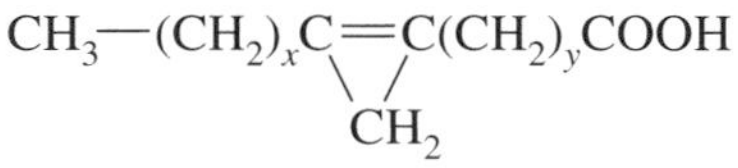

$$CH_3{-}(CH_2)_xC{=}C(CH_2)_yCOOH \quad \text{(ring: } C{=}C \text{ bridged by } CH_2\text{)}$$

Malvalic acid
$x + y = 13$

$$CH_3(CH_2)_5CH{-}CHCH{=}CH(CH_2)_7COOH \quad \text{(ring: } CH{-}CH \text{ bridged by } CH_2\text{)}$$

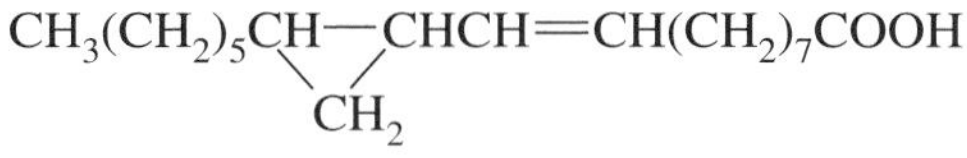

Sterculic acid

Peanut oil from the plant *Arachis hypogaea*, also known as ground nut or arachis oils, is an important world vegetable oil grown in large quantities in Africa, India, and China. World production of peanut oil in 1988 was 2.93 million metric tons (25). Peanut oil is used mainly for edible purposes in blended frying shortenings, cooking, and as a frying oil. Aflatoxin, a mycotoxin developed in moldy peanuts, does not carry through to the refined oil. Peanut oil with its low linolenic acid content has excellent oxidative stability and is considered a premium oil for frying and cooking (26). Peanut oil differs from other vegetable oils with 6% of long chain saturated fatty acids including arachidic (20:0), behenic (22:0), and lignosceric (24:0).

Olive oil is the ancient oil of the Mediterranean from the tree *Olea europa.* In 1993 about 256 million pounds had been consumed in the United States. Greece, Spain, Italy, Portugal and Tunisia account for most of the production. It is used almost entirely for edible products such as cooking or as a salad oil.

Olive oil is pressed from the fruit to obtain "virgin" oil. The greenish colored oil has a characteristic flavor and odor and may be consumed without further treatment. The oil is classified by the stage of processing. The first pressing yields the highest grade or virgin olive oil. Lower grades are produced with successive pressings or solvent extraction. Lower grade oils are frequently blended with milder flavored oils such as soybean or cottonseed oils. Olive oil is high in squalene which may be used to detect its presence in blends.

The fatty acid content of olive oil ranges from 65 to 85% oleic acid (18:1). It is very stable to oxidative deterioration with an absence of drying properties.

Rice bran oil is isolated from the bran fraction obtained during rice milling. Bran constitutes 6–10% of rough rice and contains 15 to 20% oil (27). Rice bran oil is important commercially because of the importance of rice as a cereal. Nutritional studies have shown rice bran oil to actively lower serum cholesterol (28). In Japan, more than 100 million pounds of rice bran oil are produced primarily for frying uses. Rice bran oil is also produced in India, South America and the U.S.

The production of rice bran oil has been limited by problems with oil quality. The high lipase activity of the bran quickly hydrolyzes the oil into free fatty acids causing severe flavor problems. The high wax content of the crude oil results in high refining losses.

The newest member to this group of oils is high oleic sunflower oil with as much as 80% oleic acid. Other versions have shown oleic acid up to 90%. The high oxidative stability of high oleic sunflower oil makes it suitable in applications where fully saturated lauric fats have been suitable. Coatings for cereals and raisins, and for roasting of nutmeats are prepared from high oleic oils. Frying applications also benefit from its stability. High oleic oils are also particularly adapted to synthesis of other oil products such as cocoa butter equivalents.

8.4 Linolenic Acid Oils

The most important members of the linolenic acid group are oils of annual plants. All oils in the group have good drying properties, useful in paints and coatings. Because of the high degree of unsaturation, these oils tend to oxidize readily. Flavor reversion is common even with only slight oxidation. Linolenic acid oils are not as desirable for food use as the oleic–linoleic group of oils. Most find only limited use in soap making for the same reason. Overall, linolenic acid oils are somewhat cheaper than the oleic–linoleic acid oils.

The linolenic acid oils consist of linseed, soybean, hempseed, perilla, wheat germ, and canola oil, among others. Soybean is, of course, the most widely produced oil in the world as well as the United States. Worldwide production in 1991–1992 was 16.8 million metric tons. In the U.S., more than 13 billion pounds (6 million MT) had been produced in 1992. The crude oil is obtained from the legume, *Soya max*, almost entirely by solvent extraction. The major U.S. food uses are as salad and cooking oils (4.9 billion lbs), baking and frying oils (4.1 billion lbs), and margarine (4.1 billion lbs).

Traditionally, the use of soybean oil has been limited by the tendency of the oil to revert or oxidize due to the linolenic acid content (5–9%). It has been estimated that up to 70% of the oil is hydrogenated (22). Soybean oil is a semidrying oil and, as such, is used extensively in drying oil-based products.

Crude soybean oil contains significant quantities of nontriglyceride components (1.5–2.5%) and phosphatides or phospholipids are among these. The phosphatides are removed from the crude oil by water washing. The dried phosphatides form the basis for lecithin production, an ingredient useful as an emulsifier or surface active agent in formulated food and feed products.

Efforts to improve soybean oil have generally involved the reduction of linolenic acid content. Breeding out, extracting, and reacting out the linolenic acid have all been tried. Selective hydrogenation has been most successful. Genetic development appears promising for future improvement of soybean oil (29).

Light hydrogenation of soybean oil reduces the linolenic acid (18:3) content to less than 3%. Flavor stability is improved. More recently, soybean oil marketed as a salad oil has not been hydrogenated (30). Hydrogenation of soybean oil is used for preparing shortenings for baking, frying, and margarine applications.

Linseed oil from flax (35–45% oil) has traditionally been used as a drying oil particularly in coating formulations. With the development of water-based coatings, the demand for linseed oil has decreased. Other industrial uses of linseed oil include varnishes, linoleum, oilcloth, and printing inks. Food use is limited because of the characteristic odor that redevelops, even after deodorization.

The quality of linseed oil for inedible purposes is dependent on the iodine value. Typical values are 180 to 185 with more than 45% linolenic acid present. Recently, efforts have been made to produce edible quality linseed oil having a reduced polyunsaturated fatty acid content. A linseed oil with 2% 18:3 and 72% 18:2, called Linola, has been developed (31). Perilla oil, from the plant *Perilla ocymoides*, is also a drying oil that resembles linseed oil but has a higher iodine value. It exhibits faster drying properties. Hempseed oil from *Cannabis sativa* is also similar to linseed oil.

Canola oil is a recent addition to the linolenic acid group. The low erucic acid oils are the result of selective breeding of *Brassica napus* and *Brassica campestris* that contain less than 2% erucic acid (22:1). Originally, rapeseed oil contained 25–50% erucic acid. The linolenic acid content is 9–12% and the oleic acid content is 64%. Canola oil has the lowest content of saturated fatty acids of the vegetable oils. Canola oil finds wide use as salad, frying, and cooking oils as well as inclusion in margarine formulations.

8.5 Vegetable Butters

The vegetable butters are primarily from the seeds of tropical trees. They are similar to the lauric oils in that they melt over a narrow temperature range. Most are characterized by the fatty acid distribution of the triglycerides. All contain 50% or more saturated fatty acids of C-14 to C-18 carbon chain lengths and consist of simple mixtures of triglycerides having a single or limited number of triglyceride types present. Seventy-five percent of the triglyceride types in cocoa butter are composed of oleic acid in combination with palmitic or stearic acid. Only small quantities of trisaturated glycerides are present. The unsaturated fatty acid occupies the center or *sn*-2 position of the triglyceride. Other vegetable butters are borneo tallow and mowrah fat. Both are sometimes referred to as Illipe butter. Borneo tallow is most similar to cocoa butter.

Chinese vegetable tallow, obtained from the Chinese tallow tree (*Stillingia sebifera*), is primarily a palmitic acid oil. Stillingia oil from the kernel is a drying oil. Other vegetable butters include nutmeg, ucuhuba, malabar tallow, aceituno oil and phulwara butter. The single triglyceride type imposes distinctive crystallization behaviors. For cocoa butter, polymorphism is a major consideration in its use.

Generally, the vegetable butters are higher priced fats. They are used mainly for the manufacture of chocolate coatings, confections, and pharmaceutical products (suppositories, lotions, creams).

8.6 Erucic Acid Oils

Erucic acid oils having commercial significance are mustard, ravison, and rapeseed oil. Crambe oil is being developed along with high erucic rapeseed oil. Erucic acid contents are in the 40–55% range with a small amount of linolenic and eicosenoic acid. Major producing countries are China, India, continental Europe (Sweden, Poland, France), and Canada. As a result of genetic development, both high and low erucic acid rapeseed oils are available. The low erucic acid rapeseed, formerly referred to as LEAR, has since acquired the label "canola" oil. Canola oil is intended for food applications since high erucic acid oils have been suspected to be nutritionally undesirable. The high erucic acid varieties are intended for industrial or nonfood applications for the manufacture of lubricants and fatty acid derivatives such as the amides.

8.7 Conjugated Acid Group

The commercially important members of this group are tung oil and oiticica oil. These contain 85% eleostearic (9,11,13 octadecatrienoic acid) and 70–80% licanic (4 keto-9,11,13 octadecatrienoic acid) acids respectively. The conjugated position of the double bonds favors oxidation and polymerization. These oils dry more readily than conventional drying oils and are in demand for production of varnish, enamel and the manufacture of protective coatings. They are unsuitable for production of soaps. China is the single largest source of the conjugated acid oils with price being based on the availability of the Chinese crop.

Table 2.14 Castor oil consumption, with inedible by category, United States, 1985/86–1992 (1)[a]

Year 1/	Total Consumption	Total Edible	Total Inedible	Soap	Paint or Varnish	Feed	Resins and Plastics	Lubricants 2/	Fatty Acids	Other Products
1985/86	59.9	0.0	59.9	d[b]	d	0.0	4.5	d	d	d
1986/87	70.4	0.0	70.4	d	4.6	0.0	4.2	5.6	d	53.8
1987/88	74.6	0.0	74.6	d	4.3	0.0	4.8	6.1	d	59.0
1988/89	59.2	0.0	59.2	d	4.8	0.0	4.5	6.2	0.0	43.2
1989/90	51.4	0.0	51.4	d	5.9	0.0	4.0	5.7	0.0	d
1991	48.0	0.0	46.0	d	5.9	0.0	4.0	d	0.0	31.7
1992	41.3	0.0	41.3	d	d	0.0	3.3	3.5	0.0	28.4

[a] Million pounds.
[b] d = Data withheld to avoid disclosing figures for individual companies.

8.8 Hydroxy Acid Group

Castor oil is the sole representative of this group. Consumption totaled 41.3 million pounds in the U.S. in 1992 (Table 2.14). The triglycerides contain ricinoleic acid (12 hydroxyoctadeca-9-enoic acid) and small amounts of 9,10-dihydroxy stearic acid. Up to about 90% of the castor oil is made up of these fatty acids. The oil is not edible and is not used in soap making. Castor oil may be catalytically dehydrated to form a conjugated acid oil similar to tung or oiticica oil, hence being useful for coatings. Specialty uses of castor oil are many with applications as lubricants, oils for sulfonation, and as a hydraulic fluid. Alkaline fusion of castor oil yields sebacic acid and 2-octanol. Pyrolysis yields undecylenic acid and heptaldehyde.

STATUS OF VEGETABLE OILS

Total lipids in the diet have been a major topic for nutritionists and dieticians for the last several years (32). Per capita consumption in the United States has shown some change. The production of reduced fat, low fat, and no fat foods has been promoted by most major food companies. The nutritional issues have now gone beyond total fat consumption to include saturated fatty acid content and, most recently, trans fatty acid content.

The utilization of oils and fats for industrial or nonfood applications has decreased in favor of alternatives from petroleum products. The oils industry has responded somewhat with the introduction of modified oils such as the high oleic sunflower and high erucic rapeseed varieties.

Future opportunities exist for the biological modification of oils. Using biological triglyceride modification through breeding, mutation, or genetic manipulation may result in: (1) production of high content saturated fatty acid oil from soybeans replaces the need for hydrogenation in the production of functional or stable oils for foods, (2) production of nearly pure fatty acid triglycerides suitable for industrial utilization, or (3) synthesis of modified triglycerides for a lower calorie oil for foods.

REFERENCES

1. *Oil Crops* S&O/OCS-40 (Jan. 1994).
2. Foreign Agricultural Service, "Oilseeds and Products," *Counselor and Attache Reports*, Feb. 1994.
3. American Heart Assoc., Nutrition Committee, *Circulation* **77,** 721A–724A (1988).
4. W. C. Willett, *Science* **264,** 532–536 (1994).
5. R. P. Mensink and M. B. Katan, *Arteriosal Thromb.* **12,** 911 (1992).
6. J. T. Judd and co-workers, *Am. J. Clin. Nutr.* **59,** 861 (1994).
7. J. E. Hunter and T. H. Applewhite, *Am. J. Clin. Nutr.* **44,** 707–717 (1986).

8. D. Andreas, *Am. Oil Chem. Soc., INFORM* **3**(9), 1016–1017 (1992).
9. *Am. Oil Chem. Soc., INFORM* **5**(5), 589–592 (1994).
10. T. H. Applewhite, *World Congress on Biotechnology for the Fats and Oils Industry, Am. Oil Chem. Soc.,* 1988.
11. C. Rutledge, P. Dawson, and J. Rattray, *Biotechnology for the Oils and Fats Industry, Am. Oil Chem. Soc.,* 1984.
12. H. M. Davies and F. J. Flider, *Chem. Tech.,* 33–37 (Apr. 1994).
13. *Am. Oil Chem. Soc., INFORM* **5**(2), 145–149 (1994).
14. R. F. Wilson, *Am. Oil Chem. Soc., INFORM* **4**(2), 193 (1993).
15. A. Svendsen, *Am. Oil Chem. Soc., INFORM* **5**(5), 619 (1994).
16. G. Griffiths, S. Stymne, and K. Stobart, *World Conference on Biotechnology for the Fats and Oils Industry. Am. Oil Chem. Soc.,* 1988.
17. A. G. Green, *Crop Sci.* **26**, 961 (1986).
18. J. L. Harwood, *World Conference on Biotechnology for the Fats and Oils Industry, Am. Oil Chem. Soc.,* 1988.
19. D. Swern, *Bailey's Industrial Oil and Fat Products,* vol. 1, 4th ed., John Wiley & Sons, Inc., New York, 1979.
20. N.O.V. Sonntag in D. Swern, ed., *Bailey's Industrial Oil and Fat Products,* vol. 1, 4th ed., John Wiley & Sons, Inc., New York, 1979, pp. 289–459.
21. J. J. Stanton and J. Blumenfeld, *INFORM* **3**(9), 1019–1022 (1992).
22. R. Wood in C. Chow, ed., *Fatty Acids in Foods and Their Health Implications,* Marcel Dekker, Inc., New York, 1992.
23. H. N. Bhagavan and P. Padmanabhan, in C. Chow, ed., *Fatty Acids in Foods and Their Health Implications,* Marcel Dekker, Inc., New York, 1992.
24. L. B. Wrenn, *Amer. Oil Chem. Soc., INFORM* **4**(1), 6–20 (1993).
25. B. F. Haumann, A. R. Baldwin, A. Seiffert, K. G. Berger, and A. M. Gavin, *J. Am. Oil Chem. Soc.* **65**, 702 (1988).
26. J. C. Cowan, H. Moser, G. R. List, and C. D. Evans, *J. Am. Oil Chem. Soc.* **48,** 835–839 (1971).
27. F. T. Orthoefer and R. J. Nicolosi, "Rice Bran Oil Characteristics," *84th Am. Oil Chem. Soc. Annual Meeting,* Anaheim, CA, Apr. 29, 1993.
28. R. J. Nicolosi, E. Rogers, L. Ausman, and F. T. Orthoefer, *Rice Science and Technology,* J. Wadsworth and W. Marshall, eds., Marcel Dekker, Inc., New York, 1992, pp. 421–438.
29. E. G. Hammond and W. R. Fehr, *J. Am. Oil Chem. Soc.* **61,** 1713 (1984).
30. P. J. White in C. Chow, ed., *Fatty Acids in Foods and Their Health Implications,* Marcel Dekker, Inc., New York, 1992, pp. 238–239.
31. B. F. Haumann, *Am. Oil Chem. Soc. INFORM* **1,** 934 (1990).
32. C. K. Chow, *Fatty Acids in Foods and Their Health Implications,* Marcel Dekker, Inc., New York, 1992.
33. *Am. Oil Chem. Soc., INFORM* **3**(8), 930–933 (1992).
34. B. F. Haumann, *Am. Oil Chem. Soc., INFORM* 3(10),1080–1093 (1992).

3

Oilseeds and Products Trading

World markets for fats, oils, and high protein meals have undergone substantial changes since World War II. Both the value and the complexity of the markets have been increasing as population and income growth worldwide has driven upward consumer demand for oils and livestock products. World oilseed production, however, has consistently responded to the challenge. Through increasing diversification of oilseed production and increased trade in oilseeds and their products, protein meal and oils, global demand has been met.

The growing globalization and complexity of oilseed markets have resulted in ever-growing uncertainties and risks. The implications of changing international oilseed policies and diverse world patterns of production and consumption make price discovery more difficult. The trading institutions have had to adapt to this increasingly international market and as the institutions adapt, uncertainty is reduced, serving as a catalyst to increased trade and globalization.

This chapter deals with the importance of oil crops in world trade, the role of the U.S. oilseed industry, and the changing nature of the market and forces that affect price determination. Also, the evolution of the trading institutions, which are increasingly faced with the uncertainties and risk generated by a growing global market for oilseeds, are examined.

OVERVIEW OF THE OILSEED COMPLEX

The importance of oil crops as a vital part of the world's food supply is evidenced in world agricultural trade statistics. In 1992, the value of trade in oilseeds and oilseed products, at $31 billion, or 9% of the total agricultural trade, constituted the third most valuable component in total world agricultural trade—ranked behind only meat products and cereals (1). As the world sources of fats, oils, and protein meals are becoming increasingly diverse, international trade is becoming increasingly important.

The global market for oilseeds and their products has historically been dominated by the major producers of soybeans, specifically the United States. This contrasts to production patterns prior to World War II when much of the world's oilseed production was concentrated in Asia and Africa to supply European colonial powers with fats and oils (2). However, presently the major oilseed producing areas have been in the temperate zones, with the U.S. at the forefront of world oilseed production.

On average between 1990 and 1992, soybeans accounted for nearly 50% of world oilseed production while rapeseed and sunflowerseed combined accounted for a further 22% of production. (Figure 3.1). Cottonseed, a cotton by-product, is the second largest contributor to the oilseed complex, representing 15% of total production. The most important tropical oilseeds are copra (from which coconut oil is derived) and palm kernel, together constituting less than 4% of global production. Peanuts, accounting for 10%, are also grown in tropical countries, particularly Africa, as well as in the United States, China, India, and Argentina.

1.1 U.S. Oilseed Production and Trends

The United States' commercial production of oilseeds includes soybeans, cottonseed, sunflower seed, peanuts, flaxseed, rapeseed, safflower, mustard seed and canola. Average planted acreage, production and farm gate value of these oilseeds for the 1990–1992 crops is presented in Table 3.1. An inspection of Table 3.1 reveals that, irrespective of the measure chosen, soybeans dominate the U.S. oilseed sector, comprising over the period presented, 76% of oilseed planted area, 86% of production and 83% of the farm value of production.

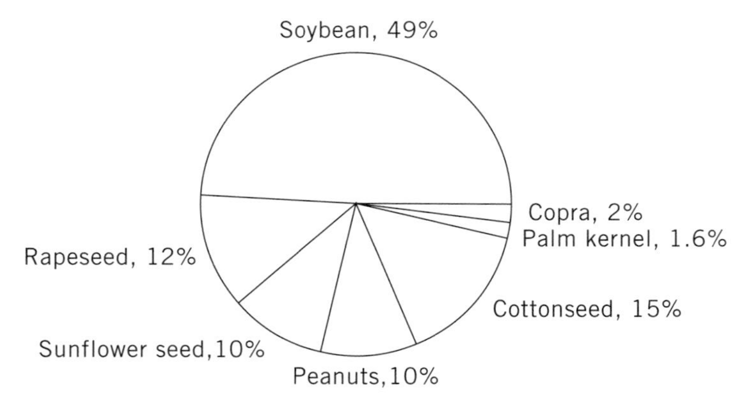

Figure 3.1 Share of world oilseed production by type, 1990–1992 average = 222 million tons. Data from U.S. Department of Agriculture.

Table 3.1 Average 1990–1992 planted acreage, production and value for U.S. oilseeds

Crop	Planted Area (1000 acres)	Production (1000 short tons)	Farm Value (Mill. $)
Soybeans	58,768	61,091	11,429
Cottonseed	13,230	6,386	608
Sunflowerseed	2,289	1,415	272
Peanuts	1,863	2,136	1,309
Flaxseed	262	124	18
Rapeseed[a]	15	8	2
Safflower[a]	282	164	42
Mustardseed	17	8	2
Canola[a]	155	91	18
Total	76,881	71,423	13,700

[a] 1991–1992 average; USDA began reporting with the 1991 crop.

In addition to their dominance within the U.S. oilseed sector, soybeans are among the leading U.S. commodities for cash receipts. Among field crops, soybeans has ranked behind only corn in cash receipts for the 1991 and 1992 seasons. Among all commodities, including livestock enterprises, soybeans usually rank fifth, following, in addition to corn; cattle and calves, dairy products, and hogs.

Soybeans. The story of U.S. soybean production can generally be divided into two periods: the first, a period of meteoric production growth culminating with the benchmark 1979–1980 season, and the more recent period, since 1980 (Figure 3.2). In 1950, U.S. soybean production had been nearly 300 million bushels, doubling by 1960 and doubling again by 1970. This rapid growth continued through the 1979 season when the U.S. soybean production record was established at 2.26 billion bushels on a record 71.4 million planted acres. Since the record 1979–1980 season, U.S. soybean acreage has retreated and stabilized recently at a much lower level, though production has been maintained at high levels owing to rising yields (3).

The principal soybean producing area in the United States has always been that part denominated as Corn Belt states, i.e., Illinois, Indiana, Ohio, Iowa, Missouri, Nebraska, and Minnesota. Agronomic factors in this geographic area have resulted in the highest soybean yields per harvested acre of any U.S. region. In their heyday of rapid area expansion, soybeans were planted in other peripheral states where yields were lower. As acreage has contracted in more recent years, soybean production has become reconcentrated in the traditional higher-yielding corn belt. Thus, record matching or record setting national soybean yields have been observed in the 1990, 1991 and 1992 seasons.

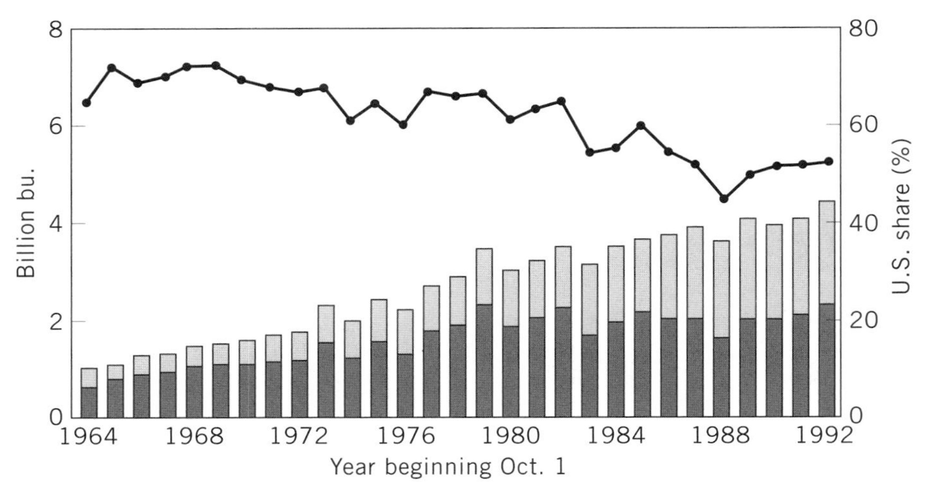

Figure 3.2 World and U.S. soybean production (bars) and % U.S. share (line with data points) 1964–1992. ▭ = world production; ■ = U.S. production.

Cottonseed. Among the principal oilseeds produced in the United States, cottonseed occupies a unique position. As a by-product of cotton fiber production, cottonseed production and supply are comparatively unresponsive to the demand side of its balance sheet. That is, U.S. cottonseed producers generally do not produce more cottonseed in response to strong seed prices, but rather focus on the prospects for profit from cotton fiber production. The value of cottonseed is, however, not unimportant to cotton producers as most count on the seed value to compensate for a large part of the cost of ginning cotton; that is, separation of lint from seed.

Cottonseed production in the United States is long-established and occurs in the area known as the Cotton Belt. This region consists of 14 states located along the southern border of the United States and includes areas of irrigated cotton production in the desert Southwest, such as California's San Joaquin Valley.

Also distinguishing cottonseed production from soybean production, but not all other oilseed production, is the impact of U.S. commodity programs. Production of cotton fiber in the United States is highly influenced by the target prices, loan rates and acreage reduction requirements of USDA programs for upland cotton and extra-long staple cotton. These programs, which are designed, among other things, to control the supply of cotton lint, also indirectly control the supply of cottonseed through their impact on cotton planted area.

Peanuts. U.S. peanut production is heavily influenced by the USDA commodity program for peanuts. The peanut program is a two-tiered price support program providing a higher support level for peanuts under a marketing quota and a lower support level for additional peanuts. Quota peanuts are those produced to satisfy the domestic edible market within the annual quota poundage, while additional peanuts are those produced in excess of the quota.

The annual level of U.S. peanut production is closely related to the demand for peanuts for domestic food purposes. Peanuts produced in excess of those eligible for food use are either exported or crushed—generally resulting in much lower-valued end uses.

Peanuts are produced in the United States in many of the same areas as cottonseed, owing to common desirable agronomic factors. Peanuts are well-suited to sandy soils, a warm climate and plentiful moisture. In addition, the intricacies of the commodity program for peanuts, which allocates marketing quota on the basis of the locale of historical peanut production, tends to geographically stabilize a planted area. There are 7 major peanut-producing states, but the most significant, by far, is Georgia, which usually accounts for one-half or more of annual U.S. peanut production (4).

Sunflowerseed. Although sunflowers are native to North America, making them unique among all the oilseed plants, commercial sunflower production is relatively new to U.S. agriculture and has yet to gain acceptance as a major crop (5). Sunflowers have ranked fifteenth among the top 30 U.S. field crops in value in 1991, yet have accounted for only 2% of U.S. oilseed production. Production is primarily concentrated in the Plains States, with North Dakota, South Dakota, Minnesota, Kansas, Nebraska, and Texas contributing 95% of the national production in 1991.

U.S. sunflower production has been erratic. Planted acreage that approached 6 million acres in the late 1970s fell precipitously thoughout much of the 1980s before rebounding from 1989 to 1991. In the past, the health of the industry has been tied almost exclusively to exports. In recent years, domestic demand for sunflowerseed and its products has become increasingly important. Recent changes in Federal farm policy, and growing domestic and international markets for sunflower products, are making sunflowers more attractive to U.S. producers.

Canola. Canola is the name given to seed, oil, and meal derived from rapeseed cultivars low in erucic acid and glucosinolates. In January of 1985, the Food and Drug Administration granted GRAS (Generally Recognized as Safe) status for low erucic acid rapeseed products, of which canola seed, oil, and meal are subsets. Since obtaining GRAS status, interest in canola has been growing in the United States (6).

A member of the mustard family, rapeseed is suitable to colder areas of the United States. However, until recently, rapeseed seed production in the United States has been small. Prior to 1985, production had been geared towards specialized industrial uses and, accordingly, acreage remained small and had been concentrated in areas of the Northern Plains, Northwest, and parts of the southern Corn Belt. Since that point, area and production have expanded significantly.

With the ongoing development of new varieties that are suited to both winter and spring plantings specific to U.S. conditions, canola production has spread throughout the rest of the country and as far south as Florida. In these areas, canola provides farmers with an alternative to fall-planted crops such as winter wheat. Interest in production has been fueled more recently by the implied health benefits associated with the lowest saturated-fat content among all major vegetable oils.

1.2 Foreign Oilseed Production and Trends

U.S. production of oilseeds, while accounting for about 50% of world oilseed production in the 1960s, has dropped to less than 30% in 1993 (Figure 3.3). Expanding competition from the major soybean producers in South America, Brazil and Argentina, and increasing production of other oilseeds, have consistently eroded the U.S. share of world production over the past three decades. Similarly, the U.S. share of global soybean production, accounting for around 70% of world totals in the 1960s has dropped to less than 60% in the early 1990s (Figure 3.2).

Soybeans. While world production of soybeans has increased by 44% to 116.5 million tons from 1980 to 1992, most of the gains have been realized by foreign producers (Table 3.2). Both Brazil and Argentina, the United States' major competitors, have rapidly expanded soybean production since the early

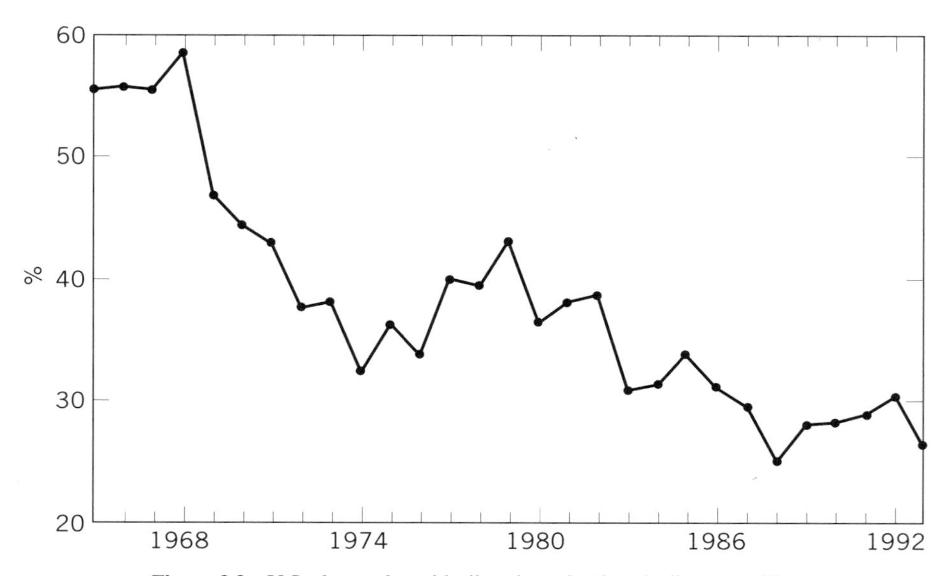

Figure 3.3 U.S. share of world oilseed production declines steadily.

Table 3.2 World and foreign oilseed production, 1980 and 1992 [a]

	World Production		Foreign Production		Change (%)	
Crop	1980	1992	1980	1992	World	Foreign
Soybeans	81.0	116.7	32.1	57.2	44	78
Cottonseed	24.8	31.6	20.8	25.9	27	25
Sunflowerseed	13.2	21.5	11.5	20.4	63	77
Peanuts	16.3	23.2	15.2	21.2	42	40
Rapeseed	11.1	25.2	11.1	25.1	127	126
Copra	4.8	4.7	4.8	4.7	−2	−3
Palm kernel	1.5	4.0	1.5	4.4	167	191
Total	154.9	226.8	99.0	158.4	46	60

[a] Million metric tons.

1980s, particularly Argentina. Their combined production jumped over 75% since 1980 and presently accounts for more than 30% of world production. Similarly both China and India have become important producers of soybeans in the past decade, jointly more than doubling their production since the 1980s, and presently accounting for about 15% of world soybean production.

Rapeseed and Sunflowerseed. Soybean's share of the world market is increasingly being challenged by other oilseeds, particularly the "soft seeds" (so called because they yield more oil), such as rapeseed and sunflowerseed. Rapeseed, in particular, has enjoyed an annual 8% growth in production since the 1960s. This compares to the 3% annual growth rate for overall world oilseed production.

Government policies in producing countries have stimulated large production of oilseeds and influenced trading patterns. In both the EU (European Union) and India, major producers of rapeseed, government policies have focused on self-sufficiency in oilseed production. A system of government supports that guaranteed producers minimum prices for production, combined with import restrictions on vegetable oils in India, have fostered annual growth rates in the EU of 12% and 8% in India over the 1980s.

World sunflowerseed production has increased 50% since 1980 to 21.4 million tons in 1992. Increases in foreign production have driven this increase with significant gains experienced in many of the producing countries including Argentina and the EU. Other major producers, such as the former Soviet Union and Eastern Europe, increased production throughout the 1980s but have recently experienced declines due to economic difficulties resulting from transitions to market economies.

Other Oilseeds. Production of other oilseeds, such as cottonseed and peanuts, has increased only slightly while copra production actually has declined

Table 3.3 Global distribution of oilseed production, 1965–1970 and 1990–1992 average

	Share of Production (%)	
	1965–1970	1990–1992
North America	54.5	31.9
South America	4.8	18.3
Europe	5.9	12.8
Asia	31.0	31.9
Africa	3.1	3.5
Other	0.7	1.6
Total	100	100

since 1980. The major and most significant exception is the strong growth in palm fruit production, more than doubling over the period. Increasing production in Malaysia and Indonesia, the two major producers, have been stimulated by government policies which diversified the agricultural economy away from rubber, previously a major export crop. Expansion in palm fruit production has had a significant impact on the world vegetable oil markets. Of all the oil-bearing plants, palm kernel has produced the most oil, both palm kernel and palm oil, per unit area. Consequently, palm oil's share of world vegetable oil production has increased dramatically, from 13% in 1980 to 21% in 1992.

Oilseed production has been dominated by developed countries or developing countries with relatively high incomes. While North America accounted for about 55% of total oilseed production between 1965 and 1970, increased production in Latin America and Europe had dropped that share to 32% by the early 1990s (Table 3.3).

Many of the industrialized countries, such as the United States, Canada, and the EU, however, have stable per capita consumption of both vegetable oil and meat. While income and demand for meat products are growing in Brazil and Argentina, high production and excess crushing capacity maintains strong availability of exportable supplies of oilseeds and products. Consequently, many of the oilseeds or their products, are exported to countries where growing income and higher demand for oils and meat have stimulated trade.

INTERNATIONAL TRADING OF OILSEEDS

There are three different groups of oil crops that have distinctive production and trading patterns: a) oilseeds that are planted and harvested annually, such as soybeans, sunflowerseed, and rapeseed; b) perennial tree-crops such as coconut and palm trees whose slow maturation constrains rapid expansion in production; and c) crops such as cotton where the cottonseed is a by-product of the cotton fiber industry (7).

Due to the nature of tree crops, coconuts and the palm fruit cannot be transported far for economic and quality reasons. Consequently, these oilseeds are typically processed locally and then exported as an oil. The Africa palm is unique among edible oil-bearing crops due to its production characteristics. The tree matures and begins to bear fruit after five years and produces two crops annually for up to 30 years.

In addition, trading patterns for oilseeds are influenced by the proportion of oil and protein meal produced by the different oilseeds during the crushing process. Soybeans have an oil content of 18% and soybean meal contains protein ranging from 44–50%. Copra, however, is 65% oil and sunflowerseed and rapeseed about 40% oil; the oilcake remaining contains significantly less protein than soybean meal.

2.1 What Drives Trade

Increasing production and trade of oilseeds and their products have occurred due both to the rising consumption of livestock products and the concurrent rapid growth in meal demand, as well as strong demand for vegetable oils. Changes in the supply and demand conditions in different producing and consuming countries create excess demand and supply, generating trade.

Part of the world's oilseed production is exported in the form of seeds or processed products. The major exporting regions are North America (United States and Canada), South America (Brazil and Argentina), and increasingly Malaysia and Indonesia (Table 3.4). The EU and Japan remain the major import markets, accounting for more than 50% of world trade in oilseeds. However, in the late 1980s, strong demand for protein meals from Southeast Asian countries, such as Malaysia, the Philippines and Thailand, is increasingly driving trade in protein meals, specifically soybean meal. This demand is partially filling in the vacuum left by the FSU and Eastern Europe as economic turmoil in the early 1990s constrained consumption and imports of most commodities.

In 1992 approximately 17% of total oilseeds were traded, with trade of

Table 3.4 Regional distribution of exports and imports of oilseeds and protein meals, 1965–1970 and 1990–1992 average (%)

	Share of Bean Exports		Share of Bean Imports		Share of Meal Exports		Share of Meal Imports	
Region	1965–1970	1990–1992	1965–1970	1990–1992	1965–1970	1990–1992	1965–1970	1990–1992
North America	76.4	58.5	5.7	8.0	31.5	14.5	7.9	3.6
South America	2.0	22.6	0.8	2.6	24.0	44.0	0.9	3.3
Europe	3.0	12.3	55.4	52.4	15.4	18.2	70.8	68.4
Asia	12.1	4.6	32.1	32.7	5.6	18.6	3.1	14.2
Africa	4.7	0.9	0.5	1.0	5.8	1.3	0.1	3.2
Other	1.9	1.1	5.5	3.2	17.7	3.4	17.3	7.4
Total	100	100	100	100	100	100	100	100

soybeans at 25% of total production, exceeding that of other oilseeds. Constraints on trade in tree crops is evidenced by the small amount of copra and palm kernel entering world markets, only 3% in 1992. Trade in soybeans is driven mainly by continuing strong demand for protein meal for livestock feeding.

A small amount of oilseeds, basically soybeans, are traded for use as food products. Consumed mainly in Asia, soybeans for food use are processed into bean curd, fermented sauces and pastes, soy milk, or other processed products. In 1992, consumption of soybeans for food purposed accounted for approximately 8% of world soybean consumption.

Issues Affecting Oilseed and Product Trade. A long history of government intervention has helped to shape the existing patterns of global oilseed production and seed and oilseed product trade. While oilseeds have historically been less affected by trade distorting policies than other commodities, intervention has altered demand for oilseeds from low-cost, efficient producers. The comparative advantage of low cost producers has been eroded by support for oilseed producers in higher cost countries, such as the EU. Shares of oilseed production and exports from countries such as the U.S. have declined while those of less efficient producers have increased (Table 3.3).

Market intervention occurs in a variety of forms, ranging from production support, such as producer payments and price supports, to trade intervention. The latter include export subsidies, differential export taxes, and import controls. In addition, individual countries may pursue macroeconomic policies which affect the agricultural sector and its comparative advantage.

Production and Trade Distorting Policies. Policies affecting production patterns include: price supports, producer payments, input subsidies, and land restrictions. The United States operates price support programs for most oilseeds; however, oilseed production patterns have historically been more affected by programs for competing crops, specifically corn and cotton.

In the EU and India, domestic policies, specifically price supports and import controls, have altered production patterns. These policies guarantee a place in the local market for high-cost producers, displacing imports from efficient, lower-cost producers. Trade-distorting policies may either restrict or promote exports. Many developing countries tax agriculture. For example, Argentina and Brazil have a system of differential export taxes which effectively tax the export of oilseeds more than products, leading to an export mix of beans versus meal favoring the processed products. Differential tax policies are largely responsible for Brazil becoming the largest exporter of soybean meal in the world.

Industrialized countries, on the other hand, tend to promote exports through export price subsidies. The United States has, since 1986, instituted three programs that have directly affected vegetable oil exports. The Export Enhancement Program (EEP) was authorized in 1986 to improve the competitiveness of U.S. agriculture exports in the face of increasing subsidized foreign exports, specifically from the EU. Two other export programs, specifically for vegetable oils, were set up in 1988: the Sunflowerseed Oil Assistance Program (SOAP) and the Cottonseed Oil Assistance Program (COAP).

Arguments abound in support of government intervention in agriculture. However, the gains must be weighed against adverse consequences. Supporters of intervention cite the importance of stable farm income, adequate farm output, and national protection from unfair foreign trade policies.

These benefits must be weighed against high costs, wasted resources, market disequilibrium, and production and trade share distortions. Inevitably, intervention results in countries which have a comparative advantage in production exporting less, often at a higher cost, than would occur in the absence of intervention policies (8).

2.2 Changing Trends in Oilseeds and Product Trading

Trends in international consumption and trade of oilseeds and products underwent a significant change in the 1980s as trade in products surpassed that of the seed themselves (Figure 3.4). In 1992 exports of protein meals and vegetable oils as a share of production significantly exceeded that of the seed: 33% versus 17%. In the 1980s increasing use of government support programs for oilseeds and products changed the composition of world oilseeds production and shifted the relative prices to favor trade in products rather than seeds. The shift in composition of trade was also caused by the significant expansion in crushing capacity in South America, stimulated by government policies, as well as increased demand from markets that have limited crushing facilities.

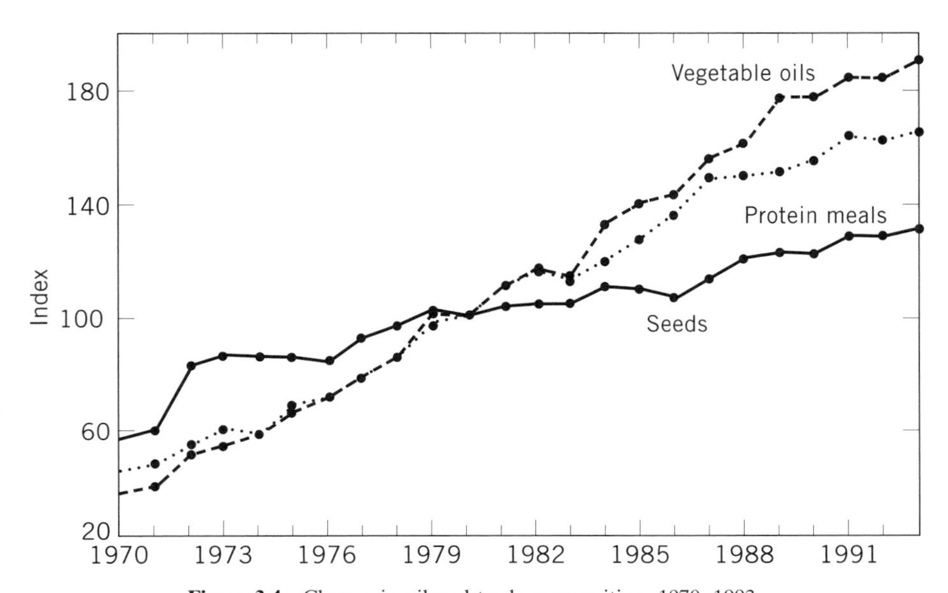

Figure 3.4 Change in oilseed trade composition, 1970–1993.

Increased availabilities of "soft seeds," i.e., oilseeds high in oil content, and strong growth in palm kernel production in the 1980s maintained low vegetable oil prices and stimulated consumption. Expansion of soybean production in South America where policies favor the export of products over seeds, led to gains in exportable supplies of soybean meal.

Protein Meals. Global demand for protein meals has almost tripled since 1970, with the bulk of demand stemming from industrialized countries, particularly the U.S. and the EU. The major importing countries are those industrialized countries with high livestock populations and, more increasingly, the developing countries in Southeast Asia where booming economies and strong meat demand are fuelling demand for imported protein meals. However, the EU continues to account for half of the total world import demand for protein meals, while accounting for 20% of the total consumption in 1992.

Protein meal demand is a function of demand for animal protein, meat and poultry products, which are dependent on factors such as population, income growth and consumer preferences. Stronger demand and extensive livestock production has resulted in farmers feeding their livestock compound feeds containing a mixture of cereals, molasses, minerals and vitamins with a balanced proportion of protein products such as oilcake or protein meals. Differences in feed rations depend on the relative prices of the various ingredients.

Soybean meal dominates the export market for protein meals due to its higher protein content as well as its prevalence in production. While about only 36% of soybean meal production is traded, it accounted for 66% of protein meal traded in 1992. The U.S. is the major producer and consumer of soybean meal. While dominating world trade in the 1960s, the U.S. share of world soybean and meal trade has been eroded by increasing exports from both Brazil and Argentina (Figure 3.5).

Soybean meal faces more competition from other protein meals as production of other oilseeds expand, generating additional supplies of competing meal. Relative price changes allow some substitutability between meals; however, this interchangeability is constrained by the high protein content of soybean meal and its suitability for feeding to all types and ages of livestock and poultry.

A small percentage of soybean meal (less than 1%) is used for human and industrial consumption. Soybean meal for food use includes high protein derivatives of meal used in cake mixes, waffles, cereals, breads, snack food, soups, and baby food. Industrial uses of soybean meal are found in the dietetic, health, and cosmetics industries as well as in the production of antibiotics (7).

Vegetable Oils. Spurred by income and population growth in developing countries—as well as rapidly expanding food processing industries in Asia and other areas—the global growth in consumption and trade of vegetable oils is outpacing that of most other agricultural products. Consumption of

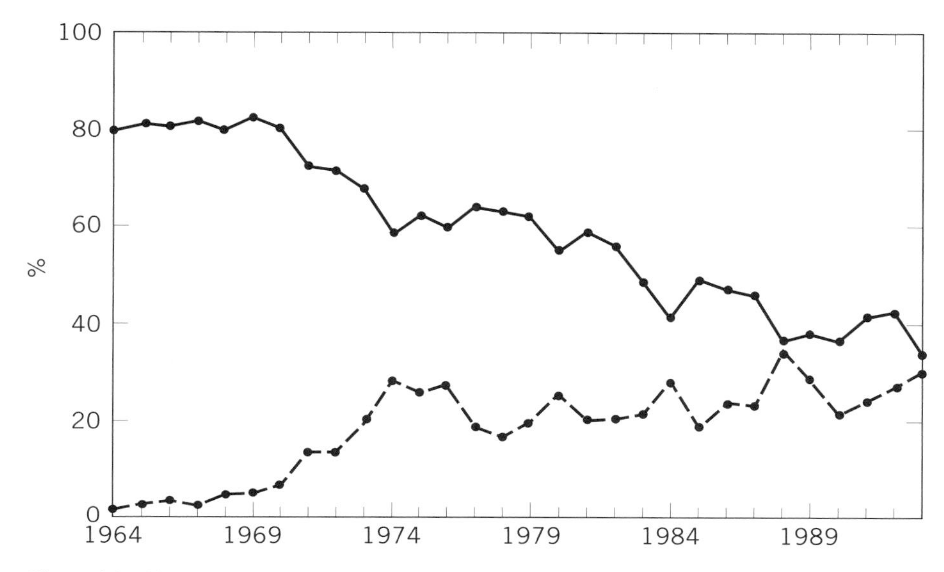

Figure 3.5 The U.S. loses market share to South America. —— = U.S. share; ---- = South America share.

vegetable oils worldwide has grown at an average annual rate of 4.2% over the past decade (9).

The United States and the EU are the largest consumers of vegetable oils, accounting for approximately one-third of world consumption. However, growth in demand is strongest in the newly industrialized countries in east Asia and developing countries, such as China, India, and Pakistan. Developing countries are consuming more vegetable oils because of rapid growth in population and income. As incomes increase, preferences shift torward more processed foods and more food prepared away from home.

Vegetable Oil Consumption. Two-thirds of total fat and oil production is supplied by oilseeds, with soybeans supplying about 28% of total vegetable oil demand. Four oilseeds—soybeans, rapeseed, sunflowerseed, and palm kernel—supply three-quarters of all vegetable oil production (Table 3.5).

World supplies of vegetable oils in 1992 are 6 times greater than in 1964 with per capita consumption increasing from 7.5 pounds per person in 1964 to over 23.8 pounds in 1992. This change implies that adequate supplies of fats available for human consumption have allowed substitution of animal fats by vegetable fats. It also implies that the growth in production and consumption of vegetable oils has far exceeded that of population growth.

While the bulk of vegetable oils are used for direct food use in the form of shortening, margarines, salad oils, and cooking oils, industrialized uses for oils continue to grow. Industrial use of oils include the manufacture of chemicals, paints, varnishes, sealants, lubricants, and adhesives.

Table 3.5 World production of major vegetable oils, averages by period[a]

Item	1970–1979	1980–1989	1990–1992
Soybean oil	9	14	17
Rapeseed oil	3	6	9
Sunflowerseed oil	3	6	8
Palm oil	3	7	12
Total	29	47	60
Share of world output	0.62	0.72	0.75

[a] Million metric tons.

Aggregate demand figures mask disparate consumption patterns between regions of the world. World per capita consumption of vegetable oil has doubled since the early 1970s. Yet, a wide disparity exists between industrialized and developing countries, with the highest per capita consumption accounted for by the United States. U.S. per capita consumption, estimated at 66.0 pounds in 1992, has almost been three times higher than the world average.

Consumption growth of vegetable oils has expanded at the brisk pace of 4% over the past decade with much of this growth in developed countries. However, with vegetable oil consumption growth stabilizing in these markets, growth prospects for vegetable oil reside in developing countries where there is a greater potential for raising vegetable oil consumption due to lower per capita consumption levels, faster population growth rates, and strong potential for income gains.

There is a significant potential for expansion in vegetable oil usage for nonfood purposes. Concerns about the environment and advancement in new technologies has fuelled interest, particularly in some European countries, in the development of alternative fuels using vegetable oils. Production of this biofuel used to replace diesel has been encouraged by economic incentives to producers and processors of rapeseed in Europe. This, given present EU oilseed policies, could significantly affect the long term outlook for the oilseed industry, affecting availabilities and prices of both protein meals and vegetable oils.

Vegetable Oil Trade. Since the 1950s, soybean oil has been the leading vegetable oil in production and use worldwide. However, world supplies of other vegetable oils, notably palm and rapeseed, have been growing, gradually reducing the relative importance of soybean oil (Figure 3.6). This growth in the importance of other oils can be attributed to country-specific policies enacted during the 1980s. Stronger availabilities of vegetable oils have maintained relatively low vegetable oil prices, thus encouraging increased consumption.

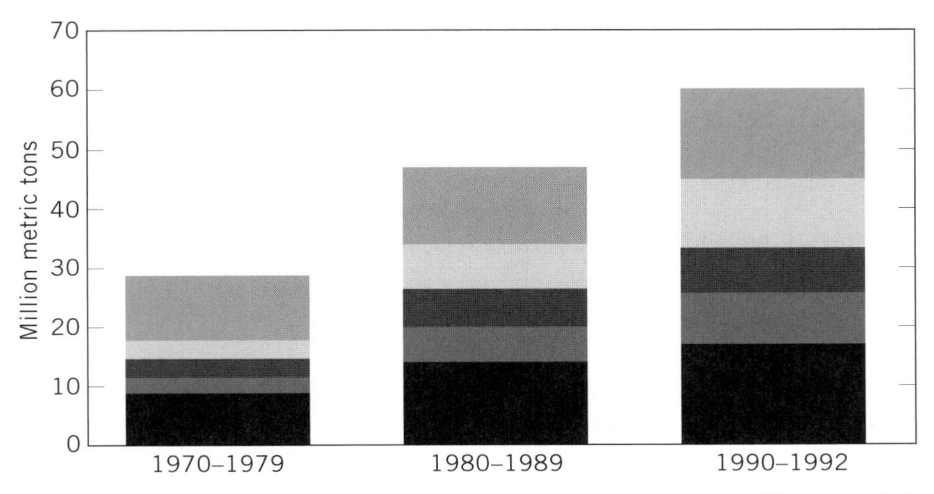

Figure 3.6 Growth in vegetable oil production driven by palm oil. ■ Soybean; ■ rapeseed; ■ sunflowerseed; □ palm; ■ other.

Tremendous growth in palm fruit production, particularly in Malaysia and Indonesia, and the physical characteristics of palm nuts which restrict trade in the fruit themselves, have stimulated trade of palm and palm kernel oil. Two-thirds of all palm oil production is typically destined for the export market. In 1992, palm oil exports, at 8.1 million tons, accounted for 40% of all vegetable oil exports. While the EU is the major importer of palm oil, increasing demand is stemming from developing countries where foreign exchange constraints generate interest in lower-priced palm oil.

Both sunflowerseed and rapeseed oils have increased their share of world trade, estimated in 1992 at 20%. Health concerns have stimulated interest in the oils with lower levels of saturated fats: soybean oil, sunflowerseed, and rapeseed oil. Argentina, the EU, and the United States are the major exporters of sunflowerseed oil, while the EU and Canada dominate the export market for rapeseed oil.

In 1992, trade in lauric oils, coconut and palm oil, accounted for only 4% of world trade with exports constituting more than 50% of lauric oil production. Destined for industrialized countries, particularly the EU, lauric oils are targeted for niche industrial uses and some edible oil markets (coconut only).

THE ECONOMICS OF THE OILSEED COMPLEX

Oilseed marketing is undertaken like that of other commodities. Most of the marketing is carried out by large international companies, rather than State trading institutions. Many of these companies, such as Cargill, Feruzzi, Archer-

Daniels Midland, and Continental Grain are also involved in oilseed processing. There are also numerous private companies which dominate trading in specific countries. Oilseed pricing, however, is much more complex than for other agricultural commodities. Prices for all commodities, in the absence of distorting policies, are determined by demand and supply. However, prices in the oilseed complex are also influenced by the joint nature of oilseeds and their products and the large amounts of protein meals and oil produced worldwide which have high, yet varying degrees of interchangeability.

3.1. The Role of Price

Resource Allocation and Trade. In market-based economies, prices play the central role in directing resources to satisfy demand, and in this respect, distinguish market-based economies from all others. In addition to organizing production, prices also serve to facilitate trade; it is this latter function with which this chapter is primarily concerned.

In order to fully develop the role of prices in the oilseed complex it is also necessary to examine the theory of derived demand, joint products, and price formulation and discovery. It is to this last process, price formulation and discovery, that formal trading institutions make their most significant contribution.

The most basic definition of price is that denoted as the market price. The market price for a commodity is the price which a willing buyer will give a willing seller, neither being under compulsion to buy or sell, and both being appraised of the commodity's highest and best use. The result of the qualitative process described above is a market price, a quantifiable value, known with certainty. Despite its numerical certainty, it is a value over which market participants may, and do, differ.

The well-known graphical illustration of market price is the price corresponding to the intersection of a supply and demand curve. It is a theoretical concept producing a single value. In actual practice there may exist many observed prices for a specific oilseed—soybeans, for example. However, the existence of many prices does not negate the concept of market price, but rather highlights the existence of many markets for soybeans—some differentiated by quality of the commodity, as well as, spatial and temporal considerations.

Spatial and Temporal Aspects of Price. In Table 3.6 are listed a small sample of soybean prices observed on an individual day, in this case, June 1, 1994. The first row of prices are settlement prices for all soybean futures contracts actively traded on the Chicago Board of Trade (CBOT). Prices are observed for soybeans up to about a year and a half in the future, illustrating the temporal aspect of commodity prices. The prices presented exhibit the

Table 3.6 U.S. soybean prices reported on June 1, 1994 in cents per bushel[a]

Region	Amount	Region	Amount
Cash-bids at country elevators, #1 yellow			
Kansas City, KS	710.00	Central MO	686.00
Ohio Valley KY	709.00	Eastern IA	682.50
Louisville, KY	701.00	Northern IL	681.50
Northern IA[b]	701.00	Lynchburg, SC	680.00
Southern IA[b]	700.50	Western KS	678.50
Southern area[c]	700.00	Central IA	677.00
Southeast MO	697.00	Beatrice, NE	676.00
Marianna, AR	697.00	Columbus, NE	676.00
Central IL	693.50	Grand Island, NE	675.00
Northeast MO	693.50	Western TN	674.50
Norfolk, VA	693.00	Blackwell, AR	673.50
Middle TN	693.00	Western IA	671.50
Southern MN[b]	692.50	Vermillion, SD	666.50
Southern IL	692.50	Mitchell, SD	658.00
Central VA	691.00	Watertown, SD	648.00
Eudora, AR	690.00	Plainview Triangle, TX	625.50

Chicago Board of Trade, futures, settlement

July '94	Aug. '94	Sept. '94	Nov. '94	Jan. '95	Mar. '95
700.50	699.75	685.00	673.75	678.00	682.00
May '95	July '95	Nov. '95			
684.00	685.50	634.50			

[a] *Wall Street Journal*, USDA-Agricultural Marketing Service.
[b] Mississippi River.
[c] Illinois River.

usual seasonal pattern, weakening into the peak harvest period for the 1994 crop (November, 1994) and then strengthening until the 1995 crop appears on the market (November, 1995).

The remaining prices in Table 3.6 are cash-bid prices at country elevators for #1, yellow soybeans. As can be readily observed, these prices exhibit a wide range, from a low of 625.5 cents per bushel to a high of 710.0 cents. Thus, for a specific grade of soybeans there exist many prices, depending upon the geographic location of the price quote. While the prices presented are numerous, they are in fact a small portion of the thousands that could prevail in the cash price market on a particular day. Despite the wide range and volume of price quotes, the price observed in any location exhibits a precise relationship to the remaining prices. An efficiently functioning market ensures this relationship, the description of which is reserved for later presentation.

Price Reporting and Forecasting. In order to provide soybean producers, government policy makers and other market participants with a consistent measure of soybean prices, the U.S. Department of Agriculture undertakes, notwithstanding the complexities outlined above, to calculate and disseminate daily, monthly and annual soybean price information. These USDA prices reflect not only the historical prices that have been observed, but also include forecasts of expected soybean prices. While USDA reports many types of soybean prices, the most prevalent and closely followed is the average price received by farmers. This reported price, whether a daily, monthly or annual average, represents the national average price for soybeans and purports to measure the price received by soybean producers at the farm gate.

In addition to reporting historical soybean prices, USDA also forecasts the future level of prices received by farmers. This USDA forecast represents, in the case of soybeans, the average price received by farmers for soybeans sold in the soybean market year, September through August. Thus in mid-May 1994, USDA publishes forecasts for the average price received for the current marketing year 1993–1994 (September 1, 1993 to August 31, 1994) and also for the 1994–1995 season—the crop that is just being planted.

The Department of Agriculture employs numerous models and the methods to forecast future soybean prices. The results of one such model, employed to estimate the U.S. average price received by farmers for soybeans, is presented in Table 3.7 (10). The data correspond to a January, 1992 forecast of the 1991–1992 season-average price received by farmers. The methodology, a more complete description of which may be found in the reference cited, is a multistep process involving estimation of the relationship between the average price received by farmers and futures prices observed from the Chicago Board of Trade. Once these relationships are determined, monthly forecasts of prices received by farmers are made and weighted by monthly marketings to forecast a season-average price. The methodology is applicable to other commodities and illustrates the value of information provided by organized futures markets.

3.2 The Intrinsic Value of Oilseeds

Derived Demand. The demand for almost all oilseeds is a derived demand. This distinction applies to commodities, the demand for which is determined by the demand for components of the commodity rather than by the attributes of the commodity in its natural or unprocessed form. Oilseeds generally derive their value from crushing into oil and meal. The oil and meal then have their own demand, reflecting its use as food or feed.

In some cases, the demand for an oilseed product may itself be a derived demand. For example, the demand for flaxseed may be derived from the demand for linseed oil, the demand for which may be derived from the demand for housing and its related uses. Peanuts represent an oilseed having a direct demand for food use, as well as a derived demand as oil and meal.

Table 3.7 Futures method of U.S. soybean producers' season average price, 1991–1992 ($/bushel)

Item	Sept.	Oct.	Nov.	Dec.	Jan.	Feb.	Mar.	Apr.	May	June	July	Aug.	Sept.
Current futures prices[a] by contract (settlement)							5.67		5.74		5.82	5.86	5.88
Monthly futures price based on nearby contract					5.67	5.67	5.74	5.74	5.82	5.82	5.86	5.88	
Plus the basis	−0.23	−0.26	−0.25	−0.21	−0.21	−0.15	−0.22	−0.14	−0.16	−0.22	−0.09	−0.02	
Forecast of monthly average farm price					5.46	5.52	5.52	5.6	5.66	5.6	5.77	5.86	
Actual monthly farm price	5.64	5.49	5.48	5.39									
Spliced actual/forecasted monthly farm	5.64	5.49	5.48	5.39	5.46	5.52	5.6	5.66	5.6	5.77	5.86		
Annual price projections: Simple average	5.58												
(Marketing weights in percents)	7.66	23.46	11.08	7.04	12.38	6.22	6.74	6.00	5.20	4.90	4.60	4.72	
Weighted average	5.54												

[a] Contract months include: September, November, January, March, May, July, and August. Futures price quotation from the Chicago Board of Trade, January 9, 1992, settlement.

Joint Products. The principal use of soybeans is for crushing into meal, which is then fed to poultry, swine or cattle. Crushing also results in the production of soybean oil, a food product. Thus soybean meal and soybean oil are examples of joint products—with production of one resulting in the production of the other and their supply affected by their own demand as well as that of their coproduct.

For oilseeds in general, the products, meals and oils, are relatively homogeneous but with unrelated end uses. Thus, separate factors influence the demand for the two products. However, a shift in the demand for one product alters the supply and price of the other product. For example, lower grain prices versus meal prices may reduce feed demand for meal and consequently lower crushing. However, in so doing, lower meal prices indirectly reduce oil supply and raise oil prices. Figure 3.7 illustrates the joint nature of the industry and the interrelationship between demand, supply and prices for products and seeds.

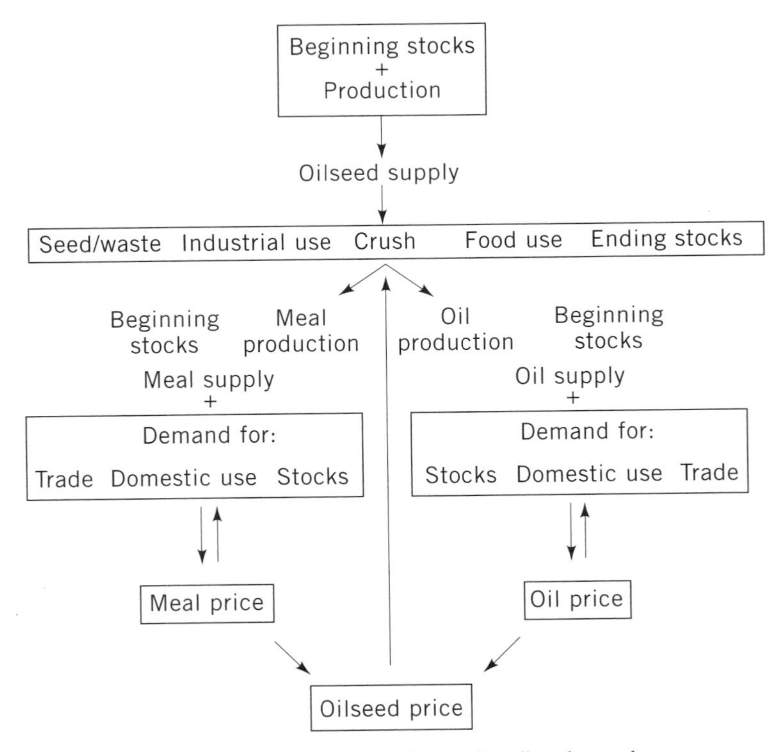

Figure 3.7 Price determination in the oilseed complex.

Demand for Soybeans. The implied value of soybeans may be determined by examination of the markets for soybean meal and soybean oil. In the United States soybean meal accounts for about 80% of all high-protein concentrates fed to livestock and over 70% of domestic soybean meal production is used in domestic animal feeding.

The demand for soybean meal depends on the profitability of livestock feeding. Over time, the relative importance of different livestock categories to soybean meal use has changed with poultry increasing in importance, beef cattle declining and hogs remaining stable.

Among these categories, soybean meal is especially important to hog and poultry feeding. In hog feeding, other high-protein feeds are not easily substituted without a loss in quality, while in poultry feeding, soybean meal is important because poultry animals require the highest ration of high-protein feeds in their diets. As ruminants, beef and dairy cattle are able to draw protein from other sources, such as chemical urea or roughage, and thus soybean meal is not as important in their rations.

The historical pattern of U.S. livestock production illustrates the increasing importance of protein feed in the livestock sector. The distribution of U.S. livestock in animal units according to principal feedstuff, i.e., grain, protein and roughage, is presented in Figure 3.8. Between 1975 and 1993, grain-consuming animal units increased 16.5%, high-protein-consuming animal units increased 32.7% and roughage-consuming animal units decreased by 20%. These patterns are expected to continue.

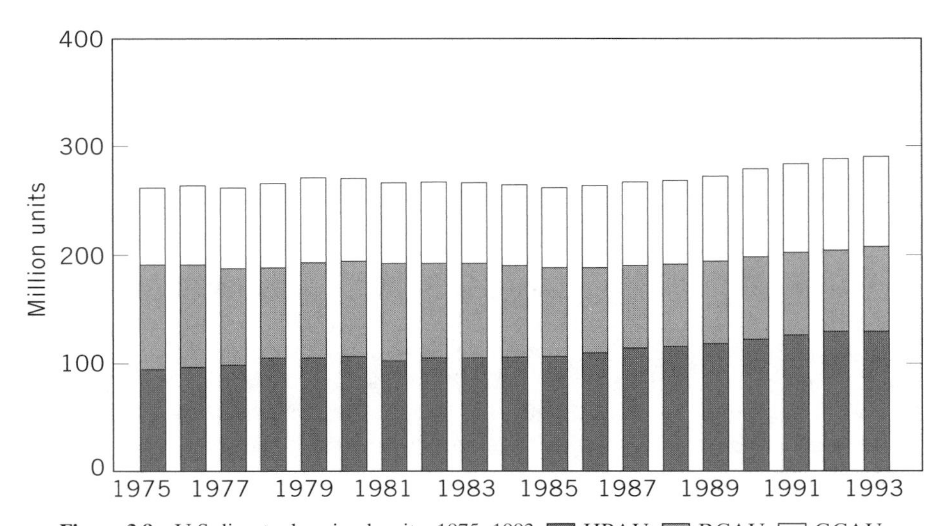

Figure 3.8 U.S. livestock animal units, 1975–1993. ■ HPAU; ▨ RCAU; □ GCAU.

The following model has been estimated to forecast quarterly soybean meal use:

$$\text{SMUDT} = \text{SMPDML1} + \text{COPFML1} + \text{PR7LVL1} + \text{HAPFC} + \text{HOSWFL1} + \text{HOSWFL2} + \text{CANPL}, \text{ where}$$

SMUDT is domestic soybean meal use in the current quarter, 1000 tons,
SMPDML1 is the average soybean meal price at Decatur, Illinois, (44% protein), in the previous quarter, $ per ton,
COPFML1 is the average farm price of corn in the previous quarter, $ per ton,
PR7LVL1 is the livestock price index of 7 meat animal prices in the previous quarter, 1977 = 100,
HAPFC is the average price received by farmers for all hay (baled) in the current quarter, $ per ton,
HOSWFL1 is sows farrowing (10 states) in the previous quarter, 1000 head,
HOSWFL2 is sows farrowing (10 states) in quarter two periods previous, 1000 head, and
CANPL is net placements of cattle into feedlots (13 states) in the current quarter, 1000 head.

Estimated model:

$$\begin{aligned} \text{SMUDT} = {} & 124.641 - \underset{\substack{(-6.05)\\ [-0.33]}}{7.451\ \text{SMPDML1}} - \underset{\substack{(-2.08)\\ [-0.18]}}{286.733\ \text{COPFML1}} \\ & + \underset{\substack{(0.70)\\ [0.07]}}{0.120\ \text{HOSWFL1}} + \underset{\substack{(2.38)\\ [0.25]}}{0.421\ \text{HOSWFL2}} + \underset{\substack{(5.78)\\ [0.57]}}{19.347\ \text{PR7LVL1}} \\ & + \underset{\substack{(2.71)\\ [0.30]}}{21.087\ \text{HAPFC}} + \underset{\substack{(4.16)\\ [0.29]}}{0.216\ \text{CANPL}} \end{aligned}$$

$$R^2 = 0.89$$
$$\text{D.W.} = 2.07$$

In this estimated model *t*-statistics for the parameters are shown in parentheses, while the associated elasticities, calculated at the variable means, are in brackets. All estimated coefficients have the expected sign (positive or negative), and all but the coefficient for HOSWFL1 are significant at the 5% level (crit. $t = 1.706$ at the 5% level for a one-tailed test). The elasticities and coefficients calculated in the model suggest among others, the following:

- a 10% rise in the price of soybean meal in one quarter will lead to a 3.3% decline in use in the following quarter

- The negative sign for the implied cross-price elasticity between meal use and lagged corn price [−0.18] suggests that corn and soybean meal are relatively poor substitutes
- Most of the boost in soybean meal use from hog feeding is attributable to sows farrowing two quarters previously
- the long-run elasticity of soybean meal demand to sows farrowing of 0.32 (obtained by adding the coefficients of farrowing of HOSWFL1 and HOSWFL2) implies that a 10% increase in sows farrowing will lead to a 3.2% total increase in meal use over the following two quarters
- a 10% increase in livestock prices will lead to a 5.7% increase in soybean meal use in the following quarter.

Thus, soybean meal use is shown to depend upon its own price, the prices of other feeds, and the prices of livestock (11). The model is particularly appealing since the data are predominantly lagged values of the variables, and thus more easily obtained than if they were concurrent with the soybean meal use forecast. Additionally, the variables are typically monitored and reported in situation and outlook work, thus readily available.

Aside from meal, the major determinant of soybean value is oil. In the United States, soybean oil represents the largest component, by far, of total vegetable oil disappearance, averaging 73% of annual use over the 1982–1986 period. However, since reaching a peak proportion of over 75% in the early 1980s, soybean oil has lost some of the market, falling to near 70% in the early 1990s. Nonetheless, among individual vegetable oils, soybean oil disappearance, estimated at 13.2 billion pounds in the 1993–1994 marketing year, simply overwhelms all others.

While soybean oil is the major component of U.S. vegetable oil disappearance, domestic disappearance is also an even larger component of total use of U.S. soybean oil. Over the 1982–1986 period, domestic use represented 85.5% of soybean oil produced in the United States, while exports represented 13.5%. Of total domestic use of soybean oil over this period, 97.4% was used in edible products. Thus the demand for soybean oil in the United States is principally food related.

The short run direct price elasticity for soybean meal presented earlier is −.33, implying that demand is relatively unresponsive to price changes in the short term. Longer term elasticities, usually larger, reflecting a stronger response over a longer period, vary widely but generally fall within the range of −.5 to −1.7.

Vegetable oils, the demand for which reflects its food use, exhibit the relatively low price elasticities associated with food products. For food as a whole, the direct-price elasticity has been estimated at −.185, the elasticity calculated for fats and oils, as a whole, is reported at −.139, lower yet than for food products in aggregate and much lower than for soybean meal (12).

These values suggest that demand for vegetable oils in general is relatively unresponsive to price. However, owing to the high degree of substitutability among vegetable oils, the demand for a particular vegetable oil usually exhibits much more responsiveness to price differences. For instance, it has been demonstrated that soybean oil and cottonseed oil will readily substitute for each other, given a significant price difference, and that this substitution may occur if the differential persists for as short a period as 2 or 3 months (13). Not surprisingly, this substitution is reflected in a strong price relationship between soybean oil and cottonseed oil over time.

Value of Oil and Meal Determine Soybean Prices. The physical quantities of soybean oil and meal contained in soybeans and prices for soybean meal and soybean oil are the major determinants of soybean prices. These attributes are transmitted primarily through the crush demand for soybeans, the largest use category. Soybean crush demand is largely determined by "crush margins," the difference between the value of the products obtained through crushing and the cost of the soybeans crushed.

The average U.S. oil yield per bushel of crushed soybeans increased from 9.5 pounds in 1947 to 11.0 pounds in 1953. Since 1953, the average yield of oil from soybeans has ranged from 10.5 to 11.3 pounds per bushel. Variability in oil yields comes almost entirely from variability in the oil content of the soybean; processing methods are not believed to affect oil yields from year to year. During 1980–1987 meal yield averaged 4.3 times that of oil (47.54 pounds of meal versus 10.98 pounds of oil per 60-pound bushel).

The oil yields from crushed soybeans have been shown to differ according to month of crushing, weather and crop yields, and region of production, among other factors. Moisture stress has been shown to raise the oil percentages of soybeans while protein was reduced. The two highest oil contents on record are associated with the 1983 and 1988 droughts in major soybean-producing areas (14).

Oil and protein contents are not yet considered in pricing and grading soybeans and thus at the farm level do not influence a producers' production decisions. Individual producers of soybeans have virtually no influence on farm-level prices in an industry comprised of 442,000 soybean farms in 1987. Thus an individual farmer is a price-taker in the market. In a system which does not specifically consider oil and meal content in pricing soybeans, producers will continue to make decisions that maximize their individual production without concern for the constituent value of their soybeans.

Crushing and Crush Margins. Soybean crush margins are closely monitored barometers of the current and future value of soybeans. The United States Department of Agriculture forecasts U.S. soybean crush as part of its supply, demand and price analysis activity for oilseeds. Forecasts are made for the total crush on a soybean crop year basis, which covers the 12-month period from September 1 to August 31. Thus the forecast soybean crush for the 1993–1994 soybean crop year, for example, would include all soybeans crushed for the period September 1, 1993 to August 31, 1994.

Soybean crush or processing margins indicate the cost, including profit, of providing crushing services. Fluctuations in soybean supply and demand cause variability in processors' crushing margins. During the 1970s, the annual average processing margin was double the average for the 1960s. In the 1980s processing margins reached a peak of $1.22 in 1987–1988 and have averaged about 90 cents since.

Table 3.8 presents data for soybean oil and meal yields, prices and values associated with soybean crushing. These data indicate that in recent years, soybean oil and meal have represented about 33 and 67%, respectively, of the value of products derived from crushing a bushel of soybeans.

Futures markets play an important role in soybean crushing. Soybean processors attempt to "lock in," through use of the futures market, as much crush as possible when processing margins are favorable. When the demand for soybean oil and meal is greater than production and the supply of soybeans is ample, soybean products sell high in relation to the cost of soybeans. This relationship means good processing margins and gives the processor a strong incentive to crush. Conversely, if the demand for soybean oil or soybean meal or both is weak or if soybeans are in short supply, the processing margins will probably be poor, reducing the incentive to crush. The seasonal pattern of soybean prices favors crush at the beginning of the crop year when soybean prices are low and the demand for millfeeds is high (due to colder weather, lack of grazing, and increased animal feed requirements). As the crop year progresses, the incentive to crush declines because of rising soybean prices and the weakening of the demand for millfeeds (3,15).

The existence of futures markets, in addition to spot markets, allows calculation of crushing margins at future dates based upon futures prices for soybeans, soybean oil and soybean meal (Table 3.9). These values are of particular importance to forecasters of future soybean crush levels. For instance, a wide margin at a future date suggests that soybean crush may strengthen in response to the enhanced profit potential.

3.3 The Structure of the Oilseed Processing Industry

Demand for Crushing Facilities. Changes in the composition of oilseed and product trade, particularly to importing countries with crushing industries, are the result of changing domestic and international demand for the products. As product prices change, the incentives facing the crushing industries change and the import mix, seeds versus meal, changes. This situation is particularly true of the EU import mix which, as the major consumer and importer of oilseeds and products, affects world prices as the import mix changes.

Changing supply and demand factors affecting the price of the products, meals and oils, have globally generated considerable investment in oilseeds

Table 3.8 Soybeans: Monthly value of products per bushel of soybeans processed, and spot price spread, U.S., 1982/83–1993/94

	Value of Products per Bushel										Price
	Soybean Oil			Soybean Meal				Percent of Value			
Date	Yield (lbs.)	Price[a] (¢)	Value ($)	Yield (lbs.)	Price[b] ($)	Value ($)	Total Value ($)	Soybean Oil	Soybean Meal	No. 1 Yellow Illinois Processor ($)	Spread Between Value of Products and Soybean Price ($)
1982/83	10.76	19.23	2.07	47.88	194.77	4.66	6.73	0.31	0.69	6.12	0.61
1983/84	11.26	31.08	3.50	47.36	210.51	4.99	8.49	0.41	0.59	7.86	0.62
1984/85	11.05	29.98	3.31	47.15	138.22	3.26	0.50	0.50	0.50	5.98	0.59
1985/86	11.01	18.73	2.06	47.27	162.49	3.84	5.90	0.35	0.65	5.30	0.60
1986/87	10.86	15.22	1.65	47.08	176.25	4.15	5.80	0.29	0.71	5.14	0.66
1987/88	11.04	22.08	2.44	47.76	231.90	5.54	7.98	0.30	0.70	6.76	1.22
1988/89	11.16	21.67	2.42	47.43	256.50	6.08	8.50	0.28	0.72	7.53	0.97
1989/90	11.17	21.50	2.40	47.55	190.22	4.53	6.93	0.35	0.65	5.96	0.97
1990/91	11.23	21.31	2.39	46.91	180.19	4.23	6.62	0.36	0.64	5.90	0.72
1991/92	11.41	19.31	2.20	47.64	190.65	4.54	6.74	0.33	0.67	5.84	0.90
1992/93	10.85	20.93	2.27	47.57	192.68	4.58	6.85	0.33	0.67	5.96	0.90
1993–1994											
September	10.95	23.61	2.58	47.98	199.90	4.80	7.38	0.35	0.65	6.43	0.95
October	10.91	22.96	2.50	47.60	194.50	4.63	7.13	0.35	0.65	6.06	1.07
November	10.74	25.43	2.73	47.47	209.40	4.97	7.70	0.35	0.65	6.64	1.06
December	10.67	28.27	3.02	47.25	206.00	4.87	7.88	0.38	0.62	6.94	0.94
January	10.77	29.91	3.22	47.55	298.30	4.71	7.94	0.41	0.59	7.01	0.93
February	10.86	28.85	3.13	47.61	198.40	4.72	7.86	0.40	0.60	6.86	1.00
March	10.87	29.03	3.15	47.60	195.40	4.65	7.80	0.40	0.60	6.92	0.88
April	10.94	27.90	3.05	47.57	188.90	4.49	7.54	0.40	0.60	6.70	0.84

[a] Crude, tanks, f.o.b. Decatur.
[b] Forty-eight percent (solvent), Decatur.

Table 3.9 Soybean and product prices and crush margins, cash and futures[a]

		Cash Markets				CBOT Futures[b]	
Item	Unit	Latest[b] June 15	Week ago[b] June 8	Month[b] May 1994	Year ago[c] June 1993	July	Sept.
Soybean oil							
Price	cents/lb	27.91	27.58	29.01	21.21	27.81	27.75
Value	$/bu	3.05	3.02	3.17	2.30	3.04	3.04
Soybean meal							
Price	$/ton	205.50	193.00	203.75	193.10	205.70	206.30
Value	$/bu	4.89	4.59	4.85	4.59	4.89	4.91
Total value of oil and meal	$/bu	7.94	7.61	8.02	6.89	7.93	7.95
Soybeans							
Price	$/bu	7.10	6.74	6.89	6.06	7.05	6.93
Crush margin	$/bu	0.84	0.87	1.13	0.83	0.88	1.02

[a] Cash price quotes from AMS based on crude oil; 48% protein meal; and soybeans at Illinois processors.
[b] Based on April 1994 extraction rates of 10.94 lbs of oil and 47.57 lbs of meal per bushel.
[c] Based on June 1993 extraction rates of 10.84 lbs of oil and 47.51 lbs of meal per bushel.

processing capacity. While the majority of processing facilities are located in the producing countries, fluctuations in prices have made it lucrative for countries that aren't major producers and exporters of oilseeds products to invest in crushing facilities, such as the EU and Japan.

Excess crushing capacity in many developed countries and the technical efficiencies inherent in modern crushing facilities, that can easily switch from crushing soybeans to other oilseeds, imply that product and oilseed prices cannot deviate significantly in the long term. Higher product prices will stimulate increased demand for the seeds for crushing, increasing oilseed imports and strengthening seed prices relative to the product prices. Consequently, equilibrium in the market between the prices of the various meals and oils is maintained by the availability of crushing facilities in importing countries.

Industry Consolidation and Integration. In general, worldwide technical innovations for improving crushing efficiencies have stimulated industry consolidation. In the United States the number of companies producing fats and oils products has declined by 4% annually from about 500 in 1982 to 390 in 1987 (the only years for which official U.S. data are available). These companies have operated 646 establishments in 1987, 391 of which had 20 employees or more each (16).

Soybean oil mills, the key industry, are located mainly in the Midwest in proximity to soybean production and hog production, which requires a large volume of soybean meal. There is also sizable soybean processing in other states, such as Arkansas, which have large poultry production.

Large oilseed mills or vegetable oil refineries are capital- and energy-intensive and can operate on a continuous basis with relatively few employees. Incorporation of computerized plant controls has further reduced the need for labor. In 1987, employment in the U.S. fats and oil industry was 39,000 persons having declined during 1982–1987 by about 3% annually.

In the last decade there has been considerable horizontal and vertical integration among U.S. soybean oil mills. In 1987, the four largest companies accounted for 71% of total industry shipments, up from 61% in 1982. Several companies have vertically integrated into further processing, thereby marketing their own feed rations.

OILSEED AND PRODUCT PRICE DETERMINATION, DISCOVERY, AND MARKETING

The prices of oilseeds and oilseed products are determined on the world market by demand and supply conditions. Unlike other agricultural commodities, trade of oilseeds and their products is not as hampered by distortions introduced by export subsidies and other trade distorting measures.

In the shorter term, prices of a specific oilseed and their products are determined by supply disruptions caused by weather, and demand issues, such as changing prices for other oilseeds and products. In the longer term, prices are a function of technology and production costs, two items that affect supplies, and also determined by issues that affect the overall demand for vegetable oil and protein meals such as changing preferences, governmental policies, and technology that allows increasing substitution.

There is a high degree of substitutability among the various vegetable oils and protein meals that allows changes in relative prices to shift both the shorter and longer term demand for the products. However, short term substitution of protein meals is constrained by the difficulty of some animals in adjusting to immediate ration changes. Consequently, there is less adjustment in response to changing short term prices than those that last at least one year.

The price of oilseeds is determined by factors affecting the production of the seeds while oilseed demand is determined by the factors affecting demand for the individual and various products. Consequently, while soybean prices will fluctuate depending on the weather conditions, the product prices will be affected by government policies changing the demand for protein meals. Changing price differentials between the various meals and oils, combined with interchangeability of the products, both shift the demand preferences and eventually translate into prices for the oilseeds.

The high degree of interchangeability between the various vegetable and animal oils and fats, as well as between the protein meals, limits the degree to which a shortage in one oilseed affects prices within the oilseed complex.

However, the importance of soybeans to the entire complex and the fact that soybean stocks account for about 85% of world oilseed totals, implies that a large production shortfall could raise product prices for the entire complex.

Prices for protein meal are determined by the level of protein content in the meal. For this reason, soybean meal typically commands a premium over other protein meals containing less protein. Also, the structure of the feeding industry in an individual country and the prices and availability of substitutes feeds such as corn gluten and citrus pulp, as well as the price of energy compliments such as grains, together determine the price of the protein meal. Demand for protein meal as an input to the livestock industry serves to make the price of protein meals dependent also on the price of meat products.

Interactions between demand and supply determine the underlying price of a commodity at a specific time and place. This process of price determination is a function of the changing factors that determine supply and demand: changing incomes; population; preferences; government policies; as well as changing technology, weather, input availabilities and prices.

Unpredictable shifts in demand and supply for commodities result in volatile prices. This volatility creates risks for producers, traders, processors, and others dependent on soybeans and their products. Consequently, a constant process of price discovery is underway in which buyers and sellers in formalized trading institutions attempt to find and use constantly changing equilibrium prices (17).

The increasing uncertainty inherent in global markets as more and more information is needed for decision-making, has strengthened the role of formalized trading institutions. Buyers and sellers are becoming increasing spatially and linguistically diverse. Consequently independent organizations have been developed to establish standard contract terms acceptable to both buyers and sellers globally.

While some formalized trading institution are in place for other oilseeds, most of the trading is limited to soybeans and takes place at the Chicago Board of Trade (CBOT). There, buyers and sellers are involved in the price discovery process. The price discovery process is the institutional process "by which buyers and sellers attempt to find and use the underlying but unknown prices for use in actual purchases and sales" (2). The commodity market provides daily cash price quotes for soybeans, soybean oil and soymeal as well as providing futures price quotes for up to 9 months into the future. The CBOT exerts considerable influence on world oilseed prices due largely to soybean's dominance in the world oilseed market, the importance of U.S. oilseeds to the world oilseed complex, and the availability of market information.

4.1 Commodity Markets and Trading

Chicago Board of Trade. Forward contracting for commodities has existed for many centuries; however, organized futures markets have a shorter history. The Chicago Board of Trade (CBOT), the oldest and largest futures exchange in the world, opened in 1848. Initially founded as a centralized marketplace for buying and selling of grains, the Board has expanded its operations to include precious metals and the financial industry. However, to this day, more than 80% of the grain futures trading in the world is done each year at the Board of Trade (17).

Soybeans have had a long history of trading activity, starting with futures contracts sold by the CBOT in 1936. However, it has been rumored that the first contracts for soybeans were actually sold in Japan several hundred years ago. Soybean–oil futures trading began at the CBOT in 1950 with soybean meal added in 1951. In 1980, soybeans were one of the most actively traded contracts on the Chicago Board of Trade, or on any other exchange. From 1987 to 1993, trading in soybeans and products average approximately 16% of total commodity trading at the exchange. The trading volume in soybeans has expanded from less than a half a billion bushels in 1936 to the 58 billion bushels traded in 1993 (18).

A variety of soybean futures contracts are traded at the CBOT and its affiliate, the Mid-America Commodity Exchange. Futures contracts are offered by soybeans (5,000 bushels per contract), soybean meal (100 tons per contract), soybean oil (60,000 pounds per contract), and an option on the soybean futures contract.

International Commodity Exchanges. Since the opening of the CBOT, futures exchanges have proliferated, both in the U.S. and internationally. The past decade has been a period of dynamic growth for futures markets as globalization of markets has increased demand for information and heightened desire for reduced risk and stable prices. Active futures markets facilitate marketing and price discovery for traded commodities.

Presently in Japan, soybeans are bought and sold by the Tokyo Grain Exchange as well as those in Osaka, Kobe, Kanmon and Nagoya. The volume traded by Tokyo dwarfs that handled by the other exchanges. In October 1952, the Tokyo Grain Exchange opened for trading in domestic soybeans, red beans, and potato starch. As trading expanded to imported soybeans and white beans, the Exchange expanded in size and is now one of the largest exchanges in Japan (19).

Other exchanges for soybeans exist in Brazil and Manila. However, in February 1994, the Brazilian commodity exchange, the Bolsa de Mercadorias & Futuros, suspended soybean futures contracts due to lack of trading volume. Soybean meal is sold at the London Commodity Exchange while the Winnipeg Commodity Exchange, the oldest and largest futures exchange in Canada, conducts the world's only spot and futures market for canola and flaxseed. In the EU, private trading and crushing firms supply daily fob price quotes for different oilseeds and their products.

In addition, numerous international exchanges exist for traded oils, including the Kuala Lumpur Commodity exchange in Malaysia. This exchange which specializes in palm oil, cocoa, rubber, and tin, commenced operations in 1980. This exchange will become increasingly important for monitoring vegetable oil prices given the increasingly important role that palm oil plays in world vegetable oil trade.

While futures trade in tropical oils has been limited to this exchange, the CBOT is proposing to introduce in 1994 an edible oils index futures. This index incorporates four vegetable oils: palm, soybean, rapeseed, and sunflower and the unit of trading shall be 100 metric tons times the price of the International Edible Oils Index. The contract standards for this specific contract was developed and submitted to the CBOT by the Federation of Oil, Seed and Fats Association LTD (FOSFA). FOSFA is responsible for publishing standard contracts for the oilseed trade and serves as a legal entity for arbitrating oilseed trading disputes. FOSFA has a long history of involvement in the oilseed sector dating back to 1863. There are presently more than fifty different contracts referring to specific products and origins. The standard contracts define quality criteria but are usually restricted to oil content, moisture content, impurities and admixtures. Any buyer or seller, however, has the option to add to the standard terms when negotiating individual contracts (7).

Futures Prices, Cash Prices, and Basis. The usefulness of prices provided by commodity futures markets has been demonstrated earlier, without discussion of how the information is generated in these markets. Price information is a major economic contribution of futures markets, and these prices have become the most widely used pricing reference in domestic and international commodity trade (15). Table 3.6 presents a sample of futures and cash prices for soybeans at a specific point in time. The cash prices, sometimes referred to as spot market prices, reflect the value of soybeans at specific locations on a given date. While the abundance, and sometimes wide divergence, of cash prices may give an initial impression that they are stochastic, these prices are related in a very specific manner.

The difference between a specific futures contract price for a commodity and its cash price is called basis, and is calculated by subtracting a specific futures price from a specific local cash price. The basis reflects, among other things, costs of storage and transportation and differences in quality, if any exist (20). Thus, the difference between the July, 1994 futures price for soybeans and the cash price at Grand Island, Nebraska (700.5 and 675.0, respectively) is the basis, and in this case, −25.5 cents per bushel, on June 1. It is generally the case that as the distance from the delivery point increases, the basis will widen to reflect higher transportation costs. Spot prices are usually lowest at harvest and tend to rise as storage costs are incurred, causing the basis to become narrower as the storage progresses. By studying the basis on a particular date over a number of years, a soybean producer in any part of the country may form a reasonable expectation of the worth of his crop from the observed futures prices.

4.2 Commodity Contracts and Marketing

Production of oilseeds and other field crops is an inherently risky venture owing to the unpredictability of weather and the potential for damage from pests. Added to the risk associated with production, is risk associated with volatile prices. As a result, farmers can be subjected to wide variations in income derived from their agricultural enterprises. One method of reducing risk, that portion associated with prices, is through marketing by use of contacts—cash, futures or options.

Cash Contracts. Cash contracts with local buyers is the most common form of contracting used by farmers. These contracts can be most closely tailored to an individual farmer's needs. Cash contracts may set a firm price or base the contract price on some cash or futures price to be observed at a later date. While offering flexibility, this pricing method has drawbacks. There may be few potential buyers offering cash contracts and there is the risk that the buyer may default.

Futures Contracts. Futures contracts are more formalized and are traded in organized exchanges. Their use requires the services of a broker and establishment of a margin account. Futures contracts are an attractive marketing tool in that they are readily available, very secure from default, widely quoted and offer flexibility to change amounts contracted or delivery dates if changes in market conditions warrant. Among their drawbacks are the need for an in-depth understanding of the futures market and the ability to meet any margin calls. Also, standardized contract sizes may not be particularly suited to the output of an individual farmer.

Options Contracts. Options contracts are another alternative available to farmers who wish to assure minimum prices for outputs. These are also purchased through a broker and give the purchaser the right, but not an obligation, to buy or sell a futures contract at a specified price during a specific time period. The farmer must pay a charge for the option, but is not subject to margin calls. The farmer retains the possibility of benefiting from favorable price changes.

Despite its attractiveness, forward pricing is used by relatively few farmers. The risk of a production shortfall and the exposure to buying back a contract at a loss is often cited as a hindrance to greater use. Cash sales at harvest or from on-farm storage is generally the most popular pricing arrangement (21).

4.3 Commodity Markets and Price Discovery

Market Participants. An efficiently functioning market incorporates all the available information about a commodity into its price and thus assures market participants of a fair exchange value. Some markets, such as that for U.S. treasury bonds, are constantly faced with new information that must be assimilated into price quotes, while other markets may be faced with a relative paucity of information. While the amount of information available may influence price movement, the number of market participants may also contribute to price movement. Markets are sometimes characterized as "thin" indicating there are relatively few market participants or trades. One market that is often characterized as thin, and which often experiences limit moves, is that for lumber.

Many market participants are neither producers nor consumers of the commodity and do not intend to deliver or take physical possession of the commodity they trade. These participants simply hope to profit from positions taken in anticipation of a future price movement. But in so doing, they provide needed market liquidity and assist in keeping markets in equilibrium. For instance, if soybean prices become relatively cheap versus the prices of its products in the soybean oil and soybean meal market, these participants will attempt to exploit the difference and bring the markets back in line. Other participants may anticipate future price movements due to weather, or other factors, and take market positions that will yield a profit if they are correct. All are necessary to keep prices in equilibrium at a specific point in time and for a particular market across time. Participants who formulate ideas of future price movements often do so based on technical analysis and/or fundamental analysis of a commodity.

Technical and Fundamental Commodity Price Analysis. Technical market analysts frequently employ methods that give the user an impression of market direction and price movements without explicitly incorporating fundamental market information. They may chart daily price ranges seeking patterns over time, or employ trend lines to make projections. One advantage of this type of analysis is that its practitioner can apply it to any market and thus does not need to make the substantial investment of time and capital required for fundamental commodity analysis. Technical analysis may require little more than a series of commodity prices and personal computer spreadsheet and graphic capability.

Fundamental commodity market analysis, on the other hand, may be extremely capital and labor intensive. The United States Department of Agriculture conducts fundamental commodity market analysis and employs thousands of persons in this activity. Fundamental market analysis attempts to assess all the supply and demand factors that are likely to influence the price of a commodity and thereby predict price levels and movement.

For soybeans in the United States these factors may include; soybean production in South America, palm oil production in Indonesia, the number of livestock placed on feed, expected consumer expenditure levels, soybean meal requirements in the EU and former Soviet Union, corn prices, and a host of other variables. Due to the interrelated markets, similar analysis must be done for other oilseeds. When USDA performs this analysis the result is the release of monthly estimates of supply, demand and prices for all major field crops and livestock. While some may question the accuracy of a particular estimate, none question their importance to commodity markets and trading. This importance has recently been illustrated as the Chicago Board of Trade has requested that the USDA move release times for estimates from after market closings on release dates to prior to market opening, in order to give U.S. markets the first opportunity to trade on this vital information.

With established commodity markets and the infrastructure to support them, the United States is a leader in commodity trading. It is through these markets that producers and consumers, both domestic and foreign, can meet the oilseed requirements of the world in an efficient manner.

THE OUTLOOK FOR OILSEEDS AND PRODUCTS

The United States Department of Agriculture (USDA) annually conducts a long term forecasting exercise for the domestic agricultural sector. The domestic outlook for U.S. commodities is supported by foreign commodity baseline projections which are a conditional, long term outlook about what would be expected to happen under current agricultural policies in the United States and other countries.

The USDA's oilseed framework forecasts foreign and U.S. soybean and soybean meal production, consumption, and trade through the year 2005. While projections for the entire oilseed complex are not generated, assumptions of relative prices and country data on competing oilseeds are incorporated into the assumptions underlying the soybean and product analysis. Country-specific and regional projections are produced by Economic Research Service (ERS) country and commodity economists using a combination of country models and judgmental analysis. The results of the country-specific projections are incorporated into a world net-trade framework which is balanced by projections of U.S. trade.

This commodity specific information is then incorporated into an interagency process which generates supply and use projections for all the major program crops in the United States. The entire process involves numerous agencies within USDA and the projections encompass statistical and econometric country and commodity models, tempered with judgmental analysis (22).

5.1 Long-Term Forecasting of the Oilseed Sector

Oilseeds: Growth in Consumption, Production, and Trade. World oilseed crush is forecast to grow initially at the same level as in the 1980s, followed by a slight drop from 2000. Stable levels of protein meal consumption growth in the next decade contrasts supply with prospects for economic recovery and higher income growth during the projected period. Consumption growth of soybeans and soybean meal is projected to slow from the rates of the 1970s and 1980s. This development is linked primarily to slower consumption growth in developed countries. For example, consumption growth in the EU will continue to slow as demand for meat products stabilizes and grain consumption increases, as a result of the newly instituted Common Agricultural Policy

(CAP) reforms. Annual consumption growth in the EC is expected to remain nearly unchanged during the projected period. EC soybean meal consumption, in the 1960s, has accounted for 50% of the total foreign meal consumption. This share is projected to decline to approximately 28% by the year 2005.

Future growth in soybean meal consumption will be driven mainly by developing countries, particularly in Southeast Asia and Latin America. For example, economic growth in China is projected to expand dramatically with stronger demand for meat products stimulating increased consumption of feed grains and protein meals.

Despite assumptions of falling real prices, world oilseed production is forecast to grow about 27% from 1993 to the year 2005. Most of this growth is driven by expanding soybean production, supplemented by slight gains from increasing production of sunflowerseed, rapeseed, and palm kernel. Much of the gains will be in developing countries such as Brazil, Argentina, India, China, and Malaysia.

Market shares for the different oilseeds are not expected to differ from the base year of 1992. However, the gains in soybean production will be significant due to soybean's dominant position within the global oilseed complex.

The market oriented reforms in U.S. agricultural policy provided in the 1990 Farm Bill gives oilseeds a competitive edge in the United States as program support has been cut back for competing crops. These policies are assumed to continue in the near future, and will lead to an annual growth of 2% in U.S. oilseed production during the next decade.

Vegetable Oils: Growth in Consumption, Production, and Trade. Continuing the trends of the 1980s (see Figure 3.9) vegetable oil demand is assumed to exceed that of protein meals through 2005, implying higher prices for the higher oil-yielding oilseeds. Limited opportunity to expand production beyond the growth rates evidenced in the 1980s, however, are expected to constrain production growth of high oil-yielding oilseeds, with the exception of palm kernels. Consequently, soybean's share of total vegetable oil and protein meal use is expected to maintain its long term share of the overall market.

World consumption will likely increase just above 2% annually during the 1990s to reach more than 80 million tons in 2005. Tropical and soybean oils will likely benefit the most from the rise in consumption. Palm oil use is forecast up 80% by 2005 and palm oil will continue to strengthen its share of total consumption, rising from 22% in 1992–1993 to 30% in 2005–2006. Demand for soybean oil will rise at an average annual rate of 2%. Consumption shares of other major oils will drop, reflecting a slow downward forecast in production.

World demand for oils depends on income and population growth. Income growth in India, China and Indonesia, which account for more than one-third of the world's population, is expected to be significant in the next decade. However, high prices and import controls in these countries limit consumption

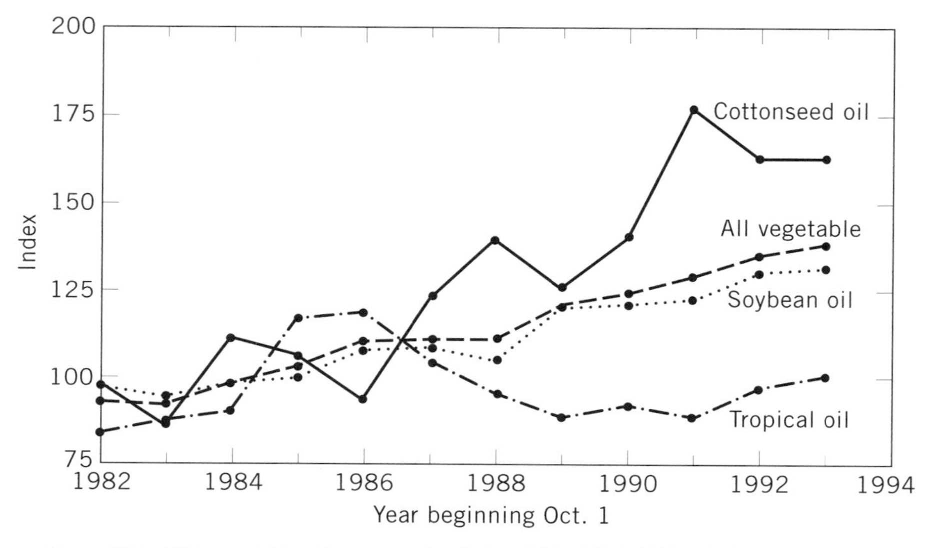

Figure 3.9 U.S. vegetable oil consumption index, 1982–1994. 1982–1986 average = 100.

and imports of vegetable oils. Despite current policies such as import restrictions that increase prices and constrain edible oil consumption in these countries, gains in income alone will be sufficient to push up edible oil use considerably.

Palm oil will likely be the primary beneficiary of this consumption growth. Indonesia, as a major producer, will consume much of its own palm oil, while China and India are expected to import palm oil because of its lower relative price and geographic proximity to producers. While soybean oil's share of world trade may decline, its share of total oil output and consumption is expected to remain at the 1992–1993 level, because consumption in producing countries, such as the U.S., Brazil, China, and India, is estimated to increase substantially.

Although demand for edible oils is relatively price sensitive in developing countries, supply is highly inelastic for most vegetable and marine oils. Tree crops are locked into a level of production regardless of short-term price signals. Oils that come from annual crops are usually joint-products and influenced as well by the coproducts (soybean meal, cotton fiber).

World edible oil production is projected to grow at an average annual rate of about 2% during the next decade. The average annual rate of supply growth is expected to slow from the 1970s and 1980s due to constrained demand for protein meals, and hence lower supplies of the residual oils. The production growth of oils coming from annual crops, especially soybean oil, could be constrained by changing demand for its joint product. If world oil stocks remain at about 10% of total production, and competition is stiff due to the variety of close substitutes, palm oil, the least expensive oil in the market and the oil with significantly less volatility in production, could become the oil with the largest production in the next century.

Total global edible oil exports will maintain a relatively high level of growth, an annual average rate of 2% per year, just under consumption growth rates. Trade in palm oil is largely driving this growth since average annual exports of palm oil will grow 2.9%. Palm oil's share of total trade will increase from an average of 41% in the period 1991–1993 to 48% in 2005. Large increases in Malaysian and Indonesian palm oil consumption, however, will prevent palm oil from growing as fast as consumption and production.

All exports of other oils besides palm oil are forecast to decline slightly over the next 10 years. Growth in soybean oil exports will likely decline as a result of a large expansion in consumption within producing countries. This growth rate is expected to expand above the 1986–1992 growth levels.

5.2 Uncertainties Underlying the Long-Term Outlook

Policy and economic uncertainties cloud the outlook for the world oilseed supply and demand and the prospects for trade over the next decade. Some of the factors affecting global demand are the agricultural reform in the European Union (EU), credit problems in the former Soviet Union (FSU), and the economic uncertainties in Eastern Europe and developing countries. Likewise, factors affecting global supply include: the price outlook for soybeans and other oilseeds; constraints on yield; area expansion by competitors; and policy reforms in major exporting countries.

Overshadowing many of the country specific uncertainties is the effect of the historic agreement concluded on December 15, 1993 which finalizes the Uruguay Round of Multilateral Trade Negotiations under the auspices of the General Agreement on Tariffs and Trade (GATT). This seven-year Trade Round has been the eighth and most ambitious round of trade negotiation under the GATT. It is unique in that it specifically addresses trade barriers, domestic support policies, and export subsidies that impede the free flow of agricultural trade.

The GATT agreement has limited implications for the global oilseed complex because world trade in oilseeds and products is relatively unencumbered by trade barriers, high domestic support, or subsidized exports. However, a slight expansion in trade opportunities for soybeans and products will occur due to higher incomes and slightly lower product tariffs in certain countries. Additional non-trade benefits for oilseeds will occur due to increased feeding of livestock as higher incomes buoy demand for meat products. The United States is expected to capture the largest share of increased trade as expansion of foreign soybean production is hampered due to increased production incentives for competing crops.

REFERENCES

1. *FAO Trade Yearbook*, Food and Agriculture Organization of the United Nations, 1993.
2. J. Houck, M. Ryan, and A. Subotnik, *Soybeans and Their Products: Markets, Models, and Policy,* University of Minnesota, 1972.
3. J. Schaub and co-workers, *The U.S. Soybean Industry,* U.S. Dept. of Agriculture, Economic Research Service, A.E.R. no. 588, Washington, D.C., May, 1988, 51pp.
4. W.C. McArthur, V. N. Grise, H. O. Doty Jr., and D. Hacklander, *U.S. Peanut Industry*, U.S. Dept. of Agriculture, Economic Research Service, A.E.R. No. 493, Washington, D.C., Nov., 1982, 45pp.
5. I. McCormick, C.W. Davison, and R.L. Hoskin, *The U.S. Sunflower Industry,* U.S. Dept. of Agriculture, Economic Research Service, A.E.R. No. 663, Washington, D.C., Oct. 1992, 58pp.
6. I. McCormick and B. Hyberg, "Prospects for Canola: Position in a Dynamic Market," *Proceedings, 1992 U.S. Canola Conference,* March 5–6, 1992, Washington, D.C., pp. 89–95.
7. G. Robbelin, R.K. Downey, and A. Ashri, *Oil Crops of the World*, McGraw-Hill Book Co., Inc., New York, 1989.
8. T.W. Bickerton and J.W. Glauber, *World Oilseed Markets–Government International and Multilateral Policy Reform,* Economic Research Service, Washington, D.C., March, 1990, pp. 36–40.
9. N.R. Morgan, Economic Research Service, Washington D.C., *Food Review*, **16**(2), 26–30 (May–Aug. 1993).
10. L.A. Hoffmann and C. Davison, "Forecasting U.S. Soybean Prices With Future Prices," in *Oil Crops Situation and Outlook Report*, OCS-32, USDA-Economic Research Service, Washington, D.C., Jan., 1992, pp. 23–27.
11. D.B. Hull, P.L. Westcott, and R.L. Hoskin, "Factors Affecting Domestic Soybean Meal Use," in *Oil Crops Situation and Outlook Report,* OCS-4, USDA-Economic Research Service, Washington, D.C., Feb., 1984, pp. 11–14.
12. K.S. Huang, "A Complete System of U.S. Demand for Food, in Tech. Bull. No. 1821, U.S. Dept. of Agriculture, Economic Research Service, Washington, D.C., Sept. 1993, 70pp.
13. E. Fryar and R. Hoskin, "U.S. Vegetable Oil Price Relationships," in *Oil Crops Situation and Outlook Report*, OCS-9, USDA-Economic Research Service, Washington, D.C., Dec., 1985, pp. 21–24.
14. B. Crowder, "Factors Affecting U.S. Oil Yields from Crushed Soybeans," in *Oil Crops Situation and Outlook Report*, OCS-24, USDA-Economic Research Service, Washington, D.C., Jan. 1990, pp. 14–19.
15. Chicago Board of Trade, *Commodity Trading Manual*, Chicago, 1985.
16. U.S. International Trade Commission, *Industry and Trade Summary—Animal and Vegetable Fats and Oils,* USITC Pub. 2631 (AG-13), Washington, D.C., May, 1993, 21pp.
17. Chicago Board of Trade, *Action in the Market Place,* Chicago, 1987.
18. Futures Industry Association, Inc., *Volume of Futures and Option Trading on U.S. Futures Exchanges*, various issues.
19. Futures Industry Association, Inc. *International Report*, various issues.
20. R.G. Heifner, B. Wright, and G. Plato, *Using Cash, Futures and Options Contracts in the Farm Business,* Economic Research Service, A.I.B. No. 665, Washington, D.C., Apr. 1993.
21. M.N. Leath, "Pricing Strategies Used by Soybean Producers," paper presented at *Southern Agricultural Economist Association Annual Meeting*, Orlando, Fl., Feb. 2–5, 1986.
22. *Long-term World Agricultural Commodity Baseline Projections*, Economic Research Service, Washington, D.C., Staff Report, in review, 1994.

4

Flavor Compounds in Fats and Oils

Most of the flavor compounds in fats and oils are produced by the reaction of oxygen with unsaturated fatty acids in triglycerides or other polar lipids. On the other hand, some flavor compounds such as those present in cocoa butter or roasted peanut oil are generated by the interaction of reducing sugars with amino compounds during the roasting process.

The development of objectionable flavor compounds by oxidation has significant detrimental effects on consumer acceptability of edible oils. In the last two decades, much progress has been made in the chemistry of volatile products of lipid oxidation, mainly as a result of advances in separation techniques and analytical methodology, particularly gas chromatography–mass spectrometry.

FREE RADICAL AUTOXIDATION OF LIPIDS

The reaction of unsaturated lipids with oxygen to form hydroperoxides is generally a free radical process involving three basic steps (1,2).

Initiation:

$$\begin{aligned} RH &\xrightarrow[\text{metal}]{\text{heat, light}} R\cdot + H\cdot \\ ROOH + M^{n+} &\longrightarrow RO\cdot + M^{(n+1)+} + OH^- \\ ROOH + M^{(n+1)+} &\longrightarrow ROO\cdot + M^{n+} + H^+ \\ 2\ ROOH &\longrightarrow ROO\cdot + RO\cdot + H_2O \end{aligned}$$

Propagation:

$$\begin{aligned} R\cdot + O_2 &\longrightarrow ROO\cdot \\ ROO\cdot + RH &\xrightarrow{\text{slow}} ROOH \\ RO\cdot + RH &\longrightarrow ROH + R\cdot \end{aligned}$$

Termination:

$$R\cdot + R\cdot \rightarrow R{-}R$$
$$R\cdot + ROO\cdot \rightarrow ROOR$$
$$ROO\cdot + ROO\cdot \rightarrow ROOR + O_2$$

RH, R·, RO·, ROO·, ROOH, and M represent an unsaturated fatty acid or ester with H attached to the allylic carbon atom, alkyl radical, alkoxy radical, peroxy radical, hydroperoxide, and transition metal, respectively.

The initiation reaction is the homolytic abstraction of hydrogen to form a carbon-centered alkyl radical in the presence of an initiator. Under normal oxygen pressure, the alkyl radical reacts rapidly with oxygen to form the peroxy radical which reacts with more unsaturated lipids to form hydroperoxide. The lipid-free radical thus formed can further react with oxygen to form a peroxy radical. Hence, the autoxidation is a free radical chain reaction. Because the rate of reaction between the alkyl radical and oxygen is fast, most of the free radicals are in the form of the peroxy radical. Consequently, the major termination takes place via the interaction between two peroxy radicals.

The rate of autoxidation increases with the degree of unsaturation. Linoleate is oxidized 10 times faster than oleate; linolenate 20–30 times faster (3).

HYDROPEROXIDES OF FATTY ACIDS OR THEIR ESTERS

It is well-known that the free radical mechanism of hydroperoxy formation involves the abstraction of the hydrogen atom from the α-methylene group of a lipid molecule. This result is favored due to the formation of a very stable allyl radical in which the electrons are delocalized over either three carbon atoms such as in the case of oleate, or five carbon atoms such as in the case of linoleate or linolenate. The mechanisms for the formation of isomeric hydroperoxides by autoxidation have been reviewed extensively (4,5).

For oleate, the hydrogen abstraction on C-8 and C-11 produces two allylic radicals. These intermediates react with oxygen to produce a mixture of 8-, 9-, 10-, and 11-allylic hydroperoxides. Autoxidation of linoleate involves hydrogen abstraction on the doubly reactive allylic C-11, with the formation of pentadienyl radical. This intermediate radical reacts with oxygen to produce a mixture of conjugated 9- and 13-diene hydroperoxides. In the case of linolenate in which there are two separate 1,4-diene systems, hydrogen abstraction will take place on the two methylene groups, C-11 and C-14. These intermediate free radicals react with oxygen to form conjugated dienes with hydroperoxides on C-9 and C-13, or C-12 and C-16, with the third double bond remaining unaffected.

2.1 Decomposition of Hydroperoxides

Hydroperoxides of unsaturated fatty acids formed by autoxidation are very unstable and break down into a wide variety of volatile flavor compounds as well as nonvolatile products. It is widely accepted that hydroperoxide decomposition involves homolytic cleavage of the —OOH group, giving rise to an alkoxy radical and a hydroxy radical (4).

$$\underset{\displaystyle\text{OOH}}{\text{R—}\overset{}{\text{CH}}\text{—R}'} \rightarrow \underset{\displaystyle\text{O}\cdot}{\text{R—}\text{CH}\text{—R}'} + \text{OH}\cdot$$

The alkoxy radical undergoes β-scission on the C—C bond, with the formation of an aldehyde and alkyl or vinyl radical. A general reaction scheme with the formation of volatile aldehyde, alkene, alkane, and alcohol is illustrated in Figure 4.1 (6).

Aldehydes. Of the volatiles produced by the breakdown of the alkoxy radicals, aldehydes are the most significant flavor compounds. Aldehydes can be produced by scission of the lipid molecules on either side of the radical. The products formed by these scission reactions depend on the fatty acids present, the hydroperoxide isomers formed, and the stability of the decomposition products. Temperature, time of heating, and degree of autoxidation are variables which affect thermal oxidation (7).

Some volatile aldehydes formed by autoxidation of unsaturated fatty acids are listed in Table 4.1. The flavors of aldehydes are generally described as green, painty, metallic, beany, and rancid, and are often responsible for the undesirable flavors in fats and oils. Hexanal has long been used as an index of oxidative deterioration in foods. Some aldehydes, particularly the unsaturated aldehydes, are very potent flavor compounds.

Hexanal and 2,4-decadienal are the primary oxidation products of linoleic acid. The autoxidation of linoleic acid generates 9- and 13-hydroperoxides of linoleic acid. Cleavage of 13-hydroperoxide will lead to hexanal and the breakdown of 9-hydroperoxide will lead to 2,4-decadienal (8). Subsequent retro-aldol reaction of 2,4-decadienal will produce 2-octenal, hexanal, and acetaldehyde (9). 2,4-Decadienal is known to be one of the most important flavor contributors to deep-fat fried foods (10).

2,4-Decadienal can undergo further oxidation to produce *trans*-4,5-epoxy-*trans*-2-decenal. This compound was recently characterized as one of the most potent odorants of soybean oil stored in the dark and has a low odor threshold of approximately 1.5 pg/L (air) (11).

Ketones. Aliphatic ketones formed by autoxidation of lipids also contribute to the flavor of oils and food products. For example, Guth and Grosch

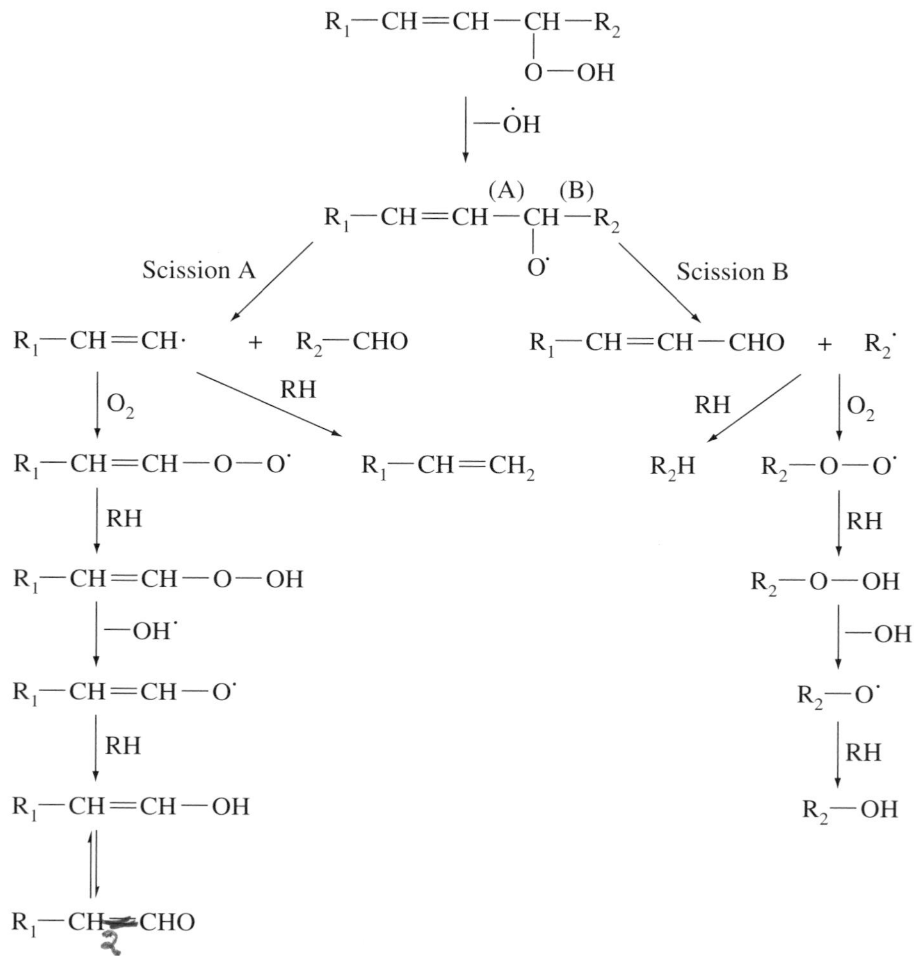

Figure 4.1 General reaction pathway for the homolytic cleavage of hydroperoxides of unsaturated fats (6).

(12) identified 1-octen-3-one as one of the odor-active compounds in reverted soybean oil. This compound was described as metallic and mushroom-like. The reaction pathway for the formation of 1-octen-3-one from the linoleate 10-hydroperoxide via the β-scission route is illustrated in Figure 4.2. 10-Hydroperoxide of linoleate was not the usual hydroperoxide formed by autoxidation of linoleate; however, it was one of the major hydroperoxides formed by the photosensitized oxidation of linoleate (13).

Furans. 2-Pentylfuran has been identified in many fats and oils. It is a well-known autoxidation product of linoleic acid and has been known as one of the compounds responsible for the reversion of soybean oil (14). Figure

Table 4.1 Some volatile aldehydes obtained from autoxidation of unsaturated fatty acids (6)

Fatty Acid	Monohydroperoxides	Aldehydes Formed
Oleate	8-OOH	2-Undecenal
		Decanal
	9-OOH	2-Decenal, nonanal
	10-OOH	Nonanal
	11-OOH	Octanal
Linoleate	9-OOH	2,4-Decadienal
		3-Nonenal
	13-OOH	Hexanal
Linolenate	9-OOH	2,4,7-Decatrienal
		3,6-Nonadienal
	12-OOH	2,4-Heptadienal
		3-Hexenal
	13-OOH	3-Hexenal
	16-OOH	Propanal
Arachidonate	8-OOH	2,4,7-Tridecarienal
		3,6-Dodecadienal
	9-OOH	3,6-Dodecadienal
Arachidonate	11-OOH	2,4-Decadienal
		3-Nonenal
	12-OOH	3-Nonenal
	15-OOH	Hexanal
Eicosapentaenoate	5-OOH	2,4,7,10,13-Hexadecapentaenal,
		3,6,9,12-Pentadecatetraenal
	8-OOH	2,4,7,10-Tridecatetraenal
		3,6,9-Dodecatrienal
	9-OOH	3,6,9-Dodecatrienal
	11-OOH	2,4,7-Decatrienal
		3,6-Nonadienal
	12-OOH	3,6-Nonadienal
	14-OOH	2,4-Heptadienal
		3-Hexenal
	15-OOH	3-Hexenal
	18-OOH	Propanal

4.3 shows the probable mechanism for its formation. The conjugated diene radical generated from the cleavage of the 9-hydroxy radical of linoleic acid may react with oxygen to produce vinyl hydroperoxide. The vinyl hydroperoxide will then undergo cyclization via the alkoxy radical to yield 2-pentylfuran (7).

Alcohols and Other Compounds. Cleavage of lipid hydroperoxides will also lead to alcohols, alkanes, alkenes, and alkynes. The mechanism for the

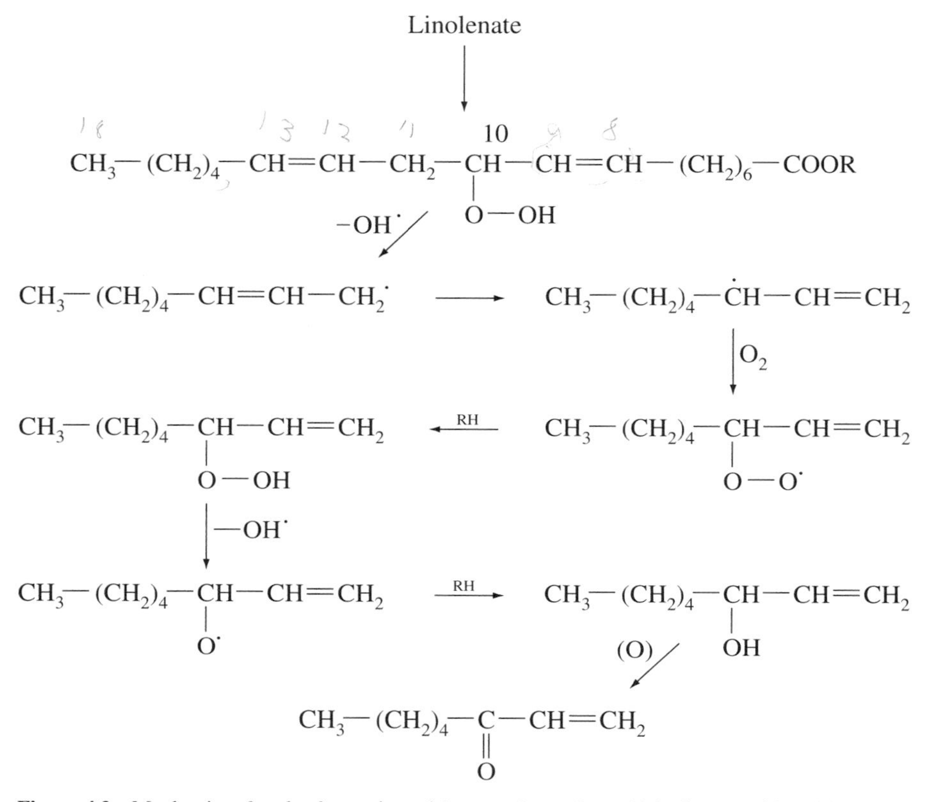

Figure 4.2 Mechanism for the formation of 1-octen-3-one from 10-hydroperoxide of linoleate.

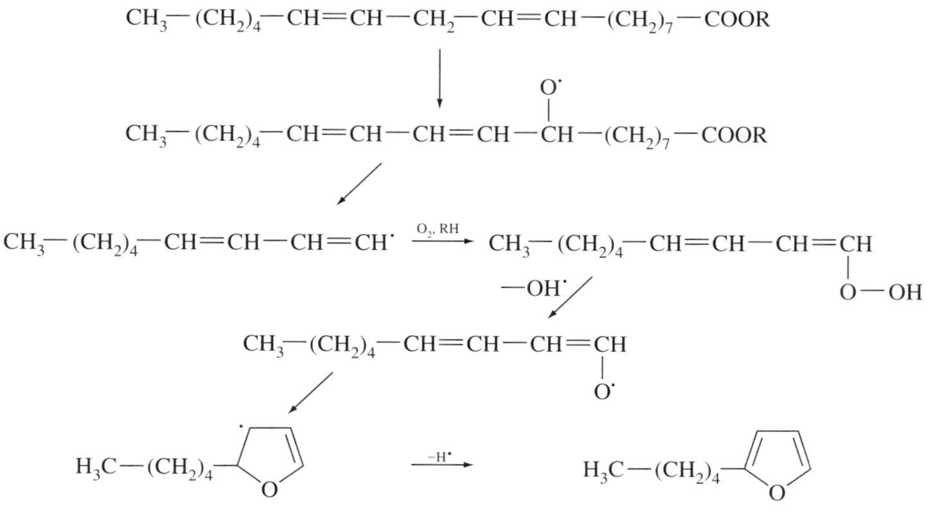

$$CH_3-(CH_2)_4-CH{=}CH-CH_2-CH{=}CH-(CH_2)_7-COOR$$

$$\downarrow$$

$$CH_3-(CH_2)_4-CH{=}CH-CH{=}CH-CH(O^\cdot)-(CH_2)_7-COOR$$

$$\swarrow$$

$$CH_3-(CH_2)_4-CH{=}CH-CH{=}CH^\cdot \xrightarrow{O_2,\ RH} CH_3-(CH_2)_4-CH{=}CH-CH{=}CH(O-OH)$$

$$\swarrow\ -OH^\cdot$$

$$CH_3-(CH_2)_4-CH{=}CH-CH{=}CH(O^\cdot)$$

$$\swarrow$$

$$H_3C-(CH_2)_4-\text{(dihydrofuranyl radical)} \xrightarrow{-H^\cdot} H_3C-(CH_2)_4-\text{(furan-2-yl)}$$

Figure 4.3 Mechanism for the formation of 2-pentylfuran (6).

formation of 1-octen-3-ol, which has a strong mushroom flavor, is also shown in Figure 4.2. Due to their relative high odor threshold, alcohols and hydrocarbons are generally not considered to be important contributors to the flavors of fats and oils and lipid-containing foods.

2.2 Singlet Oxygen Oxidation of Lipids

Oxidation of lipids occurs in the presence of molecular oxygen in both the singlet and triplet states. Atmospheric oxygen that is in the triplet state contains two unpaired electrons, while oxygen in the singlet state has no unpaired electrons (15). The electron arrangement of triplet oxygen does not allow for a direct reaction of lipid molecules that exist in the singlet state. Singlet oxygen can be generated by the interaction of light, photosensitizers, and oxygen. Singlet oxygen has been suggested to be responsible for initiating lipid oxidation of food products due to its ability to directly react with linoleic acid at least 1450 times faster than triplet oxygen (17).

The "ene" reaction of singlet oxygen is important in the photooxidation process in edible oils. In this reaction, singlet oxygen reacts with olefins to form allyl hydroperoxides through a six-membered ring transition state as shown in Figure 4.4. The lipid hydroperoxides produced from singlet oxygen are different from those generated by autoxidation. In singlet oxygen oxidation, both conjugated and nonconjugated hydroperoxides are formed; in free radical autoxidation, nonconjugated hydroperoxides are not usually formed (18).

One of the major volatile compounds formed from cottonseed oil was 1-decyne, which was reported as a predominant volatile of photooxidized cottonseed oil formed from its precursor, sterculic acid, as shown in Figure 4.5. The endogenous trace amount of chlorophyll acts as a photosensitizer to produce singlet oxygen. The singlet oxygen then attacks the cyclopropenoid fatty acid, namely sterculic acid, followed by degradative cleavage and molecular rearrangement to yield 1-decyne (19).

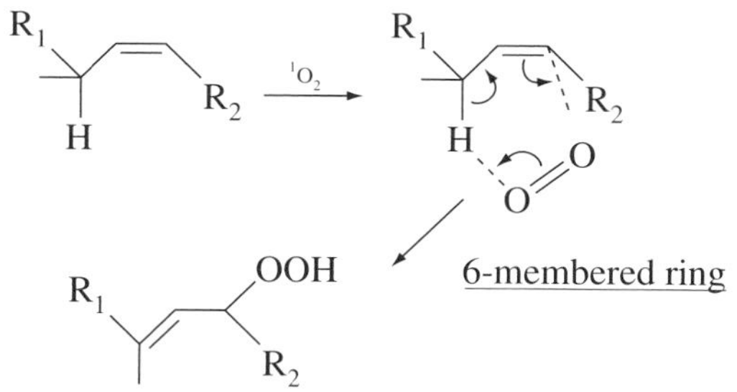

Figure 4.4 The "ene" reaction of singlet oxygen with unsaturated fatty acid.

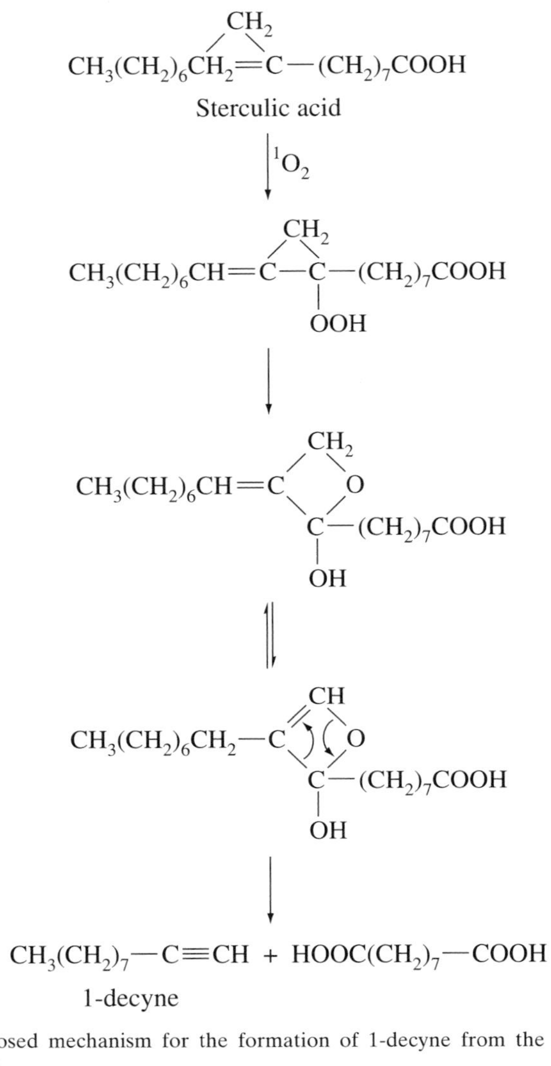

$$CH_3(CH_2)_6CH_2{=}\overset{CH_2}{C}{-}(CH_2)_7COOH$$

Sterculic acid

$$\downarrow {}^1O_2$$

$$CH_3(CH_2)_6CH{=}C{-}\underset{OOH}{\overset{CH_2}{C}}{-}(CH_2)_7COOH$$

$$\downarrow$$

$$CH_3(CH_2)_6CH{=}C\langle CH_2{-}O\rangle \underset{OH}{C}{-}(CH_2)_7COOH$$

$$\rightleftharpoons$$

$$CH_3(CH_2)_6CH_2{-}C\langle CH{-}O\rangle \underset{OH}{C}{-}(CH_2)_7COOH$$

$$\downarrow$$

$$CH_3(CH_2)_7{-}C{\equiv}CH + HOOC(CH_2)_7{-}COOH$$

1-decyne

Figure 4.5 Proposed mechanism for the formation of 1-decyne from the photooxidation of sterculic acid (19).

MAJOR VOLATILE COMPOUNDS OF COMMERCIAL FATS AND OILS

The volatile compounds of eight different vegetable oils, namely canola, corn, cottonseed, olive, peanut, safflower, soybean, and sunflowerseed oil have been analyzed and reported by Snyder and co-workers in 1985 (20). Table 4.2 shows the quantitative data of the volatile compounds identified in these vegetable oils after eight days storage at 60°C.

Table 4.2 Volatile compounds in vegetable oils after eight days storage at 60°C (20)

Volatile Compounds	GC Peak Area							
	Canola	Corn	Cottonseed	Olive	Peanut	Safflower	Soybean	Sunflowerseed
Ethane	1.0	1.0	0.8	0.2	8.0	0.2	0.9	1.0
Propane	0.7	10.6	12.6	0.6	—	10.4	9.1	12.7
Propenal	1.5	2.3	2.1	1.7	2.4	2.3	4.5	2.9
Pentene	0.2	—	—	—	—	—	0.3	—
Pentane	0.5	24.3	32.7	14.9	19.0	54.1	11.1	41.5
Propanal	1.0	—	—	0.3	0.6	—	0.7	—
Pentene	0.1	—	—	0.1	—	0.1	0.1	—
Hexane	1.1	0.4	0.3	0.3	0.2	0.4	0.3	0.1
2-Butenal	0.5	0.2	0.2	—	0.3	—	0.5	0.1
1-Penten-3-ol	0.4	—	—	—	—	—	0.8	0.1
Pentanal	0.5	1.0	1.3	1.8	2.8	1.9	2.9	2.4
Heptane	0.2	0.2	—	2.2	0.5	0.2	0.1	0.4
Pentenal	0.3	0.1	0.1	—	—	—	0.6	—
Pentanol	0.3	0.5	0.8	0.5	1.4	2.0	1.5	1.3
Octene	—	0.1	0.1	0.2	0.5	0.1	0.2	—
Hexanal	3.1	9.8	10.3	5.8	7.6	11.1	10.7	10.5
Octane	3.7	2.2	3.9	4.1	0.3	0.2	0.3	4.5
Octene	—	0.1	—	0.1	0.2	0.1	0.2	—
t-2-Hexenal	0.2	0.3	0.5	0.2	0.8	0.3	0.5	0.7
Heptanal	0.8	0.6	2.2	1.8	2.0	3.2	2.1	1.8
c-2-Heptenal	0.2	0.2	0.1	0.1	—	0.3	0.2	—
t-2-Heptenal	0.9	1.3	3.6	2.6	1.5	5.6	5.1	3.4
1-Octen-3-ol	0.1	0.3	0.2	0.3	0.3	0.1	0.3	0.3
2-Pentylfuran	0.2	0.5	0.8	0.2	0.7	0.7	1.0	0.6
t,c-2-Heptadienal	0.5	—	—	—	—	—	0.6	—
Octanal	0.5	0.3	0.4	1.1	0.6	0.4	0.5	0.3
t,t-2,4-Heptadienal	0.8	—	—	—	—	—	0.8	—
Octenal	0.2	0.5	0.5	0.6	0.7	0.6	0.4	0.5
Nonanal	1.2	0.5	0.6	2.8	0.5	0.4	0.6	0.4
t,-2-Decenal	0.2	0.2	0.3	0.7	0.2	0.2	0.3	0.3
Decenol	0.1	0.1	0.1	0.2	0.1	0.1	0.1	0.1
t,c-2,4-Decadienal	0.3	0.3	0.4	0.1	0.3	0.3	0.6	0.4
t,t-2,4-Decadienal	0.8	1.0	1.2	0.8	0.8	1.5	1.4	1.4
Undecenal	0.1	0.1	0.2	0.2	0.2	0.2	0.1	0.2
Total area	25.8	63.1	77.6	49.2	53.6	99.6	40.2	70.4

The volatile compounds in each of the stored vegetable oil samples were related to the main fatty acid components of the oil. Safflower, sunflowerseed, corn, and cottonseed oils, with the highest amount of linoleate, tended to produce the greatest amount of volatiles, especially pentane and hexanal. Canola and soybean oils, which contain linoleate, both formed measurable amounts of 2,4-heptadienal. Olive oil, with the largest quantity of oleate, produced the most octanal and nonanal.

3.1. Flavor Compounds of Selected Fats and Oils

Butter. Due to their commercial significance, the flavor of butter and butter oil has been studied extensively. More than 230 volatile compounds have been identified in different types of butter as well as butter oil (21). Using odor activity values (ratio of concentration to odor threshold), Forss and co-workers (22), Urbach and co-workers (23), and Stark and co-workers (24–26) reported δ-decalactone, δ-octalactone, decanoic acid, dodecanoic acid, skatole, and indole as important contributors to the flavor of butter oil. In addition, the data of Siek and co-workers (27) indicated that in fresh butter the levels of butanoic acid, caproic acid, dimethyl disulfide, and also δ-decalactone were above their taste thresholds.

By using aroma extract dilution analysis (AEDA) of the volatile fractions of fresh and stored butter oil, Widder and co-workers (28) recently determined diacetyl, butanoic acid, δ-octalactone, skatole, δ-decalactone, *cis*-6-dodeceno-γ-lactone, 1-octen-3-one, and 1-hexen-3-one as potent contributors to the flavor of butter oil. The concentrations of 1-octen-3-one, *trans*-2-nonenal and *cis*-1,5-octadien-3-one increased during the storage of the butter oil at room temperature.

Table 4.3 shows the sensory evaluation by Schieberle and co-workers (29) of five different kinds of butter, namely Irish sour cream (ISC), cultured butter (CB), sour cream (SC), sweet cream (SwC), and farmer sour cream (FSC). It revealed ISC butter and FSC butter with the highest overall odor intensities. Table 4.4 shows that 19 odor-active compounds were detected by AEDA in a distillate of the ISC butter. The highest flavor dilution (FD)-factors have been found for δ-decalactone, skatole, *cis*-6-dodeceno-γ-lactone and diacetyl followed by *trans*-2-nonenal, *cis*, *cis*-3,6-nonadienal, *cis*-2-nonenal, and 1-octen-3-one.

Cocoa Butter. Cocoa butter is one of the most liked and highly prized food ingredients due to its desirable flavor and unique melting behavior. As early as 1961, van Elzakker and van Zutphen (30) studied and identified 23 volatile compounds in the vacuum steam distillate of cocoa butter. Later, Rizzi (31) identified 9 alkylpyrazines including methylpyrazine, 2,5-dimethylpyrazine, 2,6-dimethylpyrazine, 2,3-dimethylpyrazine, 2-ethyl-5-methylpyrazine, trimethlypyrazine, 2,5-dimethyl-3-ethylpyrazine, 2,6-dimethyl-3-ethylpyrazine, and tetramethylpyrazine in the basic fraction of a vacuum steam distillate of cocoa butter.

The flavor of cocoa butter depends on the processing conditions to which the cocoa beans are subjected. Cocoa butter obtained from roasted cocoa

Table 4.3 Odor profiles of the butter samples (29)

Butter Samples	Odor Quality	Intensity[a]
Irish sour cream (ISC)	Buttery, creamy, sweet	3
Cultured butter (CB)	Typical butter-like, sweet	2–3
Sour cream (SC)	Mild, weakly buttery, sour	1–2
Sweet cream (SwC)	Slightly sour, mild	1
Farmer sour cream (FSC)	Rancid, like butanoic acid	3

[a] Intensity: 1, weak; 2, medium; 3, strong.

Table 4.4 Potent odorants in an Irish sour cream butter (29)

Compound	Flavor Dilution (FD) Factor	Odor Description
Diacetyl	256	Buttery
1-Penten-3-one	32	Vegetable-like
Hexanal	8	Green
1-Octen-3-one	8	Mushroom-like
cis,cis-3, 6-Nonadienal	64	Soapy
cis-2-Nonenal	64	Fatty, green
trans-2-Nonenal	128	Green, tallowy
trans, trans-2,4-Nonadienal	8	Fatty, waxy
trans, trans-2,4-Decadienal	32	Fatty, waxy
γ-Octalactone	64	Coconut-like
trans-4,5-Epoxy-*trans*-2-decenal	32	Metallic
Skatole	512	Mothball-like
δ-Decalactone	4096	Coconut-like
cis-6-Dodeceno-γ-lactone	512	Peach-like
Acetic acid	128	Pungent
Butanoic acid	512	Buttery, sweaty
Hexanoic acid	32	Pungent, musty

beans has a strong flavor reminiscent of cocoa. Cocoa butter obtained from unroasted cocoa beans which have been given a steam treatment has a considerably milder, yet distinctive, flavor.

The most comprehensive study on the flavor compounds of cocoa butter was that of Carlin and co-workers (32–35). They compared the volatile compounds of cocoa butters from roasted and unroasted cocoa beans. Pyrazines were present in greater numbers and at higher concentrations in the roasted cocoa butter. Of the 62 pyrazines identified, 57 were identified in the roasted cocoa butter and only 27 in the unroasted samples. Table 4.5 lists the comparison of the major pyrazines identified in cocoa butters from roasted and unroasted cocoa beans.

The most abundant pyrazine identified in cocoa butters was tetramethylpyrazine, which existed at an extremely high concentration in the unroasted cocoa butter but only a moderate level in the roasted cocoa butter. Tetramethylpyrazine accounted for over 90% of the pyrazine content of the unroasted cocoa butter. Besides thermal generation, tetramethylpyrazine could be formed in cocoa beans through biosynthetic reactions. Kosuge and Kamiya (36) identified tetramethylpyrazine as a metabolic product of a strain of *Bacillus subtilis.* Several species of this organism were identified in a fermenting mass of cocoa beans by Ostovar (37).

Table 4.6 lists the oxazoles and thiazoles identified in the sample of cocoa butter. They were present only in roasted cocoa butter. The sensory characteristics of these compounds shown in Table 4.6 indicated that oxazoles and

Table 4.5 Some major pyrazines identified in cocoa butters from roasted and unroasted cocoa beans (35)

Pyrazine	Relative Concentration	
	Roasted	Unroasted
Tetramethylpyrazine	270	3400
Trimethylpyrazine	2620	380
2,5-Dimethyl-3-ethylpyrazine	750	40
2,5-Dimethylpyrazine	710	60
2-Isopropyl-3-methylpyrazine	510	Trace
2,6-Dimethylpyrazine	490	—
2-Acetyl-3-methylpyrazine	390	20
2-Ethyl-3,5,6-trimethylpyrazine	310	20
2,6-Diethyl-3-methylpyrazine	10	150
6,7-Dihydro-5*H*-cyclopentapyrazine	110	—
Methylpyrazine	90	10
2-Acetyl-3-ethylpyrazine	90	10
2,5-Diethyl-3-methylpyrazine	90	—
2-Butyl-3,6-dimethylpyrazine	80	Trace
2-Methyl-6,7-dihydro-5*H*-cyclopentapyrazine	80	Trace
2-Methyl-5-vinylpyrazine	70	—
Isopropenylpyrazine	70	—
2-Methyl-3-pentylpyrazine	50	Trace
2-Methyl-6-vinylpyrazine	40	30
2,3-Dimethylpyrazine	Trace	30
2,3-Dimethyl-5-ethylpyrazine	—	30
2,3-Dimethyl-5-butylpyrazine	30	—
2,5-Dimethyl-3-isobutylpyrazine	30	—
5,6,7,8-Tetrahydroquinoxaline	30	—

thiazoles possessed interesting green, fatty, sweet, and nutty sensory qualities and were high impact flavor compounds in roasted cocoa butter (34,35). Of particular interest, 2-pentylthiazole identified had strong fatty, green, and sweet notes and may be an important contributor to cocoa butter flavor. Oxazoles could possibly be formed through the Strecker degradation of amino-ketones which result from the condensation of α-dicarbonyl compounds with amino acids (38). They might also form through reactions between amino acids (39). Maga (40) has reviewed the occurrence of thiazoles in foods and possible pathways of formation. Thiazoles could possibly form through the interaction of sulfur-containing amino acids and carbonyl-containing compounds.

Lard. Lard is a traditional edible fat for Chinese people. Lard is generally prepared either by dry-rendering or by wet-rendering. The dry-rendered lard with pork back fat as the raw material usually has better flavor than the wet-rendered lard and is used as cooking fat or shortening. The wet-rendered lard

Table 4.6 Oxazoles and thiazoles identified in roasted cocoa butters (34)

Compounds	Odor Description
2-Pentylthiazole	Strong, green, fatty, sweet
2-Acetyl-5-methylthiazole	
2-Isopropyl-4,5-dimethylthiazole	
2-Isopropyl-4-methylthiazole	Strong, camphorus
2-Acetyloxazole	
2-Isopropyl-4,5-dimethyloxazole	
2-Isopropyl-4-ethyl-5-methyloxazole	Sweet, fruity
2-Methyl-4,5-dibutyloxazole	
4,5-Dimethyloxazole	
2-Methyl-4-ethyl-5-propyloxazole	Green fatty, vegetable-like
2,5-Dimethyl-4-butyloxazole	Fresh acidic, green, pickle-like
2-Methyl-4-ethyl-5-butyloxazole	Acidic, fatty, sweet, flowery
2-Butyl-4-methyl-5-ethyloxazole	Green, sweet
2-Butyl-4-ethyl-4-methyloxazole	Green, herbal, weak, acidic, slight buttery
4,5-Dibutyloxazole	
2,5-Dibutyl-4-methyloxazole	Sweet, fruity, green

with pork belly fat as the raw material usually has an undesirable flavor and must be refined before further use.

The volatile flavor compounds of lard have been studied by Watanabe and Sato (41–46). They heated the lard at 160–170°C under a stream of air and collected the volatile compounds. They found that 2,4-decadienal and lactones contributed significantly to the flavor of lard. Recently, Hwang and Chen (47) compared the volatile flavor compounds generated by heating the crude and refined samples of both dry-rendered and wet-rendered lard at 190°C for 2 hours. Table 4.7 summarizes the amounts of some flavor-contributing volatiles in the four lards. Crude dry-rendered lard showed the highest content of these compounds followed by crude wet-rendered lard, refined dry-rendered lard and refined wet-rendered lard. Apparently, dry-rendering can yield lard with a stronger flavor.

Soybean Oil. Soybean oil is the highest volume vegetable oil produced in the world, as well as in the United States. Due to its commercial importance, the flavor chemistry of soybean oil has been extensively studied and reviewed (48).

The development of a characteristic, objectionable, beany, grassy, and hay-like flavor in soybean oil, commonly known as reversion flavor, is a classic problem of the food industry. Soybean oil tends to develop this objectionable flavor when its peroxide value is still as low as a few meq/kg while other vegetable oils, such as cottonseed, corn, and sunflower, do not (14,49). Smouse and Chang (50) identified 71 compounds in the volatiles of a typical reverted-but-not-rancid soybean oil. They reported that 2-pentylfuran formed from the

Table 4.7 The amounts of volatile flavor contributing compounds from lards of different treatments (47)

	CDL[a]	RDL	CWL	RWL
Compounds	(mg/100 g lard)			
1-Octen-3-ol	2.21	1.83	1.89	1.40
n-Hexanal	10.87	8.13	10.38	7.60
n-Octanal	2.67	2.12	2.07	1.86
trans-2-Heptenal	11.04	8.45	10.44	7.23
n-Nonanal	10.92	8.15	8.70	6.99
trans-2-Octenal	3.24	2.66	2.77	2.40
trans-2-Nonenal	3.14	2.46	2.59	1.89
2,4-Decadienal	17.01	14.34	15.06	12.10
2-Pentylfuran	0.80	0.43	0.88	0.38
γ-Octalactone	Trace	0.38	Trace	Trace
Total	61.90	48.95	54.78	41.85

[a] CDL: crude dry-rendered lard; RDL: refined dry-rendered lard; CWL: crude wet-rendered lard; RWL: refined wet-rendered lard.

autoxidation of linoleic acid, which is the major fatty acid of soybean oil, and contributes significantly to the beany and grassy flavor of soybean oil. Other compounds identified in the reverted soybean oil also have fatty acids as their precursors. For example, the "green bean" flavor is caused by *cis*-3-hexenal which is formed by the autoxidation of linolenic acid which usually constitutes 2–11% in soybean oil. Linoleic acid oxidized to 1-octen-3-ol which is characterized by its "mushroom alcohol" flavor (18).

The most significant studies on the flavor of soybean oil were those published recently by Ullrich and Grosch (51) and Guth and Grosch (12,52). By using aroma extract dilution analysis, they determined some odor compounds which strongly contributed to the off-flavor of soybean oil samples which were stored at room temperature either in daylight or in the dark. Table 4.8 lists the flavor dilution (FD)-factor values of various odor compounds in soybean oil samples. 3-Methyl-2,4-nonanedione, *cis*-3-hexenal, *cis*-2-nonenal, *cis*-1,5-octadien-3-one, 1-octen-3-hydroperoxide, 4,5-epoxy-*trans*-2-decenal, 1-octen-3-one, *cis*-1,5-octadien-3-hydroperoxide, and *trans*-2-nonenal were identified as primary odorants of soybean oil which were exposed to daylight. They also observed that the major differences in the intensity of the reversion odor of soybean oil samples were mainly due to an increase in the concentration of 3-methyl-2,4-nonanedione during storage. Guth and Grosch (53) further identified two furanoid fatty acids, namely, 10,13-epoxy-11,12-dimethyloctadeca-10,12-dienoic acid and 12,15-epoxy-13,14-dimethyloctadeca-12,14-dienoic acid in soybean oil as precursors for 3-methyl-2,4-nonanedione. The proposed mechanism for the photogeneration of 3-methyl-2,4-nonanedione is shown in Figure 4.6.

Table 4.8 Aroma extract dilution analysis of the stored soybean oils (12)

	Flavor Dilution (FD) Factor	
Compound	SBO (Daylight)[a]	SBO (Dark)[b]
cis-3-Hexenal	2048	8
Hexanal	16	<1
Pentanoic acid	32	<1
trans-2-Heptenal	64	<1
1-Octen-3-one	256	16
cis-1,5-Octadien-3-one	512	<1
2,4-Heptadienal	8	4
Octanal	16	8
2-Octenal	16	<1
trans-2-Octenal	32	8
cis-3-Nonenal	16	8
1-Octen-3-hydroperoxide	512	32
cis-1,5-Octadien-3-hydroperoxide	256	4
cis-2-Nonenal	1024	64
trans,cis-2,6-Nonadienal	16	8
trans-2-Nonenal	256	32
2,4-Nonadienal	32	16
trans, trans-2,4-Nonadienal	16	16
3-Methyl-2,4-nonandione	16348	16
4,5-Epoxy-*trans*-2-nonenal	<1	8
trans,trans-2,4-Decadienal	32	16
4,5-Epoxy-*trans*-2-decenal	512	256

[a] The oil sample was stored for 30 days at room temperature and in daylight.
[b] The oil sample stored for the same period in the dark.

The odor properties of two volatile hydroperoxides, 1-octen-3-hydroperoxide and *cis*-1,5-octadien-3-hydroperoxide are shown in Table 4.9. The odor thresholds of these hydroperoxides were 10-fold higher than those of the corresponding ketones. Precursors of both hydroperoxides are presumably the 10-hydroperoxide of linoleic and linolenic acid which is easily formed by photosensitized oxidation of linoleic and linolenic acid (48). As shown in Figure 4.7, a β-scission of the 10-hydroperoxy group, the rearrangement of the double bond and combination of the alylic radical formed with oxygen, followed by abstraction of a hydrogen atom would result in the two allyl hydroperoxides having eight carbon atoms.

As shown in Tables 4.8 and 4.9, on the basis of its high FD-factor and its odor properties, the *trans*-4,5-epoxy-*trans-2*-decenal contributed significantly to the green, hay-like overall odor in soybean oil stored in the dark. Guth and Grosch suggested (12) that an epoxyhydroperoxy fatty acid could be the precursor of such epoxy aldehydes. Figure 4.8 shows the proposed pathway for the formation of *trans*-4,5-epoxy-*trans-2*-decenal from linoleic

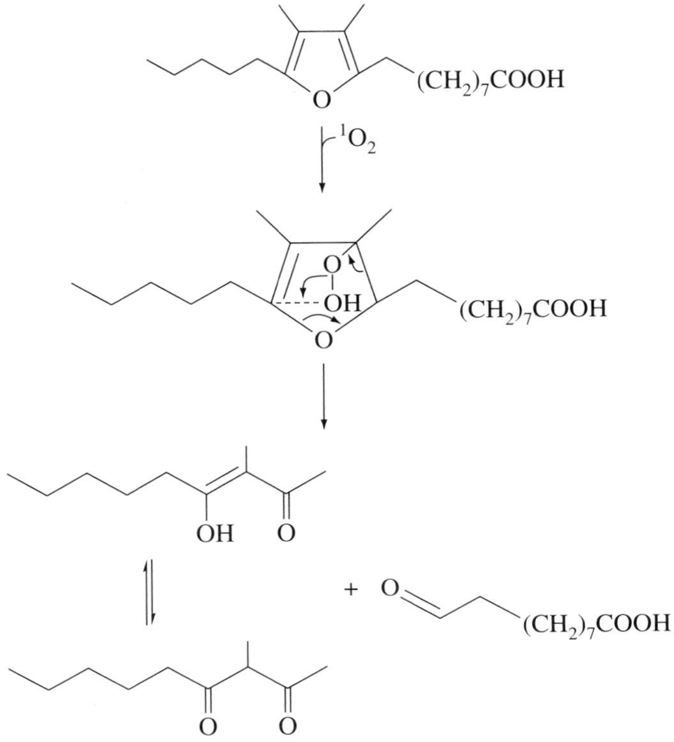

Figure 4.6 Proposed mechanism for the photo-generation of 3-methyl-2,4-nonanedione (53).

acid via the *trans*-12,13-epoxy-9-hydroperoxy-*trans*-10-octadecenoic acid intermediate.

Fish Oils. Since the use of sardine oil as a food ingredient was discontinued in the 1950s, the FDA has determined that fish oils were totally new ingredients for human foods. As a result, 90% of the fish oil produced in the United States

Table 4.9 Odor properties of the hydroperoxides and the epoxides identified in stored soybean oils (12)

Compound	Odor Description	Odor Threshold (ng/L, air)
1-Octen-3-hydroperoxide	Metallic, mushroomlike	0.6–1.2
1-Octen-3-one	Mushroomlike	0.03–0.12
cis-1,5-Octadien-3-hydroperoxide	Geraniumlike, metallic	0.03–0.06
cis-1,5-Octadien-3-one	Geraniumlike, metallic	0.003–0.006
trans-4,5-Epoxy-*trans*-2-nonenal	Metallic	0.25–1.0
trans-4,5-Epoxy-*trans*-2-decenal	Metallic, green	0.0005–0.005

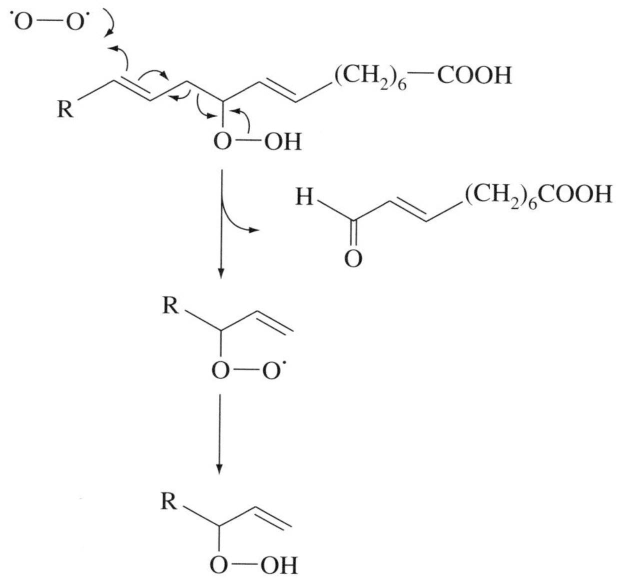

$^{\bullet}O-O^{\bullet}$

R

$(CH_2)_6-COOH$

O—OH

H

$(CH_2)_6COOH$

O

R

$O-O^{\bullet}$

R

O—OH

Figure 4.7 Reaction routes proposed for the formation of 1-octen-3-hydroperoxide ($R=CH_3-(CH_2)_4-$) and *cis*-1,5-octadien-3-hydroperoxide ($CH_3-CH_2-CH=CH-CH_2-$) (12).

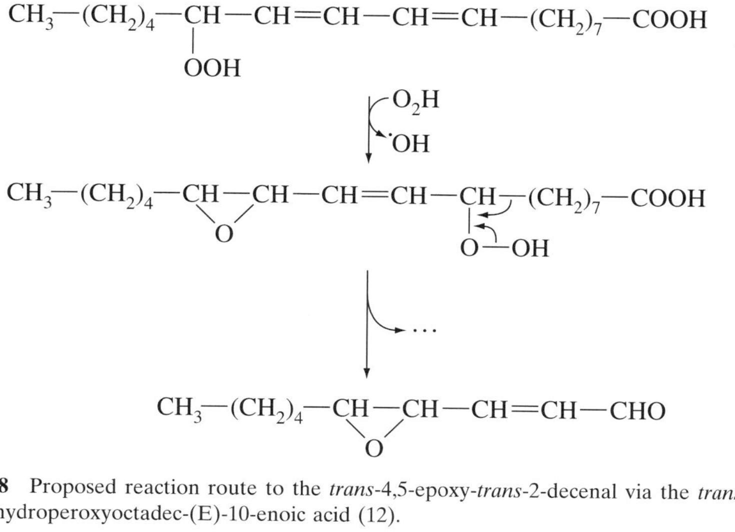

$$CH_3-(CH_2)_4-CH=CH-CH_2-CH=CH-(CH_2)_7-COOH$$

$$CH_3-(CH_2)_4-CH(OOH)-CH=CH-CH=CH-(CH_2)_7-COOH$$

O_2H

$^{\bullet}OH$

$$CH_3-(CH_2)_4-CH(-O-)CH-CH=CH-CH(O-OH)-(CH_2)_7-COOH$$

...

$$CH_3-(CH_2)_4-CH(-O-)CH-CH=CH-CHO$$

Figure 4.8 Proposed reaction route to the *trans*-4,5-epoxy-*trans*-2-decenal via the *trans*-12,13-epoxy-9-hydroperoxyoctadec-(E)-10-enoic acid (12).

was exported to Europe as food oils and 10% was used domestically as nonfood ingredients. After lengthy petition, the FDA finally affirmed the GRAS status of partially hydrogenated (PHMO) and hydrogenated menhaden oils (HMO) for direct use as human food ingredients in 1989 (54).

Fish oils are a rich source of *n*-3 polyunsaturated fatty acids such as eicosapentaenoic acid (EPA) and docosahexaenoic acid (DHA). The supplementation of Western diets with fish oils containing EPA and DHA has been recommended (55,56), and can be beneficial for ischemic heart disease and thromboembolic events.

In order to explain the reason that fish oils are much more labile than common vegetable oils, basic lipid chemistry such as oxidation rate, induction period, and oxygen uptake were determined and reported by several researchers (57,58). The relative oxidation rates of fatty esters at 36.5°C were found to be highly correlated to their molecular unsaturation, i.e., oleate (1.0), linoleate (8.0), linolenate (21.7), and EPA + DHA (39.1). Relative oxygen uptake (first two days in air) of oleate, linoleate, linolenate, EPA, and DHA were <1, 1, 99, 743, and 948, respectively. Induction periods (90 lux at 5°C) were also related to the degree of unsaturation of fatty esters. Oleate was found to be more than 100 days (estimated) and linoleate, linolenate, and EPA + DHA, 50, 20, and 4 days, respectively (55).

The high oxidation rates of EPA and DHA and the instability of their hydroperoxides caused the rapid formation of secondary products such as volatile aldehydes and other compounds which, in turn, impart flavor reversion in fish oils (53). Isomeric distribution of hydroperoxides in autoxidized EPA (59) and DHA (60) has been studied. They form eight and ten isomers respectively. Noble and Nawar (61) analyzed the volatile compounds in autoxidized DHA and identified a number of aldehydes. Most of the aldehydes identified could be explained by the β-scission of alkoxy radicals generated by the homolytic cleavage of each isomer of the hydroperoxides as shown in Figure 4.9.

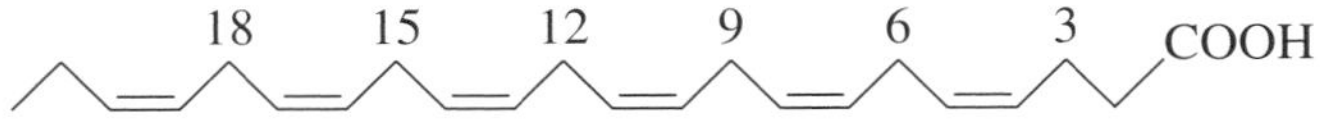

18 15 12 9 6 3 COOH

Initial H Abstractio	Hydroperoxide	Aldehydes Formed
C-18	20-OOH	Propanal
	16-OOH	2,4-Heptadienal
C-15	17-OOH	3-Hexenal
	13-OOH	2,4,7-Decatrienal
C-12	14-OOH	3,6-Nonadienal
	10-OOH	2,4,7,10-Tridecatetraenal
C-9	11-OOH	3,6,9-Dodecatrienal
	7-OOH	2,4,7,10,13-Hexadecapentaenal
C-6	8-OOH	3,6,9,12-Pentadecatetraenal
	4-OOH	2,4,7,10,13,16-Nonadecahexaenal

Figure 4.9 Aldehydes derived from autoxides decosahexaenoic acid.

Meijboom and Stroink (62) found that 2-*trans*, 4-*cis*, 7-*cis*-decatrienal was the compound responsible for the fishy off-flavors occurring in autoxidized oil containing *n*-3 fatty acids. This trienal was also found in autoxidized methyl DHA by Noble and Nawar (61) and in autoxidized mackerel oil by Ke, Ackman, and Linke (63).

The most detailed studies on the flavor of fish oil in recent years were probably those of Hsieh and co-workers (64,65) and Lin and co-workers (66). In their studies, a series of alkanals, alkenals, alkadienals, and alkatrienals were determined by dynamic headspace gas chromatography–mass spectrometry in

Table 4.10 Volatile aldehydes identified in the Gulf Menhaden Oil (54)

Compounds	GC Area (%)
n-Butanal	0.97
n-Pentanal	0.82
2-*cis*-Butenal	0.55
n-Hexanal	1.56
2-*cis*-Pentenal	1.48
n-Heptanal	1.42
2-*trans*-Hexenal	0.97
4-*trans*-Heptenal	0.24
n-Octanal	0.61
2-*trans*-Heptenal	0.31
n-Nonanal	0.51
2,4-Hexadienal	Trace
2-*cis*-Octenal	0.13
2-*trans*-Octenal	0.43
2,4-Heptadienal[a]	0.69
n-Decanal	0.15
2,4-Heptadienal[a]	1.39
Benzaldehyde	0.34
2-*trans*-Nonenal	0.14
4-*cis*-Decenal	0.14
2,4-Octadienal	0.11
2,6-Nonadienal[a]	0.06
2,4-Octadienal	0.34
4-*trans*-Decenal	0.02
2,4-Nonadienal	Trace
2,4-Nonadienal	Trace
2,4-Decadienal	Trace
2,4-Undecadienal[a]	Trace
Nonatrienal[a]	Trace
Nonatrienal[a]	Trace
Decatrienal[a]	Trace
Decatrienal[a]	Trace

[a] Configuration of geometric isomers were not determined.

crude menhaden oils (Table 4.10). Most of these aldehydes contributed to the characteristic oxidized oily odors, such as green grassy, waxy, and rancid in the crude oils. Alkatrienals, i.e., nonatrienals and decatrienals, were also found at ppb levels in the dynamic headspace of the crude oils. 2-*trans*,4-*trans*,7-*cis*-Decatrienal, 2-*trans*,4-*cis*,7-*cis*-decatrienal and 4-*cis*-heptenal impart a strong fishy odor to oils. Aldehydes have a green or plant-like note; ketones (1-octen-3-one) have a metallic off-flavor. Besides the aldehydes identified in crude menhaden oil, other compounds identified such as short-chain unsaturated alcohol (1-penten-3-ol), had a medicinal odor and others had a green unpleasant odor and may also contribute to the flavor of fish oil. Lin and co-workers (66) also reported that steam-deodorization can effectively remove a total of 99% of most aldehydes in the oils. However, reversion flavor of fish oils during storage can generate pentylfuran and aldehydes that have green and beany odors.

REFERENCES

1. H.W.S. Chan, in H.W.S. Chan, ed., *Autoxidation of Unsaturated Lipids*, Academic Press, Inc., London, 1987, p. 1.
2. S.P. Kochhar, in M.J. Saxby, ed., *Food Taints and Off-Flavours*, Blackie Academic & Professional, London, 1993, p. 151.
3. D.W.S. Wong, *Mechanism and Theory in Food Chemistry*, Van Nostrand Reinhold Co., Inc., New York, 1989, p. 1.
4. p. 17 in Ref. 1.
5. E.N. Frankel, in D.B. Min and T.H. Smouse, eds., *Flavor Chemistry of Fats and Oils*, American Oil Chemists' Society, Campaign, Ill, 1985, p. 1.
6. C.-T. Ho and Q. Chen, in C.-T. Ho and T. G. Hartman, eds., *Lipids in Food Flavors*, American Chemical Society, Washington, D.C., 1994, p. 2.
7. E.N. Frenkel, *Prog. Lipid Res.* **22,** 1 (1982).
8. C.-T. Ho and co-workers, *J. Food Rev. Int.* **5,** 253 (1989).
9. D.B. Josephson and J. Glinka, in T.H. Parliment, R.J. McGorrin, and C.-T. Ho, eds., *Thermal Generation of Aroma*, ACS Symp. Ser. No. 409, American Chemical Society, Washington, D.C., 1989, p. 242.
10. C.-T. Ho and co-workers, in M. Martens, G. A. Dalen, and H. Russwurm Jr., eds., *Flavor Science and Technology*, John Wiley & Sons, Inc., New York, 1987, p. 35.
11. P. Schieberle and W. Grosch, *Z. Lebensm. Unters. Forsch.* **192,** 130 (1991).
12. H. Guth and W. Grosch, *Lebensm.-Wiss. u. Technol.* **23,** 59 (1990).
13. E.N. Frankel, W.E. Neff, and E. Selke, *Lipids* **16,** 279 (1981).
14. C.-T. Ho, M.S. Smagula, and S.S. Chang, *J. Am. Oil Chem. Soc.* **55,** 233 (1978).
15. D.G. Bradley and D.B. Min, *Crit. Rev. Food Sci. Nutr.* **31,** 211 (1992).
16. M.B. Korycka-Dahl and T. Richardson, *Crit. Rev. Food. Sci. Nutr.* **10,** 209 (1978).
17. H.R. Rawls, and P.J. VanSanten, *J. Am. Oil Chem. Soc.* **47,** 121 (1970).
18. B.S. Mistry and D.B. Min, in G. Charalambous, ed., *Off-flavors in Foods and Beverages,* Elsevier Science Publishers, Amsterdam, 1992, p. 171.
19. L.L. Fan, J.Y. Tang and A. Wohlman, *J. Am. Oil Chem. Soc.* **60,** 1115 (1985).
20. J.M. Snyder, E.N. Frankel, and E. Selke, *J. Am. Oil Chem. Soc.* **62,** 1675 (1985).

21. H. Maarse and C.A. Visscher, *Volatile Compounds in Food. Qualitative Data, Supplement,* TNO, Zeist, the Netherlands, 1988.
22. D.A. Forss, W. Stark, and G. Urbach, *J. Dairy Res.* **34,** 131 (1967).
23. G. Urbach, W. Stark, and D.A. Forss, *J. Dairy Res.* **39,** 35 (1972).
24. W. Stark, G. Urbach, and J.S. Hamilton, *J Dairy Res.* **40,** 39 (1973).
25. W. Stark, G. Urbach, and J.S. Hamilton, *J Dairy Res.* **43,** 469 (1976).
26. W. Stark, G. Urbach, and J.S. Hamilton, *J. Dairy Res.* **43,** 479 (1976).
27. T.J. Siek, J.A. Albin, L.A. Sather, and R.C. Lindsay, *J. Food Sci.* **34,** 265 (1976).
28. S. Widder, A. Sen, and W. Grosch, *Z. Lebensm. Unters. Forsch.* **193,** 32 (1991).
29. P. Schieberle and co-workers, *Lebensm. Wissen. u.-Technol.* **26,** 347 (1993).
30. A.H.M. van Elzakker and H.J. van Zutphen, *Z. Lebensm.-Unters. Forsch.* **115,** 222 (1961).
31. G.P. Rizzi, *J. Agric. Food Chem.* **15,** 549 (1967).
32. J.T. Carlin, M.S. Thesis, Rutgers University, New Brunswick, N.J., 1980.
33. J.T. Carlin and co-workers, *Proc. 36th P.M.C.A. Production Conf.*, 1982, p. 95.
34. C.T. Ho, Q.Z. Jin, K.N. Lee, J.T. Carlin, and S.S. Chang, *J. Food Sci.* **48,** 1570 (1983).
35. J.T. Carlin and co-workers, *J. Am. Oil Chem. Soc.* **63,** 1031 (1986).
36. T. Kosuge and H. Kamiya, *Nature* **193,** 776 (1962).
37. K. Ostovar, Ph.D. Thesis, The Pennsylvania State University, 1971.
38. O.G. Vitzthum and P. Werkhoff, *Z. Lebensm.-Unters. Forsch.* **156,** 300 (1974).
39. G. Ohloff and I. Flament, *Forschr. Chem. Org. Naturst.* **36,** 231 (1979).
40. J.A. Maga, in T.E. Furia and N. Bellanca, eds., *Fenaroli's Handbook of Flavor Ingredients*, 2nd ed., vol. 1, CRC Press Inc., Cleveland, 1975, p. 228.
41. K. Watanabe and Y. Sato, *Agric. Biol. Chem.* **33,** 242 (1969).
42. K. Watanabe and Y. Sato, *Agric. Biol. Chem.* **33,** 1411 (1969).
43. K. Watanabe and Y. Sato, *Agric. Biol. Chem.* **34,** 464 (1970).
44. K. Watanabe and Y. Sato, *Agric. Biol. Chem.* **34,** 1710 (1970).
45. K. Watanabe and Y. Sato, *Agric. Biol. Chem.* **35,** 278 (1971).
46. K. Watanabe and Y. Sato, *J. Zootech. Sci.,* **42,** 393 (1971).
47. L.S. Hwang and C.W. Chen, p. 244 in Ref. 6.
48. T.H. Smouse, p. 85 in Ref. 5.
49. S.S. Chang and co-workers, *J. Am. Oil Chem. Soc.* **60,** 553 (1983).
50. T.H. Smouse and S.S. Chang, *J. Am. Oil Chem. Soc.,* **44,** 509 (1967).
51. F. Ullrich and W. Grosch, *Fat. Sci. Technol.* **90,** 332 (1988).
52. H. Guth and W. Grosch, *Fat Sci. Technol.* **91,** 225 (1989).
53. H. Guth and W. Grosch, *Fat Sci. Technol.* **93,** 249 (1991).
54. C.F. Lin, in Ref. 6.
55. K. Fujimoto, in D.B. Min and T.H. Smouse, eds., *Flavor Chemistry of Lipid Foods*, American Oil Chemists' Society, Champaign IL, 1989, p. 190.
56. W.E.M. Lands, *Fish and Human Health*, Academic Press, Inc., Orlando, Fla., 1986.
57. N. Ikeda and K.J. Fukuzumi, *Jpn. Oil Chem. Soc.* **27,** 21 (1978).
58. S.Y. Cho, K. Miyashita, T. Miyazawa, K. Fujimoto, and T. Kaneda, *J. Am. Oil Chem. Soc.,* **64,** 876 (1987).
59. R. Yamauchi, T. Yamada, K. Kato, and Y. Uevio, *Agric. Biol. Chem.* **47,** 2897 (1983).
60. M. Van Rollin and R.C. Murphy, *J. Lipids Res.* **25,** 507 (1984).
61. A.C. Noble and W.W. Nawar, *J. Am. Oil Chem. Soc.,* **48,** 800 (1971).

62. P.W. Meijboom and J.B.A. Stroink, *J. Am. Oil Chem. Soc.,* **49,** 555 (1972).
63. P.J. Ke, R.G. Ackman, and B.A. Linke, *J. Am. Oil Chem. Soc.,* **52,** 349 (1975).
64. T.C.Y. Hsieh, C.F. Lin, J.B. Crowther, A.P. Bimbo, *Tropical and Subtropical Fisheries Technological Society of the Americas*, 1988, p.154.
65. T.C.Y. Hsieh, S.S. Williams, W. Vejaphan, and S.P. Meyers, *Am. Oil Chem. Soc.,* **66,** 114 (1989).
66. C.F. Lin, T.C.Y. Hsieh, J.B. Crowther, and A.P. Bimbo, *J. Food Sci.* **55,** 1669 (1990).

5

Flavors and Sensory Evaluation

INTRODUCTION

What is sensory evaluation? Simply, it is a scientific discipline for the qualitative and quantitative evaluation of foods. Testers use their senses of smell, taste, and/or touch to identify characteristics of foods and ingredients. Testers range in background from a consumer with no training and experience to a highly skilled judge, such as a brewmaster with a vast sensory memory of odors and flavors. Large numbers of consumer testers are needed to learn more about likes and dislikes of new food products; however, one expert wine, coffee, or tea taster may be all that is needed to rate the latest beverage production. In the oil industry, the use of sensory evaluation may range from one person on the night shift tasting the oil as a quality control measure to a group of highly trained and experienced judges identifying types and intensities of sensory characteristics of oils or fat-containing foods at a research laboratory.

Sensory evaluation is the ultimate method to assess the quality and stability of fats, oils, and fat-containing foods. No instrumental or chemical methods have been developed that will take the place of our nose, taste buds, and sensory receptors. Instrumental and chemical methods measure various portions of the decomposition products that result from oxidation such as nonvolatile and volatile compounds. None of these methods integrate all factors into a total perception of overall quality or intensity that can be obtained from a sensory analysis. It is true that we are able to correlate gas chromatographic volatiles, peroxide values, and other analyses with sensory data, but correlations are not the same for all instances and in all situations. Sensory analysis is performed by humans who may not have the reproducibility of a peroxide value titration; however, trained, experienced sensory testers operating under controlled conditions can provide accurate, reliable data.

This chapter will show how sensory analysis has a place in quality control, product development, and research ranging from lab-bench facilities operating on a limited budget to state-of-the-art operations with large sensory staffs and computerized data handling. Any sensory operation, no matter how small can provide valuable information even if only to screen incoming raw materials and finished products. Also included is basic information on criteria to determine what type of sensory evaluation program is appropriate for your situation. Information on initiating a sensory program and expanding an existing program is included. Definitions will be provided for basic vocabulary in sensory analysis and for oils and fat-containing foods are in particular. Types of sensory tests that can be used for fats, oils, and oil-containing foods are presented along with examples of rating scales and score sheets. Reference standard requirements will be discussed as well as suggested standards for major oil flavor characteristics. Applications of sensory analysis for fats, oils, and fat-containing foods will be presented. This information will also be important to people wanting to collaborate or contact with a sensory operation to conduct tests.

1.1 Justification for Sensory Evalution

How do you know whether or not initiating a sensory evaluation program is necessary in your operation? There are several questions that can be asked in order to decide. First, do your existing analyses tell you what you want to know? Are your products passing quality control tests such as peroxide analysis or Lovibond color but still getting customer complaints? Do you have oils with good oxidative stability by Rancimat analysis, but the gas chromatographic volatiles analyses show high levels of degradation products? Sensory analysis may not be necessary on a routine basis as we use peroxide values in quality control; however, anytime there are discrepancies between types of tests, the final authority should be sensory analysis. The following steps can be used to determine the need for sensory analysis: First, list problems encountered in processing operations or research that affect product quality and/or stability. Common problems in production or research include a new supply source, inconsistent quality of supply, process change, ingredient change or new additive, poor storage stability or shelf life, off flavors or odors in the product, new variety of oilseeds, and alternative antioxidants. Next, list the methods available to you to address these problems. Typical instrumental, chemical, and sensory analyses to measure quality and/or stability are gas chromatographic volatiles, peroxide values, active oxygen method (AOM)/Rancimet, free fatty acids, color, conjugated diene, and oxygen uptake. Review the list of problems and the analysis methods and determine if any of the situations are not adequately covered by instrumental and chemical analyses. In an industrial setting, if you make a change in ingredients and the results of the analyses (peroxide value, gas chromatographic hexanal content) remain the same as the product with the previous ingredient, then there may be no need for sensory analysis. On the other hand, if consumer complaints appear with products that instrumental and chemical analyses gave clearance for, then sensory analyses may be needed to pinpoint problems. Also, not all off flavors and odors are detectable by instrumental and chemical analyses. The best example of this is oxidation of linolenic acid at low levels when oils develop a distinctive fishy flavor. Peroxides and volatile compound levels are usually low (<1.0 for peroxide value) in this case, but flavor quality is poor because of fishy off flavor. The flavors characteristic of hydrogenation, such as waxy, sweet, and fruity, are not detected by standard instrumental and chemical analyses but are readily identified by sensory analysis as negative attributes. Ideally, sensory evaluation should be incorporated as part of an oil or fat-containing food analysis program that also includes other methods. Sensory evaluation, gas chromatographic volatiles, and peroxide values provide the most information about the quality of salad oils (1). No test should stand alone as a measure of quality or stability. Finally, a rapid test does not mean that it is a true representation of flavor quality. Frankel reviewed tests to measure oxidation and concluded that stability tests conducted at high temperature (100°C and above) may not be indicative of shelf life stability at lower temperatures (2).

If an in-house trained, experienced panel is not an option, then contracting or collaborating with an established sensory panel may help. In a quality control situation, all personnel should be trained to recognize a good-quality product and to detect off-quality samples. Even in a research setting, scientists should smell all samples to learn differences in odor quality. In addition to measuring product quality and stability, sensory evaluation can be used for a wide variety of situations such as to monitor product quality, assess alternative processes or ingredients, monitor stability, evaluate packaging, increase product knowledge, and develop product knowledge, and develop product descriptions.

1.2 Sensory Vocabulary

Familiarity of the vocabulary used in sensory analysis is important to a better understanding of this chapter. The following definitions are typically used in sensory analysis of oils and fat-containing foods.

Sensory Evaluation. Scientific discipline used to evoke, measure, analyze, and interpret reactions to those characteristics of foods and materials as they are perceived by senses of sight, smell, taste, touch, and hearing. Sensory evaluation of oils is limited mostly to the senses of taste and smell.

Organoleptic. Archaic term meaning to sense with nose and tongue. This term was used before scientists knew that the sensory mechanisms of perception are located at higher levels of the nervous system than the level of tongue and nose. The word *sensory* is more accurate.

Odor. Sensation from stimulus of olfactory receptors in nose by volatile compounds. Thousands of volatile compounds produce simple and complex odors detected in food products.

Taste. Sensation from stimulus of taste buds on tongue; four basic tastes—sweet, salt, sour, and bitter. Taste is not usually applicable to oils because oils do not have these tastes with the possible exception of bitter.

Flavor. Integrated attributes of taste and odor. Odor is detected by smelling the volatiles of the sample, whereas flavor is detected after placing the sample in the mouth. Volatiles go to the olfactory bulb either nasally through the nose by sniffing or retronasally through the back of the oral cavity (Figure 5.1). In most cases, flavor is a more sensitive analyses of sensory characteristics because all volatiles are directed toward the olfactory epithelium and not lost between the headspace of the sample container and the nose. Panelists almost always report a stronger flavor from a sample in their mouth than they do from a sample that is only smelled.

Analytical Panel. Group of trained, experienced judges who rate products by discriminative and descriptive sensory methods. A group of 12–15 people is recommended. Judges are trained to measure intensity or quality as objectively as possible with as little subjectivity as possible.

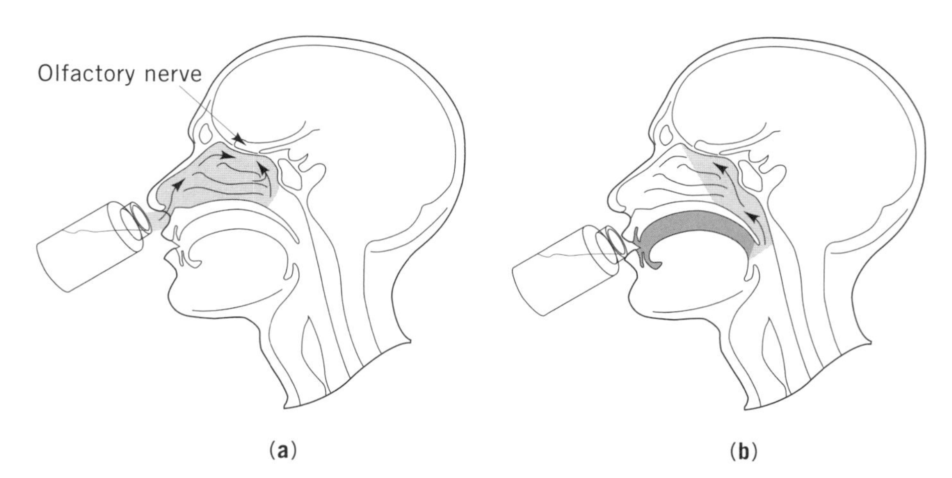

Figure 5.1 Sensory mechanisms and pathways for odor and flavor perception.

Consumer Panel. Group of untrained individuals who rate products for preference or on a like–dislike scale.

Objective. Unbiased report of stimulus from instrument or from trained, experienced sensory judges using discriminative or descriptive tests.

Subjective. Biased report from individual on preference or like–dislike scales. Consumer panels produce subjective analyses of flavor.

Intensity Scale. Method to rate the intensity or strength of an odor, taste, or flavor on a scale, usually 0–10.

Quality Scale. Method to rate the esthetic properties of a product, usually on a 1–10 scale of bad to excellent.

Bland. Total absence of odor or flavor. An oil cannot be described as "bland with a slight nutty or mild flavor."

Rancid. Typical of linoleate-containing oils. Detected in oils with peroxide value greater than 5.0.

Painty. Typical of linolenate-containing oils (soybean, low erucic acid rapeseed (canola), linseed). Detected in these oils if peroxide value is greater than 5.0. Not noted in nonlinolenic acid oils such as peanut.

Oxidized. A general term to describe a deteriorated or aged oil. Oxidation is a process of oil deterioration. An oxidized sample may have flavors ranging from buttery to painty. *Not* recommended as a flavor description.

Reverted. General term for flavor of oxidized oil that has "reverted" to flavor of partially processed oil. Reverted flavor in soybean oil can be green grassy or hay/grassy; however, each oil type has a different reverted flavor. *Not* recommended as a flavor description.

1.3 Types of Sensory Evaluation for Fats and Oils

At this point, if you have decided that sensory evaluation can play a significant role in your operation, then the following information can be used to set up a sensory evaluation program. Analyses for odor and flavor of fats, oils, and fat-containing foods are listed in Table 5.1. Salad oils can be evaluated for odor or flavor as liquid oils after deodorization as finished products. The most sensitive sensory analysis of the oil is by smelling or tasting it in liquid form without dilution or combination with any other foods. Crude and partially processed oils can also be tasted but must be diluted to make the sample

Table 5.1 Typical sensory evaluations for edible oils

Flavor/Odor Analyses	Room Odor Analyses
Salad oils	Heated oil odors
Deodorized	Frying food odors
Partially processed	
(diluted in RBD oil)	
Fat-containing foods	
Salad dressing	
Mayonnaise	
Fried food	
French fried potatoes	
Chicken	
Fish	
Potato chips	
Bread cubes	

palatable to judges (3,4). Flavor characteristics present in crude, degummed, refined and bleached stages were reliably detected by diluting the partially processed oil in good-quality oil at dilution levels of 1–5%. We were also able to use the flavor intensity and flavor characteristics of crude soybean oil to predict the storage stability of those same oils after they were deodorized and aged. Tasting bleached oils diluted to a 5:95 ratio with good-quality oil also will provide excellent reference standards for characteristic off flavors such as the sulfur flavor in canola oil (Appendix B).

Oil-containing foods can also be evaluated for flavor and storage stability by sensory methods. Procedures were developed to monitor the flavor and stability of starch-based salad dressings after aging studies (5). These methods are also applicable for mayonnaise and pourable salad dressings. Fried food flavor can be evaluated either immediately after frying and/or after aging depending on the food product. In the category of freshly fried food, the most commonly tested products are potatoes, chicken, and fish. Using potatoes has an advantage over animal products because potatoes do not contribute any fat to the frying system. A realistic frying schedule could include rotation frying of potatoes, chicken, and fish. A snack-type product such as potato chips is appropriate to test the effect of storage on fried foods. In previous research, we developed a model-type food—squares of white bread—for frying and storage stability tests (6). The fried bread was tasted immediately after frying and after aging studies under both ambient and accelerated storage conditions. Other researchers have also adopted this technique (7).

The odor of frying oil as it is heated or used to fry food is an important criteria for the consumer in judging oil quality. Consumers readily reject oil that has strong unacceptable odors. Evans and co-workers developed methods to evaluate the room odor of heated oils in the early 1970s (8,9). Subsequent research over the last 25 years has refined and expanded those procedures to evaluate blends of high and low linolenic-acid-containing oils, solid fats, liquid cooking oils, and oils from oilseeds with fatty acid composition modified by breeding (6,10–14).

ORGANIZATION OF SENSORY PANELS

2.1 Types of Sensory Panels

The two general types of panels are consumer and analytical (Table 5.2). Consumer panels are important in market research to measure preference or acceptability of a food product by those buying or consuming food products. Consumers would not taste oils directly, but they can judge oil-containing foods such as potato chips. In home-use tests, consumers can be given bottles of oil to use and asked to answer a questionnaire about their opinions on the oil quality and performance.

Trained, experienced judges—an analytical panel—are needed to evaluate oils directly for any subtle positive flavors and distinctive off flavors. The two basic types of analytical panels are difference and descriptive. A difference panel conducts simple evaluations by answering questions such as: "Is there a difference in flavor between these samples?" The level of training and experience is usually minimal; therefore, this type of panel is an excellent choice for those just wanting to start an oil panel. This panel is effective in screening incoming raw materials or a freshly processed batch of oil. On the other hand, more information is usually needed about an oil than just knowing if it is different from another sample. A descriptive panel takes the next step

Table 5.2 Types and characteristics of sensory panels

Types	Characteristics
Analytical	Measures intensity or quality differences Trained panelists with normal sensory acuity 5–20 members of panel
Difference, e.g., triangle, duo–trio, ranking, paired comparison	Measures differences between samples
Descriptive, e.g., scalar scoring, descriptive analysis	Measures intensity or quality
Consumer	Measures acceptance preference or like–dislike response 50+ untrained panelists

in oil evaluation by reporting not only if there are differences between oils, but also what those differences are. Judges are trained to rate the oils for quality or intensity based on precise scales, with reference standards that leave as little room as possible for subjective analysis. Judges can also be taught to describe the odors and flavors that they detect and to rate the intensity levels of those odors/flavors.

2.2 Requirements for Sensory Panel Operations

There are several levels of sensory panel development from the simple operation to a state-of-the-art facility. Selecting a specific level of panel depends first on the purpose or focus of the sensory analysis and, of course, the time and resources that the company or laboratory is willing to contribute. Table 5.3 presents four levels of sensory panel development with the requirements for each level. The requirements given are mainly guidelines. Some beginning operations may find need for higher level requirements.

Basic Sensory Panel. Requirements include an individual with scientific training to direct the panel and to apply scientific methodology for evaluating the oils; an odor-free, quiet room and a group of 12–16 people screened to taste oils. A training program in basic smelling and tasting techniques, followed

Table 5.3 Sensory panel requirements

	Levels			
Factors	Basic	Upgrade 1	Upgrade 2	Upgrade 3
Focus	Quality control	Effect of processing changes Storage stability	Product development Correlate instrumental and chemical analyses	Research on cause and effect
Personnel	Scientist consultant	Food technologist Technician Consultant	Sensory scientist Technicians	Sensory manager Sensory scientist Technicians
Facilities	Odor-free room Bench/table	Odor-free room Table with dividers	Odor-free room Individual booths in separate room Colored lighting Area for sample preparation	Odor-free room Individual booths in separate room Colored lighting Area for sample preparation Room odor testing area
Subjects	Screened Trained in: Mechanics of smell & taste Difference testing	Also trained in Scalar scoring	Also trained in Descriptive analysis	Also trained in Descriptive analysis Experienced
Recording method	Paper ballot	Paper ballot	Paper ballot	Computer I/O
Statistics	*t*-test Binomial distribution	Analysis of variance	Correlations	Multivariate analysis

by instructions on simple difference tests will then get a panel started. Although this panel will not solve major problems such as determining cause-and-effect relationships between volatile compounds and flavor, they will be able to determine, for example, whether or not a new supply of oil for food processing is different from the previous supply. As sensory panel operations are upgraded, panelists can be trained for scalar scoring and descriptive analysis. These techniques will not only help determine difference between oils but also provide information as to what is causing the variations. In the case of a company that uses vegetable oil in its food processing and must screen bulk oil shipments, a bench-top operation can be established in which a 500-mL sample of oil is taken to evaluate flavor and to determine peroxide value using the following protocol:

Follow American Oil Chemists Society (AOCS) method Cg 2-83 (15) for evaluating oils at 50°C.

Laboratory personnel who have been trained taste the oil and compare it with a control.

The identity of the control and the new samples should not be given.

The control should be a sample that meets the company's specifications for good oil quality or oil with weak odor/flavor intensity.

The results of the analysis of the two samples are compared to see if the new oil is of the same quality as the control.

A simple paper ballot for the triangle test is included in Appendix A.

Statistical significance of data is checked by consulting tables in sensory manuals (16,17).

As the taste tests are done, the peroxide values can be determined.

The process should take only 15 min after the oils are heated.

Upgrade 1. At this level, monitoring effects of processing changes or ingredient changes or storage stability is the primary focus. The objective here for sensory analysis is not only to determine whether or not there are differences between samples but also to identify the amount and type of difference. This panel should be trained to use a scoring scale and to rate the samples for intensity or quality. Examples of these scales are included in Appendix A and will be discussed later in this chapter. In upgrading the sensory program to this level, the panel leader should have a degree in food science with either college courses in sensory analysis or on-the-job training in sensory evaluation methodology. A sensory evaluation consultant can also be utilized to provide answers. If the volume warrants, a technician can be hired to help with the setting up of the panel and cleanup. The facilities can be upgraded to provide more comfortable surroundings by providing a table (with dividing partitions) and chairs for the judges to use while conducting their evaluations. The dividers on the table will prevent judges from talking, comparing scores, and watching the reactions of others. The data resulting from these evaluations would usually be analyzed by analysis of variance (ANOVA) to determine statistical differences.

Upgrade 2. Upgrading the sensory panel operations to the next level is usually necessary if specific work such as product development is conducted or if the research includes correlating sensory with instrumental and chemical analyses of oils. The panel should receive training in descriptive analysis to fully describe the odors and flavors detected in the products. A sensory scientist who has a degree in sensory evaluation or who has experience leading panels is invaluable in directing the operations at this level. Qualifications of a sensory scientist include knowledge of psychology, physiology, statistics, food science and chemistry. The facilities should also be upgraded to include an area specifically for the panel room with an adjoining area for the preparation of the oils. The panel room should include built-in booths. Information on the specific requirements of a taste panel room are given in sensory evaluation handbooks (16,17) and the American Oil Chemists' Society Recommended Practice Cg 2-83 (15). Techniques to mask color differences by red-colored fluorescent bulbs or red theatrical gels are effective in hiding color differences of almost all oils and oil-containing foods, unless of course there is a large difference in color intensity. Statistical analysis would include both ANOVA and correlation coefficients.

Upgrade 3 Laboratories needing this level of operations have a wide variety of oils and oil-containing foods that require several groups of trained judges. To help with the work load, a manager of the sensory operations and computer input/output (I/O) of panel data may be needed. Judges use a computer screen and keyboard or stationary mouse to record scores and descriptions. The computer I/O can be operated through a network system with a file server or by a personal computer. A separate computer screen that monitors the panelists' scores as they are entered into the computer can be installed in a room next to the panel area. This allows the panel leader to view the results and permits the panelists to check their own completed evaluations with those of other judges. A printout of the data with statistical calculations provides immediate feedback to the panel leader. An example of a printout used at our research center and explanations of computer programs to calculate, store, and retrieve data have been published (18).

2.3 Sensory Evaluation Facilities

There are comprehensive references in the literature on the design of facilities for sensory evaluation panels (16,17). The testing of salad oil for flavor requires only a few modifications not found in a standard facility, such as a method to heat oil at 50°C prior to and during panel evaluation and a method to mask color differences in the oils. An aluminum heating block (15,19) or a water bath (19) have both been used effectively to heat oils. The combination of aluminum serving blocks on an aluminum heating block is useful because the oils can be heated and served without any transfer of sample or glassware (15). This method also allows the sensory evaluation of headspace volatiles of the oil that accumulates during the heating process, since the heated oil has not been transferred from a heating container to a serving container. Red fluorescent lighting is effective in masking all but the most extreme differences in oil color (15,20,21).

Specific information on room odor facilities including floor plans and air-handling conditions have been published (10,20). One of the most important features of a room odor facility is the air lock rooms that panelists pass through before entering the actual odor room. This helps prevent significant loss of volatile odor compounds from the test room. Control of the amount of airflow into and out of the rooms must also be provided.

2.4 Sensory Judges

A major key to the success of sensory panel is the recruitment and training of a group of motivated, available judges who are interested in evaluating oils or fat-containing foods. The American Society for Testing and Materials (ASTM) has published a manual that provides excellent guidelines for recruiting and selecting panelists to judge food products (22). The five major steps in the process of developing a group of sensory judges are as follows: request volunteers for screening; screen volunters; select candidates to train; train judges; evaluate training of judges; and motivate judges.

Recruiting. The support of management in stressing the importance of sensory evaluation is needed in recruiting available personnel to be screened for the panel. A letter to all employees by a senior manager or president or director could be sent to all potential panelists requesting everyone's cooperation in the screening process.

Screening. Potential judges should be screened for their availability to participate on the panel, their health, and their level of interest in this work. Criteria for selecting potential sensory judges is in two stages: in stage one, interest and motivation, availability, and the ability to concentrate must be shown, and confidence and assertiveness as well as good health with normal sensory acuity must be proven. In the second stage, potential judges must give correct responses on preliminary odor/flavor tests. Such information can be obtained by questionnaire and/or by personal interviews. Potential panelists should have good verbal skills and be assertive in expressing their evaluation of the samples. We have found that such factors as age, educational level, job category, or gender do not make a difference in an individual's ability to evaluate foods. The triangle test is commonly used to screen panelists for their ability to evaluate products (Appendix A). To conduct the triangle test for oils, two samples of different intensity or quality levels should be selected and prepared according to AOCS method Cg 2-83 (15). The person being screened is presented three samples, two identical and one different, and is asked to identify the odd sample on the basis of flavor. The recommended number of trials that are conducted to screen judges varies, but at least six tests should be conducted. Since testers have a $33\frac{1}{3}\%$ chance of selecting the odd sample by guessing, individuals should have at least 66% correct identification before being considered as a judge. We conduct this test for a few preliminary trials to allow the candidate to be comfortable with the test. The first few sets should include samples of oil that are easy to distinguish with significant differences in intensity or quality. The actual screening samples should have only slight differences in flavor; however, the differences in the sample will depend on the sensitivity of the group being tested. Greater differences in intensity or quality may be required with individuals having no experience in sensory testing.

Training. The stages involved in training sensory judges are presented in Table 5.4. Studies involving human subjects may need review by internal or external committees for assurance that the subjects are in no way considered to be at risk. Even if no such reviews exist, the panel leader should take precautions to ensure the safety of judges by thoroughly knowing the treatments of the samples and the presence of any possible risks from the samples. Judges should sign voluntary consent forms that contain the following information: project title, treatments tested, risks and benefits of the testing, and precautions to be taken. An example of a voluntary consent form is shown in Table 5.5. Information on the treatments may be proprietary; therefore,

Table 5.4 Training sensory judges for oils and fat-containing foods

Stage 1
Protection of human objects
Chemistry of lipids and lipid oxidation
Techniques for tasting and smelling
Stage 2
Difference tests (triangle, duo-trio)
Stage 3
Scalar scoring (10-point intensity or quality scale)
Reference standards for intensity and quality levels
Control samples for weak intensity or good-quality oil
Stage 4
Reference standards for individual odors/flavors

Table 5.5 Voluntary consent form for protection of human subjects

Project Title:

Treatments Tested:

Benefits of Study:

Risks of Study:

Precautions to be taken:

Signature of Panel Leader/Date:

I, ______________________________, have been informed by ________________________ of the purpose, benefits, and risks of this study. I volunteer of my own free will to participate in this study. I understand that I may terminate my participation at any time and that this consent does not waive any of my legal rights.

Signature ______________________________ Date ________________

---

judges do not have to be given complete details, only those that may cause risks such as any non-GRAS (generally recognized as safe) additives. Examples of precautions include instructing judges not to swallow samples and to rinse their mouths thoroughly after tasting. Judges could also evaluate odor only rather than flavor if there is any question of sample safety.

During stage 1 of training, the panel leader should teach trainees the physiology of how tastes, odors, and flavors are perceived. Odors are perceived either directly through sniffing (nasal perception) or through the mouth by the retronasal pathway (Figure 5.1). Both techniques are used to evaluate oils and fats. Voirol and Daget (23) reported that the same olfactory stimulus may be perceived differently through the nose (nasally) than through the mouth (retronasally). Judges will generally perceive an odor as less intense by sniffing than by the retronasal method, possibly because there is no loss of volatiles from the oil to the olfactory receptors with the retronasal technique. Judges should be trained in the techniques of smelling and tasting that maximize odor perception. During odor evaluations, the container should always be covered and the judges should be instructed to swirl the container as it is brought to the nose, to remove the cover quickly, and to sniff the headspace of the container. The tasting technique consists of taking 5–10 mL of oil into the mouth, swishing the oil throughout the mouth, then cupping the mouth and putting the head back slightly, inhaling air over the oil, then exhaling through the nose. This procedure enhances the flow of volatiles through the back of retronasal area to the olfactory epithelium. Judges should also be instructed to repeat the procedures of pulling in air through the mouth and exhaling through the nose after the oil is expectorated so as to detect any residual flavors.

Odors interact with the olfactory system at several levels (24). First, the odor compounds enter the nasal pathways (either nasally or retronasally) and are absorbed on the mucosal lining. Next, the odor molecules react with receptor sites in the mucosa and signals from these sites are passed through the peripheral nervous system en route to the olfactory bulb. At the last stage, the electrochemical signals enter the central nervous system where the information is processed. Judges should be instructed to sniff the volatiles in the headspace of the oil container gently at first in case the odors are strong; however, repeated sniffs may be necessary. Odors disappear from the headspace quickly, so the judge should make a decision as rapidly as possible. Usually, the first impression is the best. The retronasal method for flavor perception is sensitive because volatiles from the oil are trapped in the mouth and the volatile compounds are pushed up to the olfactory epithelium through a passageway at the back of the mouth. The nasal method of odor perception is less reliable because of the loss of volatile compounds between the headspace of the oil container and the nose.

As part of training, judges should be informed of the off flavors that can occur in oils as well as the source of the flavors. Table 5.6 lists a few examples of sources of off flavor and some of the resulting off flavors. The list can be expanded depending on the types of off flavors encountered (Table 5.7). Knowing this type of information will facilitate the learning process for a judge.

At stage 2 of training, the judges should learn to use the difference tests such as triangle and duo-trio (16,17) proficiently to discriminate between different oils by odor or flavor. Judges should evaluate a wide variety of oil samples to learn the many odors/flavors that can be detected in oils of all quality levels. Training could stop at this point if the objective of the sensory panel is only to determine whether or not differences exist between oil samples.

Table 5.6 Sources of off flavors in fats and oils

Oxidation[a]	Processing	Inadequate Processing
Rancid	Hydrogenation flavor	Grassy/beany
Painty		Sulfur
Metallic		

[a] From heat, light, oxygen, metals.

Table 5.7 Flavors in edible oils[a]

Flavors	Corn	Cottonseed	Peanut	Canola/ Rapeseed	Safflower	Soy	Sunflower
Beany						D/U	
Burnt	U		X			U	X
Buttery	F	F	F	F	F	F	F
Cardboard	D	D	D	D	D	D	D
Corny	F						
Fishy				D		D	
Grassy	X		X	D	X	U/D	U/D
Hay	X		X		X	U/D	U/D
Hully							D
Nutty	F	F	F			F	
Painty				D		D	
Pine							U/D
Rancid	D	D	D	D	D	D	D
Raw	X		X	X			
Rubbery	X			X		X	
Sulfur				U/D			
Waxy	X	X			X		X
Weedy	X		X	X	X	X	X
Woody			X		X		X

[a] Other flavors may be present from contamination, processing conditions, etc.: pumpkin, melon, watermelon, petroleum, metallic, musty, etc.

[b] U = characteristic of unprocessed or partially processed oil; F = characteristic of freshly processed, good-quality oil; D = characteristic of deteriorated (from fair to bad quality) oil; X = unidentified origin.

Training should proceed to stage 3 if the objective of the panel is to determine the type and extent of differences between oil samples. Most judges quickly learn to use an intensity scale, whereas the process to learn a quality scale is usually longer. An essential part of training a panel is the use of reference standards for intensity or quality levels in the oils or food products. Appendix B contains reference standards for various flavors. Judges' training should always have a sample of the same control of either good quality or weak intensity at each testing session. Even a trained, experienced panel can benefit from receiving a control sample at each testing session if the sessions are only held infrequently. Fully trained judges should not need a control because they have odor/flavor memories that are called into use each time they evaluate a sample. A large supply of the control sample should be kept frozen in small quantities for every test. The daily testing of a control will help the trainees recalibrate their odor and flavor perceptions and recall their odor/flavor memories.

Descriptive analysis is the last and most difficult stage of panel training. Reference standards for specific odor/flavors in oils and oil-containing foods are given in Appendix B. The standards presented in Appendix B are primarily oils that have been aged or treated to produce the desired flavor. Although this is the preferred technique to achieve naturally occurring flavors, some references specify addition of chemicals to produce a flavor. Although judges may easily learn to discriminate correctly between samples using the intensity or quality scales, the process to accurately and consistently describe odors/flavors is time-consuming and may take months or years of practice. Frequent training with reference standards will enhance the learning process. Judges should be trained first to break down complex odors/flavors, such as hydrogenated, into the component parts. At our laboratory, we instruct judges to recognize the complex odor/flavor, hydrogenation, then to describe the component parts such as waxy, paraffin, sugary, flowery, fruity, and/or pastry. This information is invaluable in identifying problems of oil deterioration and in keeping judges from confusing similar odors/flavors. For example, new judges have a difficult time discriminating between a hydrogenation odor from a hydrogenated oil and the waxy odor from a high oleic sunflower oil. Judges are trained to recognize the hydrogenation odor and at the same time to describe the component parts of the odor. Therefore, a hydrogenated corn oil would be described as hydrogenated, fruity, sugary, waxy, and so forth. A high oleic sunflower oil would have a predominant description of waxy, in addition to any other minor descriptions.

The amount of time needed at each stage of training will depend on the samples and the intensity of the training. Food ingredients such as oil would probably take less time for training than a complex food such as mayonnaise. Daily training sessions of 1 h for the first 2 weeks will speed the training process. After that time, the judges should be asked to analyze actual samples that display a wide range of sensory characteristics on a daily basis. This should only take approximately 5–10 min.

Monitoring. Monitoring the judges during training is important in evaluating their progress and level of expertise. This process should continue after formal training is over to ensure that the consistency and accuracy of each judge's results are maintained. At our laboratory, we monitor judges' scoring after computer retrieval of data (18). The t test is used to determine whether or not the difference between a judge's score and the panel average is significant. If the judge has a t-test value of 2 standard deviations below or above the panel mean, that score is considered significantly different ($P < 0.05$) from the mean. Judges with a high percentage of scores significantly below or above the mean should be monitored closely to possibly change training procedures or to eliminate them from the panel.

Motivating. Keeping the judges motivated to attend panel sessions regularly and maintaining their expertise at a high level are vital parts of a sensory operation. Methods to motivate judges include providing feedback, information and education, rewards, and recognition by management. One of the greatest incentives is feedback on the results of daily tests, if possible, as well as reports of results from completed projects. We have also found that immediate feedback of results helps in the ongoing training process of judges by indicating how their evaluations compare with that of others. Rewards for panel participation vary from one laboratory to another, but range from actual money being paid or coupons for the company store to something as simple as cookies, candy, or fruit offered after each testing session. Management recognition is important and can include special luncheons, dinners, picnics, or award ceremonies for panelists.

SENSORY EVALUATION METHODS

3.1 Liquid Oils

Salad-type oils that are liquid at room temperature are evaluated for odor and flavor to determine their initial quality and/or flavor stability after storage. The AOCS has a recommended practice (Cg 2-83) (15) as does ASTM (19) for preparation and presentation of oils for sensory testing. These practices include basic information such as serving containers, oil temperature, and heating procedures for the oils.

The AOCS recommended practice also includes scoring scales based on a 1–10 system for rating the quality or intensity of liquid vegetable oils. Two scales are given, one for intensity and the other for quality in Appendix A. The intensity scale is based on a 1 = strong, 10 = bland (no odor/flavor) scale. This is recommended for rating oils that have blandness as their ultimate goal. Oils such as soy, sunflower, cottonseed, palm, low erucic acid rapeseed (canola), and safflower that have little or no flavor immediately after a proper deodorization are rated on this scale. On the other hand, a quality scale could be used for oils such as corn, peanut, and olive that have natural distinctive, although desirable flavors. Rating intensity may be less subjective than rating an oil for quality because more exact reference standards for intensity can be used to train judges. Flavorful oils can be rated for intensity of individual flavors; however, overall intensity scores for these oils are not indicative of quality since the natural flavors in addition to any off flavors would affect the score.

Reference standards for the intensity scale should include examples of oils with weak, moderate, and strong overall flavor intensity. A standard for a completely bland oil may be difficult to find; therefore, the judges should have a mental standard for blandness—no flavor. Reference standards for the quality scale should include examples of what each company defines as excellent, good, fair, poor, and bad quality. Reference standards for intensity and quality levels may be verified by instrumental and chemical analyses. For example, an oil with a peroxide value of 0–0.2 is *usually* good quality or has weak overall intensity. A volatiles profile by gas chromatographic analysis may also be used to check standards.

In using these scales, the panelists should be instructed to evaluate the overall intensity or overall flavor quality of the sample first. Then they should evaluate the oil for individual flavors and their intensity levels. In actual practice, an experienced judge will go through this process simultaneously rather than separating the two steps. The score sheet or computer screen should be set up for the judges to record their overall scores first, followed by the ratings for the individual descriptions.

The descriptions typical of oils range from no flavor (bland) to strong flavors such as rancid and painty (Table 5.7). Panelists should not only be instructed to note any odors/flavors detected in the samples but also to indicate the intensity level of each description. At our laboratory, we instruct panelists to rate the odors/flavors of salad oils on a scale of 0 = none, 1 = weak, 2 = moderate, 3 = strong (Appendix A). A weighted average of the intensity levels of each description is then calculated as follows:

$$\frac{(1 \times \text{number of weak responses}) + (2 \times \text{number of moderate responses}) + (3 \times \text{number of strong responses})}{\text{number of panelists}}$$

Quantifying the description of intensity levels provides valuable information that might not be readily seen in the overall scores. This data can also be used in statistical calculations such as correlation coefficients with levels of volatile compounds. Descriptions can also be rated on scales of other sizes such as on a 0–10 scale that we use for salad dressing or room odor (Appendix A). The range of the scale (0–3 or 0–10) chosen, depends on the ability of the judges and the extent of the evaluation that the judge is required to complete. The 0–3 scale simplifies the process of assigning intensity levels to descriptions. Judges and panel leaders should know the characteristic odors/flavors of each oil type and the source of these odors/flavors (Table 5.7). The descriptors represent flavor and odors that are present from various sources ranging from crude or partially processed oil to freshly processed oil to one that has deteriorated. For example, crude soybean oil has beany, grassy, and hay flavors. After sufficient processing and deodorization, soy oil has little or no odor/flavor and is usually described as bland or as slightly nutty or buttery. If the processing or deodorizing is not adequate, then the soybean oil may have a weak grassy or beany or hay flavor. As the oil oxidizes, odors/flavors become detectable. In the early stages of oxidation, soybean oil may have weak to moderate buttery odors/flavors. As oxidation continues, odors/flavors develop that are typical of a crude or partially processed oil. Sometimes this is referred to as a reverted flavor or retrograded flavor (25). For soybean oils, these odors/flavors are grassy, hay, or beany. Flavors detected in later stages of oxidation of soybean oil and canola oil include both rancid and painty, because both linoleic and linolenic acid are present in these oils. Oils such as peanut that contain no linolenic acid, do not develop a painty flavor since this is a flavor characteristic of linolenate-containing oils such as soybean and canola. Some inexperienced oil panelists can confuse a strong rancid flavor with painty. Oils such as cottonseed with high linoleic acid levels will develop a predominant rancid flavor in highly oxidized samples. Frankel presented reviews of fat and oil oxidation that explained both the mechanism of formation and the flavor significance of oxidation products (26,27).

3.2 Room Odor/Frying Oils and Fats

Methods to evaluate the odor characteristics of heated and frying oils were developed in the 1970s by Evans and co-workers (8,9). The initial method included simply heating a pan of oil in an empty laboratory. Judges enter the room and rated the odor in the room for overall quality and individual odor characteristics. This method was upgraded by constructing small rooms (5 × 8 × 10 ft) with controlled airflow and temperature (10,20). Judges were instructed to rate the overall intensity of room odor and the intensities of the individual odors detected. The intensity scoring scale (Appendix A) is based on a 0–10 scale (no odor to strong odor). High-oleic sunflower or low-linolenic-acid soybean oil (nonhydrogenated) heated to 190°C usually have the weakest overall odor (21). Low-erucic-acid rapeseed (canola) oil and olive oil receive ratings of 7.0 or 8.0, which are the highest mean scores given by our judges. Predominant odors for canola oil are fishy, acrid, and burnt. Hydrogenated oils have a characteristic "hydrogenation" odor that we train our panelists to recognize and to describe the individual odors that make up this complex odor. Descriptions such as fruity, flowery, and waxy are typical of hydrogenated oils. Appendix B contains standards for odors of heated oils. Details on the design of a room odor facility as well as on sensory analysis procedures are available (20,21).

3.3 Oil-Containing Foods

Methods developed to evaluate starch-based salad dressings (5) are also appropriate for mayonnaise. Pourable salad dressings can be either evaluated directly or tasted in a model system such as the emulsion method proposed by Stone and Hammond (28). Although the emulsion method was suggested as an alternative to direct tasting of oils to prevent taste fatigue, it could be an effective method to test pourable dressings. In our research, we have found that testing the dressing without a carrier is acceptable to the judges (5). Some laboratories may want to use carriers with as little flavor as possible such as unsalted soda crackers or lettuce. During panel training for these evaluations, the judges should be taught to recognize the flavor characteristics of a good-quality product and to rate the intensities of the complex variety of tastes and flavors present such as sweet, sour, stale, or rancid. The products should be evaluated with a quality scoring scale since a good-quality dressing is not bland. Individual flavors should be rated for intensity. An example of a score sheet for salad dressing analysis is given in Appendix A. A good-quality control sample should be given to the panelists at the beginning of each testing session.

The evaluation of fried foods is an important analysis because of limited methods to test these foods by instrumental and chemical analysis. Fried foods can be evaluated either immediately after frying (French fries, chicken, and fish) or after storage such as snack foods (potato chips). The judges evaluate the flavor of the food and determine whether or not there are any flavors or off flavors from the oil or fat. Since blandness is not a desired characteristic of fried foods, a quality scale is appropriate (Appendix A). Intensity of individual flavors should be rated. A good-quality control sample should be given to the panelists at the beginning of each testing session as a reference. As an alternative to frying a typical food product, we have fried 1-inch bread cubes for panel evaluation. Bread can be tested immediately after frying and/or after storage (4,13).

3.4 Data Analysis

One of the goals of sensory evaluation as well as other analytical tests is to determine whether or not there are significant differences between samples. In order for the researcher to know if the differences are significant, the data must be statistically analyzed. The statistical analysis of the simple difference tests requires no calculations, since some sensory manuals provide tables to check the significance of the data (16,17). For example, for triangle and duo-trio tests, tables in these manuals list the number of correct answers needed to establish significance at the 95, 99, and 99.9% confidence levels.

For sensory methods using scoring scales, the mean scores for the samples are usually analyzed by a two-way ANOVA test. Calculating the F ratio to determine significance is a fairly simple process (17) although statistical software packages are readily available for personal computers (PC) to conduct these calculations. Higher forms of statistical analysis, including regression analysis and correlation coefficients, are necessary to correlate data from sensory, instrumental, and chemical analyses. These programs are available as PC software along with programs to sort and rank-order data, to calculate chi-square values, and to print scatterplots and frequency distributions.

Intensity Scale for Salad Oils. Table 5.8 presents interpretations of the intensity scoring scale for salad oils. The intensity scale is set up to measure flavor changes as the oil oxidizes. As oxidation increases, the flavor intensity not only increases, but also the type of flavors detected progressively change.

Table 5.8 Interpretation of oil intensity scale[a]

Score	Intensity Level	Characteristic
10	Bland	No flavor detected
9	Trace	Detectable, but too weak to identify
8	Faint	Typical of most freshly deodorized oils
7	Slight	Typical of most commercial oils on shelf
6	Mild	Typical of oils with peroxide value (PV) <5
5	Moderate	
4	Definite	Typical of oils with PV >5
3	Strong	
2	Very strong	Typical of oils with PV >10
1	Extreme	

[a] Intensity scale: measures intensity of *overall* flavor.

Most oils, in the freshly deodorized stage, have little or no flavor; but as an oil oxidizes, the flavors change or are masked as new flavors, typical of oxidation. For example, as soybean oil oxidizes, flavors progressively change from nutty, buttery to grassy, then to rancid and painty. The last two descriptions are typical of a highly oxidized sample. The intensity scale is applicable for all oils that have no distinctive natural flavors. Freshly deodorized oils such as soybean, sunflower, canola, and cottonseed should have little or no flavor or odor if starting crude oil quality was satisfactory and if all processing was conducted properly. Judges evaluating a fresh, laboratory-processed soybean oil on an intensity scale would usually rate the oil a 9 for flavor since a trace of flavor is detectable but is too weak to be identified. A commercially processed soybean oil that is evaluated immediately after processing usually receives a score of 8 on an intensity scale, indicating a faint but detectable flavor or flavors such as nutty or buttery. After this oil has been bottled and stored on a grocery shelf for a few months, the flavor score of oil is usually 7.0, indicating a low level of oxidation. Typical flavors may be weak buttery, grassy, or beany. An intensity score of 7.0 for a fresh or slightly aged oil is usually considered the cut-off point for a good-quality oil. Oils that are scored a 6 normally have peroxide values between 1 and 5, although off flavors from processing (Table 5.11) such as hydrogenation may cause the intensity of the flavor to increase. A score of 6.0 is considered the minimum score for a good-quality oil after aging, as in the Schaal oven test (4 days at 60°C) or after 4 months at ambient temperature (25°C). Oils that are scored 5 or below are usually oxidized as indicated by peroxide values of 5.0 or above. The score would continue to decrease as the oxidation process continued.

It is important to point out that oils at an intensity level of 7 or 8 would not be described as weak rancid or painty because these flavors are typical of later stages of oxidation and therefore, generally do not exist by themselves but rather in conjunction with other off flavors such as grassy. By the time an oil reaches the peroxide level to have a rancid or painty flavor, then other flavors are present also. On the other hand, strong nutty or buttery flavors are generally not present in oxidized oils because these flavors either are masked by other off flavors or the volatile compounds producing the flavors have broken down causing other flavors.

Quality Scale for Salad Oils. The interpretation of the quality scale is given in Table 5.9. This scale is appropriate for oils with distinctive natural flavors such as corn, peanut, or olive oil. The quality characteristics used for this scale should be set by individual laboratories according to the type of oil being tested. For example, a well-processed corn oil usually has a weak corny flavor, no off flavors, and therefore can be considered good quality. Individual standards can be changed if a company considers an oil with a moderate corn flavor as indicative of good quality. The quality level of the corn oil would be rated down if the oil contained any off flavors from processing such as a weak to moderate burnt flavor. The quality would also decrease as the oil underwent oxidative deterioration and either lost some or all of its characteristic natural flavor or developed flavors indicative of oxidation such as rancid. The score would decrease further depending on the intensity of off flavors. To remind the panel of the characteristics of a good-quality product, a standard control sample can be presented during each evaluation.

Some scales used by the oil industry are based on identifying flavors and a combination of the intensity and quality scale with 10 = excellent and 1 = bad quality. Panelists are instructed to taste the oil and determine the flavors in the oil and then assign a score based on the type and intensity of the oil flavor. For example, an oil with a weak melon flavor is rated a 5, and an oil with a weak painty flavor is rated a 4. This type of scale was developed because in actual practice, this is the progression of oil deterioration. This is not an

Table 5.9 Interpretation of oil quality scale

Score	Quality Level	Characteristic
10	Excellent	
9	Very good	
8	Good	Typical of good quality oil; weak characteristic flavors, no off flavors
7		
6	Fair	Weak off flavor or some loss of characteristic flavor
5		
4	Poor	Moderate off flavors
3		
2	Bad	Strong off flavors
1		

[a] Interpretation of oil quality scale

appropriate scale for two reasons. First, a scale based on descriptions is not linear, just as a spectrum of colors is not linear. An intensity scale or a quality scale based on physical examples of specific intensity levels or of specific quality levels has equal intervals between numbers. According to Bodyfelt and co-workers (29), a linear scale is a prerequisite for statistical analysis. In the example given previously, the score difference between an oil with a weak melon flavor and a weak painty flavor is probably more than one unit; therefore, this combined scale based on descriptions is not linear and any statistical results would have little meaning. The second objection to this scale is that it depends on determining type of oil flavor in order to assign a flavor score. The validity of this scale depends on the ability of all judges to accurately and consistently identify flavors detected, which is the most difficult task for a sensory judge. A scale based on descriptions is not recommended for oil evaluations, but the use of either an intensity scale or a quality scale with reference standards for intensity or quality levels is recommended. The quality and intensity scales listed in Appendix A were used successfully in an AOCS collaborative study in 1988 (AOCS, unpublished results).

Intensity Scale for Room Odor. The characteristics of the room odor intensity levels are in Table 5.10. A room with no oil odor is rated as 0, and the score increases with increasing odor intensity. Scores of 1, 2, or 3 are indicative of weak overall odor. Cottonseed oil heated at 150°C is typical of an oil with weak overall room odor intensity. Scores of 4, 5, or 6 are indicative of moderate overall intensity. Most oils are rated in this area of the scale.

Table 5.10 Interpretation of room odor intensity scale

Score	Intensity Level	Characteristic
0	None	No odor detected
1		
2	Weak	
3		Typical of cottonseed oil heated at 150°C
4		Typical of high oleic (80%) oils heated at 190°C
5	Moderate	Typical of cottonseed oil heated at 190°C
6		Typical of hydrogenated (iodine value = 90) soybean oil heated at 190°C
7		Typical of soybean or rapeseed oils heated at 190°C
8	Strong	
9		Typical of soybean or rapeseed oils heated at 240°C
10		

[a] Room odor intensity scale: measures intensity of *overall* room odor and individual odors.

Oils with strong odor intensity would be rated as 7, 8, 9, or 10. Soybean or rapeseed (canola) oil heated to 240°C receive average scores of 8 and are our strongest odor standards.

Quality Scale for Oil-Containing Foods. The quality scale for oil-containing foods is similar to the quality scale for oils (Table 5.11). The characteristics for each type of food should be determined by each laboratory/company. For example, a good-quality potato chip with a characteristic potato flavor and no off flavors could be rated an 8. As the potato chips age and deteriorate during storage, a loss of typical potato flavor is noted and the flavor quality score decreases. As off flavors begin to develop, the score continues to decrease with increasing amounts of off flavors. With fried foods, one of the first off flavors to appear is stale.

COORDINATING SENSORY WITH INSTRUMENTAL AND CHEMICAL ANALYSES

As stated previously, no single oil analysis should stand alone. Most laboratories have the capabilities to conduct peroxide value, free fatty acid, and color analyses. These tests combined with sensory data can provide important information on initial quality of oil. If resources permit, a gas chromatograph (GC) equipped with a dynamic headspace (purge and trap) or a static headspace attachment can provide data on the volatile compounds present in fresh

Table 5.11 Interpretation of quality scale for oil-containing foods

Score	Quality Level	Characteristic
10	Excellent	
9	Very Good	
8	Good	Typical of good-quality product; weak characteristic flavors, no off flavors
7		
6	Fair	Weak off-flavor or some loss of characteristic flavor
5		No characteristic flavor and/or less desirable flavors at moderate intensity
4	Poor	Unacceptable flavors at moderate intensity
3		
2	Bad	Unacceptable flavors at strong intensity
1		

[a] Quality scale: measures quality of *overall* flavor.

and aged oil. Simple direct injection of the oil in the heated inlet liner of a GC can also be used to monitor volatiles. Several papers have been published that outline the specific procedures necessary for three volatiles methods (1,30–35). The first step in correlating GC volatiles and sensory panel results includes analyzing a variety of samples to develop a database. For example, soybean oil samples oxidized to different levels should be analyzed both by the sensory panel and with the GC to monitor how the odors/flavors change and how the volatiles change or increase. The volatile compounds that change the most rapidly as oxidation increases should be selected as "markers" to monitor oxidation. Flavor scores from the intensity or quality scales can be used for the database as well as intensity values for odor/flavor descriptions. The sensory and volatiles data can then be statistically analyzed to determine correlation coefficients. This statistical data will only show correlations between flavor and volatiles, rather than actual cause-and-effect relationships. Analyzing oils and oil-containing foods for such markers as pentane, 2,4-decadienal, and hexanal as well as total volatiles can be useful in monitoring product oxidation. Figure 5.2 shows correlation coefficients calculated between flavor intensity scores and volatiles measured by direct injection GC. This direct injection method shows high correlation coefficients between pentane

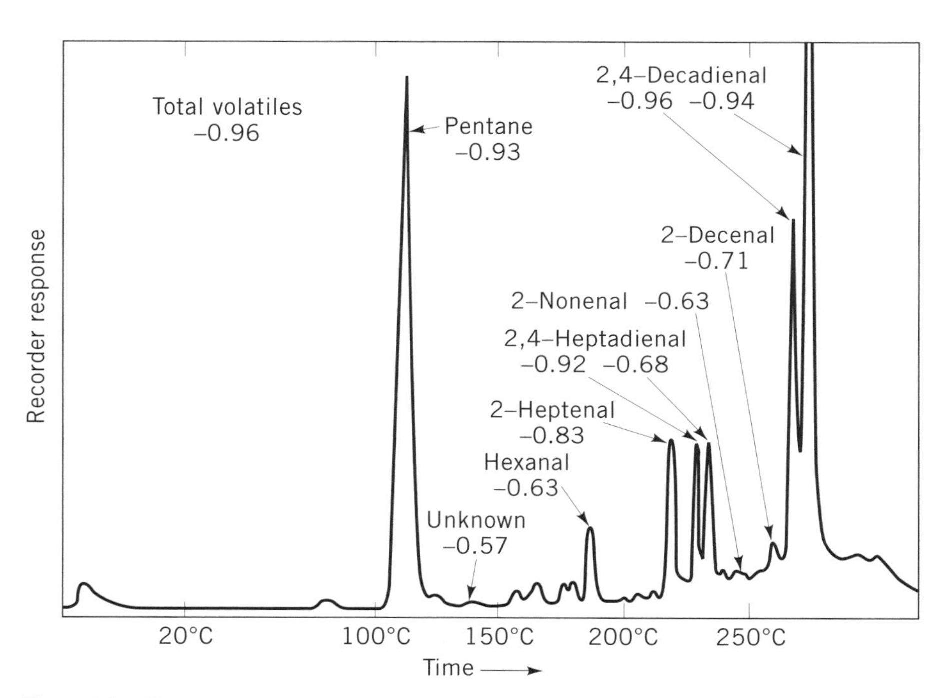

Figure 5.2 Chromatogram from direct injection analysis of oxidized soybean oil with correlation coefficients with flavor intensity scores.

and 2,4-decadienal and flavor scores. Different profiles of the same oxidized oil can be found from each of the three GC volatiles methods (35). Because of these differences, each method will show different correlation coefficients between specific volatiles and flavor results. For example, pentane and hexanal are predominant peaks in oils analyzed by static headspace; however, pentane is not easily detected by dynamic headspace.

Once the sensory results have been compared with instrumental and chemical analyses and correlation coefficients are high ($r > .90$), most future monitoring of the oil can be done by the nonsensory testing with periodic panel evaluations. Figure 5.3 presents the chromatograms of three static headspace GC analyses of fresh and oxidized soybean oils. As expected, the amounts of pentane and hexanal along with other volatile compounds increase with decreasing flavor scores. Correlations between volatiles and flavor would be high for these samples. However, flavor and volatiles do not always correlate well. As shown in Figure 5.4, the second and third chromatograms have the same flavor scores, but the volatiles profiles are different. Volatile profiles can be relied on to provide valuable information in addition to and instead of flavor evaluation, but problems apparent in Figure 5.4 show the need for sensory analysis.

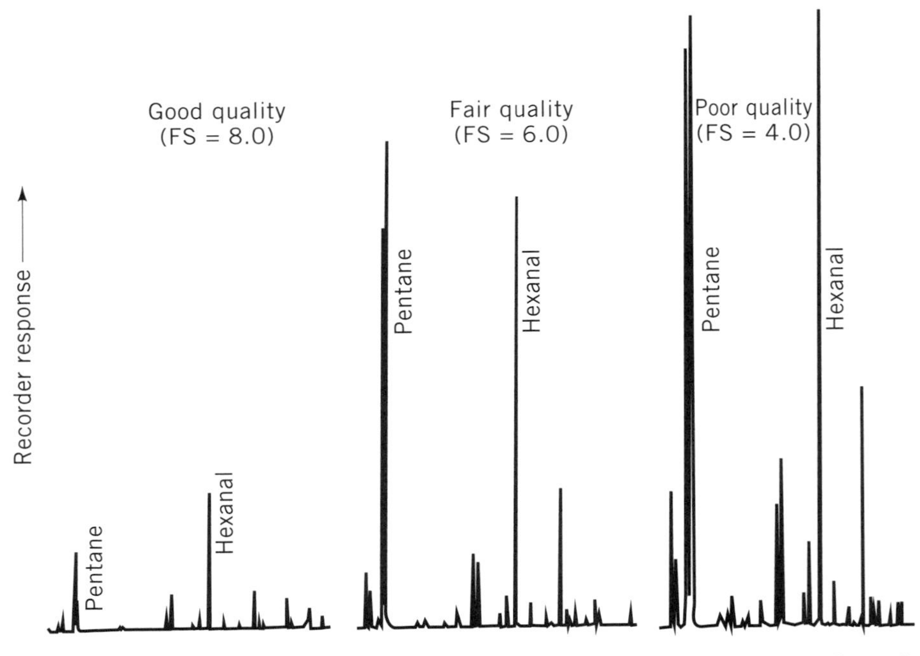

Figure 5.3 Chromatograms of three soybean oils with different levels of flavor quality and corresponding levels of volatile compounds from static headspace analysis.

FUTURE APPLICATIONS FOR SENSORY EVALUATION

Sensory evaluation has played an important role in improving and maintaining the quality of vegetable oils in the past and continues that role now. Many of the flavor and stability problems of vegetable oils have been solved, but, as new problems and situations occur, more research will be needed and sensory evaluation should play a vital part. Sensory evaluation can contribute in research on the cause-and-effect relationships between volatile compounds and flavor. Much of the previous work with flavor and volatiles was directed toward simply determining the correlations between flavor and volatiles (30–32). Carrying this work further to possibly determine the specific types and proportions of volatiles that cause specific flavors in oils would help in the research to find the sources of off flavors in oils and learn how to minimize development of off flavors. We now rely on increases in volatiles peak area to obtain measurements to correlate with flavor; however, the flavor significance of these compounds needs to be taken into consideration. For example, pentane makes an excellent marker because it changes rapidly with increasing oxidation, but it has little flavor significance because of its high odor threshold (340 ppm) (26). On the other hand, a compound such as 1-octen-3-ol, which shows only small changes in area has a very low threshold (0.0075 ppm) (26)

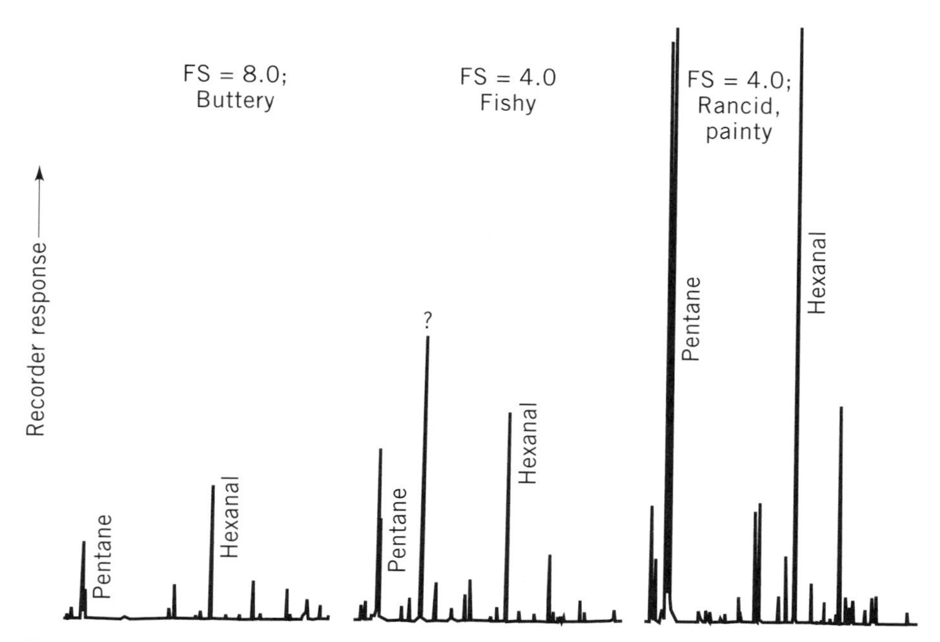

Figure 5.4 Chromatograms of three soybean oils with different volatile profiles (static headspace) but similar flavor scores for two of the oils.

and probably makes a large contribution to odor and flavor of oxidized oils. The volatiles with the most flavor significance should be used by those scientists attempting the very difficult task of determining cause-and-effect relationships between flavor and volatile compounds.

A second area of application for sensory evaluation is in the analyses of genetically modified oilseeds. Plant geneticists are developing new crops such as high oleic sunflower, low linolenate soybean and canola, and null-lipoxygenase soybeans. All of these crops were developed to provide oils with better flavor and oxidative stability. The characteristics of these oils will need to be determined by end-use evaluation. Some work as already been published on null-lipoxygenase soybeans (3) and on varieties of soybeans with genetically modified fatty acid composition (7,12,14).

The oil-processing industry is also concerned with the ability to predict the quality and stability of vegetable oils at the earliest stage in processing as possible rather than waiting for consumer complaints. Sensory evaluation along with instrumental and chemical tests may help to solve this problem. We have published some research showing that the flavor quality of the crude oil can be a predictor of the storage stability of the oil (4). We have also used a rapid oxidative stability test to determine relative differences in oil stability (1). These and other rapid methods need to be evaluated by the oil industry in the future.

SOURCES OF INFORMATION

There are several forms of assistance that are available to help establish analytical sensory panels for oils and fat-containing foods. The AOCS Recommended Practice, Cg 2-83, outlines the basic requirements for sensory testing along with rating scales for intensity and quality evaluations. ASTM Standard Practice, entitled "Bulk Sampling, Handling and Preparing of Edible Oils for Sensory Evaluation," includes specific information on how to handle oils to prevent oxidation and how to prepare the oil (temperature, heating method, glassware, etc.) for evaluation. ASTM also has published guidelines for establishing general sensory panels.

The following is a list of publications with general information about establishing and training a sensory panel, as well as specific information about oil panels.

American Oil Chemists' Society, Champaign, IL

Recommended Practice, Cg 2-83, 1989

Flavor Chemistry of Fats and Oils, 1985

Handbook for Soy Oil Processing, 1980

Methods to Assess Quality and Stability of Oils and Fat-containing Foods, 1995

American Society for Testing and Materials, Philadelphia, PA

Standard Practice for the Bulk Sampling, Handling, and Preparing Edible Vegetable Oils for Sensory Evaluation, E1346-90, 1990

Basic Principles of Sensory Evaluation, STP 433, 1968

Correlating Sensory Objective Measurements–New Methods for Answering Old Problems, STP 594, 1976

Consumer Sensory Evaluation, STP 682, 1979

Manual on Sensory Testing Methods, STP 434, 1968

Guidelines for the Selection and Training of Sensory Panel Members, STP 758, 1981

REFERENCES

1. K. Warner, E.N. Frankel, and T.L. Mounts, *J. Am. Oil Chem. Soc.* **66,** 558 (1989).
2. E.N. Frankel, *Trends Food Sci. Tech.* **41,** 220 (1993).
3. E.N. Frankel, K. Warner, and B.P. Klein, *J. Am. Oil Chem. Soc.* **65,** 147 (1988).
4. K. Warner, E.N. Frankel, and K.J. Moulton, *J. Am. Oil Chem. Soc.* **65,** 386 (1988).
5. K. Warner, E.N. Frankel, J.M. Snyder, and W.L. Porter, *J. Food Sci.* **51,** 703 (1986).
6. E.N. Frankel, K. Warner, and K.J. Moulton, *J. Am. Oil Chem. Soc.* **62,** 1354 (1985).
7. L.A. Miller and P.J. White, *J. Am. Oil Chem. Soc.* **65,** 1324 (1988).
8. C.D. Evans, H.A. Moser, G.R. List, H.J. Duttc• and J.C. Cowan, *J. Am. Oil Chem. Soc.* **48,** 711 (1971).
9. C.D. Evans, K. Warner, G.R. List and J.C. Cowan, *J. Am. Oil Chem. Soc.* **49,** 578 (1972).
10. T.L. Mounts, *J. Am. Oil Chem. Soc.* **56,** 659 (1979).
11. K. Warner, T.L. Mounts, and W.F. Kwolek, *J. Am. Oil Chem. Soc.* **62,** 1483 (1985).
12. T.L. Mounts and co-workers, *J. Am. Oil Chem. Soc.* **65,** 624 (1988).
13. K. Warner, *J. Am. Oil Chem. Soc.* **65,** 520 (1988).
14. K. Warner and T.L. Mounts, *J. Am. Oil Chem. Soc.* **70,** 983 (1993).
15. *Official and Tentative Methods of the American Oil Chemists' Society,* 4th ed., AOCS, Champaign, IL, 1989, Methods Cd 8-53, Cg 1-83, and Cg 2-83.
16. American Society for Testing and Materials, STP 434, Manual on Sensory Testing Methods, Philadelphia, 1968.
17. L.M. Poste, D.A. Mackie, G. Butler and E. Larmond, *Laboratory Methods for Sensory Analysis of Food,* Agriculture Canada, Ottawa, 1991.
18. K. Warner, J.O. Ernst, B.K. Boundy and C.D. Evans, *Food Tech.* **28,** 42 (1974).
19. American Society for Testing and Materials, Standard Practice for the Bulk Sampling, Handling and Preparing of Edible Oils for Sensory Evaluation, Philadelphia, 1990.
20. T.L. Mounts and K. Warner, "Evaluation of Finished Oil Quality," in D.R. Erickson and co-workers, eds., *Handbook of Soy Oil Processing and Utilization,* American Soybean Association and American Oil Chemists' Society, Champaign, Ill, 1980.
21. K. Warner, "Sensory Evaluation of Flavor Quality of Oils," in D. Min and T.H. Smouse, eds., *Flavor Chemistry of Fats and Oils,* American Oil Chemists' Society, Champaign, IL, 1985.

22. American Society for Testing and Materials, STP 758, Guidelines for the Selection and Training of Sensory Panel Members, Philadelphia, 1981.
23. E. Voirol and N. Daget, *Lebensm.-Wiss. u.-Technol.* **19,** 316 (1986).
24. B. Berglund, U. Berglund, and T. Lindvall, *Psychol. Rev.* **83,** 432 (1976).
25. T.H. Smouse, "Flavor Reversion of Soybean Oil," in D. Min and T.H. Smouse, eds., *Flavor Chemistry of Fats and Oils*, American Oil Chemists' Society, Champaign, Ill. 1985.
26. E.N. Frankel, "Chemistry of Autoxidation: Mechanism, Products and Flavor Significance," in D. Min and T.H. Smouse, eds., *Flavor Chemistry of Fats and Oils,* American Oil Chemists' Society, Champaign, Ill. 1985.
27. E.N. Frankel, *Prog. Lipid Res.* **22,** 1 (1982).
28. R. Stone and E.G. Hammond, *J. Am. Oil Chem. Soc.* **60,** 1277 (1983).
29. F.W. Bodyfelt, J. Tobias, and G.M. Trout, *Sensory Evaluation of Dairy Products,* Van Nostrand Reinhold, New York, 1988.
30. K. Warner, C.D. Evans, G.R. List, B.K. Boundy and W.F. Kwolek, *J. Food Sci.* **39,** 761 (1974).
31. K. Warner, C.D. Evans, G.R. List, H.P. Dupuy, J.I. Wadsworth and G.E. Goheen, *J. Am. Oil Chem. Soc.* **55,** 252 (1978).
32. K. Warner and E.N. Frankel, *J. Am. Oil Chem. Soc.* **62,** 100 (1985).
33. J.M. Snyder, E.N. Frankel, and E. Selke, *J. Am. Oil Chem. Soc.* **62,** 1675 (1985).
34. J.M. Snyder, E.N. Frankel, and K. Warner, *J. Am. Oil Chem. Soc.* **63,** 1055 (1986).
35. J.M. Snyder, E.N. Frankel, E. Selke and K. Warner, *J. Am. Oil Chem. Soc.* **65,** 1617 (1988).

APPENDIX A

Score Sheet for Triangle Test

Triangle Difference Test

Name ______________________________

Date ____________________

Instructions: Taste the three samples in the order presented.
Two of the samples are identical; one is different.
Place checkmark by number of different sample.

Code Number

391 ____________________

376 ____________________

382 ____________________

Score Sheet for Oil Intensity Evaluation

Laboratory Code No. ______

Flavor Intensity Evaluation

Name: ______________________

Directions: Take 5–10 mL of warm oil into the mouth; pull air through the oil, and exhale through the nose.
Overall Intensity Scores: Rate samples on 10–1 scale.
Descriptions: Identify flavors and rate as weak (W), moderate (M), or strong (S).

	Overall Intensity Scores *Sample Numbers*			
Intensity	______	______	______	______
10 Bland	______	______	______	______
9 Trace	______	______	______	______
8 Faint	______	______	______	______
7 Slight	______	______	______	______
6 Mild	______	______	______	______
5 Moderate	______	______	______	______
4 Definite	______	______	______	______
3 Strong	______	______	______	______
2 Very Strong	______	______	______	______
1 Extreme	______	______	______	______

	Description Intensity *Sample Numbers*			
Descriptions	______	______	______	______
Nutty	______	______	______	______
Buttery	______	______	______	______
Corny	______	______	______	______
Beany	______	______	______	______
Hydrogenated	______	______	______	______
Burnt	______	______	______	______
Weedy	______	______	______	______
Grassy	______	______	______	______
Rubbery	______	______	______	______
Melon	______	______	______	______
Rancid	______	______	______	______
Painty	______	______	______	______
Fishy	______	______	______	______
Other ______	______	______	______	______
Other ______	______	______	______	______

Score Sheet for Oil Quality Evaluation

Laboratory Code No. ______

Flavor Quality Evaluation

Name: ______________________

Directions: Take 5–10 mL of warm oil into the mouth; pull air through the oil, and exhale through the nose.

Overall Intensity Scores: Rate samples for overall flavor quality on 10-point scale.

Descriptions: Identify flavors and rate as weak (W), moderate (M), or strong (S).

Quality	Overall Quality Scores *Sample Numbers*			
	______	______	______	______
10 Excellent				
9 Good				
8				
7 Fair				
6				
5 Poor				
4				
3 Very Poor				
2				
1				

Descriptions	Description Intensity *Sample Numbers*			
	______	______	______	______
Nutty				
Buttery				
Corny				
Beany				
Hydrogenated				
Burnt				
Weedy				
Grassy				
Rubbery				
Melon				
Rancid				
Painty				
Fishy				
Other ______				
Other ______				

Score Sheet for Room Odor Evaluation

Room Odor
Date: ____________________ Name: ________________________________
Directions: Circle the number to indicate intensity level.

Overall Intensity	0	1	2	3	4	5	6	7	8	9	10
			WEAK			MODERATE			STRONG		
Fried Food	0	1	2	3	4	5	6	7	8	9	10
Fishy	0	1	2	3	4	5	6	7	8	9	10
Hydrogenated	0	1	2	3	4	5	6	7	8	9	10
Burnt	0	1	2	3	4	5	6	7	8	9	10
Doughy	0	1	2	3	4	5	6	7	8	9	10
Acrid/Pungent	0	1	2	3	4	5	6	7	8	9	10
Waxy	0	1	2	3	4	5	6	7	8	9	10
Other ________	0	1	2	3	4	5	6	7	8	9	10

Score Sheet for Salad Dressing Evaluation

Name ______________________ Date ______________

Salad Dressing Evaluation

Instructions:
Taste each sample in order presented, starting with the control (C):
Mark your intensity rating by putting the number of the sample in the boxes that represent the flavors you detect.
Rate the overall quality of the samples on the scale at the bottom of the sheet.

Intensity

Sweetness
Weak ☐☐☐☐☐☐☐☐☐☐ Strong

Sourness
Weak ☐☐☐☐☐☐☐☐☐☐ Strong

Grassiness
Weak ☐☐☐☐☐☐☐☐☐☐ Strong

Rancidity
Weak ☐☐☐☐☐☐☐☐☐☐ Strong

Paintiness
Weak ☐☐☐☐☐☐☐☐☐☐ Strong

Hydrogenated
Weak ☐☐☐☐☐☐☐☐☐☐ Strong

Staleness
Weak ☐☐☐☐☐☐☐☐☐☐ Strong

Other
Weak ☐☐☐☐☐☐☐☐☐☐ Strong

Overall Quality

☐☐☐☐☐☐☐☐☐☐
Bad Poor Fair Good Excellent

Score Sheet for Fried Food Evaluation

Fried Food Evaluation

Name ______________________ Date ______________

Instructions: Taste the samples in the order presented, beginning with the control (C). Rate the samples for individual flavors and their intensities and for overall quality. The control is an example of a good-quality product (score = 8) with moderate fried food flavor and no off flavors.

Individual Flavor Intensity

0	1	2	3	4	5	6	7	8	9	10
None		Weak			Moderate			Strong		

Intensity Scores

Sample Code	Fried Food	Stale	Fishy	Hydro-genated	Waxy	Rancid	Painty
Control	____	____	____	____	____	____	____
532	____	____	____	____	____	____	____
579	____	____	____	____	____	____	____

Overall Flavor Quality

1	2	3	4	5	6	7	8	9	10
	Bad		Poor		Fair		Good		Excellent

Sample Code	Score
Control	____
532	____
579	____

APPENDIX B: REFERENCE STANDARDS

Bacon

Definition: An aromatic reminiscent of smoked bacon
Reference: Crude undeodorized coconut oil heated to 38°C
Example: Fried smoked pork bacon

Beany

Definition: An aromatic reminiscent of raw soybeans
Reference: Crude soybean oil diluted in fresh soybean oil (5:95)
Example: Ground lima beans (dry) mixed with water (2:98 ratio)

Bitter
Definition: A basic taste stimulated by such substances as quinine, caffeine, and hop bitters
Reference: 0.2% caffeine in water
Example: Tonic water

Bland
Definition: No aromatic or taste factors perceptible
Example: Mineral oil (Fischer Scientific)

Burnt
Definition: An aromatic reminiscent of burnt popcorn or grains
Example: Air-popped popcorn

Buttery
Definition: An aromatic reminiscent of fresh, sweet, unsalted butter
Reference: Land O'Lakes fresh, sweet, unsalted butter diluted in good-quality soybean oil (1:99)
Example: Freshly processed unsalted butter

Cardboard
Definition: An aromatic associated with the odor of cardboard box packaging
Reference: Wet 1 cup unsalted, dry-roasted vacuum-packed peanuts with distilled water; place wet nuts on tray to air-dry for 24 h
Example: Wet cardboard

Corny
Definition: An aromatic from corn
Example: Raw corn: non-heat-treated corn
Cooked corn: heated or boiled corn
Toasted corn: heated enough to caramelize sugars

Deep Fried
Definition: An aromatic reminiscent of oil heated to frying temperature of 190°C.
Reference: 0.3 ppm, 2,4-decadienal in good-quality oil (odor only)
Example: Fresh cottonseed oil heated to 190°C

Fishy
Definition: An aromatic reminiscent of cod liver oil
Reference: Cod liver oil diluted in good-quality oil (1:99)
Example: Odor from canola (rapeseed) oil heated in 190°C

Fruity

Definition: An aromatic reminiscent of ripe fruit
Example: Olive oil

Grassy

Definition: An aromatic reminiscent of the green character of mowed grass
Reference: Crude soybean oil from non-heat-treated soybeans diluted in good-quality oil (5:95)
Example: Freshly cut green grass

Green

Definition: An aromatic associated with unprocessed vegetation such as fruits and grains
Example: Raw with additional characteristics of hexenals, leaves, or grass

Hay

Definition: An aromatic reminiscent of dried grass character of air-dried grain or vegetation
Reference: Crude soybean oil from heat-treated beans diluted in good-quality oil (5:95)
Example: Dried alfalfa

Hully

Definition: An aromatic associated with the outer protective coating of grain or legume
Example: Sunflower seed shells

Hydrogenated

Definition: An aromatic reminiscent of the sweet paraffinlike odor of candle wax or crayons
Reference: 10% undeodorized hydrogenated soybean oil (iodine value = 90–110) in good-quality soybean oil
Example: Crisco shortening

Lard

Definition: An aromatic associated with rendered pork fat
Example: Raw or uncooked pork fat

Light-struck

Definition: An aromatic characteristic of light-sensitive oils such as soybean that are exposed to fluorescent light or sunlight
Reference: Good-quality soybean oil exposed to fluorescent light (1076 lux) for 1 week

Melon

Definition: An aromatic reminiscent of fruits such as watermelon or cantaloupe
Reference: 0.002 ppm 2,6-nonadienal in good-quality oil (odor only)
Example: Ripe cantaloupe

Metallic

Definition: An aromatic associated with metal coins
Reference: 0.01% ferrous sulfate diluted in distilled, filtered water
Example: Copper pennies soaked in filtered water 12 h

Musty

Definition: An aromatic reminiscent of odor or moldy or dank cellar or room
Reference: Geosmin

Nutty

Definition: An aromatic reminiscent of fresh, sweet nutmeats
Reference: Freshly ground pecans
Example: Freshly processed peanut oil

Overheated

Definition: An aromatic reminiscent of oil overheated during processing
Reference: Corn oil heated to 240°C
Example: Corn oil overheated during processing

Oxidized

Definition: A general term denoting the process known as oxidative deterioration of oil. Oxidized flavors/odors range widely from buttery, grassy, rancid, to painty. Not recommended as a specific odor/flavor.

Painty

Definition: An aromatic reminiscent of linseed oil or oxidized rapeseed (canola) oil; not noted in non-linolenic acid oils such as peanut
Reference: Good-quality canola oil aged for 4–8 days at 60°C or until a peroxide value of 10.0 is reached
Example: Linseed oil

Paraffin

Definition: An aromatic reminiscent of candle wax
Reference: High oleic sunflower oil heated to 190°C for 30 min
Example: Melted paraffin

Pine

Definition: An aromatic reminiscent of pine needles that is noted in oils such as sunflower

Reference: Bleached, undeodorized sunflower oil diluted 5% in good-quality fresh sunflower oil

Example: Fresh pine needles cut in small pieces

Plastic

Definition: An aromatic reminiscent of plastic containers or food stored in plastic container

Reference: Plastic strips from polyethylene package stored 24 h in fresh, good-quality soybean oil.

Rancid

Definition: An aromatic reminiscent of odor/flavor of oxidized oils containing linoleic acid such as cottonseed or sunflower

Reference: Good-quality cottonseed oil aged for 4 days at 60°C or until a peroxide value of approximately 5.0 is reached.

Raw

Definition: An aromatic reminiscent of unripe fruits or vegetables; similar to green

Example: Soybean oil underdeodorized at too low a temperature or too short a time period

Reverted

Definition: A general term denoting the process known as oil flavor reversion. During initial stages of oxidation, most oils develop flavors/odors reminiscent of crude oil. Each oil type would have a characteristic reverted odor/flavor. Not recommended as a specific odor/flavor.

Rubbery

Definition: An aromatic reminiscent of old rubber

Example: Poorly processed corn oil; rubber stoppers from chemical laboratory

Scorched/Burnt

Definition: An aromatic reminiscent of burnt hulls or husks

Example: Poorly processed corn or sunflower oil

Soapy

Definition: An aromatic reminiscent of soap

Example: Oxidized fat containing lauric acid, such as coconut oil

Stale

Definition: General term characteristic of old, unfresh product, with loss of original flavors; usually in initial stages of oxidation

Example: Potato chips aged 2 weeks at 25°C

Sulfur

Definition: An aromatic reminiscent of oils from seeds in the sulfur-containing vegetable family

Reference: Bleached, undeodorized canola oil diluted in good-quality canola oil (5:95)

Example: Cooked brussels sprouts

Waxy

Definition: An aromatic reminiscent of candle wax

Reference: High oleic sunflower oil heated to 190°C for 30 min

Weedy

Definition: An aromatic reminiscent of freshly cut weeds

Example: Freshly cut weeds

Woody

Definition: Aromatic reminiscent of sawdust

Example: Peanut oil

6

Oleoresins and Essential Oils

OLEORESIN

An oleoresin is a material made by solvent extraction of herbs and spices followed by solvent removal. Solvents extract essential oils, fixed oils, flavor compounds, pigments, and vegetable oils from plant material. An oleoresin can be considered a concentrated mixture of natural colors, aromas, and flavors in vegetable oil. They have been available for more than 40 years (1). Most herbs and spices are commercially available as oleoresins (2). Although herbs are defined as leaves and spices are defined as all other parts of a plant, common usage often includes herbs in the general term spice (3).

The greatest advantages to the use of an oleoresin in food applications are the standardization of flavor strength, the complete utilization of flavor, the ability to blend, modify, and customize the product for specific applications, and the elimination of mold, bacteria, filth, insect fragments, and foreign matter of raw spices. In food products where the visibility of particulate matter is undesirable, oleoresins are preferred. The disadvantages of oleoresins result from the physical differences of these products from the raw spice. A liquid concentrate may require different handling and methods of addition than a ground spice. Absence of visible particles of ground or chopped spice with a liquid extract may not be acceptable in some foods. The flavor imparted by an oleoresin, due to the greater availability of the flavor compounds, is more dramatic than the spice so that sophisticated application knowledge is required. The increased solubility of the flavor compounds may also lead to losses or chemical changes in food processing and cooking. It may require a significant number of trial formulations of oleoresins to duplicate an existing flavor.

ESSENTIAL OILS

Essential oils are the volatile chemical components in plants; they are complex mixtures of aroma compounds obtained by distillation. They do not contain vegetable oil or nonvolatile flavor compounds, which are present in oleoresins. However, essential oils are often the functional fraction in oleoresins because they are responsible for the aromas. They are widely used in flavor and fragrance industries. Import statistics for essential oils are published regularly (4).

Essential oils are produced by distillation procedures, steam distillation being the most common. Although defined as volatile, the components in essential oils typically have boiling points significantly higher than that of water. To avoid thermal degradation, steam is passed through the spice (whole or ground) in a tall still, co-distilling the oil with the steam. In a variation called hydrodistillation, the plant material is soaked in water that is then heated externally to produce steam.

The distillate is collected and physically separated into two layers, essential oil and the much larger water layer. The oil layer may require further drying by centrifugation or by the use of drying agents. The water layer is recycled back to the still.

Some essential oils may be physically expressed from the plant source. Citrus oils, for example, are produced by pressing the rind of the fruit. These oils are referred to as cold-pressed (CP) oils. In cold pressing the volatile flavor compounds are not altered by heat, as may be the case with distillation.

The essential oil may be a usable product as collected or it may be subject to further purification, concentration, or separation. Vacuum distillation is the typical method of refinement for most oils. Through fractional distillation, the oil components can be separated by means of their differing boiling points. Higher quality essential oils traditionally have been characterized by the method of concentration and not by chemical composition, although they may have specifications for constituent content. Rectified, folded, and terpeneless oils are made through the selection of the various fractions obtained from redistillation.

A rectified oil is produced by redistillation with rejection of the last distillation fraction (single rectified) or of both the first and last fraction (double rectified). A folded oil is made from the higher boiling fractions. A terpeneless oil is made in an efficient distillation in which the initial distillation fractions, high in terpenes, are removed from the less volatile but more important fractions. "Terpeneless" in this case does not necessarily mean free of terpenes but rather reduced in content of hydrocarbons. Terpeneless oils are particularly important in the beverage trade because of their higher stability and solubility.

Essential oils can also be processed by extraction techniques. Citrus oils, for example, are often extracted with aqueous alcohol, a polar solvent. Partitioning occurs in which one layer contains the terpene hydrocarbons and the other, the aqueous alcohol layer, contains the polar compounds. The alcohol layer is separated and the polar compounds recovered. These oxygenated compounds are primarily responsible for the characteristic nature of the oils and supply the flavor and fragrance impact.

Adsorption is a process in which certain components of an essential oil are removed by treatment with a solid that has a selective affinity for some particular components. An example would be the use of carbon to remove colored components. Chromatography can also be done on a production scale. An essential oil is passed through a bed of adsorbent such as silica gel, then the retained components are flushed from the bed with a solvent and collected.

The term folded oil usually refers to an oil in which valued components are enriched by distillation. Generally used in reference to citrus oils, a fivefold product would have five times the flavor or aroma strength of a nonfolded oil. Current usage, however, often refers to any oil of increased flavor or aroma strength as a folded oil, not only those produced by distillation.

CHARACTERISTICS AND PROPERTIES

Density, optical rotation, refractive index, and other physical properties of essential oils are sometimes useful for quality control of essential oils. Lists of these properties are published (5–7). These properties do not determine aroma quality or natural origin in an essential oil. Their measurement is used to screen for large defects in essential oil samples or even mislabeled products.

The most widely used technique for the analysis of essential oils is gas chromatography (GC) (7–9). Although the basis of separation in this technique is boiling point, GC columns are available that modify separations by polarity and even stereochemistry. Compounds are detected as they come off the column by thermal conductivity, flame ionization, or various spectroscopic detectors. There are also detectors that are very specific and measure only sulfur- or nitrogen-containing compounds. A flame ionization detector is used most often for routine analysis. This detector is "universal" in that all volatile compounds containing carbon and hydrogen can be measured. Most detection techniques can quantify each compound. However, some, such as mass spectroscopy, can also be used to identify components (9–14). Although peaks on a chromatogram were measured graphically in the past, most analysis today is done by computer data systems. The use of computerized library searches greatly aids in chemical identification. The high temperatures used in a GC make it an unsuitable method for essential oils with thermally labile compounds.

Gas chromatography can also be used to verify the natural origin of essential oils. Essential oils may be adulterated with cheaper, synthetic compounds greatly increasing the profit margin on the sale of a "natural" oil. Verification of natural origin may be based on the presence or absence of trace components, ratios of compounds, or ratios of enantiomers (optical isomers). In some cases, techniques such as isotope ratios and radioactive carbon dating have been used to detect adulteration.

High-performance liquid chromatography (HPLC) has found limited usage with essential oils. Its major advantage is with products, such as onion oil, that contain sulfur compounds that decompose at GC temperatures (15). HPLC uses the flow of solvent through a solid adsorbent, usually at ambient temperatures, to separate compounds.

Essential oils are widely used in the food and beverage industries. They are oil soluble and are often diluted with oil or emulsified to make them water dispersible before addition to a flavor or fragrance system. In this sense, their uses are not very different from those of oleoresins, as their active components provide a significant portion of the aroma and flavor of a spice.

OLEORESIN PRODUCTION

Although details of the manufacturing process of oleoresins depend on the plant material to be processed and the solvents used, four stages are involved: grinding, extraction, desolventization, and standardization.

Grinding or a similar process (such as chopping, maceration, etc.) is used to make extraction of the plant material more rapid or more efficient. The optimal type of grinding depends on the plant material or, more specifically, its physical structure. Some herb leaves are dried and gently crumbled before extraction. Other spices, in which the oils are contained in hard-to-extract glands or in a woody matrix, are ground to very fine particles. Grinders may include a cooling mechanism to prevent degradation of heat-sensitive compounds and loss of volatile components. The grinding operation must be designed to have minimal deleterious effect on the flavor and aroma components as well as on the vegetable oils extracted.

In extraction, the ground spice is washed with a solvent to dissolve the pigments, flavor, and aroma components and separate them from the nonsoluble residue (marc or spent). Extraction equipment can be of various designs but are of two basic types: batch extractors and continuous extractors.

A batch extractor, in its simplest form, is a container that can be filled with ground spice and solvent. A stainless steel tank with a steam jacket to heat the solvent is typically used. After sufficient time has passed to dissolve the soluble materials, the mixture is separated by filtration. The oleoresin in the soluble fraction is recovered by removal of the solvent. The marc is also desolventized to recover the solvent for reuse. Extraction and filtration can be combined by use of baskets to hold the spice within the extractor. In this design the solvent flows by gravity through the bed of ground material, dissolving the flavors and oils and carrying them out of the extractor for recovery.

Variations of batch extractors include centrifugal extractors, where the solvent is forced through the ground material in a spinning container, and high-pressure extractors, where the solvent is pumped at high pressures. Efficiency of extraction is greatly influenced by adjustable parameters such as solvent temperature, contact time, ground particle size, and so forth.

The design of a continuous extractor differs in that ground material continuously enters the extractor through an air lock, passes through the solvent stream and then exits as marc through an opposing air lock. Extraction is usually a multistage process. These extractors are most efficient when flow is "counter current," with fresh solvent in contact with the ground material in the last stage of extraction.

A two-step extraction is sometimes used to obtain two fractions from a spice. This uses a nonpolar solvent to remove essential oil, followed by using a polar solvent to extract other, more polar compounds. By this procedure, black pepper, for instance, can be extracted with hexane to obtain the essential oil followed by extraction with methylene chloride to obtain a more pungent fraction.

Solvents approved for the manufacture of oleoresins in the United States are listed in Title 21 of the *Code of Federal Regulations* (CRF), part 173 (16). These include hexane, acetone, methanol, ethanol, isopropanol, and methylene chloride. Although oleoresins produced with chlorinated solvents are allowed in the United States, many other countries, such as the European Community (17) and Japan (18), do not allow food products made with chlorinated solvents—a serious consideration for exportation.

Carbon dioxide, when liquified under high pressure, is also used as an extraction solvent. Such extraction is usually done by batch extraction due to the technical difficulty of admitting ground material into a pressurized extractor. Extraction is often done at temperatures and pressures high enough to produce superfluid conditions. The extraction solvent then has properties of both a gas and a liquid. This greatly increases the solvent power and the efficiency of extraction. Because liquid and supercritical carbon dioxide are not satisfactory solvents for polar compounds or compounds of high molecular weight, traditional solvents may be added to the carbon dioxide to increase the solvent power and increase the efficiency of extraction. The disadvantages of the technical complexity of the high-pressure extraction equipment, the small extractor size, and the low efficiency of extraction must be balanced against the potentially milder extraction conditions, the ease of solvent removal, and the perceived "naturalness" of the solvent (19). The organoleptic properties of products obtained by this method are somewhat different from the equivalent products obtained from traditional solvent extract and distillation.

Desolventization is the distillation process in which the extraction solvent is removed from the oleoresin. Although this can be done at atmospheric pressure with relatively high heat, preservation of sensitive, more volatile flavor compounds usually requires the use of vacuum conditions and lower temperatures. When carbon dioxide is used as the extraction solvent, reducing the pressure to atmospheric evaporates the solvent.

If the oleoresin does not contain any desirable volatile compounds, desolventization may be vigorous, removing all volatiles. A wiped-film evaporator, in which the solvent-laden oleoresin passes as a thin film on a heated surface, can be used to strip off all the solvent as well as any other components that will evaporate. Turmeric oleoresin, in which the nonvolatile extractives are natural colorants, is often desolventized in this manner. The volatile oil contained in this oleoresin is often considered an undesirable aroma component and its loss during desolventization may improve the quality of the oleoresin.

In many oleoresins, volatile components such as terpenes contribute an important part of the flavor profile. Their loss during desolventization would lessen perceived product quality. In this case a still with reflux or fractionation capabilities is used. The less volatile essential oil is separated from the solvent and returned to the oleoresin. Alternately, a significant fraction of the essential oil can be removed with the solvent and the distillate then fractionated to recover the oil.

In the United States, the maximum residue for solvents used in extraction of oleoresins is listed in 21 CFR, part 173 (16). For the most commonly used solvents, these levels are as follows:

Solvent	Level (ppm)
Acetone	30
Methanol	50
Hexane	25
Isopropanol	50

Residual solvent level is measured by a co-distillation technique followed by GC analysis of the distillate (20,21). One of the difficulties of this technique is that it does not distinguish between the chemicals of the solvent and the same chemicals derived from plant material. The common solvents are readily detected in plant tissue. Hexane, for instance, is one of the many degradation products of vegetable oil. It would not be feasible to produce an oleoresin with zero hexane content, even if hexane was not used in extraction. Another difficulty arises from the fact that the standard assay requires the steam distillation of a sample of oleoresin. Heating oleoresin in boiling water produces many small organic compounds that may interfere with chemical identification by GC.

Standardization produces consistent flavor, color, and aroma in oleoresins; this is one of the major advantages of such products over raw spices. By blending different batches and adding recovered or additional essential oil, an oleoresin of defined characteristics can be produced. When an oleoresin or essential oil is blended with flavorants from other natural sources, the product must be described by the term WONF, or "with other natural flavors."

The addition of edible oils, such as soybean or cottonseed oil, can be used to adjust flavor intensity to within specifications. Other ingredients may be emulsifiers to improve homogeneity and ease of application and diluents to decrease viscosity. Such additives, if used, should be clearly stated on the manufacturer's label. These additives should follow Food and Drug Administration (FDA) guidelines for natural products if the oleoresin is to retain its status as a natural product.

PHYSICAL CHARACTERISTICS

Oleoresins may be liquids or solids depending on the specific nature of the flavor compounds they contain and the amount of essential oils, fixed oils, and triglycerides in them. Oleoresin cinnamon, if extracted with hexane, for example, is a thin liquid consisting of over 90% cinnamaldehyde. Only a trace of vegetable oil is extracted from the bark from which it is made. Paprika oleoresin, made from the fruit of *Capsicum annuum* L., is usually greater than 85% vegetable oil and has the viscosity of vegetable oil. The carotenoid pigments are predominantly esters of fatty acids. In contrast, nutmeg oleoresin is a pasty solid because of the high content of trimyristin, the triglyceride of myristic acid.

The flavor intensity of oleoresins is 6–40 times as strong by weight as the corresponding herb or spice. Extraction concentrates the flavor chemicals. The degree of concentration depends on the percentage of extractables in the raw spice. In food applications, oleoresins often provide a stronger flavor impact than would be predicted on the basis of a chemical comparison to the raw spice. The flavor components have been extracted from the plant matrix and are more readily incorporated into food and available to the palate.

Physical properties of oleoresins are available from the manufacturers. The *Food Chemical Index* (16) is also a source for specification data. These products are listed as GRAS, or generally recognized as safe.

The majority of oleoresins are stable when stored at room temperature. Low water activity and high content of components that are naturally bacteriostatic minimize microbiological growth. Many oleoresins are sterile. Some oleoresins may undergo oxidation due to the highly unsaturated components in them; these require refrigeration or storage under an inert atmosphere. Manufacturers' directions, such as mix before use, should obviously be followed.

CHEMICAL COMPOSITION

It is not possible to describe the chemical composition of oleoresins in general terms. Some contain various terpenes and sesquiterpenes dissolved in vegetable oil while others are terpene free. The fixed oils, the nonvolatile oils, are an important part of the chemical and flavor profile. The important components may be carotenoids, curcuminoids, phenolics, aldehydes, or many other classes of compounds. The chemical composition of oleoresins is determined by the plant material used and the solvent used for extraction. Oleoresins do not contain carbohydrates, protein, cellulose, and other materials not extracted by solvents.

QUALITY CONTROL

Physical properties of oleoresins, such as specific gravity, refractive index, optical rotation, and viscosity, are not important for determination of the quality of these materials. However, they may be very important for proper dosage and ease of addition with some process equipment. Some oleoresins require heating and stirring before use to ensure a homogenous product.

Analytical methods for quality control purposes must be chosen for the specific oleoresin or components of interest. For many oleoresins used to provide aroma, the percentage of essential oil can be determined by steam distillation with chemical composition determined by a hyphenated gas chromatographic technique: gas chromatography—mass spectroscopy (GC-MS), gas chromatography—Fourier transform infrared (GC-FTIR), etc., various methods of compound detection and identification.

For other oleoresins, specific assays have been developed to measure the components of interest. Examples are capsaicinoids in capsicum oleoresins (22), curcuminoids in turmeric oleoresin (23), piperine in black pepper (20), and others that can be found in a literature search based on the specific oleoresin (24,25).

It remains true, however, that the most important assay for a flavor material is taste. There is no substitute for sensory evaluation. Despite the high cost of maintaining flavor standards, training and testing sensory panelists, and testing all products upon receipt or before shipment, this is a necessary part of quality control (26).

APPLICATION EXAMPLES

Oleoresins are usually very soluble in vegetable oil. This makes their use in oil systems simple; they are often added to the oil before it is added to the food. Oil solubility aids in the distribution of flavor throughout a food product. Oleoresins, which are very concentrated flavors and colors, should be dissolved completely and mixed thoroughly into the food to avoid areas of excessive flavor or color. With the use of capsicum oleoresin, concentrated extractives of chili peppers, such areas would be "hot spots" in the literal sense. Capsicum oleoresins are available that are more than 1000 times hotter than hot sauces sold in grocery stores.

The oleoresins with the largest use as colorants are paprika, turmeric, and annatto. Colors of these oleoresins are limited to yellow and red-orange hues. These are listed as colors exempt from certification by the U.S. FDA as distinct from the certified colors, which are synthetics, see 21 CFR 73 (16). In other countries, such as the European Community, turmeric and paprika oleoresins are generally regarded as spices or flavors.

The chromophores in paprika oleoresin are carotenoid pigments, particularly capsanthin esters. This oleoresin is used in bread crumb coatings, processed cheese, processed meats, salad dressings, sauces, and snack food coatings where it provides a yellow-orange to red-orange hue. Method of use ranges from as simple as spraying the surface of a cracker to as complex as formulating a dry powder containing flavors, gums, cheese powders, and other ingredients to coat a cracker or chip. Paprika oleoresin is stable to aseptic, retort, and extrusion processing and not affected by variation in pH. It may be stored in closed containers at ambient temperature. Oxidative degradation can be a concern. Degradation can be catalyzed by light, by heat, and by trace metals. Paprika oleoresin stabilized with other plant extracts is commercially available.

Turmeric oleoresin is an extract of the root of *Curcuma longa* L. It contains three related pigments known as curcuminoids. It is used to provide a bright yellow hue comparable to that of FD&C Yellow No. 5. Turmeric is very heat stable but not stable at high pH (27). Curcuminoid pigments are also sensitive to ultraviolet (UV) light exposure; foods made with turmeric oleoresin usually require protective packaging. This oleoresin is widely used in bakery mixes, breading, cereal, process cheese, margarine, and soup. Most mustards and pickles contain turmeric.

The extract of annatto, *Bixa orellana* L., is sometimes termed annatto extract and not annatto oleoresin because some forms of the product do not contain a significant amount of vegetable oil. The extract may even be obtained as a powder although liquids are more common. The primary pigment, bixin, is a carotenoid containing dicarboxylic acid functional groups. It is oil soluble but can be saponified to make the water-soluble pigment, norbixin. When dissolved in oil or water, annatto provides yellow-orange to red-orange hues, depending on concentration. The precipitation or agglomeration of norbixin in cheese and cheese products gives the well-known "pinking" of the color. Low pH is usually a factor in this color change. Dairy products and snack foods are the major applications of annatto extracts, although other foods may also be colored with this natural product. Annatto is also blended with paprika and turmeric oleoresins to make other hues. Annatto has sufficient heat stability for aseptic and extrusion processing, but short heating cycles are recommended. The extracts can be sensitive to oxidation and UV degradation.

An interesting use of oleoresins is in cooking oils. Paprika and turmeric oleoresins can be used to color an oil such as popcorn oil. The colored oil is then used both for popping the popcorn and for coloring it as it pops. With the addition of spice oleoresin blends to the oil, flavor can also be added to produce a flavored, colored product. This is easily adapted to home cooking for products such as microwaved popcorn.

Oleoresins have further applications in systems in which high oil solubility is not an advantage. They can be blended with emulsifiers to make them water dispersible or, with the use of polysorbate esters, even water soluble. As an example of this, a vinaigrette, in its simplest form, would consist of vegetable oil, vinegar in water, natural color, and herb flavor. Paprika oleoresin can be used to provide an orange color and mixed oleoresins of basil and other herbs can be formulated to provide flavor. If oleoresins are used in this two-phase system, both color and flavor will dissolve in the oil layer. After mixing, the layers will quickly separate into two layers of significantly different color and flavor. If the oleoresins are first blended with emulsifiers to make water-dispersible products, the emulsified products will disperse into both layers. After the layers separate, both will contain color and herb flavor giving a more satisfactory salad dressing.

Emulsified oleoresins are effectively used in reduced-fat sauces, spreads, and dressings where oleoresins would not be soluble. The flavor evenly disperses with direct addition of the emulsified oleoresin to the food product.

ANTIOXIDANT PROPERTIES

Although oleoresins have been widely used for more than 40 years to flavor and color oils and foods, there are other applications of oleoresins in foods. The well-known preservative effects of spices and herbs also apply to oleoresins made from them. Components that have the ability to inhibit oxidation are extracted with color and flavor components, producing a spice extract that inhibits oxidation in foods.

Rancidity in fats and oils is caused by a complex set of oxidation reactions; free radical reactions are particularly important. Recent reviews of these mechanisms have been published (28–30). An antioxidant is a substance that is capable of retarding or preventing oxidation.

Various factors accelerate or inhibit oxidation. A classification system has been established to group these factors by mechanism (31). The primary inhibitors are phenolics. These compounds intercept and neutralize free radicals preventing the propagation step in oxidation.

The basic structure of a phenol is illustrated in Figure 6.1; it is a hydroxy group on an aromatic ring. Other substituents on the aromatic ring determine the efficiency of the compound as an antioxidant. The phenolic hydrogen atom of a substituted phenol readily reacts with highly reactive free radicals that develop early in the oxidation process. This quenching reaction transfers energy producing a phenolic radical, which can either lose this energy thermally or undergo further, less damaging reactions. In some cases, the antioxidant phenol is regenerated in subsequent reactions.

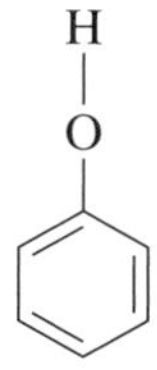

Figure 6.1 Phenol structure.

Many spices contain phenolic compounds with potential uses as antioxidants (32,33). Those most commonly cited in the literature are rosemary, sage, thyme, turmeric, oregano, mace, nutmeg, ginger, and members of the mint (Labiatae) family. Of these, rosemary has had the most widespread commercial application.

There are many phenolics in oleoresins of rosemary. Although carnosol and carnosic acid are present in highest concentration, many others have also been identified (34–36). These phenolic compounds act synergistically as antioxidants. Standardization of rosemary oleoresin by concentration of individual compounds such as carnosol is not accurate (37). Protection from oxidation is due to all the phenolic compounds present. Various commercial products based on these oleoresins have been used to protect foods against lipid oxidation, pigment fading, and flavor degradation. In a liquid oleoresin, phenolics have a much larger effect than they would have when bound within the cell structure of the ground leaf.

The type and amount of oleoresin needed in an application depends strongly on the type of fats in the system, the synergistic effects of other ingredients, and processing conditions. Effectiveness in a given fat can be measured by various techniques such as the active oxidation method and the peroxide value (20). A more rapid method, which has become common in the last few years, is the use of a Rancimat to measure the induction time for the onset of rancidity. Studies using a Rancimat have good correlation to shelf life studies with results produced in a fraction of the time (38).

A synergist increases the effectiveness of an antioxidant. Such a synergist may even be an antioxidant itself in some applications. Compounds such as ascorbic acid and citric acid may increase the effectiveness of rosemary oleoresin by an amount greater than simple additivity would predict. These acids are chelating agents capable of binding to trace metal ions, removing their prooxidant influence. Iron and copper are particularly strong prooxidants. Polyphosphates and even lecithin, which contain phosphates, can chelate metals. These materials can be incorporated into the oleoresin or added separately to help protect lipids in a food product.

Topopherols are phenolics present in most crude vegetable oils. Although they are not present in a significant amount in rosemary oleoresin, they are oil soluble and easy incorporated.

The phenolics in spice oleoresins have characteristics that make them significantly different from the synthetic phenols like BHA and BHT. With their higher molecular weights, they are much less volatile; they are not lost at frying temperatures. Rosemary and related oleoresins are flavoring extracts and may be labeled as such or, as is more common, the phrase "natural flavoring" may be used (Part 101.22 in Ref. 13).

Spice oleoresins are extracts of botanical origin. A listing of the products of this type is given in 21CFR, part 182, section 182.20 (16). They are considered GRAS (generally recognized as safe). Although they may have secondary benefits because of their antioxidant properties, they remain plant extracts.

The commercially available rosemary oleoresins are not flavorless. This flavor may limit their use in some systems where high concentrations are necessary to give the desired degree of stabilization. In other food systems, the flavor may be desirable and may improve sensory appeal.

NUTRITION

The contribution of oleoresin and essential oils to nutrition can usually be ignored; the level of usage, as a fraction of the diet, is extremely small. Obviously the vegetable oils in them are fats. Essential oils would also analyze as "fat." There may be a contribution to nutrition from oleoresins containing carotenoid pigments that have measurable vitamin A activity.

Ongoing research has indicated that diet plays a significant role in the risk of cancer and other diseases. Many herbs and spices contain antimutagens and anticarcinogens (39–41). Compounds with antiinflamatory and antioxidant properties may inhibit tumor initiation, promotion, and progression when consumed as part of the diet. The potential of phytochemicals has led to the concept of "designer foods" in which products are selected or formulated to provide for increased intake of beneficial, nonnutrient constituents (42–45).

SUMMARY

The use of oleoresins and essential oils to flavor and color foods is obvious. They can be sprayed on the surface, added to oil, dispersed with emulsifiers, spray dried with other ingredients, agglomerated, plated onto a dry carrier, or added to food in a myriad of ways. They enhance the sensory appeal of foods when incorporated into the flavor and processing systems. Oleoresin systems can also be designed to improve the stability of food products, particularly the fats and oils in them. Foods made with oleoresin can be more attractive, more flavorful, more stable, and even more healthy.

REFERENCES

1. A.D. Adamson, *Oleoresin Production and Markets with Particular Reference to the United Kingdom*, Tropical Products Institute, London, 1971.
2. D.R. Tainter and A.T. Grenis, *Spices and Seasonings: A Food Technology Handbook*, VCH Publishers, New York, 1993.
3. J.E. Simon, A.F. Chadwick, and L.E. Craker, *Herbs: An Indexed Bibliography 1971–1980*, Shoe String Press, Hamden, Conn., 1984.
4. U.S. Department of Agriculture, *U.S. Essential Oil Trade*, USDA-FAS, Washington, D.C.
5. *EOA Book of Standards and Specifications*, Essential Oil Association of USA, New York, 1975.

6. G. Fenaroli, *Fenaroli's Handbook of Flavor Ingredients*, T.E. Furia and N. Bellanca, eds. and trans., Chemical Rubber Co., Cleveland, 1971.
7. B.M. Lawrence, ed., *Essential Oils (1976–1978), (1979–1980), (1981–1987), (1988–1991)*, Allured Publishing, Wheaton, Ill.
8. E. Guenther, *The Essential Oils*, Vols 1–6, Van Nostrand Reinhold, New York, 1949.
9. S.R. Srinivas, *Atlas of Essential Oils*, ANADAMS Consulting Services, New York, 1986.
10. Y. Nasada, *Analysis of Essentials Oils by Gas Chromatography and Mass Spectroscopy*, John Wiley & Sons, New York, 1976.
11. W. Jennings and T. Shibamoto, *Qualitative Analysis of Flavor and Fragrance Volatiles by Glass Capillary Gas Chromatography*, Academic Press, Orlando, FL, 1980.
12. P. Sandra and C. Bicchi, eds., *Capillary Gas Chromatography in Essential Oil Analysis*, Heutig, New York, 1987.
13. R.P. Adams, *Identification of Essential Oils by Ion Trap Mass Spectroscopy*, Academic Press, San Diego, 1989.
14. P. Schreier, *Chromatographic Studies of Biogensis of Plant Volatiles,* Heutig, Germany, 1984.
15. E. Block, and co-workers, *J. Agric. Food Chem.* **40**, 2118 (1992).
16. Office of the Federal Register National Archives and Records Administration, *Code of Federal Regulations, Title 21 Food and Drugs*, U.S. Government Printing Office, Washington D.C., 1988.
17. European Community, *Council Directive 88/344/EEC*, 1988.
18. *Japanese Food Additive News*, **9**(9), (1989).
19. D.A. Moyler, in M.B. King and T. Reginald, eds., *Extraction of Natural Products Using Near-Critical Solvents*, Blackie, Glasgow, U.K., 1993, pp. 140–183.
20. Association of Official Analytial Chemists, *Official Methods of Analysis of the AOAC*, 15th ed., AOAC, Arlington, VA., 1990.
21. Committee on Codex Specifications, *Food Chemical Codex*, 3rd ed., National Academy Press, Washington D.C., 1981.
22. T.H. Cooper, J.A. Guzinski, and C. Fisher, *J. Agric. Food Chem.*, **39**, 2253 (1991).
23. T.H. Cooper, J.G. Clark, and J.A. Guzinski, in C.-T. Ho, T. Osawa, M.-T. Huang, and R. T. Rosen, eds., *Food Phytochemicals for Cancer Prevention II*, American Chemical Society, Washington D.C., 1994, pp. 231–236.
24. U.J. Salzer, *CRC Crit. Rev. Food Sci. Nutr.* **9**, 345 (1977).
25. U. J. Salzer, *Int. Flavours Food Addit.* **6**, 253 (1975).
26. M. Meilgaard, G. V. Civille, and B. T. Carr, *Sensory Evaluation Techniques*, 2nd ed., CRC Press, Boca Raton, FL., 1981.
27. H. H. Tønnesen and J. Karlsen, *Z. Lebensm. Unters. Forsch.* **180**, 132 (1985).
28. B.J.F. Hudson, ed., *Food Antioxidants*, Elsevier Applied Science, London and New York, 1990.
29. E.N. Frankel, *Prog. Lipid Res.* **23**, 197 (1985).
30. J.C. Allen and R.J. Hamilton, eds., *Rancidity in Foods*, 2nd ed., Elsevier Applied Science, London and New York, 1989.
31. T.P. Labuza, H. Tsuyuki, and M. Kare, *J. Am. Oil Chem. Soc.* **46**, 409 (1969).
32. K. Hermann, *Deutsche Lebensmittel-Rundschau* **77**, 134 (1981).
33. J.R. Chipault, J.R. Mizuno, and W.O. Lundberg, *Food Tech.* **10**, 209 (1952).
34. S.S. Chang, B. Osstrich-Matijaseric, O.A.L. Hsieh, and C. Huang, *J. Food Sci.* **42**, 1102 (1977).
35. K. Schwarz and W. Ternes, *Z. Lebensm. Unters. Forsch.* **195**, 95 (1992).
36. A.G. Gonzalez, C.M. Rodriguez, and J.G. Luis, *J. Chem. Res.* 114 (1988).
37. Q. Chen, H. Shi, and C.-T. Ho, *J. Am. Oil Chem. Soc.* **69**, 999 (1992).

38. J. Löliger in Ref. 30, p. 109.
39. C.-T. Ho, C. Y. Lee, and M.-T. Huang, eds., *Phenolic Compounds in Food and Their Effects on Health I*, American Chemical Society, New York, 1992.
40. G.S. Bailey and D.E. Williams, *Food Tech.* **47**(2), 105 (1993).
41. C.-T. Ho, T. Osawa, M.-T. Huang, and R.T. Rosen, eds., *Food Phytochemicals for Cancer Prevention II*, American Chemical Society, New York, 1994.
42. B.F. Haumann, *Inform* **4**, 344 (1993).
43. C.W. Boone, G.J. Kelloff, and W.E. Malone, *Cancer Res.* **50**, 2 (1990).
44. F. LaBell, *Food Process.* March, 23 (1990).
45. M.S. Kurzer, *Food Tech.* **47**(4), 80 (1993).

7

Dietary Fat and Health

NUTRITIONAL AND METABOLIC IMPORTANCE OF DIETARY LIPIDS

1.1 The Nature of Food Lipids

Fats and oils are water-insoluble, hydrophobic substances of animal and vegetable origin (1,2). The edible products of fats and oils are often referred to as food lipids. Triacylglycerols make up the majority of the lipids in edible fats and oils, and they are composed of glycerol esters of fatty acids. The structure of the triacylglycerol contains a molecule of glycerol and three molecules of fatty acids. Natural fats and oils vary in their physical properties depending on the type of fatty acids attached to the glycerol backbone (1). The reason for differences in the melting point and crystalline structure of triacylglycerols is that fats and oils contain fatty acids varying in carbon chain length and degree of unsaturation (2).

Fats of terrestrial animals contain a large proportion of saturated fatty acids in their triacylglycerols and are therefore solid at room temperature. Vegetable oils, however, generally contain unsaturated fatty acids in the triacylglycerols and are liquid at room temperature (Table 7.1). The fats of certain marine fishes can contain a large proportion of polyunsaturated fatty acids, making them a liquid at room temperature (1).

1.2 Fatty Acids in Food Lipids

Many of the fatty acids contained in food lipids have a straight chain and an even number of carbon atoms (1). The fatty acids having two hydrogens bonded to each carbon atom in the chain are saturated; they have no double bonds between carbons. The fatty acids that contain double bonds between carbon atoms are called unsaturated. The degree of unsaturation of a fatty acid is dependent on the number of double bonds in the molecule, which directly affects the unsaturation of an oil. Hence, saturated fatty acids have a higher melting point than unsaturated fatty acids of equivalent carbon chain length. All fatty acids contain a methyl end ($—CH_3$) often referred to as the omega end. The opposite end of the molecule is called the carboxyl end (—COOH).

Fatty acids found in food lipids vary in carbon chain length from 4 to 24 carbon atoms (Table 7.1). However, the primary fatty acids in foods are 16, 18, 20, and 22 carbon chain lengths (1). The unsaturated fatty acid containing one, two, and three double bonds are referred to as monoenoic, dienoic, and trienoic fatty acids, respectively (1). Usually the terms monounsaturated fatty acid and polyunsaturated fatty acid are used to describe fatty acids having one and two or more double bonds, respectively.

The more common fatty acids in food lipids are often called by their trivial name, such as lauric, palmitic, oleic, or linoleic acids (Table 7.1). The systematic name, which comes from the Geneva system of nomenclature, provides information about the fatty acid chain length given by a Greek prefix (1). Following this nomenclature the prefixes dodec-, hexadec-, and octadec- refer to fatty acids having 12, 16, and 18 carbon atoms, respectively (Table 7.1). The saturated fatty acids all contain the suffix -anoic. For double bonds, the prefix is modified to indicate the degree of unsaturation. For example, oleic acid, which contains one double bond, is an octadecenoic acid, and linoleic acid with two double bonds is an octadecadienoic acid (1). The location of the double bonds is identified by a number preceding the name. The number(s) for the double bonds is obtained by counting their position from the carboxyl carbon, which is number 1 in the chain.

Another, but more frequently used nomenclature for fatty acids is the symbol notation shown in Table 7.1. The symbol for oleic acid is 18:1*n*9, indicating a fatty acid containing 18 carbon atoms, and the number 1 following the colon indicates one double bond located at carbon number 9 (*n*-9) counting from the methyl end of the molecule. Arachidonic acid is represented by the symbol 20:4*n*6, indicating a 20-carbon fatty acid having four double bonds and the first is at the 6th position (*n*-6) counting from the methyl end. Since saturated fatty acids contain no double bonds, the number following the colon is 0. Fatty acids containing multiple double bonds are separated by a methylene ($—CH_2—$) group, so linoleic and arachidonic acids contain methylene-interrupted double bonds (1).

The unsaturated fatty acids in food lipids are usually in the cis configuration, meaning that the hydrogens are on the same side of the double bond (1). Vegetable oils, such as soybean oil and cottonseed oil, can be hydrogenated to increase the melting point of the oil. During the process of chemical hydrogenation, the content of polyunsaturated fatty acids in these oils is reduced while the amounts of monounsaturated and saturated fatty acids are increased (1). This is accomplished by adding hydrogen to the double bonds of unsaturated

Table 7.1 Fatty acids in food lipids

Symbol	Trivial Name	Systematic Name	Source
4:0	Butyric	Butanoic	Butterfat
6:0	Caproic	Hexanoic	Butterfat
8:0	Caprylic	Octanoic	Coconut oil
10:0	Capric	Decanoic	Coconut oil
10:1*n*1	Caproleic	9-Decenoic	Butterfat
12:0	Lauric	Dodecanoic	Coconut oil
12:1*n*3	Lauroleic	9-Dodecenic	Butterfat
14:0	Myristic	Tetradecanoic	Butterfat, coconut oil
15:0	Pentadecylic	Pentadecanoic	
16:0	Palmitic	Hexadecanoic	Fats and oils
16:1*n*7	Palmitoleic	9-Hexadecenoic	Fats and oils
*t*16:1*n*7	Palmitelaidic	Transhexadecenoic	Hydrogenated vegetable oils
17:0	Margaric	Heptadecanoic	Bacteria
18:0	Stearic	Octadecanoic	Fats and oils
18:1*n*9	Oleic	9-Octadecenoic	Egg yolk, olive oil
*t*18:1*n*9	Elaidic	9-Octadecenoic	Hydrogenated vegetable oils
18:1*n*7	Vaccenic	11-Octadenoic	Beef fat
18:2*n*6	Linoleic	9,12-Octadecadienoic	Vegetable oils
18:3*n*6	γ-linolenic	6,9,12-Octadecatrienoic	Evening primrose oil
18:3*n*3	α-linolenic	9,12,15-Octadecatrienoic	Soybean oil
20:0	Arachidic	Eicosanote	Lard
20:1*n*11	Gadoleic	9-Eicosanoic	Fish oil
20:1*n*9	Gondoic	11-Eicosanoic	Rapeseed oil
20:2*n*6		11,14-Eicosadienoic	
20:3*n*9	Mead	5,8,11-Eicosatrienoic	EFA-deficiency cartilage
20:3*n*6	Dihomo-γ-linolenic	8,11,14-Eicosatrienoic	Liver
20:4*n*6	Arachidonic	5,8,11,14-Eicosotetraenoic	Liver
20:5*n*3	EPA	5,8,11,14,17-Eicosapentaenoic	Fish oils
22:0	Behenic	Docosanoic	Peanut oil
22:1*n*9	Erucic	13-Docosenoic	Rapeseed oil
22:4*n*6		7,10,13,16-Docosatetraenoic	Liver
22:5*n*6		4,7,10,13,16-Docosepentaenoic	Liver
22:5*n*3		7,10,13,16,19-Docosapentaenoic	Fish oil
22:6*n*3	DHA, cervonic	4,7,10,13,16,19-Docosahexaenoic	Fish oil
24:0	Lignoceric	Tetracosanoic	
24:1	Nervonic	15-Tetracosenoic	

fatty acids. The process is usually performed with heat, pressure, and a source of hydrogen. A catalyst is also added to facilitate the hydrogenation of the oil (1). Trans fatty acid isomers are produced from the cis double bonds as a side reaction during partial hydrogenation of an oil. The trans fatty acids have a decreased bond angle compared to the *cis* isomer, which results in a higher melting point.

Most of the trans fatty acids produced from the hydrogenation of vegetable oils are trans monounsaturates of 18:1 fatty acids (1,3). Some trans fatty acids occur naturally in food lipids (4,5). The minor sources of trans fatty acids come from ruminant animal fats, such as milk, butter, and beef tallow. Ruminant animals have bacteria that produce trans fatty acids during biohydrogenation of unsaturated fatty acids in the rumen (4), and the amount in the fat is subject to seasonal variation. The largest contribution of trans fatty acids to the diet is from margarines and shortenings that contain hydrogenated vegetable oils (6,7).

New research indicates that the intake of trans fatty acids contained in hydrogenated vegetable oil compared to native vegetable oil may have a hypercholesterolemic effect. Studies with men and women demonstrated that consumption of trans fatty acids (elaidic acid substituted for oleic acid) elevated plasma levels of low-density lipoprotein cholesterol (LDL-C) but lowered those for high-density lipoprotein cholesterol (8–10). It is premature to conclude from these investigations that trans fatty acids increase the risk for cardiovascular disease. Some controversy exists on the actual amount of trans fatty acids consumed in the United States. Intakes may range from 8 to 15 grams per person per day (6,7), which is still below the levels consumed in some studies where elaidic acid elevated plasma levels of LDL-C. Perhaps the most important aspect relating trans fatty acids to health issues is the amount of total fat and hydrogenated oil consumed by children and by women during pregnancy, and its effect on essential fatty acid status.

1.3 Nutritional Importance of Food Lipids

Food lipids provide many attributes to improve the palatability of foods, such as taste, texture, and mouthfeel. Body fat, or adipose tissue, helps to protect vital organs and provides a source of energy for prolonged exercise. Dietary fats and oils provide an important source of concentrated food energy (11,12). The calories per gram of fat are 2.25 times the caloric value of carbohydrates (sugars and starches) and proteins. When dietary carbohydrates and proteins are consumed above that needed by the body, the carbon skeleton of these nutrients is converted to fatty acids and stored as triacylglycerols in the adipose tissue (11,12).

Food lipids also provide a source of fat-soluble vitamins (A, D, E, and K) and help to facilitate the digestion and absorption of these vitamins (2,12). Vegetable oils can be rich sources of tocopherols, including vitamin E, and carotenes (provitamin A); both are important natural antioxidants for the body (13). Animal fats and fish oils can be a source of vitamin D. Animal fats generally contain a higher proportion of saturated fatty acids when compared to vegetable oils. The latter are higher in unsaturated fatty acids compared to animal fats.

Young children need dietary fat and cholesterol to sustain growth and to maintain a healthy body (12). Adults, on the other hand, are advised to consume a diet containing lower total fat, especially saturated fat and cholesterol, which are linked to an increased risk of cardiovascular disease (14). The saturated fatty acids 12:0, 14:0, and 16:0, elevate the circulating levels of plasma LDL-C (15), a major risk factor for atherosclerosis. Conversely, consumption of *n*-6 and *n*-3 polyunsaturated fatty acids (PUFAs) are associated with decreased blood pressure (16). Although the American Heart Association recommends a fat intake limited to 30% of the total daily calories, it may be easier to consume foods low in fat and to balance intakes of saturated fatty acids, monounsaturated fatty acids, and PUFAs (12,14).

Certain fatty acids that must be supplied in the diet are called essential fatty acids because the body lacks enzymes to synthesize them *de novo* (1,12). Two are considered essential for humans; these are linoleic acid and α-linolenic acid (12,17,18). Linoleic acid (18:2*n*6) belongs to the *n*-6 or omega-6 family of PUFAs because the terminal double bond is six carbons from the methyl end of the molecule. Likewise, α-linolenic acid (18:3*n*3) is a member of the *n*-3 (omega-3) PUFA family since the double bond is three carbons from the methyl end. The essential fatty acids are crucial for fetal development and growth (12,17). The long-chain *n*-3 PUFAs are important for brain and retina (18) and appear to be readily digested and absorbed from dietary sources (19,20). Learning disabilities and loss of visual acuity were reported in animals consuming low levels of *n*-3 essential fatty acids (18).

1.4 Formation of Polyunsaturated Fatty Acids (PUFAs)

The essential fatty acids can follow a number of metabolic fates that include β-oxidation in mitochondria to generate adenosine 5′-triphosphate (ATP) (21), desaturation and chain elongation leading to the formation of long-chain PUFAs (12,22), and incorporation into glycerolipids (25). Total dietary fat intake and type and amounts of essential fatty acids will influence metabolic use (12,24). The liver contains desaturases and elongase enzymes to convert linoleic and α-linolenic acids to their respective PUFAs (Figure 7.1). The activities of these enzymes are influenced to some extent by changes in diet and hormones (2,12,24). The rate-limiting step is the delta-6 desaturase (Δ-6), and the preferred substrates are α-linolenic acid > linoleic acid > oleic acid (12).

The concentrations of 18:3*n*6 and 20:3*n*6 in liver are very low compared to the concentrations of linoleic and arachidonic acids because once 18:3*n*6 is formed it is rapidly elongated to 20:3*n*6 then desaturated to yield 20:4*n*6 (2,12). Recent studies conducted by Sprecher and co-workers on rat liver suggest that 22:6*n*3 is not produced directly by a putative Δ-4 desaturates (25). In the rat, the conversion involves microsomal elongation of 22:5*n*3 to 24:5*n*3

$n6$ Series

$$\begin{array}{l} 18:2n6 \xrightarrow{\Delta^6} 18:3n6 \xrightarrow{E} 20:3n6 \xrightarrow{\Delta^5} 20:4n6 \xrightarrow{E} 22:4n6 \xrightarrow{\Delta^4} 22:5n6 \end{array}$$

Linoleic; 20:3n6 PG_1; 20:4n6 Arachidonic PG_2

$n3$ Series

$$\begin{array}{l} 18:3n3 \xrightarrow{\Delta^6} 18:4n3 \xrightarrow{E} 20:4n3 \xrightarrow{\Delta^5} 20:5n3 \xrightarrow{E} 22:5n3 \xrightarrow{\Delta^4} 22:6n3 \end{array}$$

α-Linoleic; 20:5n3 Eicosapentaenoic PG_3; 22:6n3 Docosahexaenoic

Figure 7.1 Conversion of essential fatty acids to long-chain polyenoic fatty acids through steps of desaturation (Δ-6, Δ-5, Δ-4) and elongation (*E*). The PUFAs that serve as substrate for prostaglandin biosynthesis are of the 1, 2, and 3 series (PG_1, PG_2, and PG_3).

and desaturation to 24:6*n*3 by a Δ-6 desaturase. The 24:6*n*3 is then *β*-oxidized to 22:6*n*3. Although the major flux through polyene acid formation is toward long-chain PUFAs, some retroconversion does occur (12).

The nonessential fatty acids can be converted to PUFAs under certain situations. During essential fatty acid deficiency, oleic acid is converted to Mead acid 20:3*n*9 (12,24), a PUFA of the *n*-9 family (Figure 7.2). A decrease in the essential fatty acid linoleate (18:2*n*6) results in a progressive decrease in arachidonic acid (12), followed by lowered prostaglandin biosynthesis, thereby compromising normal physiologic homeostasis. However, deficiency of essential fatty acids is unlikely to occur in humans (24).

1.5 Biosynthesis of the Biologically Active Eicosanoids

Certain essential PUFAs are converted to potent biological compounds called eicosanoids (prostaglandins, leukotrienes, and lipoxins) (12,24). Biosynthesis of eicosanoids is ubiquitous in tissues and organs of the body. Many of the eicosanoids act as localized hormones, but some function as autocrine or paracrine cell-to-cell signaling agents. These compounds affect the cardiovascular, digestive, immune, nervous, reproductive, respiratory, skeletal, and urinary systems in the body (12,26–29).

Current knowledge on the metabolism of eicosanoids, including activation and release of substrate, biosynthesis, degradation, physiological effects, and nomenclature, is adequately described elsewhere (2,24,30,31). Nevertheless, to acquaint the reader, a brief description of eicosanoid metabolism will be presented. The majority of eicosanoids (prostaglandins, leukotrienes, and lipoxins) are biosynthesized from arachidonic acid, which is maintained in membrane phospholipids. Arachidonic acid is the precursor of 2-series prosta-

*n*9 Series

18:0 Stearic $\xrightarrow{\Delta^9}$ 18:1*n*9 Oleic $\xrightarrow{\Delta^6}$ 18:2*n*9 $\xrightarrow{E}$ 20:2*n*9 $\xrightarrow{\Delta^{5''}}$ 20:3*n*9 Eicosatrienoic

*n*7 Series

16:0 Palmitic $\xrightarrow{\Delta^9}$ 16:1*n*7 Palmitoleic $\xrightarrow{\Delta^6}$ 16:2*n*7 $\xrightarrow{E}$ 18:2*n*7 $\xrightarrow{\Delta^5}$ 18:3*n*7 $\xrightarrow{E}$ 20:3*n*7

Figure 7.2 Conversion of nonessential fatty acids to long-chain polyenoic fatty acids through steps of desaturation (Δ-6, Δ-5, Δ-4) and elongation (*E*).

glandins, but 20:3*n*6 and 20:5*n*3 are substrates for the 1- and 3-series prostaglandins, respectively (26) (Figure 7.3). Prior to eicosanoid formation, phospholipase A_2 cleaves arachidonic acid from the *sn*-2 position on the glycerol backbone of membrane-bound phospholipids (2,12,32,33).

Control of phospholipase activity is the principal site for regulation of eicosanoid production (12,32,33). Corticosteroids prevent prostaglandin formation by causing the synthesis of proteins (lipomodulin and macrocortin) that inhibit phospholipases. The antimalarial drug mepacrine also inhibits phospholipase-mediated release of arachidonic acid. After arachidonic acid is liberated from the phospholipid, it can undergo controlled oxidative metabolism to form a variety of eicosanoids (Figure 7.3) with differing physiologic effects (24,26,29,30). The kind and amount of eicosanoids produced are dictated by the cell type and enzymes inherent to the cell.

The cyclooxygenase pathway yields prostaglandin endoperoxides and the bioactive prostaglandins, prostacyclins, and thromboxanes. Both aspirin and indomethacin inhibit the cyclooxygenase enzyme.

The leukotrienes are another group of substances included with the eicosanoids. These compounds were originally characterized as the slow-reacting substances of anaphylaxis. Although leukotrienes were first described as products of leukocytes, it is now known that several tissues and organs produce them (28,34).

Enzymes converting substrate into eicosanoids occur in the cell membranes. After their formation, the eicosanoids exert their effect and are rapidly metabolized to inactive compounds in liver and lung and excreted by kidney or lung (26–28). The eicosanoids act as localized hormones and as autocrine/paracrine cell-to-cell regulators since most of their biological effects are limited to the site of biosynthesis. Several reviews provide details on the multitude of biological effects produced by eicosanoids (24,28,29,31,34–36). These findings suggest more detailed metabolic controls and pharmacologic actions for the eicosanoids.

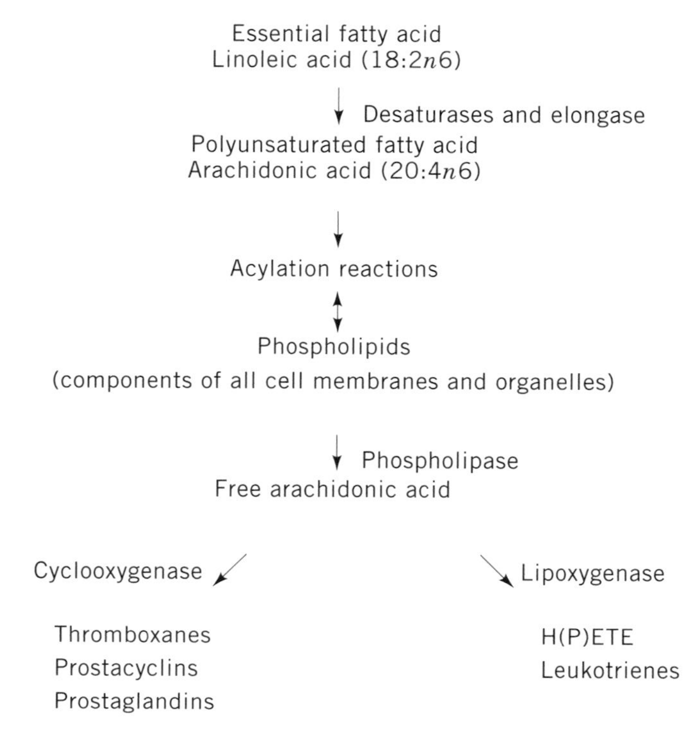

Figure 7.3 Conversion of the essential fatty acid linoleic to arachidonic acid, its incorporation into phospholipids, and subsequent biosynthesis into eicosanoids via the cyclooxygenase and lipoxygenase pathways.

Although the PUFAs exert many of their physiological effects through the biosynthesis of eicosanoids, changing the amounts and types of PUFAs in the diet can modulate eicosanoid production (16,24,36). Dietary modulation of immune response is one area of research where a change in the intake of *n*-3 PUFAs altered the eicosanoid production by immune cells (36). Such changes in immune cell activity by dietary lipids may be used to enhance or depress immune response; however, the potential impact on human health is not yet realized. Besides dietary PUFAs, the biotin status of an animal was shown to significantly effect the conversion of linoleic acid to arachidonic acid (37).

ROLE OF DIETARY FAT IN CARDIOVASCULAR DISEASE AND ATHEROSCLEROSIS

2.1 Introduction

Even though the mortality from coronary heart disease has declined recently, atherosclerosis and related vascular disorders still are the leading cause of death in the Western world. The etiology of these diseases is multifactorial, with hyperlipidemia, smoking, diabetes mellitus, hypertension, and obesity being well-established risk factors for the development of cardiovascular disease and atherosclerosis. Dietary fat affects plasma lipids and lipoproteins and thus is linked to atherosclerosis. Injury to or abnormal mechanisms of the vascular endothelium may be initiating events in the etiology of atherosclerosis.

Although epidemiological studies suggest that dietary cholesterol and saturated fatty acids increase serum cholesterol, recent evidence suggests that high intakes of polyunsaturated fats may be equally atherogenic because of their ability to convert easily to cytotoxic lipid peroxidation products. Most of the data available in the literature suggests that high-fat diets are associated with greater risks for cardiovascular disease and atherosclerosis. Conversely, low-fat diets, independent of the fat source, may be the prudent choice in prevention and treatment of atherosclerosis. Low-fat diets are usually high in dietary fiber, antioxidants, and other undefined compounds, all of which may protect against atherosclerosis.

2.2 Pathogenesis of Atherosclerosis

Theories. There are numerous theories for the pathogenesis of atherosclerosis. In spite of considerable research, the etiology of this disease is not well understood. The current trend is to consider atherosclerosis as a response of the vascular wall to a variety of initiating agents and multiple pathogenic mechanisms (e.g., hyperlipidemia), contributing to the development of atheromatous plaques. It appears that the major participants in the atherosclerotic disease process include an active vascular endothelium, smooth muscle cells, blood-borne cells such as monocytes and macrophages, and circulating lipoproteins. The result is a multifactorial sequence of events involving endothelial cell injury/dysfunction, uptake of circulating blood monocytes, and their differentiation into macrophages, coupled with smooth muscle cell migration and proliferation.

The most intensely studied current hypothesis of atherosclerosis is the response to injury hypothesis (38). The hypothesis takes into account the cellular interactions that occur during the different phases of lesion initiation, development, and progression. The initiating event appears to be injury to or dysfunction of the endothelium via lipids or lipoprotein derivatives or via mechanical, chemical, toxic, viral, or immunological agents. These events may induce growth factor secretion and changes in endothelial cell surface adhesive glycoproteins. Monocytes are attracted and attach to endothelial cells, which will contribute directly or indirectly to continued secretion of growth factors and other biologically active molecules. Subsequent subendothelial migration of monocytes may lead to fatty-streak formation and release of cytokines and growth factors. Monocytes and macrophages are major cellular components of lesions and thus are likely to play a role in their initiation and evolution.

These events provide three possible sources of cytokines and growth factors, namely from platelets, macrophages, and the endothelium. Some of the smooth muscle cells in the proliferative lesion themselves may form and secrete cytokines and growth factors such as platelet-derived growth factor. It is not clear what role dietary fat plays in the above-stated events. However, hyperlipidemia, or some component(s) of hyperlipidemic serum, as well as other risk factors, are thought to cause endothelial injury/dysfunction, resulting in endothelial cell activation, adhesion of platelets and/or monocytes, increased cytokine activity, and transmigration of monocytes into the arterial intima. Once in the subendothelial space, monocytes transform into macrophages, take up substantial amounts of lipids, and become foam cells. These foamy macrophages, as well as other cells, can also produce cytokines and growth factors, which cause migration of smooth muscle cells from the media into the intima. These interactions then lead to fibrous plaque formation and further lesion progression. In other words, once smooth muscle cells proliferate in the intima, there is further lipid accumulation as well as elaboration of the extracellular components of the atheromatous plaque.

Even though numerous risk factors, including hyperlipidemia, smoking, and hypertension, seem to contribute to the development of atherosclerosis, to date it has not been possible to link these risk factors into a common pathogenic mechanism. There is evidence, however, that modulations in the level of activity of a select set of endothelial transcription factors (e.g., endothelial nuclear factor-κB [NF-κB]) may provide a mechanism for linking these seemingly diverse processes with the generation of dysfunctional endothelium and the onset of atherosclerotic lesion formation (39). Stimuli known to activate the NF-κB complex include inflammatory cytokines, with the common denominator apparently being reactive oxygen species (40). One may speculate that oxidized lipids, when present in inappropriate levels, may induce endothelial oxidative stress and generate excess reactive oxygen species, which activate NF-κB and modulate endothelial gene expression. Antioxidants and related compounds may protect against atherosclerosis by inhibiting the activation of endothelial transcription factors such as NF-κB.

Dietary Fat. There is ample evidence demonstrating that serum cholesterol is a predictor of atherosclerosis and that serum cholesterol concentrations can be modified by varying the composition of dietary fat. Keys and coworkers (41) in 1957 and Hegsted (42) in 1965 provided the first quantitative estimates of the relative effects of the various classes of fatty acids on serum cholesterol concentrations. Both studies indicated that saturated fatty acids increased, whereas PUFAs decreased, serum cholesterol. Also, monounsaturated fatty acids had no specific effect on cholesterol concentrations. These conclusions were based on combining results from numerous studies and then describing a mathematical relationship between dietary fatty acid composition and cholesterol concentration in the serum. However, it is now known that saturated fatty acids are not equally hypercholesterolemic. For example, stearic acid (18:0) and saturated fatty acids with less than 12 carbon atoms seem to have little or no effect on raising serum cholesterol levels (43). This then would suggest that the cholesterol-raising properties of saturated fatty acids should be attributed solely to lauric (12:0), myristic (14:0), and palmitic acids (16:0). However, these three saturated fatty acids appear to have different effects on serum cholesterol concentrations as well. Numerous studies suggest that lauric acid is less, and myristic acid probably more, hypercholesterolemic than palmitic acid (42,43). Recent studies with humans suggest that lauric acid raises total serum cholesterol and LDL-C concentrations when compared to oleic acid (18:1*n*9). However, it is not as potent in increasing cholesterol concentrations as is palmitic acid (44). On the other hand, in normocholesterolemic men and women, dietary palmitic and oleic acids seemed to exert similar effects on serum cholesterol and lipoprotein profiles (45).

Although serum cholesterol appears to be a risk factor for atherosclerosis (each 1% rise in serum cholesterol is predicted to increase the risk of coronary heart disease by about 2%), the effect of dietary cholesterol on serum cholesterol concentration is not clear and far from being understood. It appears, however, that the average baseline consumption of cholesterol-containing foods can modulate the magnitude of a mathematically predicted change in serum cholesterol due to changes in dietary cholesterol (46). An increase in dietary cholesterol is expected to have the greatest effect on serum cholesterol level when the past baseline amount of dietary cholesterol was near zero. On the other hand, if the baseline cholesterol consumption is greater than 500 mg/day, additional dietary cholesterol will have a minimal impact on increasing serum cholesterol. This suggests that people who desire to achieve a maximal reduction of serum cholesterol through diet may have to reduce their cholesterol intake to minimal levels in order to observe any significant reduction in serum cholesterol. Furthermore, an individual's response to dietary cholesterol can be extremely variable. Some individuals are more responsive (hyperresponders) compared with others (normoresponders). Thus, the need to limit cholesterol intake should apply more strictly to diet-sensitive hypercholesterolemic individuals rather than to the general population. Individual variations in the response to dietary cholesterol may be mediated by differences in fat absorption efficiency, neutral sterol excretion, conversion of hepatic cholesterol to bile acids, or modulation of key enzymes involved in intracellular cholesterol metabolism, such as HMG-CoA reductase.

Even though cholesterol and saturated fatty acid intake may be primary determinants of serum cholesterol, the role of dietary fat in the development of atherosclerosis remains controversial and not well understood. The question arises whether or not dietary saturated fats should be replaced by unsaturated fats. Unsaturated fats, especially *n*-3 or omega-3 PUFAs, may be beneficial to human health (47). Some populations, such as the Greenland Eskimos who consume high levels of *n*-3 fatty acids from fish and sea mammals, have less coronary heart disease (47). In patients with hyperlipidemia, only at high doses do *n*-3 fatty acids result in a decrease of LDL-C. However, these fatty acids consistently lower serum triacylglycerols in normal subjects and in patients with hypertriacylglycerolemia. Diets high in *n*-6 and *n*-3 PUFAs may lead to a decrease in serum cholesterol, but replacing saturated with unsaturated lipids may not be desirable because of their ability to oxidize easily.

Numerous recent studies and biochemical investigations suggest that lipid oxidation products, ingested with food or produced endogenously, represent a health risk (13,48). In fact, recent evidence supports the hypothesis that LDL undergoes an oxidative modification that increases its uptake by macrophages (49). Intervention studies in the LDL receptor-deficient animal model for atherosclerosis (the Watanabe heritable hyperlipidemic rabbit), using probucol as an antioxidant, have shown that the progression of the fatty acid streak can be slowed under conditions that do not lower serum cholesterol levels (50).

Furthermore, dietary antioxidants such as vitamin E might act as an antiatherogenic agent by suppressing oxidative modification of LDL and the recruitment of monocytes into the arterial subendothelium by smooth muscle cells (51). In fact, data from subjects with varying degrees of coronary atherosclerosis support the hypothesis that high serum PUFA levels, when insufficiently protected by antioxidants (e.g., vitamin E), may indicate a higher risk of atherosclerosis (52). In particular, a positive relationship between linoleic acid (18:2*n*6) intake and coronary artery disease was seen in patients undergoing angiography (53). All these studies lead one to conclude that the type of fat becomes a less significant component in the pathogenesis of atherosclerosis, when one consumes a low-fat diet, rich in soluble fibers and natural antioxidants.

2.3 Lipoprotein Metabolism

Plasma lipoproteins are units of complex lipid and protein compositions. Lipoproteins function primarily as carriers of lipids in the blood. The apoprotein fractions of the different lipoproteins play an important role in the regulation of the metabolic fate of the different plasma lipoproteins via their role as enzymatic cofactors and their interactions with specific receptors in cell membranes. Lipoproteins can be separated by density into four major different classes: chylomicrons, very low density lipoproteins (VLDL), low-density lipoproteins (LDL), and high-density lipoproteins (HDL). Both chylomicrons and VLDL are triacylglycerol-rich particles. Chylomicrons are of mucosal cell origin, have a density of less than 0.94 g/mL, and function mainly as carriers of lipids of exogenous dietary origin to the liver and peripheral tissues. VLDL, on the other hand, are of hepatic origin, have a density of 1.006 g/mL, and transport endogenous lipids. LDL are generated primarily by the metabolism of VLDL, have a density of 1.019–1.063 g/mL, and are high in cholesterol and cholesterol esters. The main function of LDL appears to be the delivery of cholesterol to hepatic and extrahepatic tissues. HDL are the smallest particles and contain the highest relative amount of protein. These units are relatively rich in cholesterol and phospholipids and have a density of 1.063–1.21 g/mL. High levels of LDL are associated with atherosclerosis and coronary heart disease, whereas high levels of HDL provide protection from these diseases. In fact, HDL particles appear to play an important role in the removal of cholesterol from the extrahepatic tissues. The lipoproteins of greatest interest in the pathogenesis of atherosclerosis are LDL and HDL.

2.4 Low-Density Lipoproteins

There appears to be a strong association between high plasma levels of LDL and accelerated development of atherosclerosis. Furthermore, evidence supports the concept that the lipids accumulated in atherosclerotic lesions are derived primarily from plasma LDL. Receptor mediated, as well as receptor independent, mechanisms appear to regulate the internalization and deposition of cholesterol within the atherosclerotic vessel wall. In addition, several independent membrane receptors participate in the interaction between plasma lipoproteins and vascular cells. These include the classic LDL or apoB/E receptor, the chylomicron remnant or apoE receptor, the acetyl-LDL receptor, the β-VLDL receptor, and the HDL receptor.

The complicated processes that occur during atherosclerosis seem to involve the participation of modified lipoproteins. For example, modified forms of LDL are associated with increased atherogenicity (54). In contrast to native LDL, modified LDL demonstrate enhanced cellular uptake by macrophages, foam cell formation, and also cause the secretion of inflammatory cytokines and growth factors from blood-borne and arterial wall cells. Nonenzymatic modifications of LDL include the formation of LDL-glycosaminoglycan or LDL-immune complexes, as well as glycosylation of LDL apoprotein B. In addition, self-aggregated LDL are taken up more rapidly by macrophages than native, monodisperse LDL. Enzymatic modifications via lipases and oxygenases also can occur. All these modifications of LDL can affect the physicochemical (size, charge) as well as the biological (cellular uptake, secretion) properties of this lipoprotein particle.

Of special interest is the oxidative modification of LDL in the pathogenesis of atherosclerosis. Even though the *in vivo* evidence for the oxidative modification hypothesis is still weak, the contribution of lipid peroxidation to the pathogenesis of atherosclerosis is based on the following: (*1*) presence of oxidized LDL and lipid peroxides in areas of the atherosclerotic plaque, (*2*) increased susceptibility of LDL from atherosclerotic patients to undergo lipid peroxidation, and (*3*) antiatherogenicity of antioxidant therapy. In fact, cross-sectional epidemiological studies suggest an inverse relationship between intake of antioxidant vitamins and the incidence of coronary heart disease (55). Dietary lipids may play a significant role in the oxidative modification of LDL. For example, the susceptibility of LDL to undergo lipid peroxidation could include the composition and location of its PUFAs and its antioxidant content. Diets high in polyunsaturated fatty acids may greatly increase the linoleic and arachidonic acid concentrations in the core cholesterol ester and surface phospholipid fractions of LDL particles. Extrinsic factors, such as the extracellular content of iron and copper ions, vitamin E and C concentrations, the arterial matrix, and the cellular oxidative systems (e.g., oxygenases and superoxides), also can play important roles in the oxidative modification of LDL.

2.5 High-Density Lipoproteins

Epidemiological studies suggest an inverse relationship between plasma levels of HDL and the incidence of coronary heart disease. In fact, HDL may act as a preferential physiological acceptor for cholesterol from extrahepatic cells. This mechanism of "reverse cholesterol transport" was originally proposed to describe the movement of cholesterol from the extrahepatic tissues to the liver, where it may be metabolized (56). HDL, therefore, seems to prevent and/or remove cholesterol deposits within the arterial wall. This is in contrast to the apoB-containing lipoproteins (e.g., LDL), which deliver cholesterol to the cells of the arterial wall. Even though the beneficial role of HDL in atherosclerosis is not yet clearly understood, HDL appears to exhibit numerous protective effects in addition to its role in reverse cholesterol transport (57). HDL can act as an antioxidant and thus can interfere with the cytotoxicity of LDL to endothelial cells as well as with foam cell formation. Furthermore, HDL is associated with increased prostacyclin production and reduced platelet aggregation. High levels of HDL also may prevent LDL aggregation and reduce the uptake of LDL by endothelial cells by competitive inhibition for LDL receptor binding.

2.6 Dietary Fat and Lipoprotein Metabolism

There is substantial evidence to indicate that dietary fat can significantly influence not only serum levels of cholesterol and triacylglycerols but also the lipid composition of lipoproteins (58). Saturated fatty acids and cholesterol have been identified as the major nutritional factors that can raise serum LDL-C levels. However, LDL-C is only one of many risk factors for atherosclerosis, and it is not known if oxidative modification of LDL is an equally or more important factor in the pathogenesis of atherosclerosis than total LDL-C per se. More longitudinal studies are needed to answer these questions. If lipid peroxidation is a major risk factor for atherosclerosis, then excess consumption of highly unsaturated fats may not be advisable.

The quantitative relationship between cholesterol intake and cholesterol levels is still controversial, especially since in humans there appears to be a high individual variability in processing of dietary cholesterol. However, numerous animal and human studies support the concept that dietary cholesterol can raise LDL-C levels and change the size and composition of these particles. LDL particles become larger in size and enriched in cholesterol esters. Mechanisms contributing to these events include an increase in hepatic synthesis of apoB-containing lipoproteins, increased conversion of VLDL remnants to LDL, and/or a decrease in the fractional catabolic rate for LDL. Reduced LDL receptor activity due to an increase in hepatic cholesterol content, secondary to excess dietary cholesterol, may lead to a decreased uptake of both LDL and VLDL remnants.

In addition to dietary cholesterol, saturated fatty acids also are thought to raise serum LDL-C levels as well as total cholesterol concentrations. The major effect of saturated fatty acids on serum cholesterol appears to be due to a reduction in LDL receptor activity. It is likely that saturated fatty acids may contribute to a cellular redistribution of cholesterol and cholesterol oxidation derivatives, leading to a favorable environment for these lipid particles to suppress LDL receptor synthesis. In addition to their effects (directly or indirectly) on LDL receptor activity, saturated fatty acids also may promote the synthesis of apoB-containing lipoproteins.

As previously mentioned, it is now known that not all saturated fatty acids are equally hypercholesterolemic. For example, medium-chain saturated fatty acids of carbon length 8–10, as well as stearic acid (18:0), have little or no effect on serum cholesterol concentrations. In contrast, evidence indicates that palmitic acid (16:0), the principal fatty acid in most diets, can increase serum cholesterol concentrations in humans (58). However, in normocholesterolemic humans, dietary palmitic and oleic acids have been shown to exert similar effects on serum cholesterol (45), suggesting that only humans or animal species sensitive to dietary cholesterol and selected fats (hyperresponders) may exhibit significant changes in serum cholesterol in response to dietary fat intake. Myristic acid (14:0) and to a lesser extent lauric acid (12:0), which are relatively high in coconut oil, both can raise serum cholesterol and LDL-C levels. Overall, it is not clear why humans respond so differently to cholesterol or saturated fatty acids. Variations may exist at the level of fatty acid catabolism and regulation of LDL receptor activity.

In contrast to saturated fatty acids, unsaturated fatty acids may not be cholesterolemic. However, because of their ability to become oxidized and thus to contribute to oxidative stress within a cell, some unsaturated fats indirectly could be highly atherogenic. Oleic acid (18:1*n*9), the major monounsaturated fatty acid in the diet, often is called a "neutral" fatty acid because it has a neutral or cholesterol-lowering effect on serum cholesterol (58). The main classes of unsaturated fatty acids in the diet can be divided into *n*-6 and *n*-3 PUFAs. Linoleic acid (18:2*n*6) is the predominant *n*-6 fatty acid, and the parent *n*-3 fatty acid is linolenic acid (18:3*n*3). Both occur in plant oils. Fish oils of marine origin contain large amounts of long-chain PUFAs, for example, eicosapentaenoic acid (20:5*n*3) and docosahexaenoic acid (22:6*n*3). With regard to cholesterol metabolism, linoleic acid may lower serum cholesterol levels by upregulating LDL receptor activity and/or by inhibiting hepatic synthesis of apo B-containing lipoproteins. Long-chain *n*-3 PUFAs appear to have a greater influence on triacylglycerol than on cholesterol metabolism. High intake of fish-oil-derived *n*-3 fatty acids reduces triacylglycerol levels, especially when fed to individuals with hypertriacylglycerolemia (47).

The role of dietary fats on HDL metabolism is poorly understood. In general, saturated fatty acids do not reduce HDL cholesterol (HDL-C). However, dietary monounsaturated fatty acids, when substituted for saturated fatty acids, contribute to a favorable modification of the lipoprotein ratio, that is, a decrease in the LDL/HDL ratio (58). In contrast to monounsaturated fatty acids, a high intake of *n*-6 PUFAs (e.g., linoleic acid) reduces HDL-C concentrations, possibly by reducing the synthesis of apoA-I, a major HDL apoprotein. The actions of *n*-3 PUFAs on HDL-C levels are similar to those of linoleic acid. Even though unsaturated fatty acids do not appear to be hypercholesterolemic, their HDL-C lowering capacity might be of concern since HDL is directly protective against atherosclerosis. Decreased serum HDL levels would indicate reduced removal of lipids from the arterial wall.

2.7 Lipids and Endothelial Cell Dysfunction

As the knowledge of the pathogensis of atherosclerosis rapidly increases, it appears that an active vascular endothelium, smooth muscle cells, blood-borne cells, such as monocytes and macrophages, all exert metabolic/physiologic effects in the atherosclerotic disease process. Risk factors, such as elevated plasma levels of certain lipids, prooxidants and cytokines, may contribute to the chronic activation/stimulation as well as to the damage of the vascular tissues. There is evidence that supports the hypothesis that it is not only pure cholesterol and saturated fats but rather oxidation products of cholesterol and unsaturated fats (and possibly certain pure unsaturated fats) that are atherogenic, possibly by causing endothelial cell injury/dysfunction. Lipid-mediated endothelial cell dysfunction may lead to adhesion of monocytes, increased permeability of the endothelium to macromolecules (i.e., a decrease in endothelial barrier function), and disturbances in growth control of the vessel wall.

2.8 Fatty Acids

While many mechanisms for the etiology of atherosclerosis have been proposed, endothelial injury/dysfunction clearly plays a role in the atherosclerotic disease process. There is evidence to suggest that diet-derived lipids metabolically interact with the vascular endothelium and may be responsible for abnormal regulatory mechanisms and a subsequent alteration of endothelial integrity (59). For example, high levels of circulating triacylglycerol-rich lipoproteins (chylomicrons and VLDL) have been implicated in the injury process of the endothelium (59,60). Plasma chylomicron levels are elevated in humans after consuming a high-fat meal, and hepatic synthesis of VLDL is increased when caloric intake is in excess of body needs. When plasma triacylglycerol-rich lipoproteins are elevated, hydrolysis of triacylglycerols by lipoprotein lipase thus elevates the concentrations of fatty acid anions, which occurs in proximity to the endothelial surface. Such high levels of diet-derived fatty acids can cause endothelial injury or dysfunction and thus disrupt the ability of the endothelium to function as a selective barrier. This would result in lipid deposition by allowing increased penetration of cholesterol ester-rich remnant lipoproteins into the arterial wall. In support of this hypothesis, it has been shown that exposure of cultured endothelial cells to selected fatty acids, for example, oleic acid, decreased endothelial barrier function, expressed as an increased transfer of both albumin and LDL across the endothelium (61,62). In addition to oleic acid, Hennig and co-workers also investigated the effects of other fatty acids on endothelial barrier function (63). Albumin-bound palmitic and stearic acid had little effect on endothelial barrier function, but exposure of cell monolayers to linoleic acid produced an even greater increase in albumin transfer than did equal concentrations of oleic acid. Furthermore, the disruption in endothelial barrier function was exacerbated greatly in the presence of small amounts of oxidation derivatives of unsaturated fatty acids (64). This suggests that, in general, fatty acid oxidation derivatives, but not pure lipids, are extremely cytotoxic.

It is not clear why linoleic acid, and none of the saturated fatty acids studied, disrupted endothelial barrier function. The injurious effects of linoleic acid on cultured endothelial cells may be mediated in part by the induction of peroxisomes and thus by excessive hydrogen peroxide formation (65). In addition, enrichment of endothelial lipids with selective fatty acids can modify specific cellular lipid pools and alter the morphology of cultured cell monolayers. Such fatty-acid-mediated changes in membrane composition may be sufficient to alter membrane properties (e.g., fluidity and activities of membrane-bound enzymes). One may speculate from these and other data that high dietary intakes of certain unsaturated fatty acids, such as linoleic acid, might not be entirely safe.

2.9 Cholesterol

There is experimental evidence to suggest that some oxysterols, but not pure cholesterol, are the prime cause of atherosclerotic lesion formation (66). One may speculate that the deposition of pure lipids, such as cholesterol and its esters, may be merely a secondary process in response to oxysterol-induced endothelial cell injury. Cell injury/dysfunction and the subsequent disruption of endothelial barrier function by oxysterols (67,68) could initiate the early events in atherosclerosis. Such injury could allow increased uptake of cholesterol-rich lipoproteins into the arterial wall by decreasing endothelial cell prostacyclin production, thereby enhancing platelet adhesion and aggregation and by increasing monocyte adhesion and infiltration. Hennig and Boissonneault (67) have demonstrated a time- and concentration-dependent disruption of endothelial barrier function following exposure of cultured cells to the oxysterol cholestan-3β,5α,6β-triol. In contrast, pure cholesterol did not affect endothelial cell integrity with respect to endothelial barrier function (67). In addition to albumin-bound oxysterol, LDL enriched with cholestan-3β,5α,6β-triol also decreased endothelial barrier function in a dose-dependent manner (69). These studies suggest that oxysterols, and not pure cholesterol, cause dysfunction of vascular endothelial cells. Not all oxysterols, however, are equally cytotoxic, and different mechanisms of endothelial cell injury by different types of cholesterol oxidation derivatives may exist. Since relatively high concentrations of oxysterols are formed in certain processed foods, and since they are easily absorbed and transported in the blood, cholesterol oxidation derivatives may be an important dietary risk factor in cardiovascular disease.

2.10 Summary and Dietary Advice

Numerous animal and human studies suggest that dietary cholesterol and certain saturated fatty acids increase serum as well as LDL-C concentrations. Even though humans with elevated serum cholesterol levels may be at risk, evidence also is mounting to suggest that the complicated processes that occur during atherosclerosis involve the participation of modified lipoproteins. For example, oxidatively modified forms of LDL are associated with increased atherogenicity, because modified LDL, in contrast to native LDL, demonstrate enhanced cellular uptake by macrophages and foam cell formation. Modified LDL particles also cause secretion of inflammatory cytokines from blood-borne and arterial wall cells, which will lead to endothelial cell activation. The resulting disturbances in endothelial integrity possibly allow increased penetration of cholesterol-rich lipoprotein remnants into the arterial wall, a critical event in the etiology of atherosclerosis. Modulations in the level of activity of a select set of oxidative stress-responsive transcription factors (e.g., endothelial nuclear factor-B; NF-κB) may provide a common mechanism for linking these diverse processes (Figure 7.4). Reactive oxygen species appear to be the common denominator in the many stimuli known to activate the NF-κB complex. One may speculate, then, that high levels of dietary polyunsaturated lipids, which are easily oxidizable, can activate oxidative stress-responsive transcription factors that, in turn, may promote cytokine production, adhesion molecule expression, endothelial barrier dysfunction, and ultimately accelerate atherosclerosis. Interestingly, the activation of NF-κB can be inhibited by a variety of antioxidants (Figure 7.4), which suggests that certain nutrients that have antioxidant properties may protect against atherosclerosis by interfering with the proposed mechanisms of endothelial cell dysfunction. In summary, the research on cardiovascular disease and atherosclerosis is complex, but the advice for patients at risk is simple: eat less fat, independent of its source, and eat only enough food to satisfy energy needs.

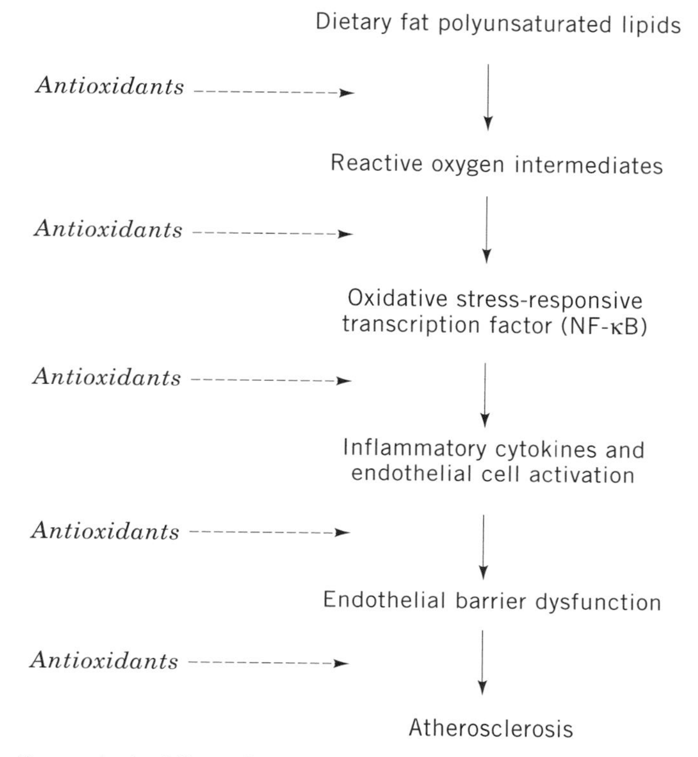

Figure 7.4 Proposed role of dietary fat, in particular unsaturated fat, in the etiology of atherosclerosis. Dietary fats, rich in certain unsaturated lipids are atherogenic by enhancing the formation of reactive oxygen intermediates. These intermediates can activate the oxidative stress-responsive transcription factor NF-κB, which in turn may promote cytokine production, adhesion molecule expression, and ultimately endothelial barrier dysfunction and atherosclerosis. Certain nutrients/chemicals, which have antioxidant properties, may protect against atherosclerosis by acting at any one of the progressive steps.

ROLE OF DIETARY FAT IN HYPERTENSION

3.1 Introduction

Hypertension is defined as an elevated arterial pressure, with a diastolic blood pressure above 90 mm Hg or a systolic blood pressure above 140 mm Hg. Hypertension remains one of the most common chronic diseases in Westernized populations. It affects approximately 30% of the adult population and is the underlying cause of 1.5 million deaths and 1.0 million disabilities annually in the United States. The incidence of hypertension is higher in men than in women, in blacks than in whites, with the highest levels in elderly, black women. It appears that several dietary factors, including dietary fats, may play a crucial role in the pathogenesis of this disease (70).

3.2 Pathogenesis of Hypertension

The pathogenesis of hypertension is complex and includes both hereditary and environmental factors. Generally, it is accepted that the increase in systemic blood pressure is secondary to vascular resistance of peripheral arterioles, which in turn may be caused by excessive contraction and/or growth of vascular smooth muscle cells (71). Several factors are postulated to participate in the functional and structural modulation of smooth muscle cells. Neurotransmitters of the sympathetic nervous system, humoral factors, abnormalities in cell membrane transport systems, and imbalances between endothelium-derived constrictor factors (e.g., endothelin), and dilator factors (e.g., prostaglandins, endothelium-derived relaxing factor) may be involved in this process (72). There is also strong evidence that the kidneys may be of primary importance in the development of hypertension. Kidney function is affected by elevated blood pressure and renal abnormalities may induce hypertension (73). Moreover, a number of dietary factors play a crucial role in the development of hypertension. The majority of epidemiological and experimental data suggest that blood pressure is positively correlated with dietary sodium and inversely related with dietary potassium, calcium, and/or magnesium (74).

Another important nutritional disorder that contributes to elevated blood pressure is obesity. It has been demonstrated clearly that obesity and hypertension are closely related. Weight loss in obese patients results in a decrease in blood pressure values. On the other hand, elevation of systolic blood pressure may be mediated by an increase in body weight. The mechanism of hypertension in overweight patients is not understood fully; however, obesity is known to lead to insulin resistance and consequently to hyperinsulinemia. Elevated insulin may do the following: (*1*) alter electrolyte balance across cellular membranes, (*2*) increase extracellular volume and cardiac output, (*3*) activate the sympathetic nervous system, (*4*) impair lipid metabolism, and (*5*) stimulate proliferation of vascular smooth muscle cells (72). All of these factors may participate in the development of hypertension (70).

3.3 Plasma Lipids in Hypertension

Dietary lipids are another factor that may influence blood pressure (Figure 7.5). However, the pathophysiological link between dietary fats and blood pressure, as well as specific hemodynamic effects of different fats, remains controversial. For example, there are no consistent data regarding the relationship between hypercholesterolemia and hypertension. Several epidemiological studies have shown no correlation between total plasma cholesterol and blood pressure (75). Therefore, these two parameters are considered to be independent but synergistic risk factors for coronary artery disease (76). However, in other epidemiological studies, the prevalence of hypercholesterolemia in patients with hypertension was approximately twice as high as that in a popula-

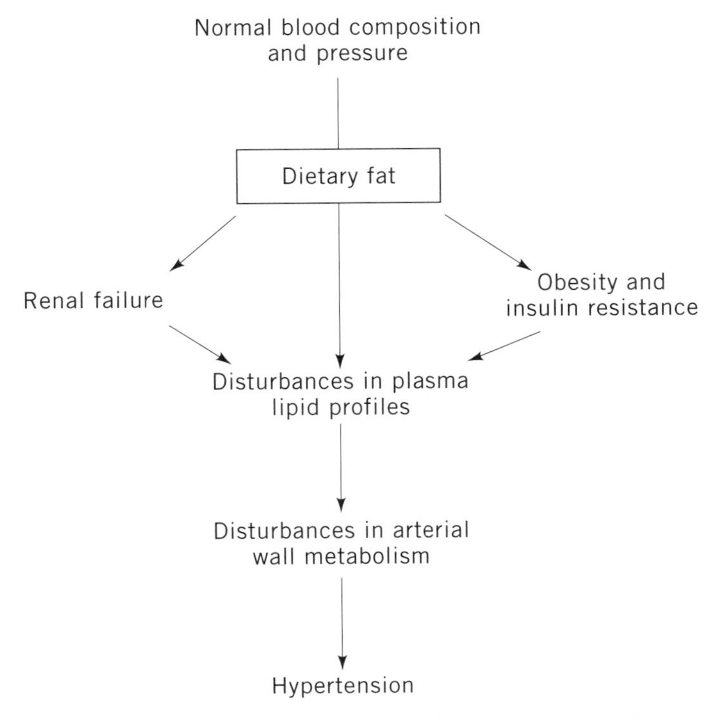

Figure 7.5. Effects of dietary fat on the development of hypertension. High-fat diets may cause changes in blood composition and/or rheology, and thus, lead to changes in metabolism of arterial wall smooth muscle cells. Additionally, high-fat diet-mediated renal injury and/or obesity may be involved in this process.

tion with normal blood pressure (77). Moreover, several factors such as fasting insulin level, physical activity, or obesity are linked with hypercholesterolemia and hypertension (78). A reduction in plasma cholesterol levels has been demonstrated to lower blood pressure (79).

Several mechanisms, including increased vascular reactivity, may be linked to increased cholesterol levels and hypertension (80). Experimental animals that were hypercholesterolemic have reduced renal blood flow, and thus cholesterol may activate renal mechanisms causing an increase in blood pressure (81). The responsiveness to noradrenaline of isolated rabbit femoral (82) and basilar (83) arteries, canine coronary arteries (84), and cynomolgus monkey hindlimb arteries (85) are enhanced in hypercholesterolemia. It is possible that elevated LDL levels are responsible for the hypertensive effects of hypercholesterolemia. There is evidence that LDL may increase calcium concentrations in arterial smooth muscle cells and, through this mechanism, exert a vasoconstrictive effect (86). Moreover, oxidized LDL were shown to produce concentration-dependent contractions in rabbit femoral arteries pretreated with noradrenaline. This effect was positively correlated with the degree of

LDL oxidation (87). In human studies, a decrease in total and LDL cholesterol levels, achieved by a low fat diet, was shown to be associated with reduced blood pressure and heart rate as well as with a selective decrease in noradrenaline cardiovascular reactivity (80).

3.4 Role of Dietary Fat in Blood Pressure Regulation

Although the correction of hypercholesterolemia by a low-fat diet enriched in PUFAs may decrease blood pressure (16), it is difficult to identify the active lipid component(s) responsible for such an effect. In several epidemiological studies, the relationship between the total fat intake and blood pressure was investigated using different populations and a large sample size. In these studies low-fat diets had no consistent effect on blood pressure (88–90). However, some of the studies, which showed no effect of the low-fat diet on blood pressure regulation, were performed on normotensive subjects. It is possible that low-fat diets have different effects in hypertensive patients. Moreover, there is no doubt that a low-fat diet may be beneficial for obese, hypertensive patients (91) and for patients with chronic renal diseases (92).

3.5 Role of Individual Dietary Fatty Acids in Blood Pressure Regulation

There is continuing controversy concerning the role of specific dietary fatty acids, especially linoleic acid, in the regulation of blood pressure (16). It was shown in experimental animals, under different experimental conditions, that dietary deprivation of linoleic acid resulted in the development of hypertension (93). On the other hand, dietary supplementation with this fatty acid was effective in decreasing blood pressure (90). In rats, dose-related inverse relationships between dietary linoleic acid and blood pressure were observed when this fatty acid was given as 0–5% of total energy consumption. In contrast, dietary supplementation with linoleic acid in excess of 5% of total energy had no effect on arterial blood pressure (94). It has been suggested that altered vascular reactivity may be involved in a linoleic-acid-mediated decrease in blood pressure in experimental animals. Diminished chronotropic and inotropic responses to noradrenaline were detected in isolated perfused hearts of rats fed a linoleic-acid-enriched diet (95). It has been postulated that these effects were mediated by alterations in adenyl cyclase activity and cyclic-adenosine 5′-monophosphate (AMP) formation. Moreover, a reduced affinity and density of postsynaptic cardiac β_1-receptors was observed in rats fed sunflower oil, which is rich in linoleic acid (96). The differential effects of linoleic acid on blood pressure regulation also might be due to a linoleic-acid-mediated influence on eicosanoid metabolism (97).

In contrast to animal studies, where linoleic acid influences blood pressure and where several mechanisms have been proposed to explain such actions, studies investigating the effect of this fatty acid on blood pressure in humans have resulted in only conflicting data. In some studies, diets enriched with linoleic acid had a lowering effect on blood pressure in normotensive and/or hypertensive patients. Studies in which the typical diet (40% energy from fat, 20% from saturated fatty acids, and a polyunsaturated to saturated fat ratio of about 0.3) of industrialized societies was changed to low-fat diets (25% energy from fat) enriched with PUFAs (polyunsaturated to saturated fat ratio of about 1.0) consistently showed that low-fat diets, high in PUFAs, decreased blood pressure. These reports concluded that the high ratios of polyunsaturated (mostly *n*-6) to saturated fatty acids were responsible for a decrease in blood pressure. Since linoleic acid was the major PUFA added to the experimental diets, the lowering of blood pressure was believed to be due to linoleic acid (90). However, these conclusions have been criticized because other dietary and nondietary factors, such as total fat, potassium, and fiber, which influence blood pressure, were not normalized during these studies (89). In several other reports no beneficial effects were observed in studies in which dietary modifications were restricted to the amount of *n*-6 PUFAs, specifically linoleic acid (88). Therefore, it appears that a low-fat diet, rather than a diet high in linoleic acid, is potent in lowering blood pressure. This hypothesis is supported by the observation that low-fat diets with either a high (0.9) or a low (0.4) ratio of polyunsaturated (mostly linoleic acid) to saturated fatty acids was equally effective in lowering blood pressure (98).

Retrospective, cross-sectional studies as well as clinical trials, in which the role of other fatty acids in blood pressure regulation was investigated, also showed contradictory results. Dietary saturated fatty acids were positively correlated with blood pressure in some studies (99), but these results were not confirmed by other reports. An inverse correlation between blood pressure and dietary oleic acid or linolenic acid also was reported. However, results regarding the effects of oleic acid were inconsistent (88,89).

The fatty acid composition of adipose tissue generally reflects the type of fat consumed when dietary fat intake is moderate to high. Therefore, the analysis of fatty acids in adipose tissue is an alternative method to measure the type of fats consumed over time. This method was employed in several studies evaluating the relationship between the type of dietary fatty acids and blood pressure. The results from these studies were contradictory. The level of palmitic acid and other saturated fatty acids were positively (100), inversely (101), or not correlated (102) with blood pressure. Several reports showed an inverse correlation between the level of adipose tissue linoleic acid and arterial blood pressure (102, 103), but other reports indicate no associations (104). In a separate report the inverse correlation between linolenic acid and blood pressure was noted (104).

3.6 Role of Dietary *n*-3 PUFAs in Blood Pressure Regulation

In contrast to saturated, monounsaturated, and *n*-6 PUFAs, which probably have little or no effect on blood pressure, the role of *n*-3 fatty acids on blood pressure regulation remains an open question. Epidemiological retrospective studies have not provided any evidence that consumption of large amounts of fish containing high levels of *n*-3 PUFAs is correlated with low blood pressure (105). For example, the Japanese who consume high amounts of fish and *n*-3 PUFAs have a very high prevalence of hypertension and incidence of stroke. However, a traditional Japanese diet is also high in salt, which increases blood pressure. This high-salt consumption possibly could mask the positive influence that fish oil has on blood pressure (106). Clinical trials evaluating the effect *n*-3 PUFAs on blood pressure provide contradictory results. In the majority of studies, a lowering effect of fish oil on systolic pressure, diastolic pressure, or both has been observed. Several reports, which showed no effects of *n*-3 PUFAs on blood pressure, can be criticized for having an inappropriate study design (105).

The mechanisms by which *n*-3 PUFAs lower blood pressure are not fully understood; however, changes in eicosanoid metabolism, blood rheology, and vascular reactivity are suggested to be responsible for such effects (107). Elevated fatty acids of the *n*-3 family may inhibit synthesis of eicosanoids from arachidonic acid, such as prostaglandin I_2 (prostacyclin) and thromboxane A_2, by competing for the same enzymatic systems. In these instances, *n*-3 fatty acids are converted to prostaglandin I_3 (PGI_3) and thromboxane A_3, respectively. While PGI_3 has the same antiaggregation and blood vessel dilatation potency as prostacyclin, thromboxane A_3 exerts much less biological activity than thromboxane A_2. In consequence, increased levels of fish oil fatty acids may decrease vasoconstriction and platelet aggregation, thus improving vasodilatation and platelet–vessel interactions (107). Moreover, fatty acids of the *n*-3 family also may exert beneficial effects on blood fluidity by changing plasma viscosity due to modifications of blood cell membranes, and to a lowering of lipid and fibrinogen levels (97). Fatty acids of the *n*-3 family also increase the sensitivity of the arterial baroreceptor reflex system, which participates in maintaining normal arterial blood pressure levels (108).

3.7 Conclusion

There is little doubt that dietary fat and plasma lipids can influence the regulation of blood pressure. However, there is an inconsistency among studies that describe blood pressure lowering effects of the different fatty acids. Furthermore, interpretation of these reports is confounded by different study designs. It appears that dietary recommendation for the prevention and/or treatment of hypertension should include: (*1*) low fat diets, (*2*) adequate dietary potassium, calcium, and magnesium, (*3*) a moderate restriction of sodium, and (*4*) an increased consumption of *n*-3 PUFAs.

ROLE OF DIETARY FAT IN THE DEVELOPMENT OF OBESITY

4.1 Introduction

More than a quarter of the U.S. population is overweight or obese, and the proportion of the population that is overweight appears to be increasing. Obesity can be expressed in terms of body mass index (BMI), that is, weight (kg)/height^2 (m). Humans with a BMI between 25 and 29 are defined as overweight, between 30 and 40 as conventional obese, and greater than 40 as morbidly obese. For other diseases, such as diabetes mellitus, hypertension, hypercholesterolemia, coronary and cardiovascular complications, and certain cancers, there is an increasing morbidity and mortality risk associated with a BMI greater than 30. Therefore, it is important to understand the etiology as well as successful treatment approaches for obesity.

Causes of obesity are multifactorial and are thought to include genetic as well as environmental factors. Nevertheless, most types of obesity can be explained, in part, by an imbalance in the classic energy balance equation, that is, a change in energy stores equals energy intake minus energy expenditure. Though great confusion in the understanding of human obesity still exists, evidence suggests that calories from fat contribute most toward the chronic energy imbalance causing obesity. As a treatment approach one might suggest that in order to lose weight, a reduction in fat intake should be considered in addition to an increase in exercise and a restriction of total calories. The following discussion will focus on the relationship between obesity and dietary fat.

4.2 Pathophysiology of Obesity

Potential mechanisms underlying different types of obesity have been clarified gradually by studying animal models, primarily rodents. Types of obesity often are subdivided into genetic obesity, hypothalamic obesity, and dietary obesity (109). In any or all of these types of obesity there may be varying degrees of imbalances in metabolic pathways, which control the sympathetic nervous system, the endocrine system, hypertrophy and hyperplasia, and the hypothalamus. Much is known about hormones, protein derivatives, and other circulating factors, such as cholecystokinin, gastrin-releasing peptide, insulin, glucagon, prolactin, satietin, and adipsin, which appear to modulate mechanisms that govern eating and regulation of body weight (110).

Even though the genetic background contributes substantially to body weight in humans, it is not well understood whether or not the genetic factors influence body weight by acting on energy expenditure and/or on food intake. For example, in studies of twins reared apart or together, it was found that about 70% of the variance in adult BMI was due to genetic influences, whereas the remaining 30% was attributed mostly to the effect of different environments unique to the individuals (111). It seems likely that a disparity in energy balance (expenditure) versus food intake can contribute significantly to the development of obesity.

Daily energy expenditure can be divided into three major components: basal metabolic rate (BMR), thermal effect of food, and energy cost of physical activity. BMR is the sum of the sleeping metabolic rate plus the energy cost of arousal and represents 50–70% of the daily energy expenditure. The thermic effect of food represents about 10%, and the energy cost of spontaneous and unrestricted (voluntary) physical activity represents betwen 20 and 40% of daily energy expenditure, respectively. These factors can vary, depending on gender, age, genetic background, and hormonal and sympathetic nervous system activity.

4.3 Risk Factors Associated with Body Weight Gain

In order to manage or treat obesity, one must understand the relationship between the long-term control of energy balance and the short-term control of food intake. It is possible that risk factors, such as low metabolic rate, physical inactivity, low rates of fat oxidation, and various socioeconomic factors, independently or in combination can modulate the relationship between energy balance and food intake (112).

The Pima Indians of Arizona, for example, who have the highest prevalence of obesity and thus high rates of body weight gain among young adults also experience a relatively low metabolic rate for their body size. Physical inactivity also may play a role in the pathogenesis of obesity in humans. There is evidence that the level of physical activity correlates with plasma norepinephrine turnover, suggesting that the activity of the sympathetic nervous system contributes to physical activity and thus to the overall energy balance (112).

A relatively low rate of fat oxidation may also be a risk factor associated with body weight gain. Studies have shown that the oxidation of protein and carbohydrate is closely tied to intake, whereas fat oxidation is not correlated as closely with intake (113). In contrast to protein and carbohydrate, fat oxidation is not driven actively by fat intake but rather occurs passively as the difference between total energy expenditure and the oxidation of protein and carbohydrate (114). Therefore, overconsumption of fat and low physical activity can contribute greatly to obestiy.

In addition to the metabolic factors mentioned above, socioeconomic and behavioral factors also can be significant determinants of the development of obesity. The prevalence of obesity often is related inversely to education level and directly to alcohol consumption. Furthermore, long-term maintenance of posttreatment weight loss appears to be better in programs that include educational parameters and behavioral treatment measures.

4.4 Dietary Fat and Obesity

Longitudinal studies clearly show that total dietary fat intake is related directly to weight gain (115). Indeed, the most consistent predictor of body mass and body weight change in both men and women is the percent of the dietary calories derived from fat (116). As mentioned earlier, oxidation rates of protein and carbohydrates are affected rapidly by changes in the intakes of protein and carbohydrates, but fat oxidation is not influenced immediately by changes in fat intake. In fact, in the postprandial metabolism of the macro-nutrients, carbohydrates promote their own oxidation by stimulating insulin release and cellular glucose uptake. Furthermore, the feeling of satiety coincides with carbohydrate, and not fat, intake. Diets with a high fat to carbohydrate ratio then would result in a higher fat intake as well as a need to ingest a certain amount of carbohydrate, that is, more fat, to obtain a feeling of satiety. This ultimately would lead to a positive energy balance and the development of obesity.

In obese subjects, the overall appetite control system appears to fail in response to high-fat foods. In fact, when eating from a range of either high-fat or high-carbohydrate foods, obese subjects voluntarily consumed twice as much energy from the fat items, thereby indicating a weak action of fat on satiety (117). Total fat, rather than type of fat, seems to be a predictor of obesity, although types of fat (e.g., saturated versus unsaturated fats) may influence the pattern of development of adipose tissue depots. Although animal studies suggest that saturated fats may have different effects than unsaturated fats on food intake, fat oxidation, adipocyte cellularity, or thermogenesis, evidence for fat effects on these metabolic parameters in humans is inconclusive and not well understood.

4.5 Obesity, Changes in Blood Lipids, and Risk for Cardiovascular Disease

Adverse health effects due to obesity are numerous. For example, obesity is associated with elevated blood pressure, lipids, and glucose. Furthermore, a 26-year follow-up of participants in the Framingham Heart Study concluded that obesity was an independent and significant risk factor for cardiovascular disease (118). Obesity is characterized by an increase in the rate of cholesterol synthesis, which can be reduced to within normal levels by weight reduction. Also, there appears to be an overproduction of lipoproteins that contain apoB, that is, VLDL and LDL. Overproduction of VLDL may be due to the increased free fatty acid flux and the increased insulin levels associated with obesity. Another factor that can lead to higher cholesterol concentrations in obese individuals is excessive intake of saturated fatty acids and cholesterol, both of which are known to suppress LDL receptor activity. All these metabolic events will accentuate the rise of LDL-C levels.

In contrast to the atherogenic lipoprotein profile associated with obesity, weight reduction, especially when linked to an exercise regimen, will result in decreased levels of apoB-rich lipoproteins and an increase in plasma concentrations of HDL (119). This suggests that for obese individuals, caloric restriction and exercise would improve lipoprotein levels and reduce the risk of cardiovascular disease.

4.6 Fat Substitutes

Long-term compliance with low-fat regimens is often difficult. This has led to a search for new food technologies, including the development of fat substitutes. An ideal fat substitute should be safe, versatile, and resemble dietary fat in its organoleptic and performance properties. It also should be able to be used in place of the fat present in foods or added during cooking without adversely affecting quality (120). Some fat substitutes that act as fat replacements (e.g., starches and gums) have been used for many years. Newer fat substitutes (e.g., microparticulated protein and olestra) more closely mimic the taste and mouth feel of fat. There are numerous experimental fat substitutes, which can be grouped into carbohydrate-, protein-, or lipid-based materials (121). Examples of carbohydrate- and protein-based materials include modified glucose polymers, modified tapioca, corn, potato, and rice starches, gums and algins, cellulose derivatives, and microparticulated proteins. Lipid-based materials are just as numerous and include fatty acid esters of sugars and sugar alcohols, polycarboxylic acid and propoxylated glycerol esters, alkyl glycerol ethers, substituted siloxane polymers, branched (sterically hindered) triacylglycerol esters, and specific naturally occurring lipids (121).

Although fat substitutes appear to pose little risk when used in small amounts, little is known about their long-term effectiveness in body weight management or obesity. Most of the carbohydrate- and protein-based materials are partially or fully digested and absorbed. In addition, the use of these materials generally results in a reduced-calorie product in foods because of their low-energy density. In contrast to carbohydrate- and protein-based materials, most lipid-based materials resist enzymatic digestion and are poorly absorbed. Thus flux of undigested hydrophobic material throughout the length of the gut potentially could result in reduced absorption of fat-soluble substances, such as some vitamins, cholesterol, and other compounds of biological importance.

4.7 Summary and Dietary Treatment of Obesity

There is evidence that dietary fat selectively can induce obesity because of its weak action on satiety. Unlike that of carbohydrates, protein, or alcohol, the oxidative metabolism of dietary fat is regulated poorly. Addition of fat to a meal does not cause increased fat oxidation, but leads to enhanced storage of this extra fat in the adipose tissue. Thus, fat is the major dietary factor related to obesity caused by a chronic positive energy balance. Furthermore, fatty foods can lead to an atherogenic lipoprotein profile, thus making obesity an independent risk factor for vascular diseases such as atherosclerosis.

Numerous treatments of obesity have been proposed. These range from behavior and diet modification to appetite-suppressing drugs and various types of invasive surgeries (122–124). However, the treatment with the greatest chance of success seems to be a combination of reduced fat intake (as well as total calories), independent of its source, with an increase in aerobic exercise. The recommendation to consume less dietary fat and to exercise will lead to a gradual and modest weight loss. Weight loss in obese subjects also reduces risk factors for developing hyperglycemia, high blood pressure, and serum triacylglycerol and cholesterol levels associated with heart disease.

ROLE OF DIETARY FAT IN DIABETES

5.1 Introduction

Diabetes is a heterogeneous metabolic disorder characterized by an elevated blood glucose level and insufficient insulin action. Common symptoms in this disease are polydipsia (frequent drinking) and polyuria (an excessive excretion of urine). It has been estimated that diabetes may affect more than 9% of the U.S. population. Several genetic, environmental, and lifestyle factors may contribute to the risk of diabetes.

On the basis of the mechanisms underlying the inadequate insulin action, diabetes may be divided into two major types, that is, insulin-dependent diabetes mellitus (IDDM) and non-insulin-dependent diabetes mellitus (NIDDM). IDDM, also called Type I diabetes, is characterized by insulin deficiency caused by autoimmune destruction of pancreatic β cells. IDDM accounts for approximately 5–10% of all cases of diabetes in the United States. The onset is usually in childhood, and most often between the ages of 8–14 years. Insulin replacement is a necessary treatment of IDDM (125,126).

NIDDM or Type II diabetes accounts for approximately 90% of all diabetes. NIDDM is associated with obesity in 80–90% of all cases. This type of disease is referred to as Type IIb diabetes in contrast to diabetes in lean patients, called Type IIa diabetes. The subjects with NIDDM have cellular defect(s) that affect insulin-mediated glucose uptake. This inability of insulin to maintain an adequate blood glucose level is called insulin resistance. Insulin secretion by the pancreas can be insufficiently sensitive to small changes in blood glucose, and insulin levels are usually normal or elevated in NIDDM. Treatment of this type of diabetes frequently does not require insulin supplementation (125,126).

There is evidence that the development of diabetes, primarly NIDDM, is strongly associated with dietary factors (Figure 7.6). NIDDM is positively correlated with calories consumed from fats and inversely related to calories from carbohydrates in the diet (127).

5.2 Role of a High-Fat Diet in the Pathogenesis of Diabetes

A key factor that links the consumption of high-fat diet with the onset of NIDDM is obesity. Excessive intake of dietary fat often results in obesity,

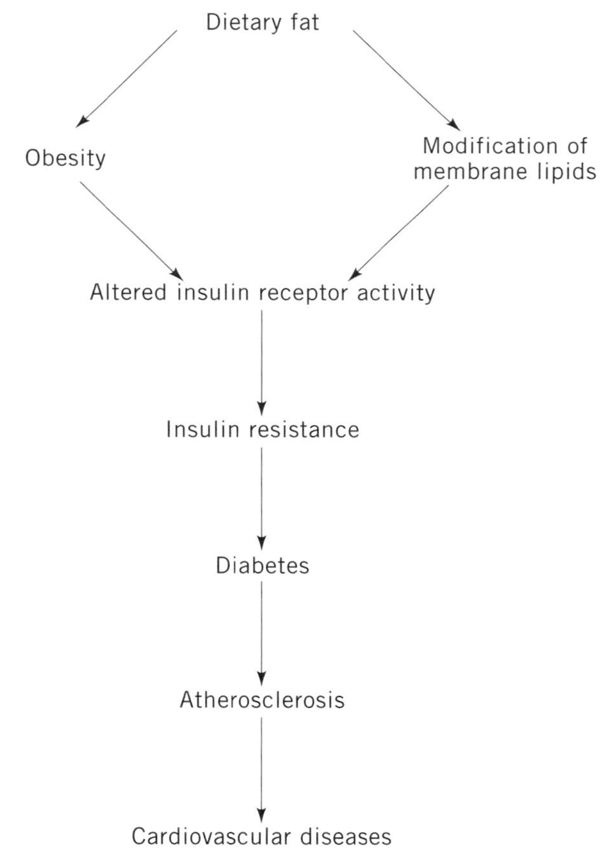

Figure 7.6. Relationship of dietary fat to obesity and the onset of non-insulin-dependent diabetes mellitus (NIDDM) and cardiovascular diseases.

which, in turn, may stimulate development of insulin resistance and Type IIb diabetes. In epidemiological studies, relative body weight is consistently associated with diabetes (128). Although several hypotheses were proposed to explain why obesity is associated with insulin resistance, this relationship remains unclear. Generally, it is known that resistance of cells to insulin action is associated with distribution of body fat. Android (chest, abdomen, arms) obesity but not gynoid (hips, thighs) obesity was related preferentially to insulin resistance. Therefore, it was suggested that android obesity and insulin resistance may be a common genetic predisposition. However, this hypothesis is criticized because weight reduction usually markedly improves insulin sensitivity (129). Additionally, there is evidence that increased blood levels of free fatty acids, a characteristic for obesity, may contribute to the development of insulin resistance. Incorporation of serum fatty acids into cellular membranes also may affect binding of insulin to its specific cellular receptor and thus diminish its hormonal activity (130).

5.3 Lipid Abnormalities in Diabetes

Diabetic persons, with IDDM or NIDDM, are at a high risk for the development of atherosclerosis and, as a result, cardiovascular diseases. This risk is expressed particularly in women because diabetes eliminates premenopausal protection against the development of atherosclerosis. Atherosclerosis and cardiovascular diseases are the cause of approximately 80% of all mortalities in diabetic persons. Coronary atherosclerosis is responsible for approximately 75%, and cerebral and peripheral vascular diseases account for the rest of the mortalities due to cardiovascular disease. Generally, it has been accepted that lipid abnormalities and glucose-mediated alterations in arterial wall structure and functions are responsible for the accelerated development of atherosclerosis in diabetes (131).

5.4 Disturbances in the Metabolism of Plasma Lipids

Disturbances in plasma lipid levels are common biochemical abnormalities in diabetic subjects, independent of the type of diabetes. Although abnormal lipid profiles are more common in NIDDM patients, untreated or poorly controlled IDDM subjects also may develop severe lipid abnormalities including elevated triacylglycerol and intermediate density lipoprotein levels. The most frequently occurring lipid abnormality in both types of diabetes is hypertriacylglycerolemia (132). It has been hypothesized that the increased formation of triacylglycerol-rich VLDL is associated with their decreased clearance, and is responsible for elevated triacylglycerol levels. VLDL synthesis is approximately 60% higher in NIDDM patients than in nondiabetic controls. The mechanisms for overproduction of VLDL in diabetes are poorly understood. However, it has been suggested that elevated levels of serum free fatty acids, which accompany the abnormal insulin action, may participate in this process. Moreover, serum lipoprotein lipase activity is decreased in diabetes, and this metabolic block may also diminish clearance of VLDL (133).

Decreased levels of HDL is another common lipid disturbance in NIDDM patients. The diminished level of HDL is caused almost entirely by a decrease in HDL_2 cholesterol, the HDL fraction responsible for reverse transport of tissue cholesterol. Moreover, the level of the HDL-specific apolipoprotein AI also may be reduced in diabetic patients (132).

Although total LDL levels are usually unchanged, LDL particles may be modified markedly in diabetes. Diabetic LDL appear to be smaller in size and denser in cholesterol content. It has been suggested that such modified LDL may exert stronger atherogenic effects (131). Moreover, glycosylation and oxidation of LDL apolipoproteins and PUFAs due to high plasma glucose levels also may result in increased LDL atherogenicity (134).

5.5 Disturbances in Cellular Lipid Metabolism

In severe diabetes both lipoprotein metabolism and tissue lipid metabolism are affected. Desaturation of fatty acids is diminished due to decreased activity of Δ-9, Δ-6, and Δ-5 desaturases (135). Impaired PUFA formation results in increased cellular concentrations of saturated fatty acids and decreased concentrations of PUFAs. In fact, the levels of highly unsaturated fatty acids such as arachidonic acid and its derivatives (e.g., prostacyclin) were shown to be decreased in diabetes (136).

Due to insufficient insulin action and reduced cellular glucose uptake, glucose oxidation is decreased in diabetes. Therefore, increased oxidation of fatty acids may be initiated as a compensatory mechanism to provide an adequate cellular energy balance. Elevated oxidation of fatty acids can result in excessive formation of ketone bodies. Similar aberrations in metabolism are expressed in IDDM subjects that are highly deficient in insulin. Moreover, catabolism of fatty acids via the alpha- (137) and omega- (138) oxidation pathways is increased in diabetes.

Activity of different lipolytic enzymes is affected in diabetes due to relative or direct insulin deficiency. For example, lipoprotein lipase, present on the endothelial surface of adipose tissue, skeletal muscle, and cardiac muscle, is an insulin-dependent enzyme. Its activity is suppressed in insulin deficiency/resistance stages. As a result, the catabolism of triacylglycerol-rich lipoproteins is reduced and the levels of VLDL are increased in diabetes as was described earlier. The activity of lipase present inside adipose cells also is dependent on insulin. However, it is regulated by mechanisms different from those that control activity of lipoprotein lipase. Intracellular lipase is stimulated by cyclic-AMP. Since insulin causes a decrease in cyclic-AMP, cellular levels of this nucleotide are elevated in insulin deficiency/resistance stages. As a consequence, cellular lipolytic processes are activated in diabetes (132). Moreover, catecholamines, cortisol, glucagon, and growth hormone, which are markedly elevated in diabetes, may enhance further cellular lipolysis by stimulation of cyclic-AMP formation (139). Such metabolic disturbances result in an increased release of fatty acids from cells and contribute to elevated levels of serum free fatty acids in diabetic subjects.

Insulin effectively stimulates cellular lipid synthesis. A lack of this mechanism in insulin deficiency or resistance stages may contribute further to an imbalance between processes of lipolysis and lipogenesis.

5.6 Role of Dietary Fatty Acids in Diabetes

Generally, it is accepted that the dietary lipid profile influences the composition of cellular polar lipids. Therefore, dietary fats may affect the cellular membrane lipid composition resulting in altered membrane function. In fact, dietary fat induced changes in fatty acid composition of the adipocyte plasma membranes in diabetic rats (140). Such changes resulted in altered insulin binding and insulin responsiveness in these animals. A diet with a high ratio of polyunsaturated to saturated fatty acids (P/S = 1.0), where linoleate (18:2*n*6) was a major PUFA, increased levels of cellular membrane PUFAs both in control and diabetic rats. Moreover, it prevented the decrease in arachidonic acid levels in diabetic animals and increased membrane insulin binding in control rats. A similar diet (i.e., high in P/S ratio) also increased insulin-stimulated glucose transport and lipogenesis in diabetic animals. Dietary PUFAs of the *n*-6 family also partially normalized glucose and glycosylated hemoglobin levels in diabetic persons (140). Moreover, *n*-6 PUFAs may have preventive effects on some diabetic complications such as diabetic retinopathy and polyneuropathy (141).

However, several reports showed deleterious effects of dietary *n*-6 PUFAs and, in contrast, beneficial influences of diets deficient in these fatty acids on the development of diabetes. Feeding rats with high-fat diets enriched with saturated, monounsaturated, or *n*-6 PUFAs equally led to severe insulin resistance in liver and peripheral tissues (142). A diet deficient in *n*-6 PUFAs decreased the incidence of spontaneous diabetes in rats. It has been hypothesized that decreased eicosanoid and leukotriene levels, and thus diminished autoimmune and/or inflammatory mechanisms, in essential fatty acid-deficient animals may be responsible for such effects (143).

A diet high in fish oil (i.e., containing *n*-3 PUFAs) has several beneficial effects on the pathogenesis of diabetes. For example, *n*-3 PUFAs were found to increase insulin-dependent glucose transport across cellular membranes. Likewise, long-chain *n*-3 PUFAs (eicosapentaenoic, 20:5*n*3, or docosahexaenoic, 22:6*n*3) prevented the development of insulin resistance in rats fed a diet enriched with fatty acids of the *n*-6 family. Similar effects were exerted by α-linolenic acid (18:3*n*3) in animals fed a diet high in saturated fat (142). Fish oil also may prevent other metabolic-disturbances in diabetes. Fish oil lowers platelet thromboxane production and platelet aggregation, decreases blood viscosity, and lowers blood pressure in diabetic persons (144). However, several studies also report that serious deleterious effects are produced when *n*-3 PUFAs are consumed by diabetic individuals. Although *n*-3 PUFAs may prevent insulin resistance, they were shown to increase glucose and glycosylated hemoglobin levels. Moreover, they may impair insulin secretion and the overall glycemic control (145).

5.7 Nutritional Recommendation in Diabetes

The main goals of dietary management of both IDDM and NIDDM include maintenance of adequate plasma glucose levels and prevention of complications (146). Since lipid abnormalities in diabetes may contribute to the development of atherosclerosis (i.e., the main cause of diabetic mortality), the management of lipid status is especially important in this disease. Dietary therapy may, at least in part, correct abnormal lipid profiles and lower the risk for cardiovascular diseases. In general, the dietary guidelines for diabetic patients should be similar to the ones for hyperlipidemic subjects. According to both the American Diabetes Association (147) and the British Diabetic Association (148), diets in the treatment of diabetes should be restricted in fat. Reduction of dietary fat to or below 30% of total caloric intake is one of the most important dietary recommendations in diabetes. A low-fat diet associated with a decrease in saturated fat not only may prevent the development of cardiovascular complications but also may improve glucose and lipid profiles in diabetic persons. Moreover, to maintain the appropriate energy requirement of a low-fat diet, dietary complex carbohydrates are increased to approximately 50–60% of total calories. However, considerable evidence has shown that diets low in fat but high in complex carbohydrates may increase triacylglycerol and VLDL and lower HDL levels. Therefore, moderate dietary enrichment with both carbohydrates and monounsaturated fatty acids has been proposed recently as an alternative dietary treatment for diabetes (149).

There are some specific recommendations for particular types of diabetes. Since IDDM affects mostly children and adolescents, one of the most important considerations in the treatment of this type of diabetes is to maintain normal growth and development. Therefore, diets for children with IDDM must meet the proper energy requirements (150). In contrast, specific dietary treatment of Type IIb diabetes includes energy restriction and weight loss as the first therapeutic step. Weight loss often results in reduction of hyperglycemia and in an improvement of sensitivity to insulin. Only when weight loss fails to improve the glucose blood levels, therapy with an oral hypoglycemic agent (preferentially biguanide, because sulfonylurea makes it difficult to lose weight) should begin (151).

5.8 Conclusion

Excessive fat intake is associated with the development of obesity, which is a risk factor for Type IIb diabetes, the most common form of this disease. Disturbances in lipid metabolism are common in all types of diabetes and play a crucial role in the development of atherosclerosis, which accounts for approximately 90% of all diabetic mortalities. Modulation of dietary fats may improve the blood lipid and glucose profiles independently of the type of diabetes. At present, the best dietary recommendations for diabetic patients include reduction in dietary fat and an increase in complex carbohydrates and soluble fiber. Low-fat diets and energy restriction are the most effective treatment approaches for controlling Type IIb diabetes.

ROLE OF DIETARY FAT IN CANCER

6.1 Introduction

Cancer is one of the major causes of death in industrialized countries. It is often equated with neoplasm (or tumor), which is defined as an autonomous, uncontrolled, and unlimited proliferation of cells. However, neoplasms may be benign or malignant. Benign neoplasms extend without invading local tissues or spreading to other sites. Malignant neoplasms are synonymous with the term cancer. They grow by invading surrounding tissues and metastasize to other tissues and organs.

Dietary factors may contribute to 35% of all cancers (152). Among them, excessive intake of fat is recognized as a crucial factor in the increased risk for cancer. There is evidence that a high-fat diet may increase the risk of several common types of cancers including breast, colon, pancreatic, and prostatic cancer. On the other hand, a low-fat diet rich in certain vitamins and minerals decreases the risk of these cancers.

6.2 Carcinogenesis

Several theories have been proposed to explain the mechanisms of carcinogenesis. They include gene mutation, viral transformation, epigenic changes, and cell selection (153). However, it is generally accepted that carcinogenesis is a two-stage process that can be divided into initiation and promotion (Figure 7.7). Initiation is the first essential step in carcinogenesis. It is a one-time, irreversible event, mediated by carcinogens, which induces some permanent mutation, such as alteration in DNA structure. Environmental carcinogens often require metabolic activation to become carcinogenic and microsomal mixed-function oxidase systems are primarily involved in this process. Since some carcinogens may provide completed neoplastic development without any

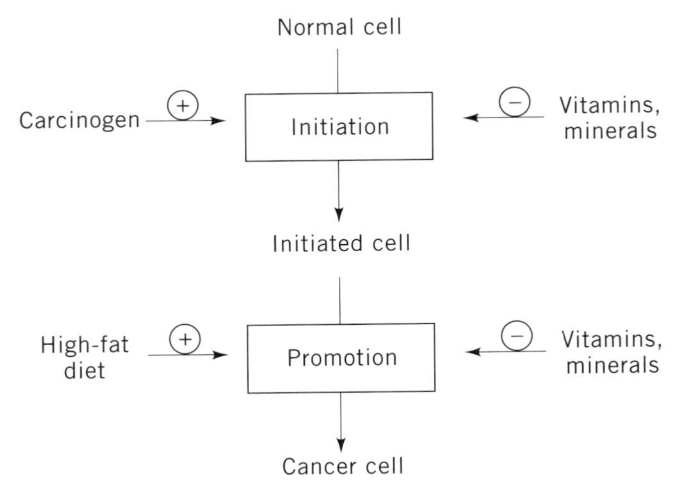

Figure 7.7 Role of nutrients, especially fat, in carcinogenesis. Dietary fat/fatty acids may influence cancer development primarily by enhancing the promotion stage of carcinogenesis. Some vitamins and minerals have anticarcinogenic properties by inhibiting the various phases of cell transformation.

other treatments, initiation is sometimes a sufficient event in carcinogenesis. However, more often, promotion is required for a full carcinogenic process. This process may be reversed partially and involves alterations in epigenic regulations that stimulate clonal proliferation of the mutant cells. Without promotion, transformed cells rarely express their neoplastic potential. To produce a cancer, promoters must act after the initiator. Factors such as dietary fats (153, 154) and lipid peroxides (13) actively may promote carcinogenesis.

6.3 Role of Dietary Fat in Breast Cancer

The hypothesis that dietary fat may contribute to the development of breast cancer began with the classical studies of Tannenbaum in 1942 (155). His original studies, showing that chemically induced mammary tumors developed more readily in rodents fed high-fat diets as compared to low-fat diets, have been proven consistently in a number of experiments (156). Higher fat intake increases mammary tumor incidence independent of enhanced caloric intake (157). The type of fat may strongly influence chemically induced mammary tumor development. In rodents, diets enriched with PUFAs of the *n*-6 family effectively shortened the latency period for tumor appearance, promoted growth and increased the incidence of mammary tumors as compared to diets with a high content of saturated fatty acids (156). It appears that these effects apply specifically to linoleic acid. When dietary linoleic acid content exceeded 4–5% of total calories, any additional fat (saturated or unsaturated) linearly increased chemically induced tumor incidence (158). It was demonstrated that mammary tumors developed more readily when a diet based on saturated fats, such as coconut oil or beef tallow, was enriched with linoleic acid (159). Dietary linoleic acid also was observed to enhance the metastasis of mammary tumors to the lung in rats (160). On the contrary, dietary fish oil, which is a source of *n*-3 PUFAs, exerted inhibitory effects on mammary tumor development. It was demonstrated that high doses of fish oil were potent in inhibiting a promoting effect of linoleic acid on mammary tumorigenesis (161). Studies examining the effects of monounsaturated fatty acids on mammary carcinogenesis have contradictory results. Moreover, saturated fatty acids alone appeared to have no effect on chemically induced mammary cancer development in experimental animals (156).

Several mechanisms have been proposed to explain the promoting effects of *n*-6 PUFAs (mainly linoleic acid) on mammary carcinogenesis. They may be related to effects exerted by these fatty acids on the endocrine system, the immune system, eicosanoid metabolism, cell membrane fluidity, cell–cell interactions, or lipid peroxidation processes (162). Also it has been suggested that the carcinogenic effects of unsaturated fatty acids may involve primarily mammary adipose tissue, which may further affect the mammary parenchyma (163).

Although the crucial role of dietary fat on mammary carcinogenesis has been proven in animal studies, the corresponding role of lipids in human breast cancer remains controversial. No consistent data has been published regarding the relationship between serum cholesterol or triacylglycerol levels and incidence of breast cancer. However, several reports found an inverse relationship between serum cholesterol levels and breast cancer in premenopausal women (164). However, it is possible that in young women high estrogen levels rather than dietary factors may be responsible for this phenomenon.

Epidemiological multicountry studies showed a strong positive correlation between the percentage of calories from dietary fat and breast cancer incidence and mortality. In countries where dietary fat accounts for less than 30% of calories, the breast cancer mortality is much lower than in countries where fat intake is higher (165). Moreover, there is evidence that migration from countries with low-fat intake to countries with high-fat intake is connected with an increased incidence of breast cancer. In addition, time–trend studies in Japan showed that enhanced fat intake during the last 20 years has been associated with increased breast cancer (166). These results are supported further by a recently published study on fat intake and mortality due to breast cancer. In this study, data was collected from 30 countries, between 1961 and 1986. It was shown that breast cancer mortality was related positively to dietary intake of animal (saturated) fat and vegetable oil (the main source of linoleic acid) but inversely correlated to fish oil consumption (167).

Obesity is associated with breast cancer incidence. A high dietary fat intake contributes to the development of obesity. On the other hand, weight loss by dietary intervention was shown to decrease the risk for developing breast cancer (164).

In contrast to results obtained in intracountry studies, prospective cohort studies and case control studies failed to provide strong evidence for a relationship between dietary lipid content and the development of breast cancer. However, methodological problems connected with these types of studies, such as small differences in dietary fat intakes, or difficulties in assessing the diets of individuals over a long period of time, may be responsible for inconsistent data. The effects of individual types of fatty acids on the incidence of breast cancer in humans also are not clear. In contrast to animal studies, human studies showed no correlation between breast cancer and PUFA intake (165). These contradictory results may be explained by the fact that the PUFA content of most human diets exceeds the threshold for maximum effect on cancer promotion. Therefore, breast cancer mortality most often is positively correlated with total fat intake as well as with the intake of saturated and monounsaturated but not with polyunsaturated fatty acids (165).

6.4 Role of Dietary Fat in Colon Cancer

Colon cancer is one of the most common cancers in the United States with more than 150,000 new cases and 60,000 deaths yearly (168). There is evidence that colon cancer is strongly associated with diet, particularly with diets high in total fat and low in fiber. Generally it is accepted that dietary fat may affect the promoting phase of cancer development, and this influence may depend on fatty acid composition. Animal studies have shown that diets high in saturated fat promote chemically induced colon cancer. Moreover, a number of studies also have shown that fat of vegetable origin, which is high in linoleic acid, may positively influence colon cancer development (169). However, such effects were not confirmed in other reports (170). This discrepancy may be explained by the fact that linoleic acid can influence colon cancer during different developmental stages of carcinogenesis. It was shown that diets enriched with linoleic acid were effective in increasing colon carcinogenesis only during the promotion phase. However, when administered in the initiation phase, a linoleic-acid-enriched diet had no influence on cancer development. On the contrary, a high level of dietary saturated fat enhanced colon carcinogenesis at both phases.

The fact that in animal studies *n*-6 PUFAs, and especially linoleic acid, were more effective in promoting colon carcinogenesis than were saturated fatty acids may be explained by the role of the individual fatty acids in prostaglandin metabolism. Fatty acids of the *n*-6 family are precursors of the series-2 prostaglandins, for example, prostaglandin E_2, which are implicated in tumor promotion (171). There is evidence that inhibition of prostaglandin E_2 biosynthesis by indomethacin inhibits colon tumor growth (172).

Various mechanisms have been proposed to explain the effects of dietary fats on colon tumorigenesis. It has been suggested that fat-stimulated development of colon cancer is mediated by alterations in bile acid metabolism. Low-fat diets maintain primarily aerobic bacterial populations in the large bowel. With a high-fat diet, the normal bacterial flora is replaced by an increased number of anaerobic bacteria. Such an exchange may facilitate a conversion of the excreted bile acids into secondary bile acids, that is, deoxycholic acid and lithocholic acid, which have tumor promoting effects. It is possible that the activation of colon epithelial ornithine decarboxylase and/or protein kinase C is involved in this process. Dietary cholesterol, which may increase bile acid and sterol excretion, also has been proposed to promote colon cancer (169). Moreover, it has been suggested that certain long-chain fatty acids, such as stearic acid and oleic acid, may promote colon carcinogenesis by formation of free ionized fatty acids, which are toxic to the colon epithelium (173). Meat/high-fat diets also may mediate colon tumorigenesis either by arylamines (from meat) and/or by promoters such as hydroxymethylfuraldehyde generated during cooking at a high temperature (174,175).

In contrast to saturated fats, evidence suggests that dietary *n*-3 PUFAs may inhibit colon tumorigenesis. Several mechanisms may be implicated in this process. Fatty acids of the *n*-3 family are known to interfere with *n*-6 PUFA formation and decrease arachidonic acid concentrations, which lowers PGE_2 production. The PGE_2 is believed to be involved in the promotion of tumor growth (176). Cancer cells, including colon tumor cells, were shown to have a high requirement for cholesterol, which is necessary for membrane synthesis (175). Therefore, it is possible that *n*-3 PUFAs may inhibit tumor growth by decreasing the cholesterol level.

An association between dietary fat and colon cancer development was examined in several case control and cohort studies. The majority of them showed a positive correlation between meat or fat intake and colon cancer incidence. Studies suggested that high intakes of beef, red meat, or animal fat increased the risk for colon cancer. However, there are several case control studies that failed to confirm the role of meat and/or saturated fat in colon carcinogenesis, and one report even showed a reverse correlation (177). Moreover, the promoting role of linoleic acid in colon cancer was not confirmed in case control and cohort studies (178). However, it is possible that methodological difficulties discussed earlier may be responsible for these results.

6.5 Role of Dietary Fat in Carcinogenesis at Other Sites

The relationship between dietary fat and skin cancer has been examined using mice. A number of studies showed a positive correlation between high-fat diets and skin tumor incidence induced by either various polycyclic hydrocarbons or ultraviolet light (179, 180). Dietary fat slightly inhibited the initiation stage of skin carcinogenesis but markedly enhanced the promotion phase (181). However, the relationship between dietary linoleic acid and skin tumor development appears to be different from those observed in mammary or colon cancers. Mouse skin tumors required approximately 4.5% dietary linoleic acid for a maximum growth. However, higher intakes of this fatty acid exerted inhibitory effects on skin carcinogenesis induced by 1,2-dimethylbenz[*a*]-anthracene (182). Moreover, no protective effects of *n*-3 PUFAs were observed in skin tumors when 12-*O*-tetradecanoylphorbol-13-acetate or benzoyl peroxide were used as promoters (183). Discrepancies between effects of linoleic acid or *n*-3 fatty acids on skin cancer as compared to those in mammary gland or colon cancers may be due to differences in deposition and/or metabolism of fatty acids in various organs.

Effects of dietary fat on the development of liver cancer were examined using different carcinogens in a number of animal studies. In general, a positive correlation between total fat intake and hepatocarcinogenesis was observed. Furthermore, PUFAs were more effective in enhancing liver cancer development than were saturated fatty acids (184). Dietary fatty acids were shown to affect primarily the initiation stage of carcinogenesis of liver tumors (185).

A majority of studies examining pancreatic carcinogenesis demonstrated a positive correlation between cancer development and high-fat diets. Such results were obtained in both animal and human studies. However, conflicting results were reported about the type of dietary fat, which may enhance pancreatic cancer. In *N*-nitrosobis(2-oxopropyl)amine-induced pancreatic cancer in hamsters, a promoting effect of both saturated and unsaturated fats was observed (186). However, diets enriched with saturated fat (beef tallow) exerted more carcinogenic effects than diets based on polyunsaturated fat (corn oil). Similar results were obtained in azaserine-induced pancreatic cancer in rats. A diet enriched with lard more markedly enhanced carcinogenesis compared to a diet supplemented with corn oil (187). Other reports have shown promoting effects of PUFAs and specifically linoleic acid on pancreatic carcinogenesis (188). The intake of several nutrients including beef, pork, poultry, seafood, and dairy products was positively related to pancreatic cancer in humans. However, other reports failed to find a significant correlation between the development of pancreatic cancer and dietary fat intake. Moreover, a decreased risk of pancreatic cancer was related to dietary consumption of butter or margarine (188).

There is less evidence that high-fat diets are related to cancer development at sites other than those discussed. However, studies suggest some correlations between fat consumption and bladder (189), prostate (190), endometrial (191), esophagus (192), lung (193), kidney (194), and ovarian (179) cancers. It was shown that subjects with high intakes of saturated fatty acids had an enhanced risk for bladder and endometrial cancer. A relationship between prostate cancer and fat intake was demonstrated in multicountry and case control studies. Prostate cancer mortality was the highest in countries with the highest fat consumption. Moreover, total fat and saturated fat intakes were found to be risks for the development of prostate cancer in case control studies (195).

6.6 Conclusion

Excessive consumption of dietary fat may enhance the risk for several cancers, particularly breast and colon. However, the potential mechanisms of the cancer-promoting effects of different fats have not been identified completely. Moreover, the influence of the various types of fat on carcinogenesis may differ depending on the organ affected. Based on recent research findings, the American Cancer Society and the National Cancer Institute have recommended that lowering fat intake to 30% of total consumed calories can prevent carcinogenesis.

However, the human diet is complex and there is no single dietary factor identified as being responsible for cancer promotion. Therefore, it may be prudent not only to reduce the total fat content of the diet but also to include sufficient amounts of potentially anticarcinogenic nutrients such as vitamins (A, C, E, and carotenes), minerals (calcium, zinc, and selenium), and fiber (13, 196, 197).

ROLE OF DIETARY FAT IN PROGRESSIVE RENAL DISEASES

7.1 Introduction

Kidneys play a crucial role in maintaining body homeostasis. Major functions of the kidneys include regulation of fluid volume and osmolality, excretion of waste products, regulation of electrolyte and acid–base balance, regulation of blood pressure, and hormonal activity. The functional unit of the kidney is the nephron, which contains the glomerulus, proximal tubule, loop of Henle, distal tubule, and collecting duct system. Filtration of blood in the glomerulus, and thus formation of the plasma ultrafiltrate, is the first step in production of urine. Since a number of kidney diseases result in glomerular dysfunction, measurement of glomerular filtration rate (GFR) is one of the most relevant tests of renal function. The most widely used measure of GFR is creatinine clearance. Except for protein content, the normal composition of the glomerular ultrafiltrate, is identical to plasma. Normal urine contains only trace amounts of protein (mainly albumin) which is not detectable by routinely used laboratory tests. Injury to the glomerulus may result in increased glomerular permeability and, consequently, in elevated protein levels in the urine.

Several kidney components are susceptible to injury mediated by a number of immune and nonimmune mechanisms. The nonimmune mechanisms involved in the damage to the glomeruli include increased glomerular capillary pressures and flow, enhanced coagulation, altered prostacyclin production, glomerular epithelial cell damage, and abnormal mesangial processing of phlogogenic macromolecules (198). Lipids also may play an essential role in the injury process as kidneys appear to be the major target organ for lipid-mediated injury. Foam cells and lipid deposition in the glomeruli are seen in the most progressive human glomerular diseases. Moreover glomerular injury develops during different forms of hyperlipidemia, both those induced by dietary fat as well as those of genetic origin (199,200).

7.2 Role of Dietary Fat in the Pathogenesis of Glomerular Injury

The most significant information concerning the role of lipids in the pathogenesis of glomerular diseases was obtained from animal studies. The role of dietary fats in the development of glomerular injury has been shown in rats (201), guinea pigs (202), and rabbits (203) fed high-cholesterol diets. Animals developed glomerular injury associated with lipid deposition when fed cholesterol-supplemented diets. The effect of hyperlipidemia on the development of glomerulosclerosis has been studied in obese Zucker rats (203). In this rat strain, obesity is an autosomal recessive trait, and the animals spontaneously develop, at an early age, hyperlipidemia, Type II diabetes, and progressive renal injury. It was documented in obese Zucker rats that glomerular damage is correlated positively with the levels of triacylglycerols and inversely with the amounts of PUFAs in the renal cortex. The hypothesis that lipids may be responsible for kidney injury was supported by the observation that pharmacological normalization of hyperlipidemia ameliorated glomerular disturbances either in obese Zucker rats (204) or in rats with unilateral nephrectomy (205).

The mechanisms of lipid-mediated glomerular injury are not clear; however, they partially resemble those in the development of atherosclerosis. Lipid deposition in the kidney is associated with the infiltration of the tissue with monocytes/macrophages (205), the process believed to be one of the first steps in atherogenesis (38). Foam cells frequently present during the progressive kidney diseases may be analogous to those present in atherosclerotic vessels. From studies of atherogenesis it is known that foam cells are formed from monocytes/macrophages and smooth muscle cells by uptake of modified LDL by scavenger receptors (38). It has been proposed that monocytes/macrophages that infiltrate the kidney may be involved in a similar process. Moreover, glomerular mesangial cells structurally resemble modified smooth muscle cells. Mesangial cells are susceptible to platelet-derived growth factor, a potent mitogen for smooth muscle cells. Stimulation of smooth muscle cell proliferation by platelet-derived growth factor is another key step in the induction of atherosclerosis in the vessel walls (199,200). Monocytes/macrophages and endothelial cells are potent sources of this growth factor (38). Furthermore, similar to the atherosclerotic vessels, glomerular lipid deposition may be not only intracellular in the form of foam cells but also extracellular. The glomerular mesangial matrix contains anionic glycosaminoglycans, which strongly bind LDL and VLDL, lipoproteins with cationic properties (199,200).

It has been hypothesized that the pathogenesis of progressive kidney injury not only resembles atherogenesis but also that these two processes may be closely related. To support this hypothesis, it was demonstrated that the degree of atherosclerosis correlated well with the amount of age-associated glomerular injury (206).

Hypercholesterolemia also may affect kidney structure by other mechanisms, including increasing glomerular capillary pressure. Glomerular hypertension is a well-known mechanism of the induction of glomerulosclerosis. It has been proposed that hypercholesterolemia may influence glomerular hemodynamics by changes in plasma viscosity or by alteration of erythrocyte membrane (207).

7.3 Disturbances in Lipid Metabolism in Chronic Renal Failure

Progressive kidney diseases, such as chronic renal failure (CRF) (208) and nephrotic syndrome (209), are associated with disturbances in lipid metabolism. CRF is characterized by the gradual and progressive decrease in renal functions manifested primarily by reduction in GFR. Treatment of the terminal stages of this disease requires dialysis and/or kidney transplantation. CRF develops as a result of almost all kidney diseases; but chronic glomerulonephritis, chronic obstructive uropathy, and hypertension are the most common causes.

Lipid abnormalities are characteristic of advanced CRF. They include hypertriacylglycerolemia associated with the enrichment of all lipoprotein fractions with triacylglycerols, and the redistribution of cholesterol from HDL to VLDL and intermediate-density lipoproteins (IDL) fractions. Moreover, βVLDL are frequently determined in patients with renal failure. Although total plasma cholesterol and LDL-C often are normal, decreased levels of HDL cause the increase in the ratio of LDL-C to HDL-C in CRF (208,210).

Even more significant than abnormal lipoprotein lipid composition are disturbances in apolipoprotein profiles in CRF patients. These changes are detected even in the early stages of CRF when no other lipid abnormalities are detected. The most characteristic changes in CRF are increased apolipoprotein C-III (apoC-III) and decreased HDL apoA-I and apoA-II. Moreover, in patients with more advanced renal failure, the elevated apoC-I and apoC-II as well as normal or slightly increased apoB and apoE are observed. Consequently, the ratios of apoA-I/apoC-III and apoA-I/apoB are reduced and the ratio of apoC-III/apoE is elevated. The changes in the total plasma apolipoprotein levels are associated with their altered distribution among lipoproteins. Apolipoproteins B, C, and E are decreased in HDL but increased in VLDL and IDL fractions. The presence of intestinal apolipoproteins, such as apoB-48 and apoA-IV are frequently increased in CRF patients. ApoB-48 is detected mainly in VLDL, and apoA-IV in VLDL, IDL, and LDL (208,210).

A growing body of evidence suggests that reduced activities of the lipolytic enzymes, lipoprotein lipase and hepatic triacylglycerol lipase, are responsible, at least in part, for lipid abnormalities in CRF patients. A decreased activity of vascular lipoprotein lipase may result in a decreased catabolism of the newly secreted chylomicrons and VLDL. A decreased activity of hepatic lipoprotein lipase may cause a reduced clearance of lipoproteins and chylomicron remnants. The activities of these enzymes are diminished even in the mild form of CRF (211). Several mechanisms are suggested for the decreased activity of plasma lipolytic enzymes. Functional deficiency of insulin, insulin resistance (212), increased parathyroid hormone (213), the presence of lipoprotein lipase inhibitor (214), and the long-term effect of heparinization in dialyzed patients (215), all are suggested as mechanisms responsible for this enzymatic defect. Moreover, the activity of lipoprotein lipase is stimulated by apoC-II and inhibited by apoC-III. Therefore, the increase in the ratio of apoC-III/apoC-II, constantly observed in CRF patients, may contribute to the reduced activity of this enzyme (208). The role of lipolytic enzymes in the pathogenesis of the development of the abnormal lipid profiles in CRF is supported by the observation that treatment with gemfibrozil, a drug that stimulates lipase activity, normalized lipid profiles in uremic patients (216).

The changes in apolipoprotein compositions and their redistribution among different lipoproteins also may influence lipoprotein catabolism in CRF. It was suggested that reduced levels of apoA-I and apoA-II in HDL fractions may decrease the activity of lecithin cholesterol acyltransferase (208). This enzyme is involved in HDL metabolism and particularly in reverse transport of cholesterol from the peripheral tissues to the plasma and liver.

In addition to the diminished catabolism of triacylglycerol-rich lipoprotein, there is evidence that patients with CRF have increased triacylglycerol production (217). In addition, the incorporation of free fatty acids into adipose tissue also is decreased in this disease (208).

7.4 Disturbances in Lipid Metabolism in Nephrotic Syndrome

Nephrotic syndrome refers to the clinical state characterized by the increased permeability of the glomerular membrane, which is detected by proteinuria in excess of 3.5 g/day. Other characteristic clinical features of nephrotic syndrome include hypoproteinemia, edema formation, and hyperlipidemia. However, in contrast to CRF, the glomerular filtration rate often is normal in this syndrome. Lipid abnormalities characteristic of nephrotic syndrome include increased levels of LDL, VLDL and IDL. Plasma total HDL levels may be normal or only slightly decreased, however, there are significant disturbances in the HDL subclasses. HDL_2, which is involved mainly in the reverse cholesterol transport from peripheral tissues, usually is markedly reduced and HDL_3, which lacks the protective effect of HDL_2, often is elevated. Moreover, triacylglycerol and cholesterol levels are elevated in all classes of lipoproteins in patients with nephrotic syndrome (209, 218).

There is evidence that the underlying mechanisms of hypercholesterolemia and hypertriacylglycerolemia in patients with nephrotic syndrome include both the increased synthesis of lipoprotein (219) in the liver and their diminished clearance from plasma (220). It was documented that a decrease in serum oncotic pressure caused by proteinuria and hypoalbuminemia may stimulate lipogenesis. Furthermore, catabolism of lipoproteins, mainly VLDL and chylomicrons, is reduced in nephrotic syndrome. It is possible that changes in apolipoprotein structures, which influence activity of plasma lipolytic enzymes, are responsible for this defect. There also is a correlation between increased plasma cholesterol or triacylglycerol levels and elevated renal clearance of albumin. Therefore, it has been hypothesized that the altered glomerular permeability may result in the loss of lipid regulatory substances (209).

7.5 Dietary Treatment of Progressive Renal Disease

Since the underlying mechanisms of lipid abnormalities in CRF and nephrotic syndrome are connected with both enhanced lipid synthesis and their decreased catabolism, even slight modifications of dietary lipids may affect dramatically the lipid profiles observed in these diseases. The beneficial effects of dietary protein restriction in the treatment of renal damage have been known for over 50 years. It has been established that low-protein diets may correct serum lipid abnormalities in CRF patients (92). The mechanism of this effect is not known; however, low-protein diets decrease cholesterol levels and increase the polyunsaturated/saturated fatty acid ratio. There is strong evidence that reduction in dietary cholesterol and carbohydrates significantly improves the overall lipid profile in progressive renal diseases. For example, such a diet was demonstrated to decrease the rate of triacylglycerol production and triacylglycerol levels in CRF patients (221).

A number of experiments, supported by some clinical trials, suggest that not only quantity but also quality of dietary lipids may influence the progression of chronic renal diseases. Several beneficial effects of diets enriched with unsaturated fatty acids were observed in animals with experimental kidney damage. Diets supplemented with linoleic acid corrected hyperlipidemia, reduced systemic hypertension, decreased proteinuria, and prevented progressive glomerulosclerosis in rats with partial nephrectomy (222). Similarly, reduction in serum total cholesterol, LDL cholesterol, and apoB levels, associated with decreased proteinuria were observed in nephrotic patients consuming soy-based vegetarian diets (223). Moreover, dietary supplementation with eicosapentaenoic acid (20 : 5*n*3) was shown to correct hyperlipidemia, ameliorate hypertension, improve blood viscosity, and decrease platelet aggregation in patients with CRF and in animals with induced progressive renal failure

(224). However, several other studies failed to confirm the beneficial effects of PUFAs on kidney functions (225). Therefore, the role of PUFAs in the treatment of progressive renal diseases remains controversial. For example, the rationale for using fish oil in kidney damage is based on the observation that *n*-3 PUFAs may decrease lipid levels, especially triacylglycerols, increase HDL-C, mainly HDL_2 cholesterol, decrease blood pressure, as well as improve blood rheology and insulin sensitivity. However, the triacylglycerol-lowering effect of fish oil is due to inhibition of hepatic triacylglycerol synthesis and the secretion of smaller VLDL molecules. The smaller VLDL are more rapidly converted to LDL, and therefore LDL-C levels may increase in patients given diets supplemented with fish oil. In addition, fish oil fatty acids interfere with *n*-6 PUFA formation and thus affect the biosynthesis of eicosanoids to change the usual kidney profile of these compounds. For example, eicosapentaenoic acid was shown to decrease kidney levels of both thromboxane A_2 and prostaglandin E_2. Although the decreased thromboxane A_2 content might improve kidney functions, the lowered prostaglandin E_2 level outweighed this potential beneficial effect. It is also known that lipid peroxidation is involved in the pathogenesis of CRF, and unsaturated fatty acids are potent stimulators of this pathway (226). These facts should not be ignored when diets supplemented with PUFAs are considered in the therapy of progressive kidney disease.

7.6 Conclusion

Progressive renal diseases are associated with disturbances in lipid metabolism, and a high-fat diet may participate in the development of these diseases. Moreover, dietary fatty acids through modification of serum lipid profiles may exert nephrotoxic effects. Some evidence would suggest that diets low in protein and total fat may be most beneficial in the treatment of patients with progressive renal diseases. However, the specific recommendations, concerning the dietary fatty acid profile, the type of PUFAs, and the ratio of polyunsaturated/saturated fatty acids in a diet need to be clarified by conducting further human studies.

OTHER HEALTH EFFECTS OF DIETARY FAT: IMMUNE FUNCTION AND BONE HEALTH

8.1 Role of Dietary Fat in Immune Function

As described in the previous sections of this chapter, the type and amount of dietary fat profoundly effects physiology, metabolism, and risk for certain diseases. To a great extent, dietary fat impacts several physiological systems in the body by altering tissue PUFA composition to modulate eicosanoid biosynthesis, influence membrane–ligand interactions, affect cell permeability, and cell-to-cell communications (16,17,227,228).

It has long been recognized that the immune system is influenced by nutritional status. For example, essential fatty acids (EFAs) play an important role in modulating immune response. EFA deficiency can result in both enhanced cell-mediated and humoral immune responses (229). These effects can also be achieved by administering cyclooxygenase inhibitors (229). In general, nutritional intervention with dietary fat to modulate eicosanoid production is primarily mediated by competition of PUFAs for desaturases/elongases, inhibition of cyclooxygenase, or production of different eicosanoids (227). Eicosanoid biosynthesis can be altered with dietary *n*-3 PUFAs to lower the concentration of arachidonic acid in membrane phospholipids, which reduces the production of 2-series prostaglandins.

More recent attention has focused on reducing arachidonic acid and elevating *n*-3 PUFAs in immune cells to produce more desirable biological responses (36,227). Immunocompetent cells are prolific synthesizers of eicosanoids, which mediate several of their effects in immune function (227). In studies with rats, dietary sources of *n*-3 PUFAs in linseed oil (230) (α-linolenate) or fish oil (229) (20:5*n*3 and 22:6*n*3) lowered the amounts of arachidonic acid in tissue phospholipids as well as the production of eicosanoids derived from 20:4*n*6. Feeding linseed oil decreased the ratio of 20:4*n*6/18:2*n*6 in serum phospholipids and depressed prostanoid production from arachidonic acid. Lokesh and Kinsella (231) found that peritoneal macrophages isolated from mice and incubated with media containing eicosapentaenoic acid (20:5*n*3) or docosahexaenoic acid (22:6*n*3) produced significantly lower amounts of prostaglandin E_2 and thromboxane B_2 than macrophages incubated without these fatty acids. A subsequent experiment demonstrated that 20:5*n*3 was more effective than 22:6*n*3 in replacing arachidonic acid in macrophages, but 22:6*n*3 was more potent in reducing leukotriene biosynthesis from arachidonic acid (232). In another study, the production of 4-series leukotrienes and prostaglandin E were reduced in peritoneal cells isolated from mice fed *n*-3 PUFAs (233).

Splenic T-cell numbers from the offspring of mice fed different lipid sources increased proportionally with the level of dietary 18:2*n*6, but the numbers were inversely proportional to the level of *n*-3 PUFAs in the diet (234). In a different study, the concentrations of 20- and 22-carbon *n*-3 and *n*-6 PUFAs in spleen phospholipids of mice reflected the dietary ratios of 18:3*n*3 to 18:2*n*6 fed; however, the dietary treatments did not affect the proportions of T-lymphocyte helper and suppressor cells (235).

Eicosanoids can be considered as regulators of macrophage activity and macrophage–lymphocyte cooperation. The eicosanoids may act as autocrine or paracrine signaling agents to facilitate cell-to-cell communications between immune cells (227). For example, the macrophage mediator interleukin-1 (IL-1) stimulates prostaglandin E_2 release from several cell types (236), but prostaglandin E_2 may also regulate IL-1 expression (237). In another immunologic reaction, complement-mediated attack on macrophages initiates cellular eicosanoid production (238). A study conducted by Lokesh, Sayers, and Kinsella (239) suggests that modulation of eicosanoid biosynthesis enhances cytokine generation. Macrophages isolated from mice fed 20 : 5*n*3 in menhaden oil had lowered prostaglandin E_2 production and higher amounts of IL-1 and tumor necrosis factor when compared to macrophages from mice fed corn oil (239).

Dietary lipids have an important role in immune function, and eicosanoids produced by immunocompetent cells participate in immunoregulation. Although the relationship between dietary lipids and eicosanoids is firmly established (16,17,24,227), the effect of dietary lipids on cellular and humoral immunity is not clearly defined. However, existing evidence would suggest that opportunities are favorable for using dietary fat to modulate immune responses. Nutritional intervention to modify eicosanoid metabolism in immunocompetent cells may potentially enhance or depress immune reactions (227), and alter inflammatory responses in humans (36,240).

8.2 Role of Dietary Fat in Bone Biology

Dietary lipids play an important role in bone biology and in bone health. In the early 1970s, research on the calcifying cartilage of bone indicated that acidic phospholipids in matrix vesicles facilitated the formation of calcium complexes to initiate mineralization (241,242). At about the same time, prostaglandin E_2 was reported to have a powerful effect on bone resorption causing calcium release from bone mineral (243). Since then, considerable evidence has demonstrated that matrix vesicles are present in all mineralizing bone tissues (241), and that prostaglandins act as local factors regulating both bone formation and resorption (29).

The events occurring during bone modeling in the young, and remodeling in the mature skeletal system, are controlled by several systemic and local factors acting in bone. The numerous growth regulatory factors in bone are prostaglandins (244), insulinlike growth factors (IGFs) (245), and cytokines (246), which are produced by chondrocytes, osteoblasts, and immune-type cells found in or associated with bone. The biosynthesis and secretion of the bone regulatory factors are induced by systemic endocrine hormones (parathyroid hormone and estrogens), vitamin $1{,}25(OH)_2D_3$ or by autocrine/paracrine signaling agents (prostaglandins and cytokines) within bone (29,244,245). The majority of these regulatory factors influence metabolic processes in bone to stimulate or inhibit matrix formation, mineralization, and resorption. Even though several localized factors act on bone cells, the prostaglandins seem to be a predominant group affecting bone since their biosynthesis in bone is induced by cytokines as well as systemic factors (29,244).

Investigations on the fatty acid composition of articular and epiphyseal cartilages of humans revealed that the concentrations of *n*-6 PUFAs are low and that Mead acid is high relative to their amounts in other tissues of the body (247). In a recent study, dietary lipids were reported to alter the fatty acid composition of articular and epiphyseal cartilages, chondrocytes, and matrix vesicles in growing chicks (248). Chicks consuming *n*-3 PUFAs had significantly elevated concentrations of 20:5*n*3 and total *n*-3 PUFAs in cartilage, chondrocytes, and matrix vesicles compared to those fed saturated fat or *n*-6 PUFA sources. However, the Mead acid content in cartilage tissues and cells of chicks was not affected by the EFA 18:2*n*6 levels in the diet (248). Since Mead acid (20:3*n*9) is normally present only during EFA deficiency, it is tempting to speculate why 20:3*n*9 is found in normal cartilages of young healthy individuals and animals. The unusual fatty acid composition of cartilage might be explained by the avascular nature of the tissue (247). On the other hand, cartilage may have a limited capacity to handle peroxidized fatty acids because supplemental vitamin E was observed to enhance bone formation rates in chicks (249).

Osteoporosis, a condition of decreased bone mass, is a major risk factor for osteoporotic fractures among the elderly. Osteoporosis affects millions of adults in the United States, and the annual expenditures for the treatment of osteoporotic fractures exceeds $10 billion. The present research on the pathogenesis of osteoporosis and prevention suggests that dietary supplementation of calcium will slow the loss of bone mineral; therefore, children are exhorted to maintain high calcium intakes to optimize consolidation of their skeleton. However, calcium is only a deterrent to osteoporosis and not a cure. A loss of coupling between bone resorption and formation in the remodeling of bone is hypothesized to contribute to osteoporosis (244,245). A loss of regulatory factors in bone is believed to cause the uncoupling (244,245), and factors produced in bone may be responsible for decreased bone formation (245). A lack of control on prostaglandin E_2 production (244) or decreased storage of IGFs in bone may contribute to the development of osteoporosis (245).

The effects of dietary fat on bone modeling revealed that saturated fat led to higher bone formation rates compared to unsaturated fat in growing chicks (250). Dietary fat appears to regulate the concentrations of IGFs in bone (250) and to effect *ex vivo* production of prostaglandin E_2 in bone organ culture (251). Prostaglandins are believed to modulate IGF production in bone cells, and prostaglandin E_2 is known to produce biphasic effects on bone formation; at low concentrations it stimulates formation, but at high concentrations prostaglandin E_2 inhibits bone formation (244). Dietary fat may prove to be useful in modulating prostaglandin and IGF concentrations in bone to benefit longitudinal growth and maintenance of bone by optimizing acquisition and preservation of skeletal mass.

Lipids perform important roles in calcifying tissues and in bone modeling/remodeling. Matrix vesicles (MV) are believed to be responsible for initiating *de novo* mineralization in most calcifying tissues (241,242), and membrane-bound acidic phospholipids located within the MV structure facilitate mineralization by forming lipid–Ca^{2+}–protein complexes (242). The events occurring during bone modeling and remodeling are orchestrated by complex interactions between systemic agents and local factors produced in bone. The prostaglandins are major participants in the local regulation of bone formation and resorption (29). The importance of dietary fat in bone health is just now being realized, and recent data suggests that bone metabolism can be modulated by dietary fat. Because dietary fat constitutes 30–40% of the food calories in Western diets, the type of fat consumed may influence metabolic and physiological processes controlling bone modeling in children and bone remodeling in adults.

REFERENCES

1. F.D. Gunstone and F.A. Norris, *Lipids in Foods,* Pergamon Press, Oxford, U.K., 1983.
2. J.F. Mead, R.B. Alfin-Slater, D.R. Howton, and G. Popjak, *Lipids: Chemistry, Biochemistry, and Nutrition,* Plenum Press, New York, 1986.
3. E.A. Emken, *Annu. Rev. Nutr.* **4,** 339 (1984).
4. J.D. Hay and W.R. Morrison, *Biochim. Biophys. Acta* **202,** 237 (1970).
5. M. Sommerfeld, *Prog. Lipid Res.* **22,** 221 (1983).
6. M.G. Enig and co-workers, *J. Am. Coll. Nutr.* **9,** 471 (1990).
7. J.E. Hunter in *Fatty Acids in Foods and Their Health Implications,* Marcel Dekker, Inc., New York, 1992.
8. P. Nestel, M. Noakes, and B. Belling, *J. Lipid Res.* **33,** 1029 (1992).
9. R.P. Mensink and M.B. Katan, *N. Eng. J. Med.* **323,** 439 (1990).
10. J.T. Judd and co-workers, *Am. J. Clin. Nutr.* **59,** 861 (1994).
11. E.A. Newsholme, P. Calder, and P. Yaqoob, *Am. J. Clin. Nutr.* **57,** 738S (1993).
12. A.J. Vergroesen and M. Crawford, *The Role of Fats in Human Nutrition,* Academic Press, New York, 1989.
13. B. Halliwell, *Nutr. Rev.* **52,** 253 (1994).
14. U.S. Department of Health and Human Services, *Surgeon General's Report on Nutrition and Health,* publication 88-50201, U.S. Government Printing Office, Washington, D.C., 1988.
15. D.K. Spady, L.A. Woollett, and J.M. Detschy, *Annu. Rev. Nutr.* **13,** 355 (1993).
16. J.M. Iacono and R.M. Dougherty, *Annu. Rev. Nutr.* **13,** 243 (1993).
17. A.P. Simopoulos, *Am. J. Clin. Nutr.* **54,** 438 (1991).
18. W.E. Connor, M. Neuringer, and S. Reisbick, *Nutr. Rev.* **50**(4), 21 (1992).
19. G.J. Nelson and R.G. Ackman, *Lipids* **23,** 1005 (1988).
20. H. Carlier, A. Bernard, and C. Caselli, *Reprod. Nutr. Dev.* **31,** 475 (1991).
21. H. Schulz, *Biochim. Biophys. Acta* **1081,** 109 (1991).
22. T. Rezanka, *Prog. Lipid Res.* **28,** 147 (1989).
23. L.B.M. Tijburg, M.J.H. Geelen, and L.M.G. van Golde, *Biochim. Biophys. Acta* **1004,** 1 (1989).
24. W.E.M. Lands, *Annu. Rev. Nutr.* **11,** 41 (1991).
25. A. Voss, M. Reinhart, S. Sankarappa, and H. Sprecher, *J. Biol. Chem.* **266,** 19995 (1991).
26. V.W. Sardesai, *J. Nutr. Biochem.* **3,** 562 (1992).
27. S. Nicosia, and C. Patrono, *FASEB J.* **3,** 1941 (1989).
28. M. Rola-Pleszczynski, *J. Lipid Mediators* **1,** 149 (1989).

29. S.C. Marks and S.C. Miller, *Endocrine J.* **1,** 337 (1993).
30. W.L. Smith and L.J. Marnett, *Biochim. Biophys. Acta* **1083,** 1 (1991).
31. S. Yamamoto, *Biochim. Biophys. Acta* **1128,** 117 (1992).
32. R.J. Mayer and L.A. Marshall, *FASEB J.* **7,** 339 (1993).
33. I. Kudo, M. Murakami, S. Hara, and K. Inoue, *Biochim. Biophys. Acta* **1170,** 217 (1993).
34. G. Feuerstein and J.M. Hallenbeck, *FASEB J.* **1,** 186 (1987).
35. W.T. Cave, *FASEB J.* **5,** 2160 (1991).
36. J.E. Kinsella, B. Lokesh, S. Broughton, and J. Whelan, *Nutrition* **6,** 24 (1990).
37. B.A. Watkins, *Nutr. Res.* **10,** 325 (1990).
38. R. Ross, *N. Engl. J. Med.* **314,** 488 (1986).
39. T. Collins, *Lab. Invest.* **68,** 499 (1993).
40. P.A. Baeuerle, *Biochim. Biophys. Acta* **1072,** 63 (1991).
41. A. Keys, J.T. Anderson, and F. Grande, *Lancet* **2,** 969 (1957).
42. D.M. Hegsted, R.B. McGandy, M.L. Myers, and F.J. Stare, *Am. J. Clin. Nutr.* **17,** 281 (1965).
43. R.P. Mensink, *Am. J. Clin. Nutr.* **57,** 711S (1993).
44. M.A. Denke and S.M. Grundy, *Am. J. Clin. Nutr.* **56,** 895 (1992).
45. T.K.W. Ng and co-workers, *J. Am. Coll. Nutr.* **11,** 383 (1992).
46. P.N. Hopkins, *Am. J. Clin. Nutr.* **55,** 1060 (1992).
47. W.E. Connor and S.L. Connor, *Adv. Intern. Med.* **35,** 139 (1990).
48. H. Esterbauer, *Am. J. Clin. Nutr.* **57,** 779S (1993).
49. D. Steinberg and J.L. Witztum, *JAMA* **264,** 3047 (1990).
50. T.E. Carew, D.C. Schwenke, and D. Steinberg, *Proc. Natl. Acad. Sci. USA* **84,** 7725 (1987).
51. D.R. Janero, *Free Rad. Biol. & Med.* **11,** 129 (1991).
52. F.J. Kok and co-workers, *Atherosclerosis* **31,** 85 (1991).
53. J.M. Hodgson, M.L. Wahlqvist, J.A. Boxall, and N.D. Balazs, *Am. J. Clin. Nutr.* **58,** 228 (1993).
54. M. Aviram, *Atherosclerosis* **98,** 1 (1993).
55. K.F. Gey and P. Puska, *Ann. NY Acad. Sci.* **570,** 268 (1989).
56. J.A. GLomset, *J. Lipid Res.* **9,** 155 (1968).
57. J.J. Badimon, V. Fuster, and L. Badimon, *Circulation* **86** [suppl III], 11186 (1992).
58. S.M. Grundy and M.A. Denke, *J. Lipid Res.* **31,** 1149 (1990).
59. P.E. DiCorleto and G.M. Chisolm, *Prog. Lipid Res.* **25,** 365 (1986).
60. D.B. Zilversmit, *Ann. NY Acad. Sci.* **275,** 138 (1976).
61. B. Hennig, D.M. Shasby, A.B. Fulton, and A.A. Spector, *Arteriosclerosis* **4,** 489 (1984).
62. B. Hennig, D.M. Shasby, and A.A. Spector, *Circ. Res.* **57,** 776 (1985).
63. B. Hennig and co-workers, *Biochem. Arch.* **6,** 409 (1990).
64. B. Hennig, C. Enoch, and C.K. Chow, *Arch. Biochem. Biophys.* **248,** 353 (1986).
65. B. Hennig and co-workers, *J. Nutr.* **120,** 331 (1990).
66. S.-K. Peng, C.B. Taylor, J.C. Hill, and R.J. Morin, *Atherosclerosis* **54,** 121 (1985).
67. B. Hennig and G.A. Boissonneault, *Atherosclerosis* **68,** 255 (1987).
68. G.A. Boissonneault, B. Hennig, and C.-M. Ouyang, *Proc. Soc. Exp. Biol. Med.* **196,** 338 (1991).
69. G.A. Boissonneault and co-workers, *Ann. Nutr. Met.* **35,** 226 (1991).
70. F.J. Haddy, *Hypertension* **18** [suppl 5], III179 (1991).
71. D.F. Bohr, A.F. Dominiczak, and R.C. Webb, *Hypertension* **18** [suppl 5], III69 (1991).
72. G.W.K. Ching and D.G. Beevers, *Postgrad. Med. J.* **67,** 230 (1991).

73. H.T. Smith, *Am. J. Hypertens.* **6,** 119S (1993).
74. A. Ascherio and co-workers, *Circulation* **86,** 1475 (1992).
75. R.M. Lauer, J. Lee, and W.R. Clarke, *Pediatrics* **82,** 309 (1988).
76. S.S. Gidding, *Pediatr. Clin. North Am.* **40,** 41 (1993).
77. G.S. Andersen, *Scand. J. Clin. Lab. Invest.* **50** [suppl 199], 25 (1990).
78. C.G. Smoak and co-workers, *Am. J. Epidemiol.* **125,** 364 (1987).
79. R. Stamler and co-workers, *JAMA* **262,** 1801 (1989).
80. N.E. Straznicky, W.J. Louis, P. McGrade, and L.G. Howes, *J. Hypertens.* **11,** 427 (1993).
81. L. Bomzon, M.C. Kew, and C. Rosendorff, *Clin. Exp. Pharmacol.* **5,** 181 (1978).
82. R. Broderick, R. Bialecki, and T. Tulenko, *Am. J. Physiol.* **257,** H170 (1989).
83. T.A. McCalden and R.G. Nath, *Stroke* **20,** 238 (1989).
84. C. Rosendorf and co-workers, *Cir. Res.* **48,** 320 (1981).
85. D.D. Heistadd and co-workers, *Cir. Res.* **54,** 711 (1984).
86. L.H. Block, *Klin. Wochenschr.* **68** [suppl 20], 60 (1990).
87. J. Galle, A. Mulsch, and E. Bassenge, *Eur. Heart J.* **10** [suppl], 37 (1989).
88. F.M. Sacks, *Nutr. Rev.* **47,** 291 (1989).
89. B.M. Margetts, *Klin. Wochenschr.* **68** [suppl 20], 11 (1990).
90. J.M. Iacono, R.M. Dougherty, and P. Puska, *Klin. Wochenschr.* **68** [suppl 20], 23 (1990).
91. A.P. Simopoulos, *Compr. Ther.* **16,** 25 (1990).
92. G. Maschio and co-workers, *Kidney Int.* **39** [suppl 31], S70 (1991).
93. J. Rosenthal, P. G. Simone, and A. Silbergleit, *Prostaglandins* **5,** 435 (1974).
94. R. Düsing, and H. Scherf, *Klin. Wochenschr.* **68** [suppl 20], 2 (1990).
95. L.C. Wince, L.E. Hugman, W.Y. Chen, R.K. Robbins, and G.M. Brenner, *J. Pharmacol. Exp. Ther.* **241,** 838 (1987).
96. L.C. Wince and C.O. Rutledge, *J. Pharmacol. Exp. Ther.* **219,** 625 (1981).
97. G. Debry and X. Pelletier, *Experientia* **15,** 172 (1991).
98. P. Puska and co-workers, *Prev. Med.* **14,** 573 (1985).
99. J.T. Salonen and co-workers, *Am. J. Clin. Nutr.* **48,** 1226 (1988).
100. P. Rubba and co-workers, *Int. J. Epidemiol.* **16,** 528 (1987).
101. R.A. Rimersma and co-workers, *Br. Med. J.* **292,** 1423 (1986).
102. D.A. Wood, S. Butler, R.A. Rimersma, M. Thompson, and M.F. Oliver, *Lancet* **2,** 117 (1984).
103. P. Oster, L. Arab, B. Schellenberg, M. Kohlmeier, and G. Schlierf, *Prog. Food Butr. Sci.* **4,** 39 (1980).
104. E.M. Berry and J. Hirsch, *Am. J. Clin. Nutr.* **44,** 336 (1986).
105. H.R. Knapp, *Nutr. Rev.* **47,** 301 (1989).
106. D.M. Reed, J.A. Resh, T. Hayashi, C. MacLean, and K. Yano, *Stroke* **19,** 820 (1988).
107. R. Salonen and co-workers, *Thromb. Haemost.* **57,** 269 (1987).
108. B. Weisser, A. Struck, B.O. Gobel, H. Vetter, and R. Düsing, *Klin. Wochenschr.* **68** [suppl 20], 49 (1990).
109. G.A. Bray, *J. Nutr.* **121,** 1146 (1991).
110. R.J. Martin, B.D. White, and M.G. Hulsey, *Am. Scient.* **79,** 528 (1991).
111. A.J. Stunkard, J.R. Harris, N.L. Pederson, and G.E. McClearn, *N. Engl. J. Med.* **322,** 1483 (1990).
112. E. Ravussin, A.M. Fontvieille, B.A. Swinburn, and C. Bogardus, *Ann. N.Y. Acad. Sci.* **683,** 141 (1993).

113. C. Bennett and co-workers, *Am. J. Clin. Nutr.* **55,** 1071 (1992).
114. J.P. Flatt, *Ann. N.Y. Acad. Sci.* **499,** 104 (1987).
115. R.C. Klesges, L.M. Klesges, C.K. Haddock, and L.H. Eck, *Am. J. Clin. Nutr.* **55,** 818 (1992).
116. A. Astrup and A. Raben, *Eur. J. Clin. Nutr.* **46,** 611 (1992).
117. J.E. Blundell, V.J. Burley, J.R. Cotton, and C.L. Lawton, *Am. J. Clin. Nutr.* **57,** 772S (1993).
118. H.B. Hubert, M. Feinleib, P.M., McNamara, and W.P. Castelli, *Circulation* **67,** 968 (1983).
119. P.D. Wood, M.L. Stefanick, P.T. Williams, and W.L. Haskell, *N. Engl. J. Med.* **325,** 461 (1991).
120. J.S. Stern and M.G. Hermann-Zaidius, *J. Am. Diet. Assoc.* **92,** 91 (1992).
121. D.J. Mela, *J. Am. Diet. Assoc.* **92,** 472 (1992).
122. G.A. Bray and D.S. Gray, *Diab./Metab. Rev.* **4,** 653 (1988).
123. J.S. Garrow, *Lancet* **340,** 409 (1992).
124. J.O. Hill, H. Drougas, and J.C. Peters, *Ann. Intern. Med.* **119,** 694 (1993).
125. National Diabetes Data Group, *Diabetes* **28,** 1039 (1979).
126. M.I. Harris, W.C. Hadden, W.C. Knowler, and P.H. Bennett, *Diabetes* **36,** 523 (1987).
127. K.M. West and J.M. Kalbfleisch, *Diabetes* **20,** 99 (1971).
128. B.A. Reeder and co-workers, *Can. Med. Assoc. J.* **146,** 2009 (1992).
129. J.E. Friedman and co-workers, *J. Clin. Invest.* **89,** 701 (1992).
130. J. Svedberg, P. Bjorntorp, W. Smith, and P. Lonuroth, *Diabetes* **39,** 570 (1990).
131. A.J. Garber, A.I. Vinik, and S.R. Crespin, *Diabetes Care* **15,** 1068 (1992).
132. P.N. Durrington, *Postgrad. Med. J.* **69** [suppl 1], S18 (1993).
133. B.V. Howard and co-workers, *Metabolism* **36,** 870 (1987).
134. C.J. Schwartz and co-workers, *Diabetes Care* **15,** 1156 (1992).
135. V. Mimouni and J.P. Poisson, *Biochim. Biophys. Acta* **1123,** 296 (1992).
136. V. Ruiz-Gutierrez and co-workers, *Diabetologia* **36,** 850 (1993).
137. T. Takahashi, H. Takahashi, H. Takeda, and M. Shichiri, *Diabetes Res. Clin. Pract.* **16,** 103 (1992).
138. H. Kozuka and co-workers, *Chem. Pharm. Bull. (Tokyo)* **39,** 1267 (1991).
139. P. Bjorntorp, *Acta Paediatr.* **383,** 59 (1992).
140. M.T. Clandinin, S. Cheema, C.J. Field, and V.E. Baracos, *Ann. N.Y. Acad. Sci.* **683,** 151 (1993).
141. D.F. Horrobin, *Am. J. Clin. Nutr.* **57,** 732S (1993).
142. L.H. Storlien and co-workers, *Diabetes* **40,** 280 (1991).
143. J. Lefkowith and co-workers, *J. Exp. Med.* **171,** 729 (1990).
144. P. Fasching and co-workers, *Diabetes* **40,** 583 (1991).
145. S.E. Kasim, *Ann. N.Y. Acad. Sci.* **683,** 250 (1993).
146. American Diabetes Association, *Diabetes Care* **10,** 126 (1987).
147. American Diabetes Association, *Diabetes Care* **12,** 573 (1989).
148. Nutrition Subcommittee of the British Diabetic Association's Professional Advisory Committee, *Diabet. Med.* **9,** 189 (1992).
149. A. Bonanome and co-workers, *Am. J. Clin. Nutr.* **54,** 586 (1991).
150. J.E. Connell and D. Thomas-Dobersen, *J. Am. Diet. Assoc.* **91,** 1556 (1991).
151. C.A. Beebe, J.G. Pastors, M.A. Powers, and J. Wylie-Rosett, *J. Am. Diet. Assoc.* **91,** 196 (1991).
152. R. Doll and R. Peto, *JNCI* **66,** 1197 (1981).
153. G. Ghourab, *Adv. Clin. Chem.* **29,** 197 (1992).
154. M.B. Roberfroid, *Mutat. Res.* **259,** 351 (1991).

155. A. Tannenbaum, *Cancer Res.* **2,** 468 (1942).
156. K.K. Carroll, *Am. J. Clin. Nutr.* **53** [suppl 4], 1064S (1991).
157. L.S. Freedman, C. Clifford, and M. Messina, *Cancer Res.* **50,** 5710 (1990).
158. C. Ip, *Am. J. Clin. Nutr.* **45,** 218 (1987).
159. C.W. Welsch, *Cancer Res.* **52** [suppl 7], 2040S (1992).
160. N.E. Hubbard and K.L. Erickson, *Cancer Lett.* **56,** 165 (1991).
161. R.A. Karmali, L. Adams, and J.R. Trout, *Prostaglandins Leukot. Essent. Fatty Acids* **48,** 309 (1993).
162. C.W. Welsh, *Am. J. Clin. Nutr.* **45,** 192 (1987).
163. K.K. Carroll and H.I. Parenteau, *Nutr. Cancer* **16,** 79 (1991).
164. R.R. Williams and co-workers, *JAMA* **245,** 247 (1981).
165. K.K. Carroll, *Lipids* **27,** 793 (1992).
166. T. Hirayama, *Prev. Med.* **7,** 173 (1978).
167. S. Sasaki, M. Horacsek, and H. Kesteloot, *Prev. Med.* **22,** 187 (1993).
168. C.C. Boring, T.S. Squires, and T. Tong, *C.A. Cancer J. Clin.* **41,** 19 (1991).
169. B.S. Reddy, *Lipids* **27,** 897 (1992).
170. K.M. Nauss, M. Locniskar, D. Sondergaard, and P.M. Newberne, *Carcinogenesis* **52,** 225–260 (1984).
171. J.S. Goodwin and J. Ceuppens, *J. Clin. Immunol.* **3,** 295 (1983).
172. T. Narisawa and co-workers, *Cancer* **56,** 1719 (1985).
173. M.F. Wargovich, V.W.S. Eng, and H.L. Newmark, *Cancer Lett.* **23,** 253 (1984).
174. T. Sugimura and S. Sato, *Cancer Res.* **43,** 2415S (1983).
175. D.E. Corpet and co-workers, *Cancer Res.* **50,** 6955 (1990).
176. M. Sakaguchi and co-workers, *Anticancer Res.* **10,** 1763 (1990).
177. R.W. Sherwin and co-workers, *J. Am. Med. Assoc.* **257,** 943 (1987).
178. P. Greenwald, *Cancer* **70** [suppl 5], 1206 (1992).
179. K.K. Carroll, *J. Environ. Pathol. Toxicol.* **3,** 253 (1980).
180. M.M. Mathews-Roth and N.I. Krinsky, *Photochem. Photobiol.* **40,** 671 (1984).
181. D.F. Birt, L.T. White, B. Choi, and J.C. Pelling, *Cancer Res.* **49,** 4170 (1989).
182. J. Leyton and co-workers, *Cancer Res.* **51,** 907 (1991).
183. M. Locniskar and co-workers, *Nutr. Cancer* **16,** 189 (1991).
184. P.M. Newberne, J. Weigert, and N. Kula, *Cancer Res.* **39,** 3986 (1979).
185. N.G. Misslbeck, T.C. Campbell, and D.A. Roe, *J. Nutr.* **114,** 2311 (1984).
186. D.F. Birt and co-workers, *Carcinogenesis* **11,** 745 (1990).
187. M.J. Appel, A. van Garderen-Hoetmer, and R.A. Woutersen, *Cancer Lett.* **55,** 239 (1990).
188. B.D. Roebuck, D.S. Longnecker, K.J. Baumgartner, and C.D. Thron, *Cancer Res.* **45,** 5252 (1985).
189. E. Riboli and co-workers, *Int. J. Cancer* **49,** 214 (1991).
190. P.M. Pour, K. Groot, K. Kazakoff, K. Anderson, and A.V. Schally, *Cancer Res.* **51,** 4757 (1991).
191. F. Levi, S. Franceschi, E. Negri, and C. La Vecchia, *Cancer* **71,** 3575 (1993).
192. S. Graham and co-workers, *Am. J. Epidemiol.* **131,** 454 (1990).
193. R. R. Beems and L. van Beek, *Carcinogenesis* **5,** 413 (1984).
194. J. Wahrendorf in A.B. Miller, ed., *Diet and the Aetiology of Cancer,* Springer-Verlag, Berlin, 1989, pp. 13–19.

195. B.S. Reddy, *Adv. Cancer Res.* **32,** 237 (1980).
196. D. Kritchevsky, *Cancer* **72** [suppl 3], 1011 (1993).
197. T. Byers, *Cancer* **72** [suppl 3], 1015 (1993).
198. A.M. El Nahas, *Nephrol. Dial. Transplant.* **4,** 843 (1989).
199. B.L. Kasiske, M.P.O'Donnell, W. Cowardin, and W.F. Keane, *Hypertension* **15,** 443 (1990).
200. W.F. Keane, B. Kasiske, and M.P. O'Donnell, *Am. J. Nephrol.* **8,** 261 (1988).
201. M. Hattori, Y. Yamaguchi, H. Kawaguchi, and K. Ito, *Nephron* **63,** 314 (1993).
202. T. Al-Shebeb, J. Frohlich, and A.B. Magil, *Kidney Int.* **33,** 498 (1988).
203. K.F. Wellmann and B.W. Volk, *Lab. Invest.* **22,** 36 (1970).
204. B.L. Kasiske, M.P. O'Donnell, and W.F. Keane, *Hypertension* **19** [suppl 1] I110 (1992).
205. H.C. Rayner, L. Ward, and J. Walls, *Nephron* **57,** 453 (1991).
206. B.L. Kasiske, *Kidney Int.* **31,** 1153 (1987).
207. B.L. Kasiske, M.P. O'Donnell, P.G. Schmitz, Y. Kim, and W.F. Keane, *Kidney Int.* **37,** 880 (1990).
208. P.O. Attman and P. Alaupovic, *Kidney Int.* **39** [Suppl 31], S16 (1991).
209. G.A. Kaysen, *Kidney Int.* **39** [Suppl 31], S8 (1991).
210. P. Cappelli and co-workers, *Nephron* **62,** 31 (1992).
211. J.-B. Roullet, B. Lacour, J.-P. Yvert, J.-J. Prat, and T. Drüeke, *Kidney Int.* **27,** 420 (1985).
212. J.-B. Roullet, B. Lacour, J.-P. Yvert, and T. Drüeke, *Am. J. Physiol.* **250,** E373 (1986).
213. M. Akmal, S.E. Kasim, A.R. Soliman, and S.G. Massry, *Kidney Int.* **37,** 854 (1990).
214. G.A. Crawford, J.F. Mahony, and J.H. Stewart, *Clin. Sci.* **60,** 73 (1981).
215. M.K. Chan, Z. Varghese, and J.F. Moorhead, *Kidney Int.* **19,** 625 (1981).
216. M.K. Chan, *Metabolism* **38,** 939 (1989).
217. P.C.K. Chan and co-workers, *Clin. Nephrol.* **31,** 88 (1989).
218. J.F. Moorhead, *Kidney Int.* **39** [Suppl 31], S35 (1991).
219. G.L. Warwick and co-workers, *Metabolism* **39,** 187 (1990).
220. I. Staprans and J.M. Felts, *Biochem. Biophys. Res. Commun.* **79,** 1272 (1977).
221. M.L. Sanfelippo, R.S. Swenson, and G.M. Reaven, *Kidney Int.* **11,** 54 (1977).
222. U.O. Barcelli and V.E. Pollak, *Contemp. Issues Nephrol.* **14,** 65 (1986).
223. G. D'Amico and co-workers, *Lancet* **339,** 1131 (1992).
224. B.J. Holub, D.J. Philbrick, A. Parbtani, and W.F. Clark, *Biochem. Cell Biol.* **69,** 485 (1991).
225. F. Thaiss, W. Schoeppe, P. Germann, and R.A. Stahl, *J. Lab. Clin. Med.* **116,** 172 (1990).
226. H.J. Bilo, J.J. Homan van der Heide, R.O. Gans, and A.J. Donker, *Nephron* **57,** 385 (1991).
227. D. Hwang, *FASEB J.* **3,** 2052 (1989).
228. J.E. Kinsella, B. Lokesh, and R.A. Stone, *Am. J. Clin. Nutr.* **52,** 1 (1990).
229. D.H. Hwang, M. Boudreau, and P. Chanmugam, *J. Nutr.* **118,** 427 (1988).
230. J.H. Lee, M. Sugano, and T. Ide, *J. Nutr. Sci. Vitaminol.* **34,** 117 (1988).
231. B.R. Lokesh and J.E. Kinsella, *Immunobiol.* **175,** 406 (1987).
232. B.R. Lokesh, J.B. German, and J.E. Kinsella, *Biochim. Biophys. Acta* **958,** 99 (1988).
233. K.S. Broughton and L.J. Morgan, *J. Nutr.* **124,** 1104 (1994).
234. A. Berger and co-workers, *J. Nutr.* **123,** 225 (1993).
235. S. Watanabe and co-workers, *J. Nutr.* **124,** 1566 (1994).
236. S.L. Kunkel, S.W. Chensue, and S.H. Phan, *J. Immunol.* **136,** 186 (1986).
237. P.J. Knudsen, C.A. Dinarello, and T.B. Strom, *J. Immunol.* **137,** 3189 (1986).
238. D.K. Imagawa and co-workers, *J. Immunol.* **136,** 4637 (1986).

239. B.R. Lokesh, T.J. Sayers, and J.E. Kinsella, *Immunol. Lett.* **23,** 281 (1989/1990).
240. J.M. Kremer and co-workers, *Arthritis Rheum.* **33,** 810 (1990).
241. H.C. Anderson, *Bone and Mineral Research,* Vol. 3, Elsevier Science Publishers, New York, 1985, pp. 109–149.
242. R.E. Wuthier, *ISI Atlas Sci. Biochem.* **1,** 231 (1988).
243. D.C. Klein and L.G. Raisz, *Endocrinology* **86,** 1436 (1970).
244. L.G. Raisz, *J. Bone Min. Res.* **8,** S457 (1993).
245. D.J. Baylink, R.D. Finkelman, and S. Mohan, *J. Bone Min. Res.* **8,** S565 (1993).
246. G.R. Mundy, *J. Bone Min. Res.* **8,** S505 (1993).
247. H.D. Adkisson and co-workers, *FASEB J.* **5,** 344 (1991).
248. H. Xu, B.A. Watkins, and H.D. Adkisson, *Lipids* **29,** 619 (1994).
249. H. Xu, B.A. Watkins, and M.F. Seifert, *J. Bone Min. Res.* **9,** B76 (1994).
250. B.A. Watkins, S.D. Bain, C. Shen, and J. McMurtry, *J. Bone Min. Res.* **8,** S322 824 (1993).
251. B.A. Watkins, C-L. Shen, M.F. Seifert, D.C. McFarland, and K.G.D. Allen, *J. Bone Min. Res.* **9,** B104 (1994).

8

Toxicity and Safety of Fats and Oils

BACKGROUND INFORMATION

Fat is considered an important component of the diet, with respect to being the principal source of energy, essential fatty acids, and fat-soluble vitamins such as vitamins A, D, E, and K (1). The classical early studies by Burr and Burr (2) demonstrated the requirement for essential fatty acids (of the *n*-6 and *n*-3 configuration) in the diet to prevent growth and development abnormalities. In addition, fats elicit important characteristic flavors, aromas, and textures in foods, as well as strong satiety signals regulating food intake and possibly eating behavior.

Epidemiological data from different countries show positive correlations between the amount of dietary fat and the fat type with the incidence of coronary artery disease (CAD); (3). Increasing total caloric intake can also result in increased total plasma cholesterol by elevating very low density lipoproteins (VLDL) concentrations, which ultimately raise low-density lipoprotein (LDL) cholesterol levels (4). Mortalities associated with CAD have been positively correlated with saturated fat intake (5), while high dietary intakes of polyunsaturated fatty acids (PUFA) may increase the risk of oxidative modification of plasma lipoproteins that lead to atherosclerosis (6). Similarly, both epidemiological (7) and experimental studies (8) have linked the level of dietary fat consumption with increases in the incidence of spontaneous, and some xenobiotic-induced cancers. Experimental evidence strongly suggests that dietary fat is a promoter, rather than an initiator, of many types of cancers such as mammary and colon cancer (9). Moreover, an association between the degree of saturation and level of cholesterol consumption exists for the incidence of a number of cancers (10). Weinstein (11), hypothesized the mechanism for the development of colon cancer from dietary fat involves the production of bile acids by the liver and the role of intestinal flora in converting dietary lipid to diacylglycerol (a modified lipid operative in stimulating protein kinase C enzyme activity) as an important factor influencing colon cancer. The alteration of colonic epithelial cells by protein kinase C, induced by diacylglycerol, has considerable appeal in explaining the interaction between specific microflora, dietary lipids, and bile salts in the development of colon cancer.

Some of the relevant toxicological responses to dietary lipids are summarized in Table 8.1. Due to the complexity of metabolic responses to various fats and oils, it is unlikely that there is a single unifying mode of action that accounts for all the changes observed. Thus, dietary modifications related to reducing the risk of developing chronic diseases such as atherosclerosis and cancer include reducing total fat intake and the proportion of fat intake comprised of saturated fat. The current North American diet, which provides

Table 8.1 Principal metabolic events associated with chronic disease and dietary fats and oils

Fat/Oil Issue	Toxicological Endpoint	Reference
High dietary fat intake	Epidemiology: colon cancer, mammary cancer	7, 9, 12
	Altered intestinal microflora: colon cancer	8, 13
	Altered immunity: colon cancer	14
	Altered acidic, neutral steroid excretion: colon cancer	15
	Initiation phase: mammary cancer	10, 16
	Epidemiology: coronary heart disease	3
High SFA intake	Aortic lesions: coronary heart disease	17
	Inhibit GST activity: inhibit xenobiotic detoxification	18, 19
High PUFA intake	Altered phospholipid metabolism: mammary cancer	15
	Co-oxidation of B(*a*)P	20
	Stimulated hepatic P-450 enzymes	21
	Deposition of mercury (*n*-3 PUFA)	22
	Genotoxic tumor promotor	23
	Altered antioxidant status	24
LOPS[a]	Reduced tissue α-tocopherol	25
	Tissue autooxidized product deposition	26
	Induction ornithine decarboxylase: colon cancer	27
Hydrogenation fats	Essential fatty acid deficiency	28
	Mammary and colon cancers	29, 30
	Neonatal growth retardation	31
	Increased LDL-cholesterol	32

[a] LOPS, lipid-oxidized products.

approximately 40% of energy as fat, with a polyunsaturated to saturate (*P*/*S*) ratio of about 0.5 and cholesterol intake of around 500 mg of cholesterol per day, has been linked to certain diseases including obesity, cardiovascular disease, hypertension, stroke, gallstones, and some forms of cancer. Current dietary recommendations include reducing fat intake to 30% of total metabolizable energy, characterized by equal amounts of saturated, monounsaturated, and polyunsaturated fatty acids ($P/S = 1.0$) and a cholesterol reduction to 300 mg or less per day.

In addition to potential toxicity manifested by naturally occurring fat constituents, there is also concern of the presence of potentially toxic oxidized unsaturated lipid in heated or cooked food. The solubility of hydrocarbons (derived or naturally occcurring) in liquid triacylglycerols is also an important topic. Although factors related to dietary fat have been suggested to be implicated with the risk of developing certain chronic diseases, the role of food preparation techniques in the development of these diseases has received less attention. Lipid peroxidation reactions occurring *in vivo* in response to the production of lipid peroxides from unsaturated fat has been associated with atherosclerosis lesions in human aortas (33,34). The significance of exposure to lipid oxidized products from diet sources and the etiology of various chronic diseases remains uncertain, however. To minimize the chemical changes associated with lipid oxidation, numerous interventions including the use of fat hydrogenation technology, addition of synthetic antioxidants, and use of high gas barrier packaging materials are employed to increase lipid stability and minimize contact with atmospheric oxygen. The benefits of employing these technologies are also frequently balanced against potential health risks.

This chapter will review the toxicity and safety issues on: *(1)* fats and accompanying natural constituents; *(2)* derived products of oxidation of fat and oils during commercial and home processing; *(3)* presence of both natural and derived toxicants; and *(4)* additives used to preserve the functionality, nutritional quality, and safety of fats and oils.

ADVERSE EFFECTS OF FATS AND ASSOCIATED NATURAL CONSTITUENTS

2.1 Fatty Acids

Saturated Fatty Acids. The effect of dietary saturated fatty acids (SFA) on the plasma cholesterol level in humans remains somewhat equivocal since both environmental and hereditary factors contribute to epidemiological results (35). Saturated fatty acids have been implicated as one dietary factor that raises serum cholesterol and triacylglycerol levels in both experimental animals (36) and humans (37). Early human studies indicated that SFA (e.g., lauric, C12:0; myristic, C14:0; palmitic, C16:0) were equally associated with increases in plasma cholesterol (38) or that myristic acid is more cholesterolemic than lauric or palmitic acids (39). Animal studies have shown that the mechanism for SFA-induced hypercholesterolemia is associated with the down regulation of hepatic LDL receptors (40). However, it remains unresolved whether all SFA contribute equally to hypercholesterolemia and atherosclerosis. For example, the frequency and severity of atherosclerotic lesions were shown to be similar in swine fed either butter (saturated) or corn (unsaturated) fats (41). Short-chain fatty acids (butyric, C4 : 0; caproic, C6 : 0; caprylic, C8 : 0; capric, C10 : 0) have no apparent effect on serum cholesterol (42). These SFA are readily absorbed and passed to the liver via the portal system for β-oxidation by hepatic mitochondria. Medium-chain fatty acids (lauric, C12 : 0; myristic, C14 : 0) are efficiently oxidized at rates similar to unsaturated acids [linolenic, C18 : 3 *n*-3; oleic C18 : 1 *n*-9; linoleic acid C18 : 2 *n*-6 (43)], but they raise serum cholesterol (44,45). Palmitic acid (C16 : 0) is a long-chain SFA that exhibits greater cholesterol-raising activity than lauric acid (45). However, as evidenced by the findings of Hayes and co-workers (46), palm oil, rich in palmitic acid, was not as effective as other saturated fats with both shorter and longer carbon chains in increasing plasma cholesterol in primates. One explanation for this finding is that palmitic acid is converted into stearic acid (C18 : 0) by the action of fatty acid elongases present in the endoplasmic reticulum and mitochondria. Stearic acid is effectively converted to oleic acid (C18 : 1 *n*-9) by the Δ^9 desaturase enzyme that incorporates a double bond between the 9th and 10th carbon from the carboxyl end of the molecule. Stearic acid has been referred to as neutral (38,39) or hypocholesterolemic under specific conditions (47). This was evident in studies where the feeding of cocoa butter, rich in stearic acid, did not result in a rise in blood cholesterol level, but rather lowered both the plasma cholesterol and the LDL cholesterol/HDL (high-density lipoprotein) cholesterol ratio equivalent to oleic acid (47). The feeding of beef fat with high levels of palmitic and oleic acids had no apparent effect on total and LDL-cholesterol, but coconut fat (high in lauric and myristic acids) increased both plasma cholesterol and LDL-cholesterol (48). These findings support the interactive effect of saturated fat and cholesterol with total plasma cholesterol and LDL-cholesterol in primates (49). Feeding saturated fat increases HDL-cholesterol as well as apoprotein A-I levels compared to feeding polyunsaturated fat (50). Differences in the chemical composition of VLDL and HDL have been reported with the ingestion of various fats with high SFA content (51). Animal studies conducted in the primate model have shown that diets with different fat types, (e.g., saturated or polyunsaturated) produced greater changes in plasma cholesterol concentrations when they were administered with high levels of dietary cholesterol (52). In contrast, studies conducted in monkeys fed cholesterol-free diets containing fat blends with *P*/*S* values ranging from 0.12 to 1.09, have indicated that incremental additions of palmitic acid at the expense of lauric and myristic acids were effective in decreasing plasma cholesterol (46). Similarly, with hamsters, low-cholesterol diets enriched with lauric and myristic acid blends were equivalent to palmitic and oleic acids in elevating plasma cholesterol.

Whereas palmitic- and oleic-enriched diets enlarged the HDL cholesterol pool, the lauric and myristic acid blend expanded the LDL cholesterol pool (53). These findings suggested significant interactions between specific fatty acids and cholesterol in eliciting changes in lipoprotein cholesterol metabolism that are associated with the induction of fibrous-plaque lesions in atherosclerosis.

Variable responses to different saturated fats by plasma triacylglycerol levels have also been reported (46). Higher plasma and VLDL triacylglycerol concentrations were found in animals fed saturated fat compared to unsaturated fat. Butter fat was shown to result in the lowest plasma triacylglycerol and HDL-cholesterol levels when compared to other saturated fats, such as beef tallow, palm oil, and coconut. These findings are of particular importance since recent epidemiology evidence indicated that plasma triacylglycerol levels are associated with coronary heart disease (CHD) (54). Results from an experimental atherosclerosis study have shown dietary coconut oil produces the greatest effect on atherosclerosis, in contrast to corn oil, which had the least effect (55). More recently, dietary cholesterol has been shown to have a more pronounced effect than dietary fat type on both plasma and hepatic cholesterol levels as well as cholesterol biosynthesis activity [3-hydroxy-3-methylglutaryl coenzyme A reductase, (HMG-CoA)]. The significance of this conclusion is emphasized by the contrasting findings in animal studies that have evaluated the development of atherosclerotic plaques in response to dietary fat intake. For example, diets containing extremely high levels of cholesterol resulted in more severe plaque progression with unsaturated fats compared to saturated fats (48), whereas an entirely opposite conclusion has been made from studies where cholesterol intake was relatively lower (17).

Monounsaturated Fatty Acids (MUFA). The monoene, oleic acid, has been found to be as effective as the polyene, linoleic acid, in lowering total serum and LDL cholesterol when compared to palmitic acid (56). Decreasing the saturated fat intake and replacing it with dietary fats containing MUFA or PUFA in hypercholesterolemic male subjects consuming diets containing 40% of total calories as fat, effectively lowered total plasma and LDL-cholesterol (56). Animal studies, however, have shown a hypercholesterolemic effect of MUFA when fed with cholesterol-free diets (57). Peanut oil, containing a relatively high concentration of oleic acid, has been reported to be atherogenic when fed with cholesterol to rodents and monkeys (58,59). However, peanut oil ingestion prevented hypercholesterolemia and actually reduced atherogenesis in monkeys fed a level of cholesterol closer to human intake (0.1%), when compared to a blend of oils that approximated the fatty acid composition of an American diet (60). This finding is noteworthy, since plasma LDL isolated from the blood of subjects fed either MUFA or PUFA containing diets are more resistant to lipid peroxidation and oxidative modification when the oleic to linoleic fatty acid ratio of the LDL particle was increased, independent of antioxidant content (47). This response would be expected to reduce potential atherogenicity, since oxidized LDLs are believed to play an important role in atherogenesis (61). Palmitoleic acid (C16:1), albeit present in only trace amounts in milk fat in particular, enhances the metabolic activity of blood vessels and prevents stroke in hypertensive, stroke-prone rats (62).

Polyunsaturated Fatty Acids. n-6 PUFA. The lowering of total plasma cholesterol by diets containing high (approximately 30% of total energy intake) proportions of *n*-6 PUFA has been associated with the reduction in HDL-cholesterol (50). HDLs (e.g., apolipoprotein A-I) affect lipoprotein metabolism by regulating the extracellular transport of cholesterol (63) and are inversely correlated with cardiovascular disease (64). Recently, an increased intake of linoleic acid has been reported to significantly enhance the risk of new atherosclerosis lesions in human coronary arteries (6). In other studies, however, human subjects fed diets containing sunflower oil (high in linoleic acid) and low erucic acid, rapeseed oil, containing a lower proportion of linoleic and markedly greater amount of α-linolenic, effectively lowered total cholesterol, VLDL, and LDL-cholesterols but had no effect on HDL-cholesterol (65). These findings indicate that fat-containing diets predominating in PUFA fat sources do not lower HDL-cholesterol when present in moderate levels (e.g., less than 15% of total energy intake).

Animal studies have demonstrated that *n*-6 PUFA containing diets enhance tumorigenesis during the promotion/progression and metastatic stages of experimental carcinogenesis. Carroll and Khor (66), reported higher yield of adenocarcinomas in 7,12-dimethylbenz(*a*)anthracene (DMBA) induced Sprague-Dawley rats fed diets containing relatively higher *n*-6 PUFA oils (e.g., cottonseed, corn, safflower, and soybean) compared to counterparts fed SFA or MUFA containing fats (e.g., butter, lard, tallow, rapeseed). The potential mechanisms for enhanced tumorigenesis attributed to dietary *n*-6 PUFA include involvement in, or effect on, arachidonic acid metabolism and prostaglandin synthesis balance (e.g., PGE_2 inhibits, while $PGF_{2\alpha}$ enhances tumor development), lipid peroxidation, and membrane fluidity (67). Moreover, PUFA intake level is an apparent factor in regulating the incidence of carcinogen-induced mammary tumors. For example, rats fed isocaloric diets containing high *n*-6 PUFA fat (25% corn oil containing *n*-6 PUFA diets), produced more tumors than counterparts fed low *n*-6 PUFA fat (5% corn oil) or a commercial nonpurified diet (10). Dietary consumption of *n*-6 PUFA from corn oil was greater than from *n*-3 supplemented diets, thus producing a potentially different effect on arachidonic acid metabolism. The affinity for Δ^6 desaturase is higher for α-linolenic (*n*-3) than linoleic (*n*-6) acid, which supports the production of prostaglandin of the 3 series. It is noteworthy that this may be more important in the rodent than the human, where linoleic acid conversion to arachidonic acid is limited (68).

n-3 PUFA. Linolenic acid (C18:3, *n*-3) is an essential fatty acid and is primarily found in plant and vegetable oils such as soybean, rapeseed, and linseed, the latter of which is not used in food products. The requirement of linolenic acid in humans can be as high as 0.5% of total caloric intake. Mammalians are able to convert linolenic acid to longer-chain *n*-3 fatty acids with greater degrees of unsaturation (e.g., C20:5, C22:5, C22:6) through elongation and desaturation pathways. Consumption of fish oil can modify phospholipid fatty acid composition by increasing long-chain *n*-3 PUFA at the expense of *n*-6 fatty acid. Marine oils contain a high content of *n*-3 PUFA (omega-3) series including eicosapentaenoic (EPA; C20:5, *n*-3) and docosahexaenoic (DHA; C22:6, n-3) acids. Omega-3 fatty acids can reduce both serum phospholipid and tissue arachidonic acid concentrations by competing for the Δ^6 desaturase enzyme, thus inducing a change in eicosanoid synthesis. Arachidonic acid is an important precursor of prostaglandins, thromboxanes, and leukotrienes; and by reducing arachidonic acid synthesis, *n*-3 fatty acids have a variety of beneficial effects in protecting against atherosclerotic disease. Diets enriched in omega-3 fatty acids have produced effects associated with lowered serum triacylglycerols and blood pressure (69), reduced platelet aggregability and altered platelet function (70,71), and lower incidences of diabetes, rheumatism, and immunological responses associated with psoriasis (72,73). The competition of EPA and DHA with arachidonic acid inhibits production of thromboxane A_2 and increases production of prostaglandin I_3. This effect results in a change in the hemostatic balance toward greater vasodilator activity with reduced platelet aggregation and has been attributed to the prolonged bleeding time observed in the Greenland Eskimos, who have characteristically high intakes of EPA and DHA from fish oils.

A chronic or high intake of *n*-3 PUFA may also represent a source of oxidative stress to cell membranes caused by lipid oxidation reactions and thus increases the requirement for *in vivo* enzymatic (e.g., superoxide dismutase, glutathione peroxidase) and nonenzymatic (e.g., tocopherol) tissue antioxidant defenses. The susceptibility of *n*-3 PUFA to autoxidation due to the high degree of unsaturation of these fatty acids has recently been evaluated as a potential risk of *in vivo* oxidative stress, as assessed by elevated tissue thiobarbituric acid reactants and urinary malondialdehyde excretion in animals fed diets containing fish oils (24).

2.2 Sterols

Sterols represent hydrogenated phenanthrene derivatives that meet the definition of a lipid but are not in the category of simple or derived lipids (Figure 8.1). Sterols are high-molecular-weight alcohols, which are present in the unsaponifiable fraction of lipids. Cholesterol ($C_{27}H_{45}OH$) is an essential constituent of all animal cells and found in the highest concentration in nervous tissue. Phytosterols are widely distributed sterols in plants, which represent end products of plant metabolism. Widely distributed phytosterols include sitosterol, from corn and stigmasterol, which is particularly rich in soybean oil. The sterol ergosterol ($C_{28}H_{43}OH$) was originally isolated from ergot and

(a) Cholesterol

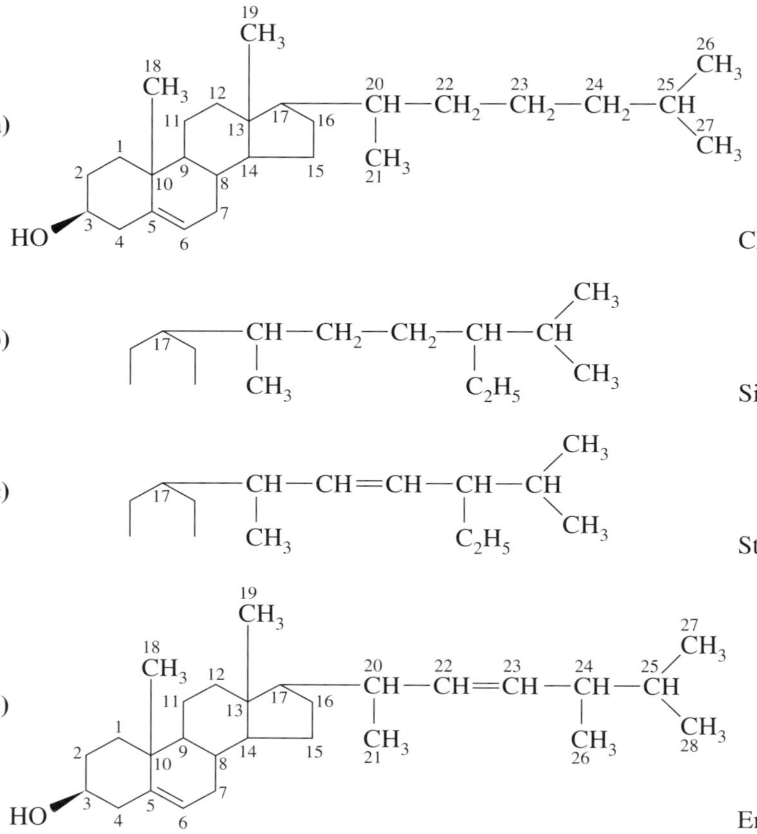

(b) Sitosterol

(c) Stigmasterol

(d) Ergosterol

Figure 8.1 Sterols found in animal and plant tissues. (**a**) Cholesterol found in animal tissues; (**b**) sitosterol found in corn; (**c**) stigmasterol found in soybean oil; and (**d**) ergosterol found in ergot and is a provitamin of vitamin D.

is obtainable from yeast. Upon irradiation this sterol has antirachitic activity and is the provitamin of vitamin D (see discussion later in this chapter).

Cholesterol. Cholesterol is the principal sterol of all animal and human body tissues. It is present in the nonsaponifiable portion of animal lipids with other sterols such as cholestanol. Circulating cholesterol is derived from either the diet or synthesized endogenously by the reduction of HMG-CoA to mevalonic acid by the enzyme HMG-CoA reductase (EC 1.1.1.34). Cholesterol 7α-hydroxylase is the rate-limiting enzyme involved with the removal of cholesterol from the liver through the synthesis of bile acids (24). Both cholesterogenesis and degradation are closely controlled by dietary constituents. Regulating enzymes involved in synthesis and degradation of cholesterol are important in suppressing endogenous cholesterol synthesis and stimulating fecal steroid excretion in response to increased dietary cholesterol intake. High dietary cholesterol intake reduces endogenous cholesterol biosynthesis but to a lesser extent in humans (74) than animals. An increase in cholesterol deposition within the hepatic cell also results in an increase in serum LDL-cholesterol (75), largely regarded as the principal atherogenic lipoprotein. This result occurs from the combined activity of enhanced conversion of VLDL fragments to LDL and decreased activity of LDL receptors to take cholesterol into the liver. The potential contributing factor of increased serum LDL in atherosclerosis is based on the susceptibility of LDL to undergo oxidation, thus producing lipid peroxidation products associated with the lipoprotein with potential atherogenic properties. In humans, an increase in plasma cholesterol in response to dietary cholesterol intake is highly variable (suggested variability = 15 mg/dL; 76) and influenced by the level of cholesterol intake. The large interindividual variability in cholesterolemic response to dietary cholesterol intake suggests that individuals are either hyporesponders or hyperresponders to dietary cholesterol (77). Noted increases in plasma cholesterol have been reported to be linear or curvilinear at intake levels up to 500 mg/day (38,39). The minimal dietary cholesterol intake required to increase plasma cholesterol is referred to as the threshold amount. Feeding cholesterol does not significantly affect plasma cholesterol until daily intakes reach around 100–300 mg/day. Furthermore, increases in plasma cholesterol reach plateau and a marked reduction in incremental change occurs at 500 mg/day (78,79).

Although there remains many unresolved issues regarding dietary cholesterol and cardiovascular disease (CVD), the importance of controlling for hypercholesterolemia for the prevention of atherosclerosis and CAD has been consistently documented in both epidemiological and experimental research. Atherosclerosis is characterized by the decrease (75–100%) in cross-section area of the major epicardial coronary arteries by plaque composed of cholesterol, mineral, and collagen. In both human and animal studies, radioactive cholesterol tracers deposit into atherosclerotic lesions in both human and animal aortas from the diet. Characteristic pathologic features of coronary atherosclerosis such as fatty streaks have been reported in young adult soldiers killed in combat (80). Elevated plasma LDL levels is the primary lipoprotein disorder in hypercholesterolemia. The presence of "foam cells," which result from the accumulation of cholesterol in macrophage and smooth muscle cells, are related to the presence of oxidized or modified LDL (mLDL). The increase in serum LDL in response to down regulation of the LDL receptor activity and decreased fractional catabolic rate of LDL caused by increased dietary cholesterol intake renders a greater probability of oxidation of LDL, leading to structural modifications in arterial intima. The development of processes to reduce cholesterol from animal fats using supercritical CO_2 processing (81) have resulted from the health concerns of dietary cholesterol intake.

Exposure of cholesterol to singlet ($^1O_2$) and triplet ($^3O_2$) dioxygens, peroxides ($O_2^{-\cdot}$, ROO·, and H_2O_2), hydroxyl radical (·OH), derived from both dietary sources and endogenous metabolism yield low levels (ppm-ppt) of cholesterol-oxidized products (COPs). A comprehensive review of the numerous biological activities of COPs has been published by Smith and Johnson (82). The content of COPs in foods or biological materials can be quantitatively determined by gas chromatography (Figure 8.2). Among the various toxicities reported for COPs, *in vivo* toxicities including cytoxicity, angiotoxicity immunosuppresive activity, and inhibition of metabolic homeostastic mechanisms pose plausible threats to human health from both low-level dietary or endogenous derived sources of cholesterol. More information on this particular topic is given later in this chapter.

Cholesterol also has a significant role in influencing serum triacylglycerol levels. The importance of triacylglycerol-rich lipoproteins in the etiology of atherosclerosis and CHD is becoming increasingly recognized. Humans exhibiting predominant hypertriglyceridemia have elevated plasma levels of VLDL particles and remnants. Increased amounts of triacylglycerol in the chylomicron also results in hypelipidemia. Hypolipidemic agents such as HMG-CoA reductase inhibitors (e.g., Lovostatin and Pravastatin) lowers both VLDL and

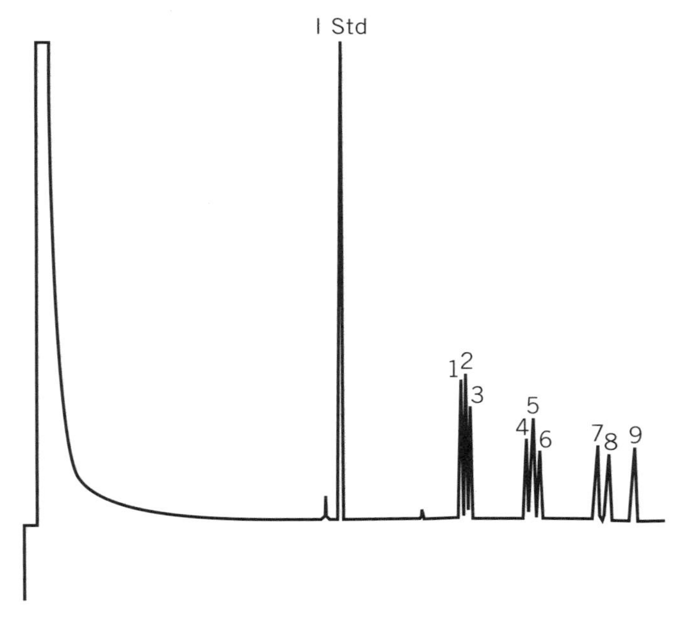

Figure 8.2 Separation of a mixture of 9 cholesterol oxides (Steraloids, Inc., Wilton, NH) using a DB-1 column (15 m × 0.25 mm i.d., 0.1 μm film thickness; J & W Scientific, Inc., Folsom, CA) on a Carlo Erba gas chromatograph with flame ionization detector (Carlo Erba Strumentazione, Italy). Conditions: Carrier gas was helium with nitrogen; injector temperature = 250°C; detector temperature = 280°C. Temperature gradient was 180–250°C at 3°C per minute. Derivatized standards, peaks (I Std) 5α-cholestane (R_T = 17.3 min); (1) cholesta-3,5,dien-7-one (R_T = 24.28 min); (4) α-epoxide (R_T = 27.30 min); (5) 7β-hydroxycholesterol (R_T = 27.63 min); (6) 4β-hydroxycholesterol (R_T = 28.01 min); (7) cholestane-triol (R_T = 31.21 min); (8) 7-ketocholesterol (R_T = 31.76 min); (9) 25 hydroxycholesterol (R_T = 33.18 min).

LDL while elevating plasma HDL in hypercholesterolemic and hypertriglyceridemic patients. Dietary cholesterol also can stimulate the secretion of VLDL-triacylglycerol, in addition to free and esterified cholesterol, in a dose-dependent manner (83). The increase in triacylglycerol synthesis with cholesterol feeding reduces fatty acid oxidation and increases incorporation of fatty acids into cholesterol ester and triacylglycerols. The subsequent accumulation of triacylglycerol in the liver that occurs with the feeding of cholesterol is concomitant with enhanced secretion of VLDL-triacylglycerol. The net effect of these actions on the metabolism of triacylglycerols is an impaired catabolism of VLDL and an increase in triacylglycerol-rich lipoproteins.

Conversely, other epidemiological studies, which have corrected for confounding variables such as lifestyle characteristics, body mass, and blood pressure, have concluded that serum cholesterol levels less than 4 mmol/L are associated with inceased death caused by cancer, digestive disorders, and respiratory disease (84). Some studies have shown a negative correlation with plasma cholesterol and death from lung cancer in men (85).

Bile Salts. The excretion of cholesterol from the body occurs both in the form of neutral steroids as well as the bile acids. In humans, gallstone formation is associated with changes in biliary bile acid composition (86). Bile acids, synthesized from cholesterol by the rate-limiting enzyme cholesterol 7α-hydroxylase in the liver, are secreted into the duodenum in amino-acid-conjugated forms. The majority of bile acids are reabsorbed from the ileum by an extrahepatic recirculation pathway when other food constituents, primarily plant materials, are not present to interfere with the reabsorption process. The nature of fatty acids and triacylglycerols in bile and the composition of biliary lipids entering the duodenum can adversely influence micellar and vesicular solubilization of cholesterol in the small intestine and enterocyte membrane permeability to cholesterol. These factors collectively contribute to the potential induced crystallization of cholesterol in the gall bladder, which can precede the formation of gallstones (87).

Unabsorbed bile acids are deconjugated and undergo epimerization by intestinal bacteria. Products of deconjugated primary and secondary bile acids are considered more toxic than conjugated bile acids on the basis of findings that both deoxycholic acid (DCA) and lithocholic acid (LCA) have carcinogenic promoter activity (88,89). The extent of mutagenicity of bile acids using the Ames/Salmonella microsome mutagenicity assay is very low (90) indicating that the bile acids may function more as comutagens than principle mutagens in affecting various aspects of cell turnover. LCA and associated glycine and taurine conjugates can enhance mutagenicity of 2-aminoantharacene and aromatic amines, and both LCA and DCA are active in enhancing the mutagenicity of 1,2-dimethylhydrazine (91).

Phytosterols. Plant sterols (e.g., sitosterol, stigmasterol, campesterol, δ-7-stigmasterol, and δ-5-avenasterol) are present in plant unsaponifiable matter. The consumption of large amounts of plant sterols from the diet interferes with cholesterol absorption to increase daily neutral steroid excretion. In animal studies, sitosterol feeding interferes with exogenous and endogenous cholesterol absorption resulting in an interruption of the enterolymphatic circulation of cholesterol, an important regulator of hepatic cholesterol synthesis. In cholesterol-containing diets supplemented with sitosterol, increases in endogenous and exogenous cholesterol, coupled with reduced excretion of fecal bile acids and decreased hepatic cholesterol content, were observed (92). These findings explain the antiatherogenic properties of plant steroids in the nonsaponifiable portion of the oil. Total sterol content of soybean oil and safflower oil is 0.29 and 0.39%, respectively. This compares to the plant sterol content of 4.1% in rice bran and 10% in rubber seed oils. Both rice bran and rubber seed oils have been shown to have a lipid-lowering effect in animals fed cholesterol-containing diets (93,94).

2.3 Vitamins (A, E, D, and K)

Free-radical-induced peroxidative damage to membrane lipids is regarded as an important step in the loss of PUFA and formation of hydroperoxides implicated in the deterioration of muscle foods as well as initiation of events leading to *in vivo* cell injury. Antioxidants such as tocopherols and retinoids quench chain propagating peroxyl radicals and thus reduce the rate of autoxidation.

Vitamin A. Vitamin A is a generic term for all ionine derivatives, excluding carotenoids. There are two dietary sources of vitamin A. Retinol, also referred to as preformed vitamin A, is present in eggs, dairy products, liver, and organ meats. Animal tissue is the source of the all-trans retinol. The fatty acid ester, retinyl palmitate, is the primary storage form of vitamin A in liver. It is released into the circulation as retinol, conjugated with a retinol-binding protein, and carried to target organs. Carotenoids, also termed provitamin A, represent a large number of naturally occurring pigments present in oils, vegetables, fruits, and animal foods (e.g., egg yolk). β-Carotene, when converted to retinyl esters, is efficiently absorbed from food but is susceptible to oxidative degradation in the gastrointestinal tract. A supposition is that 1 mmol of all-trans β-carotene should cleave to form 2 mmol of all-trans retinol. Biological functions of vitamin A include roles in vision, growth, and reproduction. Retinol and carotenoids are sensitive to pH, enzymatic activity, and photo and autooxidation reactions associated with the conjugated double-bond structure.

Deficiencies in vitamin A are well characterized to include alterations of epithelial tissue to a squamous keratinizing form, resulting in a decrease in the population of mucus-secreting cells. Vitamin A is also required for normal differentiation of tissue, and in this context the relationship between vitamin A or β-carotinoids and neoplasia has received considerable attention. Vitamin A has strong inhibitory activity on the mutagenicity of aflatoxin B_1 in the

Salmonella typhimurium Ames test (95). Carotenoids have been shown to be significantly associated with reduced risk of lung cancer, more so than retinol (96). This result indicates that carotene does not require prior metabolism into retinol for biological activity.

Hypervitaminosis A can occur both acutely or following chronic exposure. Excessive intake of carotene does not result in hypervitaminosis A but produces hypercarotemia that can lead to the skin taking on a yellowish pigment (carotenosis). Symptoms of vitamin A toxicity have resulted from the intake of polar bear liver extracts, which contain extremely high vitamin A levels (e.g., 20,000 IU/g), or following large intakes of vitamin A supplements. Symptoms range from headache, fatigue, nausea, cracked lips, dry rough skin, epistaxis, and diarrhea in adults to drowsiness, weight loss, scaly dermatitis, anorexia, and skeletal disorders in infants. Due to liver vitamin A storage, liver function abnormalities and hepatomegaly can occur. Vitamin A toxicity can also result in reduced stability of skeletal structures leading to fragile bones. There is a wide variation in individual susceptibility to vitamin A toxicity. Some individuals show toxic symptoms at levels of 50,000 IU per day, whereas other individuals exhibit reactions at daily intakes of 150,000 and 200,000 IU. Compounds such as 13-*cis*-retinoic acid are less toxic.

All *trans*-retinyl palmitate is relatively stable in commonly used storage and cooking procedures, such as boiling at 100°C for 20 min. The biological value of vitamin A palmitate in oils is, however, reduced by 50% after multiple reuses of the oil at temperatures of 170°C (97). Carotenoids in the presence of oxygen are degraded with excessive heating, photooxidation, and co-oxidation with lipid hydroperoxides. The susceptibility of carotenoids to oxidation is evidenced by the almost total loss of carotene from red palm oil following heating at 200°C for 30 min. The safety of the products of oxidation has not been ascertained. Since retinoids are sensitive to oxidation and suppress free-radical-induced oxidation of model lipid systems by quenching lipid-peroxyl radicals involved in lipid peroxidation, the biological activity of the oxidized parent compound is lost. The antioxidant activity is due to the extended polyene structure of retinoids, which are reactive toward peroxyl radicals. Antioxidant activity of carotene is highest at low O_2 tensions (98), suggesting an important role in reducing free radicals in tissues, in particular. Some of the main reaction products are 14-hydroxy-13-methoxyretinyl acetate, 5,6-epoxyretinyl acetate, 11,14-epoxyretinyl acetate, and 5,6,11,14-diepoxyretinyl acetate (99). Chemical oxidation of carotenoids, yielding 5,6-epoxide as the principle product, can undergo further isomerization to mutachrome.

Vitamin E (Tocopherols). Tocopherols are found in the nonsaponifiable fraction of fats and oils and consist of α, β, γ, and δ-tocopherols, with varying natural antioxidant activities. Commercial soybean oil is a rich source of tocopherols and contains about 1500 ppm tocopherols, of which 4, 1, 66, and 29% are α, β, γ, and δ-tocopherols, respectively (100). Wheat germ, canola, peanut, and cottonseed oils are other excellent sources of tocopherols. Vitamin

E deficiency in animals and humans can result in characteristic symptoms, involving vascular muscle and nervous system disorders and reproductive failure. In chicks, vitamin E deficiency results in changes to the central nervous system, referred to as encephalomalacia. In rodents, liver degeneration and reproductive failure are observed when tocopherols are removed from the diet. In humans, although it is rare that individuals suffer from vitamin E deficiencies, low dietary intake of this vitamin coupled with increased physiological requirements can result in erythrocyte hemolysis and hemolytic anemia.

The antioxidant activity of tocopherols is characterized by the donation of a hydrogen atom to more reactive free radicals, thus breaking the oxygen-derived free radical chain and producing a relatively more stable phenoxy radical. Although optimum concentrations of α, γ, and δ-tocopherols required to enhance oxidative stability of lipid are in the neighborhood of 100, 250, and 500 ppm, respectively, prooxidant activity of tocopherols also occur at concentrations that exceed these levels (101). The fate of tocopherols following antioxidant function has been studied. For example, substantial oxidation of tocopherol in soybean oil occurs when the oil is exposed to prolonged (e.g., 18 h) thermal (e.g., 100°C) treatment (102). Both auto- and thermal oxidation products of α-tocopherol in model lipid systems consist of dimer, trimer α-tocopherol quinone and quinone oxides, whereas α-tocopheryl quinone is the most important oxidation product of vitamin E identified in biological tissue (103). Oxidized tocopherols act as prooxidants in oils during storage, which potentially contributes to the production of lipid oxidation products. A similar reaction is likely to occur *in vivo*, as evidenced by the importance of ascorbic acid at the molecular level to reduce tocopherol radicals to tocopherol. Vitamin C can regenerate vitamin E at the radical stage prior to the irreversible conversion to tocopherol quinone (104). Animals, consuming diets deficient in vitamin C, exhibit lower vitamin E levels in liver and lung tissue compared to animals receiving adequate vitamin C. Increasing the amount of tocopherols in the food, or for that matter in lipid-containing tissues by excessive supplementation, could result in greater potential for increased oxidized tocopherol compounds and subsequent prooxidant effect in the body if present with a corresponding ascorbic acid deficiency.

Vitamin E, however, appears to be the least toxic of the fat-soluble vitamins. Recent studies have indicated that α-tocopherol is an important component in preventing oxidation of LDL (105). Oxidation of LDL has many characteristics common with lipid peroxidation in biological membranes, and there is strong evidence to suggest that the oxidation of LDL is important in the cascade of events leading to atherosclerosis. Antioxidants identified in LDL are α and γ-tocopherol, carotene, and lycopene. Oxidation of PUFA in LDL is preceded by losses of endogenous vitamin E.

Vitamin D. Vitamin D is classified as cholecalciferol (vitamin D_3) or ergocalciferol (vitamin D_2) and is obtained in the diet from animal or vegetable sources. Vitamin D_3 is an essential component of calcium homeostasis due to its role in the regulation of calcium and phosphorous for normal bone formation. During endogenous synthesis of cholecalciferol, ultraviolet irradiation of 7-dehydrocholesterol is concentrated in the skin and isomerized to form previtamin D_3. With thermal exposure, previtamin D_3 is converted to cholecalciferol. Cholecalciferol is transported to the liver by a vitamin D_3 binding protein producing 25(OH) vitamin D_3, which is the storage form of vitamin D and the often used index of vitamin D status in individuals. 25(OH) vitamin D_3 undergoes further hydroxylation in the kidney to produce 1α,25-dihydroxycholecalciferol [$1,25(OH)_2D_3$], which represents the major biologically active form of vitamin D_3. The half-life ($t_{1/2}$) of $1,25(OH)_2D_3$ is relatively low (4–6 h) compared to a longer $t_{1/2}$ for vitamins D_2 and D_3 due to storage in fat tissue. The major target organs of $1,25(OH)_2D_3$ are the intestine and bone.

Deficiencies of calcium leads to rickets, a condition of defective bone formation resulting from inadequate deposition of calcium and phosphorous. In vitamin-D-deficient individuals, the treatment of patients with rickets using vitamin D supplementation (106) or ultraviolet light (107) was shown to be successful. The American Medical Association recognized the importance of fortified milk as an important method of preventing rickets in children and recommended vitamin D fortificiation of fluid milk at a level of 400 IU or 10 μg (1 mg of vitamin D is equal to 40,000 IU)/quart of milk. The practice of fortifying milk and other food products in North America and Europe has led to the elimination of rickets.

Vitamin D is potentially the most toxic of the fat-soluble vitamins and excessive intake has been shown to contribute to several organ pathologies (Table 8.2). There are no systematic studies conducted with humans to indicate vitamin D toxicity; however, there is evidence to suggest that the chronic excessive intake of vitamin D can lead to adverse physiological effects. Calcification of soft tissues, potentially contributing to the development of atheroscle-

Table 8.2 Serum and organ changes associated with vitamin D toxicity[a]

Serum Changes	Organ Changes
Hypervitaminosis D_2	Renal calcinosis
Hypercalcemia	Glomeruli tubule damage
Hypermagnesuria	Loss of renal function
Hypercholesterolemia	Metastic calcification
	Heart
	Vascular system
	Skeletal muscle
	Osteitus fibrosa
	Osteoblast mobilization
	Excess bone calcium and absorption replacement by fibrous tissue

[a] Adapted from Refs. 108–110.

rosis and ischemic heart disease, is one example of adverse effects of chronic intakes of vitamin D that exceed the recommended daily nutrient intake (RDNI). Moreover, it has long been known from animal studies that although administering vitamin D is a corrective measure for curing rickets, the feeding of high levels of vitamin D can also induce rickets in animals (111). Increased cardiovascular degeneration has been observed at an estimated intake of 1200 IU/day (108) and renal calculi has been associated with intakes of 800 IU/day (109). These levels of intake compare to the RDNI of 100 IU/day (Canada) and 200 IU/day (U.S.) for nonpregnant, nonlactating women and adult males. The RDNI for vitamin D is higher (400 IU/day) in pregnant women and young infants. The consumption of a balanced diet containing milk (500 mL), eggs (2), cereal, meat, and a vitamin supplement adequately meets the prescribed requirements. Moreover, considering the photochemical contribution, the combined amount of vitamin D available easily mets the daily requirements.

Vitamin D concentration in bovine milk follows a temporal pattern associated with season (e.g., winter 0.06 IU/g fat versus summer 0.23 IU/g fat) and is influenced largely by the exposure of lactating animals to solar radiation (112). A recent concern of vitamin D hypervitaminosis has been the chronic low-level excesses of vitamin D attributed to fortification practices of foods. In one extreme case, an outbreak of vitamin D intoxication was reported to be caused by improper fortification of fluid milk that resulted in measured levels exceeding 230,000 IU/quart (113). There is also some concern about the limitations in vitamin D analysis that may lead to inadequate monitoring of fortified foods (114). These concerns pertain to the risks raised about vitamin D fortification of foods (110). Although acute vitaminosis D can lead to anorexia, nausea, and vomiting followed by polyuria, polydipsia, weakness, pruritus, and impaired renal function, there is growing evidence to suggest that chronic low-level excesses of dietary vitamin D can contribute to osteoporosis and atherosclerotic conditions. Individual pathologies associated with osteoporosis and atherosclerosis resulting from excessive intake of cholecalciferol and metabolites relate to increased bone resorption and soft tissue calcification, respectively. Suggestions have been made to limit or abolish vitamin D fortification in food products and modify the nutritional labeling requirements for calciferol on food products for the purpose of alleviating chronic low-level excesses.

Vitamin K. Vitamin K is a member of the quinone family of compounds and includes the biologically active vitamin K_1 (phytylmenaquinone or phylloquinone) and K_2 (multiprenyl-menaquinone). The major dietary sources of vitamin K_1 are green and leafy vegetables and oilseeds. Soybean and rapeseed oils contain the greatest amount of vitamin K_1 with almond, sunflower, safflower, walnut, and sesame oils containing lesser amount of vitamin K_1. Vitamin K_2 is synthesized by bacterial synthesis in the gastrointestinal tract. Vitamin K is required as a cofactor for the synthesis of τ-carboxyglutamic acid involved in blood coagulation. Vitamin K stimulates the conversion of the precursor of fibrinogen into fibrin and is required for the conversion of prothrombin to thrombin. The current RDNI for vitamin K is 1 μg/kg of body weight. Vitamin K_1 is affected by heat (temperatures 185–190°C), fluorescent light, and daylight. Amber glass bottles are effective in preventing photooxidation of vitamin K_1. Deficiencies in vitamin K result in bruising and increased clotting time and liver disease in adults and infants. From a medical standpoint, the relationship between vitamin K deficiency and increased blood coagulation times leads to administration of vitamin K. Oral supplementation (1–2 mg) of vitamin K from natural sources is considered acceptable due to the low toxicity. Synthetic vitamin K is not normally given orally due to a vomiting side effect.

ADVERSE EFFECTS OF SOME NATURAL CONSTITUENTS IN FATS AND OILS

3.1 Erucic Acid

13-Docosenoic acid, also known as erucic acid (*cis*-13-docosenoic acid; $C_{22}H_{42}O_2$), is found in cruciferous plants, such as mustard seed and rapeseed, and in some marine animal oils. The position of the double bond in this monounsaturated fatty acid relative to other unsaturated fatty acids is shown in Table 8.3. Erucic acid represents approximately 40% of the fatty acids in rapeseed triacylglycerol. Through the process of plant breeding, erucic acid levels in rapeseed oil (canola oil) have been reduced to only trace amounts. Low erucic acid rapeseed oil is similar to olive oil, with the exception of its minor constituents such as squalene and the content of α-linolenic acid. Rodent

Table 8.3 Structure and oil source of reactive fatty acids

Fatty Acid	Structure	Source (example)
Oleic (cis)	$CH_3{-}(CH_2)_7CH{=}CH{-}(CH_2)_7{-}COOH$ (cis: $CH_3{-}(CH_2)_7CH$ ‖ $COOH{-}(CH_2)_7CH$)	Egg, olive oil
Elaidic (trans)	$CH_3{-}(CH_2)_7CH$ ‖ $HC(CH_2)_7COOH$	Hydrogenated soybean oil
Erucic	$CH_3(CH_2)_7CH{=}CH(CH_2)_{11}COOH$	Rapeseed oil
Linoleic	$CH_3(CH_2)_4CH{=}CHCH_2CH{=}CH(CH_2)_7COOH$	Corn, cottonseed
Linolenic	$CH_3CH_2CH{=}CHCH_2CH{=}CHCH_2CH{=}CH(CH_2)_7COOH$	Soybean, linseed, rubber seed
Ricinoleic	$CH_3(CH_2)_5CH(OH){-}CH_2{-}CH{=}CH(CH_2)_7COOH$	Castor oil

and primate animals fed erucic acid develop lipidosis in heart tissues after short-term feeding. Erucic-induced myocarditis, characterized by an accumulation of erucic-acid-containing lipid in cardiac muscle, disappears when erucic acid is removed from the diet. A possible explanation for this effect is the relatively weak binding of erucic acid to albumin and low oxidation.

3.2 Unconventional Oils

Rice Bran Oil. Rice bran oil (RBO) contains a high content of unsaponifiable matter (4.1%), which limits its use as an edible oil without further refinement. High levels of free fatty acids, pigments, and waxes are also characteristic of RBO; however, these constituents are removed with modern refining technologies. Deleterious effects of RBO were detected using a three-generation rodent reproductive assay. Including RBO in rodent diets at a 10% level reduces total cholesterol, LDL and VLDL cholesterol, and triacylglycerol levels (93).

Mahua Oil. Mahua (*Madhuca latifolia*) seed oil is an important source of vegetable butters produced in India. Modern refining and hydrogenation technologies are developed to use this oil source for human consumption. Mahua unsaponifiable matter contains numerous polyenes; however, no nutritional limitations have been found with this oil. Multigeneration reproduction studies conducted with mahua oil in male rodents reported bilateral testicular atrophy and degenerative changes in seminiferous tubules that resulted in infertility. The infertility was reversed with the withdrawal of mahua oil from the diet (115).

Rubber Seed Oil. The rubber tree (*Hevea brasiliensis*) produces seeds with a higher (43% of weight) amount of oil than soybean (20% of weight) but similar content to palm kernel (44% of weight) and melon seed (49% of weight). Rubber seed oil (RSO) has a similar saponin content to soybean oil and an equivalent amount of total essential fatty acid to that of corn oil but is exceptionally high in linolenic acid (17.3%). No toxic effects of RSO have been reported in rats fed oil at dietary levels of 5 and 10%, respectively (116).

Ricinoleic Acid. The unsaturated monohydroxy fatty acid, ricinoleic ($C_{18}H_{34}O_3$), present in castor oil (Table 8.3) exhibits toxicity by cathartic activity that can be attributed to its secretagogue effect on epithelial cells of the small and large intestine (117). Castor oil has been used as an adulteration agent of edible oils.

3.3 Cyclopropenoid Fatty Acids

Cyclopropenoid fatty acids (CPFA) are present in various seed oils of plants in the Mavales order at levels varying from 0.6 to 2.5% crude oil. Deodorization of crude oil in the refining processes reduces CPFA markedly. The primary plant crop germane to Western agricultural and food industries that contains CPFA is cottonseed (*Gossypium hirsutum*). Two long-chain acids that comprise CPFA are malvalic (18 carbons; 2-octyl-1-cyclopropene-1-heptanoic acid) and sterculic (19 carbons; 2-octyl-1-cyclopropene-1-octanoic acid). The effect of CPFA on human health remains to be determined; however, results from animal studies have indicated a potential role for CPFA in adversely altering fatty acid and cholesterol metabolism. Increases in serum cholesterol in rabbits fed 0.5% CPFA have been reported (118). This reponse has been attributed to impaired clearance of cholesterol from the blood pool as a result of abnormal esterification of serum cholesterol. The lecithin–cholesterol acyltransferase (LCAT) activity regulates the source of plasma cholesterol esters *in vivo*. The alteration of the C-2 fatty acyl composition, more specifically the C182 and C20:4 fatty acids of LCAT phospholipid substrate, by CPFA has been shown (119). Moreover, reduced biliary excretion was observed in mice fed CPFA, thus resulting in impaired fecal elimination of cholesterol.

CPFA also inhibit the *in vivo* synthesis of oleic acid from stearic acid, by blocking stearoyl-CoA-Δ^9-monodesaturase activity (120). Consequently, high CPFA intake results in increased deposition of saturated fatty acids and parallel decline in monunsaturated fatty acids in body tissue lipids. Other adverse effects associated with chronic CPFA intake include growth retardation, impaired membrane functional activities, cancer promotion, and reproductive failure (121).

3.4 Structured Triacylglycerols

The interesterification of specific long-chain saturated fatty acids (e.g., stearic or behenic) and short- or medium-chain fatty acids on the glycerol backbone produces structured triacylglycerols. In addition to providing the physical properties of fat required for texture and flavor perception, structured triacylglycerols provide a lower caloric value than traditional edible oils. Two examples of structured triacylglycerols are Caprenin and the SALATRIM family of low-calorie fats. Caprenin is a randomized triacylglycerol composed of caprylic (C8 : 0), capric (C10 : 0), and behenic (C22 : 0) fatty acids. This lipid has been used as a substitute in confectionery fats due to functional and organoleptic qualities that are similar to those of cocoa butter. Due in part to the thermogenic effect of C8 : 0 and C10 : 0, as well as the limited absorption of C22 : 0 fat constituents, caprenin has an energy value of only 5 kcal/g compared to other triacylglycerols that yield 9 kcal/g (122). Short-term feeding studies in rodents have shown no apparent toxicities when consumed as a primary source of dietary fat (123).

The SALATRIM family of mixed triacylglycerols is synthesized by interesterifying hydrogenated vegetable fats (e.g., canola) with short-chain fatty acids (e.g., acetic, butyric, or propionic). SALATRIM triacylglycerols contain a high concentration of stearic acid (C18:0), esterified on positions 1 or 3 of the glycerol molecule. Caloric availability estimates of SALATRIM have been reported to range 4.5–6.0 kcal/g in rats and 4.7–5.1 kcal/g in human subjects (124). The relatively low caloric value of short-chain fatty acids coupled with the low absorption coefficient of stearic acid in SALATRIM fats (125) confer the lower caloric value of SALATRIM compared to conventional edible fat sources (e.g., corn oil, 9 kcal/g). Genetic toxicology studies have reported no evidence of mutagenic or genotoxic activity associated with SALATRIM using a battery of *in vitro* tests using *Salmonella* tester strains (126) and mammalian cell culture (127) assays. Similarly, no symptoms of toxicity could be detected in animals exposed subchronically to graded levels of SALATRIM (128).

3.5 Phytoestrogens

Phytoestrogens occur naturally in various legumes and various oils, including soybean, fennel, coffee, and anise oils (129,130). Examples of phytoestrogens are the isoflavones—genestein, formononetin, diadzein, and biochanin-A—coumesterol, and possibly the triterpenes. These compounds exert estrogenlike activity in reproductive organs of immature and ovariectomized animals (129,131). Soya products and black beans are rich sources of genestein and diadzein, whereas sprout food sources contain high levels of coumesterol and formononetin (132). These distinctly different chemical compounds possess relatively weaker estrogenic activity than natural estrogens. There is little knowledge concerning the long-term health risks of chronic human exposure to plant estrogens.

3.6 Monoterpenes

Mono- and diterpenes are found in many plants and are constituents of essential oils such as orange peel (*d*-limonene), caraway and dillweed (*d*-carvone), and lemongrass, hops, and bay (myrcene). The biological activity of these compounds has been recognized recently from both the standpoint of potential toxicity (133,134) and chemopreventative properties (135). Essential oils enter the human food supply as secondary metabolites of edible plants or as specific constituents of flavoring agents.

d-Limonene has nephrotoxic activity in mature male rodents, but no apparent toxic activity has been reported in humans. Damage to the epithelial cells of the proximal convoluted tubule were observed at 10 mg/kg body weight in male rats (134). Acute toxicity of myrcene, a flavoring additive in manufacture of alcoholic beverages, is low in rodents. Signs of fetal growth retardation and skeletal abnormalities have been reported to occur at 1.2 g/kg in the rodent (136). No embryo-fetotoxicity occurred at 0.5 g/kg myrcene.

A total of 30 essential oils have been tested for relative ability to induce the detoxifying enzyme, glutathione *S*-transferase (GST), in forestomach, liver, and small bowel mucosa in female mice (135). GST activity, which catalyzes the conjugation of electrophilic species to glutathione, resulting in the formation of less toxic, water-soluble metabolites, was significantly enhanced by essential oils obtained from dillweed, lemongrass, lemon, thyme, and caraway. Essential oils derived from oregano, thyme, and summery savory form free radicals with ultraviolet radiation and superoxide radical anion indicating potential antioxidant activity (137).

CHEMICAL REACTION IN FATS

4.1 Hydrogenation and Isomerization Reactions of Fats and Oils

Hydrogenation of vegetable fats and oils raises the melting point of oils to produce fats with enhanced firmness and plasticity and enhances the oxidative stability. Hydrogenation of vegetable fats and oils, rich in PUFAs, is a widely used process in the manufacturing of foods such as margarine and as base stock for the production of shortenings and coating fats. Subjecting hydrogenated vegetable fats to partial crystallization can lead to their use as a substitute for cacao butter in candy bars and other confectioneries. Hydrogenation of fats involves the addition of hydrogen atoms to unsaturated sites on the fatty acid molecule in the presence of a catalyst, at specific temperatures, pressures, and degree of agitation. The process translocates and isomerizes residual double bonds resulting in a cis to trans geometrical isomerization. About 80% of the double bond of the *cis*-octadecenoate isomer is located at position 9 and that of *trans*-octadecenoate isomers are positioned at carbons 9, 10, and 11 in hydrogenated vegetable oils. The extent of saturation, geometric isomerization, and the movement of double bonds along the fatty acid carbon chain are dependent on the parameters of the hydrogenation process.

Small amounts of trans fatty acids are also found in animal fats, such as butter and in some instances egg yolk. Trans fatty acids occur naturally in butter as a consequence of the biohydrogenation of fatty acids by facultative anaerobic rumen microflora in the cow. Trans unsaturated fatty acid isomers in egg yolk originate from the content in the diet of the hen and the ability to accumulate within vital embryonic lipid pools (138).

Kummerow (139) estimated that approximately 2 billion pounds of hydrogenated frying fats are used by the U.S. food industry for frying fish, chicken, potatoes, doughnuts, and other foods. Trans unsaturated fatty acids, parallel saturated fatty acids in acyl transfer reactions, oxidation, and yield of calories, cholesterol esterifying activity, and incorporation into triacylglycerol and phospholipid. For example, trans fatty acids occur mainly in position 1 of phospholipids, similar to saturated fatty acids, whereas cis unsaturated fatty acids normally occupy position 2 of the phospholipid.

Digestibility of hydrogenated fats and oils varies from 80 to 98% in humans. The triacylglycerol hydrolysis by pancreatic lipase and the absorption capabilities of the intestinal mucosal cells are not affected by the trans double bonds present in elaidic acid (trans oleic acid), as evidenced by the equivalent incorporation of both deuterium-labeled triolein and trielaidin into chylomicrons (140). However, cis–trans and trans–trans isomers of linoleic acid are devoid of essential fatty acid activity and only a small amount of these fatty acids are converted to longer chain PUFAs, thus intensifying essential fatty acid deficiency. Changes in erythrocyte and liver mitochondria membranes of rats fed hydrogenated olive oil containing a high trans fatty acid content coincide with adverse effects observed in animals afflicted with essential fatty acid deficiency (141). This observation is explained by the incorporation of trans fatty acids into phospholipids in a similar manner to that of MUFA and SFAs. The result of trans fatty acid incorporation into membrane phospholipids is a change in fluidity and permeability character of the cell membrane associated with the rigid structure and higher melting point of the trans isomer. This change in turn can influence constituent enzyme activities. Concerns of high dietary exposure to isomeric unsaturated fatty acids has been expressed for potential roles in the etiology of cancer, coronary heart disease, and effect on infant development.

Trans Isomers and Cancer. The efficacy of trans fatty acids to cause tumor initiation and promotion has undergone considerable investigation. Enig and co-workers (28) identified trans fatty acids as a significant factor in the correlation between vegetable fat intake and cancer mortality over a duration of 60 years. Many studies have examined the role of trans fatty acids in the development of colon and mammary tumors, since these two types of cancers are highly correlated with the level of dietary fat. A hypothesis has been advanced that alterations in intestinal microflora, bile acid composition, and excretion and membrane function are affected by trans fatty acids. Changes in intestinal microflora can result in the transformation of primary bile acids to secondary metabolites potentially involved in the induction or promotion of colon cancer. Reddy and co-workers (29) fed diets containing corn oil and trans fatty acids at low (5%), intermediate (13.6%), and high (23.5%) amounts to rats treated with azoxymethane, a known colon carcinogen. The results of these experiments indicated that high-fat diets resulted in a significantly greater number of colon tumors. However, the presence of trans fatty acids was shown to have no significant effect on the incidence of adenocarcinoma of the colon. Similar studies have been conducted with mammary tumors induced by the polyaromatic hydrocarbon, dimethylbenz(*a*)anthracene (DMBA). The addition of high (20%) or low (5%) levels of trans fatty acids to a diet containing a blend of fats did not result in higher incidences of chemical-induced mammary tumor yield in rats than in control animals fed a similar fat blend diet containing appropriate levels of cis fat or corn oil (30). Similarly, trans monoene isomers have no promotional effects on either liver or mammary tumorigenesis when compared to high saturated fatty acid or high cis monoene or cis diene fats (142). Results from these studies indicate that trans isomerization was not a factor in the enhancement of chemical-induced carcinogenesis in animals fed diets high in vegetable oils.

It is of particular interest that isomers *cis*-9, *trans*-11; *trans*-10, *cis*-12; *trans*-9, *trans*-11; and *trans*-10, *trans*-12 represent approximately 90% of the total heat-generated derivatives of conjugated linoleic acid (CLA). Isomeric derivatives of *cis*-9, *cis*-12-octadecadienoic acid (linoleic acid), containing a conjugated double-bond structure has been shown to have anticarcinogenic activity against the initiation of mouse skin carcinogenesis by 7,12-dimethylbenz-(*a*)anthracene (DMBA) (143) and benzo(*a*)pyrene(B(*a*)P)-induced mouse forestomach neoplasia (144). The reduced lipid peroxidation reported in mammary tissue following exposure to CLA indicates antioxidant activity (145). However, the fact that maximum concentration for antioxidant activity was shown to be related to dietary levels of 0.25% CLA, compared to 1% dietary CLA for maximum tumor inhibition, indicates that other mechanism(s) are involved in the anticarcinogenic activity of CLA.

Trans Isomers and Coronary Heart Disease. The role of trans fatty acids in dyslipidemia is controversial due to the large variability between individuals in response to trans fatty acid intake. Animal studies using swine indicate that undesirable effects associated with feeding trans fatty acids are blunted by the presence of saturated fat in the diet. Studies examining the effect of trans fatty acids on lipoprotein cholesterol levels in healthy adults have concluded that trans fatty acids have the same effect as saturated fat in raising LDL-cholesterol and decreasing HDL-cholesterol (146). A diet containing approximately 8% trans fatty acids has been shown to significantly lower HDL-cholesterol and raise LDL-cholesterol relative to the linolenic acid effect (32). Animal studies have also shown that both saturated or hydrogenated fats in combination with cholesterol, effectively decrease LDL receptor activity. These results suggest that dietary hydrogenated fats are effective in suppressing hepatic LDL receptor activity and altering lipoprotein composition. Trans fatty acids also have little effect on the development of atherosclerotic lesions in animals fed diets with varying cholesterol content.

A second mechanism for the potential undesirable effect of trans fatty acids in relation to coronary heart disease involves a role in altering essential fatty acid synthesis. *Trans* fatty acids exacerabate the deficiency of essential fatty acids (147) by effectively interfering with hepatic metabolism of cis γ-linolenic and cis α-linolenic acids by competitive inhibition of the Δ^6 desaturase enzyme. This effect reduces the synthesis of the 1-series eicosanoids because the substrate required for the synthesis of this series of eicosanoids, namely dihomo gamma-linolenic acid, is derived from γ-linolenic acid. A similar deficiency in the 3-series eicosanoids is also possible since the substrate, eicosapentaenoic acid, is also formed by successive desaturation and elongation of cis γ-linoleic acid. Since the formation of 2-series eicosanoids from arachidonic acid is independent of cis γ-linolenic acid, these compounds are not influenced by the low availability of substrates derived from elongation and desaturation reactions. However, a chronic eicosanoid imbalance results from the interferences of trans fatty acid on lipid metabolism, when the absence of essential fatty acid feedback inhibitors like dihomo-gamma-linolenic acid, docosahexaenoic acid, and the 1-series prostaglandins, lead to an excess of arachidonic acid and consequently the 2-series eicosanoids.

Taken together, the contribution of trans fatty acids to atherosclerosis and increased risk of coronary heart disease may be reduced by a decrease in dietary hydrogenated fat intake and cholesterol consumption as well as maintaining an adequate intake of essential fatty acids.

Effects of Trans Isomers on Neonatal Growth. Both omega-6 and omega-3 long-chain PUFA are required for the synthesis and deposition of eicosanoids in membrane lipids and neural and visual functional development. The effects of trans fatty acid interference on the biosynthesis of long-chain PUFA is also present in the neonate, where trans fatty acids have been inversely correlated to birth weight, but not gestational age (31). This observation raises the concern as to the safety of high intake of trans fatty acid isomers during pregnancy and the perinatal period. Infants receiving human milk ingested levels of trans fatty acids that reflect the level of dietary trans fatty acids consumed by the lactating mother (148). Animal studies that examined the effect of partially hydrogenated fat from fish and soybean oil with a trans fatty acid content of about 30%, included in diets containing linoleic acid (C18:2 *n*-6, cis–cis), reported no detrimental effects on both mother and offspring during pregnancy and the following lactation period (149). The potential adverse effects of trans fatty acids were mitigated by the intake of essential fatty acids produced beyond the Δ^6 desaturase stage of fatty acid metabolism.

4.2 Heterocyclic Amines

Browning reactions occur between oxidized unsaturated lipids and amines, amino acids and proteins in both *in vitro* and *in vivo* conditions. Carbonyl compounds and free hydroperoxides produced from lipid oxidation react with free amine groups to form Schiff's bases and resultant brown pigments. These pigments can be produced in rats fed diets containing oxidized fish oil. Similarly, hydroperoxides derived from methyl arachidonate or methyl docosahexaenoate react with phospholipids producing fluorescent pigments. Aldehydes and ketones, two examples of reaction products of lipid autooxidation, react with proteins and degraded amino acids via the Strecker degradation pathway. Heterocyclic aromatic amines can be derived in part from imidazo moieties originating from the reaction of creatinine in protein-rich foods with Maillard reaction products. The formation of these products, which include pyrazines, pyridines, and aldehydes, can be enhanced by lipids in model systems (150). The possibility that lipids are reactive with the intermediates of heterocyclic compounds was realized by Johansson and co-workers (151) in studies with both corn and olive oils. In addition to the Maillard-type reactions, the formation of free radicals from lipids is an important component of heterocyclic synthesis, since free radicals have been proposed to increase the formation of heterocyclic amines (152).

Heterocyclic aromatic amines found in fried meat and fish include IQ (2-amino-3-methylimidazo[4,5-*f*]quinoline; $C_{11}H_{10}N_4$); MeIQx (2-amino-3,8-dimethylimidazo[4,5-*f*]quinoxaline; $C_{11}H_{11}N_4$); and DiMeIQx (2-amino-3,48-trimethyllimidazo[4,5-*f*]quinoxaline $C_{12}H_{13}N_5$). Toxic potential of these compounds includes mutagenicity in the Ames *Salmonella typhimurium* assay (153) and induction of mammary and liver tumors (20). Fatty acids such as oleic and linoleic acids inhibit the mutagenicity of heterocyclic amines (154).

ADVERSE PRODUCTS OF LIPID OXIDATION DURING NORMAL PROCESSING

5.1 Lipid Oxidation Products (LOPs)

Unsaturated fatty acids undergo oxidative reactions when exposed to oxygen, radiant energy, and a number of inorganic and organic catalysts. The relative position of the unsaturated fatty acids in the triacylglycerol molecule containing palmitate, stearate, oleate, and linoleate can influence the rate of oxidation of the lipid. Oxidation of PUFA yield conjugated diene hydroperoxides, which are more toxic than the unoxidized parent compounds. Autoxidation of lipids depends on the composition of fat and the degree of unsaturation, presence of prooxidants such as transition metals, and heme compounds as well as the presence of antioxidants. The number of positional isomers from PUFA autoxidation is determined by the formula, $2n$-2, where n is the number of double bonds in the fatty acid. Thus, the number of positional isomers in linoleic, linolenic, arachidonic, and eicosapentaenoic acids are 2, 4, 6, and 8, respectively. Spontaneous autooxidation of PUFA occurs relatively slowly; however, autooxidation occurs with little or no induction period when in the presence of prooxidants. In food systems, the storage temperature, water activity (a_w), and the level of oxygen regulate the rate of oxidative reactions. Increasing temperature and decreasing a_w enhance the extent of fat oxidation. In muscle foods, the high content of phospholipids containing large amounts of PUFA are susceptible to oxidation. Peroxidative changes in muscle foods are initiated at the membrane level with lipids associated with subcellular organelles such as sarcoplasmic reticulum and mitochondria. The disruption of muscle membranes with grinding, chopping, flaking, and deboning processes thus accelerates the development of LOPs.

Ozone represents another potential agent for the development of lipid oxidation. The chemical interaction between the cyclo group of the ozone molecule with an unsaturated bond of a PUFA is referred to as the Criegee mechanism. This reaction yields hydrogen peroxide and carbonyl compounds (e.g., aldehyde or ketone). The dipolar structure of ozone renders the addition of a 1-3 cyclo group to PUFA unsaturated bonds. This reaction results in the formation of a primary ozonide, which further decomposes to a carbonyl compound and zwitterion; subsequently resulting in the formation of hydroxyl-hydroperoxides, and further decomposition products of carbonyl and hydrogen peroxide. Some of the various breakdown products of ozone-PUFA intermediates include, hydroperoxide radical, ozonide radical ion, superoxide radical ion, and hydrogen peroxide. Aldehydes, ketones, and hydrogen peroxide are the main ozonation products of PUFA and contribute to the sharp and unpleasant odors associated with lipid oxidation.

Oxidation reactions in lipids containing unsaturated fatty acids are initiated when a labile hydrogen atom is abstracted from a carbon atom next to the double bond of the unsaturated fatty acyl chain, producing a free lipid radical that reacts with oxygen to form a peroxy-free radical. Peroxy radicals are relatively stable *in vivo* ($t_{1/2}$ = approximately 8 sec) and diffuse to various cellular fractions, where they interact with the most susceptible oxidizable material of the cellular constituents. The peroxy-free radical abstracts a hydrogen from another hydrocarbon chain producing a hydroperoxide and a free radical. The principal pathway of peroxyl radical formation is therefore autooxidation (e.g., lipid peroxidation). Hydroperoxides, halogenated hydrocarbons, and transition metals are examples of effective prooxidants that initiate oxidative damage. In the presence of transition metals, such as iron, copper, and nickel, lipid hydroperoxides or organic hydroperoxides decompose rapidly, generating reactive oxygen species including peroxyl (ROO·) and hydroxyl (·OH) radicals. In addition, peroxy radicals can be further converted to alkoxyl (RO·) radicals, producing more reactive and less selective free radicals that initiate oxidation chain reactions from all autoxidizable materials. The primary oxidation products of fat are fatty acyl hydroperoxides, which subsequently degrade to peroxy, alkoxyl, and hydroxyl radicals and a large number of secondary autoxidation products, including malondialdehyde, hydrocarbons (i.e., ethane and pentane), saturated and α-β and 4-hydroxy unsaturated aldehydes, and ketones. Further oxidation may occur with the unsaturated aldehydes or in original peroxides to form cyclic peroxides, bicyclic endoperoxides, and epoxides. *In vivo,* lipid hydroperoxides can be reduced to lipid alcohols in the presence of a reductase such as the cytochrome P-450 complex, which is particularly active in the hepatic, pulmonary, and intestinal tissues. Both peroxy-radical dependent and cytochrome P-450 dependent oxidation reactions potentially contribute to a process of co-oxidation, which increases the number and complexity of potential autoxidation products derived from xenobiotic agents [e.g., benzo(*a*)pyrene and aflatoxin B_1; Figure 8.3).

In the process of *in vivo* lipid peroxidation, peroxyl radicals react with DNA, causing DNA strand breaks, protein–DNA crosslinks, and oxidative modification of DNA bases. Due to the greater stability of peroxy radicals, greater selectivity for DNA, RNA, and protein macromolecules occurs. Lipid peroxidation reactions occurring *in vivo* have been implicated in deteriorative

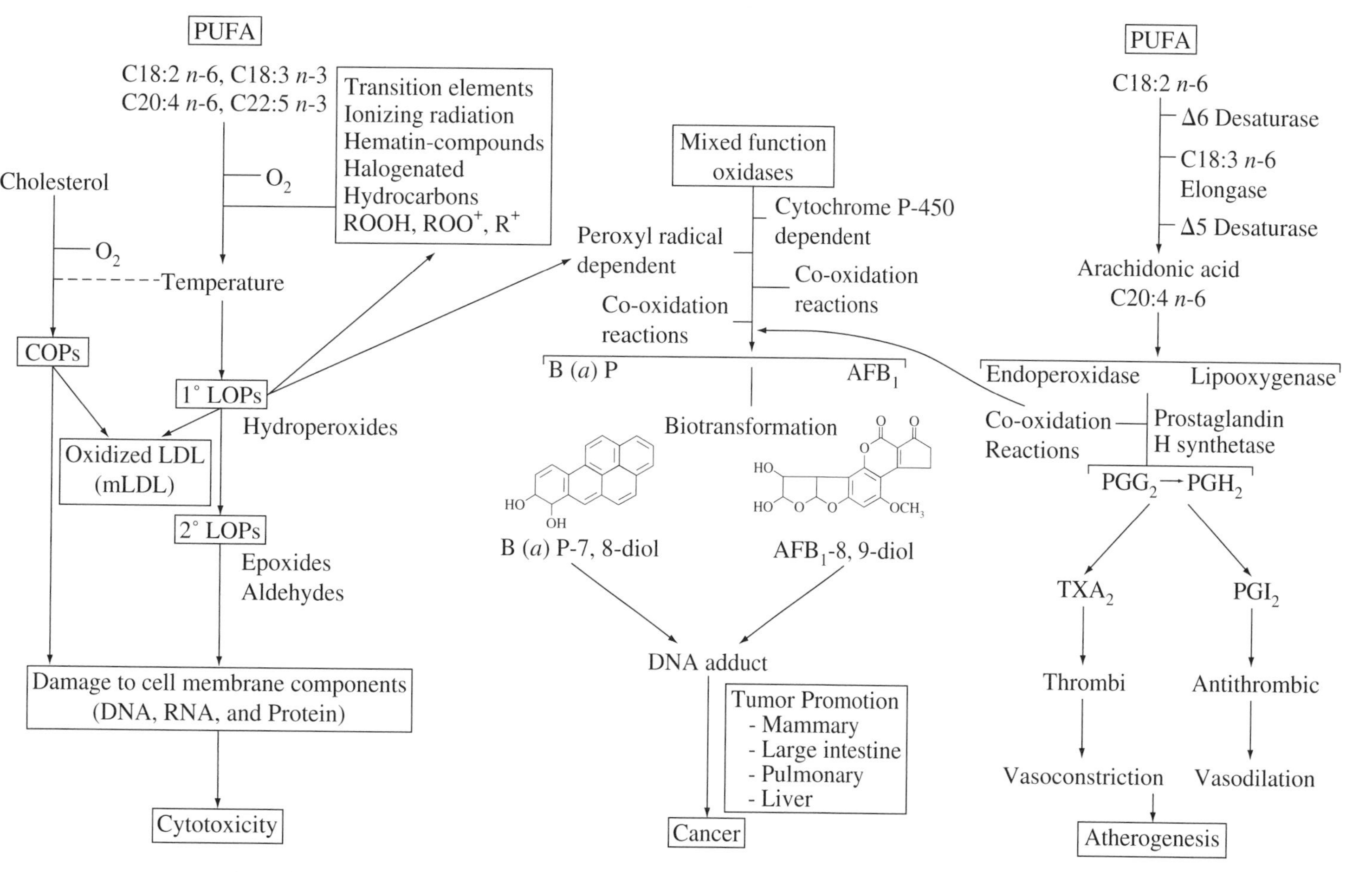

Figure 8.3 Schematic diagram showing the interaction between lipids, products of lipid metabolism and xenobiotics, and biological activities potentially leading to different chronic diseases.

processes such as aging in humans (155) and oxidation of lipoproteins. Oxidation of LDL has many attributes similar to lipid peroxidation in biological membranes. The initiation of peroxidation of LDL–PUFA leads to the production of LDL fatty acid peroxides, which in turn results in extensive scission of the fatty acids. Products of this reaction covalently attach to apoprotein B and mask the ε-amino group of lysine residues, producing oxidized LDL (mLDL). mLDL is causally associated with the formation of lipid-laden foam cells that accumulate in the arterial intima and lead to atherogenesis. Lipid peroxides have also been detected in aorta and are positively correlated with severity of aortic atherosclerosis (33). Numerous products of lipid peroxidation chain reactions have also been regarded as promoters and mutagens (156), while animal studies have documented the toxicity of lipid hydroperoxides (157). Products of lipid oxidation and their decomposition products cause significant damage to proteins, biological membranes, and cellular components that influence important cellular processes. Regardless of these findings, only a paucity of direct evidence exists that links the presence of lipid hydroperoxides in foods with a chemical hazard to human safety. Some scientists have indicated that the levels of hydroperoxides that occur normally in foods are very low and that this aspect, combined with their rapid decomposition, does not render them a significant health hazard.

Majority of LOPs absorbed from the diet are excreted as CO_2. The LD_{50} of methyl linoleate hydroperoxide has been estimated to be 1–2 mg when administered intravenously in rabbits (157). Only weak mutagencity of autooxidized linoleic and linolenic acids, oxidized methyl esters, and fatty acyl hydroperoxides has been observed in the Ames *S. typhimurium* bacterial assay (158). Part of this response may be attributed to the mutagenic activity of hydrogen peroxide, which is removed quickly by exogenous catalase. In cell culture systems, peroxyl radicals generated from lipid autooxidation can react with polynuclear aromatic hydrocarbons to produce carcinogenic arene oxides (159). Peroxyl radicals derived from unsaturated hydroperoxides initiate co-oxidation of B(*a*)P, resulting in the generation of epoxide B(a)P 7,8 diol (Figure 8.4). The conversion of benzo(*a*)pyrene to mutagenic products by LOPs has been shown *in vitro* using the sister chromatid exchange assay (160). The *in vivo* co-oxidant activity of dietary lipid with B(*a*)P resulting in B(*a*)P-DNA adducts is greater with PUFA than SFA lipid sources (161).

Secondary products of lipid autoxidation include the production of malondialdehyde (MDA), 4-hydroxyunsaturated, and α and β unsaturated aldehydes, cyclic peroxides, and saturated and unsaturated hydrocarbons. The production of MDA, a reactive three-carbon secondary product of polyunsaturated lipid oxidation, increases with the degree of unsaturation and has been a useful indicator of lipid peroxidation. MDA content in various muscle foods varies considerably and can be affected by the characteristic lipid content and the heat and time during cooking. Secondary products of LOPs are absorbed from the intestine and deposited in the liver (162). Urinary excretion of MDA following the administration of oxidized lipid to rats has been reported, thus

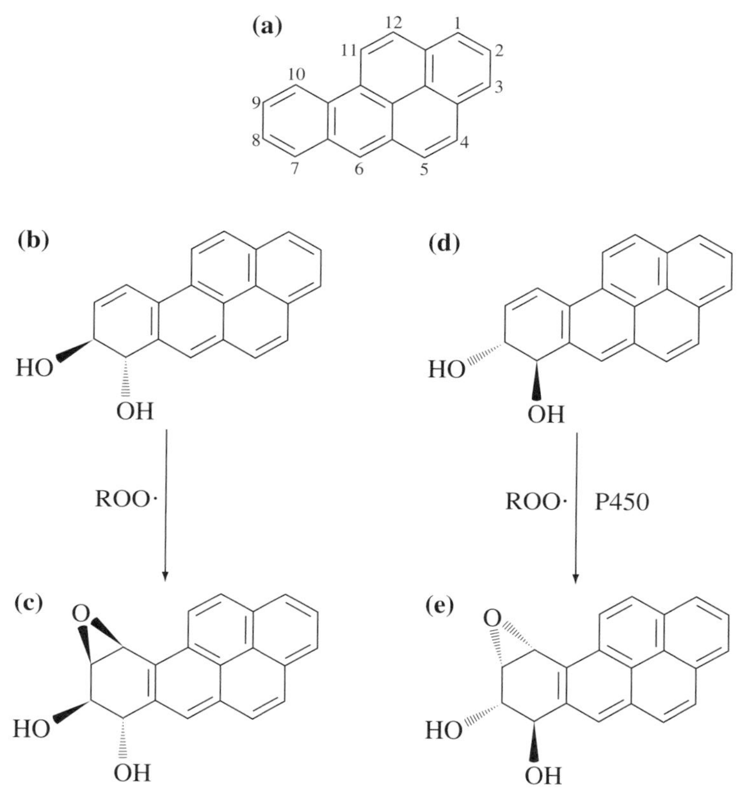

Figure 8.4 Peroxy-radical and P-450 catalyst of benzo(*a*)pyrene (B(*a*)P) diol oxidation. (**a**) B(*a*)P; (**b**) (+)-B(*a*)P diol; (**c**) (−)-anti-diolepoxide B(*a*)P enantiomer; (**d**) (−)-B(*a*)P diol; (**e**) (+)-anti-diolepoxide enantiomer. ROO· = peroxy radical; P-450 = cytochrome P-450 monoxygenase. The (−)-B(*a*)P-7,8-diol enantiomer generated from B(*a*)P *in vivo* is epoxidized by both ROO· and cytochrome P-450.

indicating *in vivo* peroxidation (163). The toxicity of MDA, expressed as LD_{50} is approximately 600 mg/kg body weight, which signifies greater toxicity than either glyoxal or formaldehyde (164). Mutagenic and genotoxic activity of MDA is associated with crosslinking with cytosine and guanidine on the DNA molecule. The prolonged application of MDA to skin of mice results in tumor formation. Model studies have indicated a co-carcinogenic action for MDA-induced carcinogenic activity involving MDA catalyst activity and the formation of *N*-nitrosamines (165). Unsaturated aldehydes and 4-hydroxyaldehydes are additional toxic secondary products of PUFA autoxidation. 4-Hydroxynonenal is particularly cytotoxic.

The reaction between LOPs with the amino acids, methionine, histidine, lysine, and tryptophan reduces the nutritive value of protein-rich foods. Methionine is readily oxidized to its sulfoxide in the presence of LOPs, whereas loss of tryptophan occurs in response to reactions of the indole ring with secondary products of fat oxidation. Interactive effects of LOPs with amino acids reduce the nutritive value of proteins by initiating protein crosslinking or by forming co-oxidative protein products. Peroxidized lipids have adverse effects on vitamins C, E, A, D and folate by reducing vitamin activity. Tocopherol loss occurs with an increase in primary and secondary oxidized products of fatty acids. Superoxide anion is involved in the decomposition of tocopherols (166).

5.2 Cholesterol Oxidation Products

Autoxidation of cholesterol to cholesterol oxidation products (COPs) occurs spontaneously when exposed to air, heat, photooxidation, and oxidative agents (167). Generation of COPs in buffer is a surface phenomenon, occurring only after prolonged exposure to light [wavelength range = 300–700 nm; (168)]. Due to the reactivity of the allylic C-7 position and the two C-20 and C-25 tertiary carbons on cholesterol, more than 70 oxidation products of cholesterol are possible. A number of reactions central to the oxidation of cholesterol occur. The primary COP is cholestanetriol (cholestan-3-5α,6β-triol), but many of the major COPs have also been identified and include epoxy derivatives [e.g., β-epoxycholesterol (cholestan-5α,6β-epoxy-3β-ol) and α-epoxycholesterol (cholestan-5α,6α-epoxy-3β-ol)]. *In vivo*, cholesterol 5,6-epoxides undergo enzymatic hydrolysis or are metabolized to bile acids (82). Oxysterols with attached hydroxyl groups at positions 7β (cholest-5-en-3β,7β-diol), 7α (cholest-5-en-3,7α-diol), 20 (cholest-5-en-3β,20-diol), and 25 (cholest-5-en-3β,25-diol) and a conjugated ketone as 7-ketocholesterol (cholest-5-en-3β-ol-7-one) have also been identified. The COP with the greatest degree of oxidation and potential toxicity is cholestane-triol (cholestan-3β,5α,6β-triol). The structural similarities of many COPs have made quantitative identification in food systems difficult. Moreover, structural alteration of some COPs (e.g., cholesta-3,5-diene-7-one derived from 7 keto-cholesterol) can occur during the purification procedure (169).

Some foodstuffs that have been identified as potential sources of COPs include spray-dried egg products (169–173), dairy (168), meat products (174,175), heated tallow (176,177), and sun-dried fish [*Spratellordes gracles* and *Decapterus maruodsi*; (178) Table 8.4]. COPs are also present in human breast milk, serum lipoproteins, and aortic tissue (Table 8.4). Estimates of dietary intake of COPs from meat sources range from 0.5 to 1.0% total cholesterol. Heating lard at 180°C results in appreciable losses (e.g., 70%) of cholesterol after 150 h and the appearance of various COPs (177). Cholestan-3β,5α-6-triol has been detected after only 20 min of heating lard, while 5,6α-epoxycholesterol and 7β-hydroxycholesterol steadily increased in concentration up to 100 min of heating. The fact that these particular COP isomers displayed a heat-labile character after 100 min of heating at 180°C, is a

Table 8.4 Presence of cholesterol-oxidized products (COPs) in foods and biologicals[a]

Source	Analysis Method	COP-Isomers								Reference
		Hydroxy				Epoxides		Keto		
		20-	25-	7α	7β	α	β	7	Triol	
Food										
Egg										
Fresh	GC, GC–MS	NR	NR	a	NR	b	—	a	—	170,171
Spray-dried	GC, GC–MS,	e	a–d	—	f	c–e	—	e	c	171,172
	NMR	—	—	d	f	d	d	d	e	170
Stored										
3 months	NMR	—	—	b	b	b	d	c	—	170
6 months	GC–MS	d	a	—	c	d	—	a	a	
12 months	GC–MS	d	a	—	d	e	—	d	c	172
Beef										
Raw	GC–MS	a	a	a	a	a	b	b	—	175
Cooked[b]	GC–MS	a	a	b	b	a	c	b	—	
Tallow[b]	GC	—	—	+	+	+	NR	+	NR	176
Potato fries[c]	GC	—	+	—	+	—	±	+	+	179
Pork										
Raw	GC–MS	—	a	a	a	a	a	b	a	175
Cooked		a	a	b	b	a	b	c	a	
Stored		—	—	g	g	e	—	g	b	174
Butter	TLC			+	+					168
Biologicals										
Modified LDL										
HDL	GC–MS	NR	—	—	a	bc	bc	c	—	180
Aortic tissue	GC–MS	—	—	g	c	e	a	a	—	
	GC–MS	NR	+3	+3	+3	NR	NR	+4	—	181

[a] Value scores: NR, not reported; (— less than 0.1 ppm, a = 0.1–0.5; b = 0.6–1.0; c = 1–5; d = 5–10; e = 11–20; f = 21–50; g = 50 ppm; + = positively reported; +3 = human aorta tissue; +4 = animal aorta tissue; modified low density lipoprotein (LDL), high density lipoprotein (HDL). GC = gas chromatography, GC–MS = gas chromatography–mass spectroscopy, NMR = nuclear magnetic resonance, TLC = thin layer chromatography.

[b] Heated samples $>$ 155°C.

[c] Heated samples $>$ 155°C, cooked in tallow fat.

plausible explanation for the observation that the amount of cholesterol lost during the thermal process did not equal the amount of COPs generated. The heating of tallow for deep fry cooking results in significant levels of COPs in French fried potatoes (179). Total COPs in dried eggs range from undetectable levels to approximately 300 ppm (182). Conversely, fresh fluid milk, butter, and shell eggs do not contain detectable levels of COPs. The application of heat to meat products during cooking increases levels of the 7-series (β and α), more so than cholesterol epoxides and cholestane-triol. Similarly, in dried whole egg and egg yolk, the application of heat during spray-drying increases cholesterol β and α-oxide content. Enhanced generation of COPs during cold storage (e.g., 3 months at −20°C) of both raw and cooked meat has also been reported (Table 8.4). It is apparent that the exposure of foodstuffs such as butter and meat products to heat prior to cold storage intensifies the generation of COPs.

The liver is a primary location for retention of COPs following ingestion of a meal rich in oxysterols. COPs are efficiently metabolized and excreted in fecal matter. Further metabolism of oxysterols following absorption from the intestine includes fatty acyl or sulfate transformation, transport in serum lipoproteins, and degradation to bile acids by hepatic tissue. Biliary secretion followed by intestinal microflora metabolism and fecal excretion occurs similarly to that of cholesterol. In human feeding studies, postprandial plasma levels of COPs have been shown to be elevated only 3 h after individuals were fed spray-dried egg powder. COPs were not detected in control subjects fed a meal containing fresh eggs, free of COPs (183). These findings confirmed earlier observations made with 25-hydroxycholesterol in animal experiments (184) and furthermore demonstrated the capacity for absorption of COPs from food sources. The absorption pattern and presence of COPs in serum, albeit highly variable between individuals, is more transient than other lipids (e.g., serum triacylglycerols). There is substantial evidence that COPs are carried by chylomicrons and LDL and may be involved to some degree in contributing to the production of modified LDL, which is an important step in facilitating atherosclerosis (167).

Cholesterol hydroperoxide and epoxide derivatives have been associated with atherogenesis (185), weak mutagenicity (186), cytotoxicity (187), and inhibition of cholesterol synthesis enzyme (HMG-CoA) activity (181). High circulating levels of cholestan-5α,6α-epoxy-3β-ol have been reported in serum collected from patients with hypercholesterolemia and high blood pressure (188). The cholesterol oxides, 25-hydroxycholesterol, 7-ketocholesterol, cholestane-3β,5α,6-triol, and 5,6-epoxide exhibit similar inhibition of cholesterol biosynthesis and LDL receptor function (189). This product of cholesterol oxidation is also believed to be an active agent in the induction of atherosclerotic plaques in rabbit aorta (184). The intravenous administration of 25-hydroxycholesterol and cholestane-3β,5α,6-triol to rabbits results in a marked angiotoxic effect (180). Angiotoxic COPs have been isolated in plasma chylomicron and LDL fractions in fasted human subjects (190).

Finally, a recent hypothesis has been put forth indicating an antioxidant role for cholesterol that resembles ascorbate and β-carotene from the standpoint that all three exhibit some degree of protection toward lipid peroxidation as well as having prooxidant activity under various circumstances (191). This suggestion is based on the reduction of lipid peroxidation reactions occurring in membranes containing cholesterol.

ADVERSE PRODUCTS FROM OVERHEATED FATS AND OILS

The occurrence of thermal and oxidative changes of fats and oils during cooking or processing of foods is a concern from both the standpoint of organoleptic and nutritional quality of foods prepared, as well as possible adverse health aspects. Thermal processes employed with fats and oils for industrial and home usage include frying, canning, and stewing. Oxidative changes in heated fats are greatly accelerated (oxidation rate doubles for every 15°C increase in temperature), compared to similar reactions occurring at ambient temperatures. The useful life of the cooking oil is dependent on the extent and nature of reactant products generated as a consequence of the source of fat used for frying, the food source undergoing this process, and the conditions used in frying. Heating of fats and oils results in the hydrolysis of triacylglycerols to fatty acids and the free radical mediated oxidative reactions of PUFA. MUFA sources, such as olive oil, are relatively thermostable compared to soybean or sunflower oils, which contain high concentrations of *n*-6 PUFA and some *n*-3 PUFA. Conversely, high concentrations of long-chain *n*-3, C:20–C:22 PUFA in fish oils are particularly susceptible to thermal oxidation. Low-fat and fatty foods also vary in their response to deep frying. Fat from the frying bath penetrates low-fat foods efficiently, thus contributing to the total lipid content. A bidirectional transfer of lipid into the frying bath from the food source, as well as uptake of lipid into the food from the frying bath occurs in fatty foods. Muscle foods rich in PUFA, such as fish, permit fewer refrying procedures due to the contribution of PUFA to the cooking oil and subsequent susceptibility of long-chain *n*-3 PUFA to thermal degradation.

Extensive heat treatment of fats can result in the production of partially transformed volatile chain scisson products, nonvolatile oxidized derivatives, cyclic substances, dimers, and polymers. Dimers and polymers both consist of oxidized and nonoxidized reactants. In the case of deep-fat frying, fats and oils are generally heated to less than 200°C. Oxygen is incorporated into the oil by agitation and water is released into the oil from the food undergoing frying. Depending on the temperature of heating, chemical reactions involving lipid oxidation (autooxidation at temperatures up to 100°C), thermal polymerization (reactions occurring between 200 and 300°C), and thermal oxidation (reactions occurring at approximately 300°C) occur. Thus, factors including the temperature and duration of heating, repeated reheating, extent of aeration and agitation, composition of triacylglycerols, and presence of catalysts and antioxidants collectively act to influence the extent of oxidation, hydrolysis, and polymerization reactions that ultimately influence the rate of fat degradation. Moreover, it is important to distinguish between thermal oxidative changes taking place in fats used in the thermal process versus the oxidative changes taking place in the food matrix. This distinction is required in part to the heat transfer afforded by the fat during the cooking process. The food surface/fat volume ratio used in frying affects both penetration kinetics of fat and the formation of potentially toxic products. For example, the temperatures reached in both the food and the fat source used in frying may not be equivalent, and thus can influence products of thermal oxidation differently in the food source. Moreover, the moisture content present in the food system undergoing a deep-frying thermal process may also act to reduce the thermal damage of lipid within the food matrix. Temperatures reached in the frying kettle are not obtained in the food until a sufficient amount of water has evaporated, thus allowing for fat penetration into food from the kettle to occur.

The type of fat (e.g., saturated versus unsaturated) is another factor regulating the composition of reactive products. The reactivity of fatty acids increases with the degree of unsaturation. Moreover, the distribution and geometry of double bonds on the fat molecule also influences reactions involved in thermal and oxidative deterioration of fats and oils. Thus, the triacylglycerol and fatty acid composition of both the fat used for frying as well as the fat source present in the food system undergoing a thermal process are important factors in evaluating thermal oxidative changes. Reactions resulting from the thermal processing of fats and oils lead to a decrease in the unsaponifiable materials in response to the formation of volatile and nonvolatile decomposition products. Of the 200 or more volatile compounds isolated from heated oil, many are released from the oil into the atmosphere, while others are incorporated into the food and contribute to the characteristic flavors associated with fried foods. Products of volatile decomposition include free fatty acids, lower alkanes, saturated and unsaturated aldehydes and ketones, and a number of different hydrocarbons (192). Nonvolatile products are absorbed into the food along with the oil. Nonvolatile compounds present in the oil cause further degradation of lipid through oxidation and give rise to different physical changes such as increased viscosity, darkening of color and characteristic foaming. Nonvolatiles derived from the formation of secondary products of oxidized unsaturated fats consist of monomeric oxygenated products, cyclic fatty acid compounds, dimers, and polymers. Monomeric oxygenated products have undergone isomerization of double bonds with the formation of shorter chain fatty acids and fatty acids with trans and conjugated double-bond systems. Epoxides are three-member ring-structured unsaturated ether derivatives, which are derived from oxygenated products of unsaturated fatty acids and are abundant in the distilled nonurea adduct fraction. Epoxides are highly reactive compounds, sensitive to nucleophilic attack. Cyclic compounds are distinguished from monomeric products by the lack of interaction with urea. They are derived from the interaction between a fatty acid radical with a carbon atom located elsewhere on the fatty acid chain. Monomeric dimers are formed by the Diels–Alder reaction, where α and β unsaturated carbonyl compounds react with conjugated dienes and carbons 1 and 4 of the conjugated diene are attached to carbon double bonds of the unsaturated carbonyl compound, forming a six-membered ring. Dimeric and polymeric triacylglycerols formed under thermal oxidative reactions are held together by carbon–carbon and carbon–oxygen bonds and contribute to the increased viscosity seen in thermally oxidized fats and oils. In the process of dimerization, highly polar compounds are also produced and account for some of the dark color characteristic of heated fat. Chain reaction and step reaction polymerization occurs between either free fatty acids, fatty acyl groups on triacylglycerol, or between fatty acids on different triacylglycerols. Bonds formed within single fatty acids result in cyclic acids, whereas bonding that occurs between two fatty acids produce dimeric acids. Polymerization reactions cause the primary deterioration of oil as a result of thermal oxidation (192). There is some uncertainty on the absorption efficiency of hydroperoxides, aldehydes, epoxides, and polymers from the intestine. Thus the toxic potential of these oxidation products in the body remains equivocal. In human subjects, fed thermally oxidized soybean oil, higher elevations in plasma thiobarbituric acid reacting substances were observed when compared to subjects fed fresh oil (193). Genotoxic activities of lipid peroxidation products have been reported (194); however, less is known about the destiny of LOPs derived from repeatedly used deep-frying oils once absorbed into the plasma (195).

Studies have also monitored the production of COPs during thermal processes. Park and Addis (176) showed in heat-treated tallow that the formation of 7-ketocholesterol follows zero-order kinetics when continuously heated at 155°C. In another study, these workers confirmed the linear increase in 7-ketocholesterol with heating time, and demonstrated that although the rate of cholesterol disappearance increased with increasing temperatures, no parallel increase in the formation of COPs occurred (196). No extraordinary difference was observed in the levels of COPs between temperatures of 135 and 150°C.

The prolonged laboratory heating of fats in the presence of air for the purpose of simulating commercial methods of thermal treatment has resulted in a myriad of toxic effects in animals fed these oils. Isolates of thermally oxidized fatty acids derived from the nonurea adduct-forming fraction containing cyclized monomers have been shown to cause distended stomach, kidney damage, multiple focal hemorrhaging of the liver and other tissues, and severe lesions in the heart (197). Oxidized fats with high carbonyl values have been associated with toxicity resulting from impaired liver thiokinase and succinic dehydrogenase activities (198). Body weight gain, feed consumption, and feed efficiency have been reported to be lower in weanling rats fed diets containing 15% laboratory-heated corn and peanut oils, compared to counterparts fed unheated oils. Animals fed heated oils also exhibited enlarged livers and kidneys, and damage to the epididymis and thymus as well as enduring symptoms of diarrhea, dermatitis, seborrhea, and loss of fur (199). Recently, the consumption of thermally oxidized PUFA by chicks has been shown to reduce free-radical scavenging activity *in vivo* by reducing tissue α-tocopherol content in the gastrointestinal tract (25).

Feeding rats laboratory-heated (100 h at 200°C) dietary olive and sunflower oils produces changes in hepatic desaturase enzyme activities that ultimately influence the fatty acid composition of hepatic microsomes. A buildup of linoleic acid resulting from a decreased activity of Δ^6 desaturase activity decreases arachidonic acid content in microsomes. An increase in microsomal linoleic acid also results in marked decreases in oleic acid content, since the desaturation and elongation of oleic acid occurs only when concentrations of linoleic and linolenic acids are relatively small. The hepatic microsomal Δ^5 desaturase enzyme system is an important pathway for the conversion of eicosadienoic to eicosatrienoic (*n*-9 series) acid. Conversion of dihomo γ-linoleic acid into arachidonic acid of the *n*-6 series is also significantly decreased in rats fed heated olive and sunflower oils. These responses in enzyme activities are similar to animals exhibiting deficiencies in essential fatty acids following the consumption of diets containing positional isomers of fatty acids produced from hydrogenation and thermal processing of cooking oils (200). It remains to be determined to what extent changes in fatty acid metabolism resulting from thermally oxidized fats have on membrane fluidity and subsequent metabolism.

Oils heated in the laboratory and used for toxicological research studies are not typical of commercially heated oils. Excessive heat treatment destroys the quality characteristics of deep-fat fried foods such as flavor, odor, and appearance and in many cases manufacturers and restaurants discard frying oil when signs of foaming, smoking, and undesirable flavors occur. Pouling and co-workers (201) reported no significant difference in food intake, growth, or liver and kidney morphology of rats fed commercially heated oils compared to animals on control diets. However, rats fed laboratory-heated oils exhibited lower initial growth rates and hepatomegaly. The reason for the differences between laboratory and commercially heated oils in these experiments has been attributed to numerous factors, such as temperature and duration of heating in repeat heating, aeration, agitation, and the source of the oil used in the commercial versus laboratory cooking.

The potential role of thermally processed oils in the incidence of various cancers has also been controversial. The potential of co-carcinogenic compounds being formed during the heating of corn oil has been suggested by findings from rodent feeding studies. Laboratory-heated corn oil or commercial frying oil included in diets caused an increase in mammary tumors induced by acetylaminofluorene. This effect was reduced dramatically when the toxic fraction of the thermally prepared oils was removed (202). The fact that fatty acyl hydroperoxides are decomposed quickly by heat has lead some investigators to conclude that the formation of mutagens in frying oils is not significant unless there has been considerable thermal abuse of the oil (203). In contrast, other workers have reported mutagenic activity in the polar fraction of repeatedly used edible oils in restaurant deep frying (204). There is a positive correlation between polar oxidation and degradation products of polyunsaturated oils and the level of thiobarbituric acid reactive substances in these fats. The primary thermal oxidation products of PUFA are linoleic acid hydroperoxides, which migrate from the food into the fat used for deep frying. This migration leads to an altered polyunsaturated/saturate fatty acid ratio for the cooking oil. Mutagenic activity of methyl linoleate hydroperoxides and oxidized linoleic acid solutions containing high peroxide values, show mutagenicity in the *S. typhimurium* assay, only when incubated in the presence of S-9 (metabolic inducer) fraction (204). These results strongly suggest that the mutagenicity of constituents present in deep-frying fat are a result of metabolites and secondary autoxidation products of linoleic hydroperoxide products. Both coconut and vegetable oil, when heated repeatedly, exhibited mutagenicity in *Salmonella* tester strains TA97 and TA100. The effects of the oils, however, did not show significant proliferation of cells in the lower gastrointestinal tissues of rats and could not be associated with a role in colon tumorigenesis when fed at a dietary level of 10% for a 4-week period (205).

TOXIC SUBSTANCES PRODUCED DURING SMOKING, CHARBROILING, AND BARBECUING OF FOODS

Polyaromatic hydrocarbons (PAHs) and extended pyrolysis products of amino acids and proteins are formed on broiling of fish and meat. Charcoal broiling and barbecuing of beef result in significant benzo(*a*)pyrene [B(*a*)P] residues (50 μg/kg) as well as other aromatic hydrocarbons (206). Malanoski and co-workers (207) reported the presence of 1.4–4.5 μg of B(*a*)P/kg in barbecued pork, beef, and reused cooking oil. The production of PAH in charcoal broiling is dependent on the fat content and the proximity of the food to the heat source during cooking (206). Production of carcinogens can be minimized by avoiding the contact of food with flame and by using longer cooking times at lower temperatures. Other workers (208) have reported negligible quantities of B(*a*)P with roasting, frying, and deep frying but higher levels following gas heat broiling. Comparatively, gas-broiled fish contain greater PAHs than electric-broiled fish (209).

Smoked food is an age-old method used to preserve food and enhance flavor. Smoke from burning wood contains at least 300 different compounds, including phenols, carbonyls, acids, alcohols, furans, lactones, and PAHs (210). Important parameters of smoke production include the fuel source, air flow, and combustion temperatures. The combustion products from fat leaving the food and coming into contact with the heat source are additional chemicals that contribute to the smoke condensate. A relatively small proportion of PAH is derived from thermal reactions occurring within the food itself. PAHs are formed in greater amounts from wood, pine, and spruce cones compared to gas and charcoal fuels. Most of the PAH derived from smoked foods are found in the surface layers, but some PAHs deposited on the surface of smoked foods migrate into the interior of the food. Factors controlling migration of xenobiotic agents into the processed foods are storage time and type of food product. Many countries have set a regulatory limit of 1 μg/kg for B(*a*)P in smoked foods. Frankfurters cooked by charcoal broiling, pan frying, or electric oven heat contain less than 1 μg B(*a*)P/kg, compared to smoked frankfurters, which can contain greater than 1 μg B(*a*)P/kg. Home-smoked mutton has been reported to contain up to 23 μg B(*a*)P/kg (211). Some Japanese smoked dried fish products contain as much as 37 μg B(*a*)P/kg fish (212). In this particular study with three types of Japanese fish products, namely *Katsubobushi, Sababushi*, and *Urumebushi*, high amounts of B(*a*)P were found in all fish. The *Katsubobushi* was prepared from broiled boned bonito fish by intermittent exposure to wood smoke and further drying in the sun for 1–2 weeks. In comparison, the *Sababushi* and *Urumebushi* were prepared from the flesh of mackerel and sardine by a similar process but with less smoking. The formation and storage of B(*a*)P in smoked bologna and bacon has also been found to be influenced by the fat composition and packaging (213). The cellulose casings containing the bologna significantly reduced the B(*a*)P content. Approximately half of the B(*a*)P in the bacon leaches into the fat drippings. In addition to muscle foods, B(*a*)P is also found in dry cereals and nuts. Indirect drying, versus direct drying of wheat and rye, results in significant difference in PAH residue content (214). Direct dried cereals such as wheat, dried over a light fuel oil flame, carry high levels of B(*a*)P (215). PAHs are also present in roasted coffee beans and in coffee soot, a by-product of the commercial roasted coffee beans (216,217).

The toxicity of wood smoke extracts are related to potential mutagenicity and carcinogenicity. Wood smoke extracts initiate tumors in mouse skin painting studies (218) and high incidences of stomach cancers have been reported in some population groups that consumed smoked foods as a major part of the diet. B(*a*)P is absorbed by mammals and aquatic species through the gastrointestinal tract, respiratory tract, and skin. Since PAHs are lipid soluble, the presence of bile secreted into the duodenum plays an important role in B(*a*)P absorption. Along with nutrient lipids, noted PAHs, B(*a*)P, 3-methylcholanthrene, and 7,12-dimethylbenz[*a*]anthracene are absorbed in the intestine and transported in the lymph as solutes in chylomicrons and very low density lipoproteins. Serum triacylglycerol concentration is a factor in the efficiency of B(*a*)P uptake by serum lipoproteins, and small changes in serum triacylglycerol concentrations can significantly affect the level of circulating carcinogen (219). Pharmacokinetic studies have indicated that at least 30% of the B(*a*)P dose is absorbed from the gut (220) and that enterohepatic circulation plays an important part in the reabsorption of B(*a*)P for reactivation by intestinal hydrolases (221).

In the process of chemical detoxification of PAHs, two types of endogenous enzymatic reactions occur; these being phase I and phase II reactions, respectively. Phase I reactions are directly related to the mixed function oxidases (MFO), which transform PAHs by oxidation, epoxidation, reduction, and hydrolysis reactions. The cytochrome P-450 dependent monooxygenases, arylhydrocarbon hydrogenase and epoxide hydrolase, are both involved in the conversion of B(*a*)P to ultimate metabolites in the phase I reactions. The metabolic activation of B(*a*)P proceeds via an initial 7,8-epoxidation by cytochrome P-450 dependent monooxygenases, hydrolysis by epoxide hydrolase to the *trans*-7,8-diol, and a monooxygenase-mediated epoxidation in the 9,10-position, producing a total of 4 enantiomeric diol-epoxides (Figure 8.4). As a result of phase I reactions, B(*a*)P dihydrodiols 7,8 and 9,10 undergo further oxidation to form diol epoxides that form adducts with DNA, RNA, and proteins and possess both mutagenic and carcinogenic activities. An alternative pathway for production of activated B(*a*)P involves co-oxidation by prostaglandin H synthetase, yielding 7,8-dihydrodiol mutagenic products (222; Figure 8.3). The enzyme prostaglandin H synthetase, known for oxygenating arachidonic acid and forming bicyclic peroxide intermediates ultimately converted to prostaglandins and prostanoids, also exhibits peroxidase activity. In addition to catalyzing the reduction of hydroperoxy endoperoxide, the peroxidase activity also causes co-oxidation reactions of chemical carcinogens [e.g., B(*a*)P], resulting in the production of mutagenic derivatives. These reactions account for the fact that mutagenesis with B(*a*)P in the *S. typhimurium* revertent assay, is dependent on prior metabolic activation (223). In addition to B(*a*)P-diol intermediates, other B(*a*)P metabolic pathways vie with activating epoxide, resulting in the production of 5 phenols, 3 dihydrodiols, and 3 quinones. Phase II reactions in B(*a*)P metabolism are conjugation reactions that directly detoxify metabolites through conjugation with glutathione and glucuronides. Products of xenobiotic conjugation are excreted from the body in urine and fecal matter.

PAHs constitute a family of lipid-soluble chemical agents that, following oxidative metabolism to reactive electrophilic compounds, exhibit a wide range of potential mutagenic, genotoxic, fetotoxic and carcinogenic properties. Although more than 70 PAHs have been identified from various sources (224), benzo(*a*)pyrene [B(*a*)P] is a major constituent and regarded as a prime indicator for the presence of PAH. The reactive metabolites of B(*a*)P following oxidative metabolism by cytochrome P-450 enzymatic complex include epoxides, quinones, and phenol derivatives. Lipid-derived peroxyl radicals also interact with B(*a*)P metabolites forming reactive intermediates that bind to vital macromolecules (167,225). Although the stereoselectivity of epoxidation of B(*a*)P-7,8-diol cytochrome P-450 and peroxyl radicals is different (Figure 8.4), the reactivity of both products is similar in chemical or biological systems. The epoxides, which include 7,8-dihydrodiol-9,10-epoxide, possess the greatest mutagenic and carcinogenic potential. Cultured human colon cells can metabolize the parent B(*a*)P to proximate epoxide carcinogen resulting in adduct formation with colonic DNA. Fecal bacteria also have the capacity to hydrolyse biliary metabolites of B(*a*)P to dihydrodiols, phenols, and quinones. B(*a*)P can potentially act as a tumor promoter by generating free radicals (226). Genotoxic reactive products other than electrophilic products of B(*a*)P, such as hydroxyl radicals, are capable of inducing single-strand breaks and small base damages in DNA.

The removal of PAHs from water and foods has had only partial success. The chlorination of water to remove B(*a*)P is unsatisfactory due to the production of mutagenic halognated hydrocarbons such as trihalomethane and chloramine. Ozonation of water to eliminate PAHs, 3-methylcholanthrene, B(*a*)P, and 7,12-dimethylbenz[*a*]anthracene have been more successful (227); albeit other workers have noted the production of mutagenic activity attributed to B(*a*)P 4,5-oxide following ozone treatment (228). In foods, the presence of specific agents and nutrients can minimize the toxicity of PAHs by inhibiting the metabolic activation of these agents to potential carcinogens. For example, the chemical antioxidants, butylhydroxytoluene (BHT) and butylhydroxyanisole (BHA) inhibit the mutagenicity of B(*a*)P in the forestomach of mice and development of pulmonary and mammary tumors in rats when administered concurrently (229,230). Vitamin A compounds (retinoids) also exhibit covalent binding of B(*a*)P to subnuclear components, thus inhibiting B(*a*)P adduct binding to DNA (231).

POTENTIAL HAZARDS FROM APPROVED ANTIOXIDANTS

Antioxidants are defined as substances that potentially delay the onset or slow down the rate of oxidation of autooxidizable materials in food, cosmetics, rubber, or plastics. Classifications of antioxidants include free radical terminators, reducing agents or oxygen scavengers, and chelating agents. Moreover, antioxidants can also be classified into groups effective in aqueous or lipid systems. Ascorbic acid and riboflavin are examples of water-soluble antioxidants. Tocopherols, BHA, BHT, and ethoxyquin (EQ) are lipid-soluble antioxidants used to prevent oxygen-induced lipid peroxidation. Synthetic antioxidants such as BHT, BHA, and tertiary butylhydroquinone (TBHQ) are examples of free radical terminators. These agents donate hydrogen from adjacent phenolic hydroxyl groups to produce a stable free radical that is not active in further propagation. This reaction is in part due to resonance delocalization and the lack of positions suitable for attack by molecular oxygen.

BHT is prepared by alkylation of *p*-cresol with isobutylene and is used in human food, animal feed, and commercial products such as plastics and soaps. BHA is an isomeric mixture of approximately 4%, 2-*tert*-butylated hydroxy anisole and 96%, 3-*tert*-butylated hydroxy anisol. Both BHT and BHA are generally regarded as safe (GRAS) substances that are used as food additives in fats and oils, dried breakfast cereals, chewing gum, margarine, dried poultry meat, essential oils, and various unstandardized foods. In muscle food systems, such as raw ground pork, a mixture of antioxidants containing BHA, BHT with citric acid (1 : 1 : 1 w/w) exemplifies effective syngerism of different antioxidants in reducing lipid oxidation (232). Shahidi and co-workers (233) reported BHA, TBHQ, and EQ to be effective antioxidants in cooked ground pork at a level of 30 ppm. BHA, not particularly effective in vegetable oils by itself, exhibits synergism with other primary antioxidants such as propyl gallate. BHA is relatively resistant to thermal decomposition, compared to propyl gallate and therefore is included in oils destined for frying and baking operations. The inability of antioxidants to completely inhibit autoxidation of dietary fats such as fish oils has been shown to result in increased *in vivo* lipid peroxidation products in heart, skeletal muscle, and mammary tissues (26).

Toxicological studies with BHA have been performed with rats, rabbits, swine, and primate animal models (234–240). Time and dose-dependent changes in forestomach epithelium include hyperplasia, papillomas, and carcinomas (237,238). Dosages used in these studies were extremely high compared with approved concentrations of 0.001–0.002% in food for BHA. Daily dosage levels of 400–500 mg/kg body weight BHA resulted in liver hypertrophy, hepatomegaly, proliferation of hepatic smooth endoplasmic reticulum, and an increase in levels of certain drug metabolizing enzymes. Primates are generally more sensitive to high levels of BHA exposure than rodents (241). More severe forestomach hyperplastic changes also occurred with the combined use of BHA and sodium ascorbate, compared to BHA alone, in chronic rodent feeding studies (242). Distinct ultrastructural changes in hepatic tissues observed in monkeys are not seen in rodents at equivalent levels of BHA exposure. No significant fetotoxic and teratogenic effects of BHA have been reported in both rabbits and swine when fed daily levels of 400 mg/kg/day (235,236). Marked proliferative and parakeratotic changes in esophagus tissue occur in swine fed BHA at subchronic levels (243). More recent studies have shown carcinogenic activity of BHA in forestomach and lung tissues in small animals (244,245).

The LD_{50} of BHT in rodents ranges between 2 and 2.5 g/kg body weight, indicating low acute toxicity (246). BHT is not mutagenic, with and without S-9 activation, in *S. typhimurium* G-46 and TA 1530 and *Saccharomyces cerevisiae* assays (247). BHT, however, produced positive chromosomal aberrations in anaphases of human WI-38 embryonic lung cells (247). Dominant lethal effects of BHT occurred with feeding 0.4% (w/w) antioxidant for 8–10 weeks to rats; however, the biological significance of these results were not ascertained (24). Similar heritable translocation studies conducted in mice with BHT produced no effect. Both BHA and BHT can promote bladder carcinogenesis in rats (234), although the same compounds also inhibit the incidence of dimethylbenz(*a*)anthracene and benzo(*a*)pyrene induced forestomach and mammary gland tumors by 50% or more (229, 249). BHA and BHT are also effective in inhibiting aflatoxin B_1 induced hepatocarcinogenesis (250). Similarly, under certain conditions, EQ promotes tumor propagation in the kidney and urinary bladder but can inhibit development of liver tumors (251,252). Although synthetic antioxidants appear to provide chemopreventation against certain xenobiotic-induced cancers, particularly relevant to tropical countries heavily exposed to mycotoxins, their potential carcinogenic activity restricts prospective intervention.

GOSSYPOL PIGMENTS

Cotton plants of the genus *Gossypium* are a primary source of the world's total oilseed producing plants, second only to soybean and exceeding that of groundnut, rapeseed, and sunflower (253). Cottonseed is a valuable by-product of the cotton plant because of the rich source of oil and proteins of high nutritional quality. In tropical and subtropical countries, in particular, cottonseed oil is used to manufacture salad and cooking oils, shortenings, and margarines. Flour made from cottonseed is a good protein supplement in human diets and the meal produced from cottonseed is an important supplement for ruminant, swine, and poultry diets. Gossypol is a yellow phenol pigment that exists in three forms in the cottonseed (Figure 8.5). The pigment is a toxic compound in its free form and accumulates in body tissues. Pigments are found in the highest concentrations in the cotton plant root and to a lesser extent in the cotton leaves and seed kernels. In the cottonseed, gossypol and other related pigments are located in discrete pigment glands that represent approximately 2.5–4.8% of the cottonseed kernel. Depending on plant variety and growing conditions of the cottonseed, gossypol content represents between 20 and 40% of the gland material. A glandless and thus gossypol-free variety of cottonseed is also available, but this variety has lower resistance to disease and insects. The current focus has been to produce varieties of cottonseed with low gossypol content, while maintaining desirable fiber content (254).

Cottonseed is processed into oils and meals for both animal and human consumption. Commercial practices employ heating the cottonseed (e.g., 90–

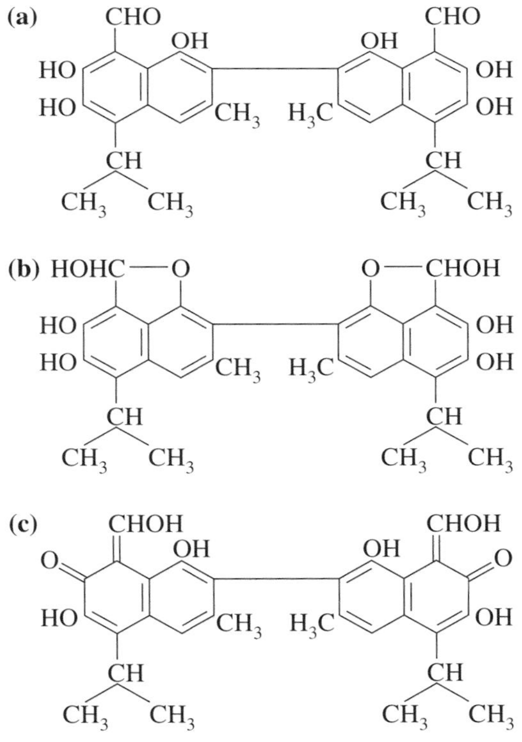

Figure 8.5 Different structures of tautomeric patterns of gossypol. (**a**) Gossypol-aldehyde, (**b**) gossypol-lactol, and (**c**) gossypol-cyclic carbonyl.

130°C), which facilitates the removal of oil, as well as reducing the amount of free gossypol to minimum levels (e.g., 0.2 g/kg dry matter). High temperatures and pressures common to extrusion processes facilitate the formation of gossypol and protein conjugates, which are responsible for the reduced amino acid digestibility and utilization of lysine in cottonseed meal. Following desolventization, cottonseed presscake retains between 0.4 and 1% oil and 0.02 and 0.07% free gossypol. The amount of residual gossypol in the meal depends on the initial gossypol content of the seed as well as the process conditions used in the oil extraction.

Toxicity of gossypol is attributed to the free gossypol, as opposed to bound gossypol. Two organizations, namely the Food and Drug Administration (FDA) and the United Nations Protein Advisory Group established limits for gossypol content in human foodstuffs to be 450 and 600 ppm, respectively. Sensitivities to gossypol toxicities vary with animal species. Ruminant species are less susceptible to gossypol toxicity due to the action of rumen microflora, which convert free gossypol to an inactive gossypol–protein bound form in the rumen. In growing lambs, cottonseed meal was incorporated safely into the diet at a level of 30% with no adverse effects observed (255). Gossypol toxicosis, however, has been reported in heifers exposed to relatively high amounts of free gossypol in whole cottonseed (256), thus indicating that gossypol detoxification by the rumen microflora can be saturated at high levels of exposure. Dogs are very sensitive to gossypol toxicity compared to mice (257), and poultry (hens and chickens), swine, rabbits, and fish are also sensitive to gossypol toxicity. Egg yolk discoloration associated with the interaction of free gossypol with yolk iron is common with laying hens exposed to gossypol contamination. The addition of ferrous sulfate to the diet of domestic animals can also reduce toxicity.

Common symptoms associated with gossypol toxicity include reduced feed intake, weight loss, hyperprothrombinemia, diarrhea, erythrocyte fragility, and associated reductions in blood hemoglobin and hematocrit values, edematous fluid accumulation in heart and lung cavities, and degenerative changes in liver and spleen. Gossypol exhibits both phenolic as well as aldehydic properties. The phenol groups form ester and ether moieties, while the aldehyde groups react with amines to form Schiff's bases. Gossypol forms Schiff's bases between carbonyl and the ε-amino groups of protein and peptide amino acids such as lysine, which results in protein crosslinking and reduced protein quality. Increasing protein intake or lysine supplementation decreases but does not prevent gossypol toxicity. The covalent binding of gossypol with proteins has been attributed to reduced mitochondrial respiratory enzyme activities (258), phospholipid sensitive Ca^{2+}-dependent protein kinases (259), and interactions with microsomal membrane proteins, which affect sterol, steroid, and fatty acid metabolism (260). In human studies, gossypol acetate has been used as a male contraceptive at levels of 20 mg/day, with no signs of toxicity. Fertility was regained following cessation of gossypol treatments and no signs of teratogenicity of subsequent offspring were found (261).

ALLYLBENZENES

Allylbenzenes are major flavor components derived from essential oils of a number of herbs and spices. The primary sources of essential oils containing allylbenzenes are camphor and sassafras (safrole), dill and parsley (anethole), and basil and tarragon oils (estragole). Safrole (4-allyl-1,2-methylenedioxybenzene) is also found in ginger, black pepper, and cinnamon leaf oil (262). The relative toxicity of allylbenzenes is influenced by chemical activation reactions in target organs, resulting in metabolic transformation of parent compounds to intermediate metabolites with increased potential for toxicity (263). The size of dose, animal species, and nutritional status are important in assessing intraspecies toxicity. Allylbenzenes exhibit a carcinogenic effect when a series of metabolic activities associated with the hydroxylation of the parent molecule at the benzylic (1′) carbon atom and subsequent sulfation reactions occur. The loss of *O*-sulfate, produces carbonium ion containing intermediates that form DNA adducts. High dosages (0.5% w/w) of safrole in the diet of rodents induce liver tumorigenesis (264). Similar high (0.46% w/w) levels of dietary intake of estragole (*p*-methoxy-allylbenzene) for extended durations (12 months) induce carcinogenicity in mice (265). Anethole, a structural isomer of estragole and important food flavor consumed in savory foods, can induce hepatocarcinomas in female rodents fed high (1% w/w) dosages of anethole over a prolonged duration of more than 25 weeks (266). Structural analogues of safrole, isosafrole, and dihydrosafrole, although carcinogenic in rats, do not exhibit similar toxicity in mice.

COLOR AGENTS USED IN EDIBLE FATS AND OILS

Colorants have important roles in the quality attributes of various food products. With the exception of some natural colors such as β-carotene and riboflavin, colorants do not contribute directly to the nutritional value of food, but rather are added to restore natural colors lost in food processing, to provide uniformity and intensify natural colors, to aid in protecting light-sensitive vitamins by absorbing light, and to provide an overall visual indicator of quality. Some food fat sources such as edible oils, butter, cheese, margarine as well as package material accessories utilize colorants. Water-insoluble colorants are naturally hydrophobic (e.g., C-1 pigment Red 3), or in the case of FD&C lakes represent aluminum hydrate emulsions that contain between 10 and 40% FD&C dyes. The insolubility of these pigments in aqueous media is an important characteristic giving prolonged color fastness and resistance to both fading and biological degradation.

Different colors used in fats and oils can be classified on the basis of (a) natural materials obtained from plant, animal, or vegetable sources; (b) synthetic colors derived from natural sources; and (c) FD&C lakes. Relative toxicities of examples of some color agents are given in Table 8.5.

11.1 Natural Colors

Carotenoids, chlorophyll, tumeric, carrot oil, and carbon black are examples of natural colors. Carotenoids (i.e., annatto) are yellow, red, and orange pigments, obtained largely from plant seeds that have found use in enhancing color in such lipid systems as red salmon and some butter products. β-Carotene, a primary example of a carotenoid colorant has an acceptable daily intake (ADI) of 2.5 mg/kg body weight. Chlorophylls provide green and olive green pigments but have limited use due to sensitivity to light and acid. Carrot oil, a solvent extract from carrots, and tumeric are additional examples of natural food colorants. Tumeric, approved for use in foods has an ADI level of 0.5 mg/kg body weight. Carbon black, also known as activated charcoal and decolorizing carbon, can be derived from vegetable material. Due to the concerns about small amounts of PAH associated with this material, carbon black has not received acceptance for direct use with foods and no ADI for carbon black is available. Use of carbon black from hydrocarbon sources is provisionally allowed in food contact materials such as wax coatings for cheese.

Table 8.5 Summary of biological effects of some colorants[a]

Color Group	Animal Species	Dose	Toxic Endpoint
Natural			
Carbon black	Mice (12–18 mo)	10% w/w diet	Adenocarcinoma, squamous cell carcinoma
	Rats	2 g/kg body weight	Enhanced DMH-induced tumor[b]
Canthaxanthin	Rats, rabbits, dogs	1 g/kg body weight	Ocular toxicity
Synthetic			
C.I. Pigment Red 3	Rats (2–12 wks) Mice	10% w/w diet	Hemolytic anemia, hematopoietic, hepatic, and spleen cell proliferation
FD & C Green No. 3	Rats	0–1 g/kg body weight/day	No effect
	Mice	0.5%	No effect
FD & C Blue No. 1	Rats (2 mo)	2%	Minor toxic effects, adenocarcinoma
	Mice	5%	Minor toxic effects, lymphoma

[a] Adapted from Refs. 267–271.
[b] DMH = dimethylhydrazine.

Carbon black is not absorbed efficiently and is likely cleared in the feces (272). Carbon black has no mutagenic activity in the *S. typhimurium* bioassay.

11.2 Natural-Identical Colors

This group of colorants are synthetic counterparts of colors/pigments derived from natural sources. The synthetic canthaxanthin (red pigment), β apo-8′-carotenal, β apo-8′-carotenoic acid ethyl ester (orange-red), and citranaxanthin (yellow) have wide use in margarines and cheese. Canthaxanthin is a diketo carotenoid pigment permitted for use in Canada, the United States, and the United Kingdom with an ADI of 25 mg/kg body weight. The ADI for β apo-8′-carotenal is 5 mg/kg body weight. It is obtained from edible mushrooms such as the Chanterellem (*Cantharellus cinnabarinus*) and various crustacea and fish sources. Accumulation in fat occurs at high dosages (250 mg/kg/day) and is slowly eliminated from adipose tissue in dogs. Citranaxanthin, another synthetic compound used primarily as an additive in animal feeds, imparts a yellow color to chicken fat and egg yolks. Like canthaxanthin, citranaxanthin is converted to vitamin A in chickens. Toxicological studies conducted in rats have failed to detect significant toxic endpoint effects (mortalities, reproductive effects) for citranaxanthin.

11.3 FD&C Lakes

FD&C lakes are color dyes extended into alumina hydrate at concentrations ranging from 10 to 40%. Lakes are provisionally listed (e.g., not approved for permanent listing for food use by FDA) and used in oil-based products, compression tablets, and icings. The number of FD&C lakes and colors available are numerous and have been comprehensively detailed (273). Some of the primary FD&C lakes of importance for this chapter include: FD&C lake Red No. 2 [Amaranth, listed (L) in 1907 and delisted (D) in 1976] was delisted by FDA on the basis of suspected carcinogenic and embryotoxic effects and insufficient evidence to confirm its safety (274). FD&C lake Red No. 3 (Erythrosine, L—1907) is presently in use. There is virtually no absorption of this colorant as indexed by its unchanged excretion in feces. No toxicity data exists for FD&C lake Red No. 3 in experimental animals and humans, albeit some concern has been raised regarding possible effects on neurotransmitter release (275). There is also no sign of carcinogenic activity from this colorant (276). FD&C lake Yellow No. 5 (Tartrazine, L—1916) is presently in use and has an ADI of 5 mg/kg body weight in dog and rat. Despite conflicting data on the mutagenicity of tartrazine (277), long-term (2-year) feeding studies have not disclosed either toxic or carcinogenic effects of tartrazine in rodents (278). FD&C yellow No. 5 has been approved for use in both Canada and the United States, but restricted usages and labeling requirements exist for this color. FD&C yellow No. 6 (sunset yellow, L—1929) is also presently in use for its reddish yellow color. No toxic effects have been seen at high consumption levels in rats (3.9 g/kg/day) and an ADI of 5 mg/kg body weight has been set for this color. FD&C green No. 3 (Fast Green, L—1927) is also nontoxic and nonmutagenic (267) and has an ADI of 2.5 mg/kg body weight. Fast Green is not absorbed from the gastrointestinal tract, as evidenced by the fact that most of the color is recovered in feces unchanged and only a small (<5%) amount is found in bile.

Colors formerly used or under investigation include Butter Yellow (4-dimethylammoazobenzene), which was used in some countries in butter and margarine products. However, due to potential causative effects on bladder and liver tumors in rats, this colorant has been delisted in many countries. Reduction of toxicity can be obtained with the substitution of the methyl group with an ethyl group. C.I. Pigment Red No. 3 [1-[(4-methyl-2-nitrophenyl)azo]-2-naphthalenol] is an organic pigment used in printing inks and therefore may come into contact with fats and oils indirectly from package materials. The structure of Pigment Red 3 resembles other phenylazonaphthols including Ponceau 3R (L—1929; D—1976), Oil Orange SS (L—1939; 3—1956) and Citrus Red 2 (L—1959; in use for coloring orange skins only). These azo compounds can undergo reductive cleavage by bacterial flora in the gastrointestinal tract or by some liver enzymes resulting in the production of aromatic amine derivatives with carcinogenic potential.

MYCOTOXINS

There are almost 100,000 different species of molds that produce secondary toxic fungal metabolites, referred to as mycotoxins. Human consumption of mycotoxins is from both animal and vegetable sources. The former reflect the transmission of mycotoxins through animal products (meat, eggs, milk) to humans from livestock fed moldy cereals. Mycotoxins can accumulate in adipose tissues at potentially high levels and exhibit noted toxicities (Table 8.6).

12.1 Aflatoxins

Aflatoxins (AF) represent a group of mycotoxins produced from the microfungi *Aspergillus flavus* and *A. parasiticus*, which grow on grains, nuts, and other crops at around 30°C and 15% moisture. AF are relatively high-molecular-weight compounds containing one or more oxygenated alicyclic rings. AFB_1, initially discovered following the accidental poisoning of turkey and ducks fed contaminated peanut meal and initially related to Turkey X disease in England, is the most potent of the naturally occurring AF. In addition to AFB_1, other fungal metabolites of AF, in order of decreasing potency are B_2 (AFB_2), G_1 (AFG_1), and G_2 (AFG_2). A hydroxylated metabolite of AFB_1, aflatoxin M_1, is present in milk and edible tissues of domestic animals that

Table 8.6 Acute oral toxicities (LD_{50} values) of major mycotoxins[a]

Mycotoxin (strain)	Source	Species	LD_{50} (mg/kg)	Toxicity Endpoint
Aflatoxin B_1 (*Aspergillus*)	Peanuts	Ducks	0.36	Liver, kidney tumor
T-2 (*Fusarium*)	Corn, barley	Mice	5–10.5	Diarrhea, vomiting, inhibition of protein and DNA synthesis, and hematopoietic and immune systems
		Rat	3–5.2	
		Guinea pig	3.1	
		Trout	6.0	
Vomitoxin (*Fusarium*)	Corn, wheat	Mice	70.0	Anorexia, fetotoxicity, teratogenicity
Ochratoxin A (*Penicillium, Aspergillus*)	Nuts, grains	Rats	22.0	Impaired kidney function
Zearalenone (*Fusarium*)	Corn, sorghum barley	Female mice	500	Estrogenic activity, reproductive failure, teratogenicity
		Male mice	5490	
		Guinea pig	2500	

[a] Adapted from Ref. 279.

have consumed AFB_1-contaminated feed. Aflatoxicosis can be acute, subacute, or chronic. Age and species differences influence the variability of AF-induced tumor susceptibility in mammalian species. The first symptoms of chronic AF poisoning in young animals are reduced feed intake and weight loss. Liver hemorrhage also occurs with chronic toxicity. There are few signs of chronic toxicity until shortly before death when symptoms include ataxia and convulsions. In rodents, such as the rat, AFB_1 has the greatest toxicity and will lead to hyperchromatic and enlarged liver parenchyma cells, which can represent initial morphological changes preceding malignancy. Kidney (renal medulla), spleen, adrenal, pancreas, colon, and bone are other organs affected by AFB_1 in animals. Human exposure of AF-contaminated foods has been associated with increased incidence of liver cancer in various countries (280) and may also act synergistically with hepatitis B virus in the induction of liver cancers. Many biochemical lesions, associated with AF toxicity in mitochondrial cells common to AF-induced toxicity, involve the inhibition of respiration and uncoupling of oxidative phosphorylation reactions. There is an inverse relationship between the LD_{50} values of AF in various mammalian species and the degree of inhibition of oxygen uptake by AF (281). This relationship may account for the different susceptibility of various mammalian species to AF-induced toxicity.

Metabolic transformation of mycotoxins by oxidation, hydrolysis, and conjugation reactions to more polar compounds facilitate excretion of AF. The metabolic activation of AF also results in products that exert biological activity.

Among the metabolites formed in these reactions are the detoxification products that are excreted from the body and activated products, which covalently bind to cellular macromolecules. There is evidence to show that environmental lipid xenobiotic agents such as polychlorinated biphenyl (PCB) stimulate both the formation of AFM_1 as well as a reactive 8,9-oxide metabolite. Monooxygenase enzymes convert AF to a 8,9-oxide intermediate, or reactive form, which in turn reacts with nucleic acids forming the adduct, *trans*-8,9-dihydro-8(N^7-guanyl)9-hydroxy AF. Similarly, co-oxidation of AFB_1 occurs during *in vivo* prostaglandin biosynthesis, which exhibits peroxidase activity from prostaglandin H synthetase activity (Figure 8.4). Mutagenicity of AF in the *Salmonella*/microsome test correlates well with carcinogenicity of various AF (282). The addition of 0.5% BHT significantly changes the deposition pattern of AFB_1 in the lactating rat (283). This effect is a result of the BHT-induced glutathione-*S*-transferase activity, which is important in the detoxification of xenobiotic agents. Fukayama and co-workers (283) also showed, however, that the enhanced metabolism of AFB_1 to AFM_1 in lactating rats resulted in the increased risk of neonates to the toxic reactions of AFM_1.

Finally, a number of lipid-soluble agents have been shown to inhibit the fungal biosynthesis of AF. Lipid-soluble substances such as methylxanthines (e.g., caffeine; 284) and chloroform extracts of carrot tissue (285) inhibit AF synthesis from *A. parasiticus*. Cinnamon and clove essential oils also inhibit growth and AF production by *A. parasiticus*.

12.2 Trichothecenes

T-2 toxin (4β,15-diacetoxy-3α-hydroxy-8α[(3-methyl-butyryl)oxy]-12,13-epoxytricholthec-9-ene) represents one of over 60 different trichothecene toxic mycotoxin metabolites generated from different species of *Fusaria* in contaminated corn and barley. Domestic animals and poultry exposed to diets containing toxic levels of T-2 exhibit typical symptoms of reduced feed intake, pharyngeal and intestinal irritations, diarrhea, and infertility. The toxin and its metabolites are eliminated primarily from the bile to excreta. A species difference exists in the relative sensitivities to the toxin. There is no evidence for mutagenic or genotoxic activity of trichothecenes attributed to short-term exposure; however, T-2 toxin has been reported to exhibit some teratogenic activity in mice (286). In humans, the common symptoms of T-2 toxicosis are vomiting and diarrhea.

12.3 Vomitoxin

Vomitoxin (2,7,15-trihydroxy-12,13-epoxytricholthec-9-ene-8-one) is another trichothecene mycotoxin produced primarily from the fungus *Fusarium graminearum*. This fungus causes scab in corn in temperate climates, which exhibit cool and wet weather conditions at the time of harvesting of corn crop. Both American and Canadian wheat crops were affected by *F. graminearum* in the early 1980s and vomitoxin levels have been reported to range from 0.1 to 4.1 ppm. Acute vomitoxin toxicity is lower than for T-2 toxin; however, common to other trichothecene mycotoxins, initial symptoms associated vomitoxin toxicosis are feed refusal and emisis. Sublethal levels of vomitoxin result in altered host resistance, humoral and cell-mediated immunities (287). Exposure to this toxin can result in abnormally elevated serum IgA levels, resulting from proliferation and differentiation of IgA-secreting cells in Peyer's patches. There is also evidence in mice that vomitoxin is both fetotoxic and teratogenic (288). These workers exposed pregnant mice on days 8–11 of gestation with varying dosages (0.5–15 mg vomitoxin/kg body weight) and reported multiple teratogenic effects at low dosages and complete conceptus resorption at the high dosages.

12.4 Ochratoxin A

Ochratoxin A (7-carboxy-5-chloro-8-hydroxy-3,4-dihydro-3-R-methyl isocoumarin) is a primary metabolite of ochratoxins derived from *Penicillium* and *Aspergillus* fungi in contaminated food, feed, and animal tissues. The primary target organ of ochratoxin A toxicosis is the kidney (proximal tubule).

12.5 Zearalenone

Zearalenone (ZEA) is a nonsteroidal phytoestrogen produced by *F. graminearum* in mold-infected corn, hay, barley, and sorghum. ZEA is a member of a rare class of natural products, referred to as β-resorcyclic acid lactones, that interact with estrogen receptors in a manner similar to natural estrogens (e.g., 17β-estradiol; 289). The acute toxicity of ZEA is less than other mycotoxins; however, a variety of toxic effects associated primarily with reproductive dysfunction occur with chronic exposure. Estrogenic syndrome (vulvar hypertrophy, vaginal eversion, and enlarged mammary glands) is common in animals exposed to high levels of ZEA. An important aspect of ZEA contamination of dairy rations is the metabolism of ZEA to the primary metabolite α-zearalenol, which has estrogenic activity that is manyfold more potent than the parent compound. Thus, the transmission of this mycotoxin metabolite into milk and edible tissues is a concern to the food industry.

12.6 Detoxification of Mycotoxins in Food

The *Aspergillus, Fusarium,* and *Penicillium* fungi produce mycotoxins including AF, T-2, vomitoxins, and zearalenone that are only partially inactivated by high-heat processing during milling and baking. Heating at high temperatures for long durations are more effective in the elimination of AF and fusarium toxins. Ochratoxin A, a secondary metabolite of the fungi *Aspergillus* and *Penicillium* is comparatively stable to heat treatment, especially when in the presence of water. Numerous chemicals including oxidizing agents, gases, bisulfite and acids, and aldehydes reduce AF levels but also form toxic residues and reduce sensory and functional properties of the food product. Ammoniation with either ammonium hydroxide or gaseous ammonium is a common and effective method for detoxifying AF and has been used in the industry to reduce AF contamination of peanut, corn, and cottonseed meals. Human exposure to various mycotoxins occurs from different dietary sources such as both cereals and baked products, as well as the toxin or its metabolite in animal products.

MANUFACTURING HAZARDS IN PROCESSING CRUDE OILS AND FATS

13.1 Refining Processes of Crude Oil

Solvent Extraction and Photooxidation Reactions. Fats derived from plant, animal, and fish without further refining are relatively crude materials with disagreeable flavor, color, and overall poor nutritional value. The oils derived from oilseeds are obtained by pressing or extracting with the solvent hexane. Hexane is also used to extract residual oil from the cake meal after the initial screw press procedure. This solvent extraction process can result in losses of hexane solvent in the cake (e.g., 5 L hexane/tonne cake). The combination of methanol/ammonia or methanol/ammonia-hexane has also been used to simultaneously extract oil and glucosinolates, as well as low-molecular-weight sugars from oilseeds and legumes. Hexane-extracted glandless cottonseed meal contains approximately 40–50% more flatulence-causing sugars than commercial soybean meals and mung beans. Reducing flatulence activity is an important consideration in improving the nutritional quality of the meal from oilseed sources, such as soybean, cottonseed, and canola.

During storage at 98°C, peroxide values used to assess the degree of autooxidation in oils, are slightly higher in soybean oil compared to canola and corn oils, but less than that in safflower oil. Olive and peanut oils and lard are the most stable fats (290). Stability of oils to photooxidation is influenced by the presence of metals and photosensitizers (e.g., carotene and chorophyll pigments), which catalyze oxidative reactions. Moreover, the stability of soybean oil to photooxidation is increased if stored in plastic bottles rather than glass containers.

Tocopherol Losses Attributed to Refining Crude Oil. Further oil refining of crude oil derived from various sources such as rapeseed, olive, sunflower, and canola require the use of degumming (e.g., phosphoric acid), neutralization (caustic soda), and decolorization (bleaching earth, bentonite) agents, as well as high temperature (e.g., 250°C at 6 mm Hg pressure for 30 min), to remove gums and volatile impurities that contribute detrimental quality attributes affecting color, odor, flavor, and stability of the oil.

The three key steps of refining, bleaching, and deodorization used to purify crude oil may reduce the native tocopherol content. Alkaline refining (e.g., heating oil with 12–15% NaOH) removes free fatty acids, color pigments, and metals that contribute to lipid oxidation reactions. This process results in losses of tocopherol that range from 14 to 20% (291,292). During the bleaching and deodorization procedures, crude oils are exposed to high temperatures and pressure conditions resulting in 10–15% losses of tocopherols and limited geometric isomerization of PUFA (293,294). Alfa-tocopherol and α-tocotrienols are more susceptible to losses associated with processing than γ-tocopherol or β-tocopherol (295). The potential loss of oxidative stability of the oil is overcome by the addition of synthetic antioxidants such as tertiary butylhydroquinone (TBHQ), BHT, or BHA at levels up to 200 ppm. Tocopherol retention is higher with TBHQ than BHA (296).

13.2 Pesticide Residues

The frequent exposure of oilseed crops to pesticides during the growth of the oil seed provides a distinct possibility of pesticide residue contamination of edible oil sources. Pesticide use and monitoring of potential pesticide residues in dietary oil sources (e.g., soybean, sunflower, cottonseed, corn oil, canola) are regulated by FDA and Agriculture Canada food safety programs. The effectiveness of crude oil refining procedures for removing pesticide contaminants has been studied extensively with numerous different pesticides (297). The purging of oil with steam at high temperatures and under vacuum, as performed during refining of crude oil during the deodorization process, substantially reduces pesticide levels.

13.3 PAH Contamination

Since PAHs are ubiquitous environmental contaminants, diesel engine emissions, incinerator and ambient air, ground and drinking water are additional potential sources of PAHs. Oil spills, industrial and domestic effluent, and incomplete combustion of fossil fuels are major sources of contributors to the contamination of mussel and fish in aquatic environments (298). The significant risk of PAH residue contamination in marine oils used for human consumption remains to be determined.

13.4 Toxic Oil Syndrome

Crude oils for industrial use undergo a denaturation process using 2% aniline or 5% ceres blue or 5% ricin oil. In the early 1980s, a syndrome referred to as toxic oil syndrome (TOS) occurred in Spain and resulted in massive reported illnesses (e.g., 2000) and over 300 deaths. The epidemic, initially referred to as atypical pneumonia, resulted in many deaths of individuals in Madrid and neighboring areas due to respiratory failure caused by pneumonitis. A primary cause for the disease was the consumption of illegally marketed edible oil, potentially contaminated with denatured industrial-grade oil containing 2% aniline (299). Three potential explanations for this intoxication put forward include: (1) the presence of unidentified contaminants in transport truck tanks, which had not been appropriately prepared for transporting food; (2) the presence of an aniline derivative or contamination of the technical-grade used in the denaturation of crude oil; and (3) the presence of a reaction product(s) derived from the normal components of the oil (e.g., oleic acid) and the aniline, or other contaminants used in the oil refining (300). Despite the numerous symptoms of toxicosis associated with the ingested contaminated oil, confirmation of aniline as the toxic agent could not be made, since the clinical manifestations of TOS were not common to aniline toxicity (299). Several suggestions regarding the possible identity of the toxic agent in TOS include derivatives formed between oleic acid and aniline, *N*-methyl-aniline and *o*-, *m*-, and *p*-toluidine. These agents have the potential to exhibit strong free radical activity, thus causing subsequent nonspecific cellular damage to vital organs (301). Gains in identifying the etiologic agent were hampered by difficulties in distinguishing the oils that contained the toxic agent(s) believed to cause TOS. A comprehensive analysis of toxic oil samples (302) resulted in identifying anilides in 62% of case oils, which confirmed the initial hypothesis set forth by Tabuenca (299), that fatty acid anilide substances contained in the edible rapeseed oil were a primary suspect for TOS in food oils.

TOS is regarded as an unique clinical condition with unknown etiology and characterized by symptoms involving both an acute stage and a chronic condition. A relatively short-lived acute phase consisted of patients exhibiting rash, lymphadenopathy, and eosinophilia. More severe symptoms included interstitial pneumonitis, pleural effusions, and pulmonary edema resulting in respiratory distress and cardiac injury. It is noteworthy that similar observations of pneumonia, intestinal irritation, and liver necrosis have been reported in ducks fed crude petroleum (303). Approximately 20% of victims inflicted with TOS displayed chronic symptoms that included neuromuscular manifestations such as severe neuropathy and myositis while others developed reversible pulmonary hypertension, scleroderma-like changes, or chronic liver disease (304,305).

13.5 Polychlorinated Biphenyls

Two very serious incidences of polychlorinated biphenyls (PCBs) contaminated rice bran cooking oils occurred in Japan in 1968 affecting over 1600 people and in Central Taiwan in 1979 affecting 2061 victims. Machine oil used to heat the rice oil during the deodorization process was determined to be the source of the PCBs in the Japanese case. In the Taiwan incident, contamination by PCBs included partial degradation of PCBs by heat and further contamination with high levels of polychlorinated dibenzofurans (PCDFs) and polychlorinated quarterphenyls (PCQs) of rice bran oil (306). The PCB poisoning in Taiwan referred to as the Yu-Cheng (oil disease) resulted in many victims having elevated serum PCB concentrations and exhibiting chloracne, hyperpigmentation, enlarged Meibomian glands, and deformities to fingernails (307). Children exposed to PCBs *in utero* in this case were called Yu-Cheng children and had lower birth weights, hyperpigmentation, neonatal teeth, delayed developmental milestones, and poor cognitive functioning (308,309).

The accumulation and/or disposition of PCBs and dioxins in bovine and human milk fat represent an additional source of exposure to PCBs. Reported concentrations of PCBs in human and bovine milk fat range from 0.11 to 65 ppm. Presence of waste oils used on roads and spills from industrial chemicals, leaching into ground water contribute to PCB contamination. Cow's milk fat, in fact, can be used as a reliable indicator of degree of PCB contamination (310). Human milk derived from older women can also be higher in PCB than milk from younger women (311). Lactating cows exposed to dioxins and furans excrete PCBs mainly in feces and the milk. Disposition of dioxins in milk of lactating cows is characterized by a distribution phase $t_{1/2}$ of 1–2 days and a longer elimination rate $t_{1/2}$ that vary between 40 and 80 days (312). Both the stage of lactation as well as the amount of milk produced have no bearing on the toxicokinetic behavior of dioxins in milk fat.

Bottom-feeding fish (e.g., carp) from the Great Lakes have been reported to contain dioxins as high as 120 ppt body weight (313). Cooking methods employing high heat such as charbroiling resulting in internal temperatures reaching 80°C have been reported to significantly reduce 2,3,7,8-tetrachloro-dibenzo-*p*-dioxin (TCDD) (314).

A variety of both biochemical and toxic responses have been reported in laboratory animals exposed to TCDD. Features of TCDD toxicity include body-weight loss and permanent retardation of body-weight gain at lethal and sublethal dosages, respectively. Adverse effects to skin, teratogenicity, and immunotoxicity have also been observed. *In vitro* studies with TCDD from lymphoid organ cells have shown direct effects on peripheral lymphocytes, which lends support to the evidence that TCDDs are also immunotoxic (315). The administration of PCB to rats also results in hypercholesterolemia (316), which may be in response to elevations in hepatic HMG-CoA activity as a result of PCB exposure (317).

PACKAGING MATERIALS: MIGRATION OF TOXIN COMPONENTS

Food packaging materials such as polyvinyl chloride (PVC), polyvinylidene chloride (PVDC), polyethylene (PE), and polyethylene terephthalate (PET) are selected for use on the basis of their ability to protect food ingredients against deterioration from light and oxygen. They also provide the consumer an opportunity to observe the quality of the product under hygienic conditions. In addition to the number of polymers used in flexible package materials, synthetic plastic additives, termed plasticizers, are incoporated into the package material to enhance physical and mechanical properties of the film (Table 8.7). There are over 100 different plasticizers used in the manufacturing of plastic materials. Plasticizers are low-molecular-weight compounds, with low melting points, or organic liquids with high boiling points. These materials are added to the plastic film to increase workability, flexibility, and extensibility properties of the resin material employed in packaging (318). Plasticizers used in the packaging of foodstuffs, which include meats, cheeses, and vegetables, are the phthalic acid esters [e.g., bis(2-ethyl hexyl) (DEHP) and butyl phenylmethyl (BBP) (319,320)], bis(2-ethyl hexyl) adipate (DEHA) (319,323), epoxide soy bean oil (ESBO) (321,324), and citrates and sebacates (321).

The low molecular weight of many plasticizers facilitate high mobility and migration into food systems. Factors that influence the migration rate of plasticizer materials into the foodstuff include the plasticizer molecular weight, the amount of plasticizer, and its solubility in the copolymer material (318,325). In addition, the fat content of the food material (325,326), storage temperature and storage time (326,327), method of heating (328), and the contact surface of the film per unit weight of food (319,326) are important factors influencing migration efficiency. Kondyli and co-workers (326) demonstrated greater migration of DEHA than phthalic acid esters in meat samples. This observation was especially true for meats that contained a high-fat content. Similarly, fresh cheese with a high moisture content contained less plasiticizer (DEHA) than other older cheeses with higher fat content (325). In other studies, the migration of vinyl chloride (VC) into water bottled with PVC was shown to increase with storage time. This result raises the suggestion that maximum storage times for PVC packaged foods should be implemented by the industry (327). Finally, specific methods of food processing, such as irradiation, can facilitate the release of radiolytic products including gases (e.g., H_2, CH_4, hydrogen chloride) and irradiated laminate materials, which interact with the food material. Killoran (329) provided evidence showing that irradiated food package

Table 8.7 Common materials used in food packaging[a]

Flexible Package Materials		Plasticizers		
Compound	Abbreviation	Compound	Abbreviation	Molecular Weight
Ethylene vinyl acetate	EVA			
Polyethylene	PE	Bis(2-ethyl hexyl)	DEHP	390.6
Polyethylene terephthalate	PET	Butyl phenylmethyl	BBP	312.0
Polystyrene	PS	Bis(2-ethylhexyl)adipate	DEHA	370.6
Polyvinyl chloride	PVC	Acetyltributylcitrate	ATBC	402.5
Polyvinylidene chloride	PVDC	Epoxized soybean oil	ESBO	—

[a] Adapted from Refs. 318–322.

materials with 60 kGy caused increased migration of plastic components derived from various PVC derivatives. A significant migration of plastic food packaging materials into foods occurs in general, and in particular, with fatty foods following irradiation that contribute to reduced organoleptic qualities (330). Irradiation of packaged food materials in the presence of oxygen produces significant levels of CO, CO_2 in addition to H_2, CH_4, and hydrocarbons. Residue levels of gaseous radiolysis products are generally related to the absorbed dose of radiation.

Daily human intakes of DEHA, acetyltribuyl citrate (ATBC), and ESBO have been estimated at 8.5, 1.5, and 1.0 mg, respectively (328). Reported levels of DEHA migration range from 0.8 to 15 mg/dm^2 in meats and cheese foods (319,323,325). Highest levels of ESBO have been reported in cooked chicken (14 ppm), pizza (16.3 ppm), microwaved cornish pastry (10.6 ppm), and sandwiches (9 ppm). Citrate ester (ATBC) residues have been detected in peanut biscuits (47 ppm) and reheated pizza (35 ppm). Vinyl chloride (VC) migration in PCV bottled water increases linearly from about 60 ppt after 25–30 days of storage to levels exceeding 150 ppt following 5–6 months of storage. The daily intake of VC has been estimated to exceed 100 ng/person/day (327).

Unlike vinyl chloride, a known human carcinogen (331), there is little evidence indicating significant toxicities associated with DEHA, ATBC, and ESBO. Long-term studies with ESBO have yielded little data indicating systemic toxicity at daily intake levels of 100 mg/kg body weight. Similarly, ATBC is not mutagenic in the Ames *S. typhimurium* assay. In 1974, the Institute of Food Technologists' Expert Panel on Food Safety and Nutrition (322) concluded that the phthalate-containing food packaging materials represented no significant hazards to human health. Studies conducted with rodents fed phthalate dibasic esters have reported reduced weight gain, liver hyperplasia, neuromuscular and skeletal abnormalities, as well as potential embryotoxicities (332–334). It has been stated that irradiation processes should not result in the production of radiolytic degradation products that are either toxic or reduce the sensory qualities of the packaged food products (330). There is, however, no toxicity data at the present time concerning irradiated food package materials, albeit high levels of volatile products if produced from radiolysis of polypropylene package materials could potentiate a concern.

RISK ASSESSMENT FACTORS IN EVALUATION OF THE SAFETY OF FATS AND OILS

Advances made in food lipid preservation and processing to provide the consumer with a nutritious and sensory appealing product have coincided with an increased public awareness of the safety of consuming food calories derived from different fat sources and exposure to lipid-soluble xenobiotic agents. The latter may be derived from either an intentional addition to the food system, (e.g., synthetic antioxidants) or as a consequence of nutrient–

nutrient interaction (e.g., LOPs, COPs, PAHs) initiated by a process. Traditionally, nutritional research has emphasized the deficiency of nutrients, and nutritional toxicology studies have also favored acute toxicities, putting lesser emphasis on the effects associated with chronic exposure. Recently, interpreting risk assessment factors concerning the safety of dietary fats has involved both the chronic and excessive intake of dietary fats, oils, and their constituents. Moreover, in addition to finding methods to evaluate the risks associated with potential toxicants found in fats and oils, a number of studies have reported the significance of "extranutritional" constituents of plant and animal fat and oil origin, which possess bioactive and potential chemopreventative properties (335). Many of these agents exhibit *in vivo* antioxidant protection [e.g., tocopherol (105), retinoids (231), and monounsaturated fatty acids (336)]. Continued research in these areas are required for comprehensive risk assessment as we evolve into an era of new biotechnologies that will certainly include a wide variety of lipid sources in our food supply.

To properly evaluate the safety of different fats and oils, or constituents thereof, a risk assessment strategy is required. In general terms, a risk can be defined as the product generated from the probability and the consequences of injury associated with the risk. The probability of the risk occurring will be influenced by the sum of factors involving the exposure level of a toxicant and the susceptibility of the individual to potential injury from the toxicant. These components are important in assessing the risk and thus safety of fats and oils. Academic inquiry that addresses risk assessment of fats and oils includes: (1) defining the *sources* of fat and oil constituents (e.g., saturated versus unsaturated versus hydrogenated fat) that may have an unsafe potential to the consumer; (2) determining the *amount* of lipid or lipid-soluble constituent consumed (e.g., mycotoxin, gossypol, colorants), which represents a realistic exposure and toxic potential to the organism; and (3) studying the *characteristics* of the potential toxicant (e.g., LOPs, COPs, plasticizer) that interfaces with the organism at the molecular level. Finally, the susceptibility of the individual (aged, pregnant, apolipoprotein E2-associated hyperlipidemic individuals) in displaying an adverse effect to the lipid, or lipid constituent, requires definition in the final assessment of risk. This chapter has attempted to identify these aspects for individual lipid constituents common to fats and oils used in our food supply.

REFERENCES

1. A.W.M. Hay and co-workers, *Lipids* **15,** 251 (1980).
2. G.O. Burr and M.M. Burr, *J. Biol. Chem.* **82,** 345 (1929).
3. T. Dwyer and B. Hetzel, *J. Epidemiol.* **9,** 65 (1980).
4. S.M. Grundy, *Am. J. Clin. Nutr.* **45,** 1168 (1987).
5. A. Nordoy and S.H. Goodnight, *Atherosclerosis* **10,** 149 (1990).
6. D.H. Blankenhorn and co-workers, *JAMA* **263,** 1646 (1990).

7. W.C. Willet, M.J. Stampfer, G.A. Colditz, B.A. Rosner and F.E. Speizer, *N. Engl. J. Med.* **323,** 1664 (1990).
8. B.S. Reddy, and co-workers, *Cancer Res.* **50,** 2562 (1990).
9. K.K. Carroll, *Cancer Res.* **35,** 3374 (1975).
10. P.C. Chan and T.L. Dao, *Cancer Res.* **41,** 164 (1981).
11. B. Weinstein, *Cancer Res. (Suppl)* **51,** 5080 (1991).
12. K.K. Carroll, in C. Ip and co-workers, eds., *Dietary Fat and Cancer*, Alan R. Liss, New York, 1986, pp. 231–248.
13. M. Morotomi, J.G. Gillem, P. LoGerfo, and I.B. Weinstein, *Cancer Res.* **50,** 3595 (1990).
14. B.R. Bansal, J.E. Rhoads, and S.C. Bansal, *Cancer Res.* **38,** 3293 (1978).
15. B.S. Reddy, S. Mangat, A. Sheinfil, J.H. Weisburger, and E.L. Wynder, *Cancer Res.* **37,** 2132 (1977).
16. P.C. Chan, K.A. Ferguson, and T.L. Dao, *Cancer Res.* **43,** 1079 (1983).
17. F.H. Mattson and S.M. Grundy, *J. Lipid Res.* **26,** 194 (1985).
18. A. Mitra, S. Govindwar, and A.P. Kulkarni, *Toxicol. Lett.* **58,** 135 (1991).
19. A. Mitra, S. Govindwar, P. Joseph, and A. Kulkarni, *Toxicol. Lett.* **60,** 281 (1992).
20. T. Tanaka, W.S. Barnes, G.M. Williams, and J.H. Weisburger, *Gann.* **76,** 570 (1985).
21. M. Sato, A. Oh-Hashi, M. Kubota, E. Nishide, and M. Yamaguchi, *Br. J. Nutr.* **63,** 249 (1990).
22. S. Hojberg, J.B. Nielson, and O. Andersen, *Food Chem. Toxic.* **30,** 703 (1992).
23. G. Bramilla and co-workers, *Mutat. Res.* **171,** 169 (1986).
24. M.R. L'Abbe, K.D. Trick, and J.L. Beare-Rogers, *J. Nutr.* **121,** 1331 (1991).
25. P.J.A. Sheehy, P.A. Morrissey, and A. Flynn, *Br. J. Nutr.* **71,** 53 (1994).
26. M.J. Gonzalez, J.I. Gray, R.A. Schemmel, L. Dugan and C.W. Welsch, *J. Nutr.* **122,** 2190 (1992).
27. A.W. Bull, N.D. Nigro, W.A. Golembieski, J.D. Crissman, and L.J. Marnett, *Cancer Res.* **44,** 4924 (1984).
28. M.G. Enig, R.J. Munn, and M. Keeney, *Fed. Proc.* **37,** 2215 (1978).
29. B.S. Reddy, T. Tanaka, and B. Simi, *J. Natl. Cancer Inst.* **75,** 791 (1985).
30. S.L. Selenskas, M.M. Ip, and C. Ip, *Cancer Res.* **44,** 1321 (1984).
31. B. Koltzko, *Acta Paediatr.* **81,** 302 (1992).
32. P.L. Zock and M.B. Katan, *J. Lipid Res.* **33,** 399 (1992).
33. W.A. Harland, J.D. Gilbert, G. Steel and C.J.W. Brooks, *Atherosclerosis* **13,** 239 (1971).
34. J. Loeper, J. Emerit, J. Goy, O. Bedu, and J. Loeper, *IRCS J. Med. Sci.* **11,** 1034 (1983).
35. A. Keys, *Circulation* **41**(suppl 1), 1 (1970).
36. P.H.E. Groot, B.C.J. DeBoer, E. Haddeman, U.M.T. Houtsmuller, W.C. Hulsmann, *J. Lipid Res.* **29,** 541 (1988).
37. E.H. Ahrens and co-workers, *Lancet* **1,** 943 (1957).
38. A. Keys, J.T. Anderson, and F. Grande, *Metabolism* **14,** 747 (1965).
39. D.M. Hegsted, R.B. McGandy, M.L. Myers and F.J. Stare, *Am. J. Clin. Nutr.* **17,** 281 (1965).
40. D.K. Spady and J. Dietschy, *Proc. Natl. Acad. Sci.* **82,** 4526 (1982).
41. J.H. Bragdon, J.M. Seller, J.W. Stevenson, *Proc. Soc. Exp. Biol. Med.* **95,** 282 (1957).
42. S.A. Hashim, A. Artega, and T.B. Van Itallie, *Lancet* **1,** 1105 (1960).
43. J. Layton, P.J. Drury, and M.A. Crawford, *Br. J. Nutr.* **57,** 383 (1986).
44. F. Grand, *J. Nutr.* **76,** 255 (1962).
45. M.A. Denke and S.M. Grundy, *Am. J. Clin. Nutr.* **56,** 895 (1992).
46. K.C. Hayes, A. Pronczuk, S. Lindsey and D. Diersen Schade, *Am. J. Clin. Nutr.* **53,** 491 (1991).

47. A. Bonanome and S.M. Grundy, *N. Engl. J. Med.* **319,** 1244 (1988).
48. R. Reiser, M.F. Sorrels, and W.C. Williams, *Circ. Res.* **7,** 833 (1959).
49. J. Shepherd, C.J. Packard, J.R. Patsch, A.M. Gotto, and O.D. Taunton, *J. Clin. Invest.* **60,** 1582 (1978).
50. D.M. Ney, H.C. Lai, J.B. Lasekan and M. Lefevre, *J. Nutr.* **121,** 1311 (1991).
51. M. Soci-Thomas, M.D. Wilson, F.L. Johnson, D.L. Williams and L.L. Rudel, *J. Biol. Chem.* **264,** 9039 (1989).
52. S. Lindsey, J. Benattar, A. Pronczuk and K.C. Hayes, *Proc. Soc. Exp. Biol. Med.* **195,** 261 (1990).
53. W.P. Castelli, *Am. Heart J.* **112,** 432 (1986).
54. D. Kritchevsky, *Am. J. Clin. Nutr.* **23,** 1105 (1970).
55. P. Nishina and co-workers, *J. Lipid Res.* **34,** 1413 (1993).
56. M.L. Fernandez and D.J. McNamara, *Metabolism* **38,** 1094 (1989).
57. D. Kritchevsky, S.A. Tepper, O. Vesselinovitch, and R.W. Wissler, *Atherosclerosis* **14,** 53 (1971).
58. D. Vesselinovitch, G.S. Getz, R.H. Hughes, R.W. Wissler, *Atherosclerosis* **20,** 303 (1974).
59. L.M. Alderson, K.C. Hayes, and R.J. Nicolosi, *Arteriosclerosis* **6,** 465 (1986).
60. A. Bonanome and co-workers, *Athero. Thromb.* **12,** 529 (1992).
61. D. Steinberg, S. Parthasaathy, T.E. Carew, J.C. Khoo, J.L. Witztum, *N. Engl. J. Med.* **320,** 915 (1989).
62. Y. Yamori and co-workers, *J. Hypertension* **4**(Suppl.), 449 (1986).
63. D. Reichel and N.E. Miller, *Arteriosclerosis* **9,** 785 (1989).
64. D.J. Gordon and co-workers, *Circulation* **79,** 8 (1989).
65. B.E. McDonald, J.M. Gerrard, V.M. Bruce and E.J. Corner, *Am. J. Clin. Nutr.* **50,** 1382 (1989).
66. K.K. Carroll and H.T. Khor, *Lipids* **6,** 415 (1971).
67. C.W. Welsch, *Am. J. Clin. Nutr.* **45,** 192 (1987).
68. D.L. Smith and A.L. Willis, in S. Abraham, ed., *Carcinogenesis and Dietary Fat*, Kluwer, Academic, Boston, 1989, pp. 53–82.
69. I. Bairati, L. Roy, and F. Meyer, *Can. J. Cardiol.* **8,** 41 (1992).
70. J. Dyerberg and K.J. Jorgensen, *Prog. Lipid Res.* **21,** 255 (1982).
71. H.R. Knapp, I.A.G. Reilly, P. Alessandrini, G.A. Fitzgerald, *N. Engl. J. Med.* **314,** 937 (1986).
72. K.K. Carroll, *Lipids* **21,** 731 (1986).
73. E. Soyland and co-workers, *N. Engl. J. Med.* **328,** 1812 (1993).
74. D.S. Lin and W.E. Connor, *J. Lipid Res.* **21,** 1042 (1981).
75. D. Applebaum-Bowden and co-workers, *Am. J. Clin. Nutr.* **39,** 360 (1984).
76. D.R. Jacobs, J.T. Anderson, P. Hannan, A. Keys, and H. Blackburn, *Arteriosclerosis* **3,** 349 (1983).
77. M.B. Katan and co-workers, *J. Lipid Res.* **29,** 883 (1988).
78. J.M.R. Beveridge, W.F. Connell, G.A. Mayer, and H.A. Haust, *J. Nutr.* **71,** 61 (1960).
79. W.E. Connor, D.B. Stone, and R.E. Hodges, *J. Clin. Invest.* **43,** 1691 (1964).
80. J.J. McNamara, M.A. Molet, J.F. Stremple, and R.T. Cutting, *J. Am. Med. Assoc.* **216,** 1185 (1971).
81. A. Arul, A. Boudreau, J. Makhlouf, R. Tardif and M.R. Sahasrabudhe, *J. Food Sci.* **52,** 1231 (1987).
82. L.L. Smith and B.H. Johnson, *Free Radical Biol. Med.* **7,** 285 (1989).
83. T.V. Fungwe, L.M. Cagen, H.G. Wilcox, and M. Heimberg, *J. Lipid Res.* **33,** 179 (1992).

84. D.R. Jacobs and co-workers, *Circulation* **86,** 1046 (1992).
85. A.J. McMichael, O.M. Jensen, D.M. Parkin, and D.G. Zaridze, *Epidemiol. Rev.* **6,** 192 (1984).
86. M. Alvaro and co-workers, in L. Barbara and co-workers, eds., *Recent Advances in Bile Acid Research,* Raven Press, New York, 1985, pp. 223–234.
87. J.L. Thislte and A.F. Hofmann, *N. Engl. J. Med.* **289,** 655 (1973).
88. B.S. Reddy, T. Narisawa, J.H. Weisburgar and E.L. Wynder, *J. Natl. Cancer Inst.* **56,** 441 (1976).
89. R.G. Cameron, K. Imaida, H. Tsuda and N. Ito, *Cancer Res.* **42,** 2426 (1982).
90. S.J. Silverman and A.W. Andrews, *J. Natl. Cancer Inst.* **59,** 1557 (1977).
91. M. Wilpart, P. Mainguet, A. Maskens and M. Roberfroid, *Carcinogenesis* **4,** 45 (1983).
92. J. Dupont, *Lipids* **15,** 133 (1980).
93. R.D. Sharma and C. Rukmini, *Lipids* **21,** 715 (1986).
94. E. Nwokolo and D.D. Kitts, *Food Chem.* **30,** 219 (1988).
95. L. Busk and U.G. Ahlborg, *Toxicol. Lett.* **6,** 243 (1980).
96. R.B. Shekelle and co-workers, *Lancet* **2,** 1185 (1981).
97. R.M.D. Favaro, C.K. Miyasaaka, I.D. Desai, J.E. Dutrade-Oliverira, *Nutr. Res.* **12,** 1357 (1992).
98. G.W. Burton and K.U. Ingold, *Science* **224,** 569 (1984).
99. R. Yamauchi, N. Miyake, K. Kato, and Y. Ueno, *BioSci. Biotech. Biochem.* **56,** 1529 (1992).
100. M.Y. Jung, S.H. Yoon, and D.B. Min, *J. Am. Oil. Chem. Soc.* **66,** 118 (1989).
101. M.Y. Jung and D.B. Min, *J. Food Sci.* **55,** 1464 (1990).
102. E.N. Frankel, C.D. Evans, and P.M. Cooney, *J. Agric. Food Chem.* **7,** 438 (1959).
103. C.K. Chow, *Free Radical Biol. Med.* **11,** 215 (1991).
104. J.E. Packer, T.E. Slater, and R.L. Willson, *Nature (London)* **278,** 737 (1979).
105. H. Esterbauer, M. Dieber-Rotheneder, G. Striegl and G. Waeg, *Am. J. Clin. Nutr.* **53,** 314S (1991).
106. E. Mellanby, Medical Research Council, Special Report Series, Number 61 (London) (1921).
107. A.F. Hess and M. Weinstock, *J. Biol. Chem.* **62,** 301 (1924).
108. V. Linden, *Br. Med. J.* **3,** 647 (1974).
109. A.W. Taylor, *Clin. Sci.* **42,** 515 (1972).
110. R.P. Holmes and F.A. Kummerow, *J. Am. Coll. Nutr.* **2,** 173 (1983).
111. A.W. Ham and D.L. Murray, *Br. J. Exp. Pathol.* **15,** 228 (1934).
112. A. Kurmann and H. Indyk, *Food Chem.* **50,** 75 (1994).
113. C.H. Jacobus and co-workers, *N. Engl. J. Med.* **326,** 1173 (1992).
114. M.F. Holick, Q. Shao, W.W. Liu and T.C. Chen, *N. Engl. J. Med.* **326,** 1178 (1992).
115. C. Rukmini, *Food Chem. Toxic.* **28,** 601 (1990).
116. E. Nwokolo, D.D. Kitts, and J. Kanhai, *Plant Foods Human Nutr.* **38,** 145 (1988).
117. T.S. Gaginella, V.S. Chadwick, J.C. Debongnie, J.C. Lewis and S.F. Phillips, *Gastroenterology* **73,** 95 (1977).
118. T.A. Eisele and co-workers, *Food Chem. Toxicol.* **20,** 407 (1982).
119. J.P. Matlock and J.E. Nixon, *Toxicol. Appl. Pharmacol.* **84,** 3 (1986).
120. A.R. Johnson, J.A. Pearson, F.S. Shenstone, and A.C. Fogerty, *Nature* **214,** 124 (1967).
121. A. Greenburg and J. Harris, *J. Chem. Ed.* **59,** 539 (1982).
122. J.C. Peters, B.N. Holcombe, L.K. Hiller, and D.R. Webb, *J. Am. Coll. Toxicol.* **10,** 357 (1991).
123. D.R. Webb, J.C. Peters, R.J. Jandacek, and N.E. Fortier, *J. Am. Coll. Toxicol.* **10,** 341 (1991).

124. J.W. Finley, L.P. Klemann, G.A. Leveille, M.S. Otterburn and G.C. Walchak, *J. Agric. Food Chem.* **42,** 495 (1994).
125. L.P. Klemann, J.W. Finley, and G.A. Leveille, *J. Agric. Food Chem.* **42,** (1994).
126. J.R. Hayes and E.S. Riccio, *J. Agric. Food Chem.* **42,** 515 (1994).
127. J.R. Hayes and co-workers, *J. Agric. Food Chem.* **42,** 521 (1994).
128. J.R. Hayes, N.H. Wilson, D.H. Pence, and K.D. Williams, *J. Agric. Food Chem.* **42,** 528 (1994).
129. D.D. Kitts, *J. Toxicol. Environ. Health* **20,** 37 (1986).
130. J.M. Concon, in *Food Toxicology, Part A: Principles and Concepts,* Marcel Dekker, New York, pp. 281–403.
131. A.A. Franke, L.J. Custer, C.M. Cerna and K.K. Narala, *J. Agric. Food Chem.* **42,** 1905 (1994).
132. F.E. Newsome and W.D. Kitts, *Can. J. Anim. Sci.* **60,** 53 (1980).
133. F.J.R. Paumgartten and co-workers, *Brazilian J. Med. Biol. Res.* **23,** 873 (1990).
134. D.R. Webb, G.M. Ridder, and C.L. Alden, *Food Chem. Toxic.* **27,** 639 (1989).
135. L.K.T. Lam and B. Zheng, *J. Agric. Food Chem.* **39,** 660 (1991).
136. I.F. Delgado and co-workers, *Food Chem. Toxic.* **31,** 31 (1993).
137. N. Deighton, S.M. Glidewell, S.G. Deans, and B.A. Goodman, *J. Sci. Food Agric.* **63,** 221 (1993).
138. A.K. Al-Athari and B.A. Watkins, *Poult. Sci.* **67,** 778 (1988).
139. F.A. Kummerow, *J. Environ. Pathol. Toxicol. Oncol.* **6**(3–4), 136 (1986).
140. E.A. Emken, in E.A. Emken and H.J. Dutton, eds., *Geometric and Positional Fatty Acid Isomer*, American Oil Chemists' Society, 1979, pp. 107–118.
141. W.J. Decker and W. Mertz, *J. Nutr.* **91,** 324 (1967).
142. R.R. Brown, *Cancer Res.* **41,** 3741 (1981).
143. Y.L. Ha, N.K. Grimm, and M.W. Pariza, *Carcinogenesis* **8,** 1881 (1987).
144. Y.L. Ha, J. Storkson, and M.W. Pariza, *Cancer Res.* **50,** 1097 (1990).
145. C. Ip, S.F. Chin, J.A. Scimeca and M.W. Pariza, *Cancer Res.* **51,** 6118 (1991).
146. R.P. Mensink and M.B. Katan, *N. Engl. J. Med.* **323,** 439 (1990).
147. E.G. Hill, S.B. Johnson, and R.T. Holman, *J. Nutr.* **109,** 1759 (1979).
148. J.E. Chappell, M.T. Clandinin, and C. Kearney-Volpe, *Am. J. Clin. Nutr.* **42,** 49 (1985).
149. J. Opstvedt, J. Pettersen, and S.J. Mork, *Lipids* **23,** 713 (1988).
150. A. Arnoldi, C. Arnoldi, O. Baldi and A. Griffini, *J. Agric. Food Chem.* **35,** 1035 (1987).
151. M. Johansson, K. Skog, and M. Jagerstad, *Carcinogenesis* **14,** 89 (1993).
152. W.S. Barnes and J.H. Weisburger, *Cancer Lett.* **24,** 221 (1984).
153. T. Sugimura, *Environ. Health Perspect.* **67,** 5 (1986).
154. H. Hayatsu and co-workers, *Mutation Res.* **81,** 287 (1981).
155. A.L. Tappel and C.J. Dillard, *Fed. Proc.* **40,** 174 (1981).
156. B. Ames, *Science* **221,** 1256 (1983).
157. G.M. Findlay, H.H. Draper, and J.G. Bergan, *Lipids* **5,** 970 (1970).
158. J.T. MacGregor and co-workers, *Food Chem. Toxic.* **23,** 1041 (1985).
159. T.A. Dix and L.J. Marnett, *Science* **221,** 77 (1983).
160. J.M. McNeil and E.D. Wills, *Chem. Biol. Interact.* **53,** 197 (1985).
161. G.Y. Kwei and L.F. Bjeldanes, *Food Chem. Toxic.* **28,** 491 (1990).
162. K. Kanazawa, E. Kanazawa, and M. Natake, *Lipids* **20,** 412 (1985).
163. H.H. Draper, L. Polensek, M. Hadley, and L.G. McGirr, *Lipids* **19,** 836 (1984).
164. D.L. Crawford, R.O. Sinnuhuber, F.M. Stout, J.E. Oldfield and J. Kaufmes, *Toxicol. Appl. Pharmacol.* **7,** 826 (1965).

165. M.H. Coleman, *J. Food Technol.* **13,** 55 (1978).
166. G. Kajimoto, K. Ikuta, H. Yoshida, and A. Shibahara, *J. Jpn. Soc. Nutr. Food Sci.* **42,** 313 (1989).
167. P.B. Addis and G.J. Warner, in O.I. Aruoma and B. Halliwell, eds., *Free Radicals and Food Additives,* Taylor & Francis, London, 1991, pp. 77–119.
168. J.M. Luby, J.I. Gray, and B.R. Harte, *J. Food Sci.* **51,** 908 (1986).
169. E. Chicoye, W.D. Powrie, and O. Fennema, *J. Food Sci.* **33,** 581 (1968).
170. A. Fontana and co-workers, *J. Food Sci.* **57,** 869 (1992).
171. P. van de Bovenkamp, T.G. Kosmeijer-Schuil, and M.B. Katan, *Lipids* **23,** 1079 (1988).
172. L. Pizzoferrato, S. Nicoli, and C. Lintas, *Chromatographia* **35,** 269 (1993).
173. B.D. Sander, P.B. Addis, S.W. Park, and D.E. Smith, *J. Food Protection* **52,** 109 (1989).
174. S.A. Park and P.B. Addis, *J. Food Sci.* **52,** 1500 (1987).
175. J.E. Pie, K. Spahis, and C. Seillan, *J. Agric. Food Chem.* **39,** 250 (1991).
176. S.W. Park and P.B. Addis, *J. Agric. Food Chem.* **34,** 653 (1986).
177. J.S. Chen and G.C. Yen, *Food Chem.* **50,** 167 (1994).
178. Y.C. Chen, C.P. Chui, and B.H. Chen. *Food Chem.* **50,** 53 (1994).
179. W.B. Zhang, P.B. Addis, and T.P. Krick, *J. Food Sci.* **56,** 716 (1991).
180. S.K. Peng, C.B. Taylor, J.C. Hill, and R.J. Morin, *Atherosclerosis* **54,** 121 (1985).
181. A.A. Kandutsch and H.W. Chen, *Lipids* **13,** 704 (1978).
182. L.S. Tsai and C.A. Hudson, *J. Food Sci.* **50,** 229 (1985).
183. H.A. Emanuel, C.A. Hassel, P.B. Addis, S.D. Bergemann and J.L. Zavoral, *J. Food Sci.* **56,** 843 (1991).
184. S.K. Peng, C.B. Taylor, W.Y. Huang, J.C. Hill and B. Mikkelson, *Atherosclerosis* **41,** 395 (1982).
185. D. Matthais, C.H. Becker, V. Godicke, R. Schmidt, and K. Ponsold, *Atherosclerosis* **63,** 115 (1987).
186. G.A.S. Ansari, R.D. Walker, V.B. Smart, and L.L. Smith, *Food Chem. Toxic.* **20,** 35 (1982).
187. S.K. Peng, P. Tham, C.B. Taylor, and B. Mikkelson, *Am. J. Clin. Nutr.* **32,** 1033 (1978).
188. M.F. Gray, T.D.V. Lawrie, and C.J.W. Brooks, *Lipids* **6,** 836 (1971).
189. G.F. Morin, B. Hu, and S.K. Peng, in S.K. Peng and R.J. Morin, eds., *CRC, Biological Effects of Cholesterol Oxides,* CRC Press, Boca Raton, FL, 1992; pp. 104–123.
190. P.B. Addis, H.A. Emanuel, S.D. Bergmann, and J.H. Zavoral, *Free Radical Biol. Med.* **7,** 179 (1989).
191. L.L. Smith, *Free Radical Biol. Med.* **11,** 47 (1991).
192. W.W. Nawar, in O.R. Fennema, eds., *Food Chemistry,* Marcel Dekker, New York, 1985, pp. 139–244.
193. M. Naruszewica, E. Wonzy, E. Mirkiewicz, G. Nowicka and W.B. Szostak, *Atherosclerosis* **66,** 45 (1987).
194. L.E. Veca, J. Wilhelm, and M. Harns-Ringdahl, *Mutation Res.* **195,** 137 (1988).
195. B. Frei, R. Stocker, and B.M. Ames, *Proc. Natl. Acad. Sci. USA,* **85,** 9748 (1988).
196. S.W. Park and P.B. Addis, *J. Food Sci.* **51,** 1380 (1986).
197. J.C. Alexander, *J. Toxicol. Environ. Health* **7,** 125 (1981).
198. J.C. Alexander, *J. Am. Oil Chem. Soc.* **55,** 711 (1977).
199. J.C. Alexander, V.E. Valli, and B.E. Chanin, *J. Toxicol. Environ. Health* **21,** 295–309 (1987).
200. M. Mahfouz, S. Johnson, and R.T. Holman, *Biochim. Biophys. Acta* **666,** 58 (1981).
201. C.E. Poling, W.D. Warner, P.E. Mone, and E.E. Rice, *J. Am. Oil Chem. Soc.* **39,** 315 (1962).

202. M. Sugai, L.A. Witting, H. Tsuchiyama, and F.A. Kummerow, *Cancer Res.* **22,** 510 (1962).
203. S.M. Taylor, C.M. Berg, N.H. Shoptaugh, and V.N. Scott, *Food Chem. Toxic.* **20,** 209 (1982).
204. G. Hageman and co-workers, *Lipids* **24,** 899 (1989).
205. G. Hageman, H. Verhagen, B. Schutte and J. Kleinjans, *Food Chem. Toxic.* **29,** 689 (1991).
206. W. Lijinksy and A.E. Ross, *Food Cosmet. Toxicol.* **5,** 343 (1967).
207. A.J. Malanoski, E.L. Greenfield, C.J. Barnes, J.M. Worthington and F.L. Joe. *J.A.O.A.C.* **51,** 114 (1968).
208. W. Fritz, *Arch. Geschwulstforsch* **40,** 81 (1972).
209. Y. Masuda, K. Mor, and M. Kuratsune, *Gann* **57,** 133 (1966).
210. A.O. Asita and co-workers, *Mutation Res.* **264,** 7 (1991).
211. T. Thorsteinsson, *Cancer* **23,** 455 (1969).
212. Y. Musada and M. Kuratsune, *Gann* **62,** 27 (1971).
213. K. Rhee and L. Bratzler, *J. Food Sci.* **35,** 146 (1970).
214. J. Fornal, L. Fornal, and L. Babukowski, *Przegl. Zbozowo. Mlyn.* **14,** 445 (1970).
215. M. Rohrlich and R. Suckow, *Getreide Mehl.* **20,** 90 (1970).
216. M. Kuratsune and W.C. Hueper, *J. Natl. Cancer Inst.* **20,** 37 (1958).
217. M. Kuratsune and W.C. Hueper, *J. Natl. Cancer Inst.* **24,** 463 (1960).
218. J. Mumford and co-workers, *Science* **231,** 217 (1987).
219. J.S.H. Yoo, J.O. Norman, and D.L. Busbee, *Proc. Soc. Exp. Biol. Med.* **177,** 434 (1984).
220. H. Foth, R. Kahl, and G.F. Kahl, *Food Chem. Toxic.* **26,** 45 (1988).
221. J.K. Chipman, P.C. Hirom, G.S. Frost, and P. Millburn, *Biochem. Pharmacol.* **30,** 937 (1981).
222. J. Guthrie, I.G.C. Robertson, E. Zeiger, J.A. Boyd, and T.E. Eling, *Cancer Res.* **42,** 1620 (1982).
223. J.J. Reiners, D. Crowe, C.M. McKeown, D.E. Nerland, and G. Sennenfeld, *Carcinogenesis* **5,** 125 (1984).
224. J.B. Andelman and M.J. Suess, *Bull. WHO* **43,** 479 (1979).
225. J.Z. Byczkowsi and A.P. Kulkarni, *Biochem. Biophys. Res. Commun.* **159,** 1199 (1989).
226. P.A. Cerutti, *Science* **227,** 375 (1985).
227. G.R. Burleson, M.J. Calfield, and M. Pollarda, *Cancer Res.* **39,** 2149 (1979).
228. J.N. Pitts and co-workers, *Science* **210,** 1347 (1980).
229. L.W. Wattenberg, *J. Natl. Cancer Inst.* **48,** 1425 (1972).
230. L.W. Wattenberg, *J. Natl. Cancer Inst.* **50,** 1541 (1973).
231. S. Nomi, T. Matsuura, H. Ueyama, and K. Ueda, *J. Nutr. Sci. Vitaminol.* **27,** 33 (1981).
232. R.S. Miles, F.K. McKeith, P.J. Bechtel, and J. Novakofski, *J. Food Protection* **49,** 222 (1986).
233. F. Shahidi, L.J. Rubin, and D.F. Wood, *Food Chem.* **23,** (1987).
234. K. Imaida and co-workers, *Carcinogenesis* **4,** 895 (1983).
235. E.V. Hansen and O. Meyer, *Toxicology* **10,** 195 (1978).
236. E.V. Hansen, O. Meye, and P. Olsen, *Toxicology* **23,** 79 (1982).
237. N. Ito, S. Fukushima, A.A. Hagiwara, M. Shibata and T. Ogiso, *J. Natl. Cancer Inst.* **70,** 343 (1983).
238. T. Masui and co-workers, *Jpn. J. Cancer Res.* **77,** 1083 (1986).
239. H. Verhagen and co-workers, *Carcinogenesis* **11,** 1461 (1990).
240. F. Iverson, E. Lok, E. Nera, K. Karpinski and D.B. Clayson, *Toxicology* **35,** (1985).
241. F. Iverson and co-workers, *Cancer Lett.* **26,** 43 (1985).
242. M. Hirose, M. Masuda, H. Tsuda, S. Uwagawa and N. Ito, *Carcinogenesis* **8,** 1731 (1987).

243. G. Wurtzen and P. Olsen, *Food Chem. Toxic.* **24,** 1229 (1986).
244. R. Abraham and co-workers, *Exp. Mol. Pathol.* **44,** 14 (1986).
245. H. Amo and co-workers, *Carcinogenesis* **11,** 151 (1990).
246. I.A. Karplyuk, *Vop. Pitan.* **18,** 24 (1959).
247. S. Green, *J. Environ. Pathol. Toxicol.* **1,** 49 (1977).
248. C.W. Sheu and co-workers, *Environ. Mutagen.* **8,** 357 (1986).
249. T.J. Slaga and W.M. Bracken, *Cancer Res.* **37,** 1631 (1977).
250. G.M. Williams, T. Tanaka, and Y. Maecura, *Carcinogenesis* **7,** 1043 (1986).
251. K. Imaida and co-workers, *Gann* **75,** 769 (1984).
252. H. Tsuda, T. Sakata, T. Masui, K. Imaida and N. Ito, *Carcinogenesis* **5,** 525 (1984).
253. FAO, *1990 FAO Production Year-book,* Vol. 44, Food and Agriculture Organization of United Nations, Rome, 1991.
254. F. Malek and P. Zandi, *Food Chem.* **37,** 289 (1990).
255. P. Nikokyris, K. Kandylis, K. Deligiannis, and D. Liamadis, *J. Dairy Sci.* **74,** 4305 (1991).
256. M.L. Gray, R.D. Randel, L.W. Greene, and G.L. Williams, *J. Anim. Sci.* **68** (Supp 1), 465 (Abst) (233) (1990).
257. V.L. Singleton and F.H. Kratzer, in *Toxicants Occurring Naturally in Foods*, Committee on Food Protection, National Academy of Sciences, Washington, D.C., 1973, Chapter 15, pp. 318–323.
258. M.B. Abou-Donia and J.W. Dieckert, *Life Sci.* **14,** 1955 (1974).
259. K. Kimura, K. Sakurada, and N. Katoh, *Biochem. Biophys. Acta* **839,** 276 (1985).
260. P.P. Moh, P.K. Li, M.V. Darby, R.W. Breggemeier and Y.C. Lin, *Res. Comm. Chem. Pathol. Pharmacol.* **76,** 305 (1992).
261. J. Frick and C. Danner, in S.J. Segal, ed., in *Gossypol—A Potential Contraceptive for Men*, Plenum Press, New York, 1985, pp. 17–23.
262. F. Homberger and E. Boger, *Cancer Res.* **28,** 2572 (1968).
263. E.W. Boberg, E.C. Miller, A. Poland, and A. Liem, *Cancer Res.* **43,** 5163 (1983).
264. E.C. Hagan and co-workers, *Toxicol. Appl. Pharmacol.* **7,** 18 (1965).
265. E.C. Miller and co-workers, *Cancer Res.* **43,** 1124 (1983).
266. R. Truhart and co-workers, *Food Chem. Toxic.* **27,** 11 (1989).
267. J.P. Brown, G.W. Roehm, and R.J. Brown, *Mutation Res.* **50,** 249 (1978).
268. J.F. Borzelleca, K. Depukat, and J.B. Hallagan, *Food Chem. Toxic.* **28,** 221 (1990).
269. D.L. Morgan, C.W. Jameson, J.H. Mennear, and J.D. Prejean, *Food Chem. Toxic.* **27,** 793 (1989).
270. C.A. Nau, G.T. Taylor, and C.H. Lawrence, *J. Occup. Med.* **18,** 732 (1976).
271. B.C. Pence and F. Buddingh, *Tox. Lett.* **25,** 237 (1985).
272. C.A. Nau, J. Neal and V.A. Stembridge, *A.M.A. Arch. Ind. Health* **17,** 21 (1958).
273. M. Baerdick, in J.N. Hathcock, ed., *Nutritional Toxicology,* Vol. 1, Academic Press, New York, 1982, pp. 343–434.
274. K.S. Khera, W. Przybylski, and W.P. McKinley, *Food Cosmet. Toxicol.* **12,** 507 (1974).
275. C.V. Vorhees, R.E. Butcher, R.L. Brunner, V. Wooten and T.J. Sobotka, *Arch. Toxicol.* **53,** 253 (1983).
276. J.F. Borzelleca and J.B. Hallagan, *Food Chem. Toxic.* **25,** 735 (1987).
277. D. Henschler and D. Wild, *Arch. Toxicol.* **59,** 69 (1986).
278. A. Maekawa and co-workers, *Food Chem. Toxic.* **25,** 891 (1987).
279. L.B. Bullerman, *J. Food Protection* **42,** 65 (1979).

280. H.G. Peers, G.A. Gilman, and C.A. Russel, *Int. J. Cancer* **17,** 167 (1976).
281. A.O. Uwaifo, *Toxicology* **31,** 33 (1984).
282. J.J. Wong and D.P.H. Hsieh, *Proc. Natl. Acad. Sci. USA* **73,** 2241 (1966).
283. M.Y. Fukayama, W.G. Helferich, and D.P.H. Hsieh, *Food Chem. Toxicol.* **22,** 857 (1984).
284. R.L. Buchanan and A.M. Fletcher, *J. Food Sci.* **43,** 654 (1978).
285. C. Batt, M. Solberg, and M. Ceponis, *J. Food Sci.* **45,** 1210 (1980).
286. G.K. Stanford, R.D. Hood, and A.W. Hayes, *Res. Commun. Chem. Pathol. Pharmacol.* **10,** 743 (1975).
287. J.J. Pestka, W. Dong, R.L. Warner, L. Rasooly and G.S. Bondy, *Food Chem. Toxic.* **28,** 693 (1990).
288. K.S. Khera, C. Whalen, G. Angers, R.F. Vesonder and T. Kuiper Goodman, *Bull. Envir. Contam. Toxicol.* **29,** 487 (1982).
289. W.D. Kitts, F.E. Newsome, and V.C. Runeckles, *Can. J. Anim. Sci.* **63,** 823 (1983).
290. J.M. de Man, T. Fan, and L. de Man, *J. Am. Oil Chem. Soc.* **64,** 993 (1987).
291. R.T. Sleeter, *Am. Oil Chem.* **58,** 283 (1981).
292. D.A. Lillard, *J. Food Protection* **46,** 61 (1983).
293. T. Gutfinger and A. Letan, *J. Sci. Food Agric.* **25,** 1143 (1974).
294. R.G. Ackman, S.N. Hooper, and D.L. Hooper, *J. Am. Oil Chem. Soc.* **51,** 42 (1974).
295. M.N. Hassapidou, G.D. Balattsouras, and A.G. Manoukas, *Food Chem.* **50,** 111 (1994).
296. G. Kajimoto, Y. Kanomi, H. Kawakami, and M. Hamatani, *J. Jpn. Soc. Nutr. Food Sci.* **45,** 291 (1992).
297. K.J. Smith, P.B. Polen, D.M. deVries and F.B. Coon, *J. Am. Oil Chem. Soc.* **45,** 866 (1968).
298. K. Rainio, R.R. Linko, and L. Ruotsila, *Bull. Environ. Contam. Toxicol.* **37,** 337 (1986).
299. J.M. Tabuenca, *Lancet* **ii,** 567 (1981).
300. M. Posada de la Paz and co-workers, *Food Chem. Toxic.* **29,** 797 (1991).
301. J.R. Spurzen and J.E. Lockey, *Arch. Intern. Med.* **144,** 249 (1984).
302. J.T. Bernert and co-workers, *J. Food Sci.* **52,** 1562 (1987).
303. R. Hartung and G.S. Hunt, *J. Wildl. Management* **30,** 564 (1966).
304. E.M. Kilbourne and co-workers, *N. Eng. J. Med.* **309,** 1408 (1983).
305. J.G. Rigau-Perez and co-workers, *Am. J. Epidemiol.* **119,** 250 (1984).
306. Y. Masuda, H. Kuroki, K. Haraguchi and J. Nagayama, *Chemosphere* **15,** 1621 (1986).
307. C.K. Wong, C.J. Chen, P.C. Cheng and P.H. Chen. *Br. J. Dermatol.* **107,** 317 (1982).
308. W.J. Rogen and co-workers, *Science* **241,** 334 (1988).
309. Y.C. Chen, Y.L. Guo, C.C. Hsu and W.J. Rogen, *J. Am. Med. Assoc.* **268,** 3213 (1992).
310. R. Frank and H.E. Braun, *Bull. Environ. Contam. Toxicol.* **42,** 666 (1989).
311. V.S. Packard, *J. Food Protection* **48,** 724 (1985).
312. L.G.M. Th Tuinstra and co-workers, *J. Agric. Food Chem.* **40,** 1772 (1992).
313. N.V. Fehringer, S.M. Walters, R.J. Kozara and L.F. Schneider, *J. Agric. Food Chem.* **33,** 626 (1985).
314. N.C. Stachiw, M.E. Zabik, A.M. Booren, and M. J. Zabick, *J. Agric. Food Chem.* **36,** 848 (1988).
315. R. Neubert, U. Jacob-Muller, H. Helge, R. Stahlmann, and D. Neubert, *Arch. Toxicol.* **65,** 213 (1991).
316. N. Kato, T. Tani, and A. Yoshida, *J. Nutr.* **111,** 123 (1981).
317. H.S. Jenko, *Biochem. Biophys. Acta* **837,** 85 (1985).
318. C.S. Giam and M.K. Wong, *J. Food Protection* **50,** 769 (1987).

319. L. Castle, A.J. Mercer, J.R. Startin and J. Gilbert, *Food Addit. Contam.* **4,** 399 (1987).
320. L. Castle, A.J. Mercer, J.R. Startin, and J. Gilbert, *Food Addit. Contam.* **5,** 9 (1988).
321. L. Castle, M. Sharman, and J. Gilbert, *J. Chromatogr.* **437,** 274 (1988).
322. IFT, *Food Technol.* **28**(3), 75 (1974).
323. J.R. Startin and co-workers, *Food Addit. Contam.* **4,** 385 (1987).
324. J. Gilbert, L. Castle, S.M. Jickells, A.J. Mercer, and M. Sharman, *Food Addit. Contam.* **5**(suppl. 1), 513 (1988).
325. C. Nerin, P. Gancedo, and J. Cacho, *J. Agric. Food Chem.* **40,** 1833 (1992).
326. E. Kondyli, P.G. Demertzis, and M.G. Kontominas, *Food Chem.* **45,** 163 (1992).
327. E. Benfenati, E. Natangelo, E. Davoli and R. Fanelli, *Food Chem. Toxic.* **29,** 131 (1991).
328. Working Party on Chemical Contaminants from Food Contact Materials: Sub-Group on Plasticisers, Plasticisers: Continuing Surveillance, The Thirtieth Report of the Steering Group on Food Survellance, MAFF Food Surveillance Paper No. 30, HMSO, London, 1990.
329. J.J. Killoran, *Fadiat. Res. Rev.* **3,** 369 (1971).
330. R. Buchalla, C. Schuttler, and K.W. Bogl, *J. Food Protection* **56,** 998 (1993).
331. IARC Working Group, Monographs on the Evaluation of the Carcinogenic Risk of Chemicals to Humans, Vol. 19, in *Some Monomers, Plastics and Synthetic Elastomers and Acrolein,* International Agency for Research on Cancer, Lyon, 1979.
332. W.H. Lawrence, *Clin. Toxicol.* **13,** 89 (1978).
333. R.N. Baker, *Toxicology* **9,** 319 (1978).
334. G.M. Pollack, R.C.K. Li, J.C. Ermer, and D.D. Shen, *Toxicol. Appl. Pharmacol.* **79,** 246 (1985).
335. D.D. Kitts, *Can. J. Physiol. Pharmacol.* **72,** 423 (1994).
336. K.L. Fritz, T.L. Nelson, V. Ruiz-Velasco, and S.D. Mercurio, *J. Nutr.* **124,** 425 (1994).

9

Fat Substitutes

INTRODUCTION

In recent years the public has been made increasingly aware of the problems associated with high levels of fat in the diet. Widely publicized reports have stressed the role of fats in cardiovascular disease and, more recently, in certain types of cancer. The health community is almost unanimous in considering the current level of fat in the diet as too high. A maximum level of no more than 30% of total calories from fat was set as a goal (1,2).

1.1 Fats and Oils in the Diet

Fats and oils in food can be divided into visible, such as margarine, salad oil, and frying fats, and invisible fats consumed in meats, dairy products, baked goods, and similar fat-containing foods. It is difficult to obtain accurate per capita consumption figures for either type. For example, a substantial percentage of spent fats used in the fast-food industry is known to be discarded or used in animal food. As to invisible fats, levels in meat are especially difficult to ascertain. The cut of meat, the method of preparing, and trimming done by the person eating the meat all influence actual fat consumed. Table 9.1 shows an estimate of current U.S. per capita fat consumption. If the daily caloric consumption is assumed to be 3700 kcal, the consumption of 57.2 kg of fat per year would represent 38% of calories from fat, well above the recommended level.

Despite this awareness of the health problems associated with a high-fat diet, there is little indication that fat consumption is decreasing. While consumption of food fats and oils has declined slightly from its peak in 1986,

Table 9.1 Estimated annual U.S. per capita fat consumption, 1993[a]

Fat	Amount (kg)
Visible fats	
Baking and frying[b]	7.6
Margarine	4.0
Salad and cooking oils	11.8
Other edible fats and oils	0.6
Butter	1.9
Total visible	25.9
Invisible fats	
Meat, poultry, and fish	18.2
Milk and other dairy products	9.2
Other invisible fats and oils	3.9
Total invisible	31.3
Total	57.2

[a] *Sources*: Fats and Oils Used in Edible Products, Bureau of the Census, 1993; Meat and Poultry Facts, 1992, American Meat Institute; Human Nutrition Information Service, USDA, unpublished data, 1993.
[b] An estimated 2.7 kg per capita is accounted for by discarded frying oil.

nevertheless the trend since 1970 has been overall rising (Figure 9.1). During the same period red meat consumption decreased but was more than replaced by poultry consumption (Figure 9.2). The effect of this is unclear since the lower fat level of the poultry may well have been offset by the high level of poultry prepared by deep frying, particularly in the rapidly growing fast-food sector.

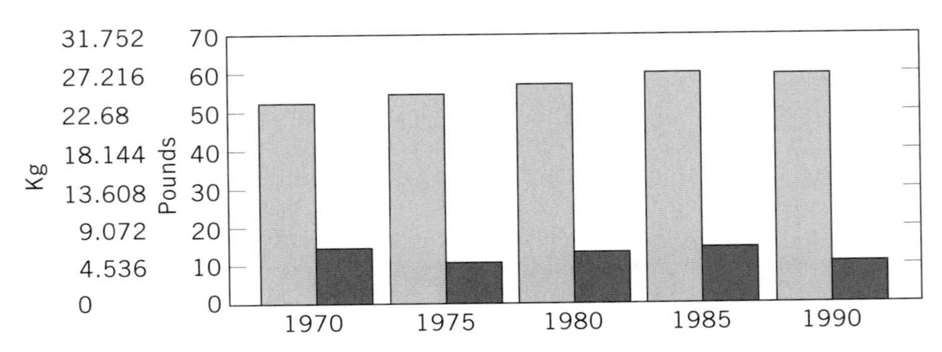

Figure 9.1 U.S. per capita consumption of fats and oils, 1970–1990. Fat content basis includes butter, margarine, direct use of lard and edible tallow, shortening, salad and cooking oils, and other fats. ▭, Vegetable; ■ animal. From Ref. 3.

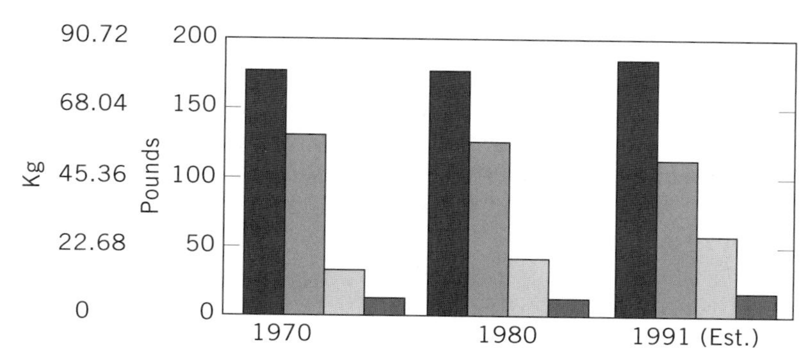

Figure 9.2 U.S. per capita consumption of meat, poultry, and fish, 1970–1991. ■, Total meat; ▨ red meat; □ poultry; ■ fish. From Ref. 3.

1.2 Functions of Fats

The most important function of fats and oils in the diet is nutritional. Fats have the highest caloric value of all foods, supplying 9 kcal per gram, over twice that of carbohydrates or proteins. In addition to the caloric value of fats, certain unsaturated fatty acids are known to be necessary in the diet. These essential fatty acids (EFAs) include 9,12-linoleic acid, which is present in most vegetable oils but not produced by the body. Fats and oils are also associated with the fat-soluble vitamins—A, D, E, and K. The antioxidant function of vitamin E has been linked to a lower incidence of cardiovascular disease.

Diets high in lipids tend to impart a sense of fullness or satiety. This is due to their high caloric value but also to the longer time it takes for them to be absorbed by the body. Dietary fats must first be split by enzymes and the resulting fatty acids and monoglycerides absorbed through the intestine. An exception to this are the medium-chain triglycerides. These C8,C10 triglycerides are absorbed directly through the portal vein to the liver and oxidized at a rate similar to that of sugars.

Most important and most difficult to replace are the flavor and physical feel that fats and oils bring to the many food products that contain lipids. These physical characteristics, or "mouthfeel," impart the smoothness needed in salads as well as the ease of mastication that high-melting-point margarines and butter bring to breads.

Fats and oils have distinctive flavors. In many cases modern processing, that is, refining, bleaching, and deodorizing, have greatly reduced these natural flavors. This is especially true for soybean oil. Other oils such as olive oil, sesame seed oil, and peanut oil are used specifically for their distinctive flavors. Lipids are also important in bringing out the flavor of other food ingredients.

Fats and oils have specific functions in particular food applications. For example, they serve as efficient heat transfer media in deep-fried food applications. In baked goods fats provide lubricating properties necessary for processing. In addition, the dispersed fat helps provide the desired texture and distribution of air. In many applications as in baked goods and salad dressings lipids provide a substantial amount of the bulk of the finished product. This can make replacement difficult and expensive. For example, in low-fat meat products, low-cost lard or tallow is replaced by higher value protein. In other applications, as in some low-fat salad dressings, fat is replaced by water. Cost, then, can be an important factor in fat replacement.

1.3 Approaches to Fat Reduction

The simplest and most effective way to reduce fat levels in the diet to the recommended 30% is by reducing fat consumption. In most cases this can be done by modest changes in the daily diet, avoiding those foods high in fat calories and replacing the lost calories primarily with increased consumption of carbohydrates. It would appear, however, that most consumers would prefer essentially the same diet with the fat content reduced but with no or minimal change in appearance or flavor. For example, a low-fat frozen yogurt would be preferred to eliminating ice cream from the diet. This has led to the development of numerous new ingredients designed to replace conventional fats and oils and the introduction of hundreds of low-fat or nonfat processed food into the market.

There are several approaches to eliminating or lowering fat calories in foods. Most obvious is the physical removal of the fat, for example, in trimming of meat. In some instances solvent extraction of fat has been suggested. Some progress has been made in lowering fat levels in pork by selective breeding. One of the most interesting is replacement of normal triglyceride lipids with "synthetic" fats, fatty-acid-based substances that have fatlike properties but are not metabolized like conventional triglycerides. The most widely used method at the present time is replacement of the fat in a specific application with nonfat materials, usually carbohydrates or proteins, that bind water so as to provide some of the bulk and mouthfeel of fats.

No matter what the approach, replacing all of the functions of fat in a specific food application is a challenge to food scientists.

NONCALORIC FATS

2.1 Theory of Nondigestible Fats

Normal triglycerides are almost completely absorbed, with the first step being the hydrolysis by pancreatic lipase followed by absorption of the resulting fatty acids and monoglycerides. In 1972, Mattson and Volpenheim demonstrated *in vitro* that the rate of hydrolysis of fatty esters by pancreatic lipase was related to the number of fatty acids in the ester (5). This was later verified with actual animal studies (6). Absorbability ranged from 100% for glycerine trioleate to 0% for sucrose octaoleate (Table 9.2). Another approach that blocks enzymatic hydrolysis is to make the ester from a fatty alcohol and a polyfunctional acid (7). In either case the normal digestion of the fatty material is prevented by interfering with the hydrolysis step.

2.2 Olestra

Background. Olestra is the name Procter & Gamble has given to its noncaloric sucrose polyester (SPE). In order to be nondigestible olestra must contain from 6 to 8 mol of fatty acid. This is not to be confused with sucrose esters containing from 1 to 3 mol, which are used in food as emulsifiers and are readily absorbed. In a 1971 patent assigned to Procter & Gamble, Mattson and Volpenheim described olestra and claimed its value as a noncaloric fat (8).

Synthesis. Because of the difficulties inherent in working with sugars that carmelize at the high temperatures needed for esterification, most approaches to making sucrose esters have focused on interesterification. The original method for production of the mono- and diesters used as emulsifiers was based on the interesterification with methyl esters (9). This was a solvent-based process using dimethylformamide. Later Procter & Gamble developed a solventless process (10). In this process a 3:1 mole ratio of fatty acid methyl esters to sucrose are reacted in the presence of potassium hydroxide to form a melt containing lower sucrose esters. A second step then adds more methyl esters to produce sucrose esters conaining six to eight fatty acids. Yields are claimed to be as high as 90%. In 1985, Volpenheim described an improved process using potassium carbonate as a catalyst with a 12:1 mole ratio of methyl esters to sucrose (11). This appears to be an extremely complex process.

Table 9.2 Potential fat reduction with Olestra[a]

Food	No Olestra		With Olestra	
	% Fat	Calories	% Fat	Calories
Foods prepared with shortening with35% Olestra				
Fried chicken	17.5	290	14.1	260
French fries	13.2	265	8.6	225
Cookie (chocolate chip)	30.1	260	12.1	235
Foodservice foods prepared with shortening with 75% Olestra				
Fried chicken	17.5	290	10.3	225
French fries (includes pan frying)	16.3	325	4.1	215
Potato chips	36.8	165	9.3	95

[a] *Source*: Olestra Food Additive Petition, The Procter & Gamble Company, Cincinnati, Ohio, 1987.

Despite the higher value of the glycerine by-product compared to sucrose, the cost of the finished olestra will be substantially higher than conventional fats and oils.

Functionality. The physical properties of sucrose polyesters are similar to normal triglycerides in that these properties are very dependent on the chain length and unsaturation of the fatty acids (12). Relative viscosity depends on the chain length and degree of unsaturation of the fatty acids, as in the case of triglycerides, but is always higher than the corresponding triglyceride, as would be expected. Likewise, melting points vary with chain length and degree of unsaturation but are higher than the comparable triglycerides. One difference between sucrose octaester and the comparable triglyceride is in the alpha phase. Compared to the triglyceride, the α-sucrose octaester is much more stable. Olestra appears to be adaptable to most applications where fats and oils are used. However, human studies with olestra based on soybean methyl esters led to complaints as to a poor mouthfeel. Blending with conventional triglycerides or using more unsaturated methyl esters may overcome this, and in fact Procter & Gamble's Food and Drug administration (FDA) petitions call for olestra as less than 100% of the fat in all applications.

Procter & Gamble has patented the use of olestra for treating hypercholesterolemia (13), although it has not petitioned for approval of olestra for pharmaceutical applications. This same mechanism probably accounts for the fact that oil-soluble vitamin levels in some subjects in human olestra studies were slightly lowered.

Another major drawback to olestra is the mineral oil effect, or "anal leakage," the result of a nondigestible fat passing through the digestive system. Early attempts to alleviate anal leakage called for the use of saturated fatty acids in the polyester. This evidently increased the mouthfeel problem. A more recent patent describes a polyester with a mix of fatty acids to provide about 20% solids at mouth temperature combined with an emulsifier (14).

Current Status. In April 1987, Procter & Gamble filed a petition with the FDA on olestra asking approval for in-home and most food service use up to 35% in most fats and oil applications and up to 75% in food service deep frying and in fried snacks. In August 1990, this petition was amended to cover only the use of olestra in snacks, but at 100% rather than 75%. Table 9.2 shows some of the fat reductions possible with olestra according to Procter & Gamble's original petition.

The evaluation of olestra has presented a unique problem to the FDA. Normally, food additives are subject to animal testing in order to establish the highest level that has no adverse affect on the test animal, usually more than 100 times the useful level. It is impossible for the test animal to consume this amount. This had led to many human clinical studies to determine the possible problems that might occur with a nondigestible material that may account for as much as 50% of the diet. The result has been a long and, as yet, not concluded process.

2.3 Other Noncaloric Fats

Polysaccharide Esters. A patent issued to Arco Chemical Technology, Inc., describes partially esterified compounds made from such polysaccharides as xanthan gum, guar gum, gum arabic, and cellulose products. The degree of esterification is sufficiently controlled to prevent lipase hydrolysis (15). Similarly, a patent issued to Curtice Burns, Inc., describes a polysaccharide comprising at least three monosaccharides esterified with at least four fatty acids and a method for producing these products by interesterification (16).

Dialkyl Dicarboxyllate Acid Esters. Esters of dialkyl dicarboxylic acids and fatty alcohols are described by Fulcher (17). Esters of malonic acid and alkylated malonic are said to be 95–98% nondigestible in *in vitro* studies. Sensory evaluation of snacks indicated that blends of soybean oil and dihexyldecylmalonate (DDM) were satisfactory compared to all soybean oil as the deep-frying media.

Trialkoxytricarballate. Esters of tricarballic and citric acids with fatty alcohols were found to resist enzymatic hydrolysis both *in vitro* and in animal studies (18). The oleyl alcohol esters were said to make a satisfactory margarine and mayonnaise.

Polyvinyl Oleate. A patent assigned to Nabisco Brands, Inc., describes the esterification of low-molecular-weight polyvinyl alcohol and fatty acids to produce an edible fat replacement for use in food products (19). A polyvinyl alcohol with a molecular weight between 1000 and 5000 is preferred. No hydrolysis of the esters was observed upon exposure to pancreated lipases *in vitro.*

Esterified Propoxylated Glycerols (EPG). A polyether polyol formed by reacting propylene oxide with glycerine is then esterified with fatty acids to produce a nondigestible fatlike substance (20). *In vitro* studies show no hydrolysis by pancreatic enzymes. By varying the number of propylene oxide units and the type of fatty acids, products for specific end uses can be formulated. Arco Chemical, the developer of EPG, is developing a version of this family for commercial use.

Tris-hydroxymethyl Lower Alkane Esters. Esters of neopentyl polyols are not fully digestible. A patent issued to Nabisco Brands, Inc. describes esters of Tris-hydroxymethylethane and Tris-hydroxymethylpropane. These esters are said to be only partially digested and have one-third the normal calories.

It is claimed that they avoid some of the undesirable side effects of other nondigestible synthetic fat materials (21).

Polyglycerol Esters. Polyglycerol is prepared by polymerization of glycerine under alkaline conditions. The glycerol molecules are joined by an ether group between the α-hydroxy groups. The polyglycerol is then esterified with varying moles of fatty acids. The moles of glycerine, degree of esterification, and types of fatty acid can be changed to formulate various end products. Polyglycerol esters up to and including decaglycerol esters are approved as food additives. Polyglycerol esters have an estimated caloric value of 6–7 kcal/g; however, the real caloric value may be as low as 2 kcal/g due to partial absorbtion (22). While polyglycerol esters are used as emulsifiers in low-fat products such as margarine-type spreads, use at the high levels needed for total fat replacement would require changes in allowable levels.

Polybasic Acid Amides. Polybasic acid amides, made by reacting dimer or trimer acids with amino acids, are described in a patent assigned to the Dow Chemical Company (23). The resulting amides are said to have organoleptic properties similar to normal triglycerides, are digestively hydrolyzable, but are much lower in available calories.

Jojoba Oil. Jojoba oil, from the jojoba plant raised in the Southwest United States, is a mixture of linear esters of fatty acids and fatty alcohols. While technically a wax, it is liquid at 10°C. Digestibility is reported to be 20–40%, which has created interest in jojoba oil as a low-calorie fat (7).

2.4 Structured Lipids

Medium-Chain Triglycerides. Medium-chain triglycerides (MCTs) are made by esterifying glycerine with fractionated coconut oil fatty acids, primarily C8 and C10, with smaller levels of C6 and C12. They have been used since the 1950s for treatment of disorders affecting lipid absorption and other clinical applications. MCTs are hydrolyzed more rapidly than conventional long-chain triglycerides (LCTs) and are absorbed from the intestine to the venal portal system to the liver. Animal studies indicate that MCTs require higher levels in the diet to achieve the same weight gain as with LCTs. While the value of MCTs as a low-calorie fat is not well established, there are indications that MCTs may help in the control of obesity (24).

Caprenin. Caprenin is a triglyceride consisting of equal moles of medium-chain caprylic and capric acids and the very long chain behenic acid. Developed by Procter & Gamble as a low-calorie alternative to cocoa butter, the caloric value is claimed to be 5 kcal/g compared to the 9 kcal/g of conventional fats (25). This is attributed in part to the lower caloric density of the medium-chain fatty acids and to the limited absorption of the behenic acid. Caprenin has been granted Generally Regarded As Safe (GRAS) status for confectionery applications. While approval for use as a cocoa butter replacement will access a limited market, potential applications for this or similar structured lipids are numerous.

A similar approach to using structured lipids as lower calorie fats is claimed in a patent assigned to Nabisco, Inc. Triglycerides containing poorly absorbed stearic acid and low caloric short-chain acetic, propionic, and butyric acids are said to have caloric utilization of about 5 kcal/g (26).

PROTEIN-BASED FAT REPLACERS

Carbohydrates such as modified starches, maltodextrins, gums, and cellulose products have been used by food formulators for some time as thickeners and stabilizers. As the need for lower fat products became apparent, it was natural for these materials to be used as fat replacers. Only recently have protein materials been developed that have value as fat substitutes.

3.1 Microparticulated Protein

Simplesse. In January 1988 the NutraSweet Company announced a protein-based fat replacer called Simplesse. The original patent covering this product (26) describes the acid denaturing of whey protein and the subsequent processing through a high shear device to obtain a particle size ranging from 0.1 to 2.0 μm. The particle size is said to be the key to creating a proper fatlike mouthfeel, since particles lower than 0.1 μm are not felt by the tongue while particles over 3.0 μm appear to be gritty.

Protein particles naturally occurring in eggs, milk, grains, and other sources do not have the proper size or shape to develop an acceptable mouthfeel. For example, casein micelles are smaller than 0.1 μm while soy protein particles can be as high as 80 μm, too large to be used as a fat replacer (27). Simplesse can be made from egg white, skim milk, or whey protein. The product is supplied as a liquid dispersion in water and should be refrigerated.

According to the Nutrasweet Company, the protein is hydrated during the manufacturing process, reducing the caloric value to only 1–2 calories per gram as compared to the normal 4 calories. This compares with the normal 9 calories per gram with conventional fats. Simplesse is recommended for a wide variety of food applications including dairy products such as yogurts, cheese products, and frozen desserts. It is also recommended for formulating low-fat baked goods such as cheesecakes and pie crusts.

3.2 Denatured Protein

Another approach to using proteins as fat replacers calls for the use of denatured protein (28). Ault Foods Limited has developed such a product based on whey protein concentrate called Dairy-Lo, which has been licensed to the Pfizer Food Science Group. Controlled heat denaturation of the protein is said to expose hydrophobic regions on the polypeptide chains resulting in amphiphilic properties that improve the emulsification characteristics of the protein. This in turn is claimed to impart water controlling properties that improve the texture of low-fat foods, especially dairy products such as frozen dairy desserts.

3.3 Gelatin

Gelatin is a mixture of water-soluble proteins obtained by boiling skin, ligaments, and bones of animals. It has been used in low-calorie foods for many years; however, its use as a fat replacer is recent. It is especially useful as a viscosity modifier and to impart a creamy texture. A recent patent describes the use of gelatin derived from fish waste as a fat replacer in a number of food applications (29).

3.4 Vegetable Protein

While vegetable proteins are not generally regarded as ingredients for low-fat food formulations, they can play an important role in the production of low-fat meat products. Soybean-based vegetable proteins are effective water binders and can form gel structures that improve the texture of processed meats. In addition, adding soy protein affects the economics. Removing all or most of the low-value fat when formulating low-fat processed meats increases the cost of the product. Extending the meat protein with lower cost yet nutritionally valuable vegetable protein makes the low-fat product more competitive.

There are three types of soy protein that can be used. Soy flour is produced from the meal left after the dehulling and oil extraction and is about 50% protein. Further protein concentration results in soy concentrate at 70% and isolated soy protein, which is 90% protein.

CARBOHYDRATE-BASED FAT REPLACERS

4.1 Starch-Based Products

Maltodextrins. The FDA defines maltodextrins as saccharide polymers consisting of D-glucose joined at α,1–4 bonds with a dextrose equivalent (DE) of less than 20. Maltodextrins have been widely used in food applications for many years due to their functional properties. They are used for their bulking properties, bland taste, ability to bind water, and dispersibility. While maltodextrins have the normal value of 4 kcal/g as fat replacers, in certain applications they may be used with a 3:1 water/maltodextrin ratio, which lowers the level to 1 kcal/g, a saving of 8 kcal/g compared to the replaced fat. Unlike protein-based fat replacers, maltodextrins are heat stable and can be used in baked goods.

Maltodextrins are produced by the partial hydrolysis of starch either by acid treatment or by enzymes. The dextrose equivalent is an indication of the number of reducing groups present after the breakdown of the starch polymer by hydrolysis, a DE of 100 indicating complete hydrolysis to 100% glucose. Low DE dextrins are produced from acid or a combination of acid with α-amylase. Properties of maltodextrins vary widely depending on the degree of depolymerization. Low DE maltodextrins have the water binding characteristics of starches and replace the bulk of the fat. The high DE products have solubility and bulking properties closer to syrups.

Commercial maltodextrins are available made from cornstarch, waxy cornstarch, potato starch, and tapioca starch. Some typical maltodextrins recommended as fat replacers are:

1. Paselli SA2 (Avebe America, Inc., Princeton, NJ) is a potato-starch based maltodextrin with a dextrose equivalent of less than 3. Gels made from Paselli SA2 are said to have a smooth, bland texture. Paselli SA2 is recommended for baked goods, spoonable salad dressings, mayonnaise-type products, low-fat spreads, and low-fat ice creams.
2. Instant N-oil (National Starch and Chemical Company, Bridgewater, NJ) is based on tapioca starch. Tapioca starch is said to have a blander flavor than cereal or potato-based starches.
3. Maltrins (Grain Processing Corporation, Muscatine, IA) are maltodextrins derived from cornstarch. Maltrin M040 has a dextrose equivalent of 5 and is recommended for low-fat spreads and salad dressings.
4. TrimChoice (A.E. Staley Manufacturing Company, Decatur, IL) is derived from oat flour. In addition to maltodextrins, it contains β-glucan, the major soluble fiber found in oats. The combination of the low DE maltodextrin and β-glucan is said to provide a fatlike mouthfeel.

Modified Starch. A unique modified starch fat replacer, called Stellar, was introduced by A.E. Staley Company in 1991. It is described as a modified food starch under 21 CFR 172.892 rather than a maltodextrin. Acid treatment depolymerizes the amylopectin structure of the starch granules. The resulting product is a dry powder that is 70% insoluble in water. The insoluble fraction has an average molecular weight of less than 20,000 as compared to a potato maltodextrin with an average molecular weight of 180,000. The maltodextrins and other starch-based products depend on the formation of polymer gels. The fatlike properties of Stellar are achieved when an aqueous slurry is processed in high shear equipment and are a result of submicron size, easily deformable aggregates, and a high degree of water immobilization. Stellar is suggested for use in most fats or oil replacement applications including salad dressings, dairy products, meats, and baked goods (30).

4.2 Polydextrose

Polydextrose, a condensation polymer of dextrose, has been widely used as a bulking agent with artificial sweeteners, but in some food applications may be used to replace a portion of the fat or oil. As such it is a "fat sparing" agent.

Polydextrose is a randomly bonded melt condensation polymer. It is prepared by vacuum polycondensation of a molten mixture of dextrose, sorbitol, and citric acid in a 89 : 10 : 1 ratio (31). The complex polymer is highly branched containing varied glucose linkages, as shown in Figure 9.3. While molecular weight appears to vary widely, the average is about 5000.

Most important for the use of polydextrose as a bulking agent with low-calorie sweeteners and in low-fat formulations is its low caloric value. This low caloric value, 1 kcal/g, is a result of the inability of human enzymes to break down such a large, complex, randomly bonded polymer. Polydextrose is soluble up to 80% at 25°C, forming a clear solution. This contrasts with the insoluble cellulosic products. At higher levels polydextrose is said to have a

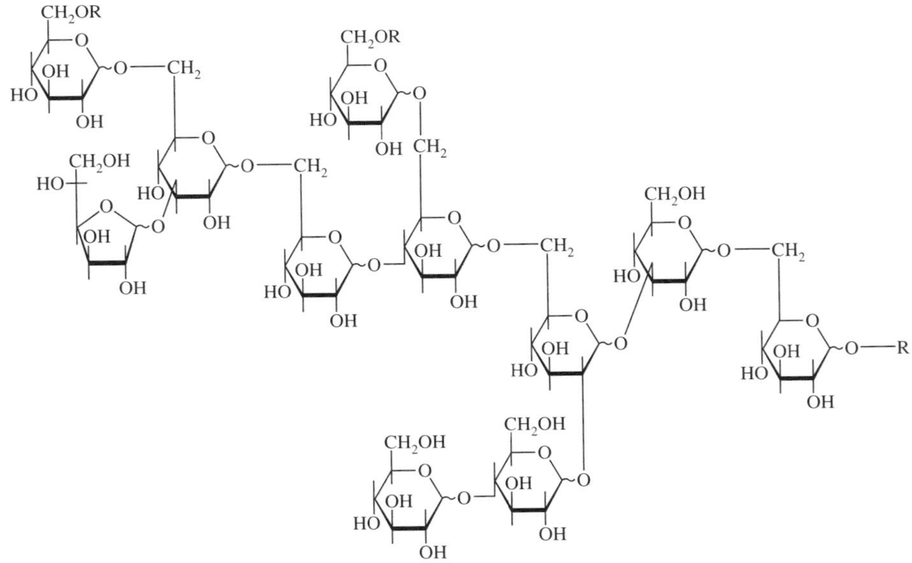

Figure 9.3 Polydextrose.

more satisfactory mouthfeel than gums when used to improve the texture of low-calorie foods (32).

In 1991 and 1993, the Pfizer Food Science Group introduced Litesse and Litesse II, blander forms of polydextrose. Litesse is suggested for use in beverages, baked goods, frozen dairy desserts, candies, and spreads.

4.3 Cellulose

Cellulose is a polymer of D-glucose units linked by β-1,4′glycosidic bonds rather than the α-1,4′ linkage seen in starches. The β-glucosidase enzyme, which is necessary to hydrolyze cellulose, is not present in humans and cellulose has no caloric value in the human diet. As a nondigestible fiber, cellulose has been used as a bulking agent in low-calorie food applications and is more recently being evaluated in fat-free and low-fat formulations. Several types of cellulose are available. Figure 9.4 shows the structure of cellulose.

Powdered cellulose is produced by grinding purified wood-derived cellulose. Particle size ranges from 20 to 300 μm. The larger micron materials have higher water absorption capacities as a result of their more open structure. Typical are the Solka Floc products made by Fiber Sales and Development Corporation. Powdered cellulose can be used in baked products. A recent patent describes the use of powdered cellulose in frying batters for low-fat foods (33).

Microcrystalline cellulose is manufactured by the FMC Corporation under the trade name Avicel. Cellulose fiber is hydrolyzed to remove the amorphous parts of the fiber thus concentrating the crystalline bundles. The resulting material is then either spray dried to a powder or further processed with cellulose gum (sodium carboxymethylcellulose) and then either spray dried or bulk dried (34). Microcrystalline cellulose dispersions with particle size as low as 0.2 μm are recommended as fat replacers in aqueous systems where they contribute a fatlike consistency and mouthfeel to low or fat-free formulations. It is recommended for low-fat salad dressings, sauces, frozen desserts, and other dairy products.

Methyl cellulose (MC) and hydroxypropyl methylcellulose (HPMC) are cellulose ethers developed by The Dow Chemical Company. They are water-soluble gums that are tasteless and have no caloric value. These cellulose

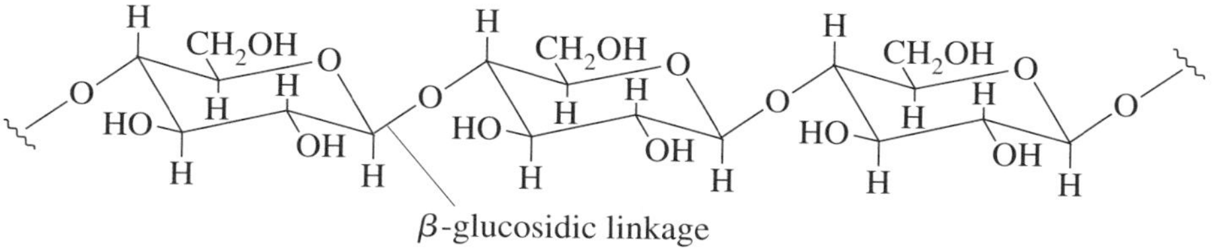

β-glucosidic linkage

Figure 9.4 Cellulose.

ethers are recommended for their ability to provide fatlike mouthfeel, control viscosity, and help retain moisture in low-fat foods. They can be used in processed poultry and meat products and in sauces and reduced-fat mayonnaise. The ability of the cellulose ethers to gel upon heating provides a barrier to moisture loss that makes them of value in low-fat baked products (35).

4.4 Natural Gums

The natural gums are carbohydrate-based hydrocolloids used historically as thickening agents in a number of food applications. They may come from plant exudates, such as tragacanth, from extracts such as carageenan, from seeds such as guar gum, or, as in the case of xanthan, be produced by fermentation. All of these gums are excellent water binders. They are often used at levels below 1.0%, replacing as much as 10 times the amount of starch needed to achieve the same viscosity.

Guar gum is the ground endosperm of the guar plant, *Cyamposis tetragonolubus*, which is cultivated in Texas and India as livestock feed. The polysaccharide guaran makes up about 85% of the gum. Guar gum has excellent water-binding qualities, especially in cold water. It is used in frozen desserts and other dairy products and in dressings and sauces.

Carageenans are sulfated hydrocolloid polysaccharides extracted from a number of related species of red seaweed. Three types of carageenan are recognized depending on the number and location of the sulfate ester groups on the galactose units. The kappa type forms brittle gels while the iota type forms more flexible gels. Lambda will not form gels. Blending the three grades offers the formulator a variety of options for the viscosity of finished products. Carageenans are recommended for dairy products, processed cheeses, and processed meats. Carageenan has been used in the formulation of a low-fat ground beef patty for a major fast-food chain (36).

Gum tragacanth is an exudate from *Astragalus gummifer*, a bush found in Iran and Syria. It is a complex mix of polysaccharides containing D-galactoronic acid and other sugars plus small amounts of starch and cellulose. Solutions of gum tragacanth are stable and high in viscosity. It is said to impart a creamy texture to salad dressings and improves the body of low-fat ice creams.

Xanthan gum is a high-molecular-weight polysaccharide produced by *Xanthomonas campestris*. It contains D-glucose, D-mannose, and D-glucuronic acid. Xanthan gum forms stable solutions in either hot or cold water. Xanthan gum solutions have good texture and flavor release.

Konjac flour is made from the tuber *Amorphophallus konjac*. It has been used for centuries in East Asia in a wide variety of foods. It is a high-molecular-weight polysaccharide made up of glucose and mannose units. The molecular weight, over 1,000,000 daltons, accounts for konjac's high pseudoplastic viscosity. Konjac flour is said to have a synergistic effect when used with carageenan or xanthan gum. It can be used with starches to form heat-stable gels useful in reduced-fat processed meat systems (37).

Pectin is a polysaccharide present in all plant tissues. The richest source of pectin is the rinds of citrus fruits. Pectin contains a mixture of methyl esterified galacturonan, galactan, and araban. It has been used for many years in the preparation of jellies and other gelled food products. Hercules Incorporated has introduced a pectin-based fat replacer, Slendid, extracted from citrus peel and apples. Slendid is recommended for most food applications including baked goods.

EMULSIFIERS

While not generally regarded as fat substitutes or replacers, nevertheless emulsifiers can play an important role in fat reduction. Emulsifiers affect the texture and mouthfeel of processed foods by modifying surface interactions. Most emulsifiers approved for food products are esters or partial esters of edible fatty acids. As such they have various caloric values, depending on their actual composition; however, they are used at low levels and disperse water so they usually have lower calories. As fat is lowered in processed foods, the role that emulsifiers perform becomes extremely important.

In many low-fat applications the water level is increased, which increases the need for emulsifiers to prevent components from separating. Emulsifiers also improve low-fat products by reducing the size of the fat globules and obtaining better dispersion of the remaining fat. Emulsifiers appear to function better than carbohydrate-based fat substitute in low-moisture baked goods such as cookies. They make the fat go farther (fat sparing action) and simulate the sensory characteristics of higher levels of fat. Emulsifiers also tenderize and provide the lubricity that the replaced fat had given. In addition they can complex with starches and proteins to improve texture and shelf life.

Selection of the proper emulsifier system is a critical factor in all areas of low-fat food development. Mono- and diglycerides appear to be the major products used, in particular distilled monoglycerides. Other widely used emulsifiers in reduced-fat foods are polysorbates, polyglycerol esters, lecithin, propylene glycol monoester, and sorbitol monoester. In the development of an emulsifier system, it is important to determine what functions are to be performed. Where more than one function is called for, it is likely that different emulsifiers will be needed. That is why many foods contain two or more emulsifiers. Low-fat food formulators must use caution, however, when dealing with multiple emulsifiers as excess emulsifiers can contribute a bitter taste to food products.

A novel use of fat emulsions in fat reduction is described in a patent assigned to the Pfizer Food Science Group. Emulsified fat particles are trapped in agar or alginate gels, providing small amounts of high-surface-area fats or oils (38).

APPLICATIONS

Interest in providing fat-free or low-fat foods has been noticeable in every segment of the food industry. The approach has varied depending on the particular target product, but two factors appear in most low-fat formulations. First, the same fat attributes—flavor, texture, and function—must be treated no matter what the end product. Second, in most cases there is no one solution but skillful formulation involving fat replacers, emulsifiers, bulking agents, flavors, and other ingredients is needed.

In most cases the flavor of the eliminated oil or fat is not a problem. Exceptions are deep-fried snacks, meat products, and foods containing high flavor oils such as olive oil and sesame oil. Rather, the problem is taste that is dependent on fat-soluble flavor ingredients. Often only a small amount of the original fat is needed to maintain the normal flavor. Most important, any added fat replacer or other ingredient must not contribute any additional taste. In many applications replacing fats causes more of a problem as to texture than flavor. This is especially true in high-fat products such as spreads and dressings.

6.1 Baked Goods

The fat content of baked goods varies widely from as low as 0% in French breads to as high as 30% in chocolate chip cookies. Table 9.3 shows the fat content of some typical bakery products.

Depending on the type of baked goods, fats and oils have several important functions. In leavening systems shortenings assist in dispersing the gas released

Table 9.3 Typical fat content of bakery products[a]

Product	% Fat	kcal/100 g	% kcal from Fat
Yeast-leavened bread	1–3	250	4–11
Muffins	9–10	285	28–32
Cakes	10–20	370	24–48
Danish pastry	23–25	425	49–53
Chocolate chip cookies	20–30	495	36–54
Doughnuts	25–30	420	54–64

[a] *Source*: Composition of Foods, Agriculture Handbook No. 8, Agricultural Research Service, USDA, Washington, D.C., 1975.

throughout the dough to provide the desired volume. Texture and mouthfeel of cookies are greatly affected by the type and level of fat in the formula. Flakiness of pie crusts is another aspect of texture. Flavor is another attribute of fats, not only the flavor of the fat used but also of any oil-soluble flavorants in the other ingredients. Replacing all of these functions can be a challenging problem.

A number of low-fat and fat-free cakes are available commercially. In most cases starch-based fat replacers, additional emulsifiers, and gums are used to hold higher moisture levels and replace the functionality of fats. A typical formula is shown in Table 9.4. A similar approach is shown in the formulation of a fat-free muffin (Table 9.5).

Low-fat formulations for baked products normally high in water are less difficult than hard products such as cookies that contain little or no water. Most low-fat commercial cookies are of the softer type, containing some moisture. A starting formulation is shown in Table 9.6.

Baking product companies have been among the leaders in producing light or healthy products, with reduced-fat products being the most recent part of this trend. New low-fat or fat-free products have been made available in every category, from crackers to cakes. Table 9.7 reviews some of the commercial products.

Table 9.4 Reduced calorie yellow cake[a]

Ingredient	%
Part A	
Sugar, granulated	18.29
Polydextrose	7.00
Nonfat milk solids	2.05
Polyglycerol monoesters and mono- and diglycerides	1.03
Modified food starch	0.82
Guar gum	0.19
Sodium bicarbonate	0.16
Artificial butter flavor	0.08
Vanilla flavor	0.08
Part B	
Cake flour	21.61
Baking powder	1.83
Glucono-delta-lactone (Pfizer)	0.20
Salt	0.35
Xanthan gum	0.10
Part C	
Whole eggs	17.77
Water	28.44

[a] *Source:* Ref. 29.

Table 9.5 Fat-free muffins[a]

Ingredients	%
Water	32.9
Cake flour	19.7
Sugar	19.0
Dried fruit	6.5
Brown sugar	6.3
Oat bran	5.4
Maltrin M 100 (maltodextrin)[b]	2.7
Nonfat dry milk	2.4
Baking powder	1.7
Dried egg whites	1.3
Emulsifier	0.6
Salt	0.5
Vanilla	0.5
SSL (Emplex, Amer. Ingred.)	0.3
Xanthan gum	0.1
Guar gum	0.1

[a] *Source*: Grain Processing Corporation, Muscatine, Iowa.
[b] Grain Processing Corporation.

6.2 Salad Dressings and Mayonnaise

Pourable Salad Dressings. The Standard of Identity for French dressing calls for a minimum of 35% oil, but many commercial products contain higher levels of oil. For some people salad dressings are a major source of fat calories. The removal of oil from a salad dressing decreases solids and reduces richness

Table 9.6 Fat-reduced cookie[a]

Ingredient	%
Flour	24.60
Chocolate chips	24.60
Shortening	4.12
Sugar	16.45
Whole eggs	8.20
Brown sugar	4.15
Water	3.81
Salt	0.50
Baking soda	0.50
Vanilla extract	0.50
Emplex	0.12
Paselli SA2 (20% gel)	12.45

[a] *Source:* AVEBE America, Inc.

Table 9.7 Calories from fat (baked goods)

Product	Fat Grams	Total Calories	% Calories from Fat
Fat-free bread	0	40	0
Fat-free crumb cake	1	80	11
Yellow cake mix	4	190	19
Pancake mix	2	130	14
Fat-free crackers	0	50	0
Carrot walnut cookies	1	60	15

as compared to the full-fat product. Some of this richness can be restored by increasing soluble solids using modified starches and adding gums. Additional sucrose or corn syrup may be added to increase the solids and offset the vinegar. Additional emsulsifier also helps to improve the mouthfeel. Table 9.8 shows a suggested starting formula.

Mayonnaise and Spoonable Dressings. Commercial mayonnaise is an oil-in-water emulsion that must contain 65% oil but usually contains about 80% oil. Such a high oil-in-water emulsion has a high viscosity providing a specific

Table 9.8 Pourable French dressings[a]

Ingredients	Control (%)	Nonfat (%)
Vegetable oil	38.20	0.00
Water	20.85	48.14
Vinegar	20.00	24.00
Sugar	11.50	11.50
Tomato paste	6.00	6.50
N-lite LP	0.00	4.00
Ultra-Tex	0.00	2.50
Salt	1.00	1.00
Mustard powder	1.00	0.30
Onion powder	0.50	0.30
MSG	0.30	0.50
Xanthan gum	0.25	0.35
PGA[b]	0.00	0.30
Garlic powder	0.20	0.20
Artificial color	0.00	0.15
Sodium benzoate	0.08	0.08
Potassium sorbate	0.08	0.08
Oleoresin paprika	0.05	0.10

[a] *Source*: National Starch and Chemical Company, Bridgewater, New Jersey.
[b] Propylene glycol alginate.

mouthfeel that is difficult to mimic. Most of the ingredients that are used in low-fat salad dressings also apply to mayonnaise and spoonable dressings. However, the ingredient list for fat-free or low-fat mayonnaise becomes even longer because of the need to replace whole eggs and egg yolks with egg whites plus additional modified starches and gums to replace body lost by the removal of eggs. A formula suggested to achieve these goals is shown in Table 9.9.

As might be expected in a food category that supplies such a high level of fat calories, a number of new low-fat or fat-free products have been introduced. In the case of the spoonable dressings, the total fat per serving has been lowered, but the percent of calories from fat remains high due to the lack of other nutritional ingredients. Table 9.10 shows the fat and calorie content of some commercial products.

6.3 Margarines and Spreads

Margarine must contain 80% fat in order to meet FDA Standards of Identity. Reduced-calorie margarines or spreads are typically divided into two categories, products containing 50–70% fat and dietary margarines in the 40–50% fat range. These types of spreads have been available since the 1960s. These lower fat products are restricted in frying and baking applications due to the high water content. As a topping on bread or other substrates more product is needed to achieve a buttery taste, increasing the actual fat consumption. In addition, problems arise as to breaking the water-in-oil emulsion at room temperature or in freeze–thaw cycles.

The formulation of low-fat margarine-type spreads involves replacing some of the triglyceride components with emulsifiers and more water. Relatively high levels of monoglycerides, polyglycerol esters, lecithin, and other fat-

Table 9.9 Imitation mayonnaise, 20% oil[a]

Ingredient	(%)
Water	46.2
Vegetable oil	20.0
Modified corn starch	9.0
Dry whole egg powder	8.4
Sugar	8.4
Vinegar	6.0
Salt	1.0
Methocel K100M premium	0.4
Mustard powder	0.4
Xanthan	0.2

[a] *Source*: The Dow Chemical Company, Midland, Michigan.

Table 9.10 Calories from fat (salad dressings)

Product	Fat Grams	Calories from Fat	% Calories from Fat
Light 1000 Island dressing	2	18	60
Light Italian dressing	4	36	90
Oil-free Italian dressing	0	0	0
Reduced-calorie mayonnaise	4	36	90
Nonfat-mayonnaise dressing	0	0	0

derived emulsifiers are used. Gelatin is also used to increase the viscosity and improve the texture. As with the spoonable dressings, while the total calories are lowered the percent of calories from fat remains high (Table 9.11).

6.4 Processed Meats

Processed meat products made from lower fat meats represents one approach to a low-fat meat diet. However, usually lower fat means lower consumer acceptance (39).

Processed meats include sausages, luncheon meats, frankfurters, and ground meats. The major problems to be overcome in replacing fat in processed meats are taste, texture, and cost. The simplest way to reduce fat is to increase the lean content. However, lean meat is more costly than fat and presents problems to food processors in items such as sausage. While spreads and salad dressings can use water to replace fat, processed meat products are more limited as to the amount of water that can be added while maintaining quality. With the reduction of fat and the increase of water, texture and flavor are changed.

In meat products that adhere to a Standard of Identity several ingredients such as starches and flours can be incorporated at levels up to 3.5%. These materials can bind water but decrease the normal meat texture. This change in texture can be compensated for by adding nonmeat proteins, for example, whey and soy protein. The proteins increase water binding and can form a gel structure to improve texture. The U.S. Department of Agriculture (USDA) has modified its regulation to allow the substitution of water for fat in frankfurters and other cooked sausages.

Table 9.11 Calories from fat (selected spreads)

Product	Fat Grams	Calories from Fat	% Calories from Fat
Low-fat butter	10	90	100
Diet margarine	6	54	100
Light spread	8	72	90
Low-fat spread	3	27	100

Ground beef products typically contain from 20 to 30% fat. While extra-lean ground beef made by trimming has been available, consumer response has been mixed probably due to taste and the higher price. A major fast-food restaurant chain has introduced a low-fat hamburger containing less than 10% fat. This is accomplished by using carageenan gum plus water with specific flavoring added (36). The apparent acceptance of this product has led to the introduction of several similar products in other fast-food restaurants and also for the retail trade. In some cases lower cost soy protein has been added to not only assist in water binding but also to supply some of the solids lost by the elimination of fat. A suggested formula for a low-fat ground meat patty is shown in Table 9.12.

Demand for lower fat products has led to the increased use of poultry in processed meats. Here again the need to improve mouthfeel and texture calls for additives that bind moisture and provide firmness. A low-fat frankfurter-type product replacing red meat with poultry is described in Table 9.13.

As with spreads, the low-fat meat products still provide a high percentage of total calories as fat. Increasing the protein content with lower cost vegetable proteins is one approach to improving this. Table 9.14 shows the caloric content of some commercial low-fat processed meat products.

6.5 Dairy Products

No sector of the food industry has achieved a greater reduction in fat content than has the dairy industry. Whole-milk consumption has been decreasing while low-fat and skim milk consumption have been increasing for several years. Likewise consumption of cottage cheese has declined as yogurt consumption increased (Table 9.15). For the most part this has been achieved by eliminating the fat from the product without the addition of any fat substitute.

Cheese. Cheese ranks high among dairy products as to fat content, and lowering fat while maintaining consumer acceptance is difficult. Controlling and optimizing flavor and physical properties of low-fat cheese has proven to

Table 9.12 Meat patty, 8–9% fat[a]

Ingredient	%
Lean meat (10% fat)	90.00
Water	9.05
Viscarine ME 389 carageenan[b]	0.45
Salt	0.35
Flavor	0.15

[a] *Source*: FMC Corporation, Philadelphia, Pennsylvania.
[b] FMC Corporation.

Table 9.13 Reduced-fat poultry cuts[a]

Ingredient	%
Deboned poultry	67.00
Water/ice	27.65
Dextrose	2.00
Salt	1.27
Spices	0.88
Gelcarin ME 421 carageenan[b]	0.50
Blend 424 phosphate	0.40
Cure salt	0.25
Sodium erythorbate	

[a] *Source*: FMC Corporation, Philadelphia, Pennsylvania.
[b] FMC Corporation.

be a challenge. A reduction of up to 35% appears to maintain acceptable flavor and texture. More than 35% reduction does not provide sufficient milk fat to impart desirable mouthfeel, lubricity, and act as a flavor precursor. Some suggested processing modifications include selection of lactic acid starter cultures with desirable characteristics, adjustment of milk clotting enzyme levels, control of acidity development during manufacturing, regulation of lactose and lactic acid levels in cheese curd, and addition of selected bacterial strains to improve flavors (38).

Yogurt. Yogurt has benefitted greatly from the trend toward healthy eating. Not only has the caloric level been cut by lowering milk fat, but the use of artificial sweeteners in flavored products has also been a major factor in lowering calories. The result has been a doubling of per capita yogurt consumption in the 1980s.

Fat-free yogurts are typically made by replacing low-fat milk with nonfat milk and adding ingredients such as modified food starch and gelatin to replace the body lost by eliminating fat. The artificially sweetened light yogurts con-

Table 9.14 Calories from fat (processed meats)

Product	Fat Grams	Calories from Fat	% Calories from Fat
Light bologna	6	54	77
Danish ham slices	1	9	36
Light wieners	7	63	70
90% fat-free hot dogs	6	54	60
Light sausage	11	99	83
Extra lean ham	1	9	30

Table 9.15 Per capita consumption change in some dairy products (in lb)[a]

Year	Whole Milk	Low-Fat Milk	Cottage Cheese	Yogurt
1970	214	30	5.2	0.8
1975	175	53	4.7	2.1
1980	142	70	4.5	2.6
1985	120	83	4.1	4.1
1990	88	98	3.4	4.2

[a] *Source*: Ref. 3.

taining aspartame have even less body due to the loss of the sugar bulk. Even with the use of modified starch and other fat replacing ingredients, they tend to have a much lower viscosity than the fat-containing products.

Frozen Desserts. The frozen dessert category has been split for several years between premium ice creams with a fat content of over 14%, compared to the normal 12%, and low-fat low-calorie frozen desserts.

Frozen low-fat yogurt containing less fat and cholesterol than conventional ice cream was introduced in the 1970s. With the advent of noncaloric artificial sweeteners, frozen low-fat yogurt became popular as a low-calorie dessert. The main problem in formulating low-fat frozen yogurt is replacing the bulk of the milk fat and the sugar. Both polydextrose and maltodextrins have been suggested for this application. Processing techniques are also important.

Most low-fat frozen desserts are in the ice milk category, with a milk fat content of 2–7%. While some of the deficiencies of low-fat or fat-free frozen desserts may be accepted in yogurts, they are more noticeable in ice cream products. Body, texture, and creaminess need to be increased in order for the product to be acceptable. The texture must be strong enough to hold the desired overun in processing. This has led to the use of fat replacers and other ingredients to improve these properties. A recent study reviewed the functionality of various ingredients suggested for improved low-fat desserts (40). Polydextrose and maltodextrins rate high as bulk replacers; however, microcrystalline cellulose has better fat substitute qualities. A fat-free frozen dessert formula is shown in Table 9.16. Some commercial low-fat dairy products are described in Table 9.17.

6.6 Snack Foods

Fat reduction in the higher fat containing snacks is difficult due to processing as well as sensory restraints. Fat content in snacks varies from a low of 7–10% in popcorn to as high as 45% in potato chips. Corn chips, tortilla chips, and cheese puffs may contain from 22 to 35% fat.

One approach being used in French fries that may have application in other snack foods is the use of batters as barrier coatings. A recent U.S. patent

Table 9.16 Nonfat frozen dessert

Ingredient	% Solids
Nonfat milk solids	11.00
Sucrose	11.00
Corn syrup solids (42 DE)	5.00
Avicel RC-581 cellulose gel	0.50
CMC-7HF[a]	0.17
Emulsifier[b]	0.10
	27.77

[a] Carboxy methylcellulose, Hercules, Inc.
[b] Tandem 100K, Witco Co.

describes the use of a powdered cellulose batter to reduce lipid retention and increase moisture content in deep-fried foods (41). High solids potatoes will also limit the fat uptake of potato chips to a limited degree. Another recent invention calls for snacks cut from sheets of farinaceous dough, coated with oil, pressed in a mold, and then cooked to produce chiplike snacks with low fat content (42).

Changes in processing rather than reformulating or using special ingredients appears to offer more promise for reduced-fat snack foods. One of the suggested approaches is the use of microwave heat to lower moisture prior to deep frying. Another method calls for partial deep frying followed by baking. In many cases better control of temperature and oil quality, especially viscosity, will lower fat content.

It is unlikely that these procedures can achieve substantial fat reduction in high-fat snacks such as potato chips. FDA approval of olestra or a similar noncaloric fat substitute holds the most promise for low-fat products in this category. Some commercial low-fat snack products are listed in Table 9.18.

Table 9.17 Calories from fat (dairy products)

Product	Fat Grams	Calories from Fat	% Calories from Fat
Lowfat yogurt	2	18	14
Light yogurt	0	0	0
Light cheddar cheese	3	27	39
Light American cheese	4	36	51
Light cream cheese	7	63	79
Nonfat cottage cheese	5	45	56
Light vanilla ice cream	3	27	27
Frozen dairy dessert	2	18	14

Table 9.18 Calories from fat (snack products)

Product	Fat Grams	Calories from Fat	% Calories from Fat
Light potato chips	6	54	42
Light cheese snack	6	54	39
Fat-free pretzels	0	0	0
Light microwave popcorn	3	27	39

GOVERNMENT REGULATIONS

The Nutrition Labeling and Education Act (NLEA) of 1990 required the Department of Health and Human Services to develop new nutritional labeling regulations by November 8, 1991. New regulations for meat and poultry products were to be developed by the Department of Agriculture. These new labeling regulations will cover over 250,000 food products and became effective in May 1994 for most products and in July 1994 for meat and poultry products. In regard to labeling, the act covers nutritional information, descriptors, serving size, and health claims.

Prior to the NLEA, regulations required manufacturers to list nutritional information on the package only if they were making nutrition claims for the product. Thus nutritional information appeared on a low-fat version of a product but not on the standard product. Under the NLEA food labels are required to list amounts for calories, calories from fat, total fat, saturated fat, and cholesterol (43). The following mandatory and optional dietary components must be listed on food label (43):

- Total calories
- Calories from Fat
  - Calories from saturated fat (optional)
- Total fat
- Saturated fat
  - Polyunsatured fat (optional)
  - Monounsaturated fat (optional)
- Cholesterol
- Sodium
  - Potassium (optional)
- Total carbohydrate
- Dietary fiber
  - Soluble fiber (optional)
  - Insoluble fiber (optional)

Sugars
  Sugar alcohol (optional)
  Other carbohydrate (optional)
Protein
Vitamin A
Vitamin C
Calcium
Iron
Other essential vitamins and minerals (optional)

Descriptors are the words or phrases used on labels to describe the level of a nutrient without making any health claim. To minimize misunderstanding, the FDA has defined these terms. Descriptors that are important in fat-containing foods are:

1. *Low:* a serving (or 50 g of food if the serving size is small) containing no more than 40 calories, 140 mg of sodium, 3 g of fat, 1 g of saturated fat, or 20 g of cholesterol.
2. *Lean:* a serving (or 100 g) of meat, poultry, seafood, and game meats containing less than 10 g of fat, 4 g of saturated fat, and 95 mg of cholesterol.
3. *Extra lean:* a serving (or 100 g) of meat, poultry, seafood, and game meats containing less than 5 g of fat, 2 g of saturated fat, and 95 mg of cholesterol.
4. *Light:* an altered product containing one-third fewer calories or 50% of the fat in the reference food; if 50% or more of the calories come from fat, the reduction must be 50% of the fat.
5. *Percent fat free*: a product must be low-fat or fat-free, and the percentage must accurately reflect the amount of fat in 100 g of the food. For example, 2.5 g of fat in 50 g of food results in a 95% fat-free claim.

Prior to the NLEA, the serving size shown on the label of a particular food product was decided by the manufacturer and did not always reflect the real amount a person would eat at one meal. The FDA and the USDA have developed serving sizes for over 150 foods, meat, and poultry product categories that must be used on the label. These serving sizes must be presented in metric and common household measures.

The NLEA states that health claims on a label can only be made where significant scientific agreement supports the legitimacy of the health claim. The NLEA directs the FDA to establish regulations on permissible health claims. If a claim has not undergone FDA review, it will be disallowed. To date a limited number of health claims have been allowed, including fat and cancer and saturated fat and cholesterol and coronary heart disease.

Under the NLEA all Standard of Identity foods will be required to have an ingredient statement on the label. Unfortunately, Standards of Identity can interfere with the marketing of more healthful food products. For example, new lower fat margarine cannot use "margarine" on the label because it lacks the level of fat needed to meet the Standard of Identity. While these Standards of Identity were developed with the intent of protecting the consumer from inferior products, they may now be standing in the way of improved nutrition. The dairy industry in particular is affected by the Standards of Identity because over 40% of foods bearing Standards of Identity are in the dairy sector. In dairy products it has been suggested that other ingredients, such as calcium and protein content, be part of the standard rather than fat content.

FUTURE CONSIDERATIONS

It is likely that the American consumer will continue to pay attention to the nutritional evidence supporting a lower fat diet. Recent history, however, would indicate that actual lowering of fat calories in the average diet will proceed at a slow pace. Continued development on the part of the food industry of more acceptable low-fat products will assist in achieving the goal of 30% fat calories in the average diet. Lower prices of low-fat products so that there is no economic barrier to good health will also be important. Of equal importance will be education, both in the schools and at the adult level.

As to fat substitutes, it is unlikely that, with one exception, any major new products will appear. The exception would be the approval of olestra or a similar nonfat or low-fat product. This would have a major effect in almost all food sectors. Structured lipids such as caprenin offer another possible approach. Meanwhile, food scientists will continue to improve the current types of fat replacers and to use all available knowledge to formulate better fat-free and low-fat food products.

REFERENCES

1. Human Health Services, *The Surgeon General's Report on Nutrition and Health,* U.S. Department of Health and Human Services, U.S. Department of Health and Human Services, U.S. Government Printing Office, Washington, D.C., 1988.
2. U.S. Department of Agriculture and the U.S. Department of Health and Human Services, *Dietary Guidelines for Americans*, Advisory Committee Report, USDA, Washington, D.C., 1990.
3. J.J. Putnam, Food Consumption, Prices, and Expenditures, 1970–1990, Statistical Bulletin No. 840, Economic Research Services, USDA, Washington, D.C. (1992).
4. Dieting and Low Calorie/Reduced Calorie Products Survey, Booth Research Services, Atlanta, Ga., 1991.
5. F.H. Mattson and R.A. Volpenheim, *J. Lipid Res.* **13,** 325 (1972).
6. F.H. Mattson and G.A. Nolen, *J. Nutr.* **102,** 1177 (1972).

7. D.J. Hamm, *J. Food Sci.* **49,** 419 (1984).
8. F.H. Mattson and R.A. Volpenheim, U.S. Pat. 3,600,186 (1971).
9. L.I. Osipow, F.D. Snell, W.C. York, and A. Finchler, *Indust. Eng. Chem.* **48,** (1956).
10. G.P. Rizzi and H.M. Taylor, *J. Am. Oil Chem. Soc.* **55,** (1978).
11. R.A. Volpenheim, U.S. Pat. 4,517,360 (1985).
12. R.J. Jandacek and M.R. Webb, *Chem. Phys. Lipids* **22,** (1976).
13. F.H. Mattson, U.S. Pat. 4,034,083 (1977).
14. C.A. Bernhardt and H.M. Taylor, U.S. Pat. 5,158,796 (1992).
15. J.F. White, U.S. Pat. 4,959,466 (1990).
16. R.S. Meyer and co-workers, U.S. Pat. 4,973,489 (1990).
17. J. Fulcher, U.S. Pat. 4,582,927 (1986).
18. D.J. Hamm, U.S. Pat. 4,508,746 (1985).
19. D.P. D'Amelia and co-workers, U.S. Pat. 4,915,974 (1990).
20. J.F. White and M.F. Pollard, U.S. Pat. 4,861,613 (1989).
21. L.P. Klemann, U.S. Pat. 4,927,659 (1990).
22. W.R. Michael and R.H. Coots, *Toxicol. Appl. Pharmacol.* **20,** 334 (1971).
23. S.B. Christensen and co-workers, U.S. Pat. 5,194,286 (1993).
24. A.C. Bach and V.K. Babayan, *Am. J. Clin. Nutr.* **950,** (1982).
25. D.R. Webb and R.A. Sanders, *J. Am. Coll. Toxicol.* **10,** 3 (1991).
26. E.L. Wheeler and co-workers, U.S. Pat. 5,258,197 (1991).
27. N.S. Singer, S. Yamamoto, and J. Latella, U.S. Pat. 4,734,287 (1988).
28. N.S. Singer and R.H. Moser, in A.M. Altschul, ed., *Low Calorie Food Hand Book*, Marcel Dekker, New York, 1993.
29. Y.J. Asher, M.A. Mollard, S. Jordan, T.J. Maurice, and K.B. Caldwell, U.S. Pat. 5,215,777 (1994).
30. S. Greenman, R. Gan, M. Bergman, A. Bakal, and P.A. Cash, U.S. Pat. 5,194,282 (1993).
31. Technical Data, Stellar Fat Replacer, TD 5513, A.E. Staley Company, Decatur, Ill., 1991.
32. H. Rennhard, U.S. Pat. 3,876,794 (1975).
33. F.K. Moppet, in L. Nabors and R.P. Gelard, eds., *Alternative Sweeteners* 2nd ed., Marcel Dekker, New York, 1991.
34. J.F. Ang, W. B. Miller, and I.M. Blais, U.S. Pat. 5,019,406 (1991).
35. Bulletin G-34, Avicel Cellulose Gel, FMC Corp., Philadelphia, Pa, 1985.
36. Methocel Food Gums, The Dow Chemical Company, Midland, Mich., 1993.
37. W.R. Egbert, and co-workers, *Food Tech.* June (1991).
38. Fu-Ning Fung, U.S. Pat. 5,082,684 (1990).
39. Nutrcol Konjac Flour, Introductory Bulletin K-1, FMC Corporation, Philadelphia, Pa., 1989.
40. A.M. Pearson and co-worker, *Proceedings of the Reciprical Meat Conference,* **40,** 105 (1987).
41. "Factors to Consider in Production of Low Fat Cheeses," IBC Conference, Fat and Cholesterol Reduced Products: Technologies and Ingredients (1991).
42. B.W. Tharp, "Use of Conventional Technology in Producing Low Fat and No Fat Frozen Desserts," IBC Conference, Fat and Cholesterol Reduced Foods: Technology and Ingredients (1991).
43. J.F. Ang, W.B. Miller, and I.M. Blais, U.S. Pat. 5,019,406 (1991).

10

Lecithins

INTRODUCTION

Phospholipids or phosphatides which include the compound phosphatidylcholine, the chemical term for lecithin, are lipids containing a phosphoric acid residue; they are nature's principal surface active agents. They are found in all living cells, whether of animal or plant origin. In humans and in animals, the phosphatides are concentrated in the vital organs, such as the brain, liver, and kidney; in vegetables they are highest in the seeds, nuts, and grains. As constituents of cell walls and active participants in metabolic processes, they are essential to life (1–9).

The commercial term "lecithin" is very general and describes a composition of lipid constituents and surface active compounds present in the product rather than the chemical entity: phosphatidylcholine (PC). Thus, in general usage, lecithin refers to a complex, naturally occurring mixture of phosphatides obtained by water-degumming crude vegetable oils, and separating and drying the hydrated gums (9). It is, however, the phospholipid portion of lecithin that is mainly responsible for giving form and function to lecithin (6,7).

Commercial lecithin is an important coproduct of edible oil processing because of its dietary significance and multifaceted functionality in food systems and industrial applications. Unless indicated otherwise, the term "lecithin" will denote the commercial designation throughout the text in this Chapter.

Lecithin has a long history of use in foods, dating back more than fifty years. The 1930s brought widespread use of commercial solvent extraction techniques for vegetable oil production, and because "degumming" of crude vegetable oil became necessary for shipping stability, a large supply of crude lecithin "gums" was produced. These gums were obtained in sufficient quantity to necessitate their becoming an item of commerce (8). In the ensuing years, there was extensive research into developing new lecithin applications, as well as product refinements and modifications. Lecithin ingredients are now recognized as valuable products which have both nutritional value and commercial, i.e., food/industrial utility (2,5).

Two of the earliest edible applications of lecithin, viscosity reduction in chocolate and confectionery products and emulsification/antispatter properties in margarine, still enjoy wide popularity and represent outlets for large volumes of lecithin products. In addition, other early uses such as in bakery goods, pasta, textiles, insecticides, paints, etc., are still active today.

SOURCES OF PHOSPHOLIPIDS

1.1 Lecithin in Human/Animal Tissues

Almost all body cells contain phospholipids. The common animal phospholipids are made of sphingomyelin, phosphatidylcholine (PC), phosphatidylethanolamine (PE), phosphatidylserine (PS), phosphatidylinositol (PI), and other glycerol phosphatides of complex fatty acid composition. Phosphatidylcholine (PC), phosphatidylethanolamine (PE), formerly also referred to as "cephalin," and phosphatidylserine (PS) are by far the most predominant phosphatides from most animal sources. As constituents of cell walls and active participants in metabolic processes, they appear to be essential to life (9).

The exact composition of human/animal phospholipids depends on the source and the method of extraction and purification. The central nervous system especially has a high phospholipid content. The liver is the site for their biosynthesis, and the lipids of the mitochondria, which are the regulators of cell metabolism and energy production in the body, consist of up to 90% of phospholipids (10).

A survey of improved fractionation and analytical methods for elucidating the molecular species of these complex animal phospholipids and the phospholipids present in primary and processed foods has been published by Kuksis (11). This survey also includes information on the quantitative analysis of phospholipids, peroxidation products of phospholipids, and the composition of selected animal phospholipids.

Only egg yolk, milk, and brain have served as major animal sources of commercial lecithin. In some instances isolated and purified lecithins have been developed for clinical nutritional uses. Weihrauch and Son (12) present a concise review of phospholipid composition in various foods.

The following brief discussion covers phospholipids from animal sources that have some commercial significance.

Egg Phospholipids. At one time, eggs, which possess a relatively high phospholipid content, served as a commercial source until soybean technology made it uneconomical to produce. The phospholipids in eggs are mainly in

Table 10.1 Composition of soy and egg lecithin (14)

Polar Lipids	Soy	Egg
Phosphatidylcholine	20–22	68–72
Phosphatidylethanolamine	21–23	12–16
Phosphatidylinositol	18–20	0–2
Phosphatidic acid	4–8	–
Sphingomyelin	–	2–4
Other phospholipids	15	10
Glycolipids	9–12	–

the yolk where at least a portion of them are combined with protein and carbohydrates.

Compositional data on commercial egg products and various lipids extracts from egg yolk have been compiled by Gornall and Kuksis (13), Kuksis (11), Schneider (14), and Satirhos and co-workers (15). Egg yolk has about 70% phosphatidylcholine, 24% phosphatidylethanolamine, 4% sphingomyelin, 1% phosphatidylserine, and 1% phosphatidylinositol, and lysophosphatidylcholine and lysophosphatidylethanolamine contribute the remaining 2% of the total phospholipids. Tables 10.1 and 10.2 (14,16) compare the composition of soy and egg lecithins and their fatty acids, and Table 10.3 (16–21) shows the distribution of phospholipid classes in egg yolk.

The main difference between plant/legume lecithin, e.g., soy, and lecithin in egg yolk is that the former has a higher unsaturated fatty acid and choline content and has no cholesterol. Egg lecithin as a commercial ingredient, with the exception of some medical feeding programs, is too expensive for routine use in food (10).

Milk Phospholipids. Milk has a phospholipid content of about 0.035%, associated with the fat by virtue of being part of a colloidal membrane which

Table 10.2 Fatty acid composition (%) (14)

Type of Acid	Soy	Egg
Saturated		
Palmitic	15–18	27–29
Stearic	3–6	14–17
Unsaturated		
Oleic	9–11	35–38
Linoleic	56–60	15–18
Linolenic	6–9	0–1
Arachidonic	0	3–5

Table 10.3 Distribution of phospholipid classes in egg yolk (16)

Phospholipid Classes	Analyses					
	[a]	[b]	[c]	[d]	[e,f]	[e,g]
			Weight %			
% Total lipid				23		
PC	66–76	69.1	77.0	69	82.6	87.1
PE	15–24	23.9	18.0	24	9.1	7.8
PS	1	2.7		Trace		
CL	1	3.2		Trace		
SPH	3–6	1.0	2.3	3	1.8	2.5
LPC + LPE	3–6		2.5	3	6.5	2.5
Unidentified						

[a] Privett and co-workers (17).
[b] Noble and Moore (18).
[c] Cook and Martin (19).
[d] Gornall and Kuksis (20).
[e] Connelly and Kuksis (21).
[f] Commercial sample.
[g] Intralipid.

surrounds the fat globule. Wittcoff (4), Morrison and co-workers (22), and Privett and co-workers (23) have reported the results of TLC analysis of the polar lipids of various milk fractions. Skim milk and milk serum have the highest portion of polar lipids as percent of the total lipids, while whole milk and cream have the least. Of the polar lipids, phosphatidylethanolamine constitutes the largest component, with phosphatidylcholine and sphingomyelin (being present in about equal proportions) at a significantly lower level (Table 10.4) (16,22–24).

Brain Phospholipids. The brain is a rich source of phospholipids and, together with the spinal cord, probably possesses the highest phospholipid content of any of the organs. There are many and different types of phospholipids in the central nervous system. Since they bypass the blood-brain barrier, adequate nutrition (biosynthesis) of the nerve cells is assured with these substances. Special enzyme systems see to it that the most efficient functioning is accomplished at all times (4).

The composition of brain phospholipids has been extensively investigated by adsorption column- and thin-layer chromatography (TLC). Table 10.5 lists the major classes of brain phospholipids from different animal species, as compiled by Kuksis (16,26–28).

Phospholipids in Liver, Kidney, Muscle, and Other Tissues. Organ meats such as liver, kidney, and muscles are a major source of dietary phospholipids.

Table 10.4 Distribution of phospholipid classes in various milk fractions (16,23)

Lipid Classes	Whole Milk	Plastic Skim	Nonfat Skim	Serum	Cream	Milk[a]	Serum[b]
				Weight %			
% Total lipid	2	19	32	22	1–2	–	–
Ceramide monohexoside	Trace	19.5	14.1	7.8	17.9	3	–
Ceramide dihexodide	Trace	10.1	–	–	–	3	–
Phosphatidylethanolamine	36.4	27.5	45.1	32.4	25.3	30	26.9
Plasmalogen	–	–	–	–	–	1	–
Phosphatidylcholine	27.0	7.3	16.4	23.2	19.9	28	29.3
Plasmalogen	–	–	–	–	–	3	–
Sphingomyelin	29.0	18.2	14.5	26.9	21.6	19	31.3
Phosphatidylserine	Trace	1.9	7.6	3.5	Trace	8	5.0
Phosphatidylinositol	Trace	Trace	Trace	1.1	Trace	5	5.9
Ganglioside + (unknown)	7.6	15.5[4]	2.3[3]	5.1[4]	15.3[4]	0	1.0

[a] Morrison and co-workers (22).
[b] Santha and Narayanan (24).

Table 10.5 Distribution of phospholipid classes in brain of different animal species (16)

	Animal Species							
Lipid Classes	Human[a,b]	Bovine[a,b]	Bovine[b,c]	Human[c,d]	Bovine[c,e]	Human[c,e]	Sheep[f]	Rat[g]
				Weight %				
% Total lipid								
PC	21.8	18.4	32.4	33.2	48.2	47.6	37.3	36.8
Plasmalogen	–	–	–	–	–	–	0.9	–
PE	35.4	36.1	23.5	25.2	24.2	17.8	7.7	36.4
Plasmalogen	–	–	–	–	–	–	16.5	–
PS	18.8	18.0	11.0	10.7	6.7	9.3	9.2	11.8
PI	1.8	1.8	4.3	4.8	7.1	5.0	2.1	3.1
PA	1.1	1.7	0.9	0.3	1.3	1.2	2.6	1.2
CL	–	–	1.2	1.0	ND[h]	0.3	2.0	2.2
PG	–	–	0.9	ND[h]	0.4	0.6	–	–
LPC (LPE)	2.0	0.2	0.9	1.0	1.0	2.5	–	–
LbisPC	1.0	0.4	2.1	0.2	0.7	0.2	–	–
SPH	16.3	15.0	20.4	17.0	4.9	10.7	12.9	5.7
Unidentified								

[a] Siakotos and co-workers (26).
[b] Myelin.
[c] Siakotos and co-workers. (25)
[d] Edothelial cells.
[e] Nuclei.
[f] Scott and co-workers. (27).
[g] Wuthier (28).
[h] ND = not detected.

The reader is referred to Kuksis (16) for the distribution of various phospholipid classes in the liver, kidney, muscles (heart and skeletal), spleen, lung, blood cells, bile, and adipose tissue of different animal species. Compositional data of fatty acids for these tissues and fluids are also given.

In blood, phosphatidylcholine is quantitatively the most important phospholipid. Sphingomyelin is present in varying amounts in perhaps all of the animal organs, most of it in the soft organs, and to a lesser degree in skeletal muscles and eggs (4). Total blood contains about 0.2 to 0.3% phospholipids. In plasma and serum, phosphatidylcholine predominates, whereas in corpuscles phosphatidylethanolamine and sphingomyelin constitute the bulk of phospholipids. Most workers have found that the phospholipid content is greater in red blood cells than in plasma, and it constitutes 60 to 65% of the total lipids in these cells (4).

1.2 Soybean

Although the highest concentrations of phospholipids occur in animal products, i.e., meat, poultry, fish, eggs, and milk/cheese, the major commercial source is the soybean, which contains 0.3 to 0.6%. Nevertheless, phospholipids from other vegetable oilseeds, i.e., corn, cottonseed, linseed, peanut, rapeseed, safflower, and sunflower, and plants have also been studied and used (5), as discussed later.

Commercial lecithin from the soybean is a complex mixture. It is comprised of phospholipids, triglycerides, with minor amounts of other substituents, i.e., phytoglycolipids, phytosterols, tocopherols, and fatty acids. The composition and molecular arrangement of this heterogenous mixture of compounds defines a product which is low in apparent polarity and has a strong tendency to promote w/o type emulsions (29).

A wide range of data has been published showing the variability in the composition of phospholipids and fatty aids in soybean lecithin (Tables 10.6 and 10.7) (30).

Soy lecithin is a coproduct of oil processing; as a result purification steps used to produce a quality oil may affect the lecithin components. Also, soybeans exposed to frost damage or subjected to a prolonged storage have reduced lecithin yields (6). Phospholipases which produce phosphatidic acid are active during storage and may reduce the yield of lecithin (31). During the maturation process, the major phospholipids, i.e., (PC), (PE), and (PI) increase, while others decrease or remain constant (30).

A change in the relative concentration of any of these components, or an alteration of its chemical structure, may cause some change in the physical or chemical properties of commercial lecithins. Lecithins can exist as a liquid, plastic, or free-flowing solid. Their color, solvent solubility, surfactancy, and chemical reactivity all can be modified. These modifications, in turn, will alter the functional properties of the lecithin in a given application (29).

Table 10.6 Components (%) of soybean lecithin (30)

Component	Range of Composition		
	Low	Intermediate	High
Phosphatidylcholine	12.0–21.0	29.0–39.0	41.0–46.0
Phosphatidylethanolamine	8.0–9.5	20.0–26.3	31.0–34.0
Phosphatidylinositol	1.7–7.0	13.0–17.5	19.0–21.0
Phosphatidic acid	0.2–1.5	5.0–9.0	14.0
Phosphatidylserine	0.2	5.9–6.3	–
Lysophosphatidylcholine	1.5	8.5	–
Lysophosphatidylinositol	0.4–1.8	–	–
Lysophosphatidylserine	1.0	–	–
Lysophosphatidic acid	1.0	–	–
Phytoglycolipids	–	14.3–15.4	29.6

The ensuing section is a brief review of phospholipid plant sources other than those from the soybean which have current or may have potential commercial applications, i.e., other oilseeds, cereals, and grains. Soybean lecithin then will be discussed in more detail later throughout the subsequent sections of the text.

1.3 Corn

Weeks and Walters (32) have found that 2.5 to 4.5% of the phosphorus in corn is in the form of phospholipids, depending upon the variety involved.

The first detailed analysis of commercial corn phosphatides was published by Scholfield and co-workers (33). In recent years, the phenomenal growth in demand for corn sweeteners and other products of the corn refining industry, such as corn lecithin, have become available and competitive with lecithin

Table 10.7 Fatty acids (%) of soybean lecithin (30)

Fatty Acid	Range of Composition		
	Low	Intermediate	High
Myristic (C14:0)	0.3–1.9	–	–
Palmitic (C16:0)	11.7–18.9	21.5–26.7	42.7
Palmitoleic (C16:1)	7.0–8.6	–	–
Stearic (C18:0)	3.7–4.3	9.3–11.7	–
Oleic (C18:1)	6.8–9.8	17.0–25.1	39.4
Linoleic (C18:2)	17.1–20.0	37.0–40.0	55.0–60.8
Linolenic (C18:3)	1.6	4.0–6.2	9.2
Arachidic (C20:0)	1.4–2.3	–	–

Table 10.8 Distribution (%) of polar lipids in corn and soybean lecithin (30)

Polar Lipid	Corn	Soybean
Sterylglycoside ester	15.0	4.3
Monogalactosyldiglyceride	1.8	0.8
Digalactosyldiglyceride	3.7	3.0
Other glycolipids	9.8	6.4
N-Acyl phosphatidylethanolamine	2.6	2.2
N-Acyl lysophosphatidylethanolamine	3.7	10.4
Phosphatidylethanolamine	3.2	14.1
Phosphatidylglycerol	1.4	1.0
Phosphatidylcholine	30.4	33.0
Phosphatidylinositol	16.3	16.8
Phosphatidic acid	9.4	6.4
Phosphatidylserine	1.0	0.4
Lysophosphatidylethanolamine	Trace	0.2
Lysophosphatidylcholine	1.7	0.9

from the soybeans (31). Tables 10.8 and 10.9 illustrate the distribution of polar lipids and fatty acids in lecithin of corn compared to those in soybeans (30).

Similar compositions were noted for corn and soy PC, PI, and phosphatidic acid. Glycolipids represent a higher proportion of polar lipids in corn than in soybean lecithin. Cherry (34) and Cherry and Kramer (30) also stated that the percentage of minor components, steryl-glycoside ester and other glycolipids, in corn are more than twice that found in soybean. The physical properties, particularly the emulsifying properties of corn lecithin, differ from those of soybean lecithin because of the higher proportion of glycolipids in the corn lecithin.

Both the glycolipids and the phospholipids of corn have lower percentages of linolenic acid (18:3) and are more saturated than those in the soybean. In general, crude corn and soybean lecithins are equal in linolenic acid (18:2) content, but linolenic acid in corn varies from 42–70% depending on the

Table 10.9 Fatty acid composition (%) of corn and soybean lecithin (30)

	Composition	
Fatty Acid	Corn	Soybean
Palmitic (C16:0)	17.7	17.4
Stearic (C18:0)	1.8	4.0
Oleic (C18:1)	25.3	17.7
Linoleic (C18:2)	54.2	54.0
Linolenic (C18:3)	1.0	6.8

variety of corn. Phytic acid, 88% of which is the corn germ, is extracted as part of the lecithin fraction (30,34). Elimination of phytic acid in corn is desirable because it binds zinc, magnesium, and calcium.

1.4 Cottonseed

The phospholipids in cottonseed are similar in many respects to those of soybeans, with the exception of their lower level of linolenic and higher level of saturated fatty acid content (35). Cherry and Kramer (30) compiled Table 10.10 to show the composition of cottonseed lecithin.

Lecithin can be fractionated from cottonseed as phospholipids and glycolipids. Cottonseed lecithin shows flavor and color deterioration when blended with other vegetable oils. The saturated/unsaturated fatty acid ratio of cottonseed phospholipids is appropriately 1:2 (36). Palmitic acid constitutes 90% of the total saturated fatty acids (36%), while linoleic acid is approximately 80% of the total unsaturated fatty acids (64%). Gossypol binds to lecithin during oil extraction from glanded cottonseed (approx. 9% in crude phospholipids); this economically negates its use as a commercial source. New cultivars of glandless or gossypol-free cottonseed, now being grown in the southwestern and western United States, however, may have some potential for providing commercial edible lecithins (34).

The composition of cottonseed lecithin and the composition of the phospholipid fraction from hexane-defatted glandless cottonseed oil are summarized in Tables 10.10 and 10.11 (30,34,37–39).

Since cottonseed lecithin contains only trace amounts of fatty acids with more than two double bonds (linolenic acid), it is more stable to oxidation and rancidity than soybean lecithin. Cottonseed phospholipids are relatively high in phosphatidylcholine which could provide good emulsifying properties in foods (30,34).

Table 10.10 Composition (%) of cottonseed lecithin (30)

Component	Extract	Phosphatidylcholine	Phosphatidylethanolamine	Phosphatidylserine
Phospholipids	1.8–2.2	34.9–35.9	13.7–20.1	7.0–26.0
Fatty acids				
Myristic (C14:0)	0.4	0.3	0.4	0.6
Palmitic (C16:0)	32.9	31.1	33.7	33.3
Palmitoleic (C16:1)	0.5	0.3	0.3	0.6
Stearic (C18:0)	2.7	2.8	2.2	0.3
Oleic (C18:1)	13.6	11.5	11.5	14.4
Linoleic (C18:2)	50.0	54.0	49.0	50.4
Total gossypol	9.13	2.34	22.43	19.90
Free gossypol	0.02	2.24	0.05	0.01

Table 10.11 Composition of phospholipid fraction from glandless cottonseed oil (34,37,38)

Phospholipid[a]	Composition (% of Total Phosphorus)
Origin	4.12
Lysophosphatidylcholine	2.56
Phosphatidylinositol	13.41
Phosphatidylserine	2.38
Phosphatidic acid	8.76
Phosphatidylcholine	23.16
Phosphatidylethanolamine	13.46
Phosphatidylglycerol	7.62
Lysophosphatidylserine	ND[b]
Lysophosphatidylethanolamine	ND
Unknown (sum: 6 TLC spots)	25.30

[a] Water (2–4%) was added to hexane-extracted glandless cottonseed oil, the resulting mixture stirred 30 min at 70°C, and centrifuged to separate the oil and phospholipid-containing fraction. The phospholipids were separated by 2-dimensional thin layer chromatography (TLC) on Silica gel-60 plates. Dimension I = chloroform: methanol : 7N NH_4OH (65 : 30 : 1); Dimension II = chloroform : methanol : acetic acid : water (170 : 25 : 25 : 4). Quantitation of the phsopholipids was according to El-Sebajy and co-workers (39).
[b] ND = not detected.

1.5 Rapeseed

Rewald (40) found approximately 20% phospholipids in rapeseed. Rapeseed lecithin has been reviewed and an extensive bibliography has been compiled (41). Table 10.12 (41) shows the composition of rapeseed and soybean gums.

The major phospholipids present in rapeseed lecithin are phosphatidylcholine, phosphatidylethanolamine, and phosphatidylinositol. The relative proportion of these components do not differ significantly from that of soybean lecithin. In lecithin from high erucic oils the long chain fatty acids (C20–C22) are present only in small amounts, while in low erucic oil lecithin (including canola) the fatty acid composition is not markedly different from that of the parent oil, except for a somewhat higher content of C16:0, leaving a slightly higher C18:1 and C18:3 fatty acid level in the degummed oil (41).

Solvent-extracted rapeseed oil has been found to contain the highest level of phosphorus. For this reason, it is common practice to degum solvent-extracted but not pressed oil. Rapeseed lecithin has found some uses in animal feeds. Further uses will depend on future quality developments (41).

Sosada (42) has determined the optimum conditions for fractionation of rapeseed lecithin with alcohols to improve purified lecithin yield and phosphatidylcholine enrichment.

Table 10.12 Composition of rapeseed and soybean gums (%) (41)

Components	Rapeseed					Soybean
	(9)[a]	(9)[b]	(10)[b]	(11)[b]	(8)[c]	(8)[c]
Water	24	—	—	—	—	—
Nonlipid	9	—	—	—	—	—
Triglycerides or neutral lipids	16	—	38.1	5.6	29	35
Phospholipids (total)	51	—	—	—	—	—
Phosphatidylcholine	—	22	16.2	24.6	20	21
Phosphatidylethanolamine	—	15	17.5	22.1	16	8
Phosphatidylinositol	—	18	7.6	14.7[d]	8	20
Lysophosphatidylcholine	—	—	—	19.4	1	0
Lysophosphatidylethanolamine	—	—	2.0	—	—	—
Sterol glycosides or glycolipids	—	9[e]	7.9	13.6	11	0
Unidentified	—	36[f]	10.7	—	15	11
Unaccounted for	—	—	—	—	—	5

[a] Complete gum sample, including water.
[b] Acetone precipitate from gum sample.
[c] Dried gum sample.
[d] Tentatively identified.
[e] Tentatively: phytoglycolipids.
[f] Tentatively: 16% acidic phospholipids (plus 20% unidentified).

1.6 Sunflower and Peanut

Although sunflower lecithin currently is not used to any great extent, as U.S. sunflower oil production is increased the availability of lecithin from this oil may be a possibility (43). Due to its high phosphatidylcholine content, sunflower lecithin can be utilized in foods and feedstuffs. Its use in the manufacture of foods and cosmetics can be increased by refining and fractionation and/or modifications (44).

The percentage of phospholipids in the oil ranges from 0.02 to 1.5%, with an average of ca. 0.75%. The composition of phospholipids varies greatly with phosphatidylcholine ranging from 12.7 to 64.2%. Use of a powdered lecithin obtained from commercial operations is hampered by its tendency to undergo oxidative deterioration (44).

Hilditch and Zaky (45) found that phosphatides in peanuts are considerably less unsaturated than those in soybean and cottonseed. Rewald (46) has shown 35.7% PC and 64.3% PE content in peanut phosphatides.

1.7 Other Plants and Microorganisms

Table 10.13 shows the phospholipid composition of selected potential plant sources (30). Parson and Price (47) have published compositional data on the phospholipids of barley grain based on thin-layer chromatography.

Table 10.13 Phospholipids (%) of selected plant sources (30)

	Phospholipids		
Source	Phosphatidylcholine	Phosphatidylethanolamine	Phosphatidylinositol
Rapeseed	16:2; 20.0–24.6	15.0–17.5; 22.1	7.6–8.0; 14.7–18.0
Sunflower seed	12.7–26.8; 42.2–64.2	9.9–29.4; 46.6	3.7–21.4; 24.0–36.6
Peanut seed	49.0	16.0	22.0
Cucurbit seed	55.8–74.9	10.5–18.7	13.7–17.2
Rice bran	20.4–23.1	17.8–20.2	5.8–6.6
Barley seed	44.3–44.4	7.6–8.8	1.1–1.3
Olive fruit	47.3–58.9	5.3–8.0	18.0–23.9
Avocado fruit	37.0–44.9	12.0–19.5	12.1–18.0
Palash seed	44.6	14.8	27.0
Jangli badam seed	30.0	23.0	40.6
Papaya seed	28.1	18.7	34.0
Coriander seed	44.0	29.3	23.1
Carrot seed	29.1	35.4	23.1

Other than in animal tissues, egg, and oilseeds, quantitative data on the phospholipids in plants are meager because of the difficulty involved in their isolation. The phospholipid content of wheat gluten is quite high (8.5 to 11.1%). The phospholipids in wheat are about 80% PC and 20% PE (4). Rye, barley, and other grains, vegetables, and fruits all contain small amounts of phosphatides. Microorganisms, especially those which are acid-fast, and lower plants also contain large amounts of lipids, including phospholipids; these entities are of interest for clinical research (4). A survey of microbiological sources of phospholipids has been published by Ratledge (48).

NOMENCLATURE, CLASSIFICATION, STRUCTURE AND COMPOSITION, AND CHEMICAL/PHYSICAL PROPERTIES

2.1 Definition

According to Wittcoff (4), three distinct polymeric alcohols provide the basic constituents for the various phospholipids. The first of these is glycerol, and the phosphatides containing it are referred to as glycerophosphatides. Included herein, in addition to phosphatidylcholine (PC), phosphatidylethanolamine (PE), and phosphatidylserine (PS), are the acetalphosphatides or plasmalogens (in body fluids, muscles, and egg), the lysophosphatides, and the phosphatidic acids. The second polyhydric alcohol is the amino–dihydroxy compound sphingosine, which is the basis for not only sphingomyelin (in the brain and spinal cord), but also for other glycolipids. All of these compounds based on sphingosine are also referred to as sphingolipids. The third polyhydric alcohol is inositol, which forms the basis of many phosphatides, e.g., phosphatidylinositol (PI), in both plant and animal tissues representing known and yet unknown chemical structures.

Compounds have been reported as having to contain both glycerol and inositol, (e.g., soybean phosphatides and many poorly defined compounds in both plant and animal sources). Phospholipids also form complexes with proteins (e.g., vitellin in egg yolk, animal and plant tissues, blood serum, and milk), carbohydrates, glycosides, alkaloids, minerals, enzymes, cholesterol, and other substances. Lysophosphatides represent a special class of compounds resulting from the chemical or enzymatic hydrolyses of phospholipids. The role of phospholipases in normal and pathological conditions as well as in cell metabolism is of great biological significance (4).

For the elucidation, synthesis, chemical properties, physical chemistry, composition, and analytical determination of the various individual phosphatide structures in animal and plant sources, the reader is referred to Wittcoff (4). Schneider (14) discusses the nomenclature used for phospholipids in more detail and provides compositional data on commercial lecithins (Table 10.14).

Because of the commercial significance of soybean lecithin, this Chapter will focus primarily on the structure, composition, analytical determination, properties, and applications of this product.

The U.S. Food Chemical Codex (49) defines lecithin as follows:

> Food-grade lecithin is obtained from soybeans and other plant sources. It is a complex mixture of acetone-insoluble phosphatides that consist chiefly of phosphatidylcholine, phosphatidylethanolamine, and phosphatidylinositol, combined with various amounts of other substances such as triglycerides, fatty acids and carbohydrates. In its oil-free form, the preponderance of triglycerides and fatty acids are removed and the product contains 90% or more of phosphatides, representing all or certain fractions of the total phosphatide complex. The consistency of both natural grades and refined grades of lecithin may vary from plastic to fluid, depending on the free fatty acid and oil content, and upon the presence or absence of other diluents. . .

2.2 Classification of Commercial Soybean Lecithin Products

The simplest method for modifying natural (crude) lecithin is by the addition of a nonreactive substance. Plastic lecithins are converted to fluid forms by

Table 10.14 Composition of commercial lecithins (%) (on oil-free basis) (14)

Lecithin	Soy	Corn	Sun-flower	Rapeseed	Egg	Bovine Brain
Phosphatidylcholine	21	31	14	37	69	18
Phosphatidyl-ethanolamine	22	3	24	29	24	36
Phosphatidylinositol	19	16	13	14	—	2
Phosphatidic acid	10	9	7	—	—	2
Phosphatidylserine	1	1	—	—	3	18
Sphingomyelin	—	—	—	—	1	15
Glycolipids	12	30	—	20	—	—

adding 2 to 5% fatty acids or carriers such as soybean oil. If the additives react with the lecithin to alter the chemical structure of one or more of the phosphatide components, the resulting product is referred to as a chemically modified lecithin. Modification can also be achieved by subjecting lecithin to partial controlled enzymatic hydrolysis. Finally, refined lecithin products can be obtained by fractionating the various phospholipid components.

A method for classifying lecithin to include modified and refined forms has been proposed by Cowell and co-workers (50). This classification distinguishes between natural (crude) lecithins and those modified by either custom blending or chemical/enzymatic treatment, e.g., hydroxylation, hydrogenation, or acetylation or refining by acetone or alcohol fractionation. These latter products reflect the state of the art regarding the availability of the various lecithin products on the market and have enhanced properties for specific uses. A listing of soybean lecithin classification follows (51).

- I. Crude commercial lecithin
  - A. Plastic
    - 1. Unbleached
    - 2. Single bleached
    - 3. Double bleached
  - B. Fluid
    - 1. Unbleached
    - 2. Single bleached
    - 3. Double bleached
- II. Compounded
- III. Chemically modified
- IV. Refined
  - A. Deoiled
    - 1. As is
    - 2. Custom blended
  - B. Fractionated
    - 1. Alcohol soluble
      - a. As is
      - b. Custom blended
    - 2. Alcohol insoluble
      - a. As is
      - b. Custom blended
  - C. Purified phosphatides

Natural (Crude) Lecithins. Specifications as defined by the *National Soybean Processors Association* (1986–1987) for natural (crude) lecithins is presented in Table 10.15 (52); similar specifications have also been published by the *Food Chemical Codex* (1981) (49).

Phospholipid content is specified in terms of acetone-insolubles (AI), product clarity, and purity in terms of hexane-insolubles (HI). The lecithins are

Table 10.15 Soybean lecithin specifications (52)

Analysis	Grade					
	Fluid Unbleached Lecithin	Fluid Bleached Lecithin	Fluid Double-Bleached Lecithin	Plastic Natural Lecithin	Plastic Bleached Lecithin	Plastic Double-Bleached Lecithin
Acetone insoluble, min.	62%	62%	62%	65%	65%	65%
Moisture, max.[a]	1%	1%	1%	1%	1%	1%
Hexane insoluble, max.	0.3%	0.3%	0.3%	0.3%	0.3%	0.3%
Acid value, max.	32	32	32	30	30	30
Color, Gardner, max.[b]	18	14	12	18	14	12
Viscosity, centipoise, at 25°C, max.[c]	15,000	15,000	15,000	—	—	—
Penetration, max.[d]	—	—	—	22 mm	22 mm	22 mm

[a] By Karl-Fischer Titration (AOCS Method Tb 2-64).
[b] Undiluted basis.
[c] By any appropriate conventional viscosimeter, or by AOCS Bubble Time Method Tq 1A-64, assuming density to be unity. Fluid lecithin having a viscosity less than 7,500 centipoises may be considered a premium grade.
[d] Using Precision cone 73525, Penetrometer 73510; sample conditioned 24 hours at 25°C.

classified as of plastic or fluid consistency and are further subdivided on the basis of manufacturing procedure as natural color, bleached, or double bleached. Acidity of phospholipids plus acidity of the carrier (i.e., the oil and fatty acids) is given by the acid value (AV), i.e., mg of potassium hydroxide required to neutralize the acids in 1 g of the lecithin sample. Crude lecithin can be filtered for utmost purity and clarity. Filtration removes hexane-insoluble (HI) matter. Such products are in demand for encapsulated nutritional supplements, for high purity pharmaceutical grades, and for high technology industrial uses requiring a high level of purity. Solvent-laden oil is the most commonly filtered, due to its low viscosity (8).

Compounded Lecithins. Compounded lecithins are blended products. Lecithin combined with selected additives can exhibit modified properties and functionalities. Lecithin may have synergistic action with some additives or, simply, compounded with ingredients for making it more compatible in a particular system. Common additives include special oils, polysorbates, monoglycerides and modified monoglycerides, lanolin derivatives, solvents, plasticizers, or other surfactants. These are added either to the wet gums prior to drying or are blended with dry fluid lecithin at elevated temperatures (8,53).

Modified Lecithins. Lecithins may be modified chemically, e.g., hydrogenation, hydroxylation, acetylation, and by enzymatic hydrolysis to produce products with improved heat resistance and emulsifying properties and increased dispersability in aqueous systems (8,53,54). One of the more important products is hydroxylated lecithin which is easily and quickly dispersed in water and, in many instances, has fat-emulsifying properties superior to the natural product. Hydroxylated lecithin is approved for food applications under Title 21 of the Code of Federal Regulations 172.814 (1977) (55).

Fractionated and Oil-free Lecithins. When crude lecithin is further refined by various fractionation methods to selectively separate its components, acetone and ethanol are the most common solvents used. Much of the recent patent activity on modified lecithins involves these fractionation procedures. Fractionation of crude lecithin yielding phosphatidylcholine of greater than 90% purity have been reported.

A commercial, nearly oil-free lecithin is prepared by acetone extraction of natural lecithin, which removes all but 3–4% of the oil and free fatty acids. Then an optional alcohol fractionation step may separate the oil-free lecithin into an alcohol-soluble lecithin enriched in phosphatidylcholine and an alcohol-insoluble fraction enriched in phosphatidylinositol. The choline fraction is an excellent emulsifier for oil-in-water emulsions and the inositol fraction for water-in-oil emulsions.

2.3 Structure of Phosphatides in Commercial Lecithins

Chemical structures for the most commonly occurring phospholipids in commercial soybean lecithin are shown in Figure 10.1 (8). Phosphatidylcholine (PC) and phosphatidylethanolamine (PE) are cationic and anionic at the same time; that is, they are zwitterions and thus they can have some buffering action for both bases and acids. Phosphatidylinositol (PI), however, is a relatively strong acid and, therefore, is anionic. The classes of compounds in commercial lecithin are as follows (29):

Phospholipids	*Glycolipids*
Anionic	Steryl glucosides
Zwitterionic	Esterified steryl glucosides
	Galactosyl glycerides

The reader is referred to Horrocks (56) for more specific discussion on the nomenclature and structure of phosphatides.

2.4 Composition

Specification ranges, chemical and fatty acid compositions for commercial natural lecithins, along with approximate compositional data for commercially refined lecithin fractions are given in Tables 10.16–10.19 (9,45,57–65), respectively.

Soybean oil contains 1.5–3.0% phospholipids (66). Crude lecithin has a soybean oil content of about 30%. Phosphatidylcholine (PC) is present at a

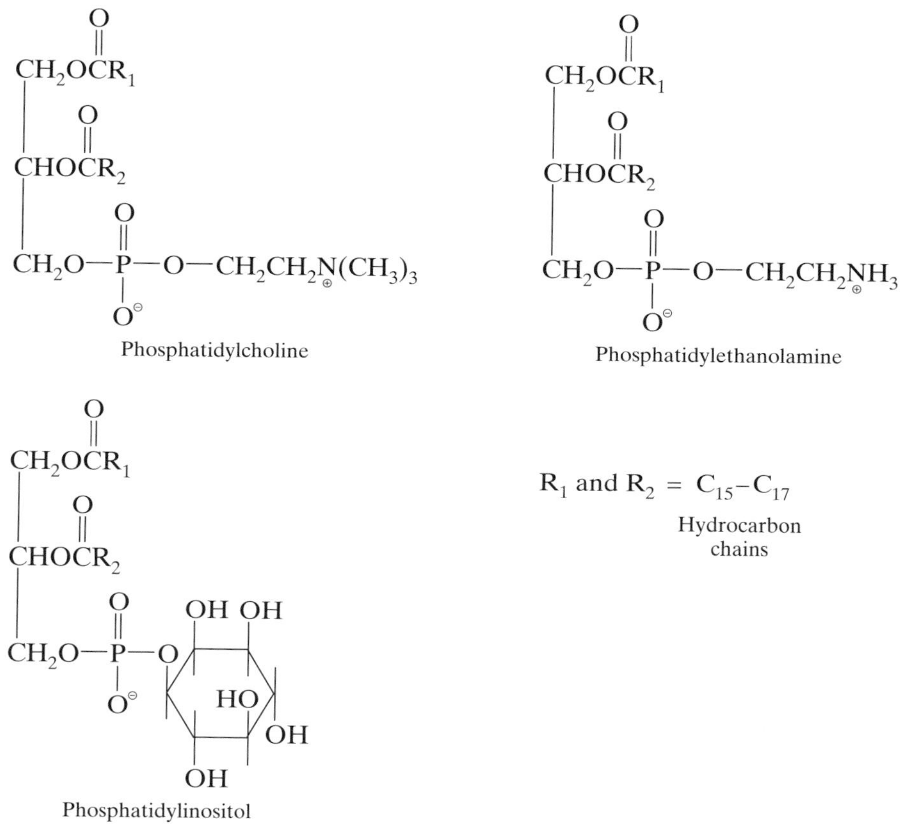

$$\begin{array}{l} CH_2OC(=O)R_1 \\ | \\ CHOC(=O)R_2 \\ | \\ CH_2O-P(=O)(O^{\ominus})-O-CH_2CH_2\overset{\oplus}{N}(CH_3)_3 \end{array}$$

Phosphatidylcholine

$$\begin{array}{l} CH_2OC(=O)R_1 \\ | \\ CHOC(=O)R_2 \\ | \\ CH_2O-P(=O)(O^{\ominus})-O-CH_2CH_2\overset{\oplus}{N}H_3 \end{array}$$

Phosphatidylethanolamine

$$\begin{array}{l} CH_2OC(=O)R_1 \\ | \\ CHOC(=O)R_2 \\ | \\ CH_2O-P(=O)(O^{\ominus})-O-C_6H_6(OH)_5 \end{array}$$

OH OH, HO, OH, OH (inositol ring substituents)

Phosphatidylinositol

R_1 and R_2 = $C_{15}-C_{17}$
Hydrocarbon chains

Figure 10.1 Three principal components of soybean lecithin (8).

Table 10.16 Specifications for commercial soybean lecithin (57,58)

	Grade					
Analysis	Fluid Natural Color Lecithin	Fluid Bleached Lecithin	Fluid Double-Bleached Lecithin	Plastic Natural Color Lecithin	Plastic Bleached Lecithin	Plastic Double-Bleached Lecithin
Acetone insoluble, % min.	62	62	62	65	65	65
Moisture, % max.[a]	1	1	1	1	1	1
Benzene insolubles, % max.	0.3	0.3	0.3	0.3	0.3	0.3
Acid value, max.	32	32	32	30	30	30
Color, Gardner, max.[b]	10	7	4	10	7	4
Viscosity, poises, at 25°C, max.	150	150	150	—	—	—
Penetration, max., in mm[c]	—	—	—	22	22	22

[a] By toluene distillation for 2 hr or less.
[b] As a 5% solution in colorless mineral oil.
[c] By specified cone penetrometer test.

Table 10.17 Approximate chemical composition of natural commercial soybean lecithin (59)

Fraction	%
Soybean oil	35
Phosphatidylcholine	16
Phosphatidylethanolamine	14
Phosphatidylinositol	10
Phytoglycolipids and other minor phosphatides and constituents	17
Carbohydrates	7
Moisture	1

level of ca. 16%., phosphatidylethanolamine (PE) ca. 14%, and phosphatidylinositol (PI) ca. 12% (8). As can be seen in Table 10.18 (9), the fatty acid compositions of soybean phospholipids are rich in polyunsaturated fatty acids. Miscellaneous low level constituents include water, phosphatidic acid, pigments, galactosyl glycerides, various glycolipids, phosphatidylserine, carbohydrates, sterols, and tocopherols. Phosphorus content of crude soybean oil extracted from flours can vary depending on extraction temperature and flour moisture (67).

2.5 Chemical/Physical Properties

Information on physical, chemical, and nutritional properties of soy lecithin products and their food and industrial uses can be obtained in references such as Sartoretto (68), Stanley (1), Szuhaj (2) Szuhaj and List (5) Wittcoff (4), and Wolf and Sessa (69).

In practice, commercial lecithin products are not marketed by phosphatide content, but under a set of unique chemical and physical properties. These properties, as indicated by product specifications, must be understood since they are used to characterize specific lecithin types.

Commercial soybean lecithin, being a mixture of three phospholipids, performs as a wetting and emulsifying agent. Stanley (1) states that in heterogeneous systems such as oil and water, the phospholipid molecules arrange themselves in monomolecular layers with the fatty acid facing the oil surface and the phosphoric acid portion facing the water surface. The arrangement lowers the interfacial tension of the oil-water boundaries with resultant benefits such as rapid wetting, lowering of viscosity, and better and more stable emulsions or dispersions.

Soybean lecithin is soluble in aliphatic and aromatic hydrocarbon solvents, partially soluble in ethyl alcohol (principally the inositol fraction), practially insoluble in acetone (less than 0.003% weight/volume at 5°C) and in water (70). When mixed with water, soybean lecithin hydrates to a thick emulsion

Table 10.18 Composition fatty acids of soybean lecithin (%) (9)

Reference	14:0	14:1	16:0	16:1	18:0	18:1	18:2	18:3	20:0	20–22 Unsaturated
Hilditch and Zaky (45)	—	—	11.7	8.6	4.0	9.8	55.0	4.0	1.4	5.5
Rzhekhin and co-workers[a] (60)	—	—	18.1	—	3.7	22.4	40.0	5.0	2.3	6.2
Vijayalakshmi and Rao[b] (61)	—	—	42.7	7.0	11.7	17.0	20.0	1.6	—	—
Daga (62)	1.9	Trace	26.7	—	9.3	25.1	37.0	—	—	—
Daga (63)	0.3	1.2	25.5	—	10.3	39.4	17.1	6.2	—	—
Rydhag and Wilton[c] (64)	—	—	21.5	—	4.3	7.2	60.9	6.1	—	—
Rydhag and Wilton[d] (64)	—	—	18.9	—	4.1	6.8	60.8	9.2	—	—

[a] Also 2.3% unidentified.
[b] $CHCl_3/CH_3OH$ extraction.
[c] Acetone-precipitated.
[d] Granulated.

Table 10.19 Approximate composition of commercially refined lecithin fractions (65)

Fraction	Oil-Free Lecithin	Alcohol-Soluble Lecithin	Alcohol-Insoluble Lecithin
Phosphatidylcholine	29%	60%	4%
Cephalin	29	30	29
Inositol and other phosphatides, including glycolipids	32	2	55
Soybean oil	3	4	4
Other constituents[a]	7	4	8
Emulsion type favored	Either oil-in-water or water-in-oil	Oil-in-water	Water-in-oil

[a] Includes sucrose, raffinose, stachyose, and about 1% moisture.

that can be thinned with water to almost any desired dilution. Acetone does dissolve readily in lecithin and will form a thin, uniform imbibition as long as the quantity of acetone is insufficient to precipitate the phosphatides. Lecithin is soluble in mineral oils and fatty acids, practically insoluble in cold vegetable and animal oils, but will disperse or melt in hot oils.

Soybean lecithin has a brown to light yellow color, depending upon the conditions used in its manufacture and the degree of bleaching.

Identification and Characterization of Phospholipids. For the topic of various techniques and methods used in the identification and characterization of phospholipids in general, the reader is referred to Kramer and co-workers (71). This reference includes: sample preparation and lipid extraction, separation of lipid classes, general approach to identify lipids, and characterization of cardiac and methanogenic lipids.

There are ongoing efforts to further identify the phospholipid components in the soybean. Hanras and Perrin (72) have reported on a preparative procedure to isolate gram quantities of phospholipid classes from soybean lecithin. Simultaneous determination of the main molecular species of soybean phosphatidylcholine or phosphatidylethanolamine and their corresponding hydroperoxides have also been obtained by lipoxygenase treatment. A single C-18 reverse-phase column was employed to separate the main molecular species and their hyperperoxides by high-performance liquid chromatography (73).

Boatright and Snyder (74) have estimated the phospholipid composition of the protein bodies in mature soybeans. Two-dimensional thin-layer chromatography revealed the ca. 38% of the phospholipid phosphorus was in the form of phosphatidylethanolamine. This phospholipid has been shown to be the result of transphosphatidylation of phosphatidylcholine with ethanol indicating extensive activity of phospholipase D within the membrane of the mature protein bodies.

Fractionation and Purification of Lecithin. Because of space limitation, it is not possible to discuss fractionation and purification processes for all vegetable and animal lecithins in this Chapter. The reader is referred to Schneider (14) who described the fractionation and purification of various vegetable lecithins and those from egg in considerable detail. Small-scale fractionation processes may include: separation of neutral oil and polar lipids (deoiling) including the use of acetone; the adsorption of a hexane solution of lecithin on a silica column; separating neutral and polar lipids from a hexane solution with the aid of membranes; treatment of lipid mixtures with supercritical gases or gas mixtures, e.g., carbon dioxide or propane–carbon dioxide; fractionation of neutral oil containing lecithins by solvent treament, e.g., aqueous methanol, ethanol, and propanol; fractionation of deoiled lecithins by solvent treatment, e.g., ethanol; solvent treatment after chemical modification, e.g., acylation prior to acetone deoiling; precipitation methods, e.g., salt; ultrafiltration methods; and many chromatographic processes, mainly for polar lipid separation but also separation focused on the degree of unsaturation.

The commercial manufacture of fractionated soybean lecithins will be covered later in this Chapter.

Synthesis and Modification of Phospholipids. Recently there has been an increase in the number of patents and publications on synthesis and modification of phospholipids. An excellent review of these developments is presented by Ghyczy (75). According to the review, depending on the starting material used, there are two ways to synthesize phospholipids. In the partial synthesis, phospholipids are isolated from natural sources and the individual constituents, fatty acids and head groups are exchanged to obtain a certain phospholipid. In the total synthesis, phopholipids are produced from fully synthesized available basic molecules which were not obtained from phospholipids. Both methods are of importance today since each manufacturing process has certain advantages with regard to definite products and fields of application.

The partial synthesis may involve several synthetic steps, depending on the basic phosphatide used, the enzyme, the final product desired, and the type and position of the phospholipid constituents to be exchanged. For example, the partial synthesis may avail of the reacylation 3-*sn*-glycerophosphorylcholine (GPC). By such a deacylation step, GPC can be obtained from phosphatidylcholine (PC) in soybean lecithin (75).

The partial synthesis may avail of PC obtained from GPC as starting material to synthesize PC with mixed fatty acids. Other types of phospholipids yield compounds after deacylation which have certain functional groups, e.g., amino groups from phosphatidylethanolamine (PE). Phosphatidylinositol (PI) can be manufactured by using the enzyme PL-D, using phospholipids from soy lecithin (75).

The most suitable starting material for the total synthesis of phospholipids are optically active derivatives from glycerol, called "chiral" C_3 building blocks. In addition to proper configuration, an early differentiation of the hydroxyl groups is also necessary to shorten the process of synthesis (75).

Recently, there have been reports on two new fields of application for modified phospholipids: change in the efficiency of biological active compounds by covalent binding to a phospholipid, and labelling of phosphatides to be able to monitor their function in biological processes.

Transesterifications have also been investigated to determine a means for preparing polyunsaturated phospholipids simply from soy phospholipids (76).

For specific examples of adducts formed by these new approaches, the reader is referred to Ghyczy (75).

Specifications for Soybean Lecithin. The following methods are routinely used for determining whether the specifications for given products are met:

Acetone-Insolubles (AI). The amount of acetone-insoluble matter (%AI) is approximately representative of these phosphatides in crude lecithin. In crude lecithin AI is also synonymous with activity, i.e., functional or nutritional. Upgraded products, however, are usually formulated to provide specific activities, in which case additional criteria may be used to define functionality in terms of viscosity control, emulsification, oxidative stability, and end-use compatability. The amount of acetone-insoluble matter is determined by the AOCS Official Method Ja-4-46 (77).

Acid Value (AV). The acid value (AV) is the titrable acidity in lecithin in mg KOH/g of sample. A product's AV is representative of the acidity contributed by phospholipids as well as from free fatty acids. The AV is usually not indicative of pH as the chemical nature of the phospholipid impart buffering qualities to most systems. Lecithins typyically exhibit a neutral pH value in aqueous media. An AV above 36 may indicate degradation of the lecithin due to improper processing or substandard quality soybeans. AV should not be confused with free fatty acid content, pH, or mineral acids. The correct method to assay for free fatty acids is to titrate only the acetone-soluble portion, whereby any contribution from the phosphatides in the AI portion is eliminated. AV is determined by the AOCS Official Method Ja 6-55 (77).

Moisture. The water content of lecithin products is usually less than 1.0%. As a consequence of lecithin's essentially moisture-free state, lecithin products have very low water activity and do not adversely contribute to the microbiological profile of most food systems. Most lecithin products are preserved well in storage. Higher moisture levels usually indicate a greater potential for spoilage or chemical degradation. Moisture is determined by the Karl Fisher aquametric titration (AOCS Official Method Tb 2-64 or Food Chemical Codex (FCC) III Official Method (1981)) (49,77). A less accurate moisture level can also be determined by azeotropic toluene distillation (AOCS Official Method Ja 2-46) (77). One cannot determine lecithin moisture by vacuum oven methods. These methods are known to degrade lecithin products and yield false moisture levels.

Hexane-Insolubles (HI). The level of hexane-insoluble (HI) matter is one measure of the purity of lecithin products. HI matter usually consists of residual fiber, but often particulate contaminants are introduced during processing (e.g. filter aids). The level of HI matter in crude lecithin should never exceed 0.3%, and rarely exceeds 0.1%. HI matter in lecithin is detrimental to clarity and usability in specific applications. This is an official Food Chemical Codex (FCC) (1981) (49) method.

Color. The color of lecithin is fundamentally an aesthetic quality standard. Many formulators instinctively dislike dark-colored ingredients. Historically, lecithins have been color graded as unbleached, single-bleached, and double-bleached. By convention, the amber color tones of lecithin are measured on the Gardner color scale. The color range covered by various lecithin products is generally in the range of Gardner 9–17 in an undiluted form. (Proposed AOCS method is based on AOCS Official Method Td-la-64) (77).

Other Physical/Chemical Properties and Quality Criteria

Consistency. Lecithins are available in both fluid and plastic (solid) forms. Fluid lecithins generally follow Newtonian flow characteristics. The viscosity profile of lecithins is a complex function of acetone-insoluble content, moisture, mineral content, acid value, and the combined effects of assorted additives such as vegetable oils and surfactants. Generally, higher AI and/or moisture content yields higher viscosity, whereas an increased AV often decreases viscosity. Certain divalent minerals, such as calcium and others, can also adjust the viscosity level.

Clarity. In some soy processing plants, high levels of hexane-insoluble matter (HI) may partition with the lecithin gums on separation from the oil. This lipid-insoluble material can cause haziness in fluid lecithins. With modern miscella and oil filtration techniques, lecithins with very low, or even no HI contents can be produced. Consequently, modern lecithins are quite clear. Additionally, moisture can also contribute to lack of clarity. Generally over 1% moisture levels can cause haziness. Haziness, besides being an aesthetic problem, if caused by HI material, can result in sediments over time; solid particles may appear on the bottom of an otherwise clear liquid product containing lecithin.

For a more detailed review of industrial methods of analysis the reader is referred to Lantz (57). A review of traditional and novel approaches to the analysis of plant phospholipids has been prepared by Marmer (78). Ackman (79) has reviewed the early developments and practical applications of GLC analysis.

Chemistry and Reactivity of Phospholipids. The chemistry of the phosphatides is generally that of their ester linkages, unsaturated fatty acids, and other reactive groups. Most of the applicable reactions of organic chemistry have been employed in their study (80). Baer and Kates (81), Brockerhoff (82), Hanahan (83), Pryde (84), Scholfield (9,80), Strickland (85), Verheij (86), and Wittcoff (4) provide major reviews of phospholipid chemistry and reactivity under various conditions. The latter covers hydrolysis, hydrogenolysis, acetolysis, hydroxylation, thermal decomposition, hydrogenation, autooxidation, browning reaction, and other reactions (e.g., bromination and complexing with various substances).

Lecithin Interaction with Other Food Ingredients. Food systems are usually heterogeneous mixes of components, in which the interaction of ingredient classes (e.g., proteins, starches, fats, surfactants) can be important to finished product quality, shelf-life, and nutritional value.

The most common modifications of lecithin and the intended physical/functional alterations are shown in Table 10.20 (29). The range of physical/functional properties available in commercial lecithins are listed in Table 10.21 (29). These changes in lecithin allow for the basic lecithin obtained from soybean oil to be converted to various emulsifier products having a wide variety of food and industrial applications. Reviews describing chemical reactions for phospholipid modification intended to obtain specific functionalities include: Eichberg (87), Hawthorn and Kemp (88), Kuksis (89), Pryde (84), Snyder (90), Strickland (85), and Van Deenen and DeHaas (91).

Model studies have given some insight into the mechanism of protein/phospholipid interactions. The interactions of soy globulins and phosphatidylcholine was reported by Kanamoto and co-workers (92). Results of these studies suggested that high energy input is necessary to the formation of stable phospholipid/protein complexes. Interacting PC vesicles with 7S and 11S soy globulins, Beckwith (93) demonstrated the extent of protein/phospholipid interaction was dependent upon both the ratio of the reactants and the specific globulin.

Table 10.20 Alteration of lecithin form/function (29)

Action	Technique	Utility
Decrease viscosity	Add special diluents	Sprayable Easier handling
Increase hydrophilic properties	Solvent fractionation Chemical modification Enzyme modification Compounding	o/w emulsifiers Wetting agents
Reduce color	Process controls Oxidative bleaching	Light-colored foods
Convert to powder form	De-oil Mix with carrier	Dry blendable

Table 10.21 Physical/functional properties of commercial lecithin (29)

Property	Commercial Range of Values
Viscosity	100 centipoise to plastic
Color	Light honey to dark amber
HLB	2–12
pH	5–8
Flavor/odor	Slightly nutty to moderate bitter, pungent
Solubility	
Nonpolar	Soluble
Lower alcohols	Partially soluble to soluble
Glycerine	Partially soluble to soluble
Water	Insoluble to dispersible

Chen and Soucie (94) showed that treatment of soy protein isolate with hydroxylated lecithin lowered the isoelectric point, increased electrophoretic mobility, and significantly increased protein dispersibility and suspension stability. Nielsen (95) investigated the interaction of peroxidized phospholipids with several proteins under N_2. His findings demonstrated a covalent attachment of phospholipids to proteins whose molecular size is increased.

While there is little doubt that proteins interact with the phospholipid components of commercial lecithins, it is the exact nature of these interactions that is important. Thus, while commercial lecithins might be expected to interact with wheat gluten, there is no data to indicate that the effect is as dramatic or functionally meaningful as gluten interaction with known dough conditioners like the steroyl-lactylates (29). The interaction of lecithin with starch can also have great functional significance in food systems. Not surprisingly, the structure of the lecithins involved determines their reactivity and hence functionality.

The absorption isotherms of several emulsifiers to fat and sugar crystals dispersed in oils have been examined (96). Unsaturated monoglycerides and phospholipids cause a decrease in adhesion for all concentrations examined. Phospholipids reduce the adhesion between sugar crystals, resulting in much denser sediments.

The browning (Maillard) reaction is the main process expected for changes in lecithin color and flavor as a result of protein–carbohydrate interactions (97). A Maillard browning reaction seems also to occur when lecithin is heated with water on paraffin at 180°C, indicating that coloration in this case is due more to aldol-condensation reactions than to Maillard reaction (98–100).

The influence of soybean lecithins on the spontaneous solidification of different model fats has been studied in the presence and absence of water (101). Lecithins added to dry fat do not affect crystallization but, in the presence of water, clearly delay it. Not only does the crystallization begin late but the surfusion phenomena are also exaggerated.

Lecithins as Antioxidants. The literature is replete with references to the antioxidant properties of lecithin: Frankel (102,103), Lundberg (104), Porter and co-workers (105), and Pryde (84). Lecithin has usually been considered an antioxidant synergist, improving or prolonging the action of primary antioxidants like tocopherols. Early literature references, however, do not always make this distinction between antioxidant and synergist. Evidence is available for the ability of the phospholipids of commercial lecithins to assist in the oxidative stabilization of fats (84). The antioxidant properties of lecithin have also been attributed to the metal-scavenging ability of the phospholipids (106,107).

The ability of phospholipids to enhance antioxidant properties of primary antioxidants vary with the type of oil being stabilized (108,109). The various phospholipids also have different synergistic tendencies. Phosphatidylethanolamine (PE) appears to have the best antioxidant activity with mixed tocopherols (108) and synthetic antioxidants (109). Dziedzic (110) has shown that the PE is consumed along with the primary antioxidant (propyl gallate in the study) during the induction period. The interpretation of this observation is that the phospholipid or its breakdown compounds regenerate the primary antioxidant by donating hydrogen radicals or protons.

Porter and co-workers (105) have studied the free radical oxidation of polyunsaturated lecithins. Two unsymmetric polyunsaturated lecithins were allowed to air oxidize and the primary products of autoxidation were isolated and characterized.

McLean and co-workers (111) have examined the role of lipid structure in the activation of phospholipase A_2 by peroxidized phospholipids. Results showed that the increase in rate of hydrolysis of peroxidized phospholipid substrates catalyzed by phospholipase A_2 is due largely to a preference for peroxidized phospholipid molecules as substrates and that peroxidation of host lipid does not significantly increase the rate of hydrolysis of nonoxidized lipids.

The reader is referred to Pryde (84) for a more thorough discussion on the kinetics of autoxidation of phospholipids; their forming metal ion, iodine, and other complexes; halogen addition; and their behavior during hydration, hydrogenation (with heterogeneous and homogeneous catalysts), hydrolysis and alcoholysis, hydroxylation, oxidation, radical and other reactions.

MANUFACTURE, FRACTIONATION, AND PURIFICATION OF LECITHINS

3.1 Manufacture of Crude Lecithin

Commercial soybean lecithin is obtained in the traditional manner by hexane extraction of the crude oil from the flake and then water degumming of the oil to yield a viscous fluid product.

The degumming of soybean oil is not an industry-wide practice. Brian (112) has estimated that only about one-third of the soybean oil produced in the

United States needs to be degummed to meet the U.S. needs for soybean lecithin production.

U.S. refiners do not find it practical to caustic refine crude soybean oil and degummed soybean oil as separate entities but prefer to blend degummed oil with the crude. The limited availability of degummed oil is a contributing factor for this practice. In Europe, however, solvent-extracted oils are usually degummed ahead of the caustic refining step. European refiners, of necessity, must process both a greater variety of oils and often oils of lower quality than do U.S. refiners (113).

Removal of all phosphatides and gums is a necessary part of the steam-refining process. However, this process has not yet been developed to the point where it can produce a refined soybean oil that is meeting U.S. competitive requirements. Studies have been carried out to fulfill this objective (114,115).

Based on a series of samples obtained from commercial processors, degumming of the oil from undamaged soybeans removed 79 to 98% of the phosphorus. Phosphorus content of the oil was lowered from 500–900 ppm in the crude to 12 to 170 ppm in the degummed oil (114). Water-degummed oil from damaged beans may have an abnormally high phosphorus level and degumming these oils poses a difficult problem.

Oil Degumming. All stages of oilseed processing affect the quality of soy lecithin. Prior to degumming, the seed quality, cleaning of the beans, extraction of the oil, handling the miscella and crude oil all have an important role in making a good quality lecithin.

Hexane extraction removes about 50% of phosphatides from the meal (116). The presence of fines in the miscella is undesirable for making lecithin and should be kept to a minimum (0.2% or less) and the crude miscella coming from the extractor should be filtered. Several miscella filtration methods are available. The crude oil also can be filtered to produce lecithin.

The method of hexane removal from the miscella is also very important in that dark colors in lecithin are believed to be due to an aldehyde–amine reaction largely formed by heating the oil during the solvent stripping operation. Most U.S. processors employ a dual-stage evaporator followed by a low pressure stripping system (117).

Hydration causes most of the phosphatides and gums present in a crude oil to become insoluble in the oil. Such hydration can come about from water added to the oil in the degumming step or from moisture picked up from the air by the oil during storage (118).

Current Commercial Practices. If the oil is recovered by solvent (hexane) extraction, some mills allow a portion of the steam blown through the oil for removal of last traces of solvent to condense and thus hydrate the gums. Sometimes operators find it difficult to closely control the moisture addition with direct steam and prefer to add hot water in controlled amounts.

Crude oil from which the lecithin is to be recovered is usually filtered prior to degumming to remove residual meal fines and seed fragments. Although more difficult, dry lecithin can also be filtered. Careful filtration results in a highly clarified lecithin with little or no residual hexane insoluble matter (6).

Brian (112) describes two methods of miscella filtration, one with and one without filter aids, but both result in a lecithin that still remains somewhat cloudy. Highly clarified lecithin products can be obtained only by filtering the crude oil, usually with the aid of plate and frame filters, wherein the dry oil is heated to 82°C and 0.1% filter aid is added (6).

Two principal degumming methods are employed: batch and continuous. In the United States, batch degumming is used while in European plants continuous degumming is more common (117). The batch degumming process is shown in Figure 10.2 (117). A flowsheet for the continuous degumming of soybean oil and production of the soybean lecithin is shown in Figure 10.3 (112). Recommended conditions for degumming of crude vegetable oils can vary greatly as shown in Table 10.22 (112,113,119–125).

In a batch degumming system, which is the principal method utilized in the United States, crude soybean oil is typically heated to about 70°C in a large tank fitted with a side-entering horizontal agitator. Water is added (about 2% by volume) and the hydrated oil is agitated for up to one hour. The hydration of soybean phosphatides proceeds rapidly, and for all practical purposes, 15 minutes is adequate for batch systems. The hydrated "gums" or "lecithin emulsion" are then removed by continuous centrifugation. This step is then followed by drying in a batch or film drier. The gums are usually dried to a moisture content of less than 1%, typically 0.3–0.75% (6,3).

In continuous systems, preheated crude oil (80°C) and water are metered into an in-dwell pipeline agitator and held only for a short period. In both systems the oil is then pumped to a centrifuge for separation of the lecithin sludge from the oil (6,68,119,126). Water with low concentration of calcium and magnesium is preferred (117).

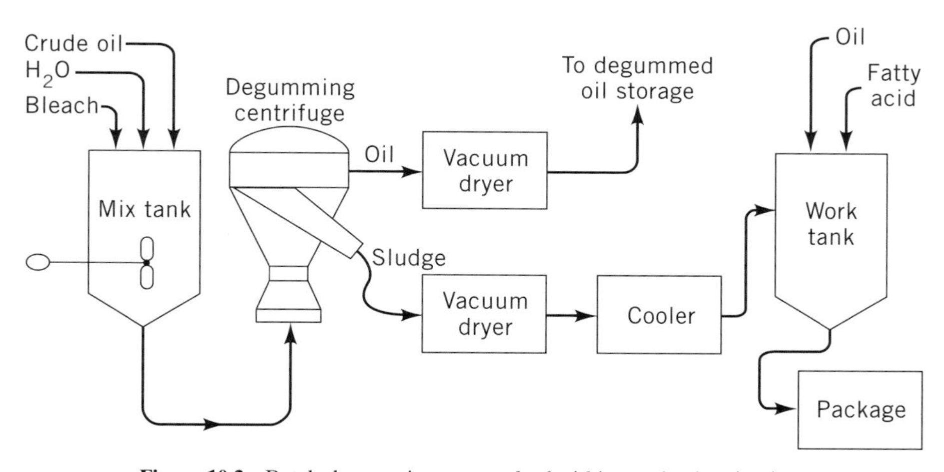

Figure 10.2 Batch degumming system for lecithin production (117).

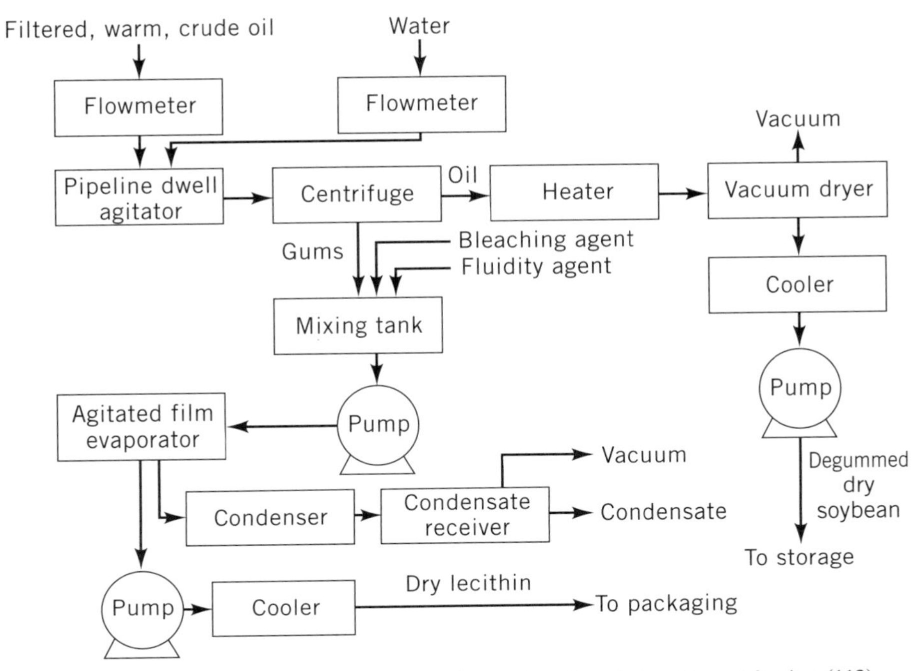

Figure 10.3 Flowsheet for degumming soybean oil and crude lecithin production (112).

Table 10.22 Degumming conditions from the literature (9,119)

Parameter	Quantity	Reference
Water	75% wt of gums	Crauer (120)
	1–2.5%	Brian (112)
	2–3%	Van Niewenhuyzen (54)
	3%	Bernardini (122)
	1%	Norris (123)
	2%	Carr (124)
	Equal to wt gums	Braae (113)
	2–5%	Andersen (125)
Temperature	32–49°C	Norris (123)
	50–70°C	Van Niewenhuyzen (54)
	65–75°C	Bernardini (122)
	70°C	Carr (124)
	95°C	Andersen (125)
Agitation	Vigorous	Bernardini (122)
	Mechanical agitation	Carr (124)
Time	30–60 min.	Carr (124)
	10–15 min.	Braae (113)

In commercial processes, the amount of degumming water required (1.5–2.0%) is roughly equivalent to the phosphatide content of the crude oil (119). Too little water will result in a dark, viscous gums phase and a hazy degummed oil that contains unhydrated phosphatides. Too much water will result in 3-phase system consisting of a free water phase, a fluid yellowish-brown gums phase, and a hazy degummed oil phase after centrifugal separation (6,119,120,126).

Flider (6) points out that the acetone-insoluble (AI) content of the gums is enhanced by raising the temperature of degumming. For example, degumming at 40°C yielded crude lecithin containing 63–65% AI, whereas at 60°C the yield increased to 68–75%. Although above 60°C some darkening of the lecithin may occur, a higher temperature, e.g., 70–80°C produces a more consistent lecithin AI on a day-to-day basis.

Agitation is an important factor in batch degumming. The AI content of the crude gums increase with agitation, presumably because at low agitation rates more oil is entrained in the gums (119).

Three types of centrifuges are in common use for oil–lecithin separations as described by Podbielniak and co-workers (127) and Sullivan (128): tubular bowl, disc bowl, and concentric plate. All of these centrifuges can be hermetically sealed, thereby protecting the process streams from the harmful effects of air (117). The newer disc-type centrifuges have a solid ejecting feature that allows the discharge of solid impurities on a regular basis. These centrifuges are also equipped with a discharge control valve that can be adjusted to vary the AI content of the gum or sludge phase. Since the hermetic centrifuges are capable of delivering a sludge of lower oil content than the conventional open bowl-types more neutral oil is available for refining, and lecithin with higher AI contents can be obtained (117).

The efficiency of commercial degumming operations is tabulated in Table 10.23 (119). Removal of phosphatides in commercial operation ranges from 75 to 96%, with an average of 87% (114).

For more specific information on the parameters of the degumming operation, the reader is referred to Brekke (129), Flider (6), List (117), List and Erickson (130), and List and co-workers (119).

Novel Degumming Approaches. List and co-workers (131) reported on a hexane-extracted crude soybean oil that had been degummed in a reactor by countercurrently contacting the oil with supercritical CO_2 at 10,000 psi at 60°C.

The phosphorus content of the crude oil was reduced from 620 ppm to less than 2 ppm. Degummed feedstocks were fed (without further processing, i.e., bleaching, directly to a batch physical refining step consisting of simultaneous deacidification–deodorization (1 h at 260°C 1–3 mm Hg) with and without 100 ppm citric acid. Flavor evaluation showed that the supercritical CO_2-processed oil had same flavor scores, both initially and after 60 days of aging and light exposure tests, as the commercially refined–bleached soybean oil control deodorized under the same conditions. These results would indicate

Table 10.23 Removal of phosphorus by commercial degumming of crude soybean oil[a] (119)

	Phosphorus (ppm)			
Processor	Crude	Degummed	Phosphorus Removed (%)	Mean (%)
A	733	167	77.2	82.8
	683	80	88.3	
B	867	53	93.8	92.3
	684	63	90.7	
C	711	89	87.5	84.8
	588	105	82.1	
D	615	40	93.4	95.9
	713	12	98.4	
E	623	102	83.6	79.7
	580	141	75.8	

[a] Plants located in Illinois, Iowa, Minnesota, Arkansas, and North Carolina. Two samples from each plant separated by at least 2 weeks.

that bleaching with adsorbent clays may be eliminated by a supercritical CO_2 countercurrent processing step, as a result of the considerable heat bleaching that takes place during deacidification–deodorization. Colors of salad oils produced under the above conditions typically ran 3Y > 0.1R.

A novel degumming process has been described by Dijkstra (132) wherein the washing water is recycled to the oil feed and used to dilute concentrated alkali. This process does not generate an aqueous effluent and can be used for both acid and alkali refining, thus allowing refiners to change gradually from alkali refining to physical refining.

Bleaching. The color of soybean lecithin can be attributed to many factors: carotenoids, melanoids, and porphyrins (68,133) in the product, age, quality of the source material, pretreatment prior to crushing of soybeans, thickness of flakes and temperature during extraction, conditions during degumming, and lecithin processing conditions (6).

Natural lecithin often has a brown color, although with advanced soybean processing technology, the color may approximate that of unbleached soybean oil (65).

Traditionally, one referred to an unbleached product as one that has not been treated with a bleach. A single-bleached product was treated with only one type of bleach (usually hydrogen peroxide) while a double-bleached product usually was treated with two types of bleach (i.e., hydrogen peroxide and formerly benzoyl peroxide). Although these grades continue to exist by name, the bleaching methods used to manufacture them are no longer uniform. High quality, relatively light-colored, unbleached lecithins are now available through modern manufacturing practices. Additionally, today's double-bleached product may have been treated with only a small quantity of one type of bleaching agent. Products are presently bleached to a color specification only, regardless of bleaching techniques or quantity (6,8).

Lecithin may also be bleached by replacing a portion of the degumming water with peroxide and carrying out the bleaching and degumming simultaneously. This method is less efficient, however, than bleaching the gums directly (6).

Although the specifications by the National Soybean Association (NSPA) *Year Book and Trading Rules, 1986–87* (52) recognizes color grades based on natural, bleached, and double-bleached lecithins, this nomenclature is technically incorrect as it is more descriptive of the process rather than of the product. Also, NSPA specifies diluted Gardner colors, but it is more common now to express lecithin colors as undiluted Gardner colors. Diluted colors of 10, 7, and 4 approximately correspond to undiluted Gardner colors of 17, 14, and 12, respectively (6). The diluted and undiluted colors usually correspond well in natural and single-bleached lecithin. In double-bleached lecithin, however, correlation can be achieved only if benzoyl peroxide is used. When only hydrogen peroxide is used to double-bleach lecithin, the diluted color is often 5 or 6 rather than the specified 4 (6).

From a regulatory point, bleached products are traditionally grouped with the unbleached forms of crude lecithin. No distinction is made between the bleached and unbleached forms as far as Generally Recognized as Safe (GRAS) status is concerned (134).

Laboratory studies by List and co-workers (119) have shown that with 1% hydrogen peroxide, complete bleaching occurs in 30 minutes at 60°C; in commercial operations, where often less efficient agitation occurs, up to 1 hr is required (6).

Bleaching with peroxides involves oxidation of the carotenes and the other color bodies within the lecithin. There is no evidence that bleaching with either hydrogen peroxide or benzoyl peroxide functionally modifies lecithins. Bleaching seems only to affect the pigments, which are not functional constituents. Scholfield and Dutton (133) reported that while hydrogen peroxide destroys all color bodies to some extent, its greatest effect is on lutein, the principle pigment (75%) found in soybean lecithin.

The color of most lecithin products will darken on prolonged heating. Color stability can be achieved, however, by avoiding exposure of lecithin to temperatures over 140°F. There are now heat-resistant lecithins on the market that maintain their light color for extended periods even at elevated temperatures (8).

Drying. After centrifugal separation and bleaching, the gums (containing 25–50% moisture) are dried (to 0.3–0.75%) as soon as possible to prevent microbial activity.

The drying operation serves not only to remove moisture, but also to lower the peroxide value as well. Peroxide destruction is rapid at or near temperatures of 100°C. Bleached products should not contain more than 50–75 meq peroxide by-product.

Two types of dryers are commonly used throughout the industry. The sludge can be dried in batch dryers operating under vaccum (20–60 mm Hg) and equipped with rotating, ball-shaped coils through which warm water is circulated to maintain the lecithin at 60–70°C (140–158°F). While these dryers require longer residence times (3–5 hrs), they are popular among European processors because less charring is apt to occur (135). In domestic lecithin processing plants, continuous, agitated film evaporators are the standard equipment. Evaporators operating on vertical and horizontal axis are available, but the horizontal type is preferred because there is less tendency for the lecithin film to break (6,121). The film evaporators operate at temperatures ranging from 80–105°C with vacuum of 25–300 mm Hg. Residence times are very short, usually 1 to 2 minutes.

Dry lecithin is highly viscous, and the viscosity increases drastically and then falls off as the moisture content increases. Comparative conditions used for drying lecithin in the two types of drying apparatus are given in Table 10.24 (54,130).

Because of the sensitivity of lecithin to heat, drying conditions are critical and the product should be cooled to 55–60°C before additional processing, and/or to 35–50°C before storage and packing (6). Shelf-life of dried lecithin products in suitable containers is more than one year at 21°C (3).

Fluidizing. Fluidizing additives such as soybean oil, fatty acids, or calcium chloride can be added to adjust the viscosity. The viscosity of dried crude lecithin can also be decreased by warming it to a maximum of 60°C. The dried crude lecithin product (unbleached or bleached) can also be used to prepare a variety of grades of lecithin by removing the oil to increase the phospholipid content, or by separating the oil-free lecithin into alcohol-soluble and alcohol-insoluble fractions.

Table 10.24 Average process conditions for drying lecithin sludge[a] (54,129)

Process Variable	Batch Dryer, Bollman Type[b]	Continuous, Agitated-Film Evaporator
Temperature		
°C	60–80	80–95
°F	140–176	176–203
Residence time, min.	180–240	1–2
Absolute pressure, mm Hg	20–60	50–300

[a] Starting product: sludge with 50% moisture. End product: lecithin with less than 1% moisture.

[b] Vacuum dryer with rotating, ball-shaped coils heated with warm water.

Besides calcium chloride, the viscosity of lecithin products may also be modified by the addition of other mono- and divalent ions, phosphoric acid, or acetic anhydride. Monovalent ions, such as sodium or potassium, are attracted to the negatively charged base groups which tend to increase the crystalline order, thereby increasing viscosity. On the other hand, divalent calcium and magnesium reduce the crystalline order and thus reduce viscosity. These techniques are used to produce fluid lecithins containing 66-70% AI without the addition of fatty acids (6).

In commercial practice, fluidized lecithins usually are made by calcium chloride addition to the gums, by the inclusion of fatty acids or vegetable oil, or with the aid of special proprietary diluents.

Fluidization with phosphoric acid is not recommended because darkening of the product and hydrolysis may occur. Degumming with acetic anhydride results in fluidized lecithins possibly because phosphatidylethanolamine (PE) is acetylated by the reagent. Nonedible lecithins may be fluidized by the addition of acidulated and dried soapstock. Degummed oil is preferred for the oil diluent. Crude or refined oils have no advantage over degummed oils (6).

Plastic lecithins are available in several forms and are typified by high AI, low AV, high moisture, or their content of certain minerals. One, or a combination of these, can produce a plastic lecithin. Oil-free lecithins are plastic, due to the removal of their nascent oil, i.e., residual soybean oil. They are generally powdered or granular in form.

Nonhydratable Phosphatides. According to Myers (136), about 90% of the phosphatides are removed from the oil by water degumming. Although most of the remaining phosphatides are removed by alkali neutralization, Braae and co-workers (137) report that soybean oil and several other types of vegetable oils often contain some phosphatides that are not removed by alkali neutralization and washing.

The impact of enzyme activity on the nonhydratable phospholipid content of crude soybean oil has been investigated by List and co-workers (138). Evaluation of flakes subjected to live steam and whole beans treated by microwave heating to inactivated phospholipase D suggests that heat, moisture, and enzyme activity are important factors contributing to the formation of nonhydratable phospholipids in extracted crude oils. Approximately 8–10 minutes of microwave heating is required to completely destroy enzymatic activity.

List and co-workers (139) later have found that four interrelated factors promote nonhydratable phosphatides (NHP): (*1*) moisture content of beans or flakes entering the extraction plant; (*2*) phospholipase D activity; (*3*) heat applied to beans or flakes prior to and during extraction; and (*4*) disruption of the cellular structure by cracking and/or flaking. Thus, NHP formation can be minimized by control of the moisture of beans and/or flakes entering the extraction process, inactivation of the phospholipase D enzyme, and optimizing temperatures during the conditioning of the cracked beans or flakes (139).

In a subsequent study, List and Mounts (140) have indicated that the adverse effects of storage conditions, excessive moisture levels, and elevated temperatures cannot be completely overcome by inactivation of phospholipase D prior to solvent extraction of the flakes.

Zhang and co-workers (141) have reported on the effects of an expander process on the phospholipids in soybean oil by comparing the differences in phosphatide compositions of the oils and the lecithins produced from expander and conventional processes by high-performance liquid chromatography (HPLC). The phosphorus content indicated that the expander-processed oil contained more phosphorus (985 ppm) than the conventionally processed oil (840 ppm). However, the phospholipids in the expander-processed oil were more hydratable than those in the conventionally processed oil. After degumming, the phosphorus content in the expander-processed oil and conventionally processed oil were reduced by 93.2% and 78.6%, respectively. The expander-processed lecithin contained 74.3% acetone-insoluble (AI) matter, and the conventionally processed lecithin contained 65.8%. There was also more phosphatidylcholine in the expander-processed lecithin (39.8%, based on AI) than in the conventionally processed lecithin (34.2%) while the phosphatidylethanolamine was lower (12.4% vs. 18.1%); and the phosphatidylinositol contents were almost the same.

Braae (142) believes that the nonhydratable phosphatides are present as calcium and magnesium salts. These phosphatides can have a deleterious effect on oil quality. They can be removed either by treatment of the oil (at 70–90°C) with a small quantity of concentrated phosphoric acid (0.25%) ahead of the neutralization step (143) or by refining the oil with a mixture of lye and sodium carbonate. The phosphoric acid pretreatment apparently also aids in the removal of deleterious iron compounds in the subsequent processing of the oil, i.e., caustic refining, bleaching, and deodorizing of the oil (115). On the other hand, while such pretreatment aids in the lowering of refinery losses and results in low phosphorus and iron content in the degummed oil, the resulting lecithin is dark and low in acetone-insolubles (AI) (6).

The use of acetic anhydride as a degumming adjunct has been described by Hayes and Wolff (144–146) and Myers (136). In this process, 0.1 wt % of acetic anhydride is mixed for 15 min. with crude soybean oil (phosphorus 750 ppm) that has been preheated to 60°C (140°F), followed by stirring the mixture for 30 min. after the addition of 1.5% water. The reaction is completed within minutes. After centrifugation and water washing, the phosphorus content of the oil ranged from 2–5 ppm.

The advantages claimed for this treatment are thought to be that the caustic refining step can be omitted and thus the loss of neutral oil because of saponification is eliminated, and higher yields are obtained from both finished deodorized oil and lecithin. On the negative side, disadvantages found were that equipment and piping must be constructed of type 316 stainless steel to handle the corrosive materials; more care is required in deodorization of the oil; and the process will not produce a satisfactory product from highly colored vegetable oil such as corn and cottonseed oils, nor from some lots of soybean oil (130). Also, according to Evans and co-workers (147), the process removes phosphorus but not iron, one of the metallic proxidants that can give soybean oil a poor flavor. Lecithin produced from this process, however, is claimed to be similar to that prepared in the conventional manner.

Other degumming agents considered include acetic, oxalic, boric and nitric acids (148), and surfactants (149). However, none of these are currently used in lecithin manufacture. Ringers (150) obtained good results in a two-step degumming process wherein an edible acid, presumably citric, was used. Soybean oil can also be degummed by heat, but this practice is confined to oils going into industrial uses.

Lunde and co-workers (151) concluded that the sequestering action of a fatty oil for metal ion depends at least in part on the oil's phospholipid content and reaches a maximum at 0.1–2 ppm of phosphorus. As a point of information, hexane extracts only about one-half of the phosphatides present in soybeans (1,116).

For further information on the nonhydratable phosphatides, the reader is referred to Hvolby (152), Letan and Yaron (153), and Nielsen (116,154).

3.2 Manufacture of Refined-Grade Lecithins

Producing High-Clarity Lecithins. Lecithin destined for certain applications may require more rigorous than the usual initial refining conditions. Clarified lecithins are carefully filtered in (*1*) the full miscella; (*2*) crude oil; (*3*) directly as lecithin. As mentioned before, the filtration is carried out on plate and frame filters with manual or automatic cleaning cycles. Filtration simply removes hexane-insoluble (HI) material producing products of utmost purity and clarity. Such ingredients can be marketed as encapsulated nutritious supplements, as high purity pharmaceutical adjuncts, and additives for high technology applications.

Although the dry lecithin can be filtered, many processors prefer to filter the crude oil before degumming (112). Usually, the crude oil is filtered through large plate-and-frame filter presses. If the degumming operation is conducted at a solvent extraction oil mill, then either the miscella, i.e., oil-solvent mixture, or the oil can be filtered. If all the fines and filter aid are not removed, the dry lecithin will appear cloudy.

Brian (112) has made suggestions for the type of pumps, centrifuges, filters, heat exchangers, and drying equipment most suitable for lecithin production. He has also provided useful engineering design data on filtration flow rates for the crude oil, miscella, and dry lecithin, quality of filter acid needed, and the overall heat transfer coefficients for agitated-film evaporators and shell-and-tube heat exchangers used for heating and cooling the oil.

Producing Compounded Lecithins. Compounded lecithins are special purpose products made by the direct addition and/or blending of functional additives, emulsifiers, diluents, surface active agents, etc. These modified lecithins represent perhaps the largest category of lecithin products.

Water-dispersible lecithins may be produced by adding a hydrophilic surfactant (5–20%) such as polysorbate or ethoxylated monoglycerides. A mixture of lecithin and nonionic surfactants (10–20%) has utility in nonedible applications where water-dispersibility is needed. Lecithins containing 50–54% phospholipids present difficulties in manufacturing. Blending of partial glycerides and lecithin followed by spray cooling results in flaked or powdered products (6).

When lecithins are diluted with soybean oil or fatty acids, they have a tendency to separate. In such cases, substituting a portion or all of the soybean oil with other oils such as peanut, cottonseed, coconut, or partially hydrogenated soybean oil will increase stability (6).

Lecithins may also exhibit synergistic actions with some compounds. Common additives include special oils, polysorbates, mono- or diglycerides, modified monoglycerides, lanolin and lanolin derivatives, solvents, plasticers, and other surfactants.

Producing Deoiled Lecithins. Deoiled lecithin represents a special category where high phospholipid content (above 90 AI) is required. When contacted with acetone, phospholipids precipitate as a fine free-flowing powder. After removing the acetone deoiled lecithins are dry powders or granules (6).

Depending on the type and efficiency of the extraction equipment, the acetone/crude lecithin ratio necessary to achieve a 95% phosphatide product is 10–20 : 1 (v/v). In batch extraction, the tank is charged with acetone prior to crude lecithin addition. Crude lecithin is then introduced into a crystallizing vessel with agitation until an acetone/crude lecithin volume ratio of ca. 5 : 1 is achieved. Only the best quality fluid crude lecithin should be used for the preparation of deoiled granular lecithins. The mixture is then agitated for 20–30 min., after which time the phospholipids are allowed to settle. The triglyceride–acetone miscella is then removed and the vessel charged again with fresh acetone for the second extraction. A single batch may be extracted 2–4 times to obtain the desired phospholipid concentration (95% minimum) (6).

In a continuous extraction, crude lecithin and acetone are simultaneously metered into a vessel. Within limits, acetone consumption can be decreased by increasing residence time in the continuous extractor, increasing raw material efficiency (6).

After extraction, the deoiled lecithin is recovered by filtration as a cake, containing 25–50% acetone. According to Flider (6), the acetone concentration of the cake is critical for optimal granulation. "Too little acetone will result in the formation of a high concentration of fines and powder. Too much acetone will result in a 'salt and pepper' effect (i.e., a mixture of coarse and fine particles) caused by agglomeration of the fines and powder during granulation." The fine and powder output is 5–50% of the total deoiled material, depending on production conditions (6).

Following granulation, the remainder of the acetone is removed by drying, preferably in a moving bed-forced air dryer. Such dryers are preferred over fluid-bed dryers because less destruction of the lecithin granules occurs. After drying, the acetone content of the product should be well below 50 ppm, preferably below 25 ppm (6). Flider (6) states that mesityl oxide, through an aldol condensation reaction, may be formed if excess acetone is present. When the lecithin is sufficiently dry, however, this is not a problem. The dried deoiled lecithin is sized by sieving through a series of screens (6).

Since the tocopherols are removed from the lecithin during the extraction process, the deoiled lecithin has less oxidative stability than the crude product. Also, the surface/volume ratio of the deoiled lecithin contributes to reduced stability. Mixed soy tocopherols are usually added back at a level of 500 ppm to prevent this. A small percent of an anticaking agent may also be added to ensure that the product remains free-flowing. A free-flowing deoiled lecithin can be easily added to other products (6).

When compared to crude lecithin, oil-free lecithin is more hydrophilic and seems to have better emulsifying activity than its AI alone would suggest. The removal of odor/flavor components with the oil also produces blander lecithins (8).

Refined deoiled lecithin can also be blended with carriers such as cocoa butter, hard butters, medium-chain triglycerides, or other diluents to obtain products with more functionality and different physical characteristics. Up to 40% phospholipids may be incorporated in these carriers without the use of solvents. These products are usually stabilized against autoxidation by the addition of antioxidants (6). Deoiled lecithin should be packaged as soon as possible to prevent moisture absorption. For more specific details on various aspects of deoiled lecithins, the reader is referred to Flider (6).

Producing Modified Lecithins. The chemistry of lecithin has been reviewed by Pryde (84) and by Wittcoff (4); Schmidt and Orthoefer (53) have discussed the manufacture and use of modified lecithin products. The latter class is represented by chemically or enzymatically modified products.

The traditional approach to the modification of phospholipid properties is by fractionation, isolation, and purification of a single component. Functions of phospholipid mixtures are also modified by partial chemical or enzymatic hydrolysis, acetylation, hydrogenation, and hydroxylation (5).

Crude lecithin contains a number of functional groups that can be successfully hydrolyzed, hydrogenated, hydroxylated, ethoxylated, halogenated, sulfonated, acylated, succinylated, ozonized, and phosphorylated, to name just a few possibilities (1). The only chemically modified food-grade products produced in significant commercial quantities at the present time are the ones obtained by hydroxylation, acetylation, and also by enzymatic hydrolysis (53). Hydroxylated or acylated lecithins represent chemical modifications to improve the functionality in water-based systems.

Acetylated Lecithin. Acetylation occurs primarily on the amino group of phosphatidylethanolamine (144–146). The amino group of phosphatidylethanolamine when acylated introduces a substituent on the positively charged portion of the zwitterionic phosphatidylethanolamine and converts it to a negatively charged lecithin with improved solubility and oil-in-water emulsifying properties (155). Lecithin can be acetylated using acetic anhydride either by adding the reagent prior to degumming or adding it to the wet gums. Acetylated lecithin products are made from natural soy lecithin hydrates by treating them with low levels of acetic anhydride (1.5 to 5.0%). Lecithin hydrates are obtained during the degumming of crude soybean oil. Following the reaction with acetic anhydride, the resulting product is neutralized with a food-grade alkali to raise the pH to 6.5 to 8.0, depending on the intended application. The amount of acetic anhydride used in the process depends on the level of phospholipids in the gum and the intended use of the final product requiring different degrees of acylation for optimum functionality. The same is true for the choice of alkali (e.g., sodium, calcium hydroxide, etc.) used. The product is then vacuum-dried (film dryer in a commercial operation) to a final moisture of less than 1.0% (156).

The active ingredients in the products sold in commerce using acetylated lecithin are the same phosphatide constituents as those found in unmodified lecithin. In an acetylated product, however, these constitutents, but more particularly the phosphatidylethanolamine (PE) fraction, which is the principal reactant with acetic anhydride in the manufacturing process, is modified. The resulting *N*-acetyl-phosphatidylethanolamine, involving nucleophilic acylsubstitution is illustrated as follows:

```
 +                         +
NH3 \/\/\/PE              NH2 \/\/\/PE
 ↓                         |                              R
R——C=O                    R——C=O⁻                          \
    \                         \                             C=O
     O        --H2O-->         O             ——→           /
    /                         /                        NH \/\/\/PE
R——C=O                    R——C=O⁻
                               |
                               OH
```

The degree of reaction is measured by determining amine nitrogen content in the resulting product (usually by formol titration). Maximum (100%) acetylation would be indicated by a zero (0) amine nitrogen value whereas a minimally acylated commercial product has approximately 1.7 mg amine nitrogen/g content. In a typical commercial operation the amine nitrogen content is usually in the range of 0.7 to 1.7 mg/g.

The total phospholipid content of commercial acetylated lecithin products can vary from about 52% (AI) to about 65% (AI), the remainder being soybean oil (or another food-grade triglyceride or fatty acid as a natural constituent or added diluent), natural pigments, carbohydrates, sterols, and other minor constituents present in crude lecithin from the soybean. The acetylated lecithin meets all the compositional requirements of the U.S. Food Chemical Codex (49) and the Codex Alimentarius (157) definition for lecithin.

Typical specifications for a minimally and maximally acetylated product are given below.

Minimally Acetylated Commercial Lecithin Specification

Acetone insolubles (%)	60.0 min.; 64.0 max.
Moisture (%)	0.75 max.
Acid value	24 max.
Viscosity (cP at 25°C)	10,000 max.
Color	17- max.
Peroxide value (meq/kg)	10 max.
Hexane insolubles (%)	0.09 max.
Amino nitrogen (mg/g)	1.65 max.
Divalent metals (%)	0.42 min.; 0.48 max.
pH	6.5 min.; 8.0 max.

Maximally Acetylated Commercial Lecithin Specification

Acetone insolubles (%)	53.0 min.; 56.0 max.
Moisture (%)	0.75 max.
Acid value	36 max.
Viscosity (cP at 25°C)	2,900 max.
Color	12+ − 14
Peroxide value (meq/kg)	100 max.
HIM (Hexane insolubles 0.8 μ-Millipore)	100 ppm max.
Amino nitrogen (mg/g)	1.0 max.
pH	7.0 min.; 7.5 max.
Visual clarity	pass
Heat resistance test (Lovibond red)	8.0 max.
Appearance	Clear and brilliant at 25°C

Acetylated lecithins have improved fluid properties, improved water dispersibility, and are effective oil-in-water emulsifiers for a wide variety of food formulations (51,53). Moderately and highly acetylated lecithins are resistant to heat and can be repeatedly heated and cooled without darkening. The intended uses for minimally acetylated products are in infant foods, coffee whiteners, meat sauces and gravies, and for oil-in-water cosmetic emulsions. Moderately and maximally acetylated products are used in cheese sauces, release agents in pumpable and aerosol formulations, and shortenings.

The following patents have been issued on the topic of making and using acetylated lecithins:

1. U.S. Pat. 3,301,881 "Process of Phosphatide Separation," 1967.
2. U.S. Pat. 3,359,201 "Lecithin Product and Method," 1967.
3. U.S. Pat. 3,499,017 "Alkaline-Hydrolyzed Phosphatides," 1970.
4. U.S. Pat. 3,823,170 "Phosphatides," 1974.
5. U.S. Pat. 3,928,056 "Pan Release Product and Process," 1975.
6. U.S. Pat. 3,962,292 "Phosphatide Preparation Process," 1976.
7. U.S. Pat. 4,479,977 "Method of Preparing Heat-Resistant Lecithin Release Agent," 1984.

Hydroxylated Lecithin. Hydrogen peroxide, in addition to bleaching, can also hydroxylate lecithin; this imparts hydrophilic properties, improves moisture retention, and contributes to the formation of stable oil-in-water emulsions.

Hydroxylated lecithin is a light colored product with increased water dispersibility and enhanced oil-in-water emulsifying properties. Hydroxylated lecithin is useful in many applications in which a water dispersible lecithin is desired. It is especially useful in baking applications where it can improve the dispersion of fats and retard staling.

Hydroxylation of lecithin is carried out by the reaction of crude lecithin with hydrogen peroxide and lactic acid or a peracid. Active sites for peroxidation appear to be double bonds as measured by IV drop and the isolation of dihydroxystearic acid from the reaction mixtures. Hydroxylation is allowed to proceed until a 10% reduction in iodine value occurs (117). The ethanolamine group is also modified during hydroxylation (53,158).

Hydrolyzed Lecithins. Crude lecithin is readily hydrolyzed in the presence of strong acids or bases. Enzymes can be used for very selective hydrolysis. Prolonged treatment leads to fatty acids, glycerophosphoric acid, or their salts with mixtures of amino compounds and carbohydrates (4,117).

In a commercial process, it is desirable to control the reaction so that just one of the fatty acids is cleaved from the phospholipid molecule. Since acid or base hydrolysis is very difficult to control, enzymes are usually preferred for most applications (53). A number of phospholipase enzymes are available.

Haas and co-workers (159) have studied enzymatic phosphatidylcholine hydrolysis in organic solvents by examining selected commercially available lipases. Enzymatic hydrolysis of oat and soy lecithins and its effect on the functional properties of lecithin were investigated by Aura and co-workers

(160). The phospholipase used was most effective at low enzyme and substrate concentrations.

Partially hydrolyzed lecithins exhibit enhanced oil-in-water emulsifying properties, particularly in the presence of calcium and magnesium ions. They do not lose their emulsifying action in the presence of calcium and magnesium ions as rapidly as do the unmodified types. Enzymatically modified lecithins have been used in calf milk replacement formulations to improve the emulsification and digestibility of fats (51).

Hydrogenated Lecithin. Lecithin can be hydrogenated to a stearin-like solid that has greater oxidative stability and is less hygroscopic than unmodified lecithin, but has reduced solubility in the usual solvents. Phospholipids are not hydrogenated as readily as soybean oil which, at lower hydrogenation pressures and temperatures, can be selectively hydrogenated (53).

Hydrogenation of lecithin is usually done under conditions to reach iodine values of 10–20 in the presence of a nickel or palladium catalyst and a suitable solvent (e.g., ethyl acetate) at 75–85°C under 70 atmospheres pressure. Bromine or chlorine also readily adds across double bonds yielding products useful in lubricant formulations. Iodine can be added by warming granular lecithin dissolved in acetic acid in the presence of iodine and magnesium or aluminum catalyst (53).

Producing Fractionated Lecithins. Finally, by fractionating crude lecithin directly or after deoiling is another example of creating a variety of products with tailormade functionality. Extensive alcohol and mixed solvent fractionation, combined with other techniques, can produce lecithin products that have been greatly enriched in particular phospholipids. Separating the acetone-solubles from crude lecithin increases the amount of phospholipids in the acetone insoluble fraction by decreasing the amount of triglycerides.

Oil-free lecithin contains nearly equal proportions of phosphatidylcholine (PC), phosphatidylethanolamine (PE), and phosphatidylinositol (PI). Such products favor both oil-in-water and water-in-oil emulsions (117). Further fractionation can separate the alcohol-soluble phosphatidylcholine from the alcohol-insoluble inositol phosphatides. Commercial products that are alcohol-soluble contain concentrated phosphatidylcholine (40–60%) and only low levels of inositol phosphatides. The alcohol insoluble products are enriched in inositol phosphatides (40–60%) whereas their phosphatidylcholine content is greatly reduced. The phosphatidylethanolamine component is approximately equally partitioned between the two fractions. The alcohol-soluble grades tend to be more oil-in-water emulsifiers, whereas the alcohol insoluble grades are more effective in water-in-oil systems (8).

The PC/PE ratios of alcohol-fractionated lecithins are largely determined by processing variables such as alcohol polarity, concentration, lecithin/alcohol ratio, temperature, and extraction time (6). By extracting natural lecithin with a PC to PE ratio of 1.2:1 with 90% ethanol, alcohol-soluble fractions with PC/PE ratio of 8:1 can be obtained (6,121). The fractions may be blended with other surfactants or carriers to obtain desired functionality.

To obtain individual phospholipids of greater than 50–60% purity, some form of selective adsorption process is usually required. Adsorption and distribution chromatography present these options. Treatment of the alcohol-soluble lecithin with alumina yields a fraction very rich in phophatidylcholine and free of phosphotidylethanolamine and inositolphosphatides (161). While these products are available only in very limited quantities for highly specialized markets, products such as a lecithin containing up to 95% phosphatidylcholine (PC) can be obtained commercially.

A continuous countercurrent process for deoiling of crude lecithin was patented by Peter (162) based on the principles of critical extraction. In this process, lecithin is obtained from vegetable oil, animal fats, or lecithin-containing mixtures by extraction with a solvent containing one or many C_3- or C_4-hydrocarbons under conditions wherein the solvent forms a low-viscosity solution with the lecithin and the extraction mixture is divided into an oil-rich fluid phase and a lecithin-rich fluid phase. Both phases are then separated to obtain, after removing the solvent, an essentially lecithin-free oil and an oil-free pure lecithin. The preferred extraction solvent used is propane or butane, or mixtures of hydrocarbons. The extraction is carried out under well-defined pressure and temperature conditions. The lecithin-rich phase obtained at the bottom of the extraction column is separated by lowering the pressure and/or elevating the temperature.

A supercritical extraction method has been applied to test the feasibility of tocopherol concentration from soybean sludge with carbon dioxide at temperatures and pressures ranging from 35–70°C and 200–400 bar, respectively (163). The supercritical solubility of the esterified soybean sludge was over 4–6 times greater than that of the original soybean sludge. By a simple batch-type one-stage method the tocopherols in the esterified soybean sludge could be concentrated up to 40 wt %. The overall result of this study shows that soybean sludge initially containing about 13–14 wt % tocopherols may require a countercurrent multistage column to be highly and effectively concentrated.

The major process developments at the present time are in the area of lecithin fractionation. Markets appear to be developing in the pharmaceutical and cosmetic industries, particularly for concentrated forms of PC. Deoiled, granular lecithins have firmly established markets and new markets are developing which should ensure the growth of these products as emulsifiers.

Storage and Handling. Liquid lecithin can be kept for years provided closed containers are used and the temperature does not exceed 20–25°C. Bleached products require more careful storage and handling. Color reversion will occur rapidly in bleached products, particularly at elevated temperatures. Decomposition of peroxide is thought to contribute to color reversion in bleached products. In order to prevent this phenomenon, low storage temperatures are recommended (117).

Very low temperatures should, however, be avoided when storing liquid lecithin products because physical separation of the phospholipids and oil may occur. Physical separation is more likely to occur in low AI products. When separation does occur, remixing at 40–60°C will redisperse the oil and lecithin phases. In bulk handling of lecithin, storage temperatures of 30–35°C are acceptable. However, prolonged storage at these temperatures may cause darkening (117).

Deoiled granular lecithin can be stored up to 2 years at temperatures below 25°C. If desired, it may be stored in a frozen state at 0°C, but due to its hygroscopic nature, the product should be allowed to come to room temperature before it is exposed to the atmosphere (6).

FOOD-GRADE LECITHIN PRODUCTS, USES

4.1 Functionality

Although the nutritional and health impact of lecithin is currently being assessed in numerous studies concerned with neurochemical disorders, atherosclerosis, and metabolic diseases, lecithin products are employed primarily for their functional benefits to food products. The combined hydrophilic and lipophilic properties of phospholipid molecules give them surface-active effects in many applications. As surfactants they can exhibit a variety of functions common to other surfactants while, at the same time, they also have some unique surfactant functionalities of their own.

Commercial lecithin products sold many decades ago for applications such as chocolate and confectionery products, margarines, bakery goods, pasta products, textiles, insecticides, and paints, are still active today because of their emulsifying, wetting, colloidal, antioxidant, and physiological properties. Lecithin's multifunctional properties and its "natural" status make it an ideal food ingredient. The major applications and functional properties of lecithin products are shown in Table 10.25 (8).

Since a variety of methods are available for modifying the emulsifying properties of crude commercial lecithin, the potential for improved, tailormade functional products is unlimited. The main functional properties are: emulsifying; antispatter; instantizing/wetting/dispersing; release/parting; and viscosity modifying.

These functional characteristics are primarily derived from the chemical structures of lecithin's major phospholipids (Figure 10.1) (8). Phospholipid molecules contain two long-chain fatty acids esterified to glycerol, as well as a phosphodiester bonding a choline, inositol, or ethanolamine group. A phosphatide's fatty acid end is nonpolar and thereby lipophilic (or fat-loving). Conversely, its phosphodiester of the above-mentioned substituents is zwitterionic (or bipolar) which explains the hydrophilic (or water-loving) properties of this portion of the molecule (65).

Because of their charged nature, the phospholipids are susceptible to the ionic environment in which they function. Based on testing in the laboratory, Dashiell (29) suggests that salt concentrations greater than 2% and pH less

Table 10.25 Functional properties (8)

Ingredient In	Function(s)
Margarine	Emulsifier, antispattering agent
Confections and snack foods	
Chocolate	Crystallization control, viscosity control, antisticking
Caramels	
Coatings	
Instant foods	
Cocoa powders	Wetting and dispersing agent, emulsifier
Instant drinks	
Instant cocoa	
Instant coffee	
Protein drinks	
Dietetic drinks	
Coffee whiteners	
Milk replacers	
Cake mixes	
Puddings	
Instant toppings	
Commercial bakery items	
Breads	Crystallization control, emulsifier, wetting agent, release agent (internal and external)
Rolls	
Donuts	
Cookies	
Cakes	
Pasta products	
Pies	
Cheese products	
Pasteurized processed	Emulsifier, release agent
Cheese and cheese food	
Imitation cheese	
Meat and poultry processing	
Meat and poultry glazes and basting compounds	Browning agent, phosphate disperant
Pet foods	Dietary supplement, release agent, emulsifier
Bacon	
Dairy and imitation dairy products	
Infant, milk formulas	Emulsifier, wetting and dispersing agent, antispattering agent, release agent
Milk and cream replacers	
Egg replacers	
Imitation eggs	
Whipped toppings	
Ice cream	
Flavored milks	
Flavored butters (garlic, etc.)	
Basting butters	

Table 10.25 (Continued)

Ingredient In	Function(s)
Miscellaneous products	
Peanut spreads	Crystallization control, emulsifier
Salad products	
Flavor and color solubilization	
Packaging aid	
Polymer package, interior coating	Release agent, sealant
Can interior coating	
Sausage casing coating	
Stocking net	
Processing equipment	
Frying surfaces	Internal (in product) and/or external release agent, lubricant
Extruders	
Conveyors	
Broilers	
Dryers	
Blenders	
Evaporators	

than 4 contribute to a detectable loss in lecithin functionality (164). Similar results have been reported elsewhere (64).

The specific composition of a commercial lecithin product defines its functionality. For instance, Rydhag (165) characterized several lecithins and found those with higher levels of anionic phospholipids gave superior emulsions. In turn, the consistency of composition defines the utility of the product to the end user.

The following product modifications and combinations with lecithin were described in a publication from Lucas Meyer Co., Hamburg, Germany (166): (*1*) improved water dispersibility through formation of *N*-actyl-phospholipids; (*2*) improved water dispersibility through use of high levels of lysophosphatides; (*3*) using lecithin as coemulsifier for monoglycerides to interact with amylose; (*4*) using deoiled soy lecithin (95% powder) as emulsifier and dispersant; (*5*) using a spray-dried combination of standard soy lecithin and milk solids (mainly lactose) as emulsifier, dispersant, wetting agent; and (*6*) using a spray-dried combination of modified partial glycerides and milk solids for flour treatment and assurance of uniformity and high quality breadmaking with wheat flours.

Weete and co-workers (167) have reported on improvement of lecithin as an emulsifier for water-in-oil emulsions by thermalization. The various forms of lecithins can be heated under certain conditions of time and temperature to greatly improve their properties as emulsifiers for water-in-oil emulsions.

Viscosity, discontinuous phase-holding capacity, stability, and water retention were greatly enhanced in emulsions containing thermalized lecithins as the emulsifier compared to those prepared with corresponding amounts of non-thermalized lecithins. The improved emulsification properties of thermalized lecithins appeared to be due, at least in part, to an increase in diglycerides and free fatty acids resulting from the thermal degradation of phosphatides.

Surfactant and Colloidal Dispersant Properties. Lecithin products function in food systems that can be classified as colloidal dispersions (8).

Emulsifying Properties. The main function of phosphatides is to emulsify fats. In an oil : water system the phospholipid components concentrate at the oil : water interface. The polar, hydrophilic parts of the molecules are directed toward the aqueous phase and the nonpolar, hydrophobic (or lipophilic) parts are directed toward the oil phase. Concentration of phosphatides at the oil : water interface lowers the surface tension and makes it possible for emulsions to form. Once the emulsion is formed, the phospholipid molecules at the surface of the oil or water droplets act as barriers that prevent the droplets from coalescing and thus stabilize the emulsion (155).

Lecithins are used in a wide assortment of food emulsions. Examples of food emulsions include milk, butter, margarine, chocolate, doughs in bakery items, cheese, cream replacers, and ice cream.

Lecithin use covers both water-in-oil (w/o) and oil-in-water (o/w) emulsions. In w/o emulsions, like margarine or ready-to-use frostings, lecithin of an oil-soluble, nonwater dispersible variety is indicated. These lecithins generally fall into the lipophilic category and are used in conjunction with monoglycerides. Lecithins used in o/w emulsions have the common property of water dispersibility (29). The use of lecithin in oil-in-water emulsions requires the modification of lecithin to increase its apparent hydrophilicity. Techniques available are somewhat limited for foodgrade lecithins. As mentioned before, the commonly employed modifications for achieving better dispersant properties are acetylation, hydroxylation, solvent fractionation, and deoiling.

The manner in which lecithin is modified to achieve increased hydrophilicity will greatly affect its emulsification properties. Two lecithins may also be modified by different means to give products having an equivalent "apparent hydrophilic : lipophilic balance (HLB)," a term used to convey the approximate degree of water dispersibility (hydrophilicity) of lecithin products (29). Techniques to estimate HLB are available (168).

When working with lecithins, one must know, however, that lecithin products emulsify on a very system-dependent basis. This is due to amphoteric (+,−) polar ends on the individual phosphatides. This makes it impossible to assign a precise general emulsifier rating (such as HLB) to lecithin products (8).

To prove this point Dashiell (29) prepared two lecithins both having an apparent HLB of about 8. One was produced by enzymatic treatment and the other by hydroxylation. When used to emulsify milk fat into water (water/

fat/lecithin, 90:9:1), the enzyme-treated lecithin gave a fine-grained stable emulsion, while the hydroxylated lecithin provided a coarse-grained emulsion that quickly separated.

As a corollary to the above observations, consideration of the type of fat to be emulsified is also important in selecting the proper lecithin emulsifier. Dashiell (29) provides a short listing of fat-types and the corresponding class of lecithin emulsifier found to give the most stable emulsion in model systems of water/fat/emulsifier.

Crude lecithins are excellent water-in-oil emulsifiers. However, modified lecithins can function to emulsify either water-in-oil or oil-in-water emulsions, depending on the type of lecithin modification and the specific parameters of the system. These system parameters can include pH, types of components, component ratios, solids content, and others. Unlike crude lecithins, hydroxylated lecithins are stable in acid systems (pH <3.5). Fractionated lecithins can be manufactured for specific emulsion types. Since lecithin emulsifying activity is dependent on its phosphatide ratios, changing the ratio can alter its emulsifying capabilities (8).

Emulsifier/stabilizer systems are normally used to make stable food emulsions. Thus, lecithin is generally not called upon to handle the entire emulsification, but rather it works in combination with other emulsifiers and stabilizing polymers such as proteins, starches, and gums (29). Lecithin will break up (emulsify) the particles, while a stabilizer (water-soluble polymer, etc.) will hold the particles in a dispersed orientation when a stable emulsion is formed. Lysolecithins, which are more hydrophilic, show stronger oil-in-water emulsifying properties. Also, in most processed food emulsions the particle size of the dispersed phase has to be much smaller for dynamic stability than is practically possible with foods (7).

Dashiell (29) reported that with high levels of good quality protein present, the selection of a system specific lecithin (one giving the best emulsion in an oil/emulsifier/water system), becomes less important. In fat creaming tests, in fat/water/protein/lecithin emulsions, results showed that with limited protein in a whey-stabilized (low-protein) system, more functional lecithins gave a clear advantage. In a casein-stabilized system (containing abundant protein), differences between products of different functionalities were less dramatic.

Although lecithin is used in a wide variety of emulsions, particular properties are often limited. For example, lecithins do not appear to function well in systems like whipped toppings because the textural properties of these products depend upon a controlled emulsion destabilization during the input of mechanical work. This behavior is generally limited to α-tending emulsifiers like lactylated or acetylated monoglycerides (29).

Dispersing/Wetting/Instantizing Properties

1. *Solubilization.* Most lecithins can aid in the production of microemulsions, an example being water-soluble dyes in oil systems.

Lecithin that contains more than 67% acetone insolubles disperse with diffculty in aqueous media but solubility can be imparted by reaction with polyhydric alcohols and epoxy compounds, hydroxylation of unsaturated fatty acid constituents, and by fractionation or compounding with dispersing agents (169).

While modified or fractionated lecithins might produce oil-in-water microemulsions, the best microemulsion can be obtained with synergistic lecithin–surfactant mixtures (e.g., ethoxylated monoglycerides). Instant dispersibility in water, on the other hand, generally requires chemical modification.

2. *Solid Particle Dispersions (Sols).* Many lecithin products are still one of the best and most effective surfactants for dispersing sols. This seems to be due to lecithin's affinity for solids–liquid surface interfaces. Phospholipids seem particularly attracted to particles containing metals and metal salts. Examples of food sols include some liquid chocolates, instant drinks, frosting mixes, pigmented foods, and others. The nonfood applications include paints, inks, and other pigmented coatings.
3. *Foams.* Refined lecithins have also been employed as effective foam control agents. Examples include whipped toppings, ice creams, and many types of candies. Refined lecithin products have also been employed as effective defoaming agents in foams caused by powdered proteins in water. This is an excellent example of the system specificity of lecithin products (8).
4. *Wetting/Instantizing Properties.* Lecithin products are effective wetting agents for a wide variety of powered or granular products. Lecithination of ingredient powders for improved wetting and control of dusting problems is widely practiced. Instantizing effects can be obtained by including the proper lecithin product in a food formulation. Specific lecithin products that are compatible with the various manufacturing techniques used for instantizing are commercially available. Agglomeration is the primary instantizing operation, and lecithin addition simply fine-tunes the instantizing process.

Lecithin products are formulated to be applied to many types of food powders to achieve rapid wetting and dissolution. Since powder compositions can vary greatly (from hydrophilic to lipophilic), proper lecithin selection is done on something of an empirical basis (29). Certain general principles apply, however. If lipophilic powders need wetting or instantizing, a hydrophilic lecithin is recommended. More lipophilic lecithins are applied also to very hydrophilic powders (such as pregelatinized starch) to slightly retard or control the wetting rate. These powders have a tendency to hydrate so quickly that they will simply ball-up on the surface of water. A very lipophilic lecithin product will work best if coated onto the powder, thereby protecting the powder from too rapid hydration, yet still hydrating. On the other hand, to wet fatty powders, a more hydrophilic lecithin is used to actively promote wetting of the surface.

Manufacturing techniques employed in producing instant products include: spray-coating dry food powders with highly fluid lecithin products, and co-spray drying powders with more hydrophilic lecithins such as the oil-free forms, or hydroxylated products.

Many lecithin wetting agents are applicable to more product-specific manufacturing techniques and there are also a few special lecithin products mixed with monoglycerides for some applications. For example, acetylated monoglycerides are more easily wetted by first coating with hydroxylated lecithin.

Examples for foods that can be instantized with lecithins include cocoa powders, instant drink mixes, powdered coffee whiteners, milk replacers, cake mixes, powdered instant puddings, and instant soups and sauce mixes.

Release/Parting Properties. Lecithin functions as the active ingredient in a wide variety of food-grade release formulations. Products for institutional and retail use are available in aerosol and nonaerosol forms containing from 0.5% to about 15% lecithin (29). Common ingredients in release formulations are as follows (29) (from a Central Soya Co. market survey, 1986).

Vegetable oil (all major vegetable oil classes are used)
Hydrogenated vegetable oil
Mineral oil
Lecithin
Flour
Amorphous silica
Artificial flavor
Artificial color
Beta-carotene
Preservatives (emulsion types only).
Antioxidants
Antifoams
Water (emulsion types only)

Lecithin usage levels in commercial release formulas are limited by a tendency of the lecithin to separate from some oils. The tendency of ordinary lecithins to darken, polymerize, and foul on heated metal baking surfaces also limits their use level in commercial release formulas. Moderately and highly acetylated lecithins, however, are resistant to heat and can be repeatedly heated and cooled without darkening (170).

Supplied as both aerosols and nonaerosols, these products require stabilization against microbiological degradation. The emulsion-type release agent is not nearly as common as the basic oil–lecithin blends.

Crystallization Control. Lecithin can control crystallization in various food systems. In foods containing sugars or fats, the presence of as little as 0.5% lecithin can produce altered crystal sizes and structures that can have positive effects on product texture and viscosity. This is important in cookie fillings, butter-containing maple syrups, ice cream toppings, and similar products (8).

4.2 Specific Food Applications

As mentioned previously, soybean lecithin is used in food because of its emulsifying, wetting, collodial, and physiological properties (68). Relatively small amounts of the lecithin are needed, often only 0.1 to 2% of the fat content in foods. These use levels are more or less consistent with those of chemical surfactants (8). At these low levels of usage, the color, flavor, and odor of the lecithin normally are not noticeable. When lecithin is used in conjunction with synthetic emulsifiers, it sometimes has a synergistic effect, and thus lesser amounts of the synthetic emulsifiers need be used.

General food applications of lecithin include margarine, confections, snack foods, soups, instant foods, bakery products, simulated dairy products, processed meat/poultry/seafood products, and dietary applications. The most widespread uses of crude lecithin products are in confections and margarine (Iveson (171), Prosise (8), Sartoretto (68), Scocca (172), and Szuhaj and Sincroft (173)).

Margarine/Shortenings. Lecithin is the classic emulsifier in margarine and is added at the 0.1 to 0.5% level to the fat. It is commonly used in conjunction with mono- and diglycerides. The lecithin prevents "weeping" or "bleeding" of the moisture present, reduces spattering and promotes browning during frying, increases the shortening effect when margarine is used in baking, and helps to protect the Vitamin A in fortified margarine from oxidation (174). More complete and uniform blending of shortening occurs when 0.5 to 1.0% lecithin is added (156).

Lecithin serves as an emulsifier in many shortenings; hydrogenated vegetable phospholipids are used in the manufacture of a liquid shortening.

In some respects, the fats and oils industry is supplying raw materials for some fat replacers, or is making fat-based products (e.g., lecithin and monoglycerides). Another example is the use of specialty lecithin (phosphatidic acid) in cooking oil, which reduces fat absorption and fosters more uniform cooking of the food (175).

Confections. There are three major specific properties for lecithin in confections: emulsification (e.g., caramels); antistick/release properties; and viscosity modification (e.g., chocolate) (176). None of these properties stand alone. For example, emulsification in caramels will influence shelflife and texture. In chocolate, viscosity modification will alter production costs and texture of the finished product.

Addition of 0.25–0.35% lecithin to the chocolate used in candy-making reduces its viscosity markedly, enables the manufacturer to apply a uniform coating and thus use lesser amounts of expensive cocoa butter, decreases the time for grinding and mixing the various ingredients, and produces a more stable chocolate. Stabilized lecithin-containing chocolate has improved handling characteristics and is more resistant to fat-and-sugar bloom or "greying". Hydrogenated lecithins are especially well-suited as emulsifiers and as inhibitors of fat bloom in chocolate formulations. Use of lecithin in other fat-containing candies also prevents graining, streaking, and greasiness.

Studies by Sinram and Schmitt (177) have shown significant improvements in dark and milk chocolates using a fractionated, phosphatidylcholine-enriched soy lecithin (ChocoTop 350) as compared to a standard soy lecithin or no lecithin at all. The effect of reduced viscosity and improved yield value in chocolate depended on the type and dosage of lecithin as well as the fat content of the chocolate.

Addition of 1 to 2% of lecithin to peanut butter gives a smoother, creamier spread. The peanut butter does not separate under wide temperature variations.

Bakery Products. Lecithin is a useful emulsifer in baked goods such as bread, cakes, sweet goods, biscuits, and crackers. It helps bring about rapid and intimate mixing of the shortening in the dough, improves the fermentation, water absorption, and handling of the dough, and gives a more tender, richer product (178). In baking formulations that call for soy flour, lecithinated (usually at a level of 15%) low-fat or high-fat flours can be used. Often lecithin is incorporated into shortenings (solid or fluid) that are used in baking.

In yeast-leavened doughs 0.1 to 0.3% commercial lecithin improves water absorption, ease of handling, fermentation tolerance, shortening value of fat, volume and uniformity, and shelf life (87). It is employed in cake formulations such as box mixes so that they will be wetted rapidly when mixed with water or milk. In biscuits and crackers, pies and cakes 1 to 3% lecithin (on shortening basis) promotes fat distribution and shortening action, facilitates mixing, and acts as a release agent (169).

More specifically, lecithin can be used in baking as (*1*) an emulsifier (alone or in conjunction with other emulsifiers) to reduce emulsifier cost, to stabilize emulsions and mixing of fats and water, to improve moisture tolerance, and to ensure uniform suspension of ingredients; (*2*) a wetting agent to provide instant wetting of powdered formulations to reduce mixing time; and (*3*) a parting agent to provide cleaner and faster release from a mold (179).

Bread and Rolls. Whereas views on the effects of lecithins on bread quality (loaf volume and crumb grain) vary, there is a general consensus about the beneficial effects on dough-handling properties (Aberham (180), Kuntze (181), Pomeranz (169), Puchkova and co-workers (182), Pyler (183), and Zapryagaeva and co-workers (184)).

Sullivan (185) reported that adding small amounts of lecithin improved dough consistency but had little effect on baking properties. Pratt (186) reported that adding lecithin did not affect bread volume and crumb color. Notable improvements were a more tender crust, a finer grain and smoother texture, a more symmetrical appearance, and longer keeping quality. A loaf containing 0.3% lecithin with 1% shortening had keeping qualities equivalent to those of a lecithin-free bread containing 4% shortening.

Pratt (186) concluded that lecithin increased the effect of shortening in bread. Pratt (187) also concluded that lecithin's contribution to increased shortening effectiveness reached a peak at 0.25–0.5% (flour weight basis). Walrod (188) confirmed that the addition of lecithin also improved the extensibility, dryness, and machinability of doughs, producing a bread having improved symmetry, grain, and texture.

Many workers have demonstrated the synergistic effects in breadmaking of lecithin in combination with mono- or diglycerides and other surface-acting agents. According to Hampl and Tvrznik (189), the use of lecithin in combination with monoglycerides (*1*) improves quality characteristics of the raw materials, (*2*) optimizes technical processing, (*3*) reduces shortening requirements, and (*4*) improves overall quality of the final product, including freshness retention and nutritive value.

Pomeranz (169) reviewed extensive published data by European and Russian authors. These authors also reported that lecithin improves dough conditioning (handling properties) and its addition to bread results in greater volume, improved crumb color and texture, and retardation of staling. Lecithin and mono- and diglycerides were the most commonly used surfactants in bread in the former East Germany.

Chung and Pomeranz (190) reported that fractionated lipids, especially phospholipids at 0.2%, provided a significant increase in loaf volume when shortening was added. Johnson and co-workers (191) reported that adding soy lecithin (PC) to chlorinated petroleum ether-extracted flour at 0.2%, flour weight, improved volume and grain beyond that obtained with the unextracted flour.

Chung and co-workers (192) found that petroleum-extracted polar lipids were required at a level of 180 mg per 100 g of flour (H.R.W. 12% protein) to produce bread of desirable volume. Polar lipids were 50 times more functional than protein in improving loaf volume.

Pomeranz and co-workers (193) have also studied the effect of 0.5% commercially available lecithins on the quality of bread made from untreated and petroleum ether-extracted flour, at three different shortening levels (0.0%, 0.5%, and 3.0%). Poorest results were obtained with hydrogenated lecithin which was unable to replace extracted free flour lipids as did other lecithins. Best results were obtained with alcohol-soluble soy phosphatides containing a 2:1 mixture of phosphatidylcholine (PC) and phosphatidylethanolamine (PE) in both untreated and petroleum ether-extracted flours. When added to petroleum ether-extracted flours, 0.5% alcohol-soluble phospholipids replaced 0.8% extracted free flour lipids and 3.0% shortening. Excellent results were also obtained with hydroxylated lecithin but only with shortening present.

Pyler (183) reports that hydroxylated lecithin improves dough extensibility. It has been suggested (169) that hydroxylated lecithins are particularly valuable in bakery products because of their apparent synergy with mono- and diglycerides in addition to their high dispersibility in water systems in contrast to the oil solubility of most lecithins.

Adler and Pomeranz (194) have shown that the addition of lecithin to soy flour-enriched bread can improve its consumer acceptability in the absence of shortening. Even in the presence of shortening, a small improvement was observed with the use of lecithin (195).

Mizrahi and co-workers (196) described the improving effect of soy lecithin on bread containing soy protein isolate. The use of soy lecithin in conjunction with sucrose esters exerted an improving effect on bread quality in high-protein breads made with soy flour (197).

In a study, two water-dispersible lecithins have been evaluated against hydrated monoglycerides as starch complexing agents to prevent staling in white pan bread (198). The staling indices demonstrated that water-dispersible deoiled soy lecithin gave no improvement in softening versus the control to which no emulsifier was added. An enzyme modified lecithin, however, gave a significant softening response. This behavior has been attributed to starch complexation by the lysophospholipids.

Acker (199) has found that adding 0.1 and 0.4% (flour weight) lysolecithin extracted from wheat to doughs increased bread volume 10 and 28%, respectively.

Knightly (200) has a patent on the synergistic effect of hydrophilic lecithins (HLB-8 or higher) on lipophilic surfactants such as GMS, used primarily in bread and other bakery foods to retard staling. Not only does the use of a hydrophilic lecithin result in improved shelflife, but also in improved dough conditioning as exemplified by increased loaf volume, improved symmetry, grain, and texture.

Shogren and co-workers (201) have found that lecithin and other dough-conditioning surfactants counteracted the deleterious effect of up to 15% wheat bran when added as a source of fiber in bread.

Other Bakery Products. In addition to bread, many bakery applications also have traditionally used crude lecithins. The only use in baking that is relatively recent is lecithin's function as a release agent.

Cookies, Crackers, Waffles. In the processing of cookies and other baked goods containing significant quantities of fat, lecithin promotes even dispersion of the fat throughout the dough (1).

Lecithins are relatively inexpensive and fairly potent emulsifiers in cookie and cracker technology (202). They are easily mixed into cracker and cookie doughs, modify the consistency, and make machining easier by reducing stickiness in the finished product. The effect on flakiness, tenderness, and other shortening-related factors is not so pronounced as in the dough. Greasiness of cookies with high shortening content is often reduced by adding small amounts of lecithin to the dough.

Cole and co-workers (203) studied the effect of phosphorus-containing lipids (polar lipids) and soy (PC) lecithin in the quality of cookies baked from defatted flour. They found that while those fractions containing lecithin completely restored cookie quality, a phosphorus-free lipid fraction did not. Clements and Donelson (204) found that lecithin at 0.5% and digalactosyl–diglycerides (DGDG) at 0.1%, but not monogalactosyl–diglycerides (MGDG), were effective in restoring the quality of defatted flour in cookies. Matz (202) also has reported on the improvements obtained with the use of lecithin in the production of cookies. Cookie dough is dryer and more machinable with the use of lecithin. Lecithin improves the dispersion of fat so that it more readily mixes with sugar, flour, and other ingredients. Improved emulsification also reduces mixing times. Over-development of the dough can result in lack of tenderness in the cookie. The release quality of lecithin improves the extrudability and release from the die, improving definition of impression.

Kissel and Yamazaki (205) studied the lipid extracts from wheat flour and soy and safflower lecithins on improving cookie spread when made from protein-fortified wheat flower. Soy lecithin was most effective. Soy lecithin also improved the quality of cookies made from weak flours such as sorghum and millet (206).

Waffles made from lecithin-containing doughs show better iron-grid release and easier handling. The waffles are stronger, crisper, did not become soggy, and retain freshness better (207). Pomeranz (169) reported that the use of lecithin in waffle formulations improves release from the grill, provides strength and crispiness, and reduces sogginess. He also cited the use of lecithin as a mold release agent (i.e., 0.3%) in low fat formulations such as ice cream cones.

Margarine. Lecithin has been used for many years in bakery margarine to improve its stability. When margarine is blended into a batter or dough, or particularly when rolled into a layered pastry, such as Danish pastry or croissant, the physical working of the margarine can cause the water phase to separate from the emulsion resulting in a nonuniform dispersion of the fat. In good quality baker's margarine, about 0.5% lecithin is used, usually in conjunction with 0.5–0.75% monoglyceride (178).

Cakes, Cake Mixes, Pancakes, Doughnuts, Sweet Rolls, Icings, etc. Incorporating lecithin in cake formulations substantially improves the quality of cakes (208). Hydroxylated products that have an intermediate degree of saturation were found to be best. Lecithins have been recommended as emulsifying agents for cake mixes to assure easy pan-release and to prevent the cakes from falling or dipping in the center (209).

Bradley (210) reported that the addition of lecithin to cake batter improved viscosity, provided a more uniform color, improved texture and grain, and extended shelf-life. Wolf and Sessa (69) have advocated the use of lecithin in cake doughnut formulations at 0.5 to 1.0% (based on mix weight) to accelerate mixing of the batter. Prolonged batter mixing results in less tender crumb in the finished product. Lecithin is also beneficial in white cakes and others which contain eggs, acting to replace the phospholipids normally coming from egg yolk in the formula.

According to Wolf and Cowan (66), the emulsifying properties of phosphatides find extensive use in cake mixes and instant foods. Adding 0.5 to 1.0% lecithin promotes wetting, thereby speeding up mixing of cake-doughnut mixes.

Seguchi and Matsuki (211) found that when the lipids of flour used in pancakes were extracted with ethyl ether, poor volume and texture resulted; quality was restored by adding 0.2% polar lipid, but not with nonpolar lipids even at the level of 0.5%.

In pancake mixes lecithin also assists in wetting and acts as a grill release agent. Lecithins in which soybean oil has been replaced by hard butters are useful in producing cake and cookie icings (166).

Adding lecithin improved keeping qualities, grain, and texture in sweet-dough products (coffee cakes, sweet rolls, etc.), and also produced "shorter" doughs in these items.

Incorporating lecithin into pie crusts reduced mixing time, produced flakier doughs, enhanced release, contributed to uniform browning, and aided as a moisture barrier to protect the crust (169).

Lecithin products act as aids for the blending of unlike ingredients. An excellent example is the formulation of cream fillings for sandwich cookies. The use of low levels of lecithin significantly improves the ease of blending and mouthfeel of these products which consist mostly of low polarity shortening and high polarity sugar. The lecithin serves as an intermediary to significantly reduce stiffness and mixing time of the filling (29).

Soy lecithin, however, has not in recent years been added to cake formulations in any significant quantity, either as a direct additive or in shortening because phospholipids find their way into cakes in significant quantities in the form of wheat flour lipids, eggs, and dairy products (178).

Instant Foods. Lecithin has been used as a wetting agent and emulsifier in instant foods. Foods including cocoa powder, instant drinks, instant cocoa and flavored coffee, powdered protein drinks, coffee whiteners, instant puddings, cake mixes, and instant toppings have widely employed application of specific lecithins. The most common method to incorporate lecithin is as an external coating on the powder particles. The particular lecithin to be employed largely depends on the hydrophilicity or lipophilicity of the powder system (8).

In recent years, instant food manufacturers have become interested in oil-free lecithins. These products are granular, but can be fluidized by the addition of water or fats. Additionally, fluid products are now available that are based on oil-free lecithin and have obvious advantages because of their blandness and handling properties for instant food manufacturers (212).

Pan and Food Release Agents. Lecithin-based release agents are employed in many applications such as frozen waffle manufacture, bakery products, pizza baking, and pasta products. Most industrial griddle frying fats are formulated with lecithin, solely for its release functionalities (8). The products may be spray- or brush-applied to achieve a thin film capable of promoting easy release of baked items from pans and belts. Bakery release agents may contain 2–6% lecithin in a variety of oil bases and may also be formulated with particular matter to provide an additional mechanical release.

The simplest of all food release agents are found in the category of pan oils or griddle greases. These products contain low levels of lecithin (0.5–1.0%). Most of the release action is provided by the mechanical barrier established by liberally coating the cooking surface.

Retail release agents for home use are marketed as aerosols and occasionally as pump sprays. In this setting, the release agent will be used for everything from simple release tasks like pancakes and fried eggs to more challenging systems like cakes or muffins. Consequently, retail products are formulated with relatively high levels of lecithin to provide extra release for difficult applications.

As stated previously, moderately and highly acetylated lecithins exhibit heat resistant properties which are very desirable to have in many release agent applications (170). A natural crude lecithin is subject to thermally induced reactions that are responsible for darkening and formation of insolubles that occur after prolonged heating. There are several viscosity grades of heat-resistant lecithins available; the viscosity varies with temperature. Low-viscosity lecithins can be easily spread without dilution, or prepared as part of the spray release system.

Typical formulas for lecithin-based release systems are: pan bread: oil 98%, lecithin 2%; aerosol spray: oil 70%, lecithin 8%, propellant 22%.

Release formulas for cakes, cookies and other difficult specialty products often include 5–15% lecithin, 1–10% particulates (flour, silica, etc.), and various types of oils (mineral, vegetable, etc.).

Lecithins can be directly applied to the surfaces of griddles, continuous oven conveyers, flame broiling equipment, and other cooking surfaces for better release and ease of cleaning. The thinnest layer possible should be used for surface release: i.e., 0.25–1% in griddle frying fats and 5–15% in spray release products. The lecithin can be atomized without dilution with a high pressure airless spraying (211 kg/cm^2) when warmed to 60°C. These application rates indicated may be too high for some applications.

Lecithin prepared according to U.S. Pat. 4,479,977 (170) is a very effective release agent when applied to a surface in a very thin film, or used in spray pan release systems. For use on a grill, grinders, extruders, pans, or skewers, spray coverage should be applied with lower misting rather than with an air sprayer. An alternative to the use of dip tanks for continuous band ovens is to blend 2–10% lecithin in liquid oil or melted shortening, and spray-apply the blend to the conveyor belt with an air spray system (213).

Refined fluid lecithins are also used to prevent the sticking of high-moisture sliced and shredded products like cheese. Specialty fluid lecithin products are sprayed or wiped onto sheets of processed cheese prior to slicing and stacking. Effective separation of the cheese product requires an even, very thin distribution of a low viscosity lecithin applied as a fine mist to a moving sheet of processed cheese. Lecithin also works well in separating the slices of certain natural cheeses where the manufacturing process allows the lecithin to be applied as the cheese is sliced. For separating cheese slices and shredded cheese products, 1 kg for approximately 500 m^2 (equivalent to 45,000 slices) of heat-resistant sprayable lecithin is used.

As a belt release for precooked food prepared in continuous cooking operations application is usually at the 10% level with room temperature water and shaking or mechanical stirring for 1–2 minutes (213).

Dairy-Type Foods. Another major application for lecithin products is in dairy and imitation dairy products. Lecithin is used in instant-type products such as dry milk powder, dry malted-milk powder, nutritional-drink supplements, and cocoa mixes to improve their dispersibility and wettability. Almost all infant formulas contain either hydrophilic or deoiled lecithins as fat emulsifiers (214). Egg replacers employ deoiled lecithin as a very bland emulsifier, while whipped toppings utilize lecithins as emulsifiers, and sometimes to assist foaming (8).

Processed Meats. A newer application employs deoiled lecithin as a key ingredient to assist in immobilizing animal fat in canned or frozen meat-containing products. When properly formulated, this application dramatically reduces or eliminates "fat-capping" in products such as canned chili, sloppy joes, gravies, and others containing high levels of animal fat (215).

Egg Replacers. Lecithins are used in conjunction with dairy and vegetable proteins in an attempt to functionally mimic the lipoprotein complex of egg yolks. A coagulable egg replacer based on whey protein, polyunsaturated fat, and lecithin have been described (29). Another formulation included soy and wheat flour blended with oil, lecithin, carrageenan, and polysorbate 60 to replace up to 75% dry or liquid eggs in a variety of mixes and prepared foods (29). Dashiell (29) also reported on a lipoprotein complex formed from soy isolate, oil, carbohydrate, and various emulsifiers which is claimed to be useful for whole or partial replacement of egg yolks in baked goods.

Nutritional and Health-Related Applications. Choline is an important constituent of lecithin. The adult human ingests 300–1000 mg choline per day (216). Healthy humans in the United States consume about 6 g of lecithin per day from various food sources in the diet; 100 mg of this amount is added to food during processing (217).

Choline is required in the diet of several animal species, including the dog, cat, rat, guinea pig, and humans (Zeisel (218), Zeisel and co-workers (219,220)). There is a growing body of evidence that some choline phospholipids (phosphatidylcholine, sphingomyelin, plasmalogens, and their metabolites) are involved in a myriad of essential metabolic reactions (e.g., vitamin B_{12}, folic acid, methionine, etc.) and are also important mediators and modulators of transmembrane signalling.

Decreased serum choline concentrations have been reported in humans when fed by parenteral nutrition using solutions low in choline (221). Choline ingestion of a diet deficient in choline has major consequences: these include hepatic, renal, memory and growth disorders, bone abnormalities, infertility, and in some species, fatty infiltration of the liver (217,222). The latter is probably due to a disturbance in the synthesis of lecithin which is needed to export triglycerides as part of the lipoproteins. Many studies also have demonstrated that feeding a choline-deficient diet to experimental animals is an effective promoter of carcinogenesis (216).

In spite of all the pathology associated with choline-deficient diets, Zeisel (217) states that it is difficult to ascertain whether choline-containing compounds are essential ingredients in the human diet. This is so because the extent of the role of methionine and other labile methyl donors in sparing or reducing choline requirement is still not exactly known. The expression of choline deficiency varies and is dependent upon the dietary calorie source, amino acid content of the diet, and upon the rate of growth of the species. Methyl-donors (such as methionine) can spare some of the choline requirement (217).

Methionine or vitamin B_{12} deficiency exacerbates the hepatic and renal damage associated with choline deficiency. It is also known that choline plays a part in a labile methyl group cycle. The complex relationships between choline, *S*-adenosylhomocystein (SAM) and lecithin make it difficult to be sure how much choline must be included in the diet (217).

Choline and choline phospholipids also have long been recognized as important structural components of membranes in plant and animal cells and tissues. Abnormalities in signal transduction and/or phospholipid metabolism appear to be related to certain disease states, including cancer and Alzheimer's disease. Phosphatidylinositol (PI) and its metabolites have become a focus of attention because they are involved in transmembrane signal transduction. Another choline phospholipid, the platelet activating factor, is an important intracellular messenger acting on varied functions involving leukocytes, platelets, immune responses, smooth muscle cells, liver, uterus, brain, and cardiovascular system (216,218–220).

Dietary fatty acid balance is important to membrane composition, membrane-bound enzyme activity, and hormone-mediated functions in a variety of physiological states (223).

High-dose lecithin administration has been reported to be successful in the treatment of certain mental and neurological disorders. This has created the assumption that the administration of lecithin as a dietary supplement may have some merit.

Phospholipids are also believed to be hypocholesterolemic agents (86,217). Lecithin decreases LDL-cholesterol and increases HDL-cholesterol in serum (224) and may also suppress intestinal cholesterol absorption (225).

O'Brien and Andrews (226) have reported on the influence of dietary egg and soybean phospholipids and triacylglycerols on human serum lipoproteins. In whole serum, cholesterol and phospholipids were significantly lower after soy phospholipids compared to egg phospholipids treatment periods. HDL_2 cholesterol was also significantly higher in phospholipids after the soy phospholipid treatment. Data suggest that soy phospholipids may selectively increase HDL_2-cholesterol and phospholipids.

Lecithin is available as a retail dietary supplement. The capsules or granules sold in this manner have usually less than 35% phosphatidylcholine. Physicians are using 20–30 g of lecithin for the treatment of neurological disorders. The administration of such large doses increases the caloric load of the diet. For this reason, deoiled purified lecithins (95% phosphatidylcholine) are commonly recommended.

The reader is referred to Zeisel (216–218) and Zeisel and co-workers (219,220) for more detailed information on the absorption, metabolism of lecithin, and the use of lecithin in the treatment of human diseases, such as nervous disorders (e.g., Altzheimer's disease), cardiovascular disease, and disturbances in lipid transport and the immune system.

4.3 Flavor in Food Systems

The flavor of lecithin products and the effect of lecithin on other flavoring materials has received some attention but much remains to be done. Some factors affecting flavor biogenesis in phospholipids have been reviewed by Sessa (227).

It has been reported, for example, that phospholipids have an antioxidant activity in vegetable oils due to the chelating effect or synergistic behavior of the phospholipids with other antioxidants. At the same time, phospholipids are also easily oxidized or discolored during heating or storage, and they may become precursors of off flavors in the finished products (Evans and co-workers (107,228) and Wilson and co-workers (229)).

The flavor profile of commercial lecithins can vary greatly. The typical odor/flavor profile of commodity grade soybean lecithins is slightly green and beany. Certain additional notes may be present in chemically modified or fractionated lecithins. Much of the green, beany character can be eliminated by removal of the vegetable oil fraction, leaving a slightly nutty flavored dry powder or granule. Through gas chromatography and mass spectrography, 79 volatile compounds were identified in soybean lecithin (230).

Producing bland fluid lecithins is possible by redispersing the deoiled lecithin in clean oils that have been properly refined and deodorized.

Lecithin flavors are not typically a problem in food applications because lecithins are used in most foods at extremely low levels. Lab tests have shown that in finished food formulations, lecithins can either accentuate or reduce the flavor intensity of salt, acids, and some spices. However, lecithin flavors can penetrate bland dairy-based systems. Studies have shown that 0.25% deoiled lecithin could be detected in cottage cheese (29).

Lecithin has found also some application for its apparent ability to modify certain flavors. Dashiell (29) has reported on the use of lecithin to modify flavors in chewing gum products, claiming that the lecithin reduced certain bitter flavor principles.

The major components of the hexane–ethanol azeotropic extracts of soybean meals consists of phospholipids, steryl glycosides, acylated steryl glycosides and nonpolar lipids, mainly tri-, di- and monoglycerides, sterols, and some hydrocarbons (231). In an assessment of phospholipids as a contributor to off-flavor in soybean, Sessa and co-workers (232) found that soybean lipid extracts possessing bitter taste had similar composition to commercial soybean phosphatides. The autoxidation products from model systems with soybean PC (233) resembled those identified by Sessa and co-workers (234) from bitter-tasting soybean PC that they isolated from hexane-defatted soybean flakes.

Phospholipids have been found to be prooxidants when present in oil systems due to their unsaturated fatty acid content and amino-containing moieties. As a result off-flavors and colored oil products can be produced (227).

Some works have provided evidence that phospholipids have sequestering properties for the prooxidant metals present in the oils. Phospholipids are also considered as synergists to other antioxidants such as tocopherols and flavonols. However, their synergistic effect has not yet been entirely clarified (227).

The effect of triglycerides and phospholipids on development of warmed-over flavor (WOF) in cooked meat was studied, using model systems from beef and chicken dark and light meats (235,236). Total phospholipids, especially phosphatidylethanolamine (PE), were shown to be the major contributors to the development of WOF in cooked meat. The triglycerides enhanced development of WOF only when combined with the phospholipids (as total lipids). Phosphatidylcholine (PC) did not influence WOF in the model system. In general, changes in the PUFAs of the phospholipids were shown to be related to the development of WOF in cooked meat. Additions of 156 ppm of nitrite significantly reduced TBA numbers and prevented development of WOF.

Willemot and co-workers (237) reported improvement of warmed-over aroma with a decrease in phospholipids (PLs) and PL-polyunsaturated fatty acids (PUFAs) in ground lean pork stored up to 16 days at 4°C after cooking to an internal temperature of 75°C. PLs were the main source of PUFA oxidation during warmed-over flavor development.

It could be misleading, however, to extrapolate from the flavor of an ingredient per se, or its flavor in a given food system to its probable flavor acceptability in a different food system. Most researchers who are experienced in food product development are aware of the phenomenon of flavor compatibility which ultimately determines the acceptability of an ingredient in food (238).

In summary, phospholipids affect the sensory properties (i.e., appearance, color, flavor, taste, and texture of foods, the key attributes that determine consumer acceptance. The flavor of phosphatides and their interaction with both desirable and undesirable flavor is extremely critical and contributes to the acceptability of foods containing these preparations. Since various phospholipids have different physical and chemical properties, they can affect food quality to different degrees. This depends on the phospholipid's nature and content in the oil, the presence of other compounds, and the oil system in which phospholipids exist (227).

Although much progress has been made in reducing flavor, some residual flavors always remain (239). In most foods, they are often diluted and masked when the lecithins are incorporated at low levels. Flavor is a particular problem, however, in bland foods, such as dairy products.

For more details on the role of lecithin in flavor development in foods, the reader is referred to reviews by Chang (240), Frankel (102), List and Erickson (130), Min (241), Sessa (227), and Sipos (238).

4.4 Phospholipids as Antioxidants/Synergists/Chelating Agents

Swift and co-workers (242) reported that purified phosphatidylethanolamine from cottonseed oil was a good antioxidant and increased the antioxygenic action of α-tocopherol. Evans and co-workers (107) studied the flavor stability of soybean oil using fractionated commercial lecithin. They found that the addition of 0.1% of crude phospholipids, the alcohol-soluble fraction or the alcohol-insoluble fraction to the soybean oil prior to deodorization improved the oil's oxidative stability and initial flavor, but caused significant darkening of the oil and the introduction of undesirable storage flavors when added at levels that improved the oxidative stability.

Chipault (243) attributed the antioxidant property of phosphatidylethanolamine to the phosphoric acid group, and explained the inactivity of phosphatidylcholine by the fact that the latter compound exists mainly in a zwitterion form involving the hydroxyl group of the phosphoric acid radical and the nitrogen of the basic choline. The nitrogen base of phosphatidylethanolamine is much weaker. As a result, the hydroxyl group of phosphoric acid remains mostly free. Chipault (243) and Lea (244) also reported that phospholipids are synergists to other antioxidants, such as tocopherols, and the antioxidant activity of phospholipids is confined to the phosphatidylethanolamine fraction.

Going (245) and Hudson and Mahgoub (246) found that refined and bleached oils are much more prone to develop peroxides than crude oil. They speculated that the phospholipids in crude oil provide an antioxidant function above that of tocopherols. They also found that even partially hydrogenated oil oxidizes more rapidly than the crude oil.

List and Erickson (130) also attributed the difference between refined-bleached oil and crude oil to crude oil's higher content of tocopherols, which serve as natural antioxidants, and to the higher content in phospholipids. Thus the complete removal of phospholipids by degumming and refining may be responsible for decreased oxidative stability of soybean oil.

Sims (247) reported that phospholipids are not antioxidants when present alone in oil, but some phospholipids are capable of reinforcing the action of tocopherols or other antioxidants of the phenolic type.

Parke (248) concluded that phospholipids are antioxidants for soybean oil at low concentrations of five to ten ppm, but high levels of phospholipids are prooxidant. He also indicated that phospholipids may possess the ability to chelate metal ions.

Evans and co-workers (228) and Cowan and co-workers (249) reported that phospholipids counteract the effect of metals added to vegetable oils, and undoubtedly act as inactivating agents for trace metals present in the original oils.

Dutton and co-workers (250) suggested that the antioxygenic action of phospholipids was due partly to the metal-scavenging effect of the fat-soluble phosphoric acid. Dutton and co-workers (251) also found that deodorized soybean oil containing 0.02% added acetone-precipitated phospholipids and 0.3 ppm of iron as ferric chloride had a superior flavor stability compared to the sample containing added 0.3 ppm of iron only. However, the addition of phospholipids at concentration higher than 0.1% was not recommended because of darkening of the oil.

Hudson and Mahgoub (246) reported that the two most common phospholipids, phosphatidylcholine and phosphatidylethanolamine, although nonantioxidants in their own right, act as powerful synergists in conjunction with representatives of at least two classes of naturally occurring antioxidants, the tocopherols and the flavonols. They suggested that phosphoric and citric acids in oil work as synergists by chelating traces of prooxidant metals. Chelation is enhanced by the presence in most phospholipids of a basic nitrogen with its ligand-forming potential.

The effectiveness of phospholipids as antioxidants is very limited as compared with phenolic oxidation inhibitors, but as natural food constituents, they have the advantage of being unobjectionable on toxicity grounds.

ANIMAL FEEDS, FOOD-RELATED, AND INDUSTRIAL USES

Utilization of lecithins has expanded beyond traditional applications in oil-based paints, chocolate, margarine, and bakery goods. Lecithin is now used in animal feeds (e.g. calf milk replacers), performs as a release agent for the baking and plastics industries, and is used as a dispersing agent in many industrial applications (65).

Lecithin serves as a dispersing agent in water-thinned paints, latex paints, and printing inks; as an antisludge additive in motor lubricants; and in combination with phenolic antioxidants it serves as an antigumming agent in gasoline. It serves as an emulsifier, penetrant, spreading agent, and antioxidant in the textile industry and as an antioxidant and dispersant in the production of rubber compounds. Lecithin has found a place in pesticide and herbicide formulations, and it is being used in cosmetics and the pharmaceutical industry. One of its latest applications is as a surfactant in magnetic recording media.

Miscellaneous functions of lecithins in industrial applications and a listing of their nonfood uses are given in Table 10.26 (51).

Table 10.26 Nonfood applications for lecithin, levels of use, and functions (51)

Application	Levels of Use (%)[a]	Miscellaneous Functions
Adhesives		Dispersing agent or mixing aid, plasticizer
Adsorbents		Adhesion aid, coupling agent, flocculant
Animal feeds	1–3	Antidusting agent, antioxidant, biodegradable additive, dispersing agent or mixing aid, emulsifier or surfactant, machining aid, nutritional supplement or vitamin source, wetting agent, release or nonstick agent
Catalysts		Catalyst, emulsifier or surfactant, modifier, wetting agent
Ceramics and glass		Dispersing agent or mixing aid, release or nonstick agent, water repellant
Cosmetics and soaps	1–5	Antioxidant, dispersing agent or mixing aid, conditioning agent, emollient or softening agent, emulsifier or surfactant, liposomal encapsulating agent, moisturizer, nutritional supplement or vitamin source, penetrating agent, stabilizer, wetting agent

Table 10.26 (Continued)

Application	Levels of Use (%)[a]	Miscellaneous Functions
Detergents		Anticorrosive, emulsifier or surfactant
Dust control		Adhesion aid, antidusting agent
Dyes	0.5–2	Dispersing agent or mixing aid, coupling agent
Explosives		Antidusting agent, emulsifier or surfactant, stabilizer
Fertilizers		Antidusting agent, conditioning agent, spreading aid
Inks	0.5–3	Color intensifier, dispersing agent or mixing aid, emulsifier or surfactant, grinding aid, stabilizer, suspending agent, wetting agent
Leather	1–2 of oil	Conditioning agent, emollient or softening agent, emulsifier or surfactant, lubricant, penetrating agent
Magnetic tapes	0.5–1.5	Antioxidant, dispersing agent or mixing aid, emulsifier or surfactant, lubricant, wetting agent
Masonry and asphalt products		Anticorrosive, dispersing agent or mixing aid, emulsifier or surfactant, plasticizer, release or antistick agent, strengthening agent, wetting agent
Metal processing		Anticorrosive, antispatter agent, flocculant, lubricant, release or antistick agent
Paints and other coatings	0.5–5 of pigment	Antioxidant, color intensifier, dispersing agent or mixing aid, emulsifier or surfactant, grinding aid, promoter, spreading aid, viscosity modifier, stabilizer, suspending agent, wetting agent
Paper		Dispersing agent or mixing aid, emollient or softening agent

Table 10.26 (Continued)

Application	Levels of Use (%)[a]	Miscellaneous Functions
Pesticides	0.1–5	Adhesion aid, antioxidant, biodegradable additive, dispersing agent or mixing aid, emulsifier or surfactant, penetrating agent, spreading aid, stabilizer, synergist, biologically active agent
Petroleum and other fuel products	0.005–2	Adhesion aid, anticorrosive, antioxidant, dispersing agent or mixing aid, emulsifier or surfactant, lubricant, stabilizer
Pharmaceuticals		Antioxidant, emulsifier or surfactant, liposomal encapsulating agent, machining aid, nutritional supplement or vitamin aid, stabilizer, wetting agent, biologically active agent
Polymers, including rubber	0.5–1.5	Antibleed agent (as in fat bloom), antioxidant, emulsifier or surfactant, dispersing agent or mixing aid, modifier, plasticizer, release or antistick agent, stabilizer, strengthening agent
Printing, photocopying, and photography		Color intensifier, dispersing agent or mixing aid, grinding aid, photosensitive agent, wetting agent
Release agents	2–10	Emulsifier or surfactant, lubricant, release or antistick agent
Textiles	0.2–0.5	Antidusting agent, conditioning agent, dispersing agent or mixing aid, emollient or softening agent, emulsifier or surfactant, lubricant, release or antistick agent, water repellant, wetting agent
Waste treatment		Adhesion aid, dispersing agent or mixing aid, flocculant

[a] Usage level is 0.05–1.5% unless indicated.

5.1 Animal Feeds

A portion of lecithin not isolated for commercial use is added to animal feeds. Additionally, perhaps 10–15% of commercial lecithin is sold for feeds in the manufacture of animal feeds such as calf milk replacers, and feeds for cattle, pig, poultry, fur-bearing animals, fish, and pet foods. In addition to helping stabilize the product and providing antioxidant properties, the lecithin promotes fat absorption in the digestive system. In some pathological conditions (e.g., neurological disorders, seizures in dogs, etc.) lecithin has been reported to have some beneficial effect in animals. Lysolecithin is effective in enhancing the absorption of orally administered substances (e.g., taurine) (252).

Lecithin serves principally to emulsify fats in the feeds, improving their digestion and thus increasing weight gain efficiency. This is especially important in calf milk replacer formulations and fish foods.

Soy lecithin is widely used in aquatic diets to enhance physical properties of feeds and impart a variety of nutritional attributes (253). These include enhancement of digestion and assimilation of dietary nutrients, protection against lipid peroxidation, improved feed palatibility, and growth enhancement and survival. Commercially deoiled lecithin provides an excellent source of concentrated phospholipids, especially phosphatidylcholine (PC). Investigations of various species of crustaceans and fish, expecially in early growth stages, have demonstrated the beneficial effect of PC, as well as phosphatidylinositol, on survivability and growth. Other work has shown PC as the active ingredient in refined soy lecithin for lobster molting.

In fish, liver dysfunction can result from fat-supplemented diets that are too low in emulsifiers (51,254). For this reason, some compounded and chemically modified lecithins are added which function better as fat emulsifiers than does the crude commercial lecithin.

Lecithin also serves as an appetite stimulant, antioxidant, and vitamin and mineral source in feeds and promotes healthy coats in fur-bearing animals (252). Patents in feed areas also cover bloat control (255) and particulate lipidprotein (256) formulations.

5.2 Agricultural and Agriproduct Processing Uses

Misato and co-workers (257,258) have found that soybean lecithin will arrest the development of powdery mildew on cucumbers, eggplants, green peppers, and strawberries. Sodium bicarbonate containing 0.1% lecithin inhibited citrus common green mold, cucumber powdery mildew, and rice blast better than did the bicarbonate alone (259). Lecithin is also effective against mites (260).

Lecithin can be used as an additive in agricultural pesticide formulations. It can function as an emulsifier and improve adhesions and penetration of insecticides (51). It has been reported that lecithin can combine with fatty substances of vegetable leaves forming a coating which aids in the rejection of plant viruses. Lecithin may improve plant tolerance to certain insecticides. In one study, the phytotoxic effect of a pesticide formulation containing oil was eliminated by using lecithin as an additive (261).

Phospholipids formulated with alcohols, glycols, and/or glycol ethers are reported to have antibacterial effects in agrichemical applications (262). Lecithin is a component of coating compositions recommended for a variety of fruits and vegetables (263). Post-harvest treatment of apples with aqueous emulsions containing lecithin and various fats and oils can reduce the incidence of soft scald that occurs during cold (−1°C) storage (264). Fungicidal preparations containing organic acids and lecithin have been reported to be useful for oranges (265).

5.3 Catalyst Applications

Lecithin has some catalytic or cocatalytic effect in multiphase systems due to its surface active properties, if only via its action as an emulsifier. Lecithin is reported to be useful as an emulsifier in the curing of aqueous dispersions of unsaturated polyesters (266). The products are more easily removed from their molds and have improved mechanical properties when lecithin is used. In a fermentation application, 1.5% soybean lecithin acts as an inducer in the preparation of cholesterol esterase using a strain of *Pseudomonas* bacteria (267). Aside from its role as a catalyst, initiator, or modifier, lecithin may have ancillary uses in catalyst systems as part of a protective coating (268).

5.4 Coatings

Paints and other types of coatings make up a major portion of nonfood applications for lecithin products. The broad range of functional properties of lecithin makes it highly suitable for many different coating formulations. These include paints, waxes, polishes, and wood preservatives. Lecithin is also used as a release agent in coating-related manufacture, and applications are currently being found in cosmetics and magnetic tape coatings. These latter two areas are covered separately.

The literature is replete with information on special lecithin products in coating applications (51). Various formulations have been published and special lecithin blends patented for improved functionality in specific areas. Examples of these applications are antioxidant; color intensifier agent; catalyst; conditioning aid modifier; dispersing aid; emulsifier/surfactant; grinding, mixing, or spreading aid; wetting agent; stabilizer; suspending agent; synergist; and viscosity regulation (269). Miscellaneous applications involving coating principles include cosmetics, mold release agents, magnetic tape coatings, printing and inks, toner formulations, photographic additives, and polyamide coatings (i.e., nylon leather substitutes) (51).

Since lecithin products can function as interfacial agents in paints, lacquers, and printing inks, they can serve as wetting, dispersing, suspending, and stabilizing agents in both oil-base and latex/resin emulsion paints. In paints, lacquers, and printing inks, the choice of wetting agents used depends on the nature of the pigment, the vehicle, and the processing procedure. As a rule, natural grades of lecithin have been recommended up to 1% (on pigment weight basis) and in some formulas containing carbon black or iron blues as much as 2% may be required. However, comparative tests have shown that in many coatings the tailormade refined-grade lecithins are among the best wetting and dispersing agents that can compete with other surfactants available on the market (Iveson (270), Kronstein (271), Schmidt and Orthoefer (51), and Stanley (1)).

Lecithin products can function in paints to facilitate pigment dispersion and redispersion, and also to regulate (generally reduce) viscosity. Lecithin serves as an emulsifier and wetting, dispersing, and stabilizing agent in both oil- and latex-based paints. It shortens mixing time, aids brushing, and increases the covering power of the paint. In oil-based paints lecithin has been traditionally used as a low-cost pigment grinding aid. Lecithin coats the particles of metal oxide pigments rendering them readily dispersible. If more than one pigment is used, lecithin coating helps maintain a uniform mixture (269).

In water-based paint systems water-dispersible lecithin products are recommended as low-cost emulsifiers, stabilizers, thickening agents, and spreading aids. It has been demonstrated that in latex paints pigments were dispersed more rapidly in the presence of lecithin. Paints based on rubber-type vehicles may also be formulated with lecithin to improve color uniformity. In addition, the advantages claimed for lecithin in oil-base paints (i.e., aiding pigment dispersion, shortening grinding time, preventing undesirable hard settling of pigments and eliminating deflocculation while contributing to the overall stability of the emulsion system) are still valid in water-based coatings (269).

Water-dispersible lecithins are made with chemical modification or mixing ordinary lecithin with nonionic surfactants. Many of the products recommended in the literature and technical brochures for water-based compositions include such chemically modified water-dispersable lecithin compounds (e.g., hydroxylated, acylated, fractionated, and refined grades) (271–274). Usually 0.5 to 1% modified lecithin is recommended in polyvinyl acetate-based paints, acrylic emulsions, and in butadiene–styrene emulsion paints, probably as a matter of tradition and habit.

In mixed pigmentations, lecithin additives help to maintain a uniform mixture in both alkyd and latex paints ensuring joint settling in storage. Latex paint redispersion is also made more effective with a water-dispersible lecithin additive than with a synthetic surfactant. Paint specimens with lecithin additives remixed after 28 months of storage maintained a very high level of uniform color stability (275).

Metal salts of lecithin have been patented as paint additives (276). These salts are reported to improve pigment dispersion and also act as drying promoters and inhibit yellowing of the product. The same patent also promotes the use of phosphorylated lecithin as a rust inhibitor in primer paints.

Lecithin as well as certain derivatized and hydrolyzed lecithins were found to be useful in coatings to inhibit the rusting of steel in salt spray tests (277). These findings are expected since lecithin functions as an antioxidant in a number of applications. Synergistic effects are also reported between phosphatides and other antioxidants (246,278,279).

An improved new process for the production of paint containing a chemically modified lecithin material has been patented and published (280–282), The process entails the mechanical treatment of lecithin with metal oxide without application of heat (by simply ball-milling a standard crude oil-carrier lecithin), producing lecithin-treated metal oxide and metal-oxide treated lecithin that can be used in a powder form as a pigment for the formulation of solvent- or water-based paints alone or in combination with other pigments as corrosion preventing compounds. The resulting paints have improved properties compared with paints formulated with untreated pigments.

Paints and inks are mature markets for lecithin applications. Among coatings, the most active patent area is in magnetic tapes and cosmetic uses of lecithin. Since tailormade lecithin products match the performance of many of the additives under consideration for the rapidly advancing technology in the coating industry, their future use should be bright in this area (51,269).

5.5 Cosmetics and Soaps

Lecithin, being a natural compound and having multifunctional properties, is a logical choice for cosmetic formulations. These preparations include skin creams and lotions, foundations and cleansing creams, sunscreens, soaps, bath oils, shampoos and hair conditioners, shaving creams, preshave and aftershave lotions, nail enamels, face powder, eye color creams, lipstick, and hair sprays.

Although lecithin can be used as an emulsifier, spreading agent and/or wetting agent, the primary reason lecithins chosen for use in cosmetics is its role in providing "skin feel." Use levels in cosmetics range 0.5–1.0%.

When lecithin is used primarily in a coating process which produces pigments and other particulates with smoother surfaces, improved adhesion to the skin and better color stability occurs. This translates to longer-wearing blushers, eyeshadows, and face powders. Lecithin can also reduce the undesirable oily feeling in cosmetics containing oils. In the latter, the product also wears longer and reduces the transfer of substances to clothing due to improved film adhesion. Lecithin alters the emulsion break ("rubout") during the application of day creams while at the same time acting as a moisturizer (283).

Use of a hydrogenated (egg yolk) lecithin and its metal salts (mica groups) has been described in cosmetics for coating of pigments (e.g. Kaolin) (284–286).

Lecithin, however, also has some disadvantages. It makes products more susceptable to microbial growth and, being multifunctional, its efficiency and ability to accommodate varying HLB's cannot always make it to compete with a host of customized synthetics in special situations (283).

5.6 Detergents

A patent describes the use of soybean lecithin to improve the detergency of anionic detergent compounds in the cleaning of dishes containing dried or baked soils (287). The use of soybean lecithin also has been described to improve fish cleaning competition containing alkali metal salts and soaps (288). Fractionated lecithin improves the detergency of *N*-acylglutamate salts (289).

5.7 Liposomes

Liposomes are formed by allowing phospholipids to swell in aqueous solutions. Alternatively, methanol, ethanol, or ether solutions of phosphatides may be injected into tris- and phosphate-buffered saline (290). Size can be controlled by various means including sonication (291) and extrusion trough polycarbonate membranes (292).

Giant unilamellar liposomes up to 100 microns in diameter can be prepared under certain conditions in the presence of methylglucoside (293). The aqueous compartments between bilayers can be used to entrap therapeutic molecules. The permeability of the liposome can be reduced by incorporating cholesterol into the bilayers or by modifying the lipid composition (by using dipalmitoyl phosphatidylcholine, for example). Liposomes are being studied as delivery and sustained release vehicles for drugs, vaccines, enzymes, and hormones to be used orally and intravenously.

A decrease in blood glucose was found in diabetic rats fed liposomes containing insulin, for example (294). This could lead to oral forms of insulin in man. Liposomes have been found to enhance antibody production against entrapped protein (295) which would make them useful antiviral agents.

Liposomes can also be used to encapsulate cosmetics and foods (296).

5.8 Paper, Printing, and Ink

Lecithin is reported to be useful as a dispersing and softening agent in paper manufacturing (297). Partially saponified lecithin has been recommended for this use (298). Lecithin has functionality in printing inks. It also increases the intensity of colors and serves as a grinding and remixing aid. Some patented products include multidetectable ink compositions (299) and magnetic ink (300). Lecithin is also useful in formulations for removing offset and printing inks (301). Most of the patent activity in this area recently has involved toner formulations for photocopying (302).

Oil-free phospholipids can be used as photosensitive agents for making positive and negative photographic images for lithographic prints, printed circuits, or similar uses (303,304). An electrosensitive film useful for duplicating designs uses lecithin (305). Lecithin can be used as a dispersing aid for oil-soluble photographic additives (306). It can also be used for pressure-sensitive sheets for transferring colors (307–309) or for correcting typewriter errors (310).

5.9 Pesticides

Lecithin is reported to be an active agent in mosquito control (311,312). The exact mechanism of action is unknown, but the lecithin forms a stable film on water surfaces which may interfere with the breathing mechanism of the mosquito pupae.

Lecithin can be used as an emulsifier for various pesticide formulations. Machine oil and phospholipid mixtures have been patented as fungicides (313). Lecithin functions to improve adhesion and penetration of the pesticide and may improve plant tolerances to these substances. Lecithin-containing formulations are active against mites (314) and against worm infestations on apples (315). Other pesticide applications have already been discussed under "Agricultural Uses."

5.10 Pharmaceuticals

Phospholipids are components of biological membranes in all living systems. They play important roles in biological processes such as cell permeability, the regulation of membrane-bound enzymes, and lipid and cholesterol transport (87,316). The neurological importance of phospholipids is emphasized by the phospholipid composition of dry brain tissue which is about 25% (317). Because of this biological activity, there is a great variety of applications for lecithin in the pharmaceutical field.

Aside from being an active pharmaceutical agent or a liposome-type transport agent, lecithin also has a more traditional function as an emulsifier, wetting, dispersing, and stabilizing agent in pharmaceutical applications. Representative formulations incorporating lecithin include β-lactam antibiotics (318), steroids, such as diethylstilbesterol (319), prostaglandins (320), vitamins (321), and antidiarrhea absorbent (322). Lecithin has been used as an aid to facilitate microencapsulation of pharmaceuticals with ethyl cellulose (323). Hydrogenated lecithin is used in emulsions for intravenous injections.

5.11 Mold Release Agents

A new natural lecithin-based, nontoxic mold release agent that is biodegradable and usable with materials that come into contact with food has been prompted. The odorless product, called Table L, was developed for use on most compression and injection molds. It can be used on all thermoplastics.

It does not interfere with heat sealing, part bonding, laminating, painting, ultrasonic welding, or silk screening, and it does not build up on the surface of the mold (324).

5.12 Magnetic Tape Coatings

Magnetic tape coatings are a relatively new area for lecithin use and much recent patent activity has been observed, especially among Japanese workers. Some of these tapes are reported to be suitable for video applications (325). Lecithin functions as a dispersing agent, lubricant, wetting agent, and antioxidant in these formulations. Lecithin destined for certain industrial applications may require more rigorous than usual initial refining conditions. Deoiled, refined lecithins used for magnetic tape, for example, should be derived from crude lecithin with very low peroxide and hexane insoluble values.

A typical coating contains iron oxides, sometimes doped with cobalt or chromium; a polymer system which acts as a binder; various additives, including up to 1.5% lecithin (dsb); and solvents such as methyl ethylketone and cyclohexanone (326–330). This material is milled into a dispersion and deposited onto a tape and dried.

Lecithin can also be used in a tape coating containing abrasive cleaning compounds for cleaning tape recorder heads (331). A magnetooptical imaging film can be made by coating a support with a photosensitive resin containing magnetic material (332). The coating is prepared using lecithin and the film is especially useful for Chinese character pattern memory.

Chagnon and Ferris (333) have reported a study on the effectiveness of soy lecithin as a surfactant in dispersing pigments for use in magnetic tapes (i.e., in recording media). The results have compared in audiotape samples prepared with various pigments using Centrolex P (Central Soya Co., Inc.), an oil-free soy lecithin, synthetic polyester (GAFAC RE-610, GAC Corp.), and a synthetic amine surfactant (Duomeen OL, Armak).

The conclusions from these experiment indicated that:

Oil-free soy lecithin is effective as a surfactant

Oil-free soy lecithin is effective in dispersing a wide range of magnetic particle types

Squareness, SFD (Switching Field Distribution), physical properties, and audio output, can all be enhanced by using Centrolex P as a dispersing agent in magnetic tape

Based on its chemical structure, Centrolex P should be an effective as a magnetic media dispersant in a wide range of solvents

5.13 Others

An interesting coatings application for lecithin is in producing nonpyrophoric Raney nickel (268). The lecithin in this application creates a lipophilic surface on the metal, aided with a wax, fat, or organic polymer.

Polyamide coatings containing lecithin are recommended as a finish coat in nylon leather substitutes (334). Lecithin is also used for its release and sealant properties in polymer coatings for food containers including can interior coatings, sausage casing coatings, and stocking nets for hams and other meats (296).

Lecithin is used as an emulsifier, wetting and dispersing agent, plasticizer and/or release agent in concrete, asphalt and tar shingling, linoleum tiles and surface sealants, and caulking compounds (4,296,335,336).

Lecithin is reported (at low levels, 0.05–0.1%) to improve the properties of water-glass-containing coatings for foundry cores and molds (337).

Lecithin functions as a pigment-dispersing agent and mold-release agent in the preparation of various polymer products. It facilitates dispersion of polymer resins in plasticizers to form plastisols (336).

Polymerization of vinyl chloride in the presence of poly(vinyl alcohol) and lecithin gave a poly(vinyl chloride) that was more thermally stable than polymer prepared without the lecithin (338). A curing agent for polyurethane contains lecithin as a dispersing agent (339).

Phosphatides are naturally occurring emulsifiers in rubber latex. They can be used as a foam stabilizer in foam fire extinguishing agents (340). Hydrogenated lecithins are used in lubricating oil additives.

Finally, lecithin–diamine compounds have been used as functional ingredients in multipurpose gasoline additives (341,342).

For additional lecithin uses for serving other functions, the reader is referred to List and Kleinsorgen (335), Meshandin and co-workers (343) Schmidt and Orthoefer (51,53) and Wittcoff (4).

AVAILABILITY AND ECONOMICS

Although lecithin may be obtained from other plant seeds and animal sources, soybeans provide the bulk of what is commercially available. Because of their highly priced functional value, lecithin products sell at many times the price of soybean oil. Some of the preparations and pharmaceutical uses command the highest prices.

Current world demand is estimated to be 120,000 to 150,000 metric tons (110,000–120,000 MT of standard products and 20,000 to 30,000 MT of value-added lecithins). Value-added products have been gaining market share because their specialized chemical and physical properties and the inherent limitations of the standard grades (156).

There is no industry-wide reporting of lecithin sales in specific application areas and so the exact percentage breakdown of uses is difficult to determine. This problem is also complicated by the fact that as much as 30–40% of the domestic market involves hundreds of small volume users who are serviced by distributors (51). Additionally, some customers may have both food and nonfood applications for lecithin.

For the most part, the greatest percentage of lecithin is used in foods. The second highest use is in animal feeds, followed by industrial applications. Paints and other coatings consume a major portion of the nonfood lecithin. Pharmaceuticals, including dietary supplements, and inks and cosmetics, however, also consume significant portions.

REGULATORY ASPECTS

In the United States lecithin is affirmed by the FDA as GRAS and meets standards set by the *Food Chemicals Codex* (49). It is also an approved ingredient in many important foods having a standard of identity.

Hydroxylated lecithin has special FDA approval with limitations for level and use [see 21 CFR ch. 1 (4-1-92 edition), page 72, sec. 172.814].

The U.S. Standards for Identity for Margarine and Bakery Products and Food Additive Regulations allow the use of lecithin; hydroxylated lecithin is allowed in breads without level limitations. Similarly, the use of lecithins is permitted in most countries.

A request for consideration is pending for establishing an "RDA" in the United States for choline phospholipids (lecithins) based on information available in the scientific literature, indicating the "essential" and "beneficial" nature of these compounds in nutrition and metabolism in certain health and disease states.

Both the Codex Alimentarius (*Food Chemicals Codex*, 3rd ed. 1981) and the Commission of the European Communities have established tentative criteria for Lecithins (Dir. 78/664/EEC) for food uses, and lecithins meeting these criteria are allowed under the category of "Generally Permitted Food Additives," unless they are used in "Foodstuffs With a Limited Number of Additives" (i.e., those having a "Standard of Identity" in U.S. terms).

FUTURE PROSPECTS

Just as in some food uses, some of the nonfood applications of lecithin are in mature market areas such as paints and inks. Judging from patent activity, however, growth in some areas seems reasonably assured. Among coatings, the most active patent area has been in pharmaceuticals and magnetic tapes (51).

In recent years, pharmaceutical patents accounted for almost 25% of the nonfood patent activity. Pharmaceutical applications, particularly those involving liposomes, should require increasing quantities of refined lecithins. An increased demand for lecithin as a dietary supplement is also anticipated. Cosmetic uses for lecithin should increase, particularly for hair care products. Lecithin has the virtue of being derived from a natural product as well as having good functionality for the cosmetics (51).

The major process developments at the present time are in the fractionation of lecithin. Markets appear to be developing in the pharmaceutical and cosmetic industries, particularly for concentrated forms of PC. Deoiled, granular lecithins have firmly established markets and new markets are developing which should ensure the growth of these products as emulsifiers.

Overall, the availability looks good for lecithin in both food and nonfood uses if good contact between commercial processors and patent developers can be maintained (8,51).

Part of this chapter has been adapted from *Lecithins*, B.F. Szuhaj and G. List, eds., 1985, American Oil Chemists' Society, and *Lecithins: Sources, Manufacture, Uses,* B.F. Szuhaj, ed., 1989, American Oil Chemists' Society. Permission has been granted by the American Oil Chemists' Society.

REFERENCES

1. J. Stanley, in K.S. Markley, ed., *Soybeans and Soybean Products*, Vol. 2, Interscience Publishers, Inc., New York, 1951, 26, pp. 593–647.
2. B.F. Szuhaj, ed., *Lecithins: Sources, Manufacture & Uses,* (AOCS monograph), American Oil Chemists' Society, Champaign, Ill., 1989.
3. B.F. Szuhaj, *J. Am. Oil Chem. Soc.* **60**(2), 258A–261A (1983).
4. H. Wittcoff, ed., *The Phosphatides,* American Chemical Society Monograph Series, Reinhold Publishing Corp., New York, 1951.
5. B.F. Szuhaj and G.R. List, eds., *Lecithins*, (AOCS monograph), American Oil Chemists' Society, Champaign, Ill., 1985.
6. F.J. Flider in Ref. 5, Chapt. 2, pp. 21–38.
7. B.F. Szuhaj and E.F. Sipos, in G. Charalambous and G. Doxastakis, eds., *Food Emulsifiers—Chemistry, Technology, Functional Properties and Applications,* Elsevier, Amsterdam, 1989, pp. 113–186.
8. W.E. Prosise in Ref. 5, Chapt. 8, pp. 163–182.
9. C.R. Scholfield in Ref. 5, Chapt. 1, pp. 1–20.
10. W. Schaefer and V. Wywiol, eds., *Lecithin Der Unvergleichliche Wirkstoff,* Alfred Strothe Press, Frankfurt, 1986.
11. A. Kuksis, in Ref. 2, Chapt. 4, pp. 32–71.
12. J.L. Weihrauch and Y-S. Son, *J. Am. Oil Chem. Soc.* **60,** 1971 (1983).
13. D.A. Gornall and A. Kuksis, *J. Lipid Res.* **14,** 197 (1973).
14. M. Schneider in Ref. 2, Chapt. 7, pp. 109–130.
15. N. Satirhos, B. Herslof and L. Kenne, *J. Lipid Res.* **27,** 386, (1986).
16. A. Kuksis in Ref. 5, Chapt. 7, pp. 105–162.
17. O.S. Privett, M.L. Blank, and J.A. Smith, *J. Food Sci.* **27,** 463 (1962).
18. R.C. Noble and J.H. Moore, *Can. J. Biochem.* **43,** 1677 (1965).
19. W.H. Cook and W.G. Martin, in E. Tria and A. Scanu, eds., *Structural and Functional Aspects of Lipoproteins in Living Systems,* Academic Press, Inc., New York, 1969, pp. 579–615.
20. D.A. Gornall and A. Kuksis, *Can. J. Biochem.* **49,** 51
21. P.W. Connely and A. Kuksis, *Can. J. Biochem.* **61,** 63 (1983).

22. W.R. Morrison, E.L. Jack, and L.M. Smith, *J. Am. Oil Chem. Soc.* **42,** 1142 (1965).
23. O.S. Privett, L.J. Nutter, and R.A. Gross, in H.F. Brink and D. Kritchevsky, eds., *Dairy Lipids and Lipid Metabolism,* Avi Publishing Co., Westport, Conn., pp. 99–115.
24. I.M. Santha and K.M. Narayanan, *J. Food Sci. Techn.* **15,** 24 (1978).
25. A.N. Siakotos and S. Fleischer, *Lipids* **4,** 234 (1969).
26. A.N. Siakotos, G. Rouser, and S. Fleischer, *Lipids* **4,** 239 (1969).
27. T.W. Scott, Setchall, and J.M. Bassett, *Biochem. J.* **104,** 1040 (1967).
28. R.E. Wuthier, *J. Lipid Res.* **7,** 544 (1966).
29. G.L. Dashiell in Ref. 2, Chapt. 14, pp. 213–224.
30. J.P. Cherry and W.H. Kramer in Ref. 2, Chapt. 3, pp. 16–31.
31. E.J. Weber in Ref. 5, Chapt. 3, pp. 39–56.
32. M.E. Weeks and A. Walters, *Soil Sci. Soc. Am., Proc.* **11,** 189 (1946).
33. C.R. Scholfield, T.A. McGuire, and H.J. Dutton, *J. Am. Oil Chem. Soc.* **27,** 352 (1950).
34. J.P. Cherry in Ref. 5, Chapt. 4, pp. 57–78.
35. H.S. Olcott, *Science* **100,** 226 (1944).
36. Y.A. El-Shattory, *Studies on Cottonseed Phospholipids*, Ph.D. dissertation, National Research Center, Cairo, 169 pages.
37. J.P. Cherry, *J. Am. Oil Chem. Soc.* **60,** 360 (1983).
38. J.P. Cherry and H.R. Leffler, in R.J. Kohel and C.F. Lewis, eds., *Cotton Monograph,* Am. Soc. Agron. Crop Sci. Soc. Amer., Madison, Wis., 1984, p. 511.
39. L.A. El-Sebaiy, A.R. El-Mahdy, E.K. Moustafa, and M.S. Mohamed, *Food Chem.* **5,** 217 (1980).
40. B. Rewald, *J. Soc. Chem. Ind.* **56,** 403T (1937).
41. F.W. Hougen, V.J. Thompson, and J.K. Daun in Ref. 5, Chapt. 5, pp. 79–103.
42. M. Sosada *J. Am. Oil Chem. Soc.* **70,** 405–410 (1993).
43. W.H. Morrison III in Ref. 5, Chapt. 6, pp. 97–104.
44. J. Hollo, *INFORM* **3**(4) 483 (1992).
45. T.P. Hilditch and Y.A.H. Zaky, *Biochem. J.* **36,** 815 (1942).
46. B. Rewald, *Biochem J.* **36,** 822 (1942).
47. J.G. Parsons and P.B. Price, *J. Agr. Food Chem.* **27,** 913–915 (1980).
48. C. Ratledge in Ref. 2, Chapt. 5, pp. 72–96.
49. *Food Chemicals Codex*, 3rd ed., National Academy Press, Washington, D.C., 1981, pp. 147–148 and 166–167.
50. R.D. Cowell, D.R. Sullivan, and B.F. Szuhaj, in *Surfactant Science Series,* Vol. 12, Marcel Dekker, Inc., New York, 1982, pp. 230–264.
51. J.C. Schmidt and F.T. Orthoefer in Ref. 5, Chapt. 9, pp. 183–202.
52. National Soybean Processors' Association (NSPA), *Yearbook and Trading Rules (1986–1987),* The Association, Washington, D.C., 1987.
53. J.C. Schmidt and F.T. Orthoefer in Ref. 5, Chapt. 10, pp. 203–212.
54. W. Van Nieuwenhuyzen, *J. Am. Oil Chem. Soc.* **53,** 425–427 (1976).
55. *Code of Federal Regulations, Title 21:172.814 (Food and Drugs),* U.S. Government Printing Office, Washington, D.C., 1977.
56. L.A. Horrocks in Ref. 2, Chapt. 1, pp. 1–6.
57. R.A. Lantz in Ref. 2, Chapt. 10, pp. 162–173.
58. National Soybean Processors' Association (NSPA), *Yearbook and Trading Rules (1976–1977),* The Association, Washington, D.C., 1977.

59. B.F. Szuhaj, private communication, Central Soya Co., Inc., Fort Wayne, Ind., 1978.
60. V.P. Rzhakhin, N.I. Pogonkina, and I.A. Sloveva, *Maslob.-Zhir. Prom.* **30,** 11 (1965); *Chem. Abstr.* **62:**8018b.
61. B. Vijayalakshmi and S.V. Rao, *Lipids* **9,** 82 (1972).
62. H.G.M. Daga, *Indian Chem. J.* **11,** 17 (1976); *Chem. Abstr.* **85:**190977g (1976).
63. H.G.M. Daga, *Paintindia* **27,** 21 (1977); *Chem. Abstr.* **90:**82543z (1979).
64. L. Rydhag and I. Wilton, *J. Am. Oil Chem. Soc.* **83,** 830 (1981).
65. D.R. Sullivan and B.F. Szuhaj, *J. Am. Oil Chem. Soc.* **52,** 125A, abstract no. 39, presented at Am. Oil Chem. Soc. Meeting, Dallas, Tex., April 1975.
66. W.J. Wolf and J.C. Cowan, *CRC Crit. Rev. Food Technol.* **2,** 81 (1971).
67. P.K. Clark and H.E. Snyder, *J. Am. Oil Chem. Soc.* **68,** 814–817 (1991).
68. P. Sartoretto in E.A. Parolla, ed., *Kirk-Othmer Encyclopedia of Chemical Technology,* Vol. 12, 2nd ed. John Wiley & Sons, Inc., New York, 1967, pp. 343–361.
69. W.J. Wolf and D.J. Sessa, in M.S. Peterson and A.H. Johnson, eds., *Encyclopedia of Food Technology and Food Sciences,* AVI Publishing Co., Inc., Westport, Conn., 1978, pp. 461–467.
70. R. Aneja, J.S. Chadha, and R.W. Yoell, *Fette Seifen, Anstrichm.* **73,** 643–651 (1971).
71. J.K.G. Kramer, F.D. Sauer, and E.R. Farnsworth in Ref. 2, Chapt. 6, pp. 97–108.
72. C. Hanras and J.L. Perrin, *J. Am. Oil Chem. Soc.* **68,** 804–808 (1991).
73. P. Therond, M. Couturier, J.F. Demelier, and F. Lemonnier, *Lipids* **28,** 245–249 (1993).
74. W.L. Boatright and H.E. Snyder, *INFORM* **3**(4), 541 (1992).
75. M. Ghyczy in Ref. 2, Chapt. 8, pp. 131–144.
76. Y. Totani and S. Hara, *J. Am. Oil Chem. Soc.* **68,** 848–851 (1991).
77. American Oil Chemists' Society (AOCS), *Official Methods and Recommended Practices,* Champaign, Ill., 1989.
78. W.N. Marmer in Ref. 5, Chapt. 12, pp. 247–288.
79. R.G. Ackman, *INFORM* **5**(10), 1119–1128 (1994).
80. C.R. Scholfield in Ref. 2, Chapt. 2, pp. 7–15.
81. E. Baer and M. Kates, *J. Biol. Chem.* **185,** 615 (1950).
82. H. Brockerhoff, *J. Lipid Res.* **4,** 96 (1963).
83. D.J. Hanahan, *Lipid Chemistry,* John Wiley & Sons, New York, 1960.
84. E.H. Pryde in Ref. 5, Chapt. 11, pp. 213–246.
85. K.P. Strickland in G.B. Ansell, J.N. Hawthorn, and R.M.C. Dawson, eds., *Form and Function of Phospholipids,* Elsevier Publishing Co., Amsterdam, 1973, p. 9.
86. H.M. Verheij, in R.M. Burton and F.C. Guerra, eds., *Fundamentals of Lipid Chemistry,* BI Science Publication Division, Webster Groves, Mo., 1974, p. 225.
87. J. Eichberg in M. Grayson, ed., *Kirk-Othmer Encyclopedia of Chemical Technology,* Vol. 14, 3rd ed., 1981, John Wiley & Sons, Inc., New York, pp. 250–262.
88. J.N. Hawthorn and P. Kemp, *Adv. Lipid Res.* **2,** 127 (1964).
89. A. Kuksis, *Prog. Chem. Fats and Other Lipids* **12,** 5 (1972).
90. F. Snyder in H. Busch, ed., *Cancer Research,* Vol. 6, Academic Press, Inc., New York, 1971, pp. 410–419.
91. L.L.M. Van Deenen and D.H. DeHaas, *Adv. Lipid Res.* **2,** 167 (1964).
92. R. Kanamoto and co-workers, *Agric. Biol. Chem.* **41,** 2021 (1977).
93. A. Beckwith, *J. Agric. Food Chem.* **32,** 1397 (1984).
94. W.S. Chen and W.G. Soucie, *J. Am. Oil Chem. Soc.* **62,** 1686 (1985).

95. H. Nielsen *Lipids* **16**(4), 215–222 (1981).
96. D. Johansson and B. Bergenstahl, *J. Am. Oil Chem. Soc.* **69,** 705–717 (1992).
97. M.F. El-Tarras, M.A. Moety, A.K.S. Ahmad, and M.M. Amer, *Oleagineux* **31,** 229 (1976).
98. F. Tomioka and T. Kaneda, *Yukagaku* **25,** 784 (1976); *Chem. Abstr.* **86:**42075 (1977).
99. F. Tomioka and T. Kaneda, *Kagaku To Seibutsu* **14,** 509 (1976); *Chem. Abstr.* **86:**15426 (1977a).
100. F. Tomioka and T. Kaneda, *Yukagaku* **23,** 777 (1974), *Chem. Abstr.* **82:**96613 (1975).
101. E. Sambuc, Z. Dirik, G. Reymond, and M. Naudet, *J. Am. Chem. Soc.* **58**(7), 645A (1981); *Rev. Fr. Crops Gras* **28**(1), 13–19 (1981).
102. E.N. Frankel in D.R. Erickson, E.H. Pryde, O.L. Brekke, T.L. Mounts, and R.A. Falb, eds., *Handbook: Soy Oil Processing and Utilization,* American Soybean Assoc., St. Louis, Mo., and American Oil Chemists' Society, Champaign, Ill., 1980, Chapt. 14, pp. 229–244.
103. E.N. Frankel in E.H. Pryde, ed., *Fatty Acids,* American Oil Chemists' Society, Champaign, Ill., 1979, Chapt. 18.
104. W.O. Lundberg, ed., Autoxidation and Antioxidants, Vol. 1, Interscience Publishers, New York, 1961.
105. N.A. Porter, R.A. Wolf, and H. Weenen, *Lipids* **15**(3), 163–167 (1980).
106. P. Brandt, *Lebensmittel Ind.* **201,** 31 (1973).
107. C.D. Evans, P.M. Cooney, C.R. Scholfield, and H.J. Hutton, *J. Am. Oil Chem. Soc.* **31,** 295 (1954).
108. S.Z. Dziedzic, *J. Am. Oil Chem. Soc.* **61,** 1042 (1984).
109. B.J.F. Hudson and co-workers, *Lebensmittel-Wissenschaft und Technol.* **17**(4), 191 (1984).
110. J. Dziedzic, *Agric. Food Chem.* **34,** 1027 (1986).
111. L.R. McLean, K.A. Hagaman, and W.S. Davidson, *Lipids* **28,** 505–509 (1993).
112. R. Brian, *J. Am. Oil Chem. Soc.* **53,** 27–29 (1976).
113. B. Braae, *J. Am. Oil Chem. Soc.* **53,** 353–357 (1976).
114. G.R. List, C.D. Evans, L.T. Black, and T.L. Mounts, *J. Am. Oil Chem. Soc.* **55,** 275–276 (1978).
115. G.R. List, T.L. Mounts, K. Warner, and A.J. Heakin, *J. Am. Oil Chem. Soc.* **55,** 277–280 (1978a).
116. K. Nielsen, *J. Am. Oil Chem. Soc.* **37,** 217–219 (1960).
117. G.R. List in Ref. 2, Chapt. 9, pp. 145–161.
118. J.P. Burkhalter, *J. Am. Oil Chem. Soc.* **53,** 332–333 (1976).
119. G.R. List, J.M. Avellenada, and T.L. Mounts, *J. Am. Oil Chem. Soc.* **58,** 892 (1981).
120. L.S. Crauer, Paper presented at *Meeting of International Society for Fat Research,* Goteborg, Sweden, The DeLaval Separator Co., Poughkeepsie, N.Y., 1972.
121. W. Van Nieuwenhuyzen, *J. Am. Oil Chem. Soc.* **58,** 886 (1981).
122. E. Bernardini, *The New Oil and Fat Technology,* Vol. 2, rev. ed., Technologie s.r.l., Rome, Part 3, Sec. 5, pp. 505–538, 1973.
123. F.A. Norris, in D. Swern, ed., *Bailey's Industrial Oil and Fat Products,* 3rd ed., Wiley-Interscience, New York, 1964, Chapt. 16, pp. 719–792.
124. R.A. Carr, *J. Am. Chem. Soc.* **53,** 346–352 (1976).
125. A.J.C. Andersen, in P.N. Williams, ed., *Refining of Oils and Fats for Edible Purposes,* rev. ed., Pergamon Press, New York, 1962, Chapt. 2, Sec. 1, pp. 28–42.
126. K. Klein and L.S. Crauer, *J. Am. Oil Chem. Soc.* **51,** 368–370 (1974).
127. W.J. Pobdielniak, A.M. Gavin, and H.R. Kaiser, *J. Am. Oil Chem. Soc.* **33,** 24–26 (1956).
128. F.E. Sullivan, *J.Am. Oil Chem. Soc.* **32,** 121–123 (1955).

129. O.L. Brekke in D.R. Erickson, E.H. Pryde, O.L. Brekke, T.L. Mounts, and R.A. Falb, eds., *Handbook of Soy Oil Processing and Utilization,* American Soybean Association, St. Louis, Mo. and American Oil Chemists' Society, Champaign, Ill., 1980, Chapt. 6, pp. 71–88.
130. G.R. List and D.R. Erickson in Ref. 129, Chapt. 16, pp. 267–354.
131. G.R. List, J.W. King, J.H. Johnson, K. Warner, and T.L. Mounts, *INFORM*, **4**(4), 510 (1993).
132. A.J. Dijkstra, *INFORM* **4**(4), 538 (1993).
133. C.R. Scholfield and H.J. Dutton, *J. Am. Oil Chem. Soc.* **31,** 258 (1954).
134. *Fed. Reg.* **48**(216), 51149–51150 (Nov. 7, 1983).
135. R. Leysen, *J. Am. Oil Chem. Soc.* **56,** 892A–894A (1979).
136. N.W. Myers, *J. Am. Oil Chem. Soc.* **34,** 93–96 (1957).
137. B. Braae, U. Brimberg, and M. Nyman, *J. Am. Oil Chem. Soc.* **34,** 293–299 (1957).
138. G.R. List, T.L. Mounts, A.C. Lanser, and R.K. Holloway, *INFORM.* **1**(12), 1091 (1990).
139. G.R. List, T.L. Mounts, and A.C. Lanser, *J. Am. Oil Chem. Soc.* **69,** 443–446 (1992).
140. G.R. List and T.L. Mounts, *J. Am. Oil Chem. Soc.* **70,** 639–641 (1993).
141. F. Zhang, S.S. Koseoglu, and K.C. Rhee, *J. Am. Oil Chem. Soc.* **71,** 1145–1148 (1994).
142. B. Braae, *Chem. Ind. (London)* **36,** 1152–1160 (1958).
143. F.E. Sullivan (to the De Laval Separator Co.), U.S. Pat. 2,702,813 (1953).
144. L.P. Hayes and H. Wolff (to A.E. Staley Manufacturing Co.), U.S. Pat. 2,881,195 (1959).
145. L.P. Hayes and H. Wolff (to A.E. Staley Manufacturing Co.), U.S. Pat. 2,792,411 (1957).
146. L.P. Hayes and H. Wolff (to A.E. Staley Manufacturing Co.), U.S. Pat. 2,782,216 (1957a).
147. C.D. Evans, G.R. List, R.E. Beal, and L.T. Black, *J. Am. Oil Chem. Soc.* **51,** 444–448 (1974).
148. R. Ohlson and C. Svenson, *J. Am. Oil Chem. Soc.* **53,** 8–11 (1976).
149. A.M. Nash, E.N. Frankel, and W.F. Kwolek, *J. Am. Oil Chem. Soc.* **61,** 921 (1984).
150. H.J. Ringers (to Lever Brothers Co.), U.S. Pat. 4,049,686 (1977).
151. G. Lunde, L.H. Landmark, and J. Gether, *J. Am. Oil Chem. Soc.* **53,** 207–210 (1976).
152. A. Hvolby, *J. Am. Oil Chem. Soc.* **48,** 503–509 (1971).
153. A. Letan and A. Yaron, *J. Am. Oil Chem. Soc.* **49,** 702 (1972).
154. K. Nielsen, *Studies on the Nonhydratable Soybean Phosphatides* Trans. V. Bonde, Maxsons & Co., Ltd., London, 1956.
155. W.J. Wolf and D.J. Sessa, in M.S. Peterson and A.H. Johnson, eds., *Encyclopedia of Food Science*, AVI Publishing Co., Westport, Conn., 1975, p. 476.
156. Central Soya Co., Inc. unpublished data, Fort Wayne, Ind., 1994.
157. *Codex Alimentarius,* FAO/WHO, Rome.
158. P.L. Julian, U.S. Pat. 2,629,662 (1953).
159. M.J. Haas, K. Scott, W. Jun, and G. Janssen, *J. Am. Oil Chem. Soc.* **71,** 483–490 (1994).
160. A.M. Aura, P. Forssell, A. Mustranta, T. Suortti, and K. Poutanen, *J. Am. Oil Chem. Soc.* **71,** 887–891 (1994).
161. H. Betzing, Ger. Pat. 1,902,607 (1969).
162. S. Peter, Ger. Pat. Appl. P4,222,153.6. (1992).
163. H. Lee, B.H. Chung, and Y.H. Park, *J. Am. Oil Chem. Soc.* **68,** 571–573 (1991).
164. R.A. Lantz in Ref. 2, Chapt. 14, pp. 213–224.
165. L. Rydhag, *Fette, Seifen, Anstrichmittel.* **81,** 169 (1979).
166. Anonymous, Lucas Meyer Co., Hamburg, Germany, 1978.
167. J.D. Weete, S. Betageri, and G.L. Griffith, *J. Am. Oil Chem. Soc.* **71,** 731–737 (1994).
168. P. Becher, *Emulsions: Theory and Practice,* 1965.

169. Y. Pomeranz in Ref. 2, Chapt. 13, pp. 289–322.
170. G.L. Dashiell and W.E. Prosise, (to Central Soya Co., Inc.), U.S. Pat. 4,479,977 (1984).
171. H.T. Iveson, *Soybean Dig.* **21**(8), 18–19 (1961).
172. P.M. Scocca, *J. Am. Oil Chem. Soc.* **53,** 428–429 (1976).
173. B.F. Szuhaj and D.E. Sincroft, Paper presented at *164th National Meeting of the American Chemical Society,* New York, Aug. 27–Sept. 1, the Society, Washington, D.C., 1972.
174. T.A.M. Fujikawa and M. Hamashima, Jpn. Kokai Tokkyo Koho, JP 83910 (1979); *Chem. Abstr.* **91:**17369p.
175. T.K. Mag, *INFORM.* **5**(8), 926–929 (1994).
176. R.C. Appl in Ref. 2, Chapt. 13, pp. 207–212.
177. R.D. Sinram and H. Schmitt, *INFORM* **4**(4), 509 (1993).
178. W.H. Knightly in Ref. 2, Chapt. 11, pp. 174–196.
179. Anonymous, *ADM Soy Products for the Baking Industry,* ADM, Decatur, Ill., 1975.
180. H. Aberham (1972). *Deutsche Mueller-Z.* **70**(11), 228 (1972).
181. R. Kuntze, *Proceedings of the 7th International Congress on the Problems of Modern Grain Processing and Cereal Chemistry,* Bergholz-Rehbruecke, 1972, pp. 73–82.
182. L.I. Puchkova and co-workers, *Izv. Vysshikh Uchebnykh Zaved. Pishchevaya Technolog.* **2,** 68 (1975); *Food Sci. Techn. Abstr.* **8,** 2mM272 (1976).
183. E.J. Pyler, *Baking Science and Technology,* Seibel Publishing Co., Chicago, 1973, pp. 464–485.
184. A.V. Zapryagaeva, L.I. Puchkova, G.S. Fedorova, and L.E. Chernenko, *Izv. Vysshikh Uchebnykh Zav. Pischchevaya Technolog.* **1,** 53 (1970).
185. B. Sullivan, *Cereal Chem.* **27,** 661 (1940).
186. D.B. Pratt, Jr., *Baker's Helper* **84,** 38, 64, 66, 99 (1945).
187. D.B. Pratt, Jr., *Food Ind.* **18,** 16 (1946).
188. F.E. Walrod in *Proceedings of the 23rd Annual Meeting of the American Society of Bakery Engineers,* 1947, p. 76.
189. J. Hampl and K. Tvrznik in Ref. 181, pp. 59–64.
190. O.K. Chung and Y. Pomeranz, *Bakers Digest* **51**(5), 32 (1977).
191. H.C. Johnson, R.C. Hosney, and E. Varriano-Marston, *Cereal Chem.* **56,** 333 (1979).
192. O.K. Chung, Y. Pomeranz, and K.F. Finney, *Cereal Chem.* **59,** 14 (1982).
193. Y. Pomeranz, M.D. Shogren, and K.F. Finney, *Food Techn.* **22,** 897 (1968).
194. L. Adler and Y. Pomeranz, *J. Sci. Food Agric.* **8,** 449 (1959).
195. K.F. Finney, Y. Pomeran, and G. Rubenthaler, *Cereal Sci. Today* **8,** 166 (1963).
196. S. Mizrahi, Z. Berk, and U. Kogan, *Cereal Sci. Today* **10,** 45 (1965).
197. Y. Pomeranz, M.D. Shogren, and K.F. Finney, *Cereal Chem.* **46,** 512 (1969).
198. S. Onata and co-workers, *Application of Enzymatic Modification of Phospholipids in Breadmaking,* Am. Assoc. Cereal Chem. 69th Annual Meeting.
199. L. Acker, *Fette, Seifen, Anstrichm.* **79**(1), 1 (1977).
200. W.H. Knightly, U.S. Pat. 4,684,526 (1987).
201. M.D. Shogren, Y. Pomeranz, and K.F. Finney, *Cereal Chem.* **58,** 142 (1981).
202. S.A. Matz, *Cookie and Cracker Technology,* AVI Publishing Co., Westport, Conn., 1968, pp. 54–55 and 121–122.
203. E.W. Cole, D.K. Mecham, and J.W. Pence, *Cereal Chem.* **37,** 109 (1960).
204. R.L. Clements and J.R. Donelson, *Cereal Chem.* **58,** 204 (1981).
205. L.T. Kissel and W.T. Yamazaki, *Cereal Chem.* **52,** 638 (1975).
206. S.M. Badi and R.C. Hosney, *Cereal Chem.* **53,** 733 (1976).

207. W. Schaefer, *Getreide, Mehl, Brot.* **27,** 132 (1973).
208. C.C. Elsesser and S.W. Bogyo U.S. Pat. 2,954,297 (1960).
209. C.F. Obenauf and C.W. Tatter, U.S. Pat. 3,060,030 (1962).
210. W.B. Bradley, *Proceedings American Society Bakery Engineers,* 1948, p. 53.
211. M. Seguchi and J. Matsuki, *Cereal Chem.* **54,** 918 (1977).
212. Anonymous, *Cereal Foods World* **24,** 52 (1979).
213. Central Soya Co., Inc., technical brochure, Fort Wayne, Ind., 1993.
214. E.H. Sander, in Ref. 2, Chapt. 12, pp. 197–206.
215. J.A. Haggerty and D.D. Corbin, (to Central Soya Co., Inc.), U.S. Pat. 4,472,448 (1984).
216. S.H. Zeisel in Ref. 2, Chapt. 15, pp. 225–236.
217. S.H. Zeisel in Ref. 5, Chapt. 14, pp. 323–373.
218. S.H. Zeisel, *J. Nutr. Biochem.* **1,** 332–349 (1990).
219. S.H. Zeisel and D.J. Canty, *J. Nutr. Biochem.* **4**(5), 258–263 (1993).
220. S.H. Zeisel and co-workers, *FASEB J.* **5,** 2093–2098 (1991).
221. M.E. Burt, I. Hanin, and M.F. Brennan, *Lancet* **2,** 638 (1980).
222. R.T. Bartus, R.L. Dean, A.J. Goas, and A.S. Lippas, *Science* **209,** 301 (1980).
223. M.T. Clandinin, *INFORM.* **3**(4), 519 (1992).
224. M.T. Childs, J.A. Bowlin, J.T. Ogilvie, W.R. Hazzard, and J.J. Albers, *Atherosclerosis* **38,** 217 (1981).
225. A.J. Rampone and C.M. Machida, *J. Lipid Res.* **22,** 744 (1981).
226. B.C. O'Brien and V.G. Andrews, *Lipids* **28,** 7–12 (1993).
227. D.J. Sessa in Ref. 5, Chapt. 15, pp. 347–374.
228. C.D. Evans, A.W. Schwab, H.A. Moser, J.E. Hawley and E.H. Melvin, *J. Am. Oil Chem. Soc.* **28,** 68 (1951).
229. B.R. Wilson, A.M. Pearson, and D.H. Shorland, *J. Agric. Food Chem.* **24,** 7 (1976).
230. K. Heasook, C.-T. Ho, and S.S. Chang, *J. Am. Oil Chem. Soc.* **61,** 1235–1238 (1984).
231. D.H. Honig, D.J. Sessa, R.L. Hoffman, and J.J. Rackis, *Food Technol.* **23,** 95 (1969).
232. D.J. Sessa, D.H. Honig, and J.J. Rackis, *Cereal Chem.* **46,** 675 (1969).
233. J.F. Mead, *J. Am. Oil Chem. Soc.* **57,** 393 (1980).
234. D.J. Sessa, H.W. Gardner, R. Kleiman, and D. Weisleder, *Lipids* **12,** 613 (1977).
235. J.O. Igene and A.M. Pearson, *J. Food Sci.* **44**(5), 1285–1290 (1979).
236. J.O. Igene and A.M. Pearson, *J. Food Sci.* **44,** 1285 (1979).
237. C. Willemot, L.M. Poste, J. Salvador, and D.F. Wood, *Can. Inst. Food Sci. Technol. J.* **18,** 316–322 (1985).
238. E.F. Sipos in D.B. Min and T.H. Smouse, eds., *Flavor Chemistry of Lipid Foods*, American Oil Chemists' Society, Champaign, Ill., 1989, Chapt. 15, pp. 265–289.
239. J.E. Kinsella and S. Damodaran in G. Charlambous, eds., *The Analysis and Control of Less Desirable Flavors in Foods and Beverages,* Academic Press, Inc., New York, 1980, p. 95.
240. S.S. Chang, K.M. Brobst, H. Tai and C.E. Ireland, *J. Am. Oil Soc.* **38,** 671 (1961).
241. D.B. Min and T.G. Stasinopoulos in Ref. 5, Chapt. 16, pp. 375–384.
242. C.E. Swift, and G.S. Jamieson, *Oil and Soap* **19,** (1942).
243. J.R. Chipault, in W.O. Lundberg, ed., *Autoxidation and Antioxidants,* Wiley-Interscience Publishers, New York, 1962.
244. C.H. Lea, *J. Sci. Food Agric.* **8,** 1 (1957).
245. L.H. Going, *J. Am. Chem. Soc.* **45,** 632 (1968).
246. B.J.F. Hudson and S.E.O. Mahgoub, *J. Sci. Agric.* **32,** 208–210 (1981).

247. R.P.A. Sims, in W.O. Lundberg, ed., *Autoxidation and Antioxidants,* Wiley-Interscience, Publishers, New York, 1962.
248. D.W. Parke, M.S. Thesis, Ohio State University, Columbus, Ohio, 1981.
249. J.C. Cowan, P.M. Cooney, and C.D. Evans, *J. Am. Oil Chem. Soc.* **39,** 6 (1962).
250. H.J. Dutton, A.W. Schwab, H.A. Moser, and J.C. Cowan, *J. Am. Oil Chem. Soc.* **25,** 385 (1948).
251. H.J. Dutton, A.W. Schwab, H.A. Moser, and J.C. Cowan, 1949. *J. Am. Oil Chem. Soc.* **26,** 441 (1949).
252. F.W. Kullenberg in Ref. 2, Chapt. 16, pp. 237–252.
253. S.P. Meyers, *INFORM.* **4**(4), 505 (1993).
254. J. Altfeld and co-workers, *50 Years of Lucas Meyer-25 Years Lecithin,* and references cited therein. Lucas Meyer Co., Hamburg, 1973, pp. 74–76.
255. R.F. Holmes and C.K. Kandel, S. Afr. Pat. ZA7,405,866 (1975).
256. A.B.M. Cloosterman and S. Van Kranenburg, Eur. Pat. 57,989 A2, 1982; CA 97:180608 (1982).
257. T. Misato and co-workers, (to Ajinomoto Co., Ltd.), U.S. Pat. 3,873,700 (1975).
258. T. Misato, Y. Homma and K. Ko, *Neth. J. Plant Pathol.* **83**(Suppl. 1), 395–402 (1977).
259. Y. Homma and co-workers, *Nippon Noyaku Gakkaishi* **6,** 145–153; *Chem. Ab.* **95:**127244.
260. Ajinomoto, Jpn Kokai Tokkyo Koho 81 41383 (1981).
261. G. Kalocsai and co-workers, *Hung. Teljes* **18,** 326 (1980).
262. M. Okauchi and co-workers, Jpn. Kokai Tokkyo Koho, JP 76104033 (1976).
263. W.L. Shillington and J.J. Liggett U.S. Pat. 3,533,810 (1970).
264. R.B.H. Wills and co-workers, *J. Sci. Food Agric.* **31,** 663–666; *Chem. Ab.* **94:**28976 (1950).
265. T. Misato and co-workers, Jpn. Kokai Tokkyo Koho JP 7622825 (1976).
266. T. Otsuki and K. Nagashima, Jpn. Kokai Tokkyo Koho JP 7545086 (1975).
267. K. Beaucamp and co-workers. Ger. Pat. DE 2,933,648 (1981).
268. P. Nayler, Eur. Pat. Appl. 20,123; *Chem. Ab.* **94:**128176 (1980).
269. E.F. Sipos in Ref. 2, Chapt. 18, pp. 261–276.
270. H.T. Iveson, Central Soya RD Report, Lc-2, Central Soya Co., Inc., Chicago, Ill., 1967.
271. M. Kronstein, *Mod. Paint Coat.* **68**(2), 59–63 (1978).
272. M. Kronstein, *Mod. Paint Coat.* 19–27 (Feb. 1975).
273. M. Kronstein, *Am. Chem. Soc., Div. Org. Coat. Plast. Chem.* **34**(2), 521–526 (1974).
274. M. Kronstein *Paint Ind. Mag.* **72,** 6, 10 (1957).
275. M. Kronstein and J. Eichberg U.S. Pat. 4,056,494 (1977).
276. M. Kronstein and J. Eichberg U.S. Pat. 2,997,398 (1961).
277. H. Pardun (1981). *Seifen, Oele, Fatte, Wasche* **107,** 409–411 (1981).
278. Asaki Denka Kogyo K.K. Jpn Kokai Tokkyo Koho, JP 82:143398 (1982).
279. B.F. Szuhaj *Lecithin in the Paint Industry, New Technology Position Paper,* Central Soya Co., Inc., Chicago, Ill., 1974.
280. Anonymous, *Mod. Paint Coat.* 62 (Aug. 1985).
281. M. Kronstein, *Mod. Paint Coat.* 31–39, (July 1984).
282. M. Kronstein and J. Eichberg (to American Lecithin Co.), U.S. Pat. 4,520,153 (1985).
283. C. Baker in Ref. 2, Chapt. 17, pp. 253–260.
284. Miyoshi Kasei Kogyo, Y.K., Jpn. Kokai Tokkyo Koho, JP 85 190,705 (1985).
285. Miyoshi Kasei Kogyio, Y.K., Jpn. Kokai Tokkyo Koho, JP 85 184,571 (1985).
286. U.S. Pat. 4,126,591 (1978).

287. L.W. Bernardino U.S. Pat. 4,297,251 (1981).
288. (to Kao Soap Co. Ltd.), Jpn. Kokai Tokkyo Koho JP 8128298 (1981).
289. Takisawa and co-workers, Jpn. Kokai Tokkyo Koho, JP 7570410 (1975).
290. D.D. Lasic and co-workers, *Farm Vestro (Ljubljana)* **31,** 187–195 (1980); **94:**117354.
291. U. Pick *Arch. Biochem. Biophys. Acta Chem. Ab.* **212,** 186–194 (1981).
292. D. Paapahadjopoulos, *Liposomes, Immunobiol. Proc. Natl. Symp.,* 151–164 (1980); *Chem. Ab.* **95:**49331.
293. N. Oku and co-workers, *Biochem. Biophys. Acta* **692,** 384–388 (1982).
294. R.M. Patel and B.E. Ryman, *FEBS Letters* **62,** 60–63 (1976).
295. A.C. Allison and G. Gregoriadis, *Nature* **252,** 252 (1974).
296. R.M. Earidjani and Vanlerberghe U.S. Pat. 4,217,344 (1978).
297. B.F. Szuhaj in F.T. Frederick, ed., *World Research Conference II: Proceedings,* Westview Press, Bolder, Colo., 1980, pp. 681–691.
298. M. Foa Fr. Pat. 2,475,053 (1981).
299. W.F. McDonuogh and co-workers U.S. Pat. 3,928,226 (1975).
300. H. Friedman and co-workers Fr Pat. 1,548,700 (1968).
301. G. Hajduk and co-workers Pol. Pat. PL 104,287; (1979) *Chem. Ab.* **92:** 112454.
302. N.E. Wolff and J.W. Weigh in Ref. 87, VOl. 8, pp. 794–826.
303. L.P. Hayes *Canada Chem. Abstr.* 945800 (1974).
304. L.P. Hayes and co-workers, S. Afr. Pat. ZA 6906125 (1970).
305. S. Shimomukai and co-workers Ger. Pat. DE 2,916,494 (1979).
306. (to Fuji Photo Film Co., Ltd.), Jpn. Kokai Tokkyo Koho JP 80 88,045 (1980).
307. P.E. Kinzli and co-workers Ger. Pat. DE 2,117,214 (1971).
308. P.N. Maniar Ger. Pat. DE 2,117,075 (1971).
309. D.A. Newman Ger. Pat. DE 2,116,902 (1971).
310. W. Dabisch U.S. Pat. 3,515,572 (1970).
311. A.I. McMullen Brit. Pat. 1,357,952 (1974).
312. D.P. Paapahadjopoulos and co-workers, Ger. Pat. DE 2,907,303 (1979).
313. (to Kumiai Chemical Industry Co., Ltd.), Jpn. Tokkyo Koho JP 8108804 (1981).
314. (to Nisshin Oil Mills, Ltd.), Jpn. Kokai Tokkyo Koho JP 81140910 (1981).
315. G. Ritzman and co-workers Ger. Pat. DF 3,010,041 (1981).
316. G.B. Ansell and co-workers, *Form and Function of Phospholipids,* Elsevier, New York, 1973.
317. H.J. Devel, Jr., *The Lipids, Their Chemistry and Biochemistry,* Vol. 2, 1955, p. 743.
318. (to Sherico Ltd.), Ger. Pat. DE 2,756,079 (1978).
319. N.H. Ludwig and W.A. White U.S. Pat. 3,808,338 (1974).
320. M. Gordon and J.A. Wiesbach, *Canadian Chem. Abstr.* 932324 (1973).
321. R. Yehara and co-workers, Jpn. Kokai Tokkyo Koho JP 8449313 (1980).
322. G.H. Bullous Brit. Pat. 1,182,463 (1970).
323. M. Samejima and co-workers Ger. Pat. DE 3,039,908 (1981).
324. Anonymous, Chem. Week, 54 (May 20, 1987).
325. K. Fujiki Ger. Pat. DE 3,100,686 (1981).
326. A. Hosaka and co-workers, Ger. Pat. DE 3,104,702 (1982).
327. (to Intermagnetics Corp.), Ger. Pat. (Demand) FR 2,448,560 (1979).
328. H. Kawakara, Ger. Pat. DE 3,114,646 (1982).
329. M. Ohlinger, Ger. Pat. DE 2,942,646 (1981).

330. H. Tamura Ger. Pat. DE 3,017,651 (1980).
331. M. Buzniak, Pol. Pat. PL 117,092 (1982).
332. (to Toray Industries, Inc.), Jpn. Kokai Tokkyo Koho JP 80 83,041 (1980).
333. M. Chagnon and J. Ferris in Ref. 2, Chapt. 19, pp. 277–283.
334. K. Kuriyama and M. Toyama, Jpn. Kokai Tokkyo Koho JP 74 481,167 (1974).
335. P.H. List and R. Von Kleinsorgen, Ger. Pat. DE 2,928,040 (1980).
336. J. Stanley in Ref. 87, pp. 250–269.
337. R.I. Buhlstein and co-workers U.S.S.R. Pat. SU 624697 (1978).
338. A. Heijkoop Ger. Pat. DE 2,213,927 (1972).
339. P.P. Caruso U.S. Pat. 4,282,344 (1981).
340. H. Tjujimito and co-workers U.S. Pat. 4,049,556 (1977).
341. D. Sincroft and E. Sipos U.S. Pat. 2,987,527 (1959).
342. E. Sipos and D. Sincroft U.S. Pat. 2,991,163 (1959).
343. A.G. Meshandin and co-workers, U.S.S.R. Pat. SU 905,236 (1982).

11

Chemistry

This chapter presents the basic chemistry of fats and oils. A variety of general books and review articles are recommended (1). Some of the latest developments are briefly mentioned and are cited throughout the text.

The term *lipids* covers a heterogeneous group of substances that are insoluble in water but soluble in organic solvents. Lipids are commonly divided into three major classes: simple lipids, which are esters of fatty acids and alcohols, compound lipids, which are simple lipids conjugated with nonlipid molecules, and derived lipids, which are products of lipid hydrolysis.

Fats and oils consist mainly of acylglycerols, that is, esters of glycerol and fatty acids. In general the acylglycerols and phospholipids are the major lipid compounds in nature. However, lipids usually contain small amounts of various minor components many of which impact significantly on their chemical and physical properties.

COMPOSITION AND STRUCTURE

1.1 Major Components

Fatty Acids. *Fatty acids* are the building blocks of several classes of lipids including acylglycerols, phosphoglycerols, glycolipids, cholesterol esters, and some waxes. All fatty acids consist of a hydrocarbon chain and a terminal carboxyl group. They vary in chain length and in the number, position, and configuration of their double bonds. The carbon chains of most naturally occurring fatty acids are even numbered, and their unsaturated linkages are in the cis form. When two or more double bonds are present, they are invariably methylene interrupted as in linoleic acid:

$$\overset{18}{C}H_3(CH_2)_4-\overset{13}{C}H=\overset{12}{C}H-CH_2-\overset{10}{C}H=\overset{9}{C}H-(CH_2)_7\cdots\overset{1}{C}OOH$$

Table 11.1 Nomenclature, melting points (MP), and neutralization values[a] (NV) of saturated fatty acids

Abbreviation	Systematic Name	Trivial Name	MP (°C)	NV
4:0	Butanoic	Butyric	−7.9	637
5:0	Pentanoic	Valeric	−33.5	549
6:0	Hexanoic	Caproic	−3.4	483
7:0	Heptanoic	Enanthic	−7.1	430
8:0	Octanoic	Caprylic	16.7	389
9:0	Nonanoic	Pelargonic	12.5	355
10	Decanoic	Capric	31.6	326
11	Undecanoic	Undecylic	28.7	301
12	Dodecanoic	Lauric	44.2	280
13	Tridecanoic	Tridecylic	41.1	262
14	Tetradecanoic	Myristic	54.4	246
15	Pentadecanoic	Pentadecylic	52.1	231
16	Hexadecanoic	Palmitic	62.9	219
17	Heptadecanoic	Margaric	61.3	207
18	Octadecanoic	Stearic	69.6	197
19	Nonadecanoic	Nonadecylic	68.6	188
20	Eicosanoic	Arachidic	75.4	180
21	Heneicosanoic		74.3	172
22	Docosanoic	Behenic	80.0	165
23	Tricosanoic		79.1	158
24	Tetracosanoic	Lignoceric	84.2	152

[a] Neutralization value is equal to milligrams of KOH required to neutralize 1 g of the acid.

The structure of a fatty acid is commonly denoted by a systematic name after the nomenclature of its parent hydrocarbon, by its common or "trivial" name, or by a convenient shorthand designation showing the number of carbon atoms and number and position of its double bonds (2). For systematic nomenclature the terminal carboxyl carbon is given the number 1. Linoleic acid can therefore be referred to as *cis*-9,12-octadecadienoic acid or 18:2, *cis*-9,12. Often biochemists find it convenient to denote unsaturated linkages in fatty acid chains by designating the location of the first double bond counting from the methyl end of the molecule, that is, the omega carbon. The shorthand description for linoleic acid can thus be 18:2 *w*6. More recently the symbol *w* has been replaced by *n*- (e.g., 18:2 *n*-6). Tables 11.1 and 11.2 list the structure and nomenclature of saturated and unsaturated fatty acids, respectively. The fatty acid composition of natural fats varies significantly depending not only on the plant or animal species but also within the same species due to variations in season, geographical location, feed, and so forth (Tables 11.3 and 11.4).

Table 11.2 Nomenclature, melting points (MP), and neutralization values[a] (NV) of some common unsaturated fatty acids

Abbreviation	Systematic Name	Common Name	MP (°C)	NV
14:1 n-5 c	*cis*-9-Tetradecenoic	Myristoleic		247.5
16:1 n-7 c	*cis*-9-Hexadecenoic	Palmitoleic	−0.5	220.5
16:1 n-7 tr	*trans*-9-Hexadecenoic	Palmitelaidic	31	220.5
18:1 n-12 c	*cis*-6-Octadecenoic	Petroselinic	30	198.6
18:1 n-9 c	*cis*-9-Octadecenoic	Oleic	13–16	198.6
18:1 n-9 tr	*trans*-9-Octadecenoic	Elaidic	43.7	198.6
18:1 n-7 tr	*tras*-11-Octadecenoic	Vaccenic	44	198.6
18:2 n-6 c	*cis*-9,12-Octadecenoic	Linoleic	−5.0	200.06
18:3 n-6 c	*cis*-6,9,12-Octadecatrienoic	Gamma linolenic	201.51	
18:3 n-3 c	*cis*-9,12,15-Octadecatetraenoic	Alpha linolenic	−11.0	201.51
18:4 n-3 c	*cis*-6,9,12,15-Octadecatetraenoic	Stearidonic		202.9
20:3 n-6 c	*cis*-8,11,14-Eicosatrienoic	Dihomo-gamma-linolenic		167.7
20:4 n-6 c	*cis*-8,11,14-Eicosatetraenoic	Arachidonic	−49.5	184.28
20:5 n-3 c	*cis*-5,8,11,14,17-Eicosapentaenoic	EPA		
22:1 n-9 c	*cis*-13-Docosenoic	Erucic	33.5	165.7
22:5 n-3 c	*cis*-7,10,13,16,19-Docosapentaenoic	Clupanodonic		
22:6 n-3 c	*cis*-4,7,10,13,16,19-Docosahexaenoic	DHA		

[a] Neutralization value is equal to milligrams of KOH required to neutralize 1 g of the acid.

Acylglycerols. Acyl esters of glycerol constitute the major storage lipids in plants and most animals. Approximately 98% of these fats are made up of mixtures of triacylglycerols, that is, glycerol molecules, each esterified with three fatty acids. The balance consists of mono- and diacyl glycerols plus various other minor components.

Since glycerol contains two primary hydroxyl groups, the central carbon acquires chirality if one of the OH groups is esterified or if both are esterified to

Table 11.3 Fatty acid composition of some vegetable oils (%)

Oil	10:0	12:0	14:0	16:0	18:0	18:1	18:2	18:3
Cocoa butter				26	3.5	37	2.5	
Coconut	7	48	17	8	3	6	2	
Corn				11	3	27	57	
Cottonseed			0.8	25	2	18	53	0.1
Olive			1	15	2.6	67	13	1
Palm			1	46	5	39	9	0.4
Palm kernel	3.6	50	16	8	2	15	1	
Peanut				10	3	52	29	
Rapeseed				3	1	25	17	8.5
Safflower				6	2	14	74	0.4
Soybean				11	3.5	22	54	8
Sunflower				7	4	71	71	0.3

Table 11.4 Fatty acid composition of stone animal and marine oils (%)

Oil	4:0–12:0	14:0	16:0	18:0	18:1	18:2	18:3	20:5	22:6
Beef tallow		3	26	22	42	2	0.2		
Butter	12	10.8	26	10.3	28	2	1		
Chicken		0.9	22	10	41	20	2		
Lard		1.3	24	15.5	46	9	0.3		
Mackerel		6	16	3	15	2	1	5	9

different fatty acids. The most common way for designating the stereospecific structure of acylglycerols is by using the Fischer planar projection of glycerol with the middle hydroxyl group positioned on the left side of the central carbon. The carbon atoms are numbered 1, 2, and 3 in the conventional top to bottom sequence. The prefix *sn-* indicates stereospecific numbering of the glycerol carbons (3). The compound

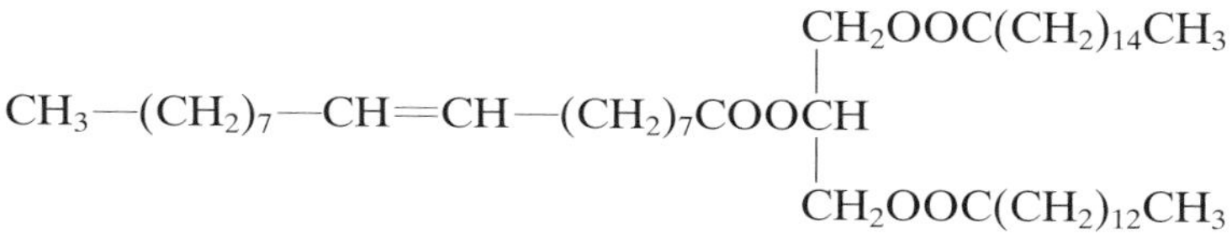

$$\begin{array}{r}
CH_2OOC(CH_2)_{14}CH_3 \\
| \qquad\qquad\quad \\
CH_3{-}(CH_2)_7{-}CH{=}CH{-}(CH_2)_7COOCH \qquad\qquad\qquad\quad \\
| \qquad\qquad\quad \\
CH_2OOC(CH_2)_{12}CH_3
\end{array}$$

can thus be named *sn*-glycerol-1-palmitate-2-oleate-3-myristate or 1-palmitoyl-2-oleyl-3-myristoyl-*sn*-glycerol. Convenient symbols are also frequently used, for example, *sn*-POM (4).

Phosphoglycerides (Glycerophospholipids). The parent compound of this series is *sn*-glycerol-3-phosphate rather than glycerol. In phosphatidic acids two fatty acids are esterified to the hydroxyl groups at carbons 1 and 2 while phosphoric acid is esterified to the hydroxyl group at position 3.

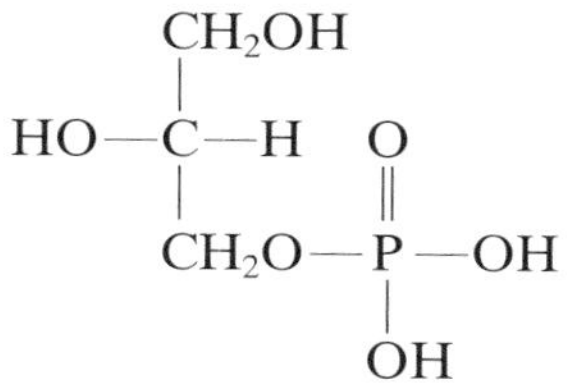

$$\begin{array}{l}
\quad\;\; CH_2OH \\
\qquad | \\
HO{-}C{-}H \quad O \\
\qquad | \qquad\; \| \\
\quad\;\; CH_2O{-}P{-}OH \\
\qquad\qquad\quad | \\
\qquad\qquad\; OH
\end{array}
\qquad\qquad
\begin{array}{l}
\qquad\qquad\;\; CH_2OCOR_1 \\
\qquad\qquad\quad | \\
R_2COO{-}C{-}H \quad O \\
\qquad\qquad\quad | \qquad\; \| \\
\qquad\qquad\;\; CH_2O{-}P{-}OH \\
\qquad\qquad\qquad\qquad | \\
\qquad\qquad\qquad\;\;\; OH
\end{array}$$

sn-Glycerol-3-phosphate 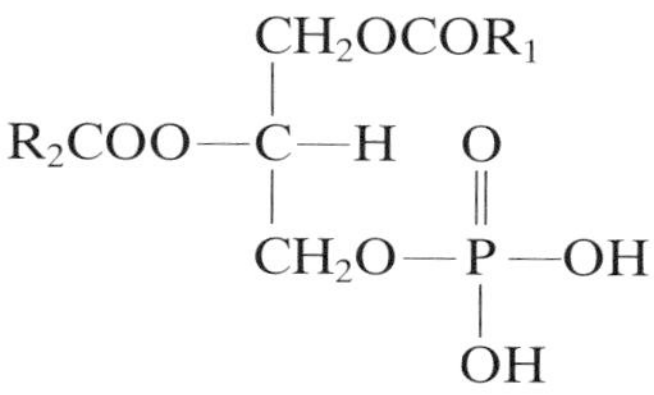 — Phosphatidic acid (PA)

The glycerol–phosphoric acid ester bond is more resistant to alkali hydrolysis than the glycerol–fatty acid linkage.

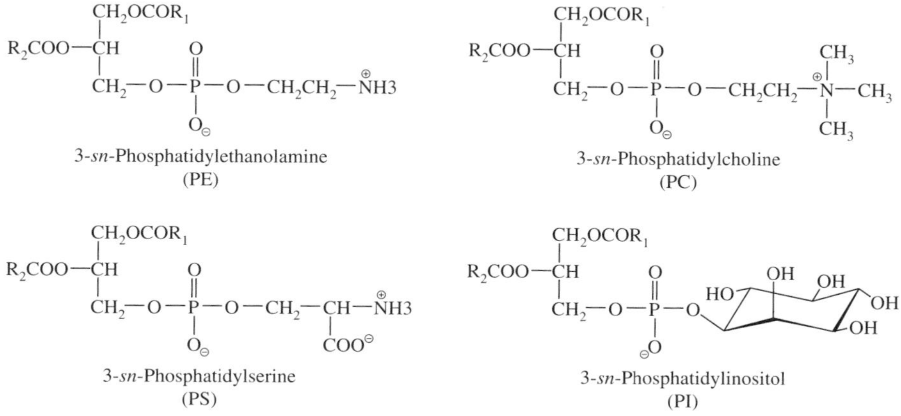

3-*sn*-Phosphatidylethanolamine
(PE)

3-*sn*-Phosphatidylcholine
(PC)

3-*sn*-Phosphatidylserine
(PS)

3-*sn*-Phosphatidylinositol
(PI)

Figure 11.1 Commonly occurring phosphatidyl esters.

The most abundant phosphoglycerides in animals and plants are the phosphatidyl esters. These are derivatives of phosphatidic acid in which the phosphate group is also esterified with another hydroxy compound. The phosphatidyl esters most commonly encountered are those containing ethanolamine, choline, serine, or inositol (Figure 11.1). Compared to triacylglycerols, phospholipids usually occur in small amounts. However, they are significant constituents of cellular membranes. All phosphoglycerides are amphiphilic since they contain a polar head, i.e., alcoholic phosphate group and two long-chain fatty acid residues.

Other derivatives of phosphatidic acid include the phosphatidylglycerols and the diphosphatidylglycerols

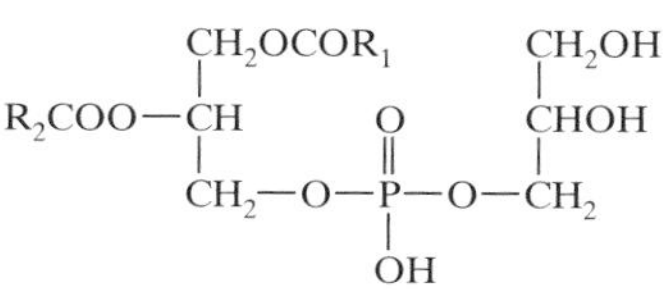

Phosphatidylglycerol

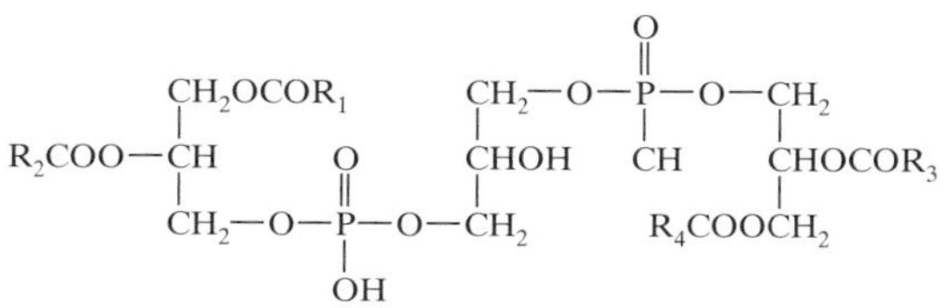

Diphosphatidylglycerol
(Cardiolipin)

Sphingolipids. Sphingolipids contain sphingosine instead of glycerol:

$$CH_3(CH_2)_{12}-CH=CH-CHOH-CHNH_2-CH_2OH$$

Sphingosine

The amino group is attached to a fatty acid by an amide bond. Examples of sphingolipids are sphingomyelin, ceramide, and cerebroside:

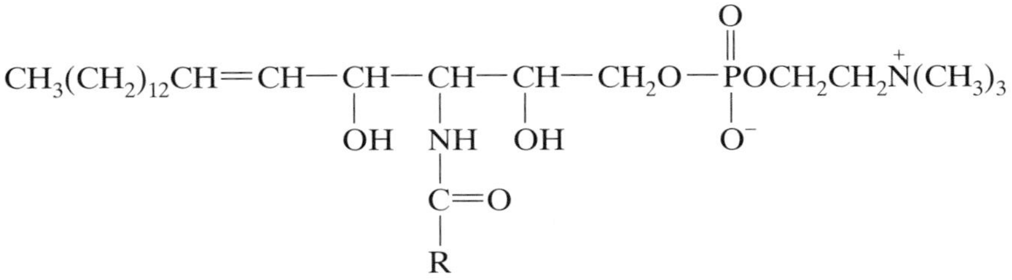

$$CH_3(CH_2)_{12}CH{=}CH{-}\underset{OH}{CH}{-}\underset{NH-C(=O)-R}{CH}{-}\underset{OH}{CH}{-}CH_2O{-}\overset{O}{\overset{\|}{P}}(O^-)OCH_2CH_2\overset{+}{N}(CH_3)_3$$

Sphingomyelin

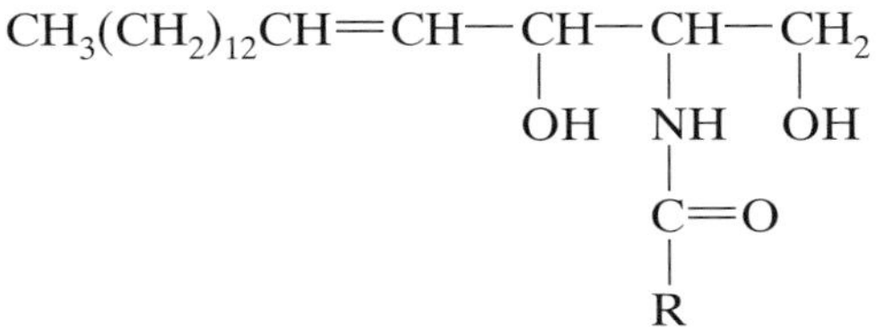

$$CH_3(CH_2)_{12}CH{=}CH{-}\underset{OH}{CH}{-}\underset{NH-C(=O)-R}{CH}{-}\underset{OH}{CH_2}$$

Ceramide

$$CH_3(CH_2)_{12}CH{=}CH{-}\underset{OH}{CH}{-}\underset{NH-C(=O)-(CH_2)_{22}-CH_3}{CH}{-}CH_2{-}O{-}[\text{sugar ring: OH, HO, O, OH, } CH_2OH]$$

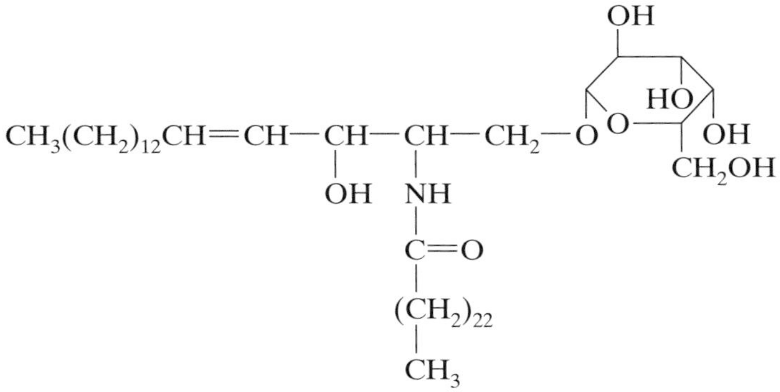

Cerebroside

1.2 Minor Components

Hydrocarbons. Trace amounts of hydrocarbons, aliphatic and branched, saturated as well as unsaturated, are commonly found in edible oils (5–8). Squalene is found in some oils in significant amounts, for example, fish oil:

Squalene

Sterols. Sterols constitute a class of steroid compounds that contain a hydroxy group at C_3 and a branched aliphatic chain of 8–10 carbon atoms at C_{17}. They occur as free alcohols or as esters of long-chain fatty acids (9–15).

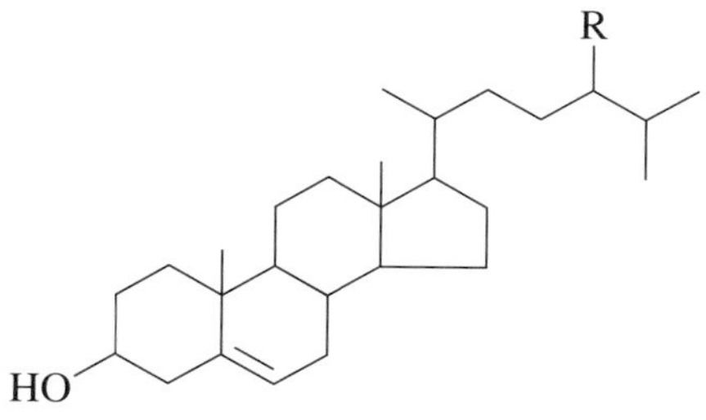

R = H, Cholesterol
R = CH_3, Campesterol
R = C_2H_5, Sitosterol

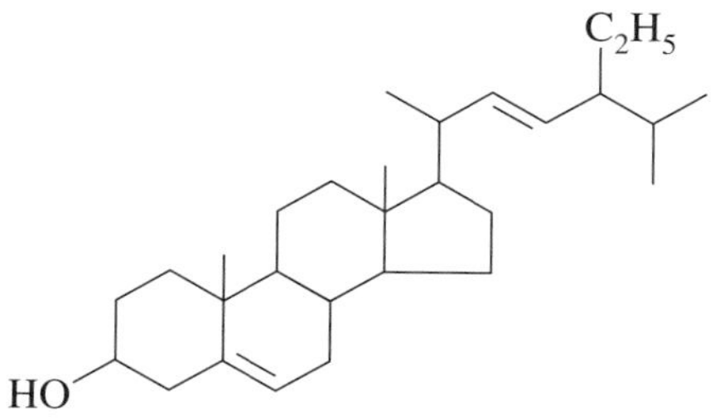

Stigmasterol

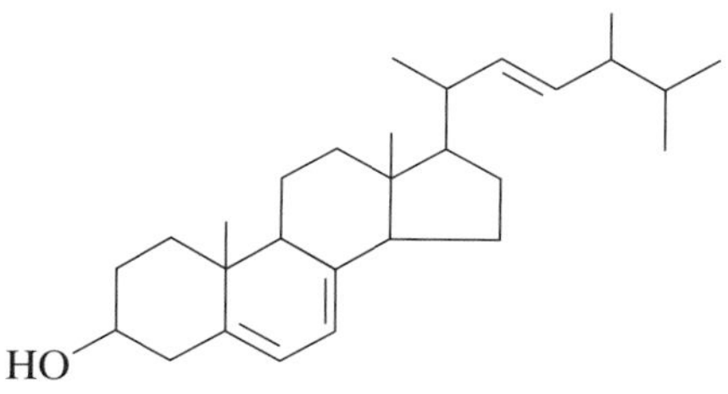

Ergosterol

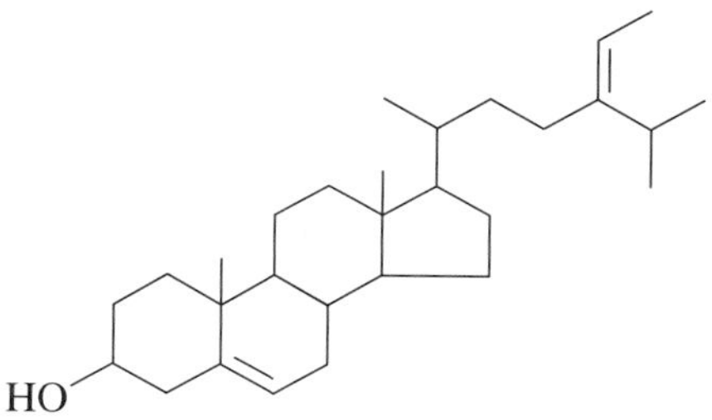

Δ^5-Ergosterol

Figure 11.2 Some common phytesterols.

Cholesterol is the most abundant sterol in animal lipids but is found in very small amounts in plants (13,16). Campesterol, stigmasterol, and sitosterol are the most abundant plant sterols (phytosterols) (Figure 11.2). Other sterols present in plants in smaller concentrations are Δ^5 and Δ^7-avenasterol, Δ^7-stigmasterol, campesterol, and fucosterol. The amounts of total sterols in some vegetable oils are shown in Table 11.5. Table 11.6 gives the distribution of individual sterols in these oils.

Chlorophylls. Detailed information regarding the structure and nomenclature of the chlorophylls and some of their breakdown products are available in various monographs (17,18). Chlorophylls are green pigments having porphyrin structures with magnesium occupying the central position. In addition to the four substituted pyrrole rings, they have a fifth ring that is not a pyrrole and a phytol side chain esterified to a carbonyl group substituent in ring IV. The four nitrogen atoms in the chlorophyll molecule are coordinated with the magnesium. The five-ring system has an extended polyene structure that is responsible for the strong absorption in the visible region of the spectrum. Chlorophyll "a" has a methyl group attached to ring II while chlorophyll "b" has an aldehyde group instead. The absorption spectra of these two compounds

Table 11.5 Unsaponifiables in some vegetable oils

Oil	Total Unsaponifiables	Sterols	Triterpene
Cocoa butter	0.3	0.2	0.05
Coconut	0.4	0.1	0.02
Corn	2	1	0.01
Cottonseed	0.6	0.4	0.02
Olive	0.8	0.1	0.2
Palm	0.4	0.03	0.02
Palm kernel	0.4	0.1	0.03
Peanut	0.9	0.3	0.04
Rapeseed	0.9	0.6	0.03
Safflower	0.6	0.6	0.07
Soybean	1.2	0.4	0.06
Sunflower	0.7	0.4	0.2

are slightly different:

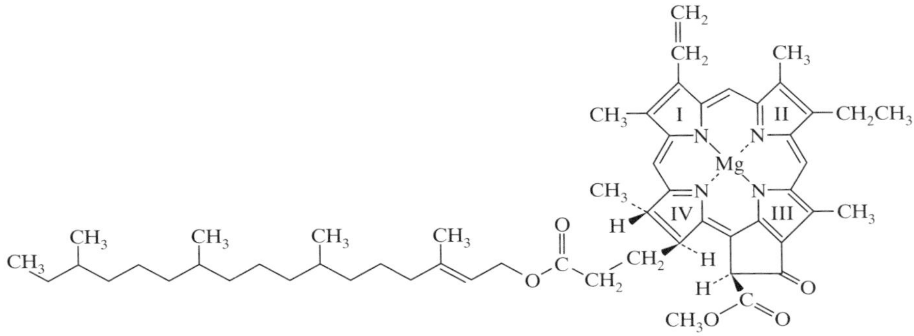

Chlorophyll

Chlorophyll a and b are easily degraded. Loss of their phytol side chains produces the corresponding chlorophyllides, which give rise to pheophorbides upon loss of the magnesium. The chlorophylls also produce the corresponding brown color compounds called pheophytins by losing the magnesium, and the

Table 11.6 Individual sterols in some vegetable oils (% of total sterols)

Oil	β-Sitosterol	Stigmasterol	Compesterol	Stigmasterol	Avenasterol	Avenasterol	Cholesterol
Cocao butter	59	26	9	1	3	tr	2
Coconut	52	13	8	4	20	1	1
Corn	66	6	22	1	4	tr	tr
Cottonseed	89	1	5	tr	2	tr	tr
Olive	87	2	2	2	3	tr	tr
Palm	74	8	14	1	2	—	1
Palm kernel	70	12	9	1	7	tr	2
Peanut	64	8	16	2	7	1	tr
Rapeseed	52	tr	30	5	2	—	tr
Safflower	52	9	13	20	1	3	—
Soybean	52	19	20	3	4	1	tr
Sunflower	60	7	8	10	6	4	tr

pheophytins can give rise to pyropheophytins by losing the carbomethoxy group at C-10.

Carotenoids. These pigments are synthesized only by plants and are the precursors of vitamin A. They may be yellow, red, or purple. Carotenoids are tetraterpenes, biosynthesized from eight isoprene units. Their most favored state is the all-trans.

Carotenoids are divided into two main classes, the carotenes, which are strictly polyene hydrocarbons, and the xanthophylls, which contain oxygen (19). The simplest carotene is lycopene. This aliphatic molecule consists of two isoprene diterpene units linked head to head:

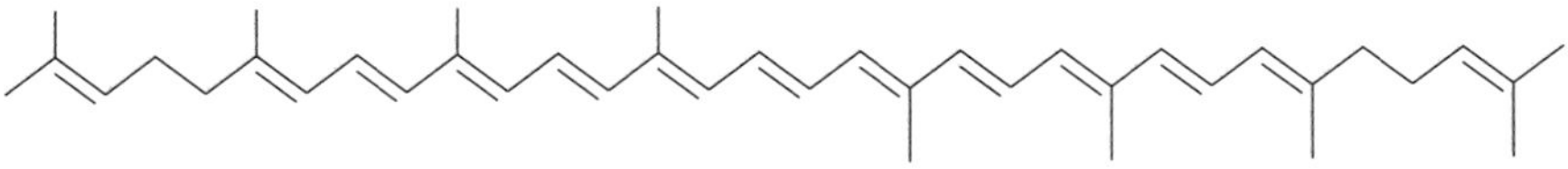

Lycopene

Carotenes may also be monocylic, for example, β- and γ-carotene or bicyclic, for example, α-carotene:

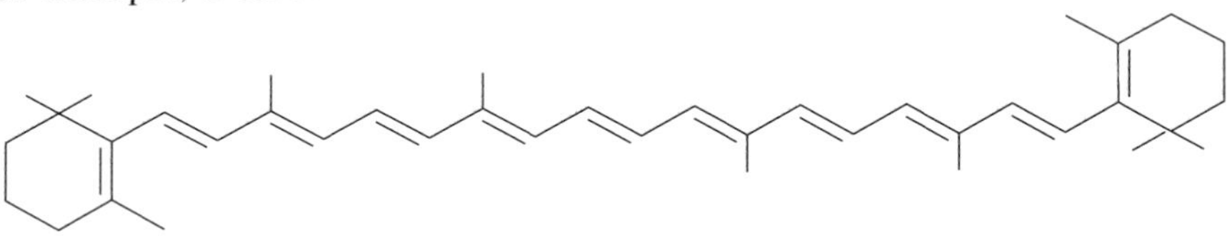

β-Carotene

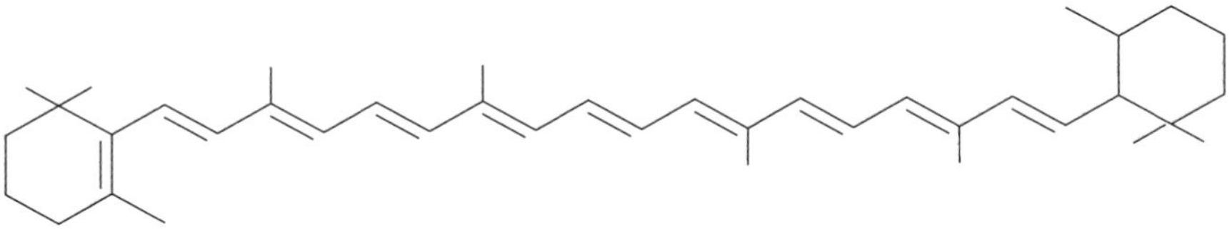

α-Carotene

The oxygen in xanthophylls may be in the form of hydroxy (e.g., zeaxanthin and lutein), keto (e.g., astaxanthin), epoxy (e.g., violaxanthin), or carboxyl (e.g., bixin) groups:

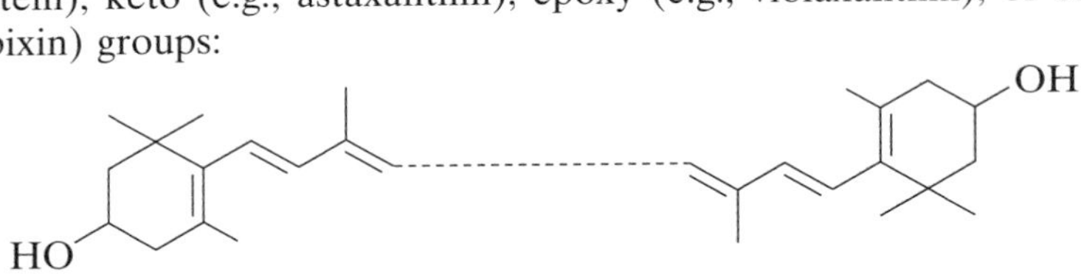

OH

HO

Zeaxanthin

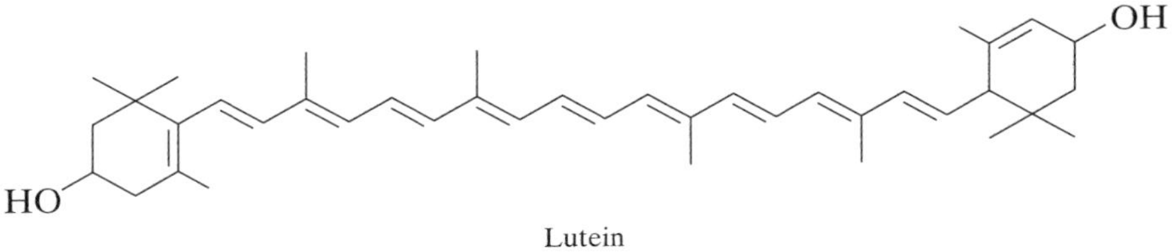

OH

HO

Lutein

HO O O OH

Violaxanthin

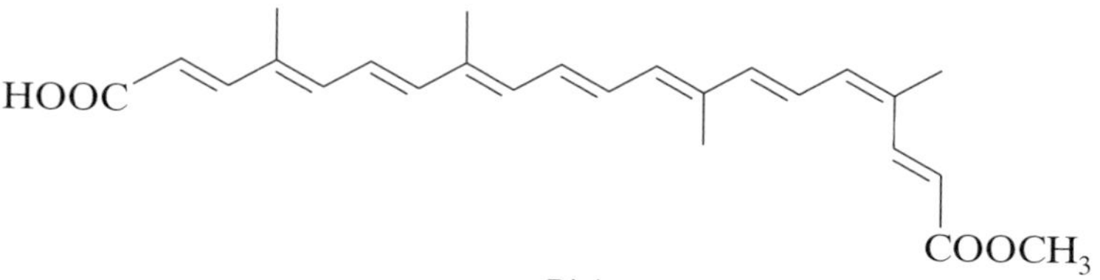

HOOC $COOCH_3$

Bixin

Carotenoids are generally stable in their natural environment but labile if extracted or heated. They are sensitive to oxygen and light. Their oxidation is accelerated by hydroperoxides generated from lipid oxidation leading to discoloration or bleaching. Among the products formed during the oxidative decomposition of carotenoids are α- and β-ionone, β-13 and β-14-apocarotenals, and β-13-apocarotenone.

Carotenoids are believed to play a major protective role against oxidation in biological systems:

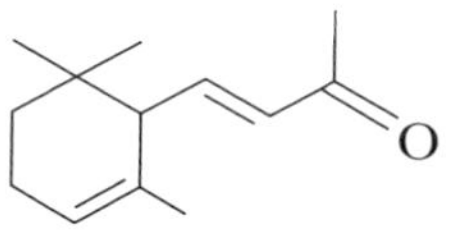

O O

α-Ionone 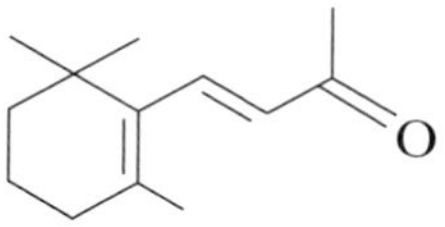 β-Ionone

Triterpene Alcohols. Terpenoid and steroid substances are mainly derived from the same origin, squalene, which upon cyclization gives rise to pentacyclic and tetracyclic triterpene alcohols (20). Demethylation of the latter compounds leads to formation of sterols:

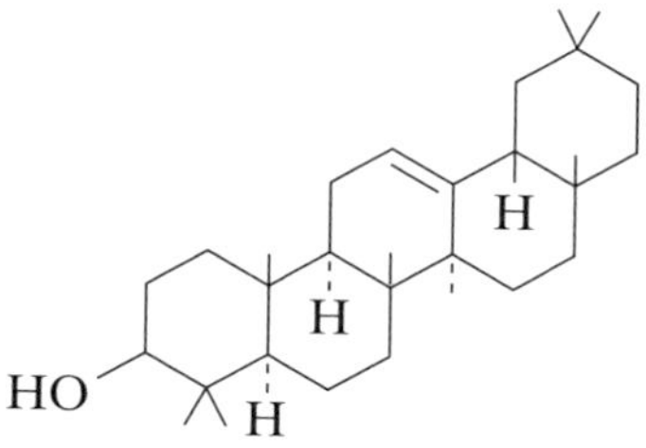

H H HO H HO H H

β -Amyrin 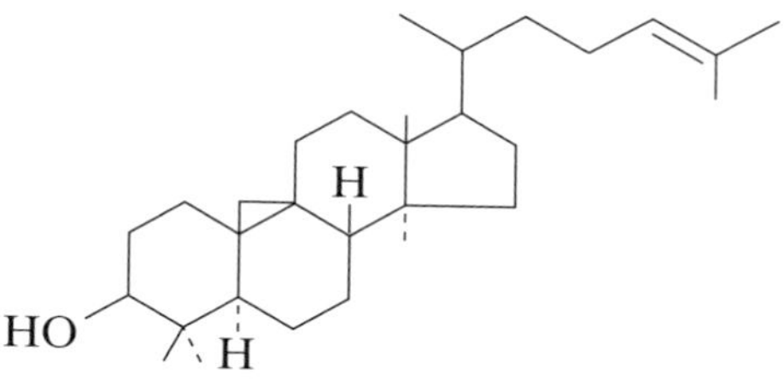 Cycloartenol

Cycloartenol and 24-methylene cycloartenol are found in all vegetable oils except soybean oil which contains only cycloartenol.

Phenolic Compounds. Fats and oils contain a variety of phenolic compounds in varying concentrations. Most of these compounds possess antioxidative properties and some are important flavor compounds. Among the phenolic compounds found in nature are caffeic, syringic, and ferulic acids:

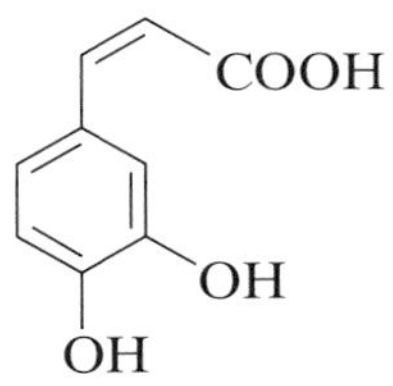

Caffeic acid 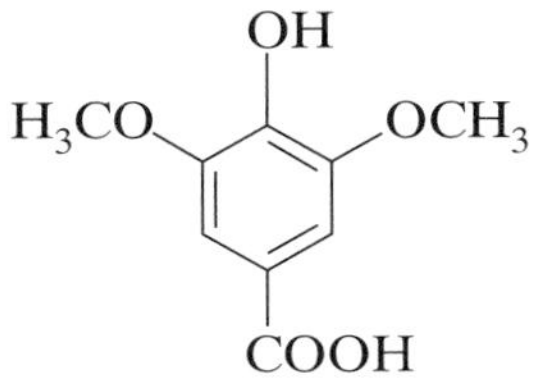   Syringic acid 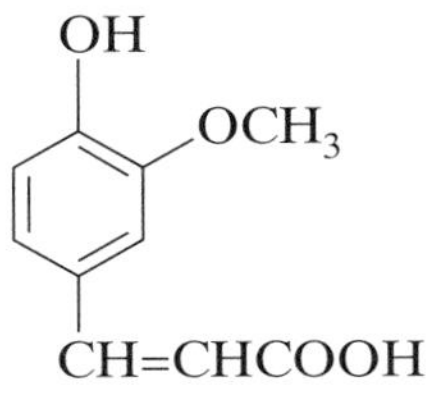   Ferulic acid

Other naturally occurring compounds include vanillic, *p*-coumaric, *o*-coumaric acids, tyrosol (4-hydroxyphenylethanol), *p*-hydroxybenzoic acid, *p*-hydroxyphenylacetic acid, hydroxytyrosol (3,4-dihydroxyphenylethanol), 3,4-dihydroxyphenylacetic acid, 3-hydroxyphenylethanol, and 3,4-dihydroxyphenylethanol (21).

REACTIONS

2.1 Hydrolysis and Saponification (4,22–24)

The ester linkages in acylglycerols can be hydrolyzed to yield free fatty acids and glycerol. The reaction proceeds in stages and is reversible. It is catalyzed by acid, alkali, enzymes, and various other substances. In industrial splitting, a high degree of hydrolysis is achieved by using a large excess of water, high pressure, and zinc, magnesium, or calcium oxides or certain sulfonated alkylbenzenes as catalysts.

If a triacylglycerol or a free acid is treated with alkali, it produces the salt of the alkali metal (soap) and glycerol or salt and water, respectively.

This is the basic reaction in the making of soap and glycerin and also serves as the basis for analytical determinations: the acid or neutralization number (i.e., milligrams of KOH required to saponify the free fatty acids in 1 g of fat); the saponification number (i.e., milligrams of KOH required to saponify 1 g of fat); the preparation of methyl esters; and calculations of fatty acid distribution on the 2-position of triacylglycerols.

With very few exceptions, vegetable oil unsaponifiables contain saturated and unsaturated hydrocarbons, sterols, aliphatic and triterpenic compounds, and phenolic compounds. Many of these compounds have been found to possess anti- or prooxidant properties.

2.2 Interesterification (25–27)

The term *interesterification* is used broadly to refer to a number of reactions in which an ester of a fatty acid may react with fatty acids, alcohols, or other fatty acid esters to produce an ester differing in composition from the original ester. Interesterification reactions can be classified as:

1. *Alcoholysis:* the displacement of the alcohol radical of an ester by another alcohol:

$$RCOOR_1 + R_2OH \rightarrow RCOOR_2 + R_1OH$$

2. *Glycerolysis:* the reaction between a triacylglycerol and an excess of glycerol. Di- and monoacylglycerol are formed:

$$\begin{array}{lclcl} CH_2OCOR_1 & & CH_2OH & & CH_2OH & & CH_2OCOR_1 \\ | & & | & & | & & | \\ CHOCOR_2 & + & CHOH & \rightarrow & CHOCOR_2 & + & CHOH \\ | & & | & & | & & | \\ CH_2OCOR_3 & & CH_2OH & & CH_2OCOR_3 & & CHOH \end{array}$$

3. *Acidolysis:* the displacement of the acid radical of an ester by another acid:

$$R_1COOR_2 + R_3COOH \rightarrow R_3COOR_2 + R_1COOH$$

4. *Ester interchange* (also termed *transesterification*): the reaction involving an exchange of the acyl group of one ester with that of another:

$$R_1COOR_2 + R_3COOR_4 \rightarrow R_3COOR_2 + R_1COOR_4$$

This exchange can also occur within the same triacylglycerol molecule:

$$\begin{array}{lclclcl} CH_2OCOR_1 & & CH_2OCOR_2 & & CH_2OCOR_2 & & CH_2OCOR_1 \\ | & & | & & | & & | \\ CHOCOR_1 & + & CHOCOR_2 & \rightleftharpoons & CHOCOR_1 & + & CHOCOR_2 \\ | & & | & & | & & | \\ CH_2OCOR_1 & & CH_2OCOR_2 & & CH_2OCOR_2 & & CH_2OCOR_1 \end{array}$$

Transesterification affords a means of improving many natural fats that otherwise possess limited applications. The process results in a redistribution of the triacylglycerol fatty acids. It is known that a fat so modified will accord more closely with that formulated on the basis of random distribution. If the mixture is submitted to ester interchange, the product will contain less fully saturated glycerides and will be more acceptable and more useful. Interesterification finds its greatest application in the technology of lard. Natural lard tends to develop fairly large crystals of disaturated glycerides (2-palmitoyloleoylstearin) that gives a grainy texture when the lard is chilled and makes it difficult to cream in butters and doughs. Rearranged lard, prepared from natural lard by redistributing the acyl groups of its triacylglycerol molecules, is a product of an overall performance superior to the premium shortenings. Beef and mutton tallows are very hard oils containing large percentages of tristearin, palmitodistearin, and other glycerides having melting points above body temperature that are not easily absorbed or digested by many animals. A product of more acceptable consistency and utility can be obtained by ester interchange between these hard fats and some liquid oils.

The process of interesterification is usually conducted by heating the fat, or mixture of fats, at a high temperature for a relatively long time, or for a much shorter time, less than one hour, in the presence of alkali metals or alkali metal alkylates as catalysts.

2.3 Hydrogenation (28,29)

Hydrogenation of oils, sometimes called "hardening," is a process in which hydrogen gas is added to unsaturated linkages in the fatty acid chains. The purpose is twofold: altering the physical characteristics of liquid oils to render them more solid and improving their stability against oxidation. Hydrogenation reactions are also applied to reduce the carbonyl group of the carboxyl of an acid or an ester to produce alcohols or to reduce keto acids to hydroxy acids. Certain nitrogen derivatives of fatty acids can be hydrogenized to produce amines. Hydrogenation involving reduction rather than addition to unsaturated linkages is commonly referred to as hydrogenolysis to distinguish the two types of reactions.

Catalytic hydrogenation involving addition of hydrogen to double bonds is by far the most important from the standpoint of industrial applications. In practice, the reaction is stopped before reaching complete saturation of all the unsaturated centers in the glyceride molecules. Depending on the experimental conditions, the reaction is more or less selective, which means that there is a pronounced tendency for polyethenoid chains to be reduced to the monoethenoids before there is appreciable reduction of monoethenoid into saturated esters. In addition, the reaction is accompanied by double-bond movement leading to both positional and geometrical isomers. Complete hydrogenation of methyl oleate gives methyl stearate only. But if the reaction is stopped before completion, it is found that stearate and oleate are accompanied by a mixture of isomeric octadecenoates often referred to as iso-oleates. These are derived from the oleate by movement of the double bond and/or by change of configuration. Double-bond migration occurs in both directions but is probably more pronounced in the direction away from the ester group. The formation of trans isomers is exensive and occurs rapidly.

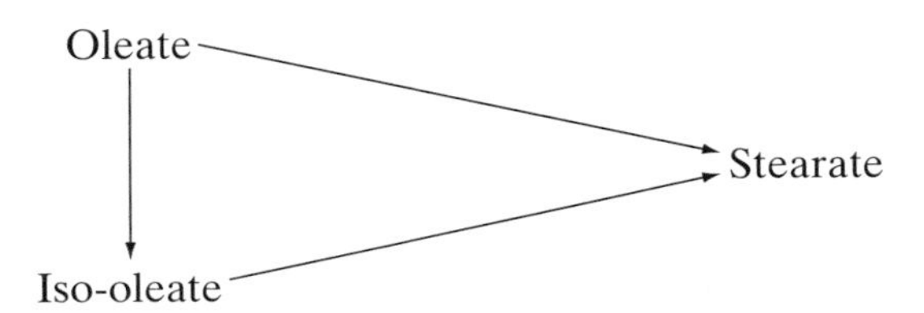

Hydrogenation of linoleates results in high concentrations of monoenes with double bonds in the 10 and 11 positions of which significant percentages are in the trans configuration:

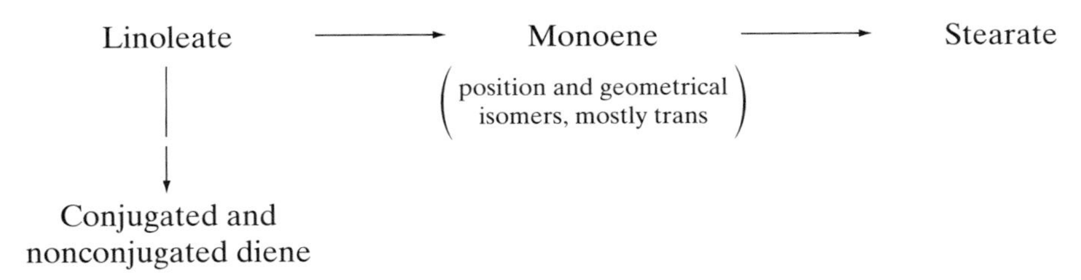

Linolenates are isomerized to conjugated trienes which are in turn hydrogenated to conjugated dienes at a high rate, and/or to nonconjugatable dienes:

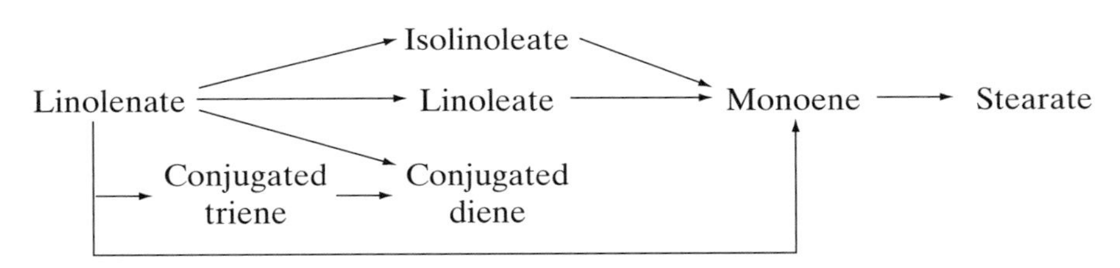

Double bond isomerization during the hydrogenation process is believed to occur via a half-hydrogenation dehydrogenation reaction sequence. The olefin is first adsorbed on the catalyst. This process, described as chemabsorption, occurs by the associated mechanism in which the double bond is broken and two carbon metal bonds are formed by reaction with adsorbed hydrogen. This is converted first to the half-hydrogenated state in which the olefin is now attached to the catalyst by only one link and is thus capable of free rotation. The species may then suffer one of two fates. Further reaction with adsorbed hydrogen followed by desorption from the catalyst produces the hydrogenated product. Alternatively, the half-hydrogenated olefin may again become attached to the catalyst by a second carbon metal bond with loss of hydrogen from either carbon atom adjacent to the existing carbon double bonds. The adsorbed olefin may then be a positional or geometrical isomer of the original substrate, which may be desorbed as such or may enter again into a similar cycle of reactions.

2.4 Halogenation (30–36)

Iodine, bromine, and chlorine can add to double bonds:

$$R'{-}CH{=}CH{-}R'' + Br_2 \rightarrow R'{-}CHBr{-}CHBr{-}R''$$

The addition of halogens to double bonds of unsaturated fatty acids is the basis for the determination of the "iodine value," which indicates the weight of iodine absorbed under standard conditions by 100 g lipid. A solution of iodine monochloride in acetic acid (Wijs' reagent) is usually used for the reaction. Alternatively, iodine monobromide in acetic acid can be used (the Hanus technique). The time of analysis can be significantly reduced with the use of mercuric acetate as a catalyst. The addition of chlorine or bromine to double bonds is not always a quantitative one since halogens tend to add incompletely or to replace hydrogen atoms in the alkyl chain.

Since the solubility of the resulting bromides in cold ether and petroleum either decreases with increase in the number of double bonds, determination of bromides has been used to detect the presence of highly unsaturated acids. For example, the tetrabromide or hexabromide number represents grams of insoluble bromides obtained from 100 g fatty acids under standard conditions.

Today, halogen additions are less commonly used for the determination of unsaturation since detailed and quantitative analyses of fatty acids can easily be done by gas chromatography of their methyl esters.

Several applications have been developed for the use of halogenated fatty acid derivatives as end products, that is, products that impart antiflammability properties to textile additions, or as reactive intermediates for the preparation of other compounds.

Halogen derivatives are prepared by addition to double bonds or by substitution of the halogen for hydrogen in the hydrocarbon chain. However, when one of these two reactions is intended, it is often difficult to exclude the other. In the bromination of oleic acid, for example, a small percentage of the bromoderivatives formed are halogenated in other than the 9 and 10 positions. Substitution of hydrogen atoms in the chain is generally random except in certain specific cases. For example, 2-position bromination of saturated fatty acids can be achieved by means of the Hell–Vohard–Zelinsky reaction:

$$R(CH_2)COOH \xrightarrow{BR_2} RCH_2COBr \xrightarrow{BR_2} RCHBrCOBr + HBr$$

$$RCHBrCOBr \xrightarrow{H_2O} RCHBrCOOH + HBr$$

Bromination at positions allyl to double bonds can be conducted with the use of *N*-bromosuccinimide (37,38):

$$C_6H_4(CO)_2NBr + -CH{=}CH-CH_2- \longrightarrow -CH{=}CH-CHBr- + C_6H_4(CO)_2NH$$

Substitution of halogen for the hydroxyl of the carboxyl group can be used in the preparation of acid chlorides:

$$3\ RCOOH + PCl_3 \rightarrow 3\ RCOCl + H_3PO_2$$

Halogen derivatives prepared by substitution may also be used to produce compounds of increased unsaturation by dehydrohalogenation reactions.

2.5 Oxidation

Chemical Oxidation The use of chemical oxidation in the investigation of fats and fatty acids has been reviewed (39,40). Reactions with ozone are used to identify positional isomers of unsaturated fatty acids. Ozonization of the double bonds is usually followed by reductive cleavage of the ozonides and chromatographic separation and quantitation of the resulting fragments (41–43). Reductive ozononlysis of glyceride oils can also be used for commercial production of certain acids (44).

Oxidative degradation by chromates, periodates, and permanganates has been used for a variety of purposes. These include the production of azelaic and pelargonic acids from oleic acid; determination of α-monoglycerides, generation of aldehydes and epoxide cleavage (45,46); location of carbon chain branching; and determination of double-bond position (47). Permanganate–periodate oxidation of a number of aliphatic acids, alcohols, aldehydes, ketones, ethers, and esters has been described by von Rudloff (48).

Hydrogen peroxide and peracids have been widely used in numerous laboratory and industrial reactions having a wide range of applications. These include epoxidation of olefins; crosslinking of polymers; curing of polyester resins, silicone rubbers, and polyurethanes; and the preparation of various intermediates (49–52).

Oxidative cleavage with nitric acid and nitrogen oxides have also been used. Such reactions are useful in catalyzing cis–trans isomerization of unsaturated fatty acids and the preparation of dibasic acids (53–56).

Reaction with Oxygen. Lipid oxidation plays a pivotal role in a number of essential biological functions, for example, as a source of energy, in prostaglandin biosynthesis, and in antibacterial cell activities. On the other hand, lipid peroxidation can cause damage to membranes, enzymes, and proteins, and has been implicated in many free radical pathologies including circulatory diseases, carcinogenesis, asthma, and aging. Similarly, oxidation of the lipids in food is of paramount importance to food quality. It may lead to the development of rancid off-flavors, cause changes in color or texture, reduce shelf life, and/or impair nutritional quality. Some products of lipid oxidation are toxic at relatively low concentratons, for example, cyclic monomers and oxycholesterols. Conversely, a limited degree of lipid oxidation is often desirable, as in the formation of typical flavors and aromas, for example, cheeses and fried food.

Basis Mechanisms. It is generally established that the autoxidation of lipids occurs largely via a self-propagating free radical mechanism (57–64). Since direct reaction of unsaturated fatty acids with oxygen is thermodynamically difficult, production of the first few radicals necessary to start the propagation reaction (i.e., initiation) must occur by some catalytic means. It has been proposed that the initiation step may take place by decomposition of preformed hydroperoxides, via metal catalysis, heat or exposure to light, or by mechanisms where singlet oxygen is the active species involved.

Upon the formation of sufficient free radicals, the chain reaction is propagated by the abstraction of hydrogen atoms at positions alpha to double bonds $RH \rightarrow R^{\cdot}$ (where RH is the substrate fatty acid and H is the α-methylenic hydrogen atom), followed by oxygen attack at these locations and resulting in the production of peroxy radicals $R^{\cdot} + O_2 \rightarrow ROO^{\cdot}$, which in turn abstract hydrogen from α-methylenic groups of other substrate molecules

$$ROO^{\cdot} + RH \rightarrow ROOH + R^{\cdot}$$

to form hydroperoxides, ROOH, and yield $R^{\cdot}$ groups, which react with oxygen, and so on.

Due to resonance stabilization of the $R^{\cdot}$ species, the reaction is usually accompanied by shifting in the position of double bonds, resulting in the formation of isomeric hydroperoxides often containing conjugated diene groups (63,65–69). Thus, abstraction of hydrogen from the two α-methylenic groups of oleate gives rise to two resonance-stabilized allkyl radicals:

$$-\overset{11}{CH_2}-\overset{10}{CH}=\overset{9}{CH}-\overset{8}{CH_2}-$$

$$\downarrow$$

$$\overset{11}{CH}\cdots\overset{10}{CH}\cdots\overset{9}{CH} + \overset{10}{CH}\cdots\overset{9}{CH}\cdots\overset{8}{CH}$$

which results in cis–trans isomers of the 8-, 9-, 10-, and 11-hydroperoxides. Linoleates are much more reactive due to the presence of 1,4-pentadiene system having a doubly allylic methylene group:

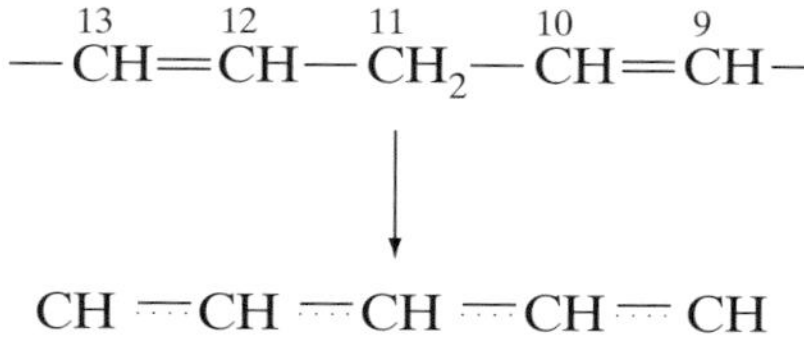

$$\overset{13}{-CH}=\overset{12}{CH}-\overset{11}{CH_2}-\overset{10}{CH}=\overset{9}{CH}-$$

$$\downarrow$$

$$CH \overset{\cdots}{=} CH \overset{\cdots}{=} CH \overset{\cdots}{=} CH \overset{\cdots}{=} CH$$

and resulting in the formation of isomeric 9- and 13-hydroperoxides.

The formation of allylic hydroperoxides from unsaturated fatty acids also occurs via photosynthesized oxidation, which involves the reaction with oxygen in the presence of light and a sensitizer (70,71). The activation of ground-state oxygen $^3O_2$ generates singlet oxygen $^1O_2$, which reacts with alkenes via a nonradical concerted process known as the "ene reaction." Oxygen attaches to either of the unsaturated carbon atoms, and the double bond is shifted resulting in the formation of trans-allylic hydroperoxides.

In the presence of trace metals or heat, the hydroperoxide oxygen–oxygen bond may break yielding hydroxy and alkoxy radicals:

$$ROOH \rightarrow RO^{\cdot} + {}^{\cdot}OH$$

Homolytic cleavage on either side of the alkoxy groups leads to the formation of typical breakdown products (Tables 11.7, 11.8, 11.9) (63,72–79). Vinylic radicals resulting from hydroperoxide cleavage, which would be expected to be reactive and unstable, are reported to react with hydroxy radicals to form 1-enols, which tautomerize to give the corresponding aldehydes (63):

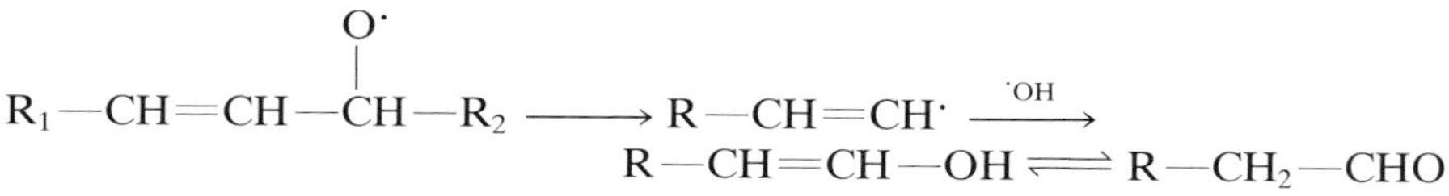

$$R_1-CH=CH-\underset{}{\overset{O^{\cdot}}{\overset{|}{CH}}}-R_2 \longrightarrow R-CH=CH^{\cdot} \xrightarrow{{}^{\cdot}OH}$$
$$R-CH=CH-OH \rightleftharpoons R-CH_2-CHO$$

Furthermore, the free radical intermediates generated in the above reactions can interact in a variety of ways to produce dimeric, polymeric, and cyclic compounds. These reactions include the combination of alkyl radicals with other alkyl and/or alkoxy radicals, Diels/Alder reactions between a double bond and a conjugated diene, and addition of free radicals to double bonds.

Table 11.7 Oleic acid hydroperoxide cleavage

Hydroperoxide Isomers	Hydroperoxide Breakdown Products
8-ROOH	2-undecenal + heptanoic acid + decanal[a] + 8-oxooctanoic acid
9-ROOH	2-decenal + octanoic acid + nonanal[a] + 9-oxononanoic acid
10-ROOH	nonanal + 9-oxononanoic acid[a] + octane + 10-oxo-8-decenoic acid
12-ROOH	octanal + 10-oxodecanoic acid[a] + heptane + 11-oxo-9-undecenoic acid

[a] From vinylic cleavage.

Table 11.8 Linoleic acid hydroperoxide cleavage

Hydroperoxide Isomers	Hydroperoxide Breakdown Products
9-ROOH	2,4-decadienal + octanoic acid + 3-nonenal[a] + 9-oxononanoic acid
13-ROOH	hexanal + 12-oxo-9-dodecanoic acid[a] + pentane + 13-oxo-9,11-tridecadienoic acid
10-ROOH[b]	3-nonenal + 9-oxononanoic acid[a] + 2-octene + 10-oxo-8-decenoic acid
12-ROOH[b]	2-heptenal + 9-undecenoic acid + hexanal[a] + 12-oxo-9-dodecenoic acid

[a] From vinylic cleavage.
[b] From photosensitized oxidation.

2.6 Antioxidation (80–82)

Antioxidants are substances that are capable of slowing the rate of oxidation in oxidizable materials. In foods, many such substances are known to occur naturally. However, antioxidants, both natural and synthetic, are frequently added as additives to retard oxidative deterioration of flavor, color, and texture.

Mechanism. In general, antioxidants function by interfering with one or more of the steps involved in autoxidation, that is, initiation, the free radical chain reaction, and termination. Furthermore, certain compounds delay oxidation by rendering ineffective factors that promote oxidation.

The majority of antioxidants function as free radical scavangers. Due to their phenolic structure, they act as hydrogen or electron donors. The phenoxy radical formed by, for example, the reaction of the antioxidant with a fatty acid peroxy radical is stabilized by delocalization of unpaired electrons around the aromatic ring:

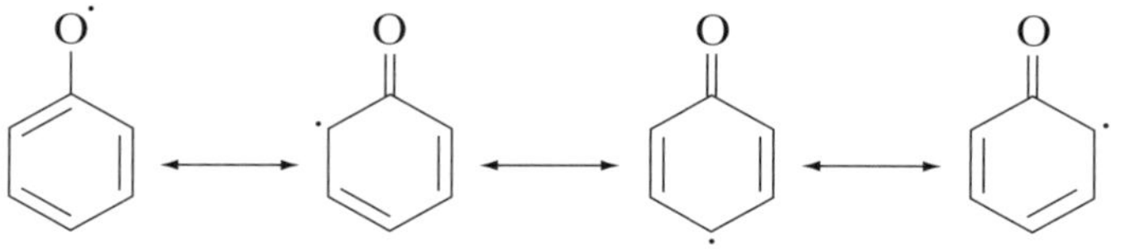

Table 11.9 Linolenic acid hydroperoxide decomposition

Hydroperoxide Isomers	Hydroperoxide Breakdown Products
9-ROOH	2,4,7-decatrienal + octanoic acid + 3,6-nonadienal[a] + 9-oxononanoic acid
12-ROOH	2,4-heptadienal + 9-undecenoic acid + 3-hexenal[a] + 12-oxo-9-dodecenoic acid
13-ROOH	3-hexenal + 12-oxo-9-dodecenoic acid[a] + 2-pentene + 13-oxo-9,11-tridecadienoic acid
16-ROOH	propanal + 15-oxo-9,12-pentadecadienoic acid[a] + ethane + 16-oxo-9,12,14-hexadecatrienoic acid
10-ROOH[b]	3,6-nonadienal + 9-oxononanoic acid[a] + 2,5-octadiene + 10-oxo-8-decenoic acid
15-ROOH[b]	butenal + 9,12-butadecadienoic acid + propanal[a] + 15-oxo-9,12-pentadecadienoic acid

[a] From vinylic cleavage.
[b] From photosensitized oxidation.

In this manner the reaction of a phenolic antioxidant (AH) with a lipid radical ($ROO^{\cdot}$)

$$AH + ROO^{\cdot} \rightarrow A^{\cdot} + ROOH$$

competes with the ability of the peroxy radical to propagate the chain reaction by abstracting a hydrogen atom from a new substrate molecule:

$$RH + ROO^{\cdot} \rightarrow R^{\cdot} + ROOH$$

Introduction of bulky side chains, for example, branched alkyl side chains, further stabilizes the phenoxy radical by increasing steric hindrance in the region of the radicals, thus reducing the chances of propagation by the phenolic radical itself.

Some phenolic compounds, amines, and thiopropionic acid can exert their antioxidative ability by acting as peroxide decomposers. Other compounds may act as quenchers of singlet oxygen (e.g., β-carotene and tocopherol), as metal chelators or reducing agents (e.g., ascorbic and citric acids).

Selection of Antioxidants. Antioxidants vary widely in their structure, properties, mode of action, and effectiveness. The choice of an antioxidant or combination of antioxidants depends on the specific requirements of the system in which they are used. Ideally, the antioxidant, its metabolites or decomposition products, should be toxicologically safe at the dosages used, should not impart off-flavors or colors upon addition, or after processing or storage, and should be active at low concentrations, easily incorporated in the oil or food product, readily available at an economic cost, and should have good stability and carry-through characteristics. Better results are often achieved by using a combination of two or more antioxidants.

Synthetic Antioxidants. The major phenolic antioxidants permitted as food additives in many countries are butylated hydroxyanisole (BHA), butylated hydroxytoluene (BHT), tertiary butylhydroquinone (THBQ), and a number of alkyl gallates.

1. Butylated Hydroxyanisole (BHA). BHA, a mixture of two isomers (85%, 2-*tert*-butyl-4-methoxyphenol + 15% 3-*tert*-butyl-4-methoxyphenol), is commercially available as white waxy flakes highly soluble in oil and insoluble in water. It has a phenolic odor, albeit not noticeable at concentrations used in food. Although its potency is low when added to vegetable oils containing relatively high concentrations of natural antioxidants, it is effectively used in combination with other primary antioxidants (e.g., gallates). BHA is stable under mild basic conditions and provides good carry-through protection but undergoes significant loss at elevated temperatures due to vaporization.

2. Butylated Hydroxytoluene (BHT). Like BHA, 2,4-di-*tert*-butyl-4-methylphenol is a hindered phenol, relatively weak when used alone for vegetable oils, but more effective if combined with other antioxidants. It is a white crystalline solid with properties similar to BHA.

3. Tert-butylhydroquinone (TBHQ). TBHQ is a white or beige-colored powder, moderately soluble in oil with very slight water solubility. It has several advantages over the other authorized antioxidants. It is more potent, more resistant to heat and unlike propyl gallates, does not cause discoloration since it does not complex with copper or iron.

4. Alkylgallates. These compounds are synthesized by esterification of gallic acid with primary alcohols. They are highly potent antioxidants due to their trihydroxyl structures. The lower gallates (e.g., propyl and butyl) have relatively low solubility in oil, but this can be overcome by increasing the alkyl chain length of the alcohol (e.g., octyl, dodecyl). These compounds have lower volatility than BHA and BHT. In spite of the relative effectiveness of gallates as antioxidants, their use has been hampered due to their tendency to form colored complexes with trace metals.

Natural Antioxidants (83–87). The presence in living systems of various compounds that protect the organism against oxidative damage has long been recognized. Understandably, a great deal of interest has been generated in screening natural materials for constituents that possess antioxidative properties. Sources available for such purpose include oils and oil seeds, grains, fruits, vegetables, bark, roots, herbs and spices, seaweeds, and animal and microbial products. Some such products have already been prepared and are produced commercially (e.g., tocopherol from palm oil). Large-scale screening for natural antioxidants is at present the subject of extensive research.

Among the naturally occurring antioxidants, plant phenolics are by far the most prevalent. These include flavonoid compounds, cinnamic acid derivatives, coumarins, tocopherols, and polyfunctional organic acids (21).

Flavonoids. The structure of these compounds is based on the flavan nucleus, which consists of two aromatic rings linked by a three-carbon aliphatic chain that has been condensed to a pyran ring:

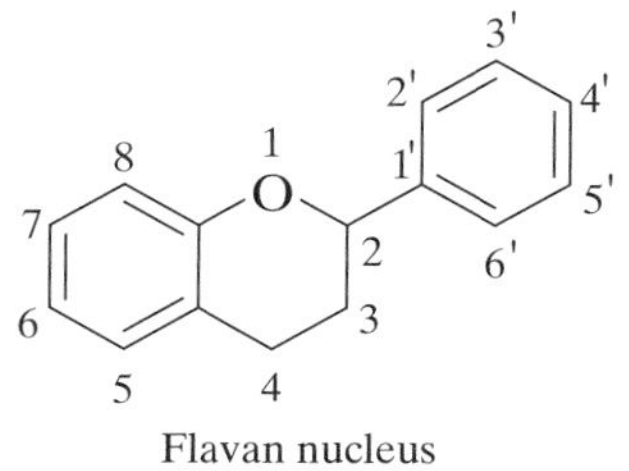

Flavan nucleus

Flavone may be considered the general type of compound of the flavonoid group (88–91). Flavonoids are generally divided in two main classes:

1. Anthoxanthins. Flavonoids that possesses a carbonyl group in the 4-position. This class may be further divided into several subgroups.

a. *Flavones:* hydroxylated or methoxylated derivatives of flavone. The 3-hydroxyflavones are commonly referred to as flavonols (e.g., quercetin):

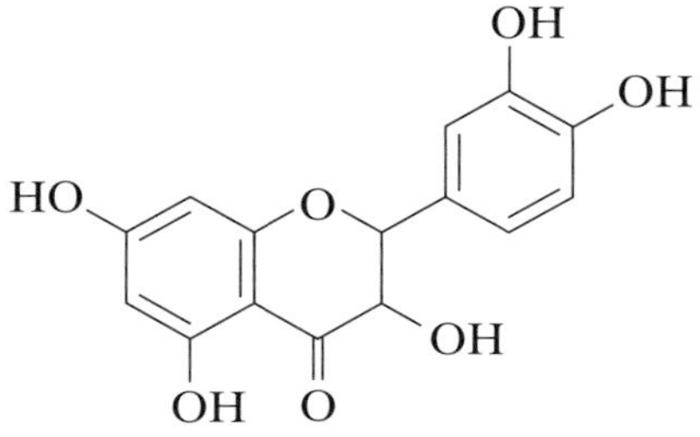

Quercetin

Flavonols chelate metal ions at the 3-hydroxy-4-keto or the 5-hydroxy-4-keto groups:

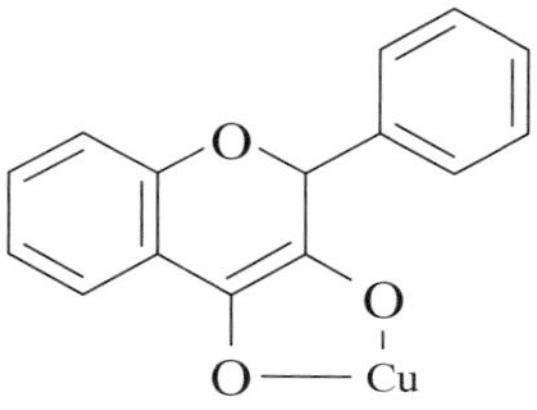

3-Hydroxyflavanone

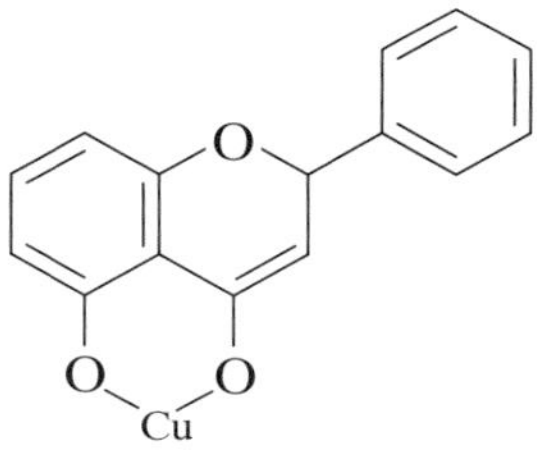

5-Hydroxyflavanone

b. Flavanones: hydroxylated or methoxylated derivatives of flavanone (e.g., eriodictyol and naringenin):

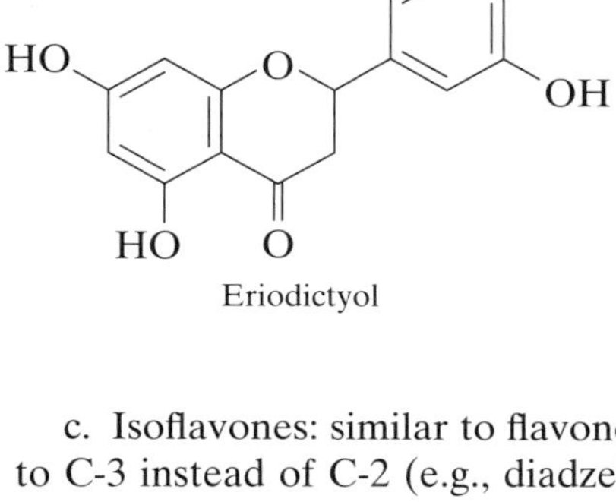

Eriodictyol

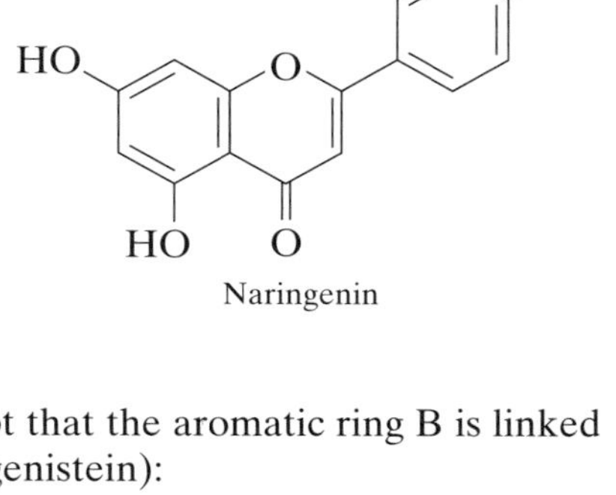

Naringenin

c. Isoflavones: similar to flavones except that the aromatic ring B is linked to C-3 instead of C-2 (e.g., diadzein and genistein):

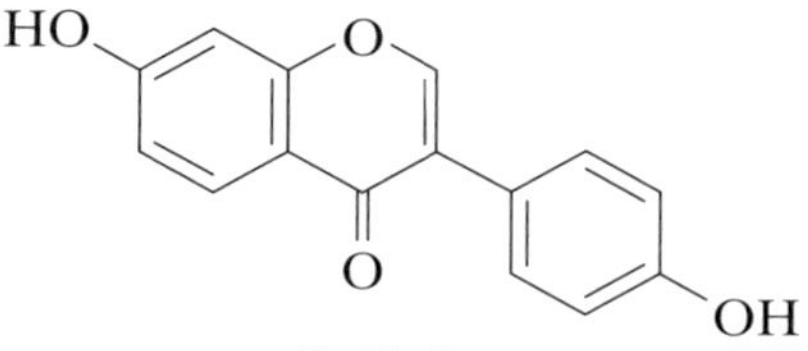

Daidzein

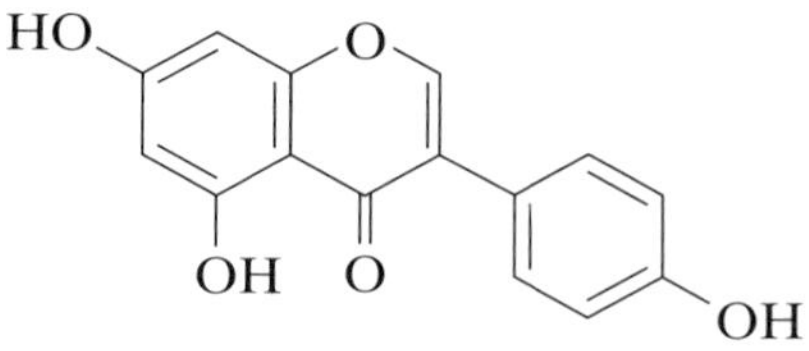

Genistein

d. Chalcones: hydroxylated or methoxylated derivatives of two aromatic rings linked by a three-carbon aliphatic fragment (the aliphatic fragment is not condensed into a ring) (e.g., butein and okanin):

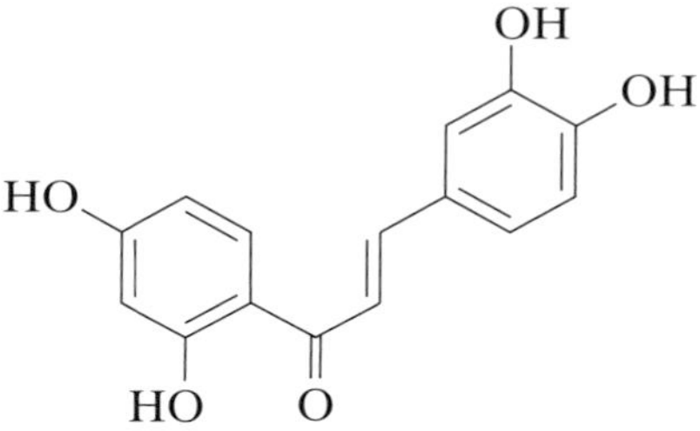

Butein

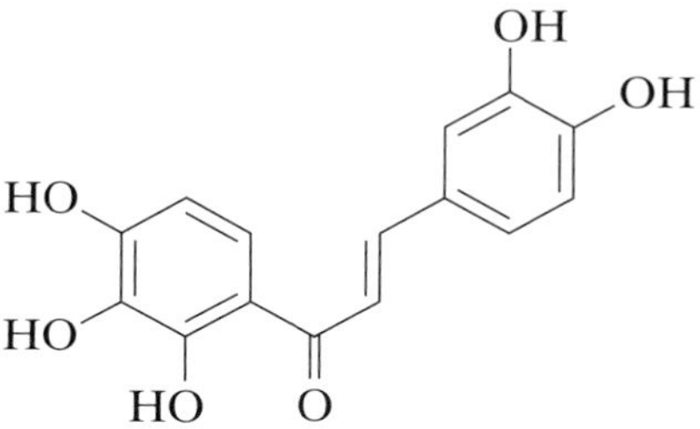

Okanin

2. Flavans. Flavonoids that do not possess a carboxyl group in the 4-position. These include the catechins (3,7,4′-hydroxyflavans) and the leucoanthocyanins, which are hydroxylated or methoxylated derivatives of 3,4-dihydroxyflavan.

Phenolic Compounds Other Than Flavonoids. These are mainly cinnamic acid and coumarin derivatives. Antioxidant activities of some flavones, flavonols, flavonones, and cinnamic acid derivatives are given in Table 11.10. Al-

Table 11.10 Antioxidant activity of some phenolic compounds

Compounds	Antioxidant Index (5×10^{-4} *M*)
Quercetin	3.8
Quercetin 3-rhamnoside	3.7
Quercetin 3-ramnoglucoside	1.6
Myricetin	4.5
Naringenin	1.6
Hesperitin	1.2
D-Catechin	3.5
Caffeic acid	3.6
Chlorogenic acid	3.7
Quinic acid	1.5
Propyl gallate	2.1

though these compounds have low solubility in lipids, they are reported to provide antioxidative protection when suspended in lipid systems (92–94) or in the aqueous phase of an aqueous lipid system (95).

Other plant phenolics that have shown significant antioxidant activities include carnosic and rosmaric acids, rosmanol, carnosol, and rosemaridiphenol, all found in rosemary extracts:

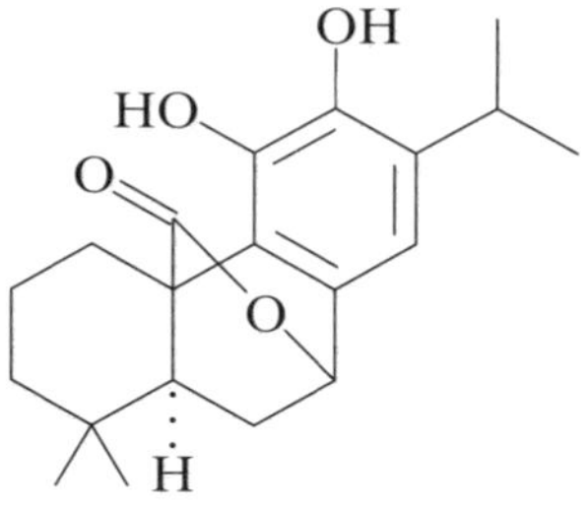

Carnosol

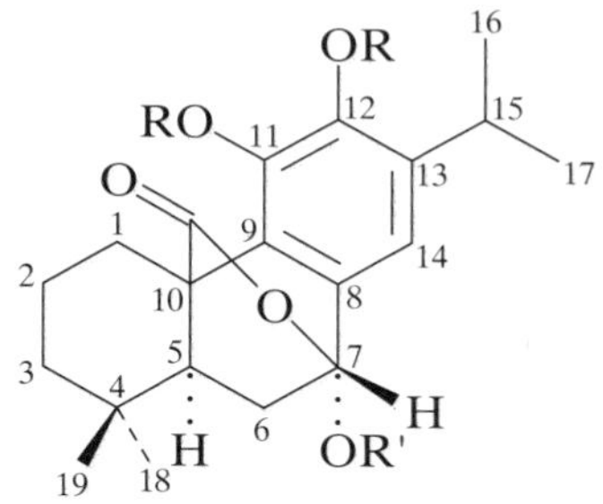

R, R' = H

Rosmanol (7α–)

and sesamol (3,4-methylenediphenoxy phenol) and sesamolinol from sesame seed oil:

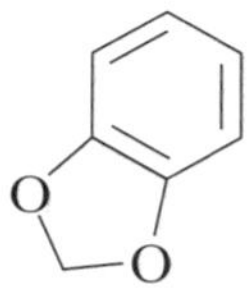

Sesamol

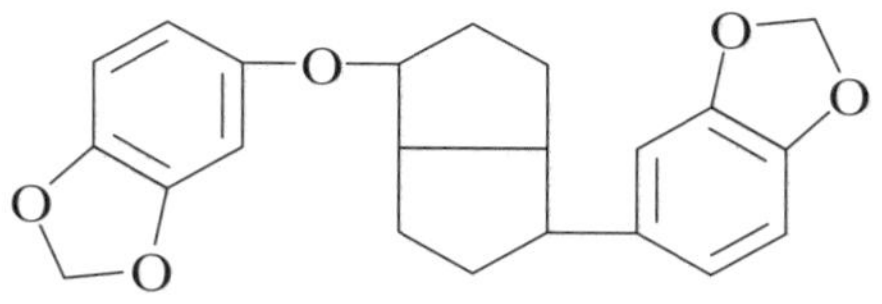

Sesamolin

RECENT ADVANCES

The following are two examples of recent trends in high-efficiency reaction techniques.

3.1 High-Intensity Ultrasound

The application of ultrasound to chemical reactions is a new approach which has been the subject of many recent investigations. The exact mechanism of sonochemistry is not well understood, but it is believed that high-intensity ultrasound exerts its effects through "cavitaion" (96,97), a phenomenon which involves the formation and the collapse of microbubbles. The collapse of the bubbles occurs in a very short time causing an extreme rise in the temperature and pressure (3000 K and 300 atm, respectively) at the center of the bubble.

This can produce dramatic enhancement in both the reactivity and selectivity of a variety of reactions (98). For example, ultrasound was found to greatly accelerate two-phase saponification of aromatic carboxylic acid esters. Reactions normally requiring 90 minutes of reflux could be accomplished after only 10 minutes of sonication (99).

3.2 Microwave Irradiation

Interest in the potential applications of microwave irradiation in lipid chemistry is rapidly growing (100). The technique has several attractive features. It is simple, practical, safe, and economical (a domestic microwave oven can be used). High pressures are avoided by using solid mineral supports in dry media or by phase transfer catalysis in the absence of solvents (101). Excellent yields and reduced reaction times are obtained.

Recently reported microwave-induced reactions include saponification of hindered esters:

$$\text{Ar}-\overset{O}{\overset{\|}{C}}-O-nC_8H_{17} \longrightarrow \text{Ar}-\overset{O}{\overset{\|}{C}}-OH + nC_8H_{17}OH$$

esterification and transesterification (100–103), alkylation, Diels-Alder, and "ene" reactions.

3.3 Hydrogenation (104–107)

Nutritional concerns regarding trans-fatty acids prompted the development of various techniques designed to reduce the amount of these isomers in the finished product. These include the selection of alternative catalysts, e.g., platinum, chromium, and precious metal catalysts, modification of reactor design and hardening conditions, and low-temperature electrocatalytic processes. In the latter technique, atomic hydrogen is directly generated on the surface of an active catalyst, Raney nickel powder cathode, by the reduction of protons from the adjacent solution. The absorbed hydrogen reacts with the unsaturated linkages in the oil. The low reaction temperature minimizes the formation of trans-isomers.

3.4 Interesterification

Recent advances in interesterification techniques have focused on the production of high-value speciality fats as well as hardened fats of low trans-acid content. In this regard, combining interesterification with hardening and/or blending has been very effective. Enzymatic interesterification is also used to produce a variety of value-added products. The high specificity of many commerically available lipases allows the production of triacylglycerols of improved properties for specific applications, e.g., confectioneries and infant formulae.

REFERENCES

1. General references: K.S. Markely, *Fatty Acids,* 5th ed., Wiley-Interscience, New York, 1968; D. Swern, ed., *Bailey's Industrial Oil and Fat Products,* Vol. 1, 4th ed., John Wiley & Sons, Inc., New York, 1979; F.D. Gunstone and F.A. Morris, *Lipids in Foods: Chemistry, Biochemistry and Technology,* Pergamon Press, Oxford, 1983; W.W. Nawar in O.R. Fennema, ed., *Food Chemistry,* Marcel Dekker, New York, 1985; B.F.J. Hudson, eds., *Food Antioxidants,* Elsevier, Amsterdam, 1990.
2. IUPAC-IUB Commission on Biochemical Nomenclature, *Lipids* **12,** 445 (1977).
3. H. Hirschmann, *J. Biol. Chem.* **235,** 2762 (1960).
4. C. Litchfield, *Analysis of Triglycerides,* Academic Press, Inc., New York, 1972.
5. J. Eisner, L.J. Iverson, A.K. Mozingo, and D. Firestone, *J. Am. Oil Chem. Soc.* **48,** 417 (1965).
6. J.A. Moura Fe, W.H. Brown, F.M. Whiting, and J.W. Stull, *J. Sci. Fd. Agric.,* 26 (1975).
7. G. Jacini and E. Fedeli, *Fette Seifen Anstr.* **77,** 1 (1975).
8. M. Bastic, L.J. Bastic, J.A. Jovanovic, and G. Spiteller, *J. Am. Oil Chem. Soc.* **55,** 886 (1978).
9. *R.P. Cook, ed., Cholesterol,* Academic Press, Inc., New York, 1958.
10. R.P. Cook, *Analyst* **86,** 373 (1961).
11. W.E. Davies, T.W. Goodwin, and E.I. Mercer, *The Identification and Determination of Plant Sterols in Determination of Sterols,* Monograph no. 2, Society of Analytical Chemistry, 1962, pp. 38–50.
12. T. Itoh, T. Tamura, and T. Mastumato, *J. Am. Oil Chem. Soc.* **50,** 123 (1973).
13. A. Kornfeldt and L.B. Croon, *Lipids* **16,** 3 (1981).
14. J. Krzynowek, *Food Technol.* **39**(2), 61 (1985).
15. D. Kritchevsky, S. Tepper, N.W. Ditullo, and W. Holmes, *J. Food Sci.* **32,** 64 (1967).
16. R.M. Feeley, P.E. Criner, and B.K. Watt, *J. Am. Dietetic Assoc.* **61,** 134 (1972).
17. J.B.S. Braverman, *Introduction to the Biochemistry of Foods,* Elsevier, New York, 1963.
18. L.P. Vernon and G.R. Seely, *The Chlorophylls,* Academic Press, Inc., New York, 1966.
19. O. Isler, *Carotenoids,* Birkhauser Verlag, Basel, 1971.
20. E. Fedeli, *Rev. Franc. Corps Gras.* **15,** 281 (1968).
21. G. Charalambous and J. Katz, eds., *Phenolic, Sulfur, and Nitrogen Compounds in Food Flavors,* ACS Series no. 26, American Chemical Society, Washington D.C., 1976.
22. R.J. Vander Wal, *J. Am. Oil Chem. Soc.* **40,** 568 (1963).
23. M.H. Coleman, *J. Am. Oil Chem. Soc.* **37,** 18 (1960).
24. M. Kates, *J. Lipid Res.* **5,** 132 (1964).
25. W.Q. Braun, *J. Am. Oil Chem. Soc.* **37,** 598 (1960).
26. E.S. Lutton, M.F. Mallery, and J. Burgers, *J. Am. Oil Chem. Soc.* **39,** 233 (1962).
27. B. Sreenivasan, *J. Am. Oil Chem. Soc.* **55,** 796 (1978).
28. L.F. Albright, *J. Am. Oil Chem. Soc.* **42,** 250 (1965).
29. K.S. Markley in K.S. Markley, ed., *Fatty Acids,* Part 5, Wiley-Interscience, New York, 1968, pp. 3611–3655.

30. N.O.V. Sonntag in Ref. 29, pp. 3589–3610.
31. A.J. Graupner and V.A. Aluise, *J. Am. Oil Chem. Soc.* **43,** 81 (1966).
32. I. Tanay and L. Madaras, *Acta Pharm. Hung.* **34**(2), 75 (1964).
33. S. Maruta and Shimizu, *Yukagaku* **13**(9), 465 (1964).
34. A. Jovtscheff, *Nahrung* **3,** 153 (1959); *Chem. Ber.* **93,** 2048 (1960).
35. A. Iovchev, *Izv. Inst. Org. Khim. Bulgar Akad. Nauk.* **2,** 77 (1965).
36. M.F. Abdel-Wahab and S.A. El-Kinawi, *Z. Anal. Chem.* **186,** 364 (1962).
37. C. Djerassi, *Chem. Revs.* **43,** 271 (1948).
38. A. Wohl, *Ber.* **52B,** 51 (1919).
39. D. Swern in K.S. Markely, eds., *Fatty Acids,* Part 2, Wiley-Interscience, New York, 1961, pp. 1307–1385.
40. L. Goldblatt in Ref. 29, pp. 3657–3684.
41. O.S. Privett and M.L. Blank, *J. Lipid Res.* **2,** 37 (1961).
42. O.S. Privett and E. Christense Nickell, *J. Am. Oil Chem. Soc.* **39,** 414 (1962).
43. E.H. Pryde, D.E. Anders, H.M. Teeter, and J.C. Cowan, *J. Org. Chem.* **25,** 618 (1960); **27,** 3055 (1962); *J. Am. Oil Chem. Soc.* **38,** 375 (1961); **40,** 497 (1963); E.H. Pryde and J.C. Cowan, *Ibid.* **39,** 496 (1962); E.H. Pryde and D.E. Anders (to See'y Agr.) U.S. Pat. 3,112,329 (Nov. 26, 1963).
44. E.H. Pryde, D.E. Anders, H.M. Teeter, and J.C. Cowan, *J. Org. Chem.* **27,** 3055 (1962); *J. Am. Oil Chem. Soc.* **38,** 375 (1961); *Ibid.* **40,** 497 (1963).
45. American Oil Chemists' Society, *Official and Tentative Methods of the American Oil Chemists' Society,* Method Cd. 11-57, ammended 1959, official 1960, revised to 1965, Chicago.
46. G. Maerker and E.T. Haeberer, *J. Am. Oil Chem. Soc.* **5,** 43, 97 (1966).
47. R.U. Lemieux and E. von Rudloff, *Can. J. Chem.* **33,** 1701 (1955).
48. E. von Rudloff, *Can J. Chem.* **43,** 1784 (1965).
49. O.L. Mageli and J.R. Kolezynski, *Papers presented at the Atlantic City Meeting Division of Organic Coatings and Plastics Chemistry,* American Chemical Society, Sept. 1965, pp. 101–124.
50. G.R. Riser, R.W. Riemenschneider, and L.P. Witnaner, *J. Am. Oill Chem. Soc.* **43,** 456 (1966).
51. *J. Am. Oil Chem. Soc.* **38,** 349 (1961); **39,** 362 (1962); **40,** 386 (1963); **41,** 574 (1964); **42,** 436A (1965).
52. Anon., *Chem. Week* **92**(14), 55, 60, 64 (Apr. 6, 1963).
53. G. Gut and A. Guyer, *Helv. Chim. Acta* **47,** 1673 (1964).
54. C. Litchfield, J.E. Lord, A.F. Isbell, and R. Reiser, *J. Am. Oil Chem. Soc.* **40,** 553 (1963).
55. M.A. McCutchon, R.T. O'Connor, E.F. DuPre, L.A. Goldblatt, and W.G. Bickford, *J. Am. Oil Chem. Soc.* **36,** 115 (1959).
56. C. Litchfield, R.D. Harlow, A.F. Isbell, and R. Reiser, *J. Am. Oil Chem. Soc.* **42,** 73 (1965).
57. J.A. Howard in J.K. Kochi, ed., *Free Radicals,* Vol. 11, John Wiley & Sons, Inc., New York, 1973.
58. K.U. Ingold, *Acc. Chem. Res.* **2,** 1 (1969).
59. C. Walling, *Free Radicals in Solution,* John Wiley & Sons, Inc., New York, 1957.
60. E.H. Farmer, G.G. Bloomfield, S. Sundralingam, and D.A. Sutton, *Trans. Faraday Soc.* **38,** 348 (1942).
61. E.H. Farmer, H.P. Koch, and D.A. Sutton, *J. Chem. Soc.,* 541 (1943).
62. E.H. Farmer and D.A. Sutton, *J. Chem. Soc.,* 119 (1945), 10 (1946).
63. E.N. Frankel, *Prog. Lipid Res.* **19,** 1 (1980).
64. L. Bateman, H. Hughes, and A.L. Morris, *Disc. Faraday Soc.* **14,** 190 (1953).

65. E.N. Frankel, R.F. Garwood, B.P.S. Khamby, G.P. Moss, and B.C. Weedon, *J. Chem. Soc. Perkin Trans.* **1,** 2233 (1984).
66. E.N. Frankel, C.D. Evans, D.G. McConnel, and E.P. Jones, *J. Am. Oil Chem. Soc.* **38,** 134 (1961).
67. H.W.S. Chan and G. Levett, *Lipids* **12,** 99 (1977).
68. H.W.S. Chan and G. Levett, *Lipids* **12,** 837 (1977).
69. E.N. Frankel, in E.H. Pryde, ed., *Fatty Acids,* Am. Oil Chem. Soc., Champaign, Ill., 1979, pp. 353–390.
70. C.S. Foote, *Acc. Chem. Soc.* **1,** 104 (1968).
71. H.W.S. Chan, *J. Am. Oil Chem. Soc.* **54,** 100 (1977).
72. E.N. Frankel in H.W. Shultz and R.O. Sinnhuber, eds., *Symposium on Foods: Lipids and their Oxidation,* Avi Publishing Co., Westport, 1962, pp. 51–78.
73. H.W.S. Chang, F.A.A. Prescott, and P.A.T. Swoboda, *J. Am. Oil Chem. Soc.* **53,** 572 (1976).
74. S. Patton, I.J. Barnes, and L.E. Evans, *J. Am. Oil Chem. Soc.* **36,** 280 (1959).
75. J. Terao, T. Ogawa, and S. Matsushita, *Agric. Biol. Chem.* **39,** 397 (1975).
76. H.T. Badings, *Neth. Milk Dairy J.* **14,** 215 (1960).
77. R. Ellis, A.M. Gaddis, and G.T. Currie, *J. Food Sci.* **26,** 131 (1961).
78. C.D. Evans, G.R. List, A. Dolev, D.G. McConnell, and R.L. Hoffman, *Lipids* **2,** 432 (1967).
79. R.J. Horvat, W.G. Lane, H. Ng, and A.D. Shepherd, *Nature* (London) **203,** 523 (1964).
80. M. Namiki, *Crit. Rev. in Food Sci. and Nutr.* **29**(4), 273 (1990).
81. F. Shahidi, P.K. Janitha, and P.D. Wanasundara, *Crit. Rev. in Food Sci. and Nutr.,* **32**(1), 67 (1992).
82. B.J.F. Hudson, ed., *Food Antioxidants,* Elsevier, Amsterdam 1990.
83. U. Bracco, J. Loliger, and J.L. Viret, *J. Am. Oil Chem. Soc.* **58,** 686 (1981).
84. G. Papadopoulos and D. Boskou, *J. Am. Oil Chem. Soc.* **68,** 669 (1991).
85. S.S. Chang, B. Matijasevic, O. Hsieh, and C. Huang, *J. Food Sci.* **42,** 1102 (1977).
86. T. Tashiro, Y. Fukuda, T. Osawa, and M. Namiki, *J.Am. Oil Chem. Soc.* **67,** 508 (1990).
87. J. Pokorny, *Trends Food Sci. Tech.* **2,** 223 (1991).
88. A.D.E. Pratt in G. Charalambons and J. Katz, eds., *Phenolic, Sulfur, and Nitrogen Compounds in Good Flavors,* ACS Symposium Series no. 26, ACS, Washington, D.C., 1976, Chapt. 1.
89. T.A. Geissman and E. Hinreiner, *Botan. Rev.* **18,** 77 (1952).
90. E.C. Bate-Smith, *Adv. Food Res.* **5,** 261 (1954).
91. T.A. Geissman in T.A. Geissman, ed., *The Chemistry of Flavonoid Compounds,* Macmillan Co., New York, 1962, Chapt. 1.
92. D.E. Pratt and B.M. Watts, *J. Food Sci.* **29,** 27 (1964).
93. D.E. Pratt, *J. Food Sci.* **30,** 737 (1965).
94. M.B. Ramsey and B.M. Watts, *Food Technol.* **17,** 1056 (1963).
95. C.H. Lea and P.A.T. Swoboda, *Chem. Ind.* 1426 (1956).
96. K.S. Suslick, ed., *Ultrasound, its Chemical, Physical, and Biological Effects,* VCH Publishers, New York, 1988.
97. T. Mason, ed., *Advances in Sonochemistry,* JAI Press, London, Vol. 1, 1990, Vol. 2, 1991, Vol. 3, 1993.
98. J.L. Luche, *OCL* **1,** 69 (1994).
99. S. Moon, L. Duchin, and J.V. Cooney, *Tetrahedron Lett.* **41,** 3917 (1979).
100. A. Loupy, *OCL* **1,** 62 (1994).

101. A. Loupy and co-workers, *Can. J. Chem.* **71,** 90 (1993).
102. A. Loupy, P. Pigeon, M. Ramdani, and P. Jacquault, *Synth. Comm.* **24,** 159 (1994).
103. A. Dasgupta, P. Banerjee, and S. Malik, *Chem. Phys. Lipids* **62,** 281 (1992).
104. B.P. Haumann, *Inform* **5,** 668 (1994).
105. P.A. Bernstein, W.F. Graydon, and L.L. Diosady, *J. Am. Oil Chem. Soc.* **66,** 680 (1989).
106. P.N. Pintauro, U.S. Pat. 5,225,581 (July 6, 1993).
107. P.N. Pintauro and J.R. Bontha, *J. Appl. Electrochem.* **21,** 799 (1991).

12

Fatty Acids in Newer Fats and Oils

FATTY ACIDS AND FOOD OILS

At one time section "I" of the *Official and Tentative Methods of the American Oil Chemists' Society* provided a tabulation of the physical and chemical characteristics of as many as 72 common or known fats, oils, and waxes of animal and vegetable origin. This information has not been included in the fourth edition (1). The famous reference book authored by T.P. Hilditch and P.N. Williams (2) provided reasonable fatty acid details and also much biological information on fats of all types. As recently as 1973 the *Handbook of Chemistry and Physics* (3) was providing a list of fatty acids of 37 plant, land animal, and marine fats and oils. However, a cautionary note says the values are "typical rather than average" and "extreme variations may occur depending on a number of variables such as source, treatment, and age of fat or oil." To illustrate this problem for a limited geographical area, Table 12.1 reproduces a quality table for flaxseed varieties grown in western Canada. This tabulation is part of the ongoing monitoring of oilseed qualities of the Canadian Grain Commission, Winnipeg, for canola, flaxseed, soybean, and to some extent sunflower crops.

This information is useful for trading purposes and for bodies such as Codex Alimentarius (4), but does not necessarily help the nutritionist or consumer. Although blended and/or interesterified oils are of increasing interest (5–7), labels and/or advertising still tend to focus on individual oils and fats.

Until recently the botanical and common names of oilseed crops provided clearly understood identifications and this system was utilized in part in the first edition of *The Lipid Handbook* (8) by F. Padley and co-workers (9), but primarily for "major" vegetable and animal fats. One or more pages of text with fatty acid tables described the history, uses, and chemical properties of these oils. Valuable support was provided in this single volume by another

Table 12.1 Quality of flaxseed varieties from a 1993 harvest survey in western Canada[a]

Variety	Number of samples	Oil content[b]	Protein content[c]	FFA	Iodine Value	Fatty Acid Composition				
						C16:0	C18:0	C18:1	C18:2	C18:3
Manitoba										
Flanders	9	45.5	21.3	0.17	196	4.8	3.9	16.0	14.9	59.6
McGregor	11	44.0	22.2	0.21	193	5.5	3.4	17.0	15.5	57.8
Norlin	39	43.6	22.8	0.20	192	5.1	2.9	20.0	13.3	57.9
Norman	32	44.9	22.2	0.18	192	5.2	2.8	19.1	14.3	57.7
Somme	7	43.6	21.1	0.15	196	5.5	2.9	16.0	14.4	60.3
Vimy	3	45.7	20.6	0.22	200	5.4	3.0	14.0	14.8	62.1
Saskatchewan										
Flanders	7	46.7	20.2	0.14	199	4.7	3.5	14.2	15.9	61.0
McGregor	5	46.0	20.0	0.11	197	5.5	3.2	14.2	17.0	59.4
Norlin	36	44.7	21.2	0.14	198	5.0	2.5	16.8	14.1	60.8
Norman	4	46.3	19.8	0.20	201	5.0	2.7	14.2	15.0	62.2
Somme	14	44.4	20.7	0.23	202	5.5	2.6	13.0	14.9	62.2
Vimy	106	46.1	20.0	0.13	198	5.4	3.0	14.5	16.0	60.4
Alberta										
Andro	4	44.5	23.2	0.15	201	4.6	2.9	14.9	14.1	62.7
McGregor	4	44.7	22.4	0.12	194	5.5	3.7	14.4	19.0	56.7
Norlin	9	44.9	21.2	0.19	199	5.0	2.7	16.3	13.5	61.7

[a] Courtesy of the Grain Research Laboratory, Canadian Grain Commission, Winnipeg.
[b] Moisture-free basis.
[c] %N × 6.25; moisture-free basis.

section with detailed descriptions of hundreds of natural and synthetic fatty acids. The rarer seed oil fatty acids, especially of Indian origin, are listed by R. Badami and K. Patil (10). These seeds include important conjugated acetylenic, epoxy, and other types of fatty acids not generally found among the fatty acids of edible oilseeds grown on a large scale. An exception is of course the cyclopropenoid acids (circa 1% of the total) of cottonseed oil.

As is well-known, the introduction of gas–liquid chromatography (GLC) has provided a vast amount of information on the fatty acid composition of edible fats and oils. The packed GLC column was not totally satisfactory because of the chain length overlap problem with the commonly used polyester liquid phases (11–13). Thus many early analyses did not specify the 20:0 or 20:1 contents of oils and fats because one of these usually coincided with 18:3*n*-3 (α-linolenic acid) on polar GLC columns. On nonpolar (methyl silicone) columns the 18:3*n*-3 coincided with 18:2*n*-6 (linoleic acid), as did most similar combinations of *n*-3 and *n*-6 fatty acids differing by two carbon atoms and one ethylenic bond (14,15). Possibly this technical defect contributed to the lack of information on, and interest in, the omega-3 fatty acids that is only now being rectified. Recent authoritative nutrition and medical reports (16,17) recognize α-linolenic acid (18:3*n*-3) as essential, a status formerly conferred primarily on linoleic acid (18:2*n*-6).

Much better data on fatty acid compositions is now provided by the widespread use of GLC analyses on capillary (wall-coated open-tubular) columns. Progress on these columns was made as early as the 1960–1970 era (12,18–20). However, shortly after 1980 the flexible fused silica GLC column was introduced (21) and in combination with the sensitive FID (flame ionization detector) is now the basis of all modern fatty acid analyses of edible plant and animal oils and fats. A vital point is the conversion of the popular digitized peak "area" percent for methyl esters of fatty acids to actual fatty acid weight percent through correction factors for FID response (Table 12.2). This conversion is not stipulated in AOAC method No. 963.22 (22) but is included in the first action AOAC method for encapsulated fish oils (23) soon to be issued in the next (16th) AOAC Official Methods as 991.39. The similar AOCS method Ce-1b-89 also provides for correction factors relative to an added internal standard, both methods being designed to give g fatty acid/100 g sample for longer chain fatty acids when applying the factors of Table 12.2. These and AOCS methods Ce-1d-91 and Ce-1e-91 are designed for capillary columns, but such methods often do not expressly focus on the need for the general use of FID correction factors.

The justification for the table is amply demonstrated in a series of papers by one group on both methyl ester preparation and quantitative gas–liquid chromatography of vegetable oil fatty acids (24–33). The marine oil fatty acids extend from 14:0 to 22:6*n*-3 (and 24:1) as do the fatty acids of many animal organ lipids, of the lean red meats (34), and even of the white meats of fowl (35). At the other end of the chromatograms the dairy products include fatty acids from 4:0 to 18:2*n*-6 and 18:3*n*-3. Both ranges of chain lengths clearly

Table 12.2 Theoretical relative response factor (TRF) for fatty acid methyl esters (FAME)[a]

FAME	TRF	FAME	TRF	FAME	TRF	FAME	TRF
4:0	1.5396	14:1	1.0354	18:2	0.9865	22:1	0.9664
5:0	1.4009	15:0	1.0308	18:3	0.9797	22:2	0.9609
6:0	1.3084	15:1	1.0227	18:4	0.9730	22:3	0.9554
7:0	1.2423	16:0	1.0193	19:0	0.9919	22:4	0.9499
8:0	1.1927	16:1	1.0117	20:0	0.9846	22:5	0.9443
9:0	1.1542	16:2	1.0041	20:1	0.9785	22:6	0.9388
10:0	1.1233	16:3	0.9965	20:2	0.9724	23:0	0.9665
11:0	1.0981	16:4	0.9889	20:3	0.9663	24:0	0.9614
12:0	1.0771	17:0	1.0091	20:4	0.9603	24:1	0.9564
12:1	1.0670	17:1	1.0019	20:5	0.9452		
13:0	1.0593	18:0	1.0000	21:0	0.9780		
14:0	1.0440	18:1	0.9932	22:0	0.9720		

[a] Atomic weights used: carbon, 12.011; hydrogen, 1.0079; oxygen, 15.994. Factors are relative to 18:0, which has a factor of 1.000 by definition. Factors for the following FAME have been verified experimentally: 4:0; 6:0; 8:0; 10:0; 12:0; 14:0; 17:0; 20:0; *c*9-18:1; *c*9,*c*12,*c*15-18:3; *c*5,*c*8,*c*11,*c*14-20:4; *c*4,*c*7,*c*10,*c*13,*c*16,*c*19-22:6. Only one factor is given for all positional and geometric isomers and for branched-chain FAME, as the factors are dependent only on the content of carbon to which hydrogen is bonded. Calculations by Craske and Bannon (32).

require the FID correction factors of Table 12.2. Modern data processing is usually capable of automatically including the FID conversion factors in any calculation of electronic "peaks" for fatty acid compositions of the type shown in Table 12.3 from area percent to weight percent. This is a list of the most common North American commercial fats and oils, all recently analyzed by one laboratory. This table is a summary of analyses of retail materials purchased in 1994 so that five or six samples of diverse origins in Canada or the U.S. are included for each oil or fat listed. It is noteworthy that these analyses reveal the C_{20}, C_{22}, and C_{24} fatty acids that are not included in most such tables of fatty acids of common fats and oils.

One flaw in such tables is that they are based on potentially obsolete oil identification information. The new varieties of oilseed plants are usually said to be based on genetic manipulation (36,37). Although canola was developed over nearly two decades by conventional plant breeding methods the newer developments have been achieved much more rapidly by modern biotechnology. This biotechnology is leading to more and more litigation and problems for patent and regulatory agencies (38).

Table 12.4 lists some new information on fatty acids of crop varieties just entering the market. The resurgence of interest in high-erucic acid rapeseed oil warrants inclusion of the new cultivar HERO. Possibly the application of correction factors to the more usual short range of C_{16}–C_{18} chain lengths in these oils has only a moderate impact on the conversion of peak area percent to fatty acid weight percent. Figure 12.1 includes a gas chromatographic record of the fatty acids of the high-oleic/low-saturated fatty acid canola oil, "Sunola," and of a more conventional but low-linolenic acid canola oil product. The points to note in this figure are the presence of *cis*-vaccenic acid immediately following the major oleic acid peak, and the obvious presence of 20:0 and 20:1 peaks. In one, between 18:2*n*-6 and 20:0, are no fewer than three peaks for α-linolenic acid (18:3*n*-3). The largest is the all-cis natural fatty acid and the adjacent peaks are the 9-*cis*,12-*cis*,15-*trans*-18:3*n*-3, and the 9-*trans*,12-*cis*,15-*cis* isomers, discussed in more detail in the next section.

The question of food label identifications of the newer varieties of edible oils in common use may thus become one of looking at minor components, particularly sterols, hydrocarbons, etc. An example of "a solution in search of a problem" has been generated by the 1993 food labels required by the U.S. Food and Drug Administration (39). The total recovery of fats from food, and their definitions as lipid classes, is extremely difficult, in both "natural" foods (40) and in the modern food combinations expected by the consumer. The solution (41–44) is to recover all the fatty acids, in most cases from total saponification, and recombine them into a "synthetic" triacylglyceride (44,45). This solution supersedes the former practice of using the known ingredients and the fatty acid data of USDA Handbook No. 18 for label calculations (34). The new analytical process would be facilitated by a fatty acid internal standard, and so the factors in Table 12.2 can be used to provide appropriate relationships to that internal standard, especially if dairy fats or tropical oils (mostly C_8–C_{16}) are included. The use of the latter has declined

Table 12.3 Summary of area % fatty acid compositions[a] and cholesterol analyses of some retail fats and oils common in 1994[b]

Fatty Acid	Canola	Safflower	Sunflower	Corn	Olive	Soybean	Peanut	Cottonseed	Coconut	Palm	Lard	Beef Tallow	Butterfat
4:0	–	–	–	–	–	–	–	–	–	–	–	–	2.3
6:0	–	–	–	–	–	–	–	–	0.6	–	–	–	1.9
8:0	–	–	–	–	–	–	–	–	7.7	0.1	–	–	1.3
10:0	–	–	–	–	–	–	–	–	6.0	0.1	0.1	0.1	3.0
12:0	–	–	–	–	–	–	–	–	46.7	0.9	0.1	0.1	3.6
14:0	0.1	0.1	0.1	–	–	0.1	–	0.8	18.3	1.3	1.4	3.3	11.6
14:1	–	–	–	–	–	–	–	–	–	–	–	1.1	1.1
15:0	–	–	–	–	–	–	–	–	–	–	0.1	0.5	1.4
15:1	–	–	–	–	–	–	–	–	–	–	–	0.2	–
16:0	3.9	6.6	6.0	9.9	11.0	10.3	10.4	22.9	9.2	43.6	25.4	24.7	32.0
16:1	0.3	0.1	0.1	0.2	0.9	0.1	0.1	0.6	–	0.2	2.7	4.0	2.1
16:2	–	–	–	–	–	–	–	–	–	–	–	0.7	–
17:0	–	–	–	–	–	–	–	–	–	0.1	0.5	1.4	0.7
17:1	–	–	–	–	–	–	–	–	–	–	0.5	1.0	0.4
18:0	1.8	2.5	4.6	2.0	3.0	3.9	2.5	2.5	2.9	4.5	15.1	17.0	10.7
18:1	58.0	13.9	15.7	28.7	73.4	22.1	45.5	17.8	6.9	38.7	43.0	42.2	24.2
18:2	21.0	75.5	71.4	56.9	9.5	54.1	32.9	54.2	1.7	9.5	8.5	2.0	3.1
18:3	11.1	0.1	0.6	1.1	0.6	8.3	0.5	0.5	–	0.3	1.1	0.5	0.6
19:0	–	–	–	–	–	–	–	–	–	–	–	0.6	–
20:0	0.7	0.4	0.3	0.5	0.4	0.3	1.3	0.3	–	0.2	0.2	0.2	–
20:1	1.7	0.2	0.2	0.4	0.3	0.3	1.5	0.1	–	0.1	0.9	0.4	–
20:2	0.1	–	–	–	–	–	–	–	–	–	0.4	–	–
22:0	0.4	0.3	0.8	0.2	0.1	0.4	3.3	0.2	–	0.1	–	–	–
22:1	0.5	–	–	–	–	–	0.2	–	–	–	–	–	–
24:0	0.2	0.1	0.2	0.2	0.1	0.1	1.8	0.1	–	0.1	–	–	–
24:1	0.3	0.2	–	–	0.6	–	–	–	–	–	–	–	–
% Saturates	7.0	9.9	12.0	12.6	14.7	15.2	19.3	26.8	91.4	51.2	42.9	47.9	68.6
% Monoenoics	60.8	14.4	16.0	29.3	75.2	22.4	47.3	18.5	6.9	39.0	47.1	48.9	27.8
% Polyenoics	32.3	75.7	72.0	58.1	10.1	62.4	33.4	54.7	1.7	9.8	10.0	3.2	3.6
Cholesterol (mg/g)	–	–	–	–	–	–	–	–	–	–	1.06	1.46	3.66

[a] Fatty acid content normalized to peak area 100%.
[b] Courtesy of the Canola Council of Canada, Winnipeg.

Table 12.4 Fatty acid profiles of oils from some recently developed oilseed varieties

	Rapeseed[a,b]	Canola	Sunflower			Flaxseed		Soybean
Fatty Acid	High-Erucic Acid Area %	InterMountain Retail Cooking Oil[c,d] Wt. %	High-Oleic/ Low-Saturated "Sunola"[e] Wt. %	High-Oleic (a)[f]	High-Oleic (b)[g]	Conventional[a] Area %	Low-Linolenic "Linola"[h] Area %	Low-Linolenic Acid[g]
16:0	2.66	4.71	3.0	3.3	4.0	5.25	5.6	10
16:1	0.20	0.32	–	–	–	0.16	–	–
18:0	0.90	2.28	2.9	4.4	4.3	3.26	4.0	4
18:1	12.52	63.13	87.7	77.1	79.2	17.28	15.9	33
18:2*n*-6	11.79	21.89	4.3	12.6	11.1	14.34	71.9	50
18:3*n*-3	8.64	3.33	0.1	0.5	0.4	59.07	2.0	3
20:0	0.78	0.57	0.3	–	–	–	–	–
20:1	6.05	1.16	0.3	–	–	–	–	–
20:2	1.31	–	–	–	–	–	–	–
22:0	0.76	–	0.9	–	–	–	–	–
22:1	50.87	0.67	–	–	–	–	–	–
24:0	0.32	–	0.3	–	–	–	–	–
24:1	1.22	0.46	–	–	–	–	–	–

[a] Courtesy of J.K. Daun.
[b] Cultivar HERO.
[c] Analysis by author.
[d] Retail oil sample.
[e] Courtesy of Western Grower Seed Corp., Saskatoon.
[f] Frankel and Huang (5).
[g] Erickson and Frey (37).
[h] Courtesy of United Grain Growers, Winnipeg. For further information see *INFORM* **4,** 907 (1993). The Flax Council of Canada has adopted the generic term "solin" for low (≤5% by weight) linolenic acid varieties of flaxseed.

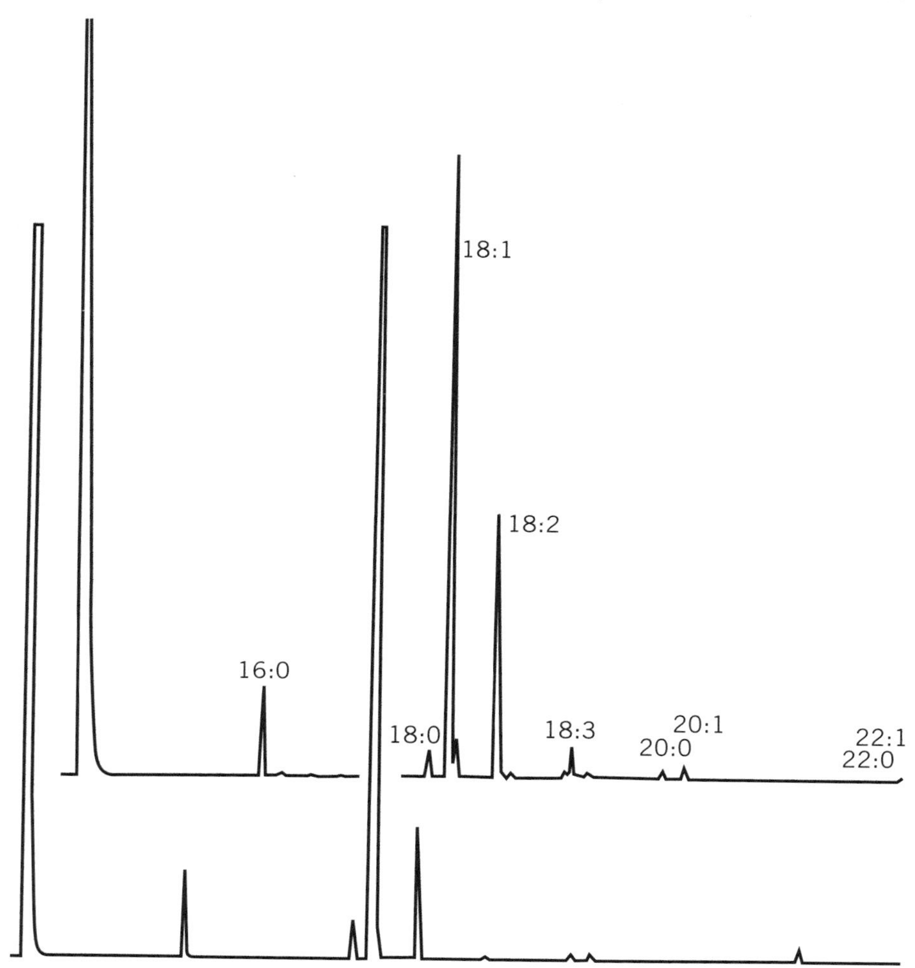

Figure 12.1 Comparison of chromatograms from the gas–liquid chromatographic analysis of methyl esters of fatty acids of two contemporary edible oils. Above, low-linolenic U.S.-grown Clear Valley retail canola oil; below, high oleic Canadian grown "Sunola" sunflower seed oil. Sample courtesy of Western Growers Seed Corporation, Saskatoon, Canada.

in the U.S., but tropical oils are major vegetable oil resources on a world basis. The objections to their use are based on their saturated fatty acids, are long-standing, and related indirectly through serum cholesterol (46,47) to health matters too complex to discuss here in any detail.

1.1 Fatty Acids and Human Health

As shown in Tables 12.3 and 12.4 there is considerable variation between tropical and temperate-zone vegetable oils, and among the latter, in their saturated fatty acid content, now an important point in marketing (47). Reducing serum cholesterol by reducing the intake of saturated fatty acids is an appealing simplification of very complex lipid biochemistry (48). The controversial topic of the "best" fatty acid mixture in our diets now motivates much of the market changes in edible fats and oils (6,37). In the opinion of some scientists the polyunsaturated fatty acids are equally if not more important in the formation of atheroma aortic plaques (49). Linolenic acid (18:3*n*-3) has traditionally been reduced to about 3% in liquid frying oils to improve stability, hence the composition of some of the new varieties of oils in Table 12.4.

In brief, there are now viewpoints that there is too much linoleic (18:2*n*-6) acid in our diets (50), and not enough α-linolenic (18:3*n*-3) acid (16,17,50–52). High oleic acid (Mediterranean) diets have been touted (53), but in all the medical controversy it is usually forgotten that flavors of fats and oils may benefit from moderate oxidation (5,37,54) of the polyunsaturated acids.

In the arguments (55) over the hazards of the slight change in serum cholesterol fractions induced by trans fatty acids (56–58), the isolated trans ethylenic bonds of food fats may be determined by an infrared spectral absorption method (forthcoming AOAC 16th edition, Method 965.34) or by a gas–liquid chromatographic method (AOAC Method 985.21). The methods vary in sensitivity (cf. AOCS Method, Cd-14-61) and a comparatively recent AOCS-GLC method (Ce-1c-89) has been criticized for underestimating the *trans*-octadecenoates in favor of the corresponding cis isomers (59). A new AOCS method (Cd-14b-93) combines gas chromatography and infrared technology for partially hydrogenated oils. Figure 12.2 is an example of the type of GLC profile encountered from a paper which explained the method in more depth (60).

What is not generally realized is that vegetable oil deodorization produces mono- and even di-trans ethylenic bonds in the C_{18} polyunsaturated fatty acids. Hydrogenation is not involved. These are structurally very similar to the original fatty acids deemed essential in man (16,17,50–52) which would naturally remain all-cis unless exposed to high processing temperatures. The problem of possible adverse biochemical effects of these geometrical isomer artifact fatty acids containing trans ethylenic bonds is almost unexplored except by J-L. Sebedio and co-workers (61,62). Analytical conditions have been defined for isomers of both linoleic (18:2*n*-6) acid (63) and linolenic (18:3*n*-3) acid (64–66).

Hydrogenation produces both geometrical and positional isomers of the ethylenic bonds in fatty acids (63). Figure 12.3 is an assessment of the incorporation of the hydrogenation artifacts of mono- and polyunsaturated fatty acids into rat liver phospholipids (67). Presumably the deodorization artifacts will undergo the same biochemical processes in our bodies. Possibly these processes are a new health problem that is not inseparable from that of blending or interesterifying liquid (refined, deodorized, but probably unhydrogenated) vegetable oils with tropic oils or highly hydrogenated C_{18}-rich oils in an attempt

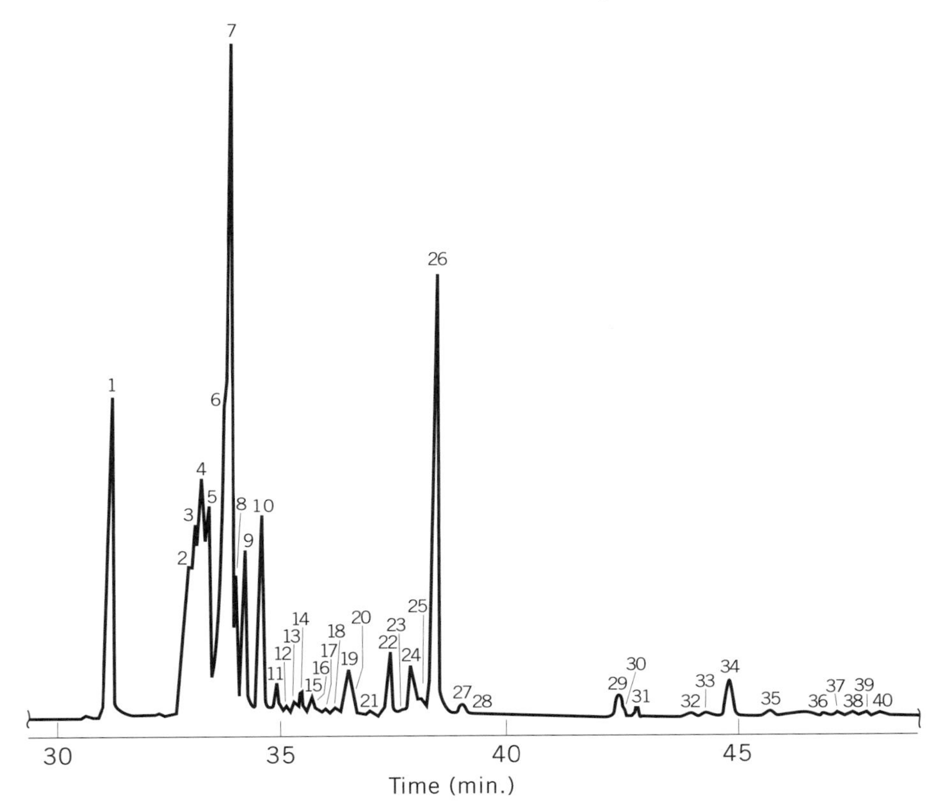

Figure 12.2 The C_{18} region of the gas chromatogram of the fatty acid methyl esters from partially hydrogenated soybean oil, using 100 m × 0.25 mm fused silica capillary column coated with SP-2560. Numbers correspond to 1, 18:0; 2, 18:1Δ6-8*t*; 3, 18:1Δ9*t*; 4, 18:1Δ10*t*; 5, 18:1Δ11*t*; 6, 18:1Δ12*t*; 7, 18:1Δ6-8*c* + 18:1Δ9*c* (major component) + 18:1Δ 13-14*t*; 8, 18:1Δ10*c* + 18:1Δ15*t*; 9, 18:1Δ11*c*; 10, 18:1Δ12*c*; 11, 18:1Δ13*c*; 12, 18:2*tt*; 13, 18:1Δ14*c* + 18:1Δ16*t*; 14, 18:2*tt*; 15, 18:1Δ15*c*; 16, 17, 18:2*tt*; 18, 18:2Δ9*t*,12*t*; 19, 20, 18:2Δ9*c*,13*t* + 18:2Δ8*t*,12*c*; 21, 18:2*ct*/*tc*; 22, 18:2Δ9*c*,12*t*; 23, 18:2Δ8*c*,13*c*; 24, 18:2Δ9*t*,12*c*; 25, 18:2Δ9*t*,15*c* + 18:2Δ10*t*,15*c* + 18:2Δ9*c*,13*c*; 26, 18:2Δ9*c*,12*c* (linoleate); 27, 18:2Δ9*c*,15*c*; 28, unknown; 29, 20:0; 30, 18:3?; 31, 18:3Δ9*c*,12*c*,15*t*; 32, 18:3Δ9*c*,12*t*,15*c*; 33, 18:3Δ9*t*,12*c*,15*c*; 34, 18:3Δ9*c*,12*c*,15*c* (linolenate); 35, 20:1*c*; 36–40, 18:2 conjugates. Reproduced by courtesy of the American Oil Chemists' Society and W.M.N. Ratnayake (60).

to reduce the hydrogenation-generated trans acid contents of foods (67). Although studies of processed fats and oils and human health are ongoing, the analytical technology for both the oils and animal tissues is obviously complex. Very few laboratories are able or willing to conduct the fatty acid radioisotope tracer studies with humans recently publicized by E.A. Emken of the U.S. Department of Agriculture. He has listed and refuted (68) a number of misconceptions popular with the public and even some scientists.

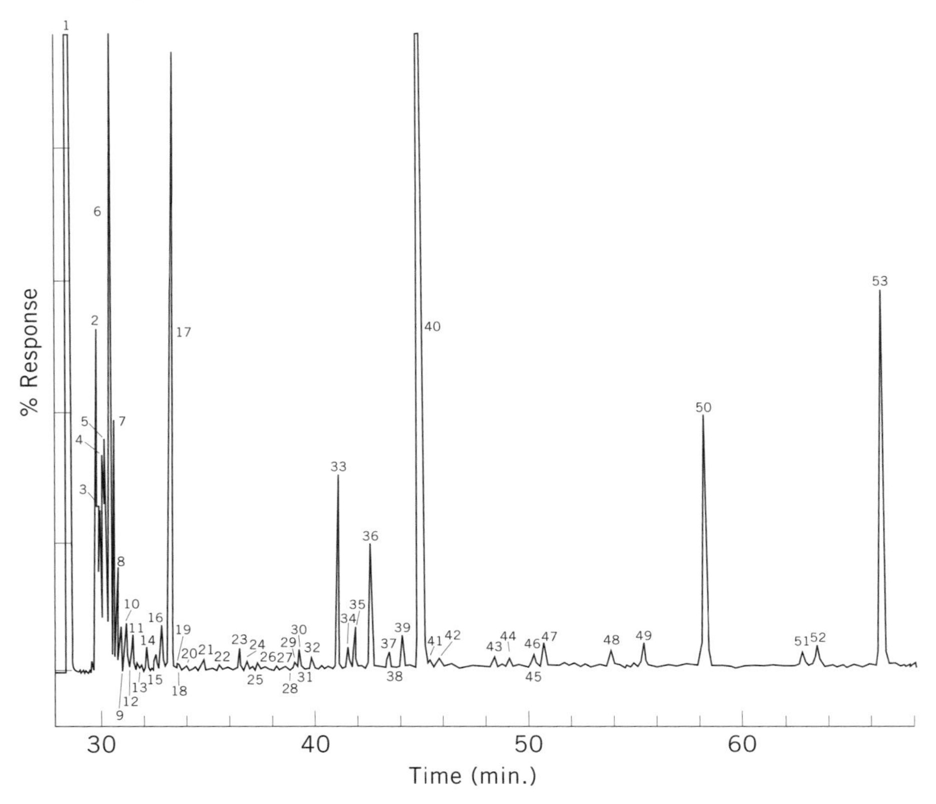

Figure 12.3 The C_{18}, C_{20}, and C_{22} regions of the gas chromatogram on SP-2560 capillary column (100 m × 0.25 mm) of the FAME from liver phospholipids of rats fed partially hydrogenated oil. Peak identifications: 1, 18:0; 2, 8*c*-9*c*-18:1; 3, 10*t*-18:1; 4, 11*t*-18:1; 5, 12*t*-18:1; 6, 13*t*-15*t*-18:1 + 6*c*-10*c*-18:1; 7, 11*c*-18:1; 8, 12*c*-18:1; 9, *tt*-18:2; 10, 13*c*-18:1; 11, 14*c*-18:1 + 6*t*-18:1; 12, *tt*-18:2 + 15*c*-18:1; 13, 9*t*,12*t*-18:2; 14, 9*c*,13*t*/8*t*,12*c*-18:2; 15, 9*c*,12*t*-18:2 + 8*c*,13*c*-18:2 + 9*t*,13*c*-18:2; 16, 9*t*,12*c*-18:2 + 10*t*,15*c*/9*t*,15*c*-18:2 + 9*c*,13*c*-18:2; 17, 18:2*n*-6 (linoleate); 18, 9*c*,14*c*-18:2; 19, 9*c*,15*c*-18:2; 20, unknown; 21, 12*c*,15*c*-18:2; 22, 20:0; 23, 11*c*-20:1; 24, 12*c*-20:1; 25, 18:2 conjugates; 26, 13*c*-20:1; 27, 18:2 conjugates; 28, 18:2 conjugates; 29, 11*c*,14*t*-20:2; 30, 8*c*,14*c*-20:2; 31, 11*t*,14*c*-20:2; 32, 20:2*n*-6; 33, 20:3*n*-9; 34, 5*c*,8*c*,14*c*-20:3; 35, 5*c*,11*c*,14*c*-20:3; 36, 20:3*n*-6; 37, 5*c*,8*c*,11*c*,15*t*-20:4; 38, unknown; 39, 5*c*,8*c*,11*c*,14*t*-20:4; 40, 20:4*n*-6 (arachidonate); 41, 22:1; 42, 22:1; 43, 22:2; 44, 22:2; 45, 20:5*n*-3; 46, unknown; 47, unknown; 48, 24:0; 49, 24:1; 50, 22:5*n*-6; 51, 22:5*n*-3; 52, *t*-22:6*n*-3; 53, 22:6*n*-3. Reproduced by courtesy of the American Oil Chemists' Society and W.M.N. Ratnayake (67).

These misconceptions are as follows:

Stearic acid is hypercholesterolemic because it is a saturated fat.

Stearic acid (18:0) is not cholesterolemic because it is poorly absorbed and rapidly converted to oleic acid.

trans Fatty acids and 18:0 have the same structural conformation, and therefore 9*t*-18:1 is metabolically equivalent to 18:0 rather than to oleic acid (9*c*-18:1).

People are not able to metabolize the trans fatty acids formed during partial hydrogenation of vegetable oils because trans fatty acids are unnatural.

Margarine is as bad as butter with regard to increasing serum cholesterol levels.

The composition of the fat in a nursing mother's diet is not very important because the mammary gland controls the fatty acid composition of human milk fat.

Metabolic conversion of linolenic acid (18:3*n*-3) to fish oil *n*-3 fatty acids (20:5*n*-3 and 22:6*n*-3) has little nutritional importance.

The cholesterol content of animal fats is responsible for the hypercholesterolemic effect of animal fats.

However, if such specific adverse biochemical effects of fatty acids can be defined, the plant breeders will probably be able to contribute to solutions.

REFERENCES

1. *Official Methods and Recommended Practices of the American Oil Chemists' Society*, 4th ed., including *Additions and Revisions 1992*, American Oil Chemists' Society, Champaign, Ill.
2. T.P. Hilditch and P.N. Williams, *The Chemical Constitution of Natural Fats,* 4th ed., Chapman and Hall, London, 1974.
3. R.C. Weast, ed., *Handbook of Chemistry and Physics,* 54th ed., The Chemical Rubber Co., 1973, pp. D-189, D-190.
4. P. Gillatt, *INFORM* **5,** 981 (1994).
5. E.N. Frankel and S-W. Huang, *J. Am. Oil. Chem. Soc.* **71,** 255 (1994).
6. B.F. Haumann, *INFORM* **5,** 668 (1994).
7. T.H. Applewhite, *INFORM* **5,** 914 (1994).
8. F.D. Gunstone, J.L. Harwood and F.B. Padley, eds., *The Lipid Handbook,* Chapman and Hall, London, 1986.
9. F.B. Padley, F.D. Gunstone, and J.L. Harwood in Ref. 8, p. 49.
10. R.C. Badami and K.B. Patil, *Prog. Lipid Res.* **19,** 119–153 (1981).
11. R.G. Ackman, *Lipids* **2,** 502 (1967).
12. R.G. Ackman in H.K. Mangold, ed., *CRC Handbook of Chromatography, Lipids,* Vol. 1, CRC Press, Boca Raton, Fla., 1984, p. 95.
13. R.G. Ackman, *Food Res. Internat.* **25,** 453 (1992).
14. R.G. Ackman, *Lipids* **29,** 445 (1994).
15. W.W. Christie, *J. Chromatogr.* **447,** 305 (1988).
16. The British Nutrition Foundation, *Unsaturated Fatty Acids: Nutritional and Physiological Significance: The Report of the British Nutrition Foundation's Task Force,* Chapman and Hall, London, 1992.

17. Committee on Medical Aspects of Food Policy, Department of Health, *Nutritional Aspects of Cardiovascular Disease: Report of the Cardiovascular Review Group,* No. 46, HMSO, London, 1994.
18. J. Flanzy, M. Boudon, D. Leger and J. Pihet, *J. Chromatogr. Sci.* **14,** 17 (1976).
19. R. G. Ackman, S. N. Hooper, and D. L. Hooper, *J. Am. Oil Chem. Soc.* **51,** 42 (1974).
20. R.G. Ackman in *Progress in the Chemistry of Fats and Other Lipids*, Vol. **12,** 1972, p. 65.
21. R.G. Ackman, *Chemistry and Industry,* 715 (Oct. 17, 1981).
22. *Official Methods of Analysis,* 15th ed., Association of Official Analytical Chemists, Inc., Arlington, VA, 1990.
23. J.D. Joseph and R.G. Ackman, *J. AOAC Internat.* **75,** 488 (1992).
24. C.D. Bannon, J.D. Craske, N.T. Hai, N.L. Harper, and K.L. O'Rourke, *J. Chromatogr.* **247,** 63 (1982).
25. C.D. Bannon and co-workers, *J. Chromatogr.* **247,** 71 (1982).
26. D.E. Albertyn and co-workers, *J. Chromatogr.* **247,** 47 (1982).
27. C.D. Bannon, J.D. Craske, and A.E. Hilliker, *J. Am. Oil Chem. Soc.* **62,** 1501 (1985).
28. C.D. Bannon, J.D. Craske, and A.E. Hilliker, *J. Am. Oil Chem. Soc.* **63,** 105 (1986).
29. C.D. Bannon and co-workers, *J. Chromatogr.* **404,** 340 (1987).
30. J.D. Craske and C.D. Bannon, *J. Am. Oil Chem. Soc.* **64,** 1413 (1987).
31. J.D. Craske, C.D. Bannon, and L.M. Norman, *J. Am. Oil Chem. Soc.* **65,** 262 (1988).
32. J.D. Craske and C.D. Bannon, *J. Am. Oil Chem. Soc.* **65,** 1190 (1988).
33. J.D. Craske, *J. Am. Oil Chem. Soc.* **70,** 325 (1993).
34. K.S. Rhee, *J. Food Lipids* **1,** 285 (1994).
35. W.M.N. Ratnayake, R.G. Ackman, and H.W. Hulan, *J. Sci. Food Agr.* **49,** 59 (1989).
36. C.R. Somerville, *Am. J. Clin. Nutr.* **58(suppl),** 270S (1993).
37. M.D. Erickson and N. Frey, *Food Technol.* **48** (11), 63 (1994).
38. B.F. Haumann, *INFORM* **5,** 1198 (1994).
39. Federal Register, U.S. Government Printing Office 58, Jan. 6, 1993, FR 2066-2964.
40. R.G. Ackman in C.A. Drevon, I. Baksaas, and H.E. Krokan, *Omega-3 Fatty Acids: Metabolism and Biological Effects,* Birkhäuser Verlag, Basel, Switzerland, 1993.
41. J.W. DeVries and A.L. Nelson, *Food Technol.* **48,** 73 (July 1994).
42. N.H. Mermelstein, *Food Technol.* **48,** 62 (July 1994).
43. W. Ellefson in D.M. Sullivan and D.E. Carpenter, *Methods of Analysis for Nutrition Labeling,* AOAC International, Arlington, VA, 1993, p. 3.
44. D.E. Carpenter, J. Ngeh-Ngwainbi, and S. Lee in D.M. Sullivan and D.E. Carpenter, *Methods of Analysis for Nutrition Labeling,* AOAC International, Arlington, VA, 1993, p. 85.
45. S.D. House, P.A. Larson, R.R. Johnson, J.W. DeVries, and D.L. Martin, *J. AOAC Internat.* **77,** 960 (1994).
46. C.E. Elson, *Critical Rev. Food Sci. Nutr.* **31,** 79 (1992).
47. B. Vessby, *INFORM* **5,** 182 (1994).
48. J.B. Allred, *J. Nutr.* **123,** 1453 (1993).
49. C.V. Felton, D. Crook, M.J. Davies, and M.F. Oliver, *Lancet* **344,** 1195 (1994).
50. W.E.M. Lands, *Fish and Human Health,* Academic Press, Inc., Orlando, Fla., 1986.
51. E.B. Schmidt and J. Dyerberg, *Drugs* **47,** 405 (1994).
52. R. Freese, M. Mutanen, L.M. Valsta and I. Salminen, *Thrombosis and Haemostasis* **71,** 73 (1994).
53. M. de Lorgeril and co-workers, *Lancet* **343,** 1454 (1994).

54. J.M. Banks and co-workers, *Food Res. Internat.* **25,** 365 (1992).
55. J. Wagner, *Food Business* **7,** 8 (1994).
56. R.P. Mensink and M.B. Katan, *Prog. Lipid Res.* **32,** 111 (1993).
57. R. Troisi, W.C. Willet, and S.T. Weiss, *Am. J. Clin. Nutr.* **56,** 1019 (1992).
58. J.E. Hunter and T.H. Applewhite, *Am. J. Clin. Nutr.* **54,** 363 (1991).
59. W.M.N. Ratnayake, *J. Am. Oil Chem. Soc.* **69,** 192 (1992).
60. W.M.N. Ratnayake, R. Hollywood, E. O'Grady and J.L. Beare-Rogers, *J. Am. Oil Chem. Soc.* **67,** 804 (1990).
61. P. Juanéda, J-L. Sebedio, and W.W. Christie, *J. High Resolution Chromatogr.* **17,** 321 (1994).
62. S.F. O'Keefe and co-workers, *J. Lipid Res.* **31,** 1241 (1990).
63. W.M.N. Ratnayake and G. Pelletier, *J. Am. Oil Chem. Soc.* **69,** 95 (1992).
64. R.L. Wolff, *J. Chromatogr. Sci.* **30,** 17 (1991).
65. R.L. Wolff, *J. Am. Oil Chem. Soc.* **70,** 425 (1993).
66. R.L. Wolff, *J. Am. Oil Chem. Soc.* **71,** 1129 (1994).
67. W.M.N. Ratnayake, Z.Y. Chen, G. Pelletier, and D. Weber, *Lipids* **29,** 707 (1994).
68. E.A. Emken, *INFORM* **5,** 906 (1994).

13

Lipid Composition of Selected Foods

Many health-risk issues are now being related to fats in the diet; and as studies emerge they continue to show that not all fats carry the same health risk. The majority of these studies are based on the nutrient detail found in nutrient databases, so it is important that the information be regularly scrutinized and updated for accurate representation. The results of the studies will affect the type of ingredients used in food products; sources of food supplies; new product formulation; business plans and marketing strategies.

BACKGROUND OF USDA INFORMATION SOURCES

The interest in information on fats has shifted dramatically over the years. In 1963, the U.S. Department of Agriculture (USDA) published a Composition of Foods Table, the original *Handbook 8*. Total fat, total saturated fat, oleic, and linoleic unsaturated fats were the measures for fats. In 1975, the same four fat categories were published in USDA Handbook 456 (*Nutritive Value of American Foods in Common Units*).

Now the USDA prints technical volumes of foods: *Handbook 8 Series—The Composition of Foods*. Each volume is dedicated to a major food group. The major fatty acid (FA) categories are listed (8 saturated fats, 4 monounsaturated fats, 7 polyunsaturated fats). Cholesterol and phytosterol are listed for each food, when available. With the advent of computers, all of the *Handbook 8* data is combined and released each year as USDA reference data. This information can be purchased and downloaded to a computer.

THE IMPORTANCE OF EXPANDED NUTRIENT DATABASES

Most nutrition software programs use the USDA nutrient databases directly. A few organizations, however, research additional, more current nutrient information from manufacturers, scientific journal articles, and other sources. By combining these additional sources with USDA data, a nutrient database can be expanded to include additional food items for a more representative, more complete list of ingredients and fewer missing values for individual nutrients.

There can be a wide variation in nutrient values, and this uncertainty is important to keep in mind. Variations result from the type of growing season, the soil, plant genetics, the laboratory technique used, sample preparation and the storage and processing methods. Therefore, the compilation of many analyses in a database can more accurately represent the overall expectations of nutrient content than a single chemical analysis.

FUTURE CONSIDERATIONS FOR NUTRIENT INFORMATION AND FOOD COMPOSITION TABLES

Here are a few issues for consideration:

3.1 Arithmetic Anomaly

Of particular visibility is an arithmetic anomaly—the values for saturated, monounsaturated, and polyunsaturated almost never add to the total fat reported. There are a variety of reasons for the difference; however, the most prevalent reason for these other fats is the presence of Trans Fatty Acids (TFAs).

Trans fatty acid data are not included here because the data are emerging and will continue to change now that TFAs are recognized as potential contributors to health risks. This recognition may bring changes in the hydrogenation process as food manufacturers try to reduce the amount of TFAs in their products. In addition to the presence of TFAs from the hydrogenation process, as hydrogenated oils get into the food supply, TFAs will begin to show up in foods that do not naturally contain trans fatty acids.

Reporting TFA data may also be complicated because there are many different forms of trans fatty acids. Yet, these data will become more and more important to researchers as the possible connections to health risks continue to emerge.

3.2 Health Risk Factors

The Health Risk factors related to fats are not entirely understood, and future research will be needed to clarify some of these issues.

Here are some examples:

1. Not all fats are to be avoided. There are some fatty acids essential to good health, and some which seem to be protective against certain health problems.
2. There are certain fats that appear to be protective against one type of health problem, yet contribute to a different health problem.
3. The antioxidant vitamins A and E cannot be transported into the body without fat in the diet, which poses the question: Will a low fat diet reduce the body's absorption of these important nutrients?

Hopefully, food nutrient databases will help the research results of tomorrow. A listing of the current available data for fats and oils is presented here. This information, from the ESHA Research data bank, is based on USDA sources, additional values from extensive research references, and food composition tables of Canada, England, and Germany.

DATA PRESENTED

In this chapter, the following data are presented:

1. Table 1: Lipid composition of vegetable oils and related products.
2. Table 2: Lipid composition of nuts, seeds, and related products.
3. Table 3: Lipid composition of animal fats and related products.
4. Table 4: Lipid composition of dairy food products.
5. Table 5: Lipid composition of fish and related products.
6. Table 6: Lipid composition of shellfish and related products.
7. Table 7: Lipid composition of miscellaneous food products with a high fat content.

In each table, the contents per 100 grams of product of the following are included:

1. Water.
2. Ash.
3. Calories (Cals).
4. Protein (Prot).
5. Carbohydrate (Carb).
6. Total fat (Fat-T).
7. Saturated fat (Fat-S).
8. Each of the saturated fatty acids: 4:0, 6:0, 8:0, 10:0, 12:0, 14:0, 15:0, 16:0, 17:0, 18:0, 20:0, 22:0, 24:0
9. Monounsaturated fat (Fat-M).
10. Each of the monounsaturated fatty acids: 14:1, 16:1, 18:1, 20:1, 22:1.

(*Text continued on page 454.*)

Table 13.1 Lipid composition of vegetable oils and related products, contents/100 g

Item	H_2O	Ash	Cals.	Prot.	Carb.	Fat-T	Fat-S	4:0	6:0	8:0	10:0	12:0	14:0	15:0	16:0	17:0	18:0
Almond oil	–	–	884	–	–	100	8.2	–	–	–	–	–	–	–	6.5	–	1.7
Apricot kernel oil	–	–	884	–	–	100	6.3	–	–	–	–	–	–	–	5.8	–	0.5
Avocado oil	–	–	884	–	–	100	11.6	–	–	–	–	–	–	–	10.9	–	0.66
Babassu oil	–	–	884	–	–	100	81.2	–	0.2	6	5.5	43.5	15	–	8.2	–	2.8
Balsam pear seed oil	–	–	884	–	–	100	–	–	–	–	–	–	–	–	2.68	–	20.7
Black henbane seed oil	–	–	884	–	–	100	11.1	–	–	–	–	–	–	–	8.71	–	2.39
Canola oil	–	–	884	–	–	100	7.1	–	–	–	–	–	–	–	4	–	1.8
Carob seed oil	–	–	884	–	–	100	22.2	–	–	–	–	–	–	–	17.7	–	4.45
Cocoa butter oil	–	–	884	–	–	100	59.7	–	–	–	–	–	0.1	–	25.4	–	33.2
Coconut oil	–	–	862	–	–	100	86.5	–	0.6	7.5	6	44.6	16.8	–	8.2	–	2.8
Coconut oil refined	0.09	0.1	878	0.8	0.01	99	86.5	–	–	7.6	5.7	45	17.2	–	8.6	–	2.4
Corn oil	–	–	884	–	–	100	13.5	–	–	–	–	–	0.3	–	10.9	–	1.8
Cottonseed oil	–	–	884	–	–	100	26.4	–	–	–	–	3.8	0.8	–	22.7	–	2.3
Cupu assu oil	–	–	884	–	–	100	53.2	–	–	–	–	–	0.9	–	9.6	–	32.7
Date pit oil	–	–	884	–	–	100	45.9	–	–	–	–	20.4	12.6	–	10	–	1.08
Eggplant seed oil	–	–	884	–	–	100	25.5	–	–	–	–	–	–	–	18.8	–	6.69
Fenugreek seed oil	–	–	884	–	–	100	40.1	–	–	–	–	–	–	–	25.8	–	10
Grapeseed oil	–	–	884	–	–	100	9.6	–	–	–	–	–	0.1	–	6.7	–	2.7
Groundnut seed oil	–	–	884	–	–	100	15.4	–	–	–	–	–	0.46	–	9.32	–	2.38
Hazelnut oil	–	–	884	–	–	100	7.4	–	–	–	–	–	0.1	–	5.2	–	2
Illipe fat	0.5	–	879	–	–	99.5	61.9	–	–	0.2	0.2	0.3	–	–	16.3	–	43.3
Linseed oil	0.5	–	879	–	–	99.5	9.6	–	–	–	–	–	–	–	6.2	–	3.4
Mono + diglycerides, coconut oil	–	–	862	–	–	100	86.5	–	0.6	7.5	6	44.6	16.8	–	8.2	–	2.8
Mono + diglycerides, corn oil	–	–	850	–	–	100	12.7	–	–	–	–	–	–	–	10.9	–	1.8
Mono + diglycerides, cottonseed oil	–	–	850	–	–	100	25.9	–	–	–	–	–	0.8	–	22.7	–	2.3
Mono + diglycerides, palm kernel oil	–	–	850	–	–	100	81.5	–	0.2	3.3	3.7	47	16.4	–	8.1	–	2.8
Mono + diglycerides, palm oil	–	–	850	–	–	100	49.3	–	–	–	–	0.1	1	–	43.5	–	4.3
Mono + diglycerides, peanut oil	–	–	850	–	–	100	16.9	–	–	–	–	–	0.1	–	9.5	–	2.2
Mono + diglycerides, soybean oil	–	–	850	–	–	100	14.4	–	–	–	–	–	0.1	–	10.3	–	3.8
Mono + diglycerides, high linole safflower oil	–	–	850	–	–	100	9.1	–	–	–	–	–	0.1	–	6.2	–	2.2
Mono + diglycerides, high oleic safflower oil	–	–	850	–	–	100	6.1	–	–	–	–	–	–	–	4.8	–	1.3
Mono + diglycerides, hydrogenate soybean oil	–	–	850	–	–	100	14.9	–	–	–	–	–	0.1	–	9.8	–	5
Mustard oil	–	–	884	–	–	100	11.6	–	–	–	–	–	1.39	–	3.75	–	1.12
Nigerseed oil	–	–	884	–	–	100	17.6	–	–	–	–	–	–	–	9.08	–	8.51
Nutmeg butter oil	–	–	884	–	–	100	90	–	–	–	–	3.1	82.6	–	4.3	–	–
Oat oil	–	–	884	–	–	100	19.6	–	–	–	–	0.387	0.243	–	16.7	–	1.06
Olive oil	–	–	884	–	–	100	13.5	–	–	–	–	–	–	–	11	–	2.2
Palm kernel oil	–	–	862	–	–	100	81.5	–	0.2	5.8	3.88	46	15.6	–	7.8	–	2.45
Palm oil	–	–	884	–	–	100	49.3	–	–	–	–	0.1	1	–	43.5	–	4.3
Palm oil refined	0.3	1	891	–	–	98.7	46.6	–	–	–	–	0.095	0.945	–	41.1	–	4.06
Peach kernel oil	–	–	884	–	–	100	36.1	–	–	–	–	–	–	–	33.2	–	2.96
Peanut oil	–	–	884	–	–	100	16.5	–	–	–	–	–	0.1	–	8.9	–	3
Pimiento seed oil	–	–	884	–	–	100	20.6	–	–	–	–	–	–	–	15.1	–	5.45
Pistachio nut kernel oil	–	–	884	–	–	100	13.8	–	–	–	–	–	–	–	12.8	–	0.96
Poppyseed oil	–	–	884	–	–	100	13.5	–	–	–	–	–	–	–	10.6	–	2.9
Pumpkin seed oil	0.5	–	879	–	–	99.5	19.8	–	–	–	–	–	–	–	15	–	4.8
Red sorrel seed oil	–	–	884	–	–	100	18.3	–	–	–	–	–	–	–	15.6	–	2.68
Rice bran oil	–	–	884	–	–	100	19.7	–	–	–	–	–	0.7	–	16.9	–	1.6
Safflower oil, linoleic >70%	–	–	884	–	–	100	9.1	–	–	–	–	–	0.1	–	6.6	–	2.2
Sesame oil	–	–	884	–	–	100	14.2	–	–	–	–	–	–	–	8.9	–	4.8
Sheanut oil	–	–	884	–	–	100	46.6	–	–	0.2	0.2	1.4	0.3	–	4.4	–	38.9

20:0	22:0	24:0	Fat-M	14:1	16:1	18:1	20:1	22:1	Fat-P	18:2	18:3	18:4	20:4	20:5	22:5	22:6	Chol., mg	Omeg3	Omeg6
–	–	–	69.9	–	0.6	69.4	–	–	17.4	17.4	–	–	–	–	–	–	–	–	17.4
–	–	–	60	–	1.5	58.5	–	–	29.3	29.3	–	–	–	–	–	–	–	–	29.3
–	–	–	70.6	–	2.67	67.9	–	–	13.5	12.5	0.957	–	–	–	–	–	–	0.957	12.5
–	–	–	11.4	–	–	11.4	–	–	1.6	1.6	–	–	–	–	–	–	–	–	1.6
–	–	–	–	–	–	28.7	–	–	–	–	–	–	–	–	–	–	–	–	–
–	–	–	23.2	–	–	23.2	–	–	61.3	61.3	–	–	–	–	–	–	–	–	61.3
0.7	0.4	0.2	58.9	–	0.2	56.1	1.7	0.6	29.6	20.3	9.3	–	–	–	–	–	–	9.3	20.3
–	–	–	32.1	–	–	32.1	–	–	41.3	39.7	1.61	–	–	–	–	–	–	1.61	39.7
1	–	–	32.9	–	0.3	32.6	–	–	3	2.8	0.1	–	–	–	–	–	–	0.1	2.8
–	–	–	5.81	–	–	5.81	–	–	1.8	1.8	–	–	–	–	–	–	–	–	1.8
–	–	–	6.7	–	–	6.7	–	–	1.4	1.4	–	–	–	–	–	–	–	–	1.4
0.5	–	–	24.2	–	–	24.2	–	–	58.7	58	0.7	–	–	–	–	–	–	–	58
0.19	–	–	17.8	–	0.8	17	–	–	51.9	51.5	0.3	–	0.1	–	–	–	–	0.2	51.6
10	–	–	38.7	–	–	38.7	–	–	3.8	3.8	–	–	–	–	–	–	–	–	3.8
–	–	–	42.5	–	–	42.5	–	–	7.23	6.56	0.67	–	–	–	–	–	–	0.67	6.56
–	–	–	13.1	–	–	13.1	–	–	57	54.5	2.44	–	–	–	–	–	–	2.44	54.5
–	–	–	31.5	–	0.2	29.9	–	–	24.1	6.35	16.7	–	–	–	–	–	–	16.7	6.35
–	–	–	16.1	–	0.3	15.8	–	–	69.9	69.6	0.1	–	–	–	–	–	–	0.1	69.6
–	–	–	49.7	–	–	48.5	1.23	–	30.6	28.9	–	–	1.73	–	–	–	–	–	30.6
–	–	–	78	–	0.2	77.8	–	–	10.2	10.1	–	–	–	–	–	–	–	–	10.1
1.6	–	–	31.1	–	0.2	30.9	–	–	1.6	1.5	0.1	–	–	–	–	–	–	0.1	1.5
–	–	–	17.2	–	–	17.2	–	–	68.7	13.4	55.3	–	–	–	–	–	–	55.3	13.4
–	–	–	5.81	–	–	5.81	–	–	1.8	1.8	–	–	–	–	–	–	–	–	1.8
–	–	–	24.2	–	–	24.2	–	–	58.7	58	–	–	–	–	–	–	–	–	58
–	–	–	17.8	–	0.8	17	–	–	51.9	51.5	0.2	–	0.1	–	–	–	–	0.2	51.6
–	–	–	11.4	–	–	11.4	–	–	1.6	1.6	–	–	–	–	–	–	–	–	1.6
–	–	–	37	–	0.3	36.6	0.1	–	9.3	9.1	0.2	–	–	–	–	–	–	0.2	9.1
1.4	2.8	–	46.2	–	0.1	44.8	1.3	–	32	32	–	–	–	–	–	–	–	–	32
–	–	–	23.3	–	0.2	22.8	0.2	–	57.9	51	6.8	–	–	–	–	–	–	6.8	51
–	–	–	12.1	–	0.4	11.7	–	–	74.5	74.1	0.4	–	–	–	–	–	–	0.4	74.1
–	–	–	75.3	–	–	75.3	–	–	14.2	14.2	–	–	–	–	–	–	–	–	14.2
–	–	–	43	–	0.4	42.5	–	–	37.6	34.9	2.6	–	–	–	–	–	–	2.6	34.9
4.6	0.4	–	59.2	–	0.216	11.6	6.19	41.2	21.2	15.3	5.9	–	–	–	–	–	–	5.9	15.3
–	–	–	49.7	–	–	48.5	1.23	–	30.6	28.9	–	–	1.73	–	–	–	–	–	30.6
–	–	–	4.81	–	–	4.81	–	–	–	–	–	–	–	–	–	–	–	–	–
–	–	–	35.1	–	0.202	34.9	–	–	40.9	39.1	1.79	–	–	–	–	–	–	1.79	39.1
0.3	–	–	73.7	–	0.8	72.5	0.3	–	8.4	7.9	0.6	–	–	–	–	–	–	0.6	7.9
–	–	–	13.4	–	–	13.4	–	–	2.4	1.6	–	–	–	–	–	–	–	–	1.6
0.4	–	–	37	–	0.3	36.6	0.1	–	9.3	9.1	0.2	–	–	–	–	–	–	0.2	9.1
0.378	–	–	37.9	–	0.3	37.5	0.1	–	11	10.8	0.2	–	–	–	–	–	–	0.2	10.8
–	–	–	30.2	–	–	30.2	–	–	29.3	29.3	–	–	–	–	–	–	–	–	29.3
1.4	2.7	0.998	52.3	–	0.1	49.4	1.1	–	26.3	26.1	–	–	–	–	–	–	–	–	32
–	–	–	19.5	–	–	19.5	–	–	55.5	55.5	–	–	–	–	–	–	–	–	55.5
–	–	–	49.3	–	1.91	47.4	–	–	32.5	30.4	–	–	2.1	–	–	–	–	–	32.5
–	–	–	19.7	–	–	19.7	–	–	66.4	66.4	–	–	–	–	–	–	–	–	62.4
–	–	–	23.5	–	0.48	23	–	–	51.5	51	0.48	–	–	–	–	–	–	0.48	51
–	–	–	26.7	–	–	26.7	–	–	50.7	50.7	–	–	–	–	–	–	–	–	50.7
–	–	–	39.3	–	0.2	39.1	–	–	35	33.4	1.6	–	–	–	–	–	–	1.6	33.4
0.2	–	–	12.1	–	0.4	11.6	0.1	–	74.5	74.1	0.4	–	–	–	–	–	–	0.4	74.1
0.5	–	–	39.7	–	0.2	39.3	0.2	–	41.7	41.3	0.3	–	–	–	–	–	–	0.3	41.3
1.08	–	–	44.5	–	0.1	44.4	–	–	6.1	5.8	0.3	–	–	–	–	–	–	0.3	4.9

Table 13.1 (*Continued*) (hy = hydrogenated)

Item	H_2O	Ash	Cals.	Prot.	Carb.	Fat-T	Fat-S	4:0	6:0	8:0	10:0	12:0	14:0	15:0	16:0	17:0	18:0
Soybean and cottonseed oil	–	–	884	–	–	100	18.1	–	–	–	–	–	0.3	–	14	–	3.7
Soybean (hydrogenated) and cottonseed oil	–	–	884	–	–	100	18.1	–	–	–	–	–	0.3	–	14	–	3.7
Soybean lecithin oil	–	–	763	–	–	100	15.6	–	–	–	–	–	0.101	–	12	–	2.92
Soybean oil (Crisco/ Wesson)	–	–	884	–	–	100	14.4	–	–	–	–	–	0.1	–	10.2	–	3.8
Soybean oil, hydrogenated	–	–	884	–	–	100	14.9	–	–	–	–	–	0.1	–	9.8	–	5
Sugar apple seed oil	–	–	884	–	–	100	21.3	–	–	–	–	–	–	–	10.7	–	10.6
Sunflower oil (Wesson, Sunlite) lin >60%	–	–	884	–	–	100	11.7	–	–	–	–	–	0.08	–	5.9	–	4.5
Sunflower oil, linoleic, hydrogenated	–	–	884	–	–	100	13	–	–	–	–	–	–	–	7.1	–	5.5
Sunflower oil, low linoleic <60%	–	–	884	–	–	100	10.1	–	–	–	–	–	–	–	5.4	–	3.5
Sunflower oil, over 70% oleic	–	–	886	–	–	100	9.79	–	–	–	–	–	–	–	3.68	–	4.32
Teaseed oil	–	–	884	–	–	100	21.1	–	–	–	–	0.1	0.1	–	17.5	–	3.1
Tomatoseed oil	–	–	884	–	–	100	19.7	–	–	–	–	–	0.2	–	15	–	4.4
Ucuhuba butter oil	–	–	884	–	–	100	85.2	–	–	–	–	12.2	63.4	–	8.6	–	1
Walnut oil	–	–	884	–	–	100	9.1	–	–	–	–	–	–	–	7	–	2
Watermelon seed oil	–	–	884	–	–	100	24.6	–	–	–	–	–	–	–	12.8	–	11.7
Wheat germ oil	–	–	884	–	–	100	18.9	–	–	–	–	–	0.1	–	17.1	–	0.918
Baking shortening, hy soy + palm + cotton	–	–	884	–	–	100	28.8	–	–	–	–	–	0.45	–	23.4	–	5
Bread shortening, hy soy + cottonseed	–	–	884	–	–	100	22	–	–	–	–	–	–	–	10.4	–	11.6
Cake mix shortening, hy soy + hy cottonseed	–	–	884	–	–	100	27.2	–	–	–	–	0.1	0.4	–	18	–	8.8
Confectionary shortening, cottonseed/palm kernel	–	–	884	–	–	100	91.3	–	–	4.3	4.3	35.8	14.4	–	10.7	–	21.9
Confectionary shortening, fractioned palm	–	–	884	–	–	100	65.5	–	–	–	–	1.1	0.9	–	36	–	26.2
Frying shortening, hy palm	–	–	884	–	–	100	47.5	–	–	–	–	–	–	–	40.6	–	6.9
Frying shortening, hy soy, linoleic <1%	–	–	884	–	–	100	21.1	–	–	–	–	–	–	–	10.9	–	10.2
Frying shortening, linoleic 30%	–	–	884	–	–	100	18.4	–	–	–	–	–	–	–	10.2	–	8.3
Household shortening, hy soybean + palm	–	–	884	–	–	100	30.6	–	–	0.1	0.1	0.4	0.4	–	19.3	–	9.9
Industrial shortening, hy soy + cottonseed	–	–	884	–	–	100	25.6	–	–	–	–	–	0.5	–	13	–	12.1
Multipurpose shortening, hy soy + hy palm	–	–	884	–	–	100	30.4	–	–	0.101	0.101	0.403	0.403	–	19.4	–	9.97
Regular frying shortening, hy soy + hy cottonseed	–	–	884	–	–	100	15.4	–	–	–	–	–	–	–	9.6	–	5.8
Vegetable shortening (Crisco/Fluffo)	–	–	884	–	–	100	25	–	–	–	–	–	0.4	–	14.1	–	10.6
Butter, vegetable oil blend (Blue Bonnet spread)	15.8	2	718	0.88	0.65	80.7	28.4	1.07	0.631	0.367	0.825	0.927	3.38	–	14.3	–	6.94
Cocoa butter	0.5	–	879	–	–	99.5	59.3	–	–	–	–	–	–	–	24.8	–	33.5
Touch of Butter spread, stick	27.5	1.4	625	0.697	–	70.4	16	–	–	–	–	–	–	–	–	–	–
Touch of Butter spread, tub	56	1.28	375	0.423	–	42.3	7.08	–	–	–	–	–	–	–	4.68	–	2.39
Blue Bonnet Margarine spread, tub	46.8	2.76	439	0.634	0.507	49.3	8.26	–	–	–	–	–	–	–	5.46	–	2.8
Blue Bonnet Margarine, stick	15.7	2	634	0.9	0.9	80.5	16.7	–	–	–	–	–	–	–	9.6	–	6.9
Diet Fleischmann's Margarine, tub	58.1	2.2	345	0.5	0.4	38.8	6.4	–	–	–	–	0.1	0.1	–	4.3	–	1.9

20:0	22:0	24:0	Fat-M	14:1	16:1	18:1	20:1	22:1	Fat-P	18:2	18:3	18:4	20:4	20:5	22:5	22:6	Chol., mg	Omeg3	Omeg6
0.14	–	–	29.5	–	0.2	29.3	–	–	48.1	45.3	2.8	–	–	–	–	–	–	2.8	45.3
0.14	–	–	29.5	–	0.2	29.3	–	–	48.1	45.3	2.8	–	–	–	–	–	–	2.8	45.3
0.58	–	–	11	–	0.403	10.6	–	–	45.3	40.2	5.14	–	–	–	–	–	–	5.14	40.2
0.28	–	–	23.3	–	0.2	22.8	0.2	0.2	57.9	51	6.8	–	–	–	–	–	–	6.8	51
–	–	–	43	–	0.4	42.5	–	–	37.6	34.9	2.6	–	–	–	–	–	–	2.6	34.9
–	–	–	51	–	–	51	–	–	23.2	23.2	–	–	–	–	–	–	–	–	23.2
0.44	0.59	0.168	20.7	–	0.44	19.5	0.8	–	66.1	65.7	0.384	–	–	–	–	–	–	–	65.7
0.567	–	–	46.2	–	0.2	46	–	–	36.4	35.3	0.9	–	–	–	–	–	–	0.9	35.3
–	–	–	45.4	–	0.2	45.3	–	–	40.1	39.8	0.2	–	–	–	–	–	–	0.2	39.8
–	1	–	83.6	–	–	82.9	0.964	–	3.8	3.61	0.193	–	–	–	–	–	–	0.193	3.61
–	–	–	51.5	–	0.5	49.9	1	–	23	22.2	0.7	–	–	–	–	–	–	0.7	22.2
–	–	–	22.8	–	0.5	21.9	–	–	53.1	50.8	2.3	–	–	–	–	–	–	2.3	50.8
–	–	–	6.71	–	–	6.71	–	–	2.9	2.9	–	–	–	–	–	–	–	–	2.9
–	–	–	22.8	–	0.1	22.2	0.4	–	63.6	52.9	10.4	–	–	–	–	–	–	10.4	52.9
–	–	–	11.4	–	–	11.4	–	–	59.6	59.6	–	–	–	–	–	–	–	–	59.6
0.734	–	–	15.6	–	0.5	14.6	0.47	–	61.7	54.8	6.9	–	–	–	–	–	–	6.9	54.8
–	–	–	29.6	–	0.23	29.4	–	–	37.2	34	3.19	–	–	–	–	–	–	3.19	34
–	–	–	33	–	–	33	–	–	40.6	36.7	4	–	–	–	–	–	–	4	36.7
–	–	–	54.2	–	–	54.2	–	–	14.1	13.1	1.1	–	–	–	–	–	–	1.1	13.1
–	–	–	2.21	–	–	2.21	–	–	1	1	–	–	–	–	–	–	–	–	1
–	–	–	29.6	–	–	29.3	–	–	0.5	0.5	–	–	–	–	–	–	–	–	0.5
–	–	–	40.6	–	–	40.6	–	–	7.5	7.5	–	–	–	–	–	–	–	–	7.5
–	–	–	73.7	–	–	73.7	–	–	0.4	0.3	0.1	–	–	–	–	–	–	0.1	0.3
–	–		43.7	–	–	43.7	–	–	33.5	31.1	2.4	–	–	–	–	–	–	2.4	31.1
–	–		50.8	–	–	50.6	–	–	14.2	13.5	0.6	–	–	–	–	–	–	0.6	13.5
–	–		58	–	–	58	–	–	12	11.4	0.6	–	–	–	–	–	–	0.6	11.4
–	–	–	51	–	–	51	–	–	14.2	13.6	0.604	–	–	–	–	–	–	0.604	13.6
–	–	–	58.2	–	–	58.2	–	–	22	21.2	0.8	–	–	–	–	–	–	0.8	21.2
–	–	–	44.5	–	–	44.5	–	–	26.1	24.5	1.6	–	–	–	–	–	–	1.6	24.5
–	–	–	32.8	–	0.743	32	–	–	15.9	15.2	0.723	–	–	–	–	–	88	0.723	15.2
1	–	–	33.1	–	0.5	32.6	–	–	1.7	1.3	0.4	–	–	–	–	–	–	0.4	1.3
–	–	–	35.8	–	–	–	–	–	15.5	–	–	–	–	–	–	–	7.54	–	–
–	–	–	18.2	–	–	18.2	–	–	15	14	0.986	–	–	–	–	–	4.52	0.986	14
–	–	–	21.2	–	–	21.2	–	–	17.5	16.4	1.14	–	–	–	–	–	–	1.14	16.4
–	–	–	39.3	–	0.2	39.1	–	–	20.9	19.4	1.5	–	–	–	–	–	–	1.3	17
–	–	–	14.5	–	–	14.5	–	–	16.3	16.1	0.2	–	–	–	–	–	–	0.2	16.1

Table 13.1 (*Continued*)

Item	H_2O	Ash	Cals.	Prot.	Carb.	Fat-T	Fat-S	4:0	6:0	8:0	10:0	12:0	14:0	15:0	16:0	17:0	18:0
Diet Imperial Margarine, tub	58.1	2.2	345	0.5	0.4	38.8	6.5	–	–	–	–	–	–	–	4.3	–	2.2
Diet margarine spread	58.1	2.2	345	0.5	0.4	38.8	6.4	–	–	–	–	0.1	0.1	–	4.3	–	1.9
Fleischmann's Corn Oil Margarine, stick	15.7	2	719	0.9	0.9	80.5	13.2	–	–	–	–	–	–	–	9	–	4.2
Fleischmann's Corn Oil Margarine, tub	16.2	2	716	0.8	0.5	80.4	14.1	–	–	–	–	–	0.1	–	9.3	–	4.7
Fleischmann's Light, Margarine spread, tub	43.8	3.71	456	0.662	0.528	51.3	8.46	–	–	–	–	0.134	0.134	–	5.68	–	2.51
Fleischmann's Light Margarine, stick	45.6	1.96	458	0.57	0.57	51.3	8.41	–	–	–	–	–	–	–	5.73	–	2.68
Hard margarine w/salt, unspecified oils	15.7	2	719	0.9	0.9	80.5	15.8	–	–	–	–	–	0.2	–	9.7	–	5.9
Hard margarine, corn and hy corn, stick	15.7	2	719	0.9	0.9	80.5	14	–	–	–	–	–	0.1	–	9.2	–	4.8
Hard margarine, corn + hy soy + hy cottonseed + salt	15.7	2	719	0.894	0.894	80.4	15	–	–	–	–	0.298	0.191	–	9.6	–	5
Hard margarine, hy and regular soy + hy cottonseed	15.7	2	719	0.9	0.9	80.5	15.6	–	–	–	–	0.3	0.2	–	9.1	–	6
Hard margarine, hy safflower and soy + hy cottonseed	15.7	2	719	0.9	0.9	80.5	13.3	–	–	–	–	–	0.2	–	7.6	–	5.5
Hard margarine, hy soy + corn + hy cottonseed	15.7	2	719	0.9	0.9	80.5	19.8	–	–	–	–	–	0.3	–	10.9	–	8.6
Hard margarine, hy soy + hy and regular palm	15.7	2	719	0.9	0.9	80.5	17.5	–	–	–	–	–	–	–	13.6	–	3.9
Hard margarine, hy soybean + cottonseed	15.7	2	719	0.9	0.9	80.5	16.3	–	–	–	–	–	0.1	–	10	–	6.2
Hard margarine, hy soybean + hy cottonseed	15.7	2	719	0.9	0.9	80.5	15.1	–	–	–	–	–	0.1	–	9.1	–	5.9
Hard margarine, hy sunflower and soy + hy cottonseed	15.7	2	719	0.9	0.9	80.5	11.9	–	–	–	–	–	0.1	–	7.2	–	4.6
Hard margarine, hy and regular coconut + hy safflower palm	15.7	2	719	0.9	0.9	80.5	56.9	–	0.4	4.7	3.2	28.3	8	–	7.4	–	3.9
Hard margarine, hy corn and soy + hy cottonseed, unsalt	18.5	0.2	714	0.5	0.5	80.3	15	–	–	–	–	0.3	0.2	–	9.5	–	5
Hard margarine, hydrogen corn oil, stick	15.7	2	719	0.9	0.9	80.5	13.2	–	–	–	–	–	–	–	9	–	4.2
Hard margarine, pat (unspecified)	15.7	2	719	0.9	0.9	80.5	15	–	–	–	–	0.3	0.2	–	9.6	–	5
Hard margarine, pat, soy and hy soy	15.7	2	719	0.9	0.9	80.5	13.1	–	–	–	–	–	0.1	–	8.1	–	4.9
Hard margarine, safflower + soy + hy soy + hy cottonseed	15.7	2	719	0.9	0.9	80.5	14.4	–	–	–	–	–	0.2	–	7.7	–	6.5
Hard margarine, safflower and hy soy, stick	15.7	2	719	0.9	0.9	80.5	13.8	–	–	–	–	0.4	0.2	–	7	–	6.1
Hard margarine, soybean + hy palm	15.7	2	719	0.9	0.9	80.5	15.1	–	–	–	–	–	0.2	–	10.2	–	4.6
Hard margarine, stick-soy and hy soy	15.7	2	719	0.9	0.9	80.5	13.1	–	–	–	–	–	0.1	–	8.1	–	4.9
Imitation margarine, hy soy + cottonseed 40% fat	58.1	2.2	345	0.5		38.8	8.3	–	–	–	–	–	0.1	–	6.1	–	2.2
Imitation margarine, hy soy + hy cottonseed 40% fat	58.1	2.2	345	0.5	0.4	38.8	7.2	–	–	–	–	–	–	–	4.4	–	2.8

20:0	22:0	24:0	Fat-M	14:1	16:1	18:1	20:1	22:1	Fat-P	18:2	18:3	18:4	20:4	20:5	22:5	22:6	Chol., mg	Omeg3	Omeg6
–	–	–	16.7	–	–	16.7	–	–	13.8	12.9	0.9	–	–	–	–	–	–	0.9	12.9
–	–	–	14.5	–	–	14.5	–	–	16.3	16.1	0.2	–	–	–	–	–	–	0.2	16.1
–	–	–	45.8	–	–	45.8	–	–	18	17.7	0.2	–	–	–	–	–	–	0.2	17.7
–	–	–	31.6	–	–	31.6	–	–	31.2	30.3	0.9	–	–	–	–	–	–	0.9	30.3
–	–	–	19.2	–	–	19.2	–	–	21.6	21.3	0.268	–	–	–	–	–	–	0.268	21.3
–	–	–	29.2	–	–	29.2	–	–	11.5	11.3	0.127	–	–	–	–	–	–	0.127	11.3
–	–	–	35.8	–	–	35.8	–	–	25.4	24.3	1.1	–	–	–	–	–	–	1.1	24.3
–	–	–	38.8	–	–	38.8	–	–	24.1	23.7	0.4	–	–	–	–	–	–	0.4	23.7
–	–	–	36.8	–	–	36.8	–	–	25.1	24.9	0.298	–	–	–	–	–	–	0.298	24.9
–	–	–	36.1	–	–	35.9	0.2	–	25.3	22.5	2.8	–	–	–	–	–	–	2.8	22.5
–	–	–	23	–	–	23	–	–	40.7	40.6	0.1	–	–	–	–	–	–	0.1	40.6
–	–	–	32	–	–	32	–	–	25.1	23.6	1.5	–	–	–	–	–	–	1.5	23.6
–	–	–	31.2	–	–	31.2	–	–	28.2	25.9	2.3	–	–	–	–	–	–	2.3	25.9
–	–	–	40.7	–	–	40.7	–	–	19.9	19.2	0.7	–	–	–	–	–	–	0.7	19.2
–	–	–	47.3	–	–	47.3	–	–	14.6	13.7	0.8	–	–	–	–	–	–	0.8	13.7
–	–	–	28.5	–	–	28.5	–	–	36.6	36.5	0.1	–	–	–	–	–	–	0.1	36.5
–	–	–	8.31	–	–	8.31	–	–	11.7	11.5	0.3	–	–	–	–	–	–	0.3	11.5
–	–	–	36.7	–	–	36.7	–	–	25	24.7	0.3	–	–	–	–	–	–	0.3	24.7
–	–	–	45.8	–	–	45.8	–	–	18	17.7	0.2	–	–	–	–	–	–	0.2	17.7
–	–	–	36.8	–	–	36.8	–	–	25.1	24.8	0.3	–	–	–	–	–	–	0.3	24.8
–	–	–	37.6	–	–	37.6	–	–	26.2	24.3	1.9	–	–	–	–	–	–	1.9	24.3
–	–	–	30.3	–	0.2	30.1	–	–	32.3	32	0.2	–	–	–	–	–	–	0.2	32
–	–	–	31.7	–	0.1	31.7	–	–	31.4	31.3	0.2	–	–	–	–	–	–	0.2	31.3
–	–	–	32	–	0.1	31.9	–	–	29.8	26.8	3	–	–	–	–	–	–	3	26.8
–	–	–	37.6	–	–	37.6	–	–	26.2	24.3	1.9	–	–	–	–	–	–	1.9	24.3
–	–	–	14.2	–	0.1	14.2	–	–	14.5	14.3	0.2	–	–	–	–	–	–	0.2	14.3
–	–	–	18.4	–	–	18.4	–	–	11.4	10.7	0.7	–	–	–	–	–	–	0.7	10.7

Item	H_2O	Ash	Cals.	Prot.	Carb.	Fat-T	Fat-S	4:0	6:0	8:0	10:0	12:0	14:0	15:0	16:0	17:0	18:0
Imitation margarine, hy soy + hy and regular palm 40% fat	58.1	2.2	345	0.5	0.4	38.8	10.1	–	–	–	–	–	0.1	–	7.7	–	1.6
Imitation margarine, 40% fat, hy soybean	46.8	2.76	439	0.634	0.507	49.3	8.26	–	–	–	–	–	–	–	5.46	–	2.8
Imitation margarine, 40% fat, unspecified oils	58.1	2.2	345	0.5	0.4	38.8	7.7	–	–	–	–	0.1	0.1	–	5.4	–	2.1
Imperial Light Margarine, tub	53	1.61	396	0.493	0.493	44.4	9.2	–	–	–	–	–	0.113	–	5.29	–	3.8
Imperial Margarine, stick	15.7	2	719	0.9	0.9	80.5	16.7	–	–	–	–	–	0.2	–	9.6	–	6.9
Imperial Margarine, tub	16.2	2	716	0.8	0.5	80.4	13.5	–	–	–	–	–	0.1	–	8.7	–	4.7
Liquid margarine	15.8	2	721	1.91	–	80.6	13.2	–	–	–	–	–	0.1	–	8.6	–	4.6
Margarine spread, hy soy + hy cottonseed, 60% fat, tub	37	1.7	540	0.6	–	60.8	12	–	–	–	–	0.1	0.4	–	7.7	–	3.8
Margarine spread, 60% fat, hy soy + hy palm	37	1.7	540	0.6	–	60.8	14.1	–	–	–	–	0.1	0.2	–	10.6	–	3.2
Margarine spread, 60% fat, tub, unspecified oils	37	1.7	540	0.6	–	60.8	12.8	–	–	–	–	0.1	0.3	–	8.9	–	3.6
Margarine spread, hy soy + hy and regular palm, tub	37	1.7	540	0.6	–	60.8	13.5	–	–	–	–	–	0.1	–	10.1	–	3.4
Margarine, unsalted	18.5	0.2	714	0.5	0.5	80.3	15	–	–	–	–	0.3	0.2	–	9.5	–	5
Nucoa Margarine, stick	15.7	2	719	0.9	0.9	80.5	16.7	–	–	–	–	–	0.2	–	9.6	–	6.9
Parkay Light Margarine spread, tub	58.1	2.2	345	0.5	0.4	38.8	7.7	–	–	–	–	–	–	–	–	–	–
Parkay Margarine, stick	15.7	2	719	0.9	0.9	80.5	16.7	–	–	–	–	–	0.2	–	9.6	–	6.9
Parkay Soft Margarine, tub	16.2	2	716	0.8	0.5	80.4	13.5	–	–	–	–	–	0.1	–	8.7	–	4.7
Parkay Squeeze Liquid margarine	15.8	2	721	1.91	–	80.6	13.2	–	–	–	–	–	0.1	–	8.6	–	4.6
Promise Extra Light Margarine spread, tub	54.5	2	352	0.8	0.5	42.3	6.73	–	–	–	–	–	–	–	4.01	–	2.72
Promise Margarine Spread, tub	26.3	2	634	0.8	0.5	70.4	7	–	–	–	–	–	–	–	–	–	–
Regular hard margarine, hy soybean	15.7	2	634	0.9	0.9	80.5	16.7	–	–	–	–	–	0.2	–	9.6	–	6.9
Saffola Margarine, stick	15.7	2	719	0.9	0.9	80.5	13.3	–	–	–	–	–	0.2	–	7.6	–	5.5
Saffola Margarine, tub	16.2	5	704	0.8	0.5	77.5	10	–	–	–	–	–	–	–	6.1	–	3.8
Shedd's Spread Margarine spread, tub	58.1	2.2	414	0.5	0.4	38.8	5.5	–	–	–	–	–	–	–	–	–	–
Soft margarine w/salt, unspecified oils	16.2	2	716	0.8	0.5	80.4	13.8	–	–	–	–	–	0.1	–	8.8	–	4.8
Soft margarine, corn and hy corn	16.2	2	716	0.8	0.5	80.4	14.1	–	–	–	–	–	0.1	–	9.3	–	4.7
Soft margarine, hy and regular soy + hy cottonseed	16.2	2	716	0.8	0.5	80.4	16.1	–	–	–	–	0.3	0.2	–	9.8	–	5.8
Soft margarine, hy and regular soybean, no salt	17.9	0.2	716	0.8	0.9	80.3	13.5	–	–	–	–	–	0.1	–	8.6	–	4.7
Soft margarine, hy soy + safflower	16.2	2	716	0.8	0.5	80.4	10.4	–	–	–	–	–	–	–	6.3	–	3.9
Soft margarine, hy soy + hy and regular palm	16.2	2	716	0.8	0.5	80.4	17.1	–	–	–	–	0.2	0.2	–	12.4	–	4.2
Soft margarine, hy soy + hy cottonseed	16.2	2	716	0.8	0.5	80.4	14.2	–	–	–	–	–	0.1	–	8.7	–	5.3
Soft margarine, hy soy + hy cottonseed + soy	16.2	2	716	0.8	0.5	80.4	15.3	–	–	–	–	–	–	–	9.9	–	5.4
Soft margarine, hy soy + hy cottonseed, no salt	17.9	0.2	716	0.8	0.9	80.3	14.1	–	–	–	–	–	0.1	–	8.7	–	5.3

20:0	22:0	24:0	Fat-M	14:1	16:1	18:1	20:1	22:1	Fat-P	18:2	18:3	18:4	20:4	20:5	22:5	22:6	Chol., mg	Omeg3	Omeg6
–	–	–	14.1	–	–	14.1	–	–	12.8	12.1	0.7	–	–	–	–	–	–	0.7	12.1
–	–	–	21.2	–	–	21.2	–	–	17.5	16.4	1.14	–	–	–	–	–	–	1.14	16.4
–	–	–	15.7	–	0.1	15.6	–	–	13.8	13.2	0.5	–	–	–	–	–	–	0.5	13.2
–	–	–	21.7	–	0.113	21.5	–	–	11.5	10.7	0.824	–	–	–	–	–	–	0.824	10.7
–	–	–	39.3	–	0.2	39.1	–	–	20.9	19.4	1.5	–	–	–	–	–	–	1.5	19.4
–	–	–	36.5	–	–	36.5	–	–	26.9	26	0.9	–	–	–	–	–	–	0.9	26
–	–	–	28.1	–	–	28.1	–	–	35.8	33.4	2.4	–	–	–	–	–	–	2.4	33.4
–	–	–	38.9	–	–	38.9	–	–	7.1	7	0.2	–	–	–	–	–	–	0.2	7
–	–	–	26	–	–	26	–	–	18.1	16.5	1.6	–	–	–	–	–	–	1.6	16.5
–	–	–	31.5	–	–	31.5	–	–	13.8	12.9	0.9	–	–	–	–	–	–	0.9	12.9
–	–	–	24.1	–	–	24.1	–	–	20.4	18.8	1.6	–	–	–	–	–	–	1.6	18.8
–	–	–	36.7	–	–	36.7	–	–	25	24.7	0.3	–	–	–	–	–	–	0.3	24.7
–	–	–	39.3	–	0.2	39.1	–	–	20.9	19.4	1.5	–	–	–	–	–	–	1.5	19.4
–	–	–	15.7	–	–	–	–	–	13.8	–	–	–	–	–	–	–	–	–	–
–	–	–	39.3	–	0.2	39.1	–	–	20.9	19.4	1.5	–	–	–	–	–	–	1.5	19.4
–	–	–	36.5	–	–	36.5	–	–	26.9	26	0.9	–	–	–	–	–	–	0.9	26
–	–	–	28.1	–	–	28.1	–	–	35.8	33.4	2.4	–	–	–	–	–	–	2.4	33.4
–	–	–	8.46	–	–	8.46	–	–	25.2	25	0.21	–	–	–	–	–	–	0.21	25
–	–	–	31	–	–	–	–	–	35	–	–	–	–	–	–	–	–	–	–
–	–	–	39.3	–	0.2	39.1	–	–	20.9	19.4	1.5	–	–	–	–	–	–	1.3	17
–	–	–	23	–	–	23	–	–	40.7	40.6	0.1	–	–	–	–	–	–	0.1	40.6
–	–	–	29.9	–	–	29.9	–	–	34.2	33.1	1.1	–	–	–	–	–	–	1.1	33.1
–	–	–	16.7	–	–	16.7	–	–	16.6	–	–	–	–	–	–	–	–	–	–
–	–	–	28.5	–	–	28.5	–	–	34.6	33.5	1.1	–	–	–	–	–	–	1.1	33.5
–	–	–	31.6	–	–	31.6	–	–	31.2	30.3	0.9	–	–	–	–	–	–	0.9	30.3
–	–	–	30.7	–	0.1	30	0.6	–	30.1	27.3	2.8	–	–	–	–	–	–	2.8	27.3
–	–	–	36.4	–	–	36.4	–	–	26.8	25.9	0.9	–	–	–	–	–	–	0.9	25.9
–	–	–	31	–	–	31	–	–	35.4	34.3	1.1	–	–	–	–	–	–	1.1	34.3
–	–	–	25..2	–	–	25.1	–	–	34.6	32.6	1.9	–	–	–	–	–	–	1.9	32.6
–	–	–	38.2	–	–	38.2	–	–	24.6	23.3	1.3	–	–	–	–	–	–	1.3	23.3
–	–	–	35.8	–	–	35.8	–	–	25.8	24.8	1	–	–	–	–	–	–	1	24.8
–	–	–	38.1	–	–	38.1	–	–	24.5	23.2	1.3	–	–	–	–	–	–	1.3	23.2

Table 13.1 (*Continued*)

Item	H_2O	Ash	Cals.	Prot.	Carb.	Fat-T	Fat-S	4:0	6:0	8:0	10:0	12:0	14:0	15:0	16:0	17:0	18:0
Soft margarine, hy soybean + cottonseed	16.2	2	716	0.8	0.5	80.4	16.5	–	–	–	–	–	0.3	–	10.5	–	5.8
Soft margarine, safflower + hy cottonseed + hy peanut	16.2	2	716	0.8	0.5	80.4	13.4	–	–	–	–	–	0.1	–	8.3	–	5
Soft margarine, safflower + hy safflower	16.2	2	716	0.8	0.5	80.4	9.2	–	–	–	–	–	0.1	–	5.8	–	3.3
Soft margarine, sunflower + hy cottonseed + hy peanut	16.2	2	716	0.8	0.5	80.4	12.8	–	–	–	–	–	–	–	7.2	–	4.9
Soft margarine, tub	16.2	2	716	0.8	0.5	80.4	13.5	–	–	–	–	–	0.1	–	8.7	–	4.7
Soft margarine, unspecified oil, no salt	17.9	0.2	716	0.8	0.9	80.3	13.8	–	–	–	–	–	0.1	–	8.7	–	5
Unsalted Saffola Margarine, stick	15.7	2	719	0.9	0.9	80.5	13.3	–	–	–	–	–	0.2	–	7.6	–	5.5
Weight Watcher's Extra Light Spread, tub	58.1	2.2	345	0.5	0.4	38.8	7.7	–	–	–	–	–	–	–	–	–	–
1000 island dressing	46.1	2.1	377	0.9	15.2	35.7	6	–	–	–	–	0.1	0.1	–	4.3	–	1.3
Blue cheese dressing	32.3	3.2	504	4.81	7.41	52.3	9.9	–	–	–	–	0.1	0.6	–	6.5	–	2.4
Blue cheese/Roquefort dressing, no salt	32.3	0.5	504	4.81	7.41	52.3	9.9	–	–	–	–	–	–	–	–	–	–
Caesar's salad dressing	35.7	5.13	455	12	2.34	44.9	7.89	0.134	0.05	0.027	0.068	0.095	0.394	–	5.63	–	1.31
Cholesterol Free Dressing (Miracle Whip Light)	57	0.1	321	–	14.3	28.6	7.14	–	–	–	–	–	–	–	–	–	–
Creamy bacon salad dressing	38.7	3.15	486	0.343	5.2	52.6	7.73	–	–	–	–	–	–	–	–	–	–
Creamy cucumber salad dressing	38.7	3.15	486	0.343	5.2	52.6	7.73	–	–	–	–	–	–	–	–	–	–
Creamy italian salad dressing	38.7	3.15	486	0.343	5.2	52.6	7.73	–	–	–	–	–	–	–	–	–	–
Filled sour dressing, non butterfat	74.8	0.71	178	3.26	4.69	16.6	13.3	–	–	0.319	0.824	2.26	2.79	–	2.82	–	4.31
French dressing	38.1	3	430	0.6	17.5	41	9.5	–	–	–	–	–	0.3	–	7.7	–	0.8
French salad dressing with cottonseed oil	24.2	1.8	631	0.1	3.41	70.2	18.2	–	–	–	–	–	–	–	–	–	–
French salad dressing, homemade	24.2	1.8	631	0.1	3.41	70.2	12.6	–	–	–	–	–	0.2	–	9.8	–	2.6
French salad dressing, no salt	38.1	0.5	430	0.6	17.5	41	9.5	–	–	–	–	–	–	–	–	–	–
Imitation sour cream (IMO)	71.2	0.3	208	2.41	6.64	19.5	17.8	–	–	1.03	1.01	8.14	3.36	–	1.98	–	2.24
Italian salad dressing	38.4	2.5	467	0.7	10.2	48.3	7	–	–	–	–	–	–	–	5	–	1.8
Italian salad dressing (Seven Seas Viva)	38.4	2.5	467	0.7	10.2	48.3	7	–	–	–	–	–	–	–	5	–	1.8
Italian salad dressing, no salt	38.4	0.5	467	0.7	10.2	48.3	7	–	–	–	–	–	–	–	–	–	–
Light buttermilk dressing	68.1	2	220	3.2	6.7	20	2.9	–	–	–	–	–	–	–	–	–	–
Low Calorie bacon and tomato dressing	75.2	–	200	1.8	2	21	3.53	–	–	–	–	–	–	–	–	–	–
Low Calorie creamy cucumber salad dressing	68.4	3.7	208	0.413	7.22	20.3	3.77	–	–	–	–	–	–	–	–	–	–
Low Calorie creamy dressing, oil free	68.4	3.7	208	0.413	7.22	20.3	3.77	–	–	–	–	–	–	–	–	–	–
Mayonnaise	15.3	1.5	717	1.11	2.71	79.4	11.8	–	–	–	–	0.1	0.1	–	8.5	–	3.1
Mayonnaise type dressing (Miracle Whip)	39.9	1.9	390	0.9	23.9	33.4	4.9	–	–	–	–	0.1	–	–	3.5	–	1.4
Mayonnaise, safflower and soy oil (Saffola)	15.3	1.5	717	1.11	2.71	79.4	8.6	–	–	–	–	–	0.1	–	6.1	–	2.4
Mayonnaise, soybean oil, no salt	15.3	0.3	717	1.11	2.71	79.4	11.8	–	–	–	–	–	–	–	–	–	–
NoCholesterol imitation mayonnaise, soybean	34.6	1.8	482	0.1	15.8	47.7	7.5	–	–	–	–	–	–	–	5.5	–	2
Oil and vinegar dressing	47.4	–	449	–	2.51	50.1	9.1	–	–	–	–	–	0.2	–	7	–	1.9
Ranch salad dressing	35	19.5	366	3.02	4.62	37.9	5.65	–	–	–	–	–	–	–	–	–	–

20:0	22:0	24:0	Fat-M	14:1	16:1	18:1	20:1	22:1	Fat-P	18:2	18:3	18:4	20:4	20:5	22:5	22:6	Chol., mg	Omeg3	Omeg6
–	–	–	31.3	–	–	31.3	–	–	29.1	27.5	1.6	–	–	–	–	–	–	1.6	27.5
–	–	–	14	–	–	14	–	–	49.5	49.5	–	–	–	–	–	–	–	–	49.5
–	–	–	23.2	–	–	23.2	–	–	44.5	44.5	–	–	–	–	–	–	–	–	44.5
–	–	–	16.1	–	–	16.1	–	–	48	47.6	0.4	–	–	–	–	–	–	0.4	47.6
–	–	–	36.5	–	–	36.5	–	–	26.9	26	0.9	–	–	–	–	–	–	0.9	26
–	–	–	37.3	–	–	37.3	–	–	25.7	24.6	1.1	–	–	–	–	–	–	1.1	24.6
–	–	–	23	–	–	23	–	–	40.7	40.6	0.1	–	–	–	–	–	–	0.1	40.6
–	–	–	15.7	–	–	–	–	–	13.8	–	–	–	–	–	–	–	–	–	–
–	–	–	8.31	–	0.2	7.71	–	–	19.8	16.5	2.5	–	–	–	–	–	26	2.5	16.5
–	–	–	12.3	–	0.3	11.8	–	–	27.8	23.5	3.7	–	–	–	–	–	17	3.7	23.5
–	–	–	–	–	–	11.8	–	–	–	23.5	–	–	–	–	–	–	17	–	23.5
0.002	0.002	–	30.5	0.001	0.521	29.8	0.122	0.033	4.11	3.36	0.275	0.016	0.027	0.155	0.008	0.267	99.7	0.694	3.38
–	–	–	7.14	–	–	–	–	–	14.3	–	–	–	–	–	–	–	–	–	–
–	–	–	12.8	–	–	–	–	–	29.7	–	–	–	–	–	–	–	4.49	–	–
–	–	–	12.8	–	–	–	–	–	29.7	–	–	–	–	–	–	–	4.49	–	–
–	–	–	12.8	–	–	–	–	–	29.7	–	–	–	–	–	–	–	4.49	–	–
–	–	–	1.96	–	0.87	1.87	–	–	0.468	0.468	–	–	–	–	–	–	5.4	–	0.468
–	–	–	8.01	–	0.4	7.11	–	–	21.7	20.3	–	–	–	–	–	–	58	–	20.3
–	–	–	–	–	–	11.9	–	–	–	36.2	–	–	–	–	–	–	–	–	36.2
–	–	–	20.7	–	0.2	20.5	–	–	33.7	31.7	1.9	–	–	–	–	–	–	1.9	31.7
–	–	–	–	–	–	7.11	–	–	–	20.3	–	–	–	–	–	–	–	–	20.3
–	–	–	0.588	–	–	0.589	–	–	0.056	0.056	–	–	–	–	–	–	–	–	0.056
–	–	–	11.2	–	0.1	11	0.1	–	28	24.6	3.3	–	–	–	–	–	–	3.3	24.6
–	–	–	11.2	–	0.1	11	0.1	–	28	24.6	3.3	–	–	–	–	–	–	3.3	24.6
–	–	–	–	–	–	11	–	–	–	24.6	–	–	–	–	–	–	67	–	24.6
–	–	–	4.7	–	–	–	–	–	11.6	–	–	–	–	–	–	–	33	–	–
–	–	–	5.59	–	–	–	–	–	11	–	–	–	–	–	–	–	4	–	–
–	–	–	5.97	–	–	–	–	–	9.63	–	–	–	–	–	–	–	1.14	–	–
–	–	–	5.97	–	–	–	–	–	9.63	–	–	–	–	–	–	–	1.14	–	–
–	–	–	22.7	–	0.1	22.5	0.1	–	41.3	37.1	4.2	–	–	–	–	–	59	4.2	37.1
–	–	–	9.01	–	–	9.01	–	–	18	16	2	–	–	–	–	–	26	2	16
–	–	–	13	–	–	13	–	–	55	52	3	–	–	–	–	–	59	3	52
–	–	–	–	–	–	22.5	–	–	–	37.1	–	–	–	–	–	–	59	–	37.1
–	–	–	10.5	–	–	10.5	–	–	27.6	23	4.6	–	–	–	–	–	–	4.6	23
–	–	–	14.8	–	0.1	14.7	–	–	24.1	22.7	1.4	–	–	–	–	–	–	1.4	22.7
–	–	–	16.3	–	–	–	–	–	14.3	–	–	–	–	–	–	–	39.2	–	–

Table 13.1 *(Continued)*

Item	H_2O	Ash	Cals.	Prot.	Carb.	Fat-T	Fat-S	4:0	6:0	8:0	10:0	12:0	14:0	15:0	16:0	17:0	18:0
Rancher's Choice Creamy Dressing	–	–	600	–	6.67	66.7	6.67	–	–	–	–	–	–	–	–	–	–
Russian dressing	34.5	2.7	494	1.61	10.4	50.8	7.3	–	–	–	–	–	–	–	5.2	–	1.9
Sandwich spread with chopped pickle	40.8	1.9	389	0.9	22.4	34	5.1	–	–	–	–	–	–	–	–	–	–
Sandwich spread with pickles, unspecified oil	40.8	2	389	0.9	22.4	34	5.1	–	–	–	–	–	–	–	3.9	–	1.2
Sesame seed salad dressing	39.2	3.9	443	3.11	8.61	45.2	6.2	–	–	–	–	–	–	–	4.2	–	1.9
Vinegarette salad dressing	38.4	2.5	467	0.7	10.2	48.3	7	–	–	–	–	–	–	–	5	–	1.8
Bean cake	23.3	0.905	406	5.41	49.1	21.3	3.19	–	–	–	–	–	–	–	–	–	–
California avocado	72.6	1.1	177	2.12	6.92	17.3	2.59	–	–	–	–	–	–	–	2.56	–	0.021
California avocado, mashed	72.6	1.1	177	2.12	6.92	17.3	2.59	–	–	–	–	–	–	–	2.56	–	0.021
Dried tofu, koyadofu, frozen with calcium sulfate	5.78	1.38	480	47.9	14.6	30.3	4.42	–	–	–	–	0.018	0.084	–	3.22	–	1.08
Dry roasted soybeans	0.8	5.29	450	39.6	32.7	21.6	3.13	–	–	–	–	–	0.06	–	2.3	–	0.772
Non dairy topping, pressurized	60.4	0.28	264	0.98	16.1	22.3	18.9	–	–	0.449	0.642	8.01	3.25	–	2.41	–	4.17
Soybeans, roasted with salt added	1.96	3.89	471	35.2	33.6	25.4	3.67	–	–	–	–	0.01	0.028	–	2.67	–	0.907
Tofu with calcium sulfate, fried	50.5	1.62	271	17.2	10.5	20.2	2.94	–	–	–	–	0.012	0.056	–	2.14	–	0.72
Tofu with nigari, koya-dofu, dried, frozen	5.78	1.32	480	47.9	14.6	30.4	4.42	–	–	–	–	0.018	0.084	–	3.22	–	1.08

11. Polyunsaturated fat (Fat-P).
12. Each of the polyunsaturated fatty acids: 18:2, 18:3, 18:4, 20:4, 20:5, 22:5, 22:6.
13. Mg (not g) cholesterol (Chol).
14. Omega 3 fatty acid (Omeg3).
15. Omega 6 fatty acid (Omeg6).

AN IMPORTANT PREMISE

The extent of coverage in the data presented in this chapter is mandated by the editor of this five volume text. The data are useful as a general reference guide on the fat contents of selected food. However, if they are to be used in a commercial application, certain considerations are important. Some examples are cited here:

1. There will be many new results available by the time these data are in print. Therefore, routine search for new data is advised.
2. In view of the limited amount of data presented, additional sources of information are needed for an appropriate evaluation.

20:0	22:0	24:0	Fat-M	14:1	16:1	18:1	20:1	22:1	Fat-P	18:2	18:3	18:4	20:4	20:5	22:5	22:6	Chol., mg	Omeg3	Omeg6
–	–	–	–	–	–	–	–	–	40	–	–	–	–	–	–	–	33.3	–	–
–	–	–	11.8	–	0.1	11.6	0.1	–	29.4	25.9	3.5	–	–	–	–	–	18	3.5	25.9
–	–	–	7.4	–	–	–	–	–	20	–	–	–	–	–	–	–	76	–	–
–	–	–	7.41	–	–	7.41	–	–	20	18.1	1.9	–	–	–	–	–	76	1.9	18.1
–	–	–	11.9	–	–	11.9	–	–	25.1	23.2	2	–	–	–	–	–	–	2	23.2
–	–	–	11.2	–	0.1	11	0.1	–	28	24.6	3.3	–	–	–	–	–	–	3.3	24.6
–	–	–	9.02	–	–	–	–	–	8.04	–	–	–	–	–	–	–	–	–	–
0.09	–	–	11.2	–	0.553	10.7	–	–	2.04	1.92	0.115	–	0.005	–	–	–	–	0.115	1.93
0.009	–	–	11.2	–	0.553	10.7	–	–	2.04	1.92	0.115	–	0.005	–	–	–	–	0.115	1.93
0.018	–	–	6.74	–	0.085	6.65	–	–	17.1	15.1	2.02	–	–	–	–	–	–	2.02	15.1
–	–	–	4.78	–	0.06	4.72	–	–	12.2	10.8	1.44	–	–	–	–	–	–	1.44	10.8
–	–	–	1.93	–	–	1.93	–	–	0.241	0.241	–	–	–	–	–	–	–	–	0.241
0.05	0.01	–	5.61	–	0.061	5.5	0.05	–	14.3	12.6	1.69	–	–	–	–	–	–	1.69	12.6
0.012	–	–	4.48	–	0.056	4.42	–	–	11.4	10	1.35	–	–	–	–	–	–	1.35	10
0.018	–	–	6.74	–	0.085	6.65	–	–	17.1	15.1	2.02	–	–	–	–	–	–	2.02	15.1

3. Standard principles applicable to using nutritional data for specific products must be applied. These include: samples, part of plant or animal product used, equipment, techniques, statistics, etc.
4. Any application involving legal compliance is not advised. For such circumstances, a current database with the most updated information should be used.

The reader is advised that there are other considerations not stated here. If further details are needed, assistance should be obtained from professionals specializing in this area.

Editor's note on ESHA research and the nutrient data bank

ESHA Research is located in Salem, Oregon. ESHA researches and compiles nutrient data for foods and ingredients to be used in nutrition analysis software for the food and health industries. Currently, 165 nutrient factors (including trans, *n*-3, and *n*-6 polyunsaturated fats) and over 22,000 foods are in the ESHA Nutrient Data Bank. This data bank is compiled from over 1,000 sources of information worldwide. The research and updating is ongoing to ensure accuracy for the users. The primary software systems are called Genesis R & D and the Food Processor.

Table 13.2 Lipid composition of nuts, seeds, and related products, contents/100 g

Item	H_2O	Ash	Cals.	Prot.	Carb.	Fat-T	Fat-S	4:0	6:0	8:0	10:0	12:0	14:0	15:0	16:0	17:0	18:0
Acorns, raw	27.9	1.36	369	6.16	40.8	23.9	3.1	–	–	–	–	–	–	–	2.85	–	0.252
Dried acorn nuts	5.07	1.79	509	8.11	53.7	31.4	4.03	–	–	–	–	–	–	–	3.75	–	0.332
Blanched almonds, sliced	5.43	3.12	586	20.4	18.5	52.5	4.93	–	–	–	–	0.005	0.299	–	3.48	–	1.03
Dried almonds, chopped	4.43	3.04	589	20	20.4	52.2	4.98	–	–	–	–	0.005	0.298	–	3.46	0.03	1.02
Dried almonds, each measure	4.43	3.04	589	20	20.4	52.2	4.98	–	–	–	–	0.005	0.298	–	3.46	0.03	1.02
Dried almonds, slices	4.43	3.04	589	20	20.4	52.2	4.98	–	–	–	–	0.005	0.298	–	3.46	0.03	1.02
Dried almonds, slivered, packed	4.43	3.04	589	20	20.4	52.2	4.98	–	–	–	–	0.005	0.298	–	3.46	0.03	1.02
Dried almonds, whole	4.43	3.04	589	20	20.4	52.2	4.98	–	–	–	–	0.005	0.298	–	3.46	0.03	1.02
Dried blanched almonds, whole	5.43	3.12	586	20.4	18.5	52.5	4.98	–	–	–	–	0.005	0.299	–	3.48	–	1.03
Dry roasted almonds, salted	3.01	4.91	587	16.3	24.2	51.6	4.92	–	–	–	–	0.005	0.294	–	3.41	0.03	1.01
Dry roasted almonds, whole, unsalted	3.01	4.91	587	16.3	24.2	51.6	4.92	–	–	–	–	0.005	0.294	–	3.41	0.03	1.01
Honey roasted almonds, unblanched	1.71	2.34	594	18.2	27.9	49.9	4.73	–	–	–	–	0.005	0.287	–	3.33	–	0.984
Oil roasted almonds	3.09	2.99	618	20.4	15.9	57.7	5.47	–	–	–	–	0.005	0.329	–	3.82	–	1.13
Oil roasted almonds, unsalted	3.09	2.99	618	20.4	15.9	57.7	5.47	–	–	–	–	0.005	0.329	–	3.82	–	1.13
Oil roasted almonds, blanched w/salt	3.52	2.91	613	19	18	56.5	5.36	–	–	–	–	0.005	0.322	–	3.74	–	1.11
Oil roast blanched almonds, no salt	3.52	2.91	613	19	18	56.5	5.36	–	–	–	–	0.005	0.322	–	3.74	–	1.11
Toasted almonds, whole	2.61	3.35	589	20.4	22.9	50.8	4.81	–	–	–	–	0.005	0.289	–	3.36	–	0.993
Dried beechnuts	6.61	3.71	576	6.21	33.5	50	5.72	–	–	–	–	–	0.063	–	3.57	–	1.18
Dried brazilnuts, unsalted	3.35	3.31	656	14.3	12.8	66.2	16.2	–	–	–	–	–	0.62	–	9.52	–	5.68
Large dried brazilnuts	3.35	3.31	656	14.3	12.8	66.2	16.2	–	–	–	–	–	0.62	–	9.52	–	5.68
Dried butternuts	3.35	2.74	612	24.9	12.1	57	1.31	–	–	–	–	–	–	–	0.872	–	0.425
Dry roasted cashews	1.71	3.96	574	15.3	32.7	46.4	9.16	–	–	0.132	0.132	0.784	0.347	–	4.35	–	2.97
Dry roasted cashews, unsalted	1.71	3.96	574	15.3	32.7	46.4	9.16	–	–	0.132	0.132	0.784	0.347	–	4.35	–	2.97
Oil roasted cashews	3.92	3.22	576	16.2	28.5	48.2	9.53	–	–	0.137	0.137	0.816	0.361	–	4.53	–	3.09
Oil roasted cashews, unsalted	3.92	3.22	576	16.2	28.5	48.2	9.53	–	–	0.137	0.137	0.816	0.361	–	4.53	–	3.09
Small oil roasted cashews, no salt	3.92	3.22	576	16.2	28.5	48.2	9.53	–	–	0.137	0.137	0.816	0.361	–	4.53	–	3.09
Large oil roasted cashews, no salt	3.92	3.22	576	16.2	28.5	48.2	9.53	–	–	0.137	0.137	0.816	0.361	–	4.53	–	3.09
Coconut, dried, creamed	1.68	2.45	684	5.31	21.5	69.1	61.3	–	0.393	4.84	3.85	30.6	12.1	–	5.86	–	3.58
Dried coconut	3.01	1.95	660	6.89	24.4	64.5	57.2	–	0.367	4.52	3.59	28.6	11.3	–	5.47	–	3.34
Dried shredded coconut, sweetened	12.6	1.43	501	2.89	47.7	35.5	31.5	–	0.202	2.49	1.98	15.7	6.22	–	3.01	–	1.84
Flaked coconut, sweetened, canned	23.3	0.78	443	3.36	40.9	31.7	28.1	–	0.18	2.22	1.76	14.1	5.55	–	2.69	–	1.64
Flaked coconut, sweetened, package	15.6	1.38	474	3.29	47.6	32.2	28.5	–	0.183	2.25	1.79	14.3	5.63	–	2.73	–	1.67
Fresh coconut, 2.5 × 2 inch	47	0.97	354	3.34	15.2	33.5	29.7	–	0.191	2.35	1.86	14.9	5.87	–	2.84	–	1.73
Fresh coconut, grated	47	0.97	354	3.34	15.2	33.5	29.7	–	0.191	2.35	1.86	14.9	5.87	–	2.84	–	1.73
Fresh grated coconut, packed	47	0.97	354	3.34	15.2	33.5	29.7	–	0.191	2.35	1.86	14.9	5.87	–	2.84	–	1.73
Shredded coconut, sweetened, package	12.6	1.43	501	2.89	47.7	35.5	31.5	–	0.202	2.49	1.98	15.7	6.22	–	3.01	–	1.84
Toasted coconut	1.01	2.31	592	5.31	44.4	47	41.7	–	0.267	3.29	2.62	20.9	8.23	–	3.98	–	2.43
Blanched filberts/hazelnuts, dried	1.91	2.11	672	12.7	16	67.3	4.95	–	–	–	–	–	125	–	3.35	–	1.37
Dried filberts/hazelnuts, chopped	5.43	3.62	632	13	15.3	62.6	4.6	–	–	–	–	–	0.117	–	3.12	–	1.27
Dried filberts/hazelnuts, whole	5.43	3.62	632	13	15.3	62.6	4.6	–	–	–	–	–	0.117	–	3.12	–	1.27
Dried ground filberts/hazelnuts	5.43	3.62	632	13	15.3	62.6	4.6	–	–	–	–	–	0.117	–	3.12	–	1.27

20:0	22:0	24:0	Fat-M	14:1	16:1	18:1	20:1	22:1	Fat-P	18:2	18:3	18:4	20:4	20:5	22:5	22:6	Chol., mg	Omeg3	Omeg6
–	–	–	15.1	–	–	15.1	–	–	4.6	4.6	–	–	–	–	–	–	–	–	4.6
–	–	–	19.9	–	–	19.9	–	–	6.05	6.05	–	–	–	–	–	–	–	–	6.05
0.166	–	–	34.1	–	0.306	33.5	0.05	–	11	10.6	0.376	–	–	–	–	–	–	0.376	10.6
0.167	–	–	33.9	–	0.304	33.3	0.05	–	11	10.5	0.374	–	–	–	–	–	–	0.374	10.5
0.167	–	–	33.9	–	0.304	33.3	0.05	–	11	10.5	0.374	–	–	–	–	–	–	0.374	10.5
0.167	–	–	33.9	–	0.304	33.3	0.05	–	11	10.5	0.374	–	–	–	–	–	–	0.374	10.5
0.167	–	–	33.9	–	0.304	33.3	0.05	–	11	10.5	0.374	–	–	–	–	–	–	0.374	10.5
0.167	–	–	33.9	–	0.304	33.3	0.05	–	11	10.5	0.374	–	–	–	–	–	–	0.374	10.5
0.166	–	–	34.1	–	0.306	33.5	0.05	–	11	10.6	0.376	–	–	–	–	–	–	0.376	10.6
0.171	–	–	33.5	–	0.3	32.9	0.049	–	10.8	10.4	0.369	–	–	–	–	–	–	0.369	10.4
0.171	–	–	33.5	–	0.3	32.9	0.49	–	10.8	10.4	0.369	–	–	–	–	–	–	0.369	10.4
0.124	–	–	32.4	–	0.293	32.1	0.048	–	10.5	10.1	0.36	–	–	–	–	–	–	0.36	10.1
0.186	–	–	37.4	–	0.336	36.8	0.055	–	12.1	11.6	0.413	–	–	–	–	–	–	0.413	11.6
0.186	–	–	37.4	–	0.336	36.8	0.055	–	12.1	11.6	0.413	–	–	–	–	–	–	0.413	11.6
0.183	–	–	36.7	–	0.329	36	0.054	–	11.9	1.14	0.405	–	–	–	–	–	–	0.405	11.4
0.183	–	–	36.7	–	0.329	36	0.054	–	11.9	11.4	0.405	–	–	–	–	–	–	0.405	11.4
0.163	–	–	33	–	0.295	32.4	0.048	–	10.7	10.2	0.364	–	–	–	–	–	–	0.364	10.2
–	–	–	21.9	–	0.287	18.8	2.76	–	20.1	18.4	1.7	–	–	–	–	–	–	1.7	18.4
–	–	–	23	–	0.324	22.4	0.062	–	24.1	23.8	0.062	–	–	–	–	–	–	0.062	23.8
–	–	–	23	–	0.324	22.4	0.062	–	24.1	23.8	0.062	–	–	–	–	–	–	0.062	23.8
–	–	–	10.4	–	–	10.4	–	–	42.7	33.7	8.72	–	–	–	–	–	–	8.72	33.7
0.445	–	–	27.3	–	0.318	26.8	0.139	–	7.84	7.66	0.161	–	–	–	–	–	–	0.161	7.66
0.445	–	–	27.3	–	0.318	26.8	0.139	–	7.84	7.66	0.161	–	–	–	–	–	–	0.161	7.66
0.459	–	–	28.4	–	0.331	27.9	0.144	–	8.15	7.97	0.167	–	–	–	–	–	–	0.167	7.97
0.459	–	–	28.4	–	0.331	27.9	0.144	–	8.15	7.97	0.167	–	–	–	–	–	–	0.167	7.97
0.459	–	–	28.4	–	0.331	27.9	0.144	–	8.15	7.97	0.167	–	–	–	–	–	–	0.167	7.97
0.459	–	–	28.4	–	0.331	27.9	0.144	–	8.15	7.97	0.167	–	–	–	–	–	–	0.167	7.97
0.077	–	–	2.94	–	–	2.94	–	–	0.756	0.756	–	–	–	–	–	–	–	–	0.756
0.013	–	–	2.75	–	–	2.75	–	–	0.706	0.706	–	–	–	–	–	–	–	–	0.706
0.058	–	–	1.51	–	–	1.51	–	–	0.388	0.388	–	–	–	–	–	–	–	–	0.388
–	–	–	1.35	–	–	1.35	–	–	0.347	0.347	–	–	–	–	–	–	–	–	0.347
–	–	–	1.37	–	–	1.37	–	–	0.352	0.352	–	–	–	–	–	–	–	–	0.352
–	–	–	1.43	–	–	1.43	–	–	0.366	0.366	–	–	–	–	–	–	–	–	0.366
–	–	–	1.43	–	–	1.43	–	–	0.366	0.366	–	–	–	–	–	–	–	–	0.366
–	–	–	1.43	–	–	1.43	–	–	0.366	0.366	–	–	–	–	–	–	–	–	0.366
0.058	–	–	1.51	–	–	1.51	–	–	0.388	0.388	–	–	–	–	–	–	–	–	0.388
–	–	–	2	–	–	2	–	–	0.514	0.514	–	–	–	–	–	–	–	–	0.514
0.105	–	–	52.7	–	0.226	52.3	0.105	–	6.45	6.27	0.163	–	–	–	–	–	–	0.163	6.27
0.093	–	–	49.1	–	0.21	48.6	0.098	–	6	5.83	0.152	–	–	–	–	–	–	0.152	5.83
0.093	–	–	49.1	–	0.21	48.6	0.098	–	6	5.83	0.152	–	–	–	–	–	–	0.152	5.83
0.093	–	–	49.1	–	0.21	48.6	0.098	–	6	5.83	0.152	–	–	–	–	–	–	0.152	5.83

Table 13.2 (Continued)

Item	H_2O	Ash	Cals.	Prot.	Carb.	Fat-T	Fat-S	4:0	6:0	8:0	10:0	12:0	14:0	15:0	16:0	17:0	18:0
Dry roasted filberts/hazelnuts, salted	1.91	3.91	662	10	17.9	66.3	4.87	–	–	–	–	–	0.123	–	3.3	–	1.35
Dry roasted filberts/hazelnuts, unsalted	1.91	3.91	662	10	17.9	66.3	4.87	–	–	–	–	–	0.123	–	3.3	–	1.35
Oil roasted filberts/hazelnuts, salted	1.21	1.81	660	14.3	19.2	63.6	4.68	–	–	–	–	–	0.118	–	3.17	–	1.29
Oil roasted filberts/hazelnuts, unsalted	1.21	1.81	660	14.3	19.2	63.6	4.68	–	–	–	–	–	0.118	–	3.17	–	1.29
Formed nuts (wheat), salted	2.51	2.31	622	13.8	23.7	57.7	8.7	–	–	–	–	–	0.094	–	5.96	–	2.69
Formed nuts, macadamia flavor	3.11	1.31	619	11.2	27.9	56.5	8.48	–	–	–	–	–	0.079	–	5.72	–	2.7
Formed nuts, other flavors	2.01	1.81	647	13.1	20.8	62.3	9.37	–	–	–	–	–	0.093	–	6.36	–	2.95
Hickory nuts, dried	2.66	2.01	657	12.7	18.3	64.4	7.04	–	–	–	–	–	–	–	5.42	–	1.42
Dried macadamia nuts	2.89	1.37	702	8.31	13.7	73.7	11	0.49	–	–	–	–	0.468	–	6.25	–	2.7
Oil roasted macadamia nuts	1.68	1.66	718	7.27	12.9	76.5	11.5	0.49	–	–	–	–	0.486	–	6.48	–	2.81
Oil roast macadamia nuts, each, no salt	1.68	1.66	718	7.27	12.9	76.5	11.5	0.49	–	–	–	–	0.486	–	6.48	–	2.81
Oil roast macadamia nuts, chopped, no salt	1.68	1.66	718	7.27	12.9	76.5	11.5	0.49	–	–	–	–	0.486	–	6.48	–	2.81
Oil roast macadamia nuts, halves, no salt	1.68	1.66	718	7.27	12.9	76.5	11.5	0.49	–	–	–	–	0.486	–	6.48	–	2.81
Dry roasted mixed nuts	1.76	4.16	594	17.3	25.4	51.5	6.9	–	–	0.045	0.045	0.271	0.216	–	4.21	–	1.75
Dry roasted mixed nuts, unsalted	1.76	4.16	594	17.3	25.4	51.5	6.9	–	–	0.045	0.045	0.271	0.216	–	4.21	–	1.75
Oil roasted mixed nuts	2.04	3.48	617	16.8	21.4	56.3	8.73	–	–	0.05	0.05	0.296	0.255	–	5.24	–	2.42
Oil roasted mixed nuts, no peanuts	3.16	2.9	615	15.5	22.3	56.2	9.09	–	–	0.078	0.078	0.467	0.342	–	4.93	–	2.78
Oil roasted mixed nuts, no peanuts, unsalted	3.16	2.9	615	15.5	22.3	56.2	9.09	–	–	0.078	0.078	0.467	0.342	–	4.93	–	2.78
Oil roasted mixed nuts, unsalted	2.04	3.48	617	16.8	21.4	56.3	8.73	–	–	0.05	0.05	0.296	0.255	–	5.24	–	2.42
Dry roasted peanuts, salted	1.56	3.61	585	23.7	21.5	49.7	6.89	–	–	–	–	–	0.025	–	5.2	–	1.11
Dry roasted peanuts, unsalted	1.55	3.55	585	23.7	21.5	49.7	6.92	–	–	–	–	–	0.025	–	5.2	–	1.11
Oil roasted peanuts	1.96	3.49	581	26.4	18.9	49.3	6.84	–	–	–	–	–	0.025	–	5.16	–	1.1
Oil roasted peanuts, unsalted	1.95	3.45	581	26.3	19	49.3	6.84	–	–	–	–	–	0.025	–	5.16	–	1.1
Oil roasted spanish peanuts, salted	1.79	3.73	579	28	17.5	49	7.56	–	–	–	–	–	0.028	–	5.61	–	1.27
Oil roasted spanish peanuts, no salt	1.78	3.72	579	28	17.5	49	7.55	–	–	–	–	–	0.028	–	5.61	–	1.27
Oil roasted valencia peanuts, salted	2.13	3.3	589	27	16.3	51.2	7.89	–	–	–	–	–	0.029	–	5.86	–	1.33
Oil roasted valencia peanuts, no salt	2.12	3.3	589	27	16.3	51.2	7.89	–	–	–	–	–	0.029	–	5.86	–	1.33
Oil roasted virginia peanuts, salted	2.18	3.49	578	25.9	19.9	48.6	6.35	–	–	–	–	–	0.012	–	4.5	–	1.15
Oil roasted virginia peanuts, no salt	2.17	3.48	578	25.9	19.9	48.6	6.34	–	–	–	–	–	0.012	–	4.5	–	1.15
Parched peanuts	4.5	2.2	571	26.3	18.8	48.2	6.71	–	–	–	–	–	0.024	–	5.04	–	1.08
Peanuts, boiled	41.8	1.46	318	13.5	21.3	22	3.06	–	–	–	–	–	0.011	–	2.3	–	0.492
Peanuts, fried without skin	6.4	2.2	569	18.1	24.9	48.4	6.71	–	–	–	–	–	0.025	–	5.07	–	1.08
Peanuts, raw	6.51	2.3	562	25.8	16.4	48.4	8	–	–	–	–	–	–	–	4.3	–	1.4
Spanish peanuts, raw	6.4	2.04	570	26.2	15.8	49.6	7.64	–	–	–	–	–	0.028	–	5.67	–	1.29
Valencia peanuts, raw	4.27	2.18	570	25.1	20.9	47.6	7.33	–	–	–	–	–	0.027	–	5.44	–	1.24
Virginia peanuts, raw	6.92	2.62	563	25.2	16.5	48.8	6.36	–	–	–	–	–	0.012	–	4.51	–	1.15
Dried ground pecans	4.83	1.57	667	7.76	18.2	67.6	5.42	–	–	–	–	–	–	–	4.14	–	1
Dried pecans, chopped	4.83	1.57	667	7.76	18.2	67.6	5.42	–	–	–	–	–	–	–	4.14	–	1
Dried pecans, halves	4.83	1.57	667	7.76	18.2	67.6	5.42	–	–	–	–	–	–	–	4.14	–	1
Dried pecans, mammoth	4.83	1.57	667	7.76	18.2	67.6	5.42	–	–	–	–	–	–	–	4.14	–	1
Dry roasted pecans	1.11	4.01	659	7.98	22.3	64.6	5.18	–	–	–	–	–	–	–	3.95	–	0.957

20:0	22:0	24:0	Fat-M	14:1	16:1	18:1	20:1	22:1	Fat-P	18:2	18:3	18:4	20:4	20:5	22:5	22:6	Chol., mg	Omeg3	Omeg6
0.097	–	–	52	–	0.223	51.5	0.103	–	6.35	6.17	0.16	–	–	–	–	–	–	0.16	6.17
0.097	–	–	52	–	0.223	51.5	0.103	–	6.35	6.17	0.16	–	–	–	–	–	–	0.16	6.17
0.102	–	–	49.8	–	0.214	49.4	0.099	–	6.1	5.92	0.154	–	–	–	–	–	–	0.154	5.92
0.102	–	–	49.8	–	0.214	49.4	0.099	–	6.1	5.92	0.154	–	–	–	–	–	–	0.154	5.92
–	–	–	23.5	–	0.23	23.3	–	–	22.7	21	1.69	–	–	–	–	–	–	1.69	21
–	–	–	23.5	–	0.225	23.2	–	–	21.9	20.2	1.58	–	–	–	–	–	–	1.58	20.2
–	–	–	25.7	–	0.248	25.4	–	–	24.3	22.4	1.78	–	–	–	–	–	–	1.78	22.4
–	–	–	32.6	–	0.277	32	–	–	21.9	20.6	1.05	–	–	–	–	–	–	1.05	20.6
–	–	–	58.2	–	16	41.2	1.05	–	1.27	1.27	–	–	–	–	–	–	–	–	1.27
–	–	–	60.4	–	16.6	42.7	1.09	–	1.32	1.32	–	–	–	–	–	–	–	–	1.32
–	–	–	60.4	–	16.6	42.7	1.09	–	1.32	1.32	–	–	–	–	–	–	–	–	1.32
–	–	–	60.4	–	16.6	42.7	1.09	–	1.32	1.32	–	–	–	–	–	–	–	–	1.32
–	–	–	60.4	–	16.6	42.7	1.09	–	1.32	1.32	–	–	–	–	–	–	–	–	1.32
0.143	0.165	0.055	31.4	–	0.223	30.8	0.259	–	10.8	10.5	0.19	–	–	–	–	–	–	0.19	10.5
0.143	0.165	0.055	31.4	–	0.223	30.8	0.259	–	10.8	10.5	0.19	–	–	–	–	–	–	0.19	10.5
0.166	0.19	0.063	31.7	–	0.238	31	0.299	–	13.3	13.1	0.176	–	–	–	–	–	–	0.176	13.1
0.164	0.188	0.063	33.1	–	0.329	32.5	0.149	–	11.4	11.1	0.249	–	–	–	–	–	–	0.249	11.1
0.164	0.188	0.063	33.1	–	0.329	32.5	.0149	–	11.4	11.1	0.249	–	–	–	–	–	–	0.249	11.1
0.166	0.19	0.063	31.7	–	0.238	31	0.299	–	13.3	13.1	0.176	–	–	–	–	–	–	0.176	13.1
0.117	0.331	0.108	24.6	–	0.009	24	0.667	–	15.7	15.7	0.003	–	–	–	–	–	–	0.003	15.7
0.123	0.349	0.114	24.7	–	0.009	24	0.667	–	15.7	15.7	0.003	–	–	–	–	–	–	0.003	15.7
0.117	0.331	0.108	24.5	–	0.009	23.8	0.662	–	15.6	15.6	0.003	–	–	–	–	–	–	0.003	15.6
0.114	0.325	0.106	24.5	–	0.009	23.8	0.662	–	15.6	15.6	0.003	–	–	–	–	–	–	0.003	15.6
0.137	0.389	0.076	22.1	–	0.044	21.5	0.517	–	17	17	0.01	–	–	–	–	–	–	0.01	17
0.135	0.384	0.126	22.1	–	0.044	21.5	0.517	–	17	17	0.01	–	–	–	–	–	–	0.01	17
0.141	0.4	0.078	23.1	–	0.046	22.5	0.541	–	17.8	17.8	0.01	–	–	–	–	–	–	0.01	17.8
0.142	0.402	0.131	23.1	–	0.046	22.5	0.541	–	17.8	17.8	0.01	–	–	–	–	–	–	0.01	17.8
0.144	0.41	0.134	25.2	–	0.053	24.6	0.542	–	14.7	14.6	0.019	–	–	–	–	–	–	0.019	14.6
0.144	0.409	0.134	25.2	–	0.053	24.6	0.542	–	14.7	14.6	0.019	–	–	–	–	–	–	0.019	14.6
0.119	0.339	0.12	24	–	0.009	23.3	0.647	–	15.2	15.2	0.003	–	–	–	–	–	–	0.003	15.2
0.054	0.153	0.05	10.9	–	0.04	10.6	0.296	–	6.96	6.95	0.001	–	–	–	–	–	–	0.001	6.95
0.112	0.319	0.12	24.1	–	0.009	23.4	0.65	–	15.3	15.3	0.003	–	–	–	–	–	–	0.003	15.3
0.07	1.3	0.102	25.3	–	–	23.9	0.5	–	12.7	12.7	–	–	–	–	–	–	–	0.003	15.6
0.137	0.389	0.127	22.3	–	0.045	21.8	0.523	–	17.2	17.2	0.01	–	–	–	–	–	–	0.01	17.2
0.131	0.371	0.072	21.4	–	0.043	20.9	0.502	–	16.5	16.5	0.01	–	–	–	–	–	–	0.01	16.5
0.144	0.41	0.134	25.3	–	0.053	24.7	0.544	–	14.7	14.7	0.019	–	–	–	–	–	–	0.019	14.7
–	–	–	42.2	–	0.31	41.2	0.465	–	16.7	16	0.677	–	–	–	–	–	–	0.677	16
–	–	–	42.2	–	0.31	41.2	0.465	–	16.7	16	0.677	–	–	–	–	–	–	0.677	16
–	–	–	42.2	–	0.31	41.2	0.465	–	16.7	16	0.677	–	–	–	–	–	–	0.677	16
–	–	–	42.2	–	0.31	41.2	0.465	–	16.7	16	0.677	–	–	–	–	–	–	0.677	16
–	–	–	40.3	–	0.296	39.3	0.444	–	16	15.3	0.647	–	–	–	–	–	–	0.647	15.3

Table 13.2 (*Continued*)

Item	H_2O	Ash	Cals.	Prot.	Carb.	Fat-T	Fat-S	4:0	6:0	8:0	10:0	12:0	14:0	15:0	16:0	17:0	18:0
Dry roasted pecans, unsalted	1.11	4.01	659	7.98	22.3	64.6	5.18	–	–	–	–	–	–	–	3.95	–	0.957
Large dried pecans	4.83	1.57	667	7.76	18.2	67.6	5.42	–	–	–	–	–	–	–	4.14	–	1
Oil roasted pecans	4.21	1.61	685	6.96	16.1	71.2	5.7	–	–	–	–	–	–	–	4.36	–	1.06
Oil roasted pecans, halves	4.21	1.61	685	6.96	16.1	71.2	5.7	–	–	–	–	–	–	–	4.36	–	1.06
Oil roasted pecans, unsalted	4.21	1.61	685	6.96	16.1	71.2	5.7	–	–	–	–	–	–	–	4.36	–	1.06
Dried pilinuts/canarytree	2.78	2.92	719	10.8	3.99	79.6	31.2	–	–	–	–	–	–	–	22	–	9.15
Dried pine nuts (Pignoia)	6.7	4.41	515	24	14.2	50.7	7.8	–	–	–	–	–	–	–	3.67	–	1.67
Dried pine nuts (pinon/pinyon)	5.91	2.27	568	11.6	19.3	61	9.38	–	–	–	–	–	–	–	4.42	–	2.01
Dried pistachio nuts	3.88	2.41	577	20.6	24.8	48.4	6.13	–	–	–	–	–	0.045	–	5.35	–	0.628
Dry roasted pistachio nuts	2.1	2.64	606	14.9	27.5	52.8	6.69	–	–	–	–	–	0.05	–	5.84	–	6.85
Dry roasted pistachio nuts, unsalted	2.1	2.64	606	14.9	27.5	52.8	6.69	–	–	–	–	–	0.05	–	5.84	–	0.685
Soy nuts/peanuts	4.24	4.26	453	37	30.6	24	3.3	–	–	–	–	0.012	0.024	–	2.27	–	0.91
Dried black walnuts, chopped	4.37	2.62	607	24.4	12.1	56.6	3.63	–	–	–	–	–	–	–	2.13	–	1.39
Dried ground black walnuts	4.37	2.62	607	24.4	12.1	56.6	3.63	–	–	–	–	–	–	–	2.13	–	1.39
Dried English walnuts, halves	3.66	1.87	642	14.3	18.3	61.9	5.59	–	–	–	–	–	0.189	–	4.24	–	1.08
Dried English walnuts, chopped	3.66	1.87	642	14.3	18.3	61.9	5.59	–	–	–	–	–	0.189	–	4.24	–	1.08
Dried English walnuts, halves, cup meas	3.66	1.87	642	14.3	18.3	61.9	5.59	–	–	–	–	–	0.189	–	4.24	–	1.08
Chia seeds, dried	5.31	3.97	472	16.6	47.9	26.3	10.5	–	–	–	–	–	–	–	6.88	–	3.09
Cottonseed kernels, roasted	4.66	4.59	506	32.6	21.9	36.3	9.7	–	–	–	–	–	0.341	–	8.43	–	0.853
Ethiopian flaxseed cake	4.1	3.4	524	18.6	36.9	37	–	–	–	–	–	–	–	–	–	–	–
Flaxseed (Canada), flaxseed council	5	4.5	450	21	–	45	–	–	–	–	–	–	–	–	–	–	–
Flaxseed, dried	6.3	4.5	498	18	37.2	34	–	–	–	–	–	–	–	–	–	–	–
Linseed, raw SFK	6.1	–	343	24.4	38.6	30.9	–	–	–	–	–	–	–	–	–	–	–
Poppyseed	6.78	6.78	533	18	23.7	44.7	4.87	–	–	–	–	–	–	–	4.11	–	0.67
Dry pumpkin kernels	6.93	4.89	541	24.5	17.8	45.9	8.67	–	–	–	–	0.044	0.052	–	5.61	–	2.81
Roasted pumpkin kernels	7.11	4.38	522	33	13.4	42.1	7.97	–	–	–	–	0.04	0.048	–	5.16	–	2.58
Roasted pumpkin kernels, unsalted	7.11	4.38	522	33	13.4	42.1	7.97	–	–	–	–	0.04	0.048	–	5.16	–	2.58
Roasted pumpkin seeds, unsalted	4.51	3.81	446	18.6	53.8	19.4	3.67	–	–	–	–	0.019	0.022	–	2.37	–	1.19
Roasted pumpkin/squash seeds	4.51	3.81	446	18.6	53.8	19.4	3.67	–	–	–	–	0.019	0.022	–	2.37	–	1.19
Safflower seed kernels, dried	5.63	5.48	517	16.2	34.3	38.5	3.68	–	–	–	–	–	0.035	–	2.57	–	0.921
Dried sesame seeds, kernels	4.82	4.65	588	26.4	9.4	54.8	7.67	–	–	–	–	–	0.136	–	4.9	–	2.31
Dried sesame seeds, whole	4.7	4.46	573	17.7	23.5	49.7	6.96	–	–	–	–	–	0.124	–	4.44	–	2.09
Dry sesame seeds, hulled	4.81	4.64	588	26.4	9.4	54.8	7.67	–	–	–	–	–	0.136	–	4.9	–	2.31
Roasted toasted sesame seeds, whole	3.31	6.01	565	17	25.7	48	6.72	–	–	–	–	–	0.12	–	4.29	–	2.02
Sesame seeds, mechanically hulled AM	5	12.6	575	20.1	13.4	48.9	6.96	–	–	–	–	–	–	–	–	–	–
Toasted sesame seeds, salted	5.01	4.01	567	17	26	48	6.72	–	–	–	–	–	0.12	–	4.29	–	2.02
Toasted sesame seeds, no salt added	5.01	4.01	567	17	26	48	6.72	–	–	–	–	–	0.12	–	4.29	–	2.02
Roasted soybeans, unsalted kernels	4.24	4.26	453	37	30.6	24	3.3	–	–	–	–	0.012	0.024	–	2.27	–	0.91
Roasted toasted soybeans	4.24	4.26	453	37	30.6	24	3.3	–	–	–	–	0.012	0.024	–	2.27	–	0.91
Dry roasted sunflower seeds	1.21	5.61	582	19.3	24.1	49.8	5.22	–	–	–	–	–	0.051	–	2.81	–	2.21
Dry sunflower seeds	5.37	3.54	570	22.8	18.8	49.6	5.2	–	–	–	–	–	0.051	–	2.81	–	2.2
Dry roasted sunflower seed kernels, salted	1.21	5.61	582	19.3	24.1	49.8	5.22	–	–	–	–	–	0.051	–	2.81	–	2.21

20:0	22:0	24:0	Fat-M	14:1	16:1	18:1	20:1	22:1	Fat-P	18:2	18:3	18:4	20:4	20:5	22:5	22:6	Chol., mg	Omeg3	Omeg6
–	–	–	40.3	–	0.296	39.3	0.444	–	16	15.3	0.647	–	–	–	–	–	–	0.647	15.3
–	–	–	42.2	–	0.31	41.2	0.465	–	16.7	16	0.677	–	–	–	–	–	–	0.677	16
–	–	–	44.4	–	0.326	43.3	0.489	–	17.6	16.8	0.713	–	–	–	–	–	–	0.713	16.8
–	–	–	44.4	–	0.326	43.3	0.489	–	17.6	16.8	0.713	–	–	–	–	–	–	0.713	16.8
–	–	–	44.4	–	0.326	43.3	0.489	–	17.6	16.8	0.713	–	–	–	–	–	–	0.713	16.8
–	–	–	37.2	–	–	37.2	–	–	7.61	7.61	–	–	–	–	–	–	–	–	7.61
–	–	–	19.1	–	0.208	17.9	0.968	–	21.3	20.7	0.654	–	–	–	–	–	–	0.654	20.7
–	–	–	22.9	–	0.25	21.5	1.16	–	25.7	24.9	0.787	–	–	–	–	–	–	0.787	24.9
0.107	–	–	32.7	–	0.286	32	0.22	–	7.32	7.02	0.252	–	–	–	–	–	–	0.252	7.02
0.115	–	–	35.7	–	0.313	34.9	0.24	–	7.99	7.67	0.275	–	–	–	–	–	–	0.275	7.66
0.115	–	–	35.7	–	0.313	34.9	0.24	–	7.99	7.67	0.275	–	–	–	–	–	–	0.275	7.67
0.07	0.012	–	5.58	–	0.048	5.49	0.044	–	12.7	12.2	1.5	–	–	–	–	–	–	1.5	12.2
–	–	–	12.7	–	0.077	11.9	0.548	–	37.5	33.5	3.31	–	–	–	–	–	–	3.31	33.4
–	–	–	12.7	–	0.077	11.9	0.548	–	37.5	33.5	3.31	–	–	–	–	–	–	3.31	33.4
0.081	–	–	14.2	–	0.161	13.3	0.52	–	39.1	31.8	6.81	–	0.024	–	–	–	–	6.81	31.8
0.081	–	–	14.2	–	0.161	13.3	0.52	–	39.1	31.8	6.81	–	0.024	–	–	–	–	6.81	31.8
0.081	–	–	14.2	–	0.161	13.3	0.52	–	39.1	31.8	6.81	–	0.024	–	–	–	–	6.81	31.8
–	–	–	7.26	–	1.73	5.32	–	–	7.28	3.21	3.87	–	–	–	–	–	–	3.87	3.21
–	–	–	6.92	–	0.256	6.61	–	–	18.1	17.9	0.069	–	–	–	–	–	–	0.069	17.9
–	–	–	–	–	–	–	–	–	–	–	–	–	–	–	–	–	–	–	–
–	–	–	–	–	–	–	–	–	–	–	–	–	–	–	–	–	–	–	–
–	–	–	–	–	–	–	–	–	–	–	–	–	–	–	–	–	–	–	–
–	–	–	–	–	–	–	–	–	–	–	–	–	–	–	–	–	–	–	–
–	–	–	6.34	–	0.13	6.17	0.04	–	30.8	30.5	0.33	–	–	–	–	–	–	0.33	30.5
–	–	–	14.3	–	0.099	14.1	–	–	20.9	20.7	0.181	–	–	–	–	–	–	0.181	20.7
–	–	–	13.1	–	0.091	13	–	–	19.2	19	0.166	–	–	–	–	–	–	0.166	19
–	–	–	13.1	–	0.091	13	–	–	19.2	19	0.166	–	–	–	–	–	–	0.166	19
–	–	–	6.03	–	0.042	5.99	–	–	8.84	8.76	0.077	–	–	–	–	–	–	0.077	8.76
–	–	–	6.03	–	0.042	5.99	–	–	8.84	8.76	0.77	–	–	–	–	–	–	0.077	8.76
–	–	–	4.85	–	0.037	4.81	–	–	28.2	28.1	0.111	–	–	–	–	–	–	0.111	28.1
0.324	–	–	20.7	–	0.164	20.4	0.077	–	24	23.6	0.415	–	–	–	–	–	–	0.415	23.6
0.302	–	–	18.8	–	0.149	18.5	0.07	–	21.8	21.4	0.376	–	–	–	–	–	–	0.376	21.4
0.324	–	–	20.7	–	0.164	20.4	0.077	–	24	23.6	0.415	–	–	–	–	–	–	0.415	23.6
0.29	–	–	18.1	–	0.144	17.9	0.067	–	21	20.7	0.363	–	–	–	–	–	–	0.363	20.7
–	–	–	18.8	–	–	–	–	–	21.8	–	–	–	–	–	–	–	–	–	–
0.29	–	–	18.1	–	0.144	17.9	0.067	–	21	20.7	0.363	–	–	–	–	–	–	0.363	20.7
0.29	–	–	18.1	–	0.144	17.9	0.067	–	21	20.7	0.363	–	–	–	–	–	–	0.363	20.7
0.07	0.012	–	5.58	–	0.048	5.49	0.044	–	12.7	12.2	1.5	–	–	–	–	–	–	1.5	12.2
0.07	0.012	–	5.58	–	0.048	5.49	0.044	–	12.7	12.2	1.5	–	–	–	–	–	–	1.5	12.2
0.149	–	–	9.51	–	0.049	9.4	0.048	–	32.9	32.8	0.069	–	–	–	–	–	–	0.069	32.8
0.14	–	–	9.46	–	0.049	9.36	0.048	–	32.7	32.6	0.069	–	–	–	–	–	–	0.069	32.6
0.149	–	–	9.51	–	0.049	9.4	0.048	–	32.9	32.8	0.069	–	–	–	–	–	–	0.069	32.8

Table 13.2 *(Continued)*

Item	H_2O	Ash	Cals.	Prot.	Carb.	Fat-T	Fat-S	4:0	6:0	8:0	10:0	12:0	14:0	15:0	16:0	17:0	18:0
Oil roasted sunflower seeds	2.61	3.88	615	21.4	14.7	57.5	6.02	–	–	–	–	–	0.059	–	3.24	–	2.55
Oil roasted sunflower seed kernels, no salt	2.61	3.88	615	21.4	14.7	57.5	6.02	–	–	–	–	–	0.059	–	3.24	–	2.55
Toasted sunflower seed kernels, salted	1.01	4.41	619	17.2	20.6	56.8	5.95	–	–	–	–	–	0.058	–	3.2	–	2.52
Toasted sunflower seed kernels, no salt	1.01	4.41	619	17.2	20.6	56.8	5.95	–	–	–	–	–	0.058	–	3.2	–	2.52
Watermelon seeds, dried kernels	5.06	3.95	557	28.3	15.3	47.4	9.78	–	–	–	–	–	–	–	5.41	–	4.3
Almond butter, honey and cinnamon with salt	2.01	3.01	602	15.8	27	52.2	4.95	–	–	–	–	0.005	0.298	–	3.45	–	1.02
Almond butter with honey and cinnamon	2.01	3.01	602	15.8	27	52.2	4.95	–	–	–	–	0.005	0.298	–	3.45	–	1.02
Almond butter, plain	1.01	3.61	633	15.1	21.2	59.1	5.6	–	–	–	–	0.005	0.337	–	3.91	–	1.16
Almond butter, salted	1.01	3.61	633	15.1	21.1	59.1	5.6	–	–	–	–	0.005	0.337	–	3.91	–	1.16
Almond paste, packed	15.5	1.91	446	11.9	43.6	27.2	2.58	–	–	–	–	0.002	0.155	–	1.8	–	0.532
Cashew butter, salted	2.97	2.51	587	17.6	27.6	49.4	9.76	–	–	0.14	0.14	0.836	0.37	–	4.64	–	3.17
Cashew butter, unsalted	2.97	2.51	587	17.6	27.6	49.4	9.76	–	–	0.14	0.14	0.836	0.37	–	4.64	–	3.17
Chunky peanut butter	1.14	3.31	589	24.1	21.6	49.9	9.58	–	–	–	–	0.022	0.05	–	5.5	–	2.14
Chunky peanut butter, unsalted	1.13	3.67	589	24	21.6	49.6	9.57	–	–	–	–	0.022	0.05	–	5.5	–	2.14
Natural peanut butter, salted	1.5	3.3	584	24	21.5	49.7	6.91	–	–	–	–	0.02	0.025	–	5.2	–	1.11
Natural peanut butter, unsalted	1.5	3.3	584	24	21.5	49.7	6.89	–	–	–	–	0.02	0.025	–	5.2	–	1.11
Smooth peanut butter	1.43	3.3	588	24.6	20.7	50	9.59	–	–	–	–	0.022	0.05	–	5.5	–	2.14
Smooth peanut butter, unsalted	1.2	3.3	598	27.1	16	52.4	9.59	–	–	–	–	0.022	0.05	–	5.5	–	2.14
Sesame butter paste	1.61	5.41	595	18.1	25.5	50.9	7.12	–	–	–	–	–	0.127	–	4.55	–	2.14
Sunflower seed butter, salted	1.24	3.96	579	19.7	27.4	47.7	5	–	–	–	–	–	0.049	–	2.69	–	2.12
Sunflower seed butter	1.24	3.96	579	19.7	27.4	47.7	5	–	–	–	–	–	0.049	–	2.69	–	2.12
Tahini (sesame butter) fresh unroasted kernels	3.01	4.74	607	18	17.9	56.4	7.9	–	–	–	–	–	0.141	–	5.05	–	2.37
Tahini fresh raw and stone ground kernels	3.01	5.01	570	17.8	26.2	48	6.72	–	–	–	–	–	0.12	–	4.29	–	2.02
Tahini (fresh roasted and toasted kernels)	3.06	5.01	595	17	21.2	53.8	7.53	–	–	–	–	–	0.134	–	4.81	–	2.26
Coconut cream, canned	71.2	0.04	192	2.7	8.36	17.7	15.7	–	0.101	1.24	0.986	7.86	3.1	–	1.5	–	0.918
Coconut cream, raw	53.9	1.16	330	3.64	6.66	34.7	30.8	–	0.197	2.43	1.93	15.4	6.08	–	2.94	–	1.8
Coconut milk, canned	72.9	0.97	197	2.03	2.82	21.3	18.9	–	0.121	1.49	1.19	9.46	3.74	–	1.81	–	1.1
Coconut milk, frozen	71.4	0.59	202	1.62	5.59	20.8	18.4	–	0.118	1.46	1.16	9.23	3.64	–	1.76	–	1.08
Coconut milk, raw	67.6	0.72	230	2.3	5.55	23.8	21.1	–	0.136	1.67	1.33	10.6	4.18	–	2.02	–	1.23
Acorn flour, full fat	6.04	1.7	501	7.5	54.7	30.2	3.92	–	–	–	–	–	–	–	3.61	–	0.319
Almond meal, partially defatted with salt	7.21	6.11	408	39.5	28.9	18.3	1.74	–	–	–	–	0.002	0.104	–	1.21	–	0.358
Almond nut meal, partially defatted	7.21	6.11	408	39.5	28.9	18.3	1.74	–	–	–	–	0.002	0.104	–	1.21	–	0.358
Almond powder, full fat, not packed	3.34	2.84	592	19.8	22.4	51.7	4.9	–	–	–	–	0.005	0.295	–	3.42	–	1.01
Almond powder, partially defatted, not packed	9.41	5.31	393	37.5	31.8	16	1.52	–	–	–	–	0.001	0.091	–	1.06	–	0.313
Low Fat peanut flour	7.81	5.24	428	33.8	31.8	21.9	3.04	–	–	–	–	–	0.011	–	2.29	–	0.489
High fat sesame flour	0.9	4.61	526	30.8	26.6	37.1	5.2	–	–	–	–	–	0.092	–	3.32	–	1.56
Sesame meal, partially defatted	5.01	4.01	567	17	26	48	6.72	–	–	–	–	–	0.12	–	4.29	–	2.02

20:0	22:0	24:0	Fat-M	14:1	16:1	18:1	20:1	22:1	Fat-P	18:2	18:3	18:4	20:4	20:5	22:5	22:6	Chol., mg	Omeg3	Omeg6
0.171	–	–	11	–	0.057	10.8	0.055	–	37.9	37.8	0.08	–	–	–	–	–	–	0.08	37.8
0.171	–	–	11	–	0.057	10.8	0.055	–	37.9	37.8	0.08	–	–	–	–	–	–	0.08	37.8
0.172	–	–	10.8	–	0.056	10.7	0.055	–	37.5	37.4	0.079	–	–	–	–	–	–	0.079	37.4
0.172	–	–	10.8	–	0.056	10.7	0.055	–	37.5	37.4	0.079	–	–	–	–	–	–	0.079	37.4
–	–	–	7.41	–	0.09	7.32	–	–	28.1	28.1	–	–	–	–	–	–	–	–	28.1
0.177	–	–	33.9	–	0.304	33.3	0.05	–	11	10.5	0.374	–	–	–	–	–	–	0.374	10.5
0.177	–	–	33.9	–	0.304	33.3	0.05	–	11	10.5	0.374	–	–	–	–	–	–	0.374	10.5
0.188	–	–	38.4	–	0.344	37.7	0.056	–	12.4	11.9	0.423	–	–	–	–	–	–	0.423	11.9
0.188	–	–	38.4	–	0.344	37.7	0.056	–	12.4	11.9	0.423	–	–	–	–	–	–	0.423	11.9
0.091	–	–	17.7	–	0.158	17.4	0.026	–	5.71	5.47	0.195	–	–	–	–	–	–	0.195	5.47
0.464	–	–	29.1	–	0.339	28.6	0.148	–	8.35	8.17	0.171	–	–	–	–	–	–	0.171	8.17
0.464	–	–	29.1	–	0.339	28.6	0.148	–	8.35	8.17	0.171	–	–	–	–	–	–	0.171	8.17
–	0.1	0.1	23.6	–	–	22.9	0.617	–	14.4	14.1	0.078	–	0.187	–	–	–	–	0.078	14.3
–	0.1	–	23.6	–	–	22.9	0.617	–	14.4	14.1	0.078	–	0.187	–	–	–	–	0.078	14.3
–	0.1	0.4	24.7	–	0.009	24	0.667	–	15.7	15.7	0.003	–	–	–	–	–	–	0.003	15.7
–	0.1	0.4	24.7	–	0.009	24	0.667	–	15.7	15.7	0.003	–	–	–	–	–	–	0.003	15.7
–	–	–	23.6	–	–	23	0.618	–	14.4	14.1	0.078	–	0.187	–	–	–	–	0.078	14.3
–	0.1	–	23.6	–	–	23	0.618	–	14.4	14.1	0.078	–	0.187	–	–	–	–	0.078	14.3
0.303	–	–	19.2	–	0.153	19	0.071	–	22.3	21.9	0.385	–	–	–	–	–	–	0.385	21.9
0.141	–	–	9.11	–	0.047	9.01	0.046	–	31.5	31.4	0.066	–	–	–	–	–	–	0.066	31.4
0.141	–	–	9.11	–	0.047	9.01	0.046	–	31.5	31.4	0.066	–	–	–	–	–	–	0.066	31.4
0.339	–	–	21.3	–	0.169	21	0.079	–	24.7	24.3	0.427	–	–	–	–	–	–	0.427	24.3
0.29	–	–	18.1	–	0.144	17.9	0.067	–	21	20.7	0.363	–	–	–	–	–	–	0.363	20.7
0.326	–	–	20.3	–	0.161	20	0.075	–	23.6	23.1	0.407	–	–	–	–	–	–	0.407	23.1
–	–	–	0.754	–	–	0.754	–	–	0.194	0.194	–	–	–	–	–	–	–	–	0.194
0.023	–	–	1.48	–	–	1.48	–	–	0.379	0.379	–	–	–	–	–	–	–	–	0.379
–	–	–	0.907	–	–	0.907	–	–	0.233	0.233	–	–	–	–	–	–	–	–	0.233
–	–	–	0.885	–	–	0.885	–	–	0.228	0.228	–	–	–	–	–	–	–	–	0.228
–	–	–	1.01	–	–	1.01	–	–	0.261	0.261	–	–	–	–	–	–	–	–	0.261
–	–	–	19.1	–	–	19.1	–	–	5.81	5.81	–	–	–	–	–	–	–	–	5.81
0.066	–	–	11.9	–	0.107	11.7	0.017	–	3.84	3.68	0.131	–	–	–	–	–	–	0.131	3.68
0.066	–	–	11.9	–	0.107	11.7	0.017	–	3.84	3.68	0.131	–	–	–	–	–	–	0.131	3.68
0.17	–	–	33.5	–	0.301	32.9	0.049	–	10.8	10.4	0.37	–	–	–	–	–	–	0.37	10.4
0.055	–	–	10.4	–	0.093	10.2	0.015	–	3.36	3.22	0.115	–	–	–	–	–	–	0.115	3.22
–	–	–	10.9	–	0.004	10.6	0.294	–	6.92	6.92	0.001	–	–	–	–	–	–	0.001	6.92
0.228	–	–	14	–	0.111	13.8	0.052	–	16.3	16	0.281	–	–	–	–	–	–	0.281	16
0.29	–	–	18.1	–	0.144	17.9	0.067	–	21	20.7	0.363	–	–	–	–	–	–	0.363	20.7

Table 13.3 Lipid composition of animal fats and related products, contents/100 g

Item	H_2O	Ash	Cals.	Prot.	Carb.	Fat-T	Fat-S	4:0	6:0	8:0	10:0	12:0	14:0	15:0	16:0	17:0	18:0
Bacon/pork fat, cooked	–	–	902	–	–	100	45.4	–	–	–	0.116	0.232	1.51	–	27.6	0.206	15.6
Beef fat drippings	–	–	902	–	–	100	50.4	–	–	–	–	0.9	3.7	0.668	24.9	1.34	18.9
Beef fat, all cuts, raw	20.3	0.28	674	8.25	–	71.2	29.5	–	–	–	0.31	0.21	2.4	0.37	17.7	0.74	8.82
Beef fat, retail cuts, cooked	18.6	0.45	680	10.7	–	70.3	28.5	–	–	–	0.27	0.25	2.36	–	17.8	–	7.87
Beef suet raw	4.01	0.1	854	1.51	–	94	52.3	–	–	–	–	0.07	2.81	–	22.6	–	24.7
Beef tallow	–	–	902	–	–	100	49.8	–	–	–	–	0.9	3.71	–	24.9	–	18.9
Chicken fat	28.9	0.28	629	3.74	–	68	20.3	–	–	–	–	0.04	0.6	–	14.7	–	4.8
Duck fat	0.2	–	900	–	–	99.8	33.2	–	–	–	–	–	0.7	–	24.7	–	7.8
Fresh pork fat, cooked	23.2	0.75	629	12.2	–	63.9	24.5	–	–	–	0.05	0.05	0.85	–	14.9	–	8.17
Fresh pork fat, raw	24.9	0.25	638	6.4	–	68.4	23.5	–	–	–	0.05	0.07	0.84	–	14.5	0.107	7.87
Frying shortening, beef tallow and cottonseed	–	–	900	–	–	100	44.9	–	–	–	–	–	3.4	–	24.5	–	17
Goose fat	0.2	–	900	–	–	99.8	27.7	–	–	–	–	–	0.5	–	20.7	–	6.2
Hard margarine, lard (hydrogenated)	15.7	2	733	0.9	0.9	80.5	31.6	–	–	–	0.1	0.2	0.1	–	19.1	–	10.8
Household shortening, lard and vegetable oil	–	–	900	–	–	100	40.3	–	–	–	0.1	0.1	1.6	–	23	–	15.2
Industrial shortening, lard and vegetable oil	–	–	900	–	–	100	35.7	–	–	–	–	0.2	1.2	–	19.2	–	14.6
Lamb fat, choice, raw	22.5	0.36	665	6.65	–	70.5	32.2	–	–	–	0.2	0.34	3.08	–	16	–	10.1
Lamb fat, cooked	26.1	0.68	586	12.2	–	59.2	27	–	–	–	0.17	0.29	2.59	–	13.4	–	8.44
Lard (pork fat)	–	–	902	–	–	100	39.2	–	–	–	0.1	0.2	1.3	–	23.8	0.178	13.5
Mono + diglycerides, beef tallow	–	–	850	–	–	100	49.8	–	–	–	–	0.9	3.7	–	24.9	–	18.9
Mono + diglycerides, pork lard	–	–	850	–	–	100	39.2	–	–	–	0.1	0.2	1.3	–	23.8	–	13.5
Mutton tallow/fat	–	–	902	–	–	100	47.6	–	–	–	–	–	4.3	0.541	21.9	0.906	20
NZ Lamb fat, cooked	26.1	0.71	586	9.73	–	60.4	31.5	–	–	–	0.19	0.26	2.76	–	13.5	–	12.4
NZ Lamb fat, raw	25.6	0.42	640	6.93	–	67.6	35.3	–	–	–	0.21	0.29	3.09	–	15.1	–	13.9
Rendered turkey fat	0.2	–	900	–	–	99.8	29.4	–	–	–	–	–	0.9	–	20.6	–	7.5
Schmaltz, rendered chicken fat	0.2	–	900	–	–	99.8	29.8	–	–	–	–	0.1	0.9	–	21.6	–	7
Separable fat from ham and ArmPicnic, roasted	28.9	1.69	591	7.65	–	61.9	22.7	–	–	–	–	0.11	0.82	–	14.1	–	7.52
Separable fat from ham and ArmPicnic, unheated	31.1	1.78	579	5.69	0.09	61.4	22.5	–	–	–	–	0.11	0.82	–	14	–	7.47
Turkey fat	30	0.556	538	17.6	–	51.3	13.4	–	–	–	–	–	0.389	–	9.67	–	2.57
Veal fat, cooked	21.7	0.82	642	9.43	–	66.7	32.4	–	–	–	0.09	0.26	3.65	–	16.6	–	10.5
Veal fat, raw	25	0.44	638	6.09	–	68.4	32.9	–	–	–	0.09	0.26	3.71	–	16.9	–	10.6
Beef armroast, prime, lean + fat, 1/2″ trim, braised	43	0.91	391	26.1	–	31	12.7	–	–	–	0.09	0.08	0.97	–	7.5	–	3.53
Beef blade, choice, lean + fat, 1/2″ trim, braised	43	0.84	388	25.4	–	31	12.9	–	–	–	0.08	0.08	0.99	–	7.55	–	3.69
Beef blade, prime, lean + fat, 1/2″ trim, braised	41.4	0.85	417	25.5	–	34.1	14.2	–	–	–	0.08	0.08	1.09	–	8.24	–	4.1
Beef blade, prime, lean + fat, 1/2″ trim, raw	54.7	0.77	328	16.3	–	28.6	12.2	–	–	–	0.09	0.06	0.9	–	6.9	–	3.56
Beef blade roast, all, lean + fat, 1/2″ trim, braised	43.3	0.84	383	25.4	–	30.4	12.7	–	–	–	0.08	0.08	0.97	–	7.4	–	3.61
Beef blade, select, lean + fat, 1/2″ trim, braised	44.5	0.85	366	25.8	–	28.4	11.8	–	–	–	0.08	0.07	0.91	–	6.91	–	3.37
Beef breakfast strips, raw	45.2	2.81	406	12.5	0.7	38.8	16	–	–	–	0.16	0.11	1.32	–	8.49	–	5.04
Beef brisket, all, lean + fat, 1/2″ trim, raw	54.5	0.77	325	16.6	–	28.2	11.9	–	–	–	0.1	0.07	0.91	–	6.82	–	3.36
Beef brisket, all, lean + fat, 1/4″ trim, braised	44.8	0.85	385	23.5	–	31.6	12.4	–	–	–	0.09	0.08	1.04	–	7.71	–	3.48
Beef brisket, flat half, lean + fat, 1/2″ trim, braised	44.9	0.89	366	24.7	–	28.9	11.8	–	–	–	0.09	0.09	0.93	–	7.09	–	3.16
Beef brisket, flat, all, lean + fat, 1/4″ trim, braised	46.4	0.9	364	25.1	–	28.5	11	–	–	–	0.08	0.08	0.9	–	6.88	–	3.11
Beef brisket, lean + fat, 1/2″ trim, braised	43	0.83	391	23	–	32.4	13.2	–	–	–	0.1	0.09	1.06	–	7.87	–	3.53

20:0	22:0	24:0	Fat-M	14:1	16:1	18:1	20:1	22:1	Fat-P	18:2	18:3	18:4	20:4	20:5	22:5	22:6	Chol., mg	Omeg3	Omeg6
–	–	–	42.5	0.005	2.64	38.8	0.94	–	7.67	6.99	0.685	–	–	–	–	–	102	0.685	6.99
–	–	–	41.8	1.23	4.2	36	0.3	–	4	3.1	0.6	–	0.3	–	–	–	109	0.6	3.1
–	–	–	30.9	–	3.89	27	0.11	–	2.57	1.5	1.08	–	–	–	–	–	99	1.08	1.5
–	–	–	30.5	–	3.15	27.1	0.17	–	2.67	1.63	1.05	–	–	–	–	–	95	1.05	1.63
–	–	–	31.5	–	2.19	28.9	–	–	3.18	2.16	0.86	–	–	–	–	–	68	0.86	2.16
–	–	–	41.8	–	4.21	36	0.3	–	4.01	3.11	0.6	–	–	–	–	–	109	0.6	3.11
–	–	–	30.3	–	3.87	25.3	0.73	–	14.2	13.3	0.7	–	0.04	–	–	–	58	0.7	13.3
–	–	–	49.3	–	4	44.2	1.1	–	12.9	12	1	–	–	–	–	–	100	1	12
–	–	–	27.4	–	1.67	25	0.45	–	6.38	5.79	0.22	–	0.12	–	–	–	93	0.22	5.91
–	–	–	29.9	0.003	1.82	27.6	0.52	–	7.21	6.11	0.59	–	0.17	–	–	–	93	0.59	6.28
–	–	–	38.5	–	4.3	34.2	–	–	8.8	8.3	0.5	–	–	–	–	–	100	0.5	8.3
–	–	–	56.7	–	2.8	53.8	0.1	–	11	9.8	1.2	–	–	–	–	–	100	0.5	9.8
–	–	–	37.8	–	2.2	34.7	0.8	–	7.5	7.1	0.4	–	–	–	–	–	51	0.4	7.1
–	–	–	44.4	–	3.1	40.9	–	–	10.9	9.7	1.1	–	–	–	–	–	56	1.1	9.7
–	–	–	40.7	–	1.4	38	0.6	–	19.2	18.1	1	–	–	–	–	–	56	1	18.1
–	–	–	29.1	–	2.04	26.1	–	–	5.36	3.9	1.35	–	0.12	–	–	–	90	1.35	4.02
–	–	–	24.4	–	1.71	21.9	–	–	4.49	3.27	1.13	–	0.1	–	–	–	114	1.13	3.37
–	–	–	45.1	0.005	2.8	41.2	1	–	11.2	10.2	1	–	0.17	–	–	–	95	1	10.2
–	–	–	41.8	–	4.2	36	0.3	–	4	3.1	0.6	–	–	–	–	–	109	0.6	3.1
–	–	–	45.1	–	2.7	41.2	1	–	11.2	10.2	1	–	–	–	–	–	95	1	10.2
–	–	–	41.6	–	2.3	37.6	–	–	7.8	5.5	2.3	–	–	–	–	–	102	2.3	5.5
–	–	–	23.2	–	1.04	21.6	0.11	–	2.53	1.4	1.14	–	–	–	–	–	109	1.14	1.4
–	–	–	25.9	–	1.16	24.2	0.13	–	2.84	1.57	1.28	–	–	–	–	–	87	1.28	1.57
–	–	–	42.9	–	5	37.9	–	–	23.1	21.2	1.4	–	0.3	–	–	–	102	1.4	21.5
–	–	–	44.7	–	5.7	37.3	1.1	–	20.9	19.5	1	–	0.1	–	–	–	85	1	19.6
–	–	–	29.7	–	1.83	27.5	–	–	6.58	5.77	0.64	–	0.14	–	–	–	86	0.64	5.91
–	–	–	29.5	–	1.81	27.3	–	–	6.53	5.73	0.63	–	0.13	–	–	–	68	0.63	5.86
–	–	–	21.8	–	3.79	17.6	–	–	11.8	10.6	0.8	–	0.14	–	–	–	126	0.8	10.8
–	–	–	28	–	3.41	23.7	–	–	3.24	2.67	0.57	–	–	–	–	–	73	0.57	2.67
–	–	–	28.4	–	3.47	24.1	–	–	3.29	2.71	0.58	–	–	–	–	–	73	0.58	2.71
–	–	–	13.9	–	1.32	11.9	0.06	–	1.21	0.79	0.35	–	0.04	–	–	–	99	0.35	0.83
–	–	–	14	–	1.28	12.1	0.06	–	1.14	0.76	0.32	–	0.03	–	–	–	103	0.32	0.79
–	–	–	15.4	–	1.38	13.3	0.06	–	1.25	0.85	0.32	–	0.05	–	–	–	103	0.32	0.9
–	–	–	13.1	–	1.4	11.1	0.04	–	1.06	0.69	0.3	–	0.04	–	–	–	74	0.3	0.73
–	–	–	13.7	–	1.26	11.8	0.06	–	1.12	0.75	0.31	–	0.03	–	–	–	103	0.31	0.78
–	–	–	12.8	–	1.17	11	0.05	–	1.05	0.69	0.3	–	0.03	–	–	–	103	0.3	0.72
–	–	–	18.7	–	2.48	16.2	–	–	2.2	1.68	0.52	–	–	–	–	–	82	0.52	1.68
–	–	–	12.8	–	1.5	10.8	0.04	–	1.01	0.61	0.36	–	0.02	–	–	–	74	0.36	0.63
–	–	–	13.9	–	1.43	12.4	0.06	–	1.14	0.74	0.36	–	0.03	–	–	–	94	0.36	0.77
–	–	–	13.1	–	1.32	11.1	0.06	–	1.1	0.69	0.35	–	0.03	–	–	–	95	0.35	0.72
–	–	–	12.4	–	1.28	11.1	0.05	–	1.09	0.7	0.34	–	0.04	–	–	–	95	0.34	0.74
–	–	–	14.7	–	1.46	12.5	0.07	–	1.18	0.75	0.37	–	0.03	–	–	–	93	0.37	0.78

Table 13.3 (*Continued*)

Item	H_2O	Ash	Cals.	Prot.	Carb.	Fat-T	Fat-S	4:0	6:0	8:0	10:0	12:0	14:0	15:0	16:0	17:0	18:0
Beef brisket, point, all, lean + fat, 1/4″ trim, braised	43.4	0.8	404	22.1	–	34.3	13.6	–	–	–	0.1	0.09	1.15	–	8.45	–	3.82
Beef brisket, point, all, lean + fat, 1/4″ trim, raw	53.5	0.74	331	16.1	–	29.1	11.8	–	–	–	0.11	0.08	0.98	–	7.21	–	3.49
Beef brisket, point, lean + fat, 0″ trim, braised	46.1	0.84	358	23.5	–	28.5	11.2	–	–	–	0.08	0.07	0.95	–	6.99	–	3.17
Beef brisket, point half, lean + fat, 1/2″ trim, braised	41.5	0.79	408	22	–	34.9	14.5	–	–	–	0.1	0.1	1.17	–	8.6	–	3.88
Beef brisket, point half, lean + fat, 1/2″ trim, raw	52	0.72	353	15.8	–	31.6	13.5	–	–	–	0.11	0.08	1.06	–	7.8	–	3.78
Beef composite, choice, lean + fat, 1/2″ trim, cooked	46.7	1.01	351	24.9	–	27.1	11.2	–	–	–	0.08	0.07	0.86	–	6.63	–	3.11
Beef composite, prime, lean + fat, 1/2″ trim, cooked	42.6	0.94	405	23.4	–	33.8	14	–	–	–	0.1	0.09	1.09	–	8.28	–	3.87
Beef composite, prime, lean + fat, 1/2″ trim, raw	52.6	0.77	344	16.6	–	30.3	13	–	–	–	0.11	0.08	0.98	–	7.43	–	3.73
Beef large rib, choice, lean + fat, raw	51.6	0.71	356	15.5	–	32.1	14	–	–	–	0.12	0.08	1.08	–	8.03	–	4.07
Beef large rib, choice, lean + fat, 1/2″ trim, broiled	46	0.87	384	20	–	33.1	14	–	–	–	0.09	0.09	1.09	–	8.2	–	3.92
Beef large rib, choice, lean + fat, 1/2″ trim, roasted	45.8	0.9	372	22.7	–	30.5	12.9	–	–	–	0.08	0.08	1.01	–	7.57	–	3.66
Beef large rib, lean + fat, 1/2″ trim, broiled	46.7	0.88	378	20.1	–	32.3	13.7	–	–	–	0.09	0.09	1.07	–	8.05	–	3.85
Beef large rib, lean + fat, 1/2″ trim, raw	52.2	0.72	350	15.6	–	31.4	13.7	–	–	–	0.11	0.08	1.05	–	7.82	–	3.96
Beef large rib, lean + fat, 1/2″ trim, roasted	46.1	0.9	368	22.7	–	30	12.7	–	–	–	0.08	0.08	0.98	–	7.42	–	3.58
Beef large rib, prime lean + fat, 1/2″ trim, broiled	41.5	0.87	425	19.9	–	37.7	16	–	–	–	0.1	0.09	1.25	–	9.37	–	4.55
Beef large rib, prime, lean + fat, 1/2″ trim, raw	48.9	0.71	383	15.4	–	35.2	15.3	–	–	–	0.12	0.09	1.17	–	8.75	–	4.46
Beef large rib, prime, lean + fat, 1/2″ trim, roasted	43.2	0.88	407	22.2	–	34.6	14.7	–	–	–	0.09	0.09	1.14	–	8.59	–	4.18
Beef large rib, select, lean + fat, 1/2″ trim, broiled	49.1	0.89	354	20.5	–	29.6	12.6	–	–	–	0.09	0.08	0.97	–	7.37	–	3.52
Beef large rib, select, lean + fat, 1/2″ trim, raw	54.3	0.74	328	15.9	–	28.8	12.6	–	–	–	0.11	0.07	0.95	–	7.18	–	3.64
Beef large rib, select, lean + fat, 1/2″ trim, roasted	46.8	0.9	357	22.8	–	28.8	12.2	–	–	–	0.08	0.08	0.95	–	7.17	–	3.45
Beef large end rib, all, lean + fat, 0″ trim, roasted	47.7	0.92	353	23.1	–	28.2	11.4	–	–	–	0.08	0.07	0.92	–	6.96	–	3.37
Beef large end rib, choice, lean + fat, 0″ trim, roasted	46.1	0.9	372	22.8	–	30.5	12.3	–	–	–	0.08	0.08	0.99	–	7.52	–	3.64
Beef short rib, choice, lean + fat, braised	35.7	0.74	471	21.6	–	42	17.8	–	–	–	0.13	0.12	1.39	–	10.5	–	4.93
Beef short rib, choice, lean + fat, raw	48.3	0.68	388	14.4	–	36.2	15.8	–	–	–	0.14	0.09	1.21	–	9.01	–	4.55
Beef small rib, lean + fat, 1/2″ trim, roasted	47.5	0.91	359	22.3	–	29.3	12.4	–	–	–	0.08	0.08	0.96	–	7.28	–	3.4[illegible]
Beef small rib, prime, lean + fat, 1/2″ trim, broiled	46.5	1.05	364	23.7	–	29.1	12.3	–	–	–	0.07	0.07	0.95	–	7.21	–	3.51
Beef small rib, prime, lean + fat, 1/2″ trim, raw	51.9	0.78	349	16.4	–	30.9	13.3	–	–	–	0.1	0.07	1.01	–	7.59	–	3.88
Beef small rib, prime, lean + fat, 1/2″ trim, roasted	41.1	0.89	420	21.8	–	36.3	15.3	–	–	–	0.09	0.09	1.19	–	8.97	–	4.36

20:0	22:0	24:0	Fat-M	14:1	16:1	18:1	20:1	22:1	Fat-P	18:2	18:3	18:4	20:4	20:5	22:5	22:6	Chol., mg	Omeg3	Omeg6
–	–	–	15.2	–	1.56	13.6	0.07	–	1.18	0.77	0.37	–	0.03	–	–	–	92	0.37	0.3
–	–	–	13	–	1.58	11.4	0.04	–	1.01	0.61	0.37	–	0.02	–	–	–	76	0.37	0.63
–	–	–	12.7	–	1.3	11.4	0.06	–	0.96	0.64	0.29	–	0.03	–	–	–	92	0.29	0.67
–	–	–	16.2	–	1.59	13.8	0.07	–	1.23	0.79	0.38	–	0.03	–	–	–	92	0.38	0.82
–	–	–	14.7	–	1.71	12.3	0.05	–	1.09	0.66	0.39	–	0.02	–	–	–	76	0.39	0.68
–	–	–	12.2	–	1.15	10.5	0.05	–	1.04	0.66	0.31	–	0.03	–	–	–	91	0.31	0.69
–	–	–	15.2	–	1.44	13	0.07	–	1.29	0.83	0.39	–	0.04	–	–	–	92	0.39	0.87
–	–	–	13.7	–	1.57	11.5	0.04	–	1.14	0.7	0.38	–	0.03	–	–	–	73	0.38	0.73
–	–	–	14.5	–	1.67	12.1	0.05	–	1.2	0.74	0.41	–	0.02	–	–	–	74	0.41	0.76
–	–	–	14.8	–	1.39	12.7	0.06	–	1.18	0.77	0.36	–	0.03	–	–	–	87	0.36	0.8
–	–	–	13.7	–	1.27	11.8	0.06	–	1.09	0.71	0.32	–	0.03	–	–	–	85	0.32	0.74
–	–	–	14.5	–	1.36	12.5	0.06	–	1.16	0.75	0.36	–	0.03	–	–	–	87	0.36	0.78
–	–	–	14.1	–	1.63	11.8	0.05	–	1.17	0.72	0.4	–	0.02	–	–	–	73	0.04	0.74
–	–	–	13.4	–	1.25	11.6	0.06	–	1.07	0.7	0.32	–	0.03	–	–	–	85	0.32	0.73
–	–	–	17	–	1.57	14.6	0.07	–	1.34	0.88	0.39	–	0.04	–	–	–	87	0.39	0.92
–	–	–	15.7	–	1.79	13.2	0.05	–	1.32	0.82	0.43	–	0.03	–	–	–	74	0.43	0.85
–	–	–	15.5	–	1.44	13.4	0.06	–	1.23	0.81	0.35	–	0.03	–	–	–	85	0.35	0.84
–	–	–	13.3	–	1.25	11.5	0.06	–	1.07	0.69	0.33	–	0.02	–	–	–	86	0.33	0.71
–	–	–	13	–	1.5	10.9	0.04	–	1.07	0.66	0.37	–	0.02	–	–	–	72	0.37	0.68
–	–	–	13	–	1.21	11.2	0.05	–	1.03	0.67	0.31	–	0.02	–	–	–	85	0.31	0.69
–	–	–	12.1	–	1.17	10.9	0.05	–	0.98	0.66	0.29	–	0.03	–	–	–	85	0.29	0.69
–	–	–	13	–	1.27	11.7	0.06	–	1.06	0.71	0.31	–	0.03	–	–	–	85	0.31	0.74
–	–	–	18.9	–	1.79	16.2	0.08	–	1.54	0.98	0.5	–	0.03	–	–	–	94	0.5	1.01
–	–	–	16.4	–	1.89	13.7	0.05	–	1.33	0.81	0.48	–	0.02	–	–	–	76	0.48	0.83
–	–	–	13.1	–	1.21	11.3	0.05	–	1.05	0.68	0.31	–	0.03	–	–	–	85	0.31	0.71
–	–	–	13	–	1.18	11.3	0.05	–	1.03	0.68	0.23	–	0.04	–	–	–	84	0.28	0.72
–	–	–	14	–	1.53	11.9	0.04	–	1.11	0.71	0.35	–	0.03	–	–	–	71	0.35	0.74
–	–	–	16.2	–	1.47	14	0.06	–	1.28	0.85	0.35	–	0.04	–	–	–	85	0.35	0.89

Table 13.3 (Continued)

Item	H_2O	Ash	Cals.	Prot.	Carb.	Fat-T	Fat-S	4:0	6:0	8:0	10:0	12:0	14:0	15:0	16:0	17:0	18:0
Beef small end rib, choice, lean + fat, 1/2″ trim, roast	46.7	0.91	367	22.2	–	30.2	12.8	–	–	–	0.08	0.08	0.99	–	7.5	–	3.6
Beef tenderloin, prime, lean + fat, 1/2″ trim, roast	46.9	0.99	358	23.4	–	28.7	11.7	–	–	–	0.07	0.07	0.87	–	6.8	–	3.44
Beef tenderloin, prime, lean + fat, 1/4″ trim, roasted	47.5	1.01	353	23.7	–	27.9	11.1	–	–	–	0.07	0.06	0.85	–	6.6	–	3.35
Beef top loin, prime, lean + fat, 1/2″ trim, raw	54.4	0.72	322	17.9	–	27.3	11.7	–	–	–	0.09	0.06	0.89	–	6.73	–	3.33
Beef whole rib, choice, lean + fat, 1/2″ trim, raw	53	0.75	338	16	–	29.9	13	–	–	–	0.11	0.07	0.99	–	7.43	–	3.78
Beef whole rib, choice, lean + fat, 1/2″ trim, broiled	46.8	0.93	368	21.4	–	30.7	13	–	–	–	0.09	0.08	1.01	–	7.61	–	3.64
Beef whole rib, choice, lean + fat, 1/2″ trim, roasted	44.7	0.88	386	21.8	–	32.5	13.8	–	–	–	0.09	0.09	1.08	–	8.09	–	3.86
Beef whole rib, choice, lean + fat, 1/4″ trim, roasted	45.9	0.97	376	22.2	–	31.2	12.6	–	–	–	0.09	0.08	1.03	–	7.74	–	3.68
Beef whole rib, lean + fat, 1/2″ trim, broiled	47.4	0.94	362	21.5	–	30	12.7	–	–	–	0.09	0.08	0.98	–	7.46	–	3.56
Beef whole rib, lean + fat, 1/2″ trim, raw	53.6	0.75	331	16	–	29.1	12.7	–	–	–	0.1	0.07	0.96	–	7.23	–	3.67
Beef whole rib, prime, lean + fat, 1/2″ trim, raw	50	0.73	370	15.8	–	33.6	14.6	–	–	–	0.13	0.09	1.11	–	8.31	–	4.23
Beef whole rib, prime, lean + fat, 1/2″ trim, broiled	42.9	0.92	408	21.1	–	35.2	14.9	–	–	–	0.09	0.09	1.16	–	8.74	–	4.23
Beef whole rib, prime, lean + fat, 1/2″ trim, roasted	41.2	0.86	425	21.5	–	36.9	15.7	–	–	–	0.1	0.09	1.22	–	9.17	–	4.42
Beef whole rib, select, lean + fat, 1/2″ trim, broiled	49.5	0.95	340	21.8	–	27.4	11.6	–	–	–	0.08	0.07	0.91	–	6.82	–	3.26
Beef whole rib, select, lean + fat, 1/2″ trim, roasted	47.1	0.89	360	22.3	–	29.3	12.4	–	–	–	0.08	0.08	0.96	–	7.31	–	3.48
Beef whole rib, all, lean + fat, 1/2″ trim, roasted	45.2	0.88	381	21.9	–	31.8	13.4	–	–	–	0.09	0.09	1.05	–	7.9	–	3.76
Beef, large end rib, all, lean + fat, 1/4″ trim, raw	53.8	0.75	323	16.1	–	28.2	11.7	–	–	–	0.1	0.07	0.93	–	7.02	–	3.56
Beef, large end rib, all, lean + fat, 1/4″ trim, broiled	47.9	1.03	347	21.3	–	28.4	11.6	–	–	–	0.09	0.08	0.92	–	7.11	–	3.37
Beef, large end rib, all, lean + fat, 1/4″ trim, roasted	46.9	0.9	365	22.6	–	29.8	12	–	–	–	0.09	0.08	0.97	–	7.36	–	3.53
Beef, large end rib, choice, lean + fat, 1/4″ trim, raw	51.8	0.73	345	15.8	–	30.8	12.8	–	–	–	0.11	0.08	1.03	–	7.67	–	3.89
Beef, large end rib, choice, lean + fat, 1/4″ trim, broiled	46.2	1.01	367	21	–	30.8	12.5	–	–	–	0.09	0.09	1.01	–	7.7	–	3.65
Beef, large end rib, choice, lean + fat, 1/4″ trim, roasted	45.4	0.88	383	22.3	–	32	12.9	–	–	–	0.09	0.08	1.05	–	7.9	–	3.79
Beef, large end rib, prime, lean + fat, 1/4″ trim, raw	49.4	0.72	377	15.5	–	34.5	14.5	–	–	–	0.12	0.08	1.15	–	8.57	–	4.37

20:0	22:0	24:0	Fat-M	14:1	16:1	18:1	20:1	22:1	Fat-P	18:2	18:3	18:4	20:4	20:5	22:5	22:6	Chol., mg	Omeg3	Omeg6
–	–	–	13.5	–	1.25	11.7	0.05	–	1.08	0.71	0.32	–	0.03	–	–	–	85	0.32	0.74
–	–	–	12.2	–	1.08	10.6	0.04	–	1.12	0.75	0.28	–	0.05	–	–	–	88	0.28	0.8
–	–	–	11.6	–	1.05	10.3	0.04	–	1.08	0.73	0.27	–	0.05	–	–	–	88	0.27	0.78
–	–	–	12.4	–	1.36	10.5	0.04	–	0.99	0.63	0.31	–	0.03	–	–	–	70	0.31	0.66
–	–	–	13.5	–	1.54	11.3	0.04	–	1.11	0.69	0.37	–	0.02	–	–	–	72	0.37	0.71
–	–	–	13.7	–	1.28	11.8	0.06	–	1.1	0.71	0.34	–	0.03	–	–	–	86	0.34	0.74
–	–	–	14.6	–	1.36	12.6	0.06	–	1.17	0.76	0.36	–	0.03	–	–	–	85	0.36	0.79
–	–	–	13.4	–	1.34	12	0.06	–	1.12	0.73	0.35	–	0.02	–	–	–	85	0.35	0.75
–	–	–	13.5	–	1.26	11.6	0.06	–	1.07	0.7	0.33	–	0.03	–	–	–	86	0.33	0.73
–	–	–	13.1	–	1.5	11	0.04	–	1.07	0.67	0.37	–	0.02	–	–	–	72	0.37	0.69
–	–	–	15	–	1.71	12.7	0.02	–	1.27	0.78	0.43	–	0.03	–	–	–	73	0.43	0.81
–	–	–	15.8	–	1.46	13.6	0.06	–	1.24	0.82	0.36	–	0.04	–	–	–	86	0.36	0.86
–	–	–	16.6	–	1.53	14.3	0.07	–	1.32	0.86	0.39	–	0.04	–	–	–	86	0.39	0.09
–	–	–	12.3	–	1.15	10.6	0.05	–	0.97	0.64	0.3	–	0.02	–	–	–	85	0.3	0.66
–	–	–	13.1	–	1.13	11.4	0.06	–	1.06	0.68	0.33	–	0.02	–	–	–	85	0.33	0.7
–	–	–	14.2	–	1.33	12.3	0.06	–	1.14	0.74	0.35	–	0.03	–	–	–	85	0.35	0.77
–	–	–	12.1	–	1.45	10.6	0.04	–	1.04	0.65	0.35	–	0.03	–	–	–	71	0.35	0.68
–	–	–	12	–	1.24	10.8	0.06	–	1.08	0.71	0.32	–	0.04	–	–	–	81	0.32	0.75
–	–	–	12.8	–	1.25	11.5	0.06	–	1.05	0.69	0.32	–	0.03	–	–	–	85	0.32	0.72
–	–	–	13.2	–	1.59	11.6	0.04	–	1.13	0.71	0.38	–	0.03	–	–	–	73	0.38	0.74
–	–	–	13	–	1.34	11.7	0.06	–	1.17	0.78	0.35	–	0.04	–	–	–	81	0.35	0.82
–	–	–	13.7	–	1.34	12.3	0.06	–	1.13	0.74	0.34	–	0.03	–	–	–	85	0.34	0.77
–	–	–	14.9	–	1.75	13	0.05	–	1.28	0.81	0.42	–	0.03	–	–	–	73	0.42	0.84

Table 13.3 (Continued)

Item	H_2O	Ash	Cals.	Prot.	Carb.	Fat-T	Fat-S	4:0	6:0	8:0	10:0	12:0	14:0	15:0	16:0	17:0	18:0
Beef, large end rib, prime, lean + fat, 1/4″ trim, broiled	42.6	0.88	413	20.3	–	36.2	15	–	–	–	0.09	0.09	1.19	–	8.97	–	4.39
Beef, large end rib, prime, lean + fat, 1/4″ trim, roasted	43.7	0.89	402	22.5	–	33.9	14.1	–	–	–	0.09	0.08	1.12	–	8.41	–	4.1
Beef, small end rib, all, lean + fat, 1/4″ trim, roast	48.4	1.11	347	22.3	–	28	11.3	–	–	–	0.09	0.08	0.91	–	6.95	–	3.26
Beef, small end rib, choice, lean + fat, 1/4″ trim, broil	47.2	1.04	349	23.5	–	27.6	11.2	–	–	–	0.08	0.07	0.9	–	6.86	–	3.28
Beef, small end rib, choice, lean + fat, 1/4″ trim, roast	46.5	1.09	367	22	–	30.2	12.2	–	–	–	0.1	0.08	0.98	–	7.53	–	3.52
Beef, small end rib, prime, lean + fat, 1/4″ trim, broil	46.8	1.05	361	23.9	–	28.7	11.8	–	–	–	0.07	0.07	0.94	–	7.1	–	3.46
Beef, small end rib, prime, lean + fat, 1/4″ trim, raw	52.5	0.79	342	16.6	–	30.1	12.5	–	–	–	0.1	0.07	0.97	–	7.38	–	3.78
Beef, small end rib, prime, lean + fat, 1/4″ trim, roast	41.3	0.9	417	21.9	–	35.9	14.8	–	–	–	0.09	0.08	1.18	–	8.88	–	4.32
Beef, whole rib, all lean + fat, 1/4″ trim, broiled	48	1.04	342	22.2	–	27.4	11.1	–	–	–	0.08	0.08	0.89	–	6.85	–	3.25
Beef, whole rib, all, lean + fat, 1/4″ trim, roasted	47.5	0.98	358	22.5	–	29	11.7	–	–	–	0.09	0.08	0.94	–	7.19	–	3.42
Beef, whole rib, choice, lean + fat, 1/4″ trim, broil	46.6	1.02	360	22	–	29.5	12	–	–	–	0.09	0.08	0.96	–	7.36	–	3.51
Beef, whole rib, choice, lean + fat, 1/4″ trim, raw	52.7	0.75	333	16.1	–	29.4	12.1	–	–	–	0.1	0.07	0.96	–	7.29	–	3.7
Beef, whole rib, prime, lean + fat, 1/4″ trim, raw	50.6	0.74	364	15.9	–	32.9	13.7	–	–	–	0.11	0.08	1.08	–	8.13	–	4.15
Beef, whole rib, prime, lean + fat, 1/4″ trim, broiled	44.3	0.95	392	21.7	–	33.2	13.7	–	–	–	0.08	0.08	1.09	–	8.21	–	4.02
Beef, whole rib, prime, lean + fat, 1/4″ trim, roasted	42.6	0.89	409	22.2	–	34.8	14.4	–	–	–	0.09	0.08	1.15	–	8.63	–	4.2
Beef chuck, broiled roast, choice, lean + fat, 1/4″ trim, braised	46	0.86	363	26.2	–	27.8	11.1	–	–	–	0.07	0.07	0.89	–	6.76	–	3.32
Formed beef bacon (Sizzlean)	26.2	6.71	449	31.3	1.41	34.4	14.4	–	–	–	0.13	0.1	1.23	–	7.67	–	4.49
Broiler/fryer chicken back with skin, raw	58.1	0.64	319	14.1	–	28.7	8.35	–	–	–	–	0.03	0.24	–	6.07	–	1.7
Broiler/fryer chicken skin, batter fry	36.1	1.6	394	10.3	23.2	28.8	7.62	–	–	–	–	0.01	0.16	–	5.06	–	2.31
Broiler/fryer chicken skin, flour fry	28.5	0.45	502	19.1	9.35	42.6	11.7	–	–	–	–	0.03	0.31	–	8.5	–	2.61
Broiler/fryer chicken skin, raw	54.2	0.41	349	13.3	–	32.4	9.09	–	–	–	–	0.03	0.27	–	6.96	–	1.61
Broiler/fryer chicken skin, roasted	40.3	0.5	454	20.4	–	40.7	11.4	–	–	–	–	0.04	0.33	–	8.75	–	2.03
Broiler/fryer chicken skin, stewed	53.3	0.43	363	15.2	–	33	9.29	–	–	–	–	0.03	0.27	–	7.11	–	1.65
Lamb rib, choice, lean + fat, broiled	47.1	1.15	361	22.1	–	29.6	12.7	–	–	–	0.07	0.12	1.15	–	6.44	–	4.03
Lamb rib, choice, lean + fat, raw	50.8	0.74	372	14.5	–	34.4	15.2	–	–	–	0.09	0.15	1.41	–	7.61	–	4.78

20:0	22:0	24:0	Fat-M	14:1	16:1	18:1	20:1	22:1	Fat-P	18:2	18:3	18:4	20:4	20:5	22:5	22:6	Chol., mg	Omeg3	Omeg6
–	–	–	15.8	–	1.49	14	0.07	–	1.27	0.85	0.36	–	0.03	–	–	–	86	0.36	0.88
–	–	–	14.8	–	1.41	13.1	0.06	–	1.19	0.79	0.34	–	0.03	–	–	–	85	0.34	0.82
–	–	–	12.1	–	1.22	10.8	0.06	–	1.03	0.67	0.33	–	0.01	–	–	–	83	0.33	0.68
–	–	–	11.9	–	1.14	10.7	0.04	–	0.96	0.65	0.29	–	0.02	–	–	–	84	0.29	0.67
–	–	–	13.1	–	1.34	11.7	0.06	–	1.12	0.72	0.37	–	0.01	–	–	–	84	0.37	0.73
–	–	–	12.5	–	1.17	11.1	0.05	–	1.01	0.67	0.27	–	0.04	–	–	–	84	0.27	0.71
–	–	–	13.2	–	1.48	11.6	0.04	–	1.08	0.7	0.33	–	0.03	–	–	–	71	0.33	0.73
–	–	–	15.6	–	1.46	13.9	0.06	–	1.26	0.84	0.35	–	0.04	–	–	–	84	0.35	0.88
–	–	–	11.7	–	1.17	10.5	0.05	–	1.01	0.67	0.3	–	0.03	–	–	–	82	0.3	0.7
–	–	–	12.5	–	1.24	11.2	0.06	–	1.04	0.68	0.33	–	0.02	–	–	–	84	0.33	0.7
–	–	–	12.6	–	1.26	11.3	0.05	–	1.09	0.72	0.32	–	0.03	–	–	–	82	0.32	0.75
–	–	–	12.7	–	1.51	11.1	0.04	–	1.07	0.67	0.36	–	0.02	–	–	–	72	0.36	0.69
–	–	–	14.3	–	1.65	12.4	0.05	–	1.21	0.77	0.39	–	0.03	–	–	–	73	0.39	0.8
–	–	–	14.5	–	1.36	12.8	0.06	–	1.16	0.78	0.32	–	0.04	–	–	–	85	0.32	0.82
–	–	–	15.2	–	1.43	13.5	0.06	–	1.22	0.81	0.34	–	0.04	–	–	–	85	0.34	0.85
–	–	–	12	–	1.14	10.8	0.05	–	1.01	0.68	0.28	–	0.04	–	–	–	103	0.28	0.72
–	–	–	16.9	–	2.28	14.6	–	–	1.59	1.26	0.33	–	–	–	–	–	119	0.33	1.26
–	–	–	12.3	–	1.62	10.2	0.29	–	6.15	5.59	0.28	–	0.1	0.01	0.02	0.03	79	0.32	5.69
–	–	–	12.4	–	0.76	11.4	0.16	–	6.84	6.35	0.35	–	0.04	0.01	–	0.01	74	0.37	6.39
–	–	–	18	–	2	15.5	0.43	–	9.45	8.69	0.41	–	0.09	0.02	0.01	0.03	73	0.46	8.73
–	–	–	13.5	–	1.92	11.1	0.42	–	6.82	6.23	0.26	–	0.09	0.02	0.01	0.03	109	0.31	6.32
–	–	–	17	–	2.41	13.9	0.53	–	8.58	7.84	0.33	–	0.11	0.02	0.01	0.04	83	0.39	7.95
–	–	–	13.8	–	1.96	11.3	0.43	–	6.97	6.37	0.27	–	0.09	0.02	0.01	0.03	63	0.32	6.46
–	–	–	12.1	–	0.86	10.9	–	–	2.38	1.75	0.52	–	0.11	–	–	–	99	0.52	1.86
–	–	–	14.1	–	1.01	12.7	–	–	2.7	1.98	0.62	–	0.1	–	–	–	76	0.62	2.08

Table 13.3 *(Continued)*

Item	H_2O	Ash	Cals.	Prot.	Carb.	Fat-T	Fat-S	4:0	6:0	8:0	10:0	12:0	14:0	15:0	16:0	17:0	18:0
NZ Lamb rib, lean + fat, raw	53.3	0.81	346	14.9	–	31.3	16	–	–	–	0.09	0.13	1.35	–	6.93	–	6.32
NZ Lamb rib, lean + fat, roasted	50.2	1.08	340	19	–	28.8	14.5	–	–	–	0.08	0.12	1.21	–	6.34	–	5.63
Duck, roasted	51.8	0.82	337	19	–	28.4	9.68	–	–	–	–	0.04	0.17	–	6.81	–	2.44
Duck, whole with skin, raw	48.5	0.68	404	11.5	–	39.3	13.2	–	–	–	–	0.05	0.25	–	9.59	–	3.21
Goose with skin, raw	49.7	0.87	371	15.9	–	33.6	9.79	–	–	–	–	0.03	0.17	–	6.96	–	2.34
Bacon, cooked	12.9	6.79	576	30.5	0.59	49.2	17.4	–	–	–	0.08	0.07	0.62	–	11	–	5.67
BBQ flavor pork skins rinds	2.1	7	538	57.9	1.6	31.8	11.6	–	–	–	0.02	0.07	0.4	–	6.97	–	4.11
Cured pork, bacon, raw (med slices)	31.6	2.14	556	8.67	0.09	57.5	21.3	–	–	–	0.23	0.22	0.88	–	13.2	–	6.7
Cured pork, breakfast strips, raw	47.4	3.01	388	11.7	0.7	37.2	12.9	–	–	–	0.21	0.21	0.72	–	7.52	–	4.27
Fresh pork carcass, lean + fat, raw	49.8	0.72	376	13.9	–	35.1	12.4	–	–	–	0.03	0.05	0.44	–	7.65	–	4.2
Fresh pork chitterlings, simmered	62.4	0.3	303	10.3	–	28.8	10.1	–	–	–	–	0.05	0.33	–	5.8	–	3.92
Fresh pork jowl, raw	22.2	0.32	655	6.39	–	69.6	25.3	–	–	–	0.05	0.15	0.88	–	15.2	–	8.94
Fresh pork tail, raw	46.1	0.5	378	17.8	–	33.5	11.6	–	–	–	–	–	0.53	–	7.6	–	3.51
Fresh pork tail, simmered	46.7	0.4	396	17	–	35.8	12.5	–	–	–	–	–	0.57	–	8.13	–	3.75
Fresh pork, wholesale cut, backfat, raw	7.7	0.7	812	2.93	–	88.7	32.2	–	–	–	0.06	0.19	1.12	–	19.5	–	11.4
Fresh pork, wholesale cut, belly, raw	36.7	0.49	518	9.35	–	53	19.3	–	–	–	0.04	0.06	0.7	–	11.9	–	6.49
Ham patties, unheated	54.4	2.94	315	12.8	1.7	28.2	10.1	–	–	–	0.02	0.06	0.37	–	6.32	–	3.31
Ham patty, grilled	51.4	2.81	342	13.3	1.71	30.9	11.1	–	–	–	0.03	0.06	0.41	–	6.92	–	3.62
Pork breakfast strips, cooked	26.9	6.38	459	29	1.06	36.7	12.8	–	–	–	0.1	0.08	0.63	–	7.39	–	4.56
Pork cracklings, cooked	12.9	6.79	576	30.5	0.59	49.2	17.4	–	–	–	0.08	0.07	0.62	–	11	–	5.67
Pork loin blade chop, lean + fat, fried	50	1.15	342	21.5	–	27.7	10.2	–	–	–	0.02	0.02	0.35	–	6.2	–	3.39
Pork loin country back-ribs, lean + fat, roasted	45.4	1.1	370	24.3	–	29.6	11	–	–	–	0.01	0.01	0.39	–	6.75	–	3.65
Pork skin rinds, deep fried, plain	1.8	5.4	545	61.3	–	31.3	11.4	–	–	–	0.02	0.07	0.39	–	6.86	–	4.05
Pork spareribs, lean + fat, braised	40.4	1.13	397	29.1	–	30.3	11.1	–	–	–	0.01	0.01	0.36	–	6.8	–	3.82
Salt pork, raw	11	3.66	748	5.06	–	80.5	29.4	–	–	–	–	0.16	1.05	–	17.7	–	10.3
All turkey skin, raw	49.6	0.4	387	12.7	–	36.9	9.64	–	–	–	–	–	0.28	–	6.96	–	1.85
All turkey skin, roasted	39.7	0.66	442	19.7	–	39.7	10.3	–	–	–	–	–	0.31	–	7.47	–	1.98
Hen turkey, skin only, raw	46.8	0.4	417	11.8	–	40.6	10.6	–	–	–	–	–	0.31	–	7.66	–	2.03
Hen turkey, skin only, roasted	35.5	0.63	482	19	–	44.5	11.6	–	–	–	–	–	0.34	–	8.38	–	2.22
Tom turkey, skin only, raw	51.3	0.39	368	13.2	–	34.6	9.03	–	–	–	–	–	0.27	–	6.52	–	1.73
Tom turkey, skin only, roasted	41.5	0.67	422	20.1	–	37.3	9.73	–	–	–	–	–	0.29	–	7.02	–	1.86
Beef and pork bologna	54.3	2.98	316	11.7	2.8	28.3	10.7	–	–	–	0.06	0.04	0.62	–	6.28	–	3.71
Beef and pork cheese smokie/cheesefurter	52.5	3.04	327	14.1	1.51	29	10.5	–	–	–	0.09	0.1	0.56	–	6.4	–	3.34
Beef bologna	55.3	3.21	312	12.2	0.8	28.5	12.1	–	–	–	0.08	0.04	0.87	–	6.65	–	4.06
Beef frankfurter/hotdog, 8 per pkg	54.7	2.91	315	12	1.81	28.5	12.1	–	–	–	0.08	0.06	0.94	–	6.53	–	3.97
Beef pastrami	46.7	3.81	349	17.2	3.06	29.2	10.4	–	–	–	0.13	0.09	0.76	–	5.94	–	3.02
Beef + pork frankfurter/hotdog, 10 per pkg	53.9	3.16	320	11.3	2.56	29.2	10.8	–	–	–	0.08	0.06	0.53	–	6.46	–	3.66
Beef + pork frankfurter/hotdog, 8 per pkg	53.9	3.16	320	11.3	2.56	29.2	10.8	–	–	–	0.08	0.06	0.53	–	6.46	–	3.66
Beerwurst/beer salami, beef	52.9	3.11	329	12.4	1.71	29.9	13	–	–	–	0.04	0.03	1.02	–	7.36	–	4.06
Black pudding (blutwurst) SFK	45.5	2.37	404	13.3	–	38.5	–	–	–	–	–	–	–	–	–	–	–
Blood sausage/pudding, slice	47.3	2.32	378	14.6	1.29	34.5	13.4	–	–	–	–	0.03	0.62	–	8.31	–	4.42

20:0	22:0	24:0	Fat-M	14:1	16:1	18:1	20:1	22:1	Fat-P	18:2	18:3	18:4	20:4	20:5	22:5	22:6	Chol., mg	Omeg3	Omeg6
–	–	–	12	–	0.53	11.2	0.05	–	1.35	0.75	0.58	–	0.02	–	–	–	81	0.58	0.77
–	–	–	11.1	–	0.51	10.3	0.04	–	1.32	0.76	0.53	–	0.02	–	–	–	100	0.53	0.78
–	–	–	12.9	–	1.12	11.5	0.26	–	3.66	3.37	0.29	–	–	–	–	–	84	0.29	3.36
–	–	–	18.7	–	1.55	16.7	0.4	–	5.09	4.7	0.39	–	–	–	–	–	76	0.39	4.7
–	–	–	17.8	–	0.98	16.7	0.04	–	3.77	3.35	0.21	–	–	–	–	–	80	0.21	3.35
–	–	–	23.7	–	1.73	22	–	–	5.81	4.89	0.79	–	0.13	–	–	–	85	0.79	5.02
–	–	–	15	–	0.99	13.8	0.25	–	3.46	3.42	–	–	0.04	–	–	–	115	–	3.45
–	–	–	26.3	–	1.81	24.5	–	–	6.75	6	0.75	–	–	–	–	–	67	0.75	6
–	–	–	16.8	–	1.84	15	–	–	5.55	4.66	0.9	–	–	–	–	–	69	0.9	4.66
–	–	–	15.9	–	0.99	14.7	0.27	–	3.8	3.3	0.29	–	0.11	–	–	–	74	0.29	3.41
–	–	–	9.68	–	0.82	8.86	–	–	7.22	2.66	0.19	–	1.87	–	–	–	143	0.19	4.53
–	–	–	32.9	–	2.16	30.2	0.56	–	8.11	7.45	0.58	–	0.08	–	–	–	90	0.58	7.53
–	–	–	15.8	–	1.22	14.6	–	–	3.68	3.19	0.28	–	0.21	–	–	–	97	0.28	3.4
–	–	–	16.9	–	1.31	15.6	–	–	3.94	3.41	0.3	–	0.23	–	–	–	129	0.3	3.64
–	–	–	42	–	2.75	38.5	0.72	–	10.4	9.5	0.74	–	0.11	–	–	–	57	0.74	9.61
–	–	–	24.7	–	1.5	22.8	0.43	–	5.65	5.03	0.48	–	0.14	–	–	–	72	0.48	5.17
–	–	–	13.3	–	0.89	12.4	–	–	3.04	2.62	0.29	–	0.13	–	–	–	70	0.29	2.75
–	–	–	14.7	–	0.97	13.5	–	–	3.33	2.86	0.32	–	0.14	–	–	–	72	0.32	3
–	–	–	16.4	–	1.67	14.7	–	–	5.65	4.86	0.79	–	–	–	–	–	105	0.79	4.86
–	–	–	23.7	–	1.73	22	–	–	5.81	4.89	0.79	–	0.13	–	–	–	85	0.79	5.02
–	–	–	11.7	–	0.72	10.7	0.21	–	3.1	2.77	0.15	–	0.06	–	–	–	85	0.61	3.53
–	–	–	13.5	–	1	12.1	0.23	–	2.32	2.05	0.09	–	0.08	–	–	–	118	0.09	2.13
–	–	–	14.8	–	0.97	13.6	0.25	–	3.65	3.36	0.26	–	0.03	–	–	–	95	0.26	3.38
–	–	–	13.5	–	0.72	12.5	0.24	–	2.73	2.51	0.11	–	0.11	–	–	–	121	0.88	2.63
–	–	–	37.9	–	2.5	34.9	–	–	9.4	8.63	0.65	–	0.08	–	–	–	86	0.65	8.71
–	–	–	15.7	–	2.73	12.7	–	–	8.46	7.65	0.57	–	0.1	–	–	–	91	0.57	7.75
–	–	–	16.9	–	2.94	13.6	–	–	9.09	8.22	0.61	–	0.11	–	–	–	113	0.61	8.33
–	–	–	17.3	–	3.01	13.9	–	–	9.31	8.42	0.62	–	0.11	–	–	–	81	0.62	8.53
–	–	–	18.9	–	3.29	15.3	–	–	10.2	9.21	0.69	–	0.12	–	–	–	106	0.69	9.33
–	–	–	14.7	–	2.56	11.9	–	–	7.93	7.17	0.53	–	0.09	–	–	–	95	0.53	7.26
–	–	–	15.9	–	2.76	12.8	–	–	8.54	7.72	0.57	–	0.1	–	–	–	117	0.57	7.82
–	–	–	13.4	–	1.39	12	–	–	2.41	2	0.41	–	–	–	–	–	55	0.41	1.99
–	–	–	13.7	–	1.15	12.5	–	–	3.03	2.63	0.4	–	–	–	–	–	68	0.4	2.62
–	–	–	13.8	–	1.65	12.2	–	–	1.1	0.85	0.24	–	–	–	–	–	58	0.24	0.85
–	–	–	13.6	–	1.64	12	–	–	1.39	1.12	0.27	–	–	–	–	–	61	0.27	1.11
–	–	–	14.5	–	1.9	12.6	–	–	0.99	0.79	0.2	–	–	–	–	–	93	0.2	0.79
–	–	–	13.7	–	1.32	12.4	–	–	2.74	2.35	0.39	–	–	–	–	–	50	0.39	2.34
–	–	–	13.7	–	1.32	12.4	–	–	2.74	2.35	0.39	–	–	–	–	–	50	0.39	2.34
–	–	–	14	–	2.33	11.7	–	–	1.12	0.81	0.29	–	–	–	–	–	61	0.29	0.81
–	–	–	–	–	–	–	–	–	–	–	–	–	–	–	–	–	–	–	–
–	–	–	15.9	–	0.88	15	–	–	3.47	3.21	0.26	–	–	–	–	–	120	0.26	3.2

Table 13.3 (Continued)

Item	H_2O	Ash	Cals.	Prot.	Carb.	Fat-T	Fat-S	4:0	6:0	8:0	10:0	12:0	14:0	15:0	16:0	17:0	18:0
Bockwurst sausage, link	56.1	2.51	307	13.3	0.48	27.6	10.1	0.01	0.01	–	0.01	0.01	0.53	–	6.29	–	3.08
Brotwurst, link	51.3	3.68	323	14.3	2.98	27.8	9.94	–	–	–	0.04	0.02	0.35	–	6.12	–	3.41
Brown and serve sausage, unheated	44.6	0.7	393	13.5	2.7	38.5	13	–	–	–	–	–	0.602	–	8.34	–	3.88
Caribou sausage, slice	55.3	3.21	312	12.2	0.8	28.5	12.1	–	–	–	0.08	0.04	0.87	–	6.65	–	4.06
Chinese pork sausage	39.1	3.7	355	21.3	6.6	29.3	–	–	–	–	–	–	–	–	–	–	–
Chorizo sausage, link	31.9	3.93	455	24.1	1.87	38.3	14.4	–	–	–	–	–	0.87	–	9.16	–	4.13
Deer/venison bologna 2 × 1/4″ slice	55.3	3.21	312	12.2	0.8	28.5	12.1	–	–	–	0.08	0.04	0.87	–	6.65	–	4.06
Dry salami, beef and pork	34.7	5.48	418	22.9	2.6	34.4	12.2	–	–	–	0.06	0.04	0.51	–	7.61	–	4.01
Fresh pork and beef sausage, cooked link	44.6	2.71	396	13.8	2.71	36.3	13	–	–	–	–	–	0.6	–	8.32	–	3.88
Fresh pork and beef sausage, cooked patty	44.6	2.71	396	13.8	2.71	36.3	13	–	–	–	–	–	0.6	–	8.32	–	3.88
Italian sausage, raw, link	51.1	2.71	346	14.3	0.65	31.3	11.3	–	–	–	0.08	0.06	0.56	–	6.65	–	3.94
Kielbasa sausage, beef and pork	54	3.51	310	13.3	2.15	27.2	9.92	–	–	–	0.03	0.02	0.36	–	6.14	–	3.37
Knockwurst sausage link, beef and pork	55.5	3.11	308	11.9	1.77	27.8	10.2	–	–	–	0.11	0.09	0.58	–	5.98	–	3.47
Liverwurst, pork	52.1	3.11	326	14.1	2.21	28.5	10.6	–	–	–	–	–	0.43	–	6.86	–	3.32
Luncheon meat, canned	51.6	3.65	334	12.5	2.07	30.3	10.8	–	–	–	0.08	0.06	0.37	–	6.69	–	3.6
Pepperoni sausage, beef and pork	27.1	5.18	497	21	2.85	44	16.1	–	–	–	–	–	0.86	–	10.3	–	4.7
Polish sausage, pork	53.2	2.41	326	14.1	1.64	28.7	10.3	–	–	–	0.07	0.07	0.4	–	6.41	–	3.39
Pork and beef lunchmeat	49.3	3.65	353	12.6	2.347	32.2	11.6	–	–	–	0.08	0.05	0.48	–	7.1	–	3.9
Pork lunchmeat, canned	51.5	3.61	334	12.5	2.11	30.3	10.8	–	–	–	0.08	0.06	0.37	–	6.69	–	3.6
Pork salami, dry/hard, slice	36.2	5.93	407	22.6	1.61	33.7	11.9	–	–	–	–	–	0.52	–	7.65	–	3.57
Pork sausage patty, cooked	44.6	3.61	369	19.7	1.04	31.2	10.8	–	–	–	0.12	0.09	0.45	–	6.53	–	3.62
Pork sausage patty, raw	44.5	2.5	417	11.7	1.03	40.3	14.5	–	–	–	0.11	0.08	0.57	–	8.88	–	4.86
Pork sausage, cooked	44.6	3.61	369	19.7	1.04	31.2	10.8	–	–	–	0.12	0.09	0.45	–	6.53	–	3.62
Pork sausage, cooked link	44.6	3.61	369	19.7	1.04	31.2	10.8	–	–	–	0.12	0.09	0.45	–	6.53	–	3.62
Small smoked sausage link, beef and pork	52.2	2.7	336	13.4	1.44	30.3	10.6	–	–	–	0.05	0.04	0.41	–	6.58	–	3.56
Smoked link sausage, pork + beef + NFDM	53.9	3.3	313	13.3	1.93	27.6	9.73	–	–	–	0.09	0.06	0.36	–	5.93	–	3.29
Smoked meat sticks	19.1	4.4	550	21.5	5.4	49.6	20.8	–	–	–	0.01	0.33	1.44	–	11.8	–	7.27
Smoked sausage, beef and pork, link, large	52.2	2.7	336	13.4	1.44	30.3	10.6	–	–	–	0.05	0.04	0.41	–	6.58	–	3.56
Smoked sausage, pork, link	39.3	4.71	389	22.2	2.11	31.7	11.3	–	–	–	0.04	0.04	0.4	–	7.08	–	3.77
Thuringer summer sausage, beef and pork	50.8	3.67	335	15.8	0.26	29.5	12	–	–	–	0.11	0.08	0.8	–	6.36	–	4.18
Venison sausage, slice	55.3	3.21	312	12.2	0.8	28.5	12.1	–	–	–	0.08	0.04	0.87	–	6.65	–	4.06

20:0	22:0	24:0	Fat-M	14:1	16:1	18:1	20:1	22:1	Fat-P	18:2	18:3	18:4	20:4	20:5	22:5	22:6	Chol., mg	Omeg3	Omeg6
–	–	–	13	–	1.02	11.8	–	–	2.99	2.57	0.25	–	0.09	–	–	–	59	0.25	2.65
–	–	–	13.3	–	1.05	12.3	–	–	2.84	2.55	0.29	–	–	–	–	–	63	0.29	2.54
–	–	–	16.9	–	1.33	15.5	–	–	3.85	3.42	0.32	–	0.12	–	–	–	62	0.32	3
–	–	–	13.8	–	1.65	12.2	–	–	1.1	0.85	0.24	–	–	–	–	–	58	0.24	0.85
–	–	–	–	–	–	–	–	–	–	–	–	–	–	–	–	–	–	–	–
–	–	–	18.4	–	1.66	16.3	–	–	3.47	2.94	0.38	–	0.09	–	–	–	88	0.38	3.02
–	–	–	13.8	–	1.65	12.2	–	–	1.1	0.85	0.24	–	–	–	–	–	58	0.24	0.85
–	–	–	17.1	–	1.7	15.4	–	–	3.22	2.88	0.33	–	–	–	–	–	79	0.33	2.87
–	–	–	17.2	–	1.36	15.8	–	–	3.91	3.47	0.32	–	0.12	–	–	–	71	0.32	3.58
–	–	–	17.2	–	1.36	15.8	–	–	3.91	3.47	0.32	–	0.12	–	–	–	71	0.32	3.58
–	–	–	14.3	–	1.5	12.9	–	–	4.04	3.63	0.41	–	–	–	–	–	76	0.41	3.62
–	–	–	12.9	–	0.97	12	–	–	3.09	2.69	0.4	–	–	–	–	–	67	0.4	2.68
–	–	–	12.8	–	1.29	11.5	–	–	2.93	2.49	0.44	–	–	–	–	–	58	0.44	2.48
–	–	–	13.3	–	1.07	12.1	–	–	2.61	2.46	0.14	–	–	–	–	–	158	0.14	2.45
–	–	–	14.3	–	1.14	13.2	–	–	3.55	3.09	0.47	–	–	–	–	–	62	0.47	3.09
–	–	–	21.1	–	1.78	18.9	–	–	4.38	3.75	0.41	–	0.14	–	–	–	79	0.41	3.88
–	–	–	13.5	–	0.95	12.6	–	–	3.09	2.8	0.29	–	–	–	–	–	70	0.29	2.79
–	–	–	15.1	–	1.36	13.7	–	–	3.75	3.17	0.58	–	–	–	–	–	55	0.58	3.16
–	–	–	14.3	–	1.15	13.1	–	–	3.57	3.09	0.47	–	–	–	–	–	62	0.47	3.09
–	–	–	16	–	1.23	14.7	–	–	3.75	3.28	0.28	–	0.16	–	–	–	79	0.28	3.43
–	–	–	13.9	–	1.09	12.8	–	–	3.81	3.28	0.54	–	–	–	–	–	83	0.54	3.28
–	–	–	18.5	–	1.46	17.1	–	–	5.25	4.4	0.85	–	–	–	–	–	68	0.85	4.39
–	–	–	13.9	–	1.09	12.8	–	–	3.81	3.28	0.54	–	–	–	–	–	83	0.54	3.28
–	–	–	13.9	–	1.09	12.8	–	–	3.81	3.28	0.54	–	–	–	–	–	83	0.54	3.28
–	–	–	14.2	–	1.14	13.1	–	–	3.27	2.89	0.38	–	–	–	–	–	71	0.38	2.88
–	–	–	12.6	–	0.98	11.7	–	–	3.03	2.64	0.39	–	–	–	–	–	65	0.39	2.64
–	–	–	20.5	–	2.31	17.9	0.25	–	4.43	4.04	0.38	–	0.01	–	–	–	133	0.38	4.04
–	–	–	14.2	–	1.14	13.1	–	–	3.27	2.89	0.38	–	–	–	–	–	71	0.38	2.88
–	–	–	14.6	–	1.02	13.6	–	–	3.77	3.4	0.37	–	–	–	–	–	68	0.37	3.39
–	–	–	13	–	1.77	11.2	–	–	1.21	0.99	0.21	–	–	–	–	–	75	0.21	0.99
–	–	–	13.8	–	1.65	12.2	–	–	1.1	0.85	0.24	–	–	–	–	–	58	0.24	0.85

Table 13.4 Lipid composition of Dairy food products, contents/100 g

Item	H_2O	Ash	Cals.	Prot.	Carb.	Fat-T	Fat-S	4:0	6:0	8:0	10:0	12:0	14:0	15:0	16:0	17:0	18:0
100% Milkfat	–	–	900	–	–	100	62.3	3.24	1.92	1.12	2.5	2.81	10.1	1.02	26.3	0.925	12.1
Butter	15.9	2.11	717	0.85	0.06	81.1	50.5	2.63	1.56	0.906	2.03	2.28	8.16	0.828	21.3	0.75	9.83
Butter oil (Ghee)	0.24	–	876	0.28	–	99.5	61.9	3.23	1.91	1.11	2.5	2.79	10	1.01	26.2	0.919	12.1
Butter, pat	15.9	2.11	717	0.85	0.06	81.1	50.5	2.63	1.56	0.906	2.03	2.28	8.16	0.828	21.3	0.75	9.83
Butter, unsalted	17.9	0.04	717	0.85	0.16	81.1	50.5	2.63	1.56	0.906	2.03	2.28	8.16	0.828	21.3	0.75	9.83
Lightly salted butter	15.9	2.1	717	0.9	0.1	81.1	50.5	2.63	1.56	0.906	2.03	2.28	8.16	0.828	21.3	0.75	9.83
Whipped butter	15.9	2.12	717	0.85	0.06	81.1	50.5	2.63	1.56	0.906	2.03	2.28	8.16	0.828	21.3	0.75	9.83
Whipped butter, pat	15.9	2.12	717	0.85	0.06	81.1	50.5	2.63	1.56	0.906	2.03	2.28	8.16	0.828	21.3	0.75	9.83
Whole dry milk powder	2.48	6.09	496	26.3	38.4	26.7	16.7	0.866	0.24	0.269	0.596	0.614	2.82	0.463	7.52	0.462	2.85
Chocolate dip vanilla ice cream cone with nuts	47.1	1.24	279	5.76	29	16.9	7.26	–	–	–	–	–	–	–	–	–	–
Chocolate ice cream bar with chocolate coating (Dove)	37.4	1.07	344	3.39	35	23.1	14.1	–	–	–	–	–	–	–	–	–	–
Chocolate coated ice cream bar with nuts	43.1	1.4	317	4.4	30.9	20.2	11.8	–	–	–	–	–	–	–	–	–	–
Coffee/table cream	73.8	0.58	195	2.71	3.67	19.3	12	0.626	0.371	0.216	0.484	0.542	1.94	–	5.08	–	2.34
Cream, medium (25%) fat	68.5	0.55	244	2.48	3.49	25	15.6	0.811	0.48	0.279	0.627	0.702	2.51	–	6.58	–	3.03
Cultured sour cream	71	0.66	214	3.17	4.28	21	13	0.68	0.402	0.234	0.526	0.588	2.11	–	5.51	–	2.54
English toffee ice cream bar + chocolate coated (Heath)	48.1	0.893	307	3.04	25.7	22.3	17.1	–	–	–	–	–	–	–	–	–	–
Heavy whipping cream, liquid	57.7	0.45	345	2.06	2.8	37	23	1.2	0.71	0.413	0.928	1.04	3.72	–	9.73	–	4.48
Heavy whipping cream, whipped	57.7	0.45	345	2.06	2.8	37	23	1.2	0.71	0.413	0.928	1.04	3.72	–	9.73	–	4.48
Light whipping cream, liquid	63.5	0.46	292	2.18	2.97	30.9	19.3	1.08	0.302	0.309	0.629	0.366	3.29	–	8.84	–	3.37
Light whipping cream, whipped	63.5	0.46	292	2.18	2.97	30.9	19.3	1.08	0.302	0.309	0.629	0.366	3.29	–	8.84	–	3.37
Nutty Buddy ice cream cone	47.1	1.24	279	5.76	29	16.9	7.26	–	–	–	–	–	–	–	–	–	–
Rich vanilla ice cream	57.2	0.8	241	3.5	22.4	16.2	9.98	0.52	0.31	0.18	0.41	0.45	1.67	–	4.5	–	1.97
Rich vanilla ice cream, medium scoop/slice	57.2	0.8	241	3.5	22.4	16.2	9.98	0.52	0.31	0.18	0.41	0.45	1.67	–	4.5	–	1.97
Sour cream dip, buttermilk/onion	65.7	2.58	221	3.82	8.12	19.8	12.1	–	–	–	–	–	–	–	–	–	–
Whipped cream, pressurized	61.3	0.76	257	3.21	12.5	22.2	13.8	0.721	0.427	0.248	0.557	0.624	2.24	–	5.84	–	2.69
Dried egg yolk	4.66	3.17	687	30.5	0.39	61.3	18.5	–	–	0.01	0.01	0.01	0.168	–	13.5	–	4.71
Dried egg-glucose reduced	1.88	3.64	615	48.2	2.39	44	13.2	–	–	–	–	–	0.12	–	9.7	–	3.37
Dried whole egg	4.15	3.46	594	45.8	4.78	41.8	12.6	–	–	–	–	–	0.115	–	9.23	–	3.21
Egg salad sandwich, firm white	40.8	1.7	340	8.19	27.3	22	3.86	–	–	–	–	0.011	0.011	–	0.933	–	0.34
Egg salad sandwich, soft white	41.6	1.78	342	7.98	25.5	23.2	4.03	–	–	–	–	0.012	0.018	–	1.21	–	0.51
Egg yolk, cooked	48.6	2.23	357	16.7	1.77	30.7	9.51	–	–	0.009	0.009	0.009	0.103	–	6.81	–	2.4
Egg yolk, raw, frozen sugared	49.5	1.43	317	14.2	11.5	23.4	7.24	–	–	0.007	0.007	0.007	0.079	–	5.2	–	1.83
Frozen egg yolk, raw	55	1.59	309	15.8	1.66	26	8.05	–	–	0.008	0.008	0.008	0.087	–	5.78	–	2.04
Large egg yolk, fresh	48.8	1.78	358	16.8	1.79	30.9	9.55	–	–	0.009	0.009	0.009	0.104	–	6.86	–	2.42
Raw egg yolk, frozen, salted	49.5	11.4	278	14.2	1.49	23.4	7.24	–	–	0.007	0.007	0.007	0.079	–	5.2	–	1.83
American cheese food	43.1	4.44	331	19.7	8.33	24.5	15.4	0.793	0.47	0.358	0.591	0.728	2.51	0.165	6.81	0.165	2.81
American cheese food, slice	43.2	5.36	328	19.6	7.3	24.6	15.4	0.798	0.472	0.361	0.594	0.732	2.53	0.268	6.85	0.274	2.82
American cheese spread (Cheez Whiz)	47.7	5.98	290	16.4	8.74	21.2	13.3	0.688	0.408	0.311	0.513	0.632	2.18	0.119	5.91	0.109	2.43
American processed cheese	39.2	5.84	375	22.2	1.61	31.3	19.7	1.04	0.356	0.375	0.642	0.484	3.21	0.343	9.1	0.35	3.8
Asiago cheese, diced	37.2	3.54	376	28.4	3.39	27.5	17.8	1.1	0.488	0.289	0.625	0.521	3.06	0.346	7.79	0.331	3.25

20:0	22:0	24:0	Fat-M	14:1	16:1	18:1	20:1	22:1	Fat-P	18:2	18:3	18:4	20:4	20:5	22:5	22:6	Chol., mg	Omeg3	Omeg6
0.37	–	–	30.1	1.2	2.31	25.2	–	–	3.71	2.26	1.46	–	–	–	–	–	260	1.46	2.26
0.3	–	–	24.4	0.97	1.87	20.4	–	–	3.01	1.83	1.18	–	–	–	–	–	219	1.18	1.83
0.368	–	–	23.7	–	2.23	25	–	–	3.69	2.25	1.45	–	–	–	–	–	256	1.45	2.25
0.3	–	–	24.4	0.97	1.87	20.4	–	–	3.01	1.83	1.18	–	–	–	–	–	219	1.18	1.83
0.3	–	–	24.4	0.97	1.87	20.4	–	–	3.01	1.83	1.18	–	–	–	–	–	219	1.18	1.83
0.3	–	–	24.4	0.97	1.87	20.4	–	–	3.01	1.83	1.18	–	–	–	–	–	219	1.2	1.8
0.3	–	–	24.4	0.97	1.87	20.4	–	–	3.01	1.83	1.18	–	–	–	–	–	219	1.18	1.83
0.3	–	–	24.4	0.97	1.87	20.4	–	–	3.01	1.83	1.18	–	–	–	–	–	219	1.18	1.83
–	–	–	7.92	0.221	1.2	6.19	–	–	0.665	0.46	0.204	–	–	–	–	–	97.1	0.204	0.46
–	–	–	7.05	–	–	–	–	–	1.68	–	–	–	–	–	–	–	34	–	–
–	–	–	7.26	–	–	–	–	–	0.764	–	–	–	–	–	–	–	37.5	–	–
–	–	–	4.79	–	–	–	–	–	1.4	–	–	–	–	–	–	–	1	–	–
–	–	–	5.58	–	0.432	4.86	–	–	0.717	0.436	0.281	–	–	–	–	–	66.1	0.281	0.436
–	–	–	7.22	–	0.56	6.29	–	–	0.928	0.565	0.364	–	–	–	–	–	87.5	0.364	0.565
–	–	–	6.05	–	0.469	5.27	–	–	0.778	0.473	0.305	–	–	–	–	–	44.4	0.305	0.473
–	–	–	3.33	–	–	–	–	–	0.567	–	–	–	–	–	–	–	35.3	–	–
–	–	–	10.7	–	0.829	9.31	–	–	1.37	0.836	0.538	–	–	–	–	–	137	0.538	0.836
–	–	–	10.7	–	0.829	9.31	–	–	1.37	0.836	0.538	–	–	–	–	–	137	0.538	0.836
–	–	–	9.09	–	1.01	7.66	–	–	0.884	0.618	0.266	–	–	–	–	–	111	0.266	0.618
–	–	–	9.09	–	1.01	7.66	–	–	0.884	0.618	0.266	–	–	–	–	–	111	0.266	0.618
–	–	–	7.05	–	–	–	–	–	1.68	–	–	–	–	–	–	–	34	–	–
–	–	–	4.67	–	0.36	4.31	–	–	0.6	0.37	0.24	–	–	–	–	–	61	0.24	0.37
–	–	–	4.67	–	0.36	4.31	–	–	0.6	0.37	0.24	–	–	–	–	–	61	0.24	0.37
–	–	–	5.85	–	–	–	–	–	0.771	–	–	–	–	–	–	–	41.3	–	–
–	–	–	6.42	–	0.498	5.59	–	–	0.825	0.502	0.323	–	–	–	–	–	76	0.323	0.502
0.053	0.064	–	24.5	–	2.04	22.5	–	–	7.97	6.81	0.173	–	0.514	–	–	–	2928	0.173	7.32
–	–	–	17.6	–	1.46	16.1	–	–	5.71	4.88	0.124	–	0.368	–	–	–	2016	0.124	5.25
0.005	0.006	–	16.7	–	1.39	15.3	–	–	5.43	4.64	0.118	–	0.35	–	–	–	1918	0.118	4.99
–	–	–	6.75	–	0.011	3.24	0.011	–	9.83	4.52	0.461	–	–	–	–	–	124	0.461	4.52
–	–	–	7.14	–	0.017	3.35	0.012	–	10.5	4.82	0.519	–	–	–	–	–	133	0.515	4.74
0.032	0.038	–	11.7	0.024	0.918	10.7	0.086	0.009	4.19	3.52	0.103	–	0.436	0.011	–	0.113	1275	0.227	3.96
0.024	0.029	–	8.9	0.019	0.696	8.11	0.066	0.007	3.19	2.68	0.078	–	0.332	0.008	–	0.086	973	0.172	3.01
0.027	0.032	–	9.89	0.021	0.773	9.01	0.073	0.008	3.54	2.98	0.086	–	0.369	0.009	–	0.096	1081	0.191	3.35
0.032	0.038	–	11.7	0.024	0.918	10.7	0.086	0.009	4.2	3.54	0.103	–	0.438	0.011	–	0.114	1281	0.228	3.98
0.024	0.029	–	8.9	0.018	0.696	8.11	0.066	0.007	3.19	2.68	0.078	–	0.332	0.008	–	0.086	973	0.172	3.02
–	–	–	7.17	0.219	0.646	6.02	–	–	0.719	0.461	0.257	–	–	–	–	–	63.6	0.257	0.461
–	–	–	7.21	0.143	0.649	6.05	–	–	0.723	0.464	0.259	–	–	–	–	–	63.9	0.259	0.464
–	–	–	6.22	0.191	0.56	5.22	–	–	0.624	0.4	0.223	–	–	–	–	–	55.2	0.223	0.4
–	–	–	8.95	0.178	1.03	7.51	–	–	0.99	0.607	0.383	–	–	–	–	–	94.4	0.383	0.607
–	–	–	7.27	0.158	0.879	6.02	–	–	0.972	0.62	0.352	–	–	–	–	–	91.7	0.352	0.62

Table 13.4 *(Continued)*

Item	H_2O	Ash	Cals.	Prot.	Carb.	Fat-T	Fat-S	4:0	6:0	8:0	10:0	12:0	14:0	15:0	16:0	17:0	18:0
Asiago cheese, shredded	37.2	3.54	376	28.4	3.39	27.5	17.8	1.1	0.488	0.289	0.625	0.521	3.06	0.346	7.79	0.331	3.25
Beer cheese, diced	41.1	3.19	371	23.2	2.8	29.7	18.8	0.914	0.373	0.299	0.585	0.482	3.23	0.416	8.66	0.381	3.46
Blue cheese	42.4	5.11	353	21.4	2.35	28.7	18.7	0.658	0.361	0.247	0.601	0.491	3.3	0.337	9.15	0.315	3.24
Bondost cheese, diced	41.1	3.19	371	23.2	2.8	29.7	18.8	0.914	0.373	0.299	0.585	0.482	3.23	0.416	8.66	0.381	3.46
Brick cheese	41.1	3.19	371	23.2	2.8	29.7	18.8	0.914	0.373	0.299	0.585	0.482	3.23	0.416	8.66	0.381	3.46
Brick cheese with salami, shredded	43	3.2	359	22.3	2.74	28.7	17.7	0.82	0.335	0.269	0.537	0.441	2.95	0.375	8.23	0.343	3.4
Brick cheese, shredded	41.1	3.19	371	23.2	2.8	29.7	18.8	0.914	0.373	0.299	0.585	0.482	3.23	0.416	8.66	0.381	3.46
Brie cheese	48.4	2.7	334	20.8	0.45	27.7	17.4	0.564	0.323	0.297	0.673	0.504	3.07	0.438	8.25	0.401	2.88
Brie cheese, melted	48.4	2.7	334	20.8	0.45	27.7	17.4	0.564	0.323	0.297	0.673	0.504	3.07	0.438	8.25	0.401	2.88
Caciocavallo cheese, diced	41.5	4.13	357	25	1.67	27.7	17.6	1	0.515	0.338	0.687	0.771	2.97	0.323	7.73	0.296	2.96
Camembert cheese	51.8	3.68	300	19.8	0.46	24.3	15.3	0.494	0.283	0.26	0.59	0.442	2.69	0.413	7.23	0.378	2.52
Caraway cheese	39.3	3.29	376	25.2	3.07	29.2	18.6	0.922	0.466	0.246	0.529	0.477	2.93	–	8.64	–	3.53
Cheddar cheese	36.8	3.93	403	24.9	1.29	33.1	21.1	1.05	0.529	0.279	0.6	0.541	3.33	0.502	9.8	0.459	4.01
Cheddar cheese, diced	36.8	3.93	403	24.9	1.29	33.1	21.1	1.05	0.529	0.279	0.6	0.541	3.33	0.502	9.8	0.459	4.01
Cheddar cheese, melted	36.8	3.93	403	24.9	1.29	33.1	21.1	1.05	0.529	0.279	0.6	0.541	3.33	0.502	9.8	0.459	4.01
Cheddar cheese, shredded	36.8	3.93	403	24.9	1.29	33.1	21.1	1.05	0.529	0.279	0.6	0.541	3.33	0.502	9.8	0.459	4.01
Cheshire cheese	37.7	3.61	387	23.4	4.79	30.6	19.5	0.966	0.488	0.258	0.554	0.5	3.08	–	9.05	–	3.7
Brie cheese, sliced	48.4	2.7	334	20.8	0.45	27.7	17.4	0.564	0.323	0.297	0.673	0.504	3.07	0.438	8.25	0.401	2.88
Colby cheese	38.2	3.36	394	23.8	2.58	32.1	20.2	1.04	0.259	0.324	0.597	0.422	3.83	2.97	8.23	2.72	3.63
Colby cheese, diced	38.2	3.36	394	23.8	2.58	32.1	20.2	1.04	0.259	0.324	0.597	0.422	3.83	2.97	8.23	2.72	3.63
Colby cheese, shredded	38.2	3.36	394	23.8	2.58	32.1	20.2	1.04	0.259	0.324	0.597	0.422	3.83	2.97	8.23	2.72	3.63
Cream cheese	53.8	1.18	349	7.56	2.67	34.9	22	0.998	0.294	0.335	0.667	0.465	3.6	0.569	10.5	0.522	4.05
Danbo cheese, diced	37.2	3.54	376	28.4	3.39	27.5	17.8	1.1	0.488	0.289	0.625	0.521	3.06	0.346	7.79	0.331	3.25
Delft cheese, diced	41.5	4.13	357	25	1.67	27.7	17.6	1	0.515	0.338	0.687	0.771	2.97	0.323	7.73	0.296	2.96
Edam cheese/ball cheese	41.6	4.23	357	25	1.44	27.8	17.6	1	0.461	0.3	0.589	0.496	2.94	0.399	8.07	0.365	2.98
Edam cheese/ball cheese, diced	41.6	4.23	357	25	1.44	27.8	17.6	1	0.461	0.3	0.589	0.496	2.94	0.399	8.07	0.365	2.98
Elbinger cheese, diced	41.1	3.19	371	23.2	2.8	29.7	18.8	0.914	0.373	0.299	0.585	0.482	3.23	0.416	8.66	0.381	3.46
Emmentaler cheese, diced	37.2	3.54	376	28.4	3.39	27.5	17.8	1.1	0.488	0.289	0.625	0.521	3.06	0.346	7.79	0.331	3.25
European emmental cheese	35.7	5.45	385	28.7	0.45	29.7	18.5	1.15	0.51	0.3	0.65	0.54	3.19	0.28	8.14	0.28	3.39
European gorgonzola cheese	42.4	6.38	359	19.4	0.62	31.2	19.1	0.868	0.653	0.666	2.14	1.29	3.22	0.017	6.5	0.016	3.11
Feta cheese	55.2	5.21	264	14.2	4.1	21.3	14.9	0.775	0.573	0.546	1.98	1.16	2.76	0.233	5.15	0.233	1.49
Feta cheese, shredded	55.2	5.21	264	14.2	4.1	21.3	14.9	0.775	0.573	0.546	1.98	1.16	2.76	0.233	5.15	0.233	1.49
Fontina cheese	37.9	3.8	389	25.6	1.56	31.1	19.2	0.715	0.489	0.314	0.651	0.804	2.96	–	8.46	–	2.66
Fontina cheese, diced	37.9	3.8	389	25.6	1.56	31.1	19.2	0.715	0.489	0.314	0.651	0.804	2.96	–	8.46	–	2.66
Fontina cheese, shredded	37.9	3.8	389	25.6	1.56	31.1	19.2	0.715	0.489	0.314	0.651	0.804	2.96	–	8.46	–	2.66
Gjetost cheese	13.4	4.76	466	9.66	42.7	29.5	19.2	0.957	0.566	0.33	0.576	0.831	3.27	–	8.66	–	3.48
Gjetost cheese, piece	13.4	4.76	466	9.66	42.7	29.5	19.2	0.957	0.566	0.33	0.576	0.831	3.27	–	8.66	–	3.48
Goat cheese, hard	29	3.73	452	30.5	2.18	35.6	24.6	1.78	0.782	0.961	3.43	1.57	3.61	–	9.31	–	3.16
Goat cheese, soft type	60.8	1.59	268	18.5	0.89	21.1	14.6	1.06	0.463	0.569	2.03	0.931	2.14	–	5.52	–	1.87
Gorgonzola cheese	39	4.7	391	24.6	–	31.7	19.4	0.882	0.663	0.677	2.17	1.31	3.27	0.345	6.6	0.322	3.16
Gouda cheese	41.5	3.95	356	24.9	2.23	27.4	17.6	0.999	0.64	0.427	0.918	1.21	3.04	0.311	6.85	0.285	2.92
Gouda cheese, diced	41.5	3.95	356	24.9	2.23	27.4	17.6	0.999	0.64	0.427	0.918	1.21	3.04	0.311	6.85	0.285	2.92
Gruyere cheese	33.2	4.31	413	29.8	0.36	32.3	18.9	1.05	0.621	0.355	0.751	0.911	3.37	0.394	8.75	0.378	2.32
Gruyere cheese, diced	33.2	4.31	413	29.8	0.36	32.3	18.9	1.05	0.621	0.355	0.751	0.911	3.37	0.394	8.75	0.378	2.32
Gruyere cheese, shredded	33.2	4.31	413	29.8	0.36	32.3	18.9	1.05	0.621	0.355	0.751	0.911	3.37	0.394	8.75	0.378	2.32
Havarti cheese, diced	41.1	3.19	371	23.2	2.8	29.7	18.8	0.914	0.373	0.299	0.585	0.482	3.23	0.416	8.66	0.381	3.46
Jarsburg cheese, diced	37.2	3.54	376	28.4	3.39	27.5	17.8	1.1	0.488	0.289	0.625	0.521	3.06	0.346	7.79	0.331	3.25
Liederkranz cheese	53	1.2	306	17.6	0.4	28.2	18.7	0.899	0.548	0.308	0.539	0.974	3.12	–	8.29	–	3.42
Light neufchatel cheese (Philadelphia)	59.7	1.03	286	10.7	3.57	25	14.3	0.649	0.191	0.218	0.434	0.302	2.34	0.37	6.83	0.339	2.63
Limburger cheese	48.4	3.8	327	20.1	0.49	27.3	16.7	0.803	0.489	0.275	0.481	0.87	2.79	–	7.4	–	3.05
Low Fat cream cheese	63.6	1.2	231	10.6	7	17.6	11.1	0.503	0.148	0.169	0.336	0.234	1.81	0.292	5.29	0.268	2.04
Low Fat mozzarella cheese, shredded	48.6	3.71	280	27.5	3.15	17.1	10.9	0.555	0.107	0.125	0.27	0.177	1.72	0.33	5.22	0.316	2.08
Low Sodium cheddar cheese, shredded	39	2.2	398	24.3	1.9	32.6	20.7	1.06	0.265	0.331	0.611	0.432	3.92	0.998	8.42	0.913	3.71
Low Sodium cheddar cheese, diced	39	2.2	398	24.3	1.9	32.6	20.7	1.06	0.265	0.331	0.611	0.432	3.92	0.998	8.42	0.913	3.71
Low Sodium colby cheese, diced	39	2.2	398	24.3	1.9	32.6	20.7	1.06	0.265	0.331	0.611	0.432	3.92	0.998	8.42	0.913	3.71

20:0	22:0	24:0	Fat-M	14:1	16:1	18:1	20:1	22:1	Fat-P	18:2	18:3	18:4	20:4	20:5	22:5	22:6	Chol., mg	Omeg3	Omeg6
–	–	–	7.27	0.158	0.879	6.02	–	–	0.972	0.62	0.352	–	–	–	–	–	91.7	0.352	0.62
–	–	–	8.6	0.171	0.817	7.4	–	–	0.784	0.491	0.293	–	–	–	–	–	94.4	0.293	0.491
–	–	–	7.78	0.153	0.816	6.62	–	–	0.8	0.536	0.264	–	–	–	–	–	75.2	0.264	0.536
–	–	–	8.6	0.171	0.817	7.4	–	–	0.784	0.491	0.293	–	–	–	–	–	94.4	0.293	0.491
–	–	–	8.6	0.171	0.817	7.4	–	–	0.784	0.491	0.293	–	–	–	–	–	94.4	0.293	0.491
–	–	–	8.66	0.135	0.824	7.49	–	–	0.908	0.603	0.305	–	–	–	–	–	91.5	0.305	0.603
–	–	–	8.6	0.171	0.817	7.4	–	–	0.784	0.491	0.293	–	–	–	–	–	94.4	0.293	0.491
–	–	–	8.01	0.195	1.01	6.56	–	–	0.826	0.513	0.313	–	–	–	–	–	100	0.313	0.513
–	–	–	8.01	0.195	1.01	6.56	–	–	0.826	0.513	0.313	–	–	–	–	–	100	0.313	0.513
–	–	–	8.01	0.186	0.833	6.75	–	–	0.663	0.372	0.291	–	–	–	–	–	96.6	0.291	0.372
–	–	–	7.02	0.171	0.883	5.75	–	–	0.724	0.45	0.274	–	–	–	–	–	72	0.274	0.45
–	–	–	8.28	–	0.885	6.97	–	–	0.83	0.508	0.322	–	–	–	–	–	93	0.322	0.508
–	–	–	9.39	0.213	1	7.91	–	–	0.942	0.577	0.365	–	–	–	–	–	105	0.365	0.577
–	–	–	9.39	0.213	1	7.91	–	–	0.942	0.577	0.365	–	–	–	–	–	105	0.365	0.577
–	–	–	9.39	0.213	1	7.91	–	–	0.942	0.577	0.365	–	–	–	–	–	105	0.365	0.577
–	–	–	9.39	0.213	1	7.91	–	–	0.942	0.577	0.365	–	–	–	–	–	105	0.365	0.577
–	–	–	8.67	–	0.927	7.3	–	–	0.87	0.533	0.337	–	–	–	–	–	103	0.337	0.533
–	–	–	8.01	0.195	1.01	6.56	–	–	0.826	0.513	0.313	–	–	–	–	–	100	0.313	0.513
–	–	–	9.28	0.17	0.979	7.81	–	–	0.953	0.676	0.277	–	–	–	–	–	94.9	0.277	0.676
–	–	–	9.28	0.17	0.979	7.81	–	–	0.953	0.676	0.277	–	–	–	–	–	94.9	0.277	0.676
–	–	–	9.28	0.17	0.979	7.81	–	–	0.953	0.676	0.277	–	–	–	–	–	94.9	0.277	0.676
–	–	–	9.84	0.21	0.984	8.38	–	–	1.27	0.774	0.491	–	–	–	–	–	110	0.491	0.774
–	–	–	7.27	0.158	0.879	6.02	–	–	0.972	0.62	0.352	–	–	–	–	–	91.7	0.352	0.62
–	–	–	8.01	0.186	0.833	6.75	–	–	0.663	0.372	0.291	–	–	–	–	–	96.6	0.291	0.372
–	–	–	8.13	0.177	0.813	6.91	–	–	0.665	0.418	0.247	–	–	–	–	–	89.2	0.247	0.418
–	–	–	8.13	0.177	0.813	6.91	–	–	0.665	0.418	0.247	–	–	–	–	–	89.2	0.247	0.418
–	–	–	8.6	0.171	0.817	7.4	–	–	0.784	0.491	0.293	–	–	–	–	–	94.4	0.293	0.491
–	–	–	7.27	0.158	0.879	6.02	–	–	0.972	0.62	0.352	–	–	–	–	–	91.7	0.352	0.62
–	–	–	7.21	0.001	0.92	6.29	–	–	1.05	0.65	0.37	–	0.028	–	–	–	92	0.37	0.678
–	–	–	8.67	0.371	0.716	7.3	–	–	1.73	0.808	0.924	–	–	–	–	–	87	0.924	0.808
–	–	–	4.62	0.106	0.385	3.98	–	–	0.591	0.326	0.265	–	–	–	–	–	89	0.265	0.326
–	–	–	4.62	0.106	0.385	3.98	–	–	0.591	0.326	0.265	–	–	–	–	–	89	0.265	0.326
–	–	–	8.69	–	0.89	7.1	–	–	1.65	0.864	0.79	–	–	–	–	–	116	0.79	0.864
–	–	–	8.69	–	0.89	7.1	–	–	1.65	0.864	0.79	–	–	–	–	–	116	0.79	0.864
–	–	–	8.69	–	0.89	7.1	–	–	1.65	0.864	0.79	–	–	–	–	–	116	0.79	0.864
–	–	–	7.88	–	0.534	6.96	–	–	0.938	0.508	0.429	–	–	–	–	–	94	0.429	0.508
–	–	–	7.88	–	0.534	6.96	–	–	0.938	0.508	0.429	–	–	–	–	–	94	0.429	0.508
–	–	–	8.12	–	0.846	7.27	–	–	0.845	0.845	–	–	–	–	–	–	105	–	0.845
–	–	–	4.81	–	0.501	4.31	–	–	0.501	0.501	–	–	–	–	–	–	46	–	0.501
–	–	–	8.45	0.134	0.727	7.42	–	–	1.76	0.821	0.939	–	–	–	–	–	88	0.939	0.821
–	–	–	7.75	0.205	0.889	6.39	–	–	0.657	0.263	0.394	–	–	–	–	–	114	0.394	0.263
–	–	–	7.75	0.205	0.889	6.39	–	–	0.657	0.263	0.394	–	–	–	–	–	114	0.394	0.263
–	–	–	10	0.296	0.724	8.58	–	–	1.73	1.3	0.433	–	–	–	–	–	110	0.433	1.3
–	–	–	10	0.296	0.724	8.58	–	–	1.73	1.3	0.433	–	–	–	–	–	110	0.433	1.3
–	–	–	10	0.296	0.724	8.58	–	–	1.73	1.3	0.433	–	–	–	–	–	110	0.433	1.3
–	–	–	8.6	0.171	0.817	7.4	–	–	0.784	0.491	0.293	–	–	–	–	–	94.4	0.293	0.491
–	–	–	7.27	0.158	0.879	6.02	–	–	0.972	0.62	0.352	–	–	–	–	–	91.7	0.352	0.62
–	–	–	7.75	–	0.911	6.48	–	–	0.81	0.554	0.255	–	–	–	–	–	73.9	0.255	0.554
–	–	–	7.05	0.15	0.705	6	–	–	0.9	0.55	0.35	–	–	–	–	–	89.3	0.35	0.55
–	–	–	8.61	–	1.01	7.13	–	–	0.495	0.339	0.156	–	–	–	–	–	90	0.156	0.339
–	–	–	5	0.542	0.501	4.27	–	–	0.64	0.39	0.247	–	–	–	–	–	56	0.247	0.39
–	–	–	4.85	0.88	0.473	4.17	–	–	0.508	0.36	0.147	–	–	–	–	–	54	0.147	0.36
–	–	–	9.34	0.204	0.983	7.84	–	–	0.94	0.666	0.273	–	–	–	–	–	100	0.273	0.666
–	–	–	9.34	0.204	0.983	7.84	–	–	0.94	0.666	0.273	–	–	–	–	–	100	0.273	0.666
–	–	–	9.34	0.204	0.983	7.84	–	–	0.94	0.666	0.273	–	–	–	–	–	100	0.273	0.666

Table 13.4 *(Continued)*

Item	H_2O	Ash	Cals.	Prot.	Carb.	Fat-T	Fat-S	4:0	6:0	8:0	10:0	12:0	14:0	15:0	16:0	17:0	18:0
Low Sodium colby cheese, shredded	39	2.2	398	24.3	1.9	32.6	20.7	1.06	0.265	0.331	0.611	0.432	3.92	0.998	8.42	0.913	3.71
Low Sodium Low Fat cheddar cheese, diced	64.6	2.2	173	24.3	1.9	7	4.44	0.229	0.057	0.071	0.131	0.093	0.842	0.214	1.81	0.195	0.798
Low Sodium Low Fat cheddar cheese, shredded	64.6	2.2	173	24.3	1.9	7	4.44	0.229	0.57	0.71	0.131	0.93	0.842	0.214	1.72	0.195	0.798
Low Sodium mozzarella cheese, shredded	49.9	2.4	280	27.5	3.1	17.1	10.9	0.556	0.108	0.126	0.27	0.177	1.72	0.303	5.24	0.29	2.09
Low Sodium muenster cheese, diced	43.3	2.2	368	23.4	1.1	30	19.1	1.04	0.243	0.275	0.586	0.367	3.07	0.378	9.22	0.362	3.57
Low Sodium parmesan cheese, grated	22.2	2.5	456	41.6	3.7	30	19.1	1.52	0.567	0.303	0.756	1.02	3.39	0.864	8.13	0.332	2.68
Low Sodium string cheese/stick cheese	49.9	2.4	280	27.5	3.1	17.1	10.9	0.556	0.108	0.126	0.27	0.177	1.72	0.303	5.24	0.29	2.09
Low Sodium swiss cheese, diced	37.8	3	376	28.4	3.4	27.4	17.8	1.1	0.487	0.289	0.624	0.52	3.06	0.342	7.78	0.328	3.25
Mennonite cheese	36.8	3.93	403	24.9	1.28	33.1	21.1	–	–	–	–	–	–	–	–	–	–
Montery jack cheese	41	3.56	373	24.5	0.68	30.3	19.1	0.982	0.244	0.306	0.563	0.398	3.61	0.949	7.76	0.868	3.42
Montery jack cheese, diced	41	3.56	373	24.5	0.68	30.3	19.1	0.982	0.244	0.306	0.563	0.398	3.61	0.949	7.76	0.868	3.42
Montery jack cheese, shredded	41	3.56	373	24.5	0.68	30.3	19.1	0.982	0.244	0.306	0.563	0.398	3.61	0.949	7.76	0.868	3.42
Mozzarella cheese, part skim, shredded	53.8	3.28	254	24.3	2.78	15.9	10.1	0.516	0.1	0.117	0.251	0.164	1.6	0.282	4.86	0.27	1.94
Mozzarella cheese, part skim, low moist, shredded	48.6	3.71	280	27.5	3.15	17.1	10.9	0.555	0.107	0.125	0.27	0.177	1.72	0.33	5.22	0.316	2.08
Mozzarella cheese, whole milk, shredded	54.1	2.63	281	19.4	2.23	21.6	13.2	0.805	0.446	0.257	0.582	0.69	2.19	0.235	5.33	0.22	2.44
Mozzarella cheese, part skim	53.8	3.28	254	24.3	2.78	15.9	10.1	0.516	0.1	0.117	0.251	0.164	1.6	0.282	4.86	0.27	1.94
Mozzarella cheese, part skim, low moisture	48.6	3.71	280	27.5	3.15	17.1	10.9	0.555	0.107	0.125	0.27	0.177	1.72	0.33	5.22	0.316	2.08
Mozzarella cheese, whole milk	54.1	2.63	281	19.4	2.23	21.6	13.2	0.805	0.446	0.257	0.582	0.69	2.19	0.235	5.33	0.22	2.44
Mozzerella cheese, whole, milk low moisture	48.4	2.92	318	21.6	2.48	24.6	15.6	0.844	0.177	0.203	0.412	0.255	2.68	0.337	7.49	0.322	2.88
Muenster cheese	41.8	3.67	368	23.4	1.13	30	19.1	1.04	0.243	0.275	0.586	0.367	3.07	0.372	9.22	0.357	3.57
Muenster cheese, diced	41.8	3.67	368	23.4	1.13	30	19.1	1.04	0.243	0.275	0.586	0.367	3.07	0.372	9.22	0.357	3.57
Muenster cheese, shredded	41.8	3.67	368	23.4	1.13	30	19.1	1.04	0.243	0.275	0.586	0.367	3.07	0.372	9.22	0.357	0.357
Neufchatel cheese	62.2	1.47	260	9.97	2.95	23.4	14.8	0.609	0.21	0.215	0.435	0.286	2.35	0.43	6.88	0.395	2.99
Mozzarella cheese, whole milk, low moisture, shredded	48.4	2.92	318	21.6	2.48	24.6	15.6	0.844	0.177	0.203	0.412	0.255	2.68	0.337	7.49	0.322	2.88
Oka cheese, diced	41.1	3.19	371	23.2	2.8	29.7	18.8	0.914	0.373	0.299	0.585	0.482	3.23	0.416	8.66	0.381	3.46
Parmesan cheese, grated	17.7	7.03	456	41.6	3.75	30	19.1	1.51	0.565	0.302	0.753	1.02	3.38	0.976	8.1	0.375	2.68
Paremesan cheese, hard, cube	29.2	6.05	392	35.8	3.23	25.8	16.4	1.3	0.487	0.26	0.648	0.873	2.91	0.806	6.97	0.309	2.3
Parmesan cheese, purchased shredded	25	6.4	415	37.9	3.42	27.3	17.4	1.51	0.565	0.302	0.753	1.02	3.38	1.12	8.1	1.19	2.68
Pimento processed cheese	39.1	5.85	375	22.1	1.74	31.2	19.7	1.04	0.355	0.374	0.641	0.483	3.21	0.343	9.09	0.351	3.79
Pimento processed cheese, shredded	39.1	5.85	375	22.1	1.74	31.2	19.7	1.04	0.355	0.374	0.641	0.483	3.21	0.343	9.09	0.351	3.79
Port du Salut cheese	45.5	2.01	352	23.8	0.57	28.2	16.7	1.03	0.556	0.336	0.681	0.768	2.72	–	6.64	–	3.35
Port du Salut cheese, diced	45.5	2.01	352	23.8	0.57	28.2	16.7	1.03	0.556	0.336	0.681	0.768	2.72	–	6.64	–	3.35
Port du Salut cheese, shredded	45.5	2.01	352	23.8	0.57	28.2	16.7	1.03	0.556	0.336	0.681	0.768	2.72	–	6.64	–	3.35
Processed American cheese	39.2	5.85	375	22.2	1.61	31.3	19.7	1.04	0.356	0.375	0.642	0.484	3.21	0.347	9.1	0.347	3.8
Process swiss cheese	42.3	5.86	334	24.7	2.11	25	16	0.815	0.489	0.275	0.572	0.566	2.79	0.355	7.19	0.348	2.6
Provolone cheese	41	4.72	351	25.6	2.15	26.6	17.1	0.981	0.37	0.256	0.482	0.355	2.75	1.18	8.16	1.13	3.06
Provolone cheese, diced	41	4.72	351	25.6	2.15	26.6	17.1	0.981	0.37	0.256	0.482	0.355	2.75	1.18	8.16	1.13	3.06

20:0	22:0	24:0	Fat-M	14:1	16:1	18:1	20:1	22:1	Fat-P	18:2	18:3	18:4	20:4	20:5	22:5	22:6	Chol., mg	Omeg3	Omeg6
–	–	–	9.34	0.204	0.983	7.84	–	–	0.94	0.666	0.273	–	–	–	–	–	100	0.273	0.666
–	–	–	2.01	0.043	0.212	1.69	–	–	0.2	0.142	0.058	–	–	–	–	–	21	0.058	0.142
–	–	–	2.01	0.043	0.212	1.69	–	–	0.2	0.142	0.058	–	–	–	–	–	21	0.058	0.142
–	–	–	4.85	0.092	0.473	4.16	–	–	0.51	0.362	0.148	–	–	–	–	–	54	0.148	0.362
–	–	–	8.71	0.169	0.973	7.34	–	–	0.66	0.431	0.23	–	–	–	–	–	96	0.23	0.431
–	–	–	8.73	0.232	0.458	7.74	–	–	0.66	0.316	0.344	–	–	–	–	–	79	0.344	0.316
–	–	–	4.85	0.092	0.473	4.16	–	–	0.51	0.362	0.148	–	–	–	–	–	54	0.148	0.362
–	–	–	7.27	0.167	0.877	6	–	–	0.97	0.62	0.35	–	–	–	–	–	92	0.35	0.62
–	–	–	9.39	–	–	–	–	–	0.942	–	–	–	–	–	–	–	105	–	–
–	–	–	8.75	0.18	0.923	7.37	–	–	0.899	0.637	0.261	–	–	–	–	–	89	0.261	0.637
–	–	–	8.75	0.18	0.923	7.37	–	–	0.899	0.637	0.261	–	–	–	–	–	89	0.261	0.637
–	–	–	8.75	0.18	0.923	7.37	–	–	0.899	0.637	0.261	–	–	–	–	–	89	0.261	0.637
–	–	–	4.51	0.102	0.44	3.87	–	–	0.472	0.335	0.137	–	–	–	–	–	57.8	0.137	0.335
–	–	–	4.85	0.088	0.473	4.17	–	–	0.508	0.36	0.147	–	–	–	–	–	54	0.147	0.36
–	–	–	6.57	0.138	0.597	5.65	–	–	0.765	0.393	0.372	–	–	–	–	–	78.4	0.372	0.393
–	–	–	4.51	0.102	0.44	3.87	–	–	0.472	0.335	0.137	–	–	–	–	–	57.8	0.137	0.335
–	–	–	4.85	0.088	0.473	4.17	–	–	0.508	0.36	0.147	–	–	–	–	–	54	0.147	0.36
–	–	–	6.57	0.138	0.597	5.65	–	–	0.765	0.393	0.372	–	–	–	–	–	78.4	0.372	0.393
–	–	–	7.03	0.118	0.822	5.93	–	–	0.773	0.566	0.212	–	–	–	–	–	89.4	0.212	0.566
–	–	–	8.71	0.169	0.973	7.34	–	–	0.661	0.431	0.23	–	–	–	–	–	95.6	0.23	0.431
–	–	–	8.71	0.169	0.973	7.34	–	–	0.661	0.431	0.23	–	–	–	–	–	95.6	0.23	0.431
–	–	–	8.71	0.169	0.973	7.34	–	–	0.661	0.431	0.23	–	–	–	–	–	95.6	0.23	0.431
–	–	–	6.77	0.164	0.737	5.66	–	–	0.65	0.448	0.202	–	–	–	–	–	76.1	0.202	0.448
–	–	–	7.03	0.118	0.822	5.93	–	–	0.778	0.566	0.212	–	–	–	–	–	89.4	0.212	0.566
–	–	–	8.6	0.171	0.817	7.4	–	–	0.784	0.491	0.293	–	–	–	–	–	94.4	0.293	0.491
–	–	–	8.73	0.23	0.458	7.74	–	–	0.661	0.316	0.345	–	–	–	–	–	78.7	0.345	0.316
–	–	–	7.51	0.198	0.394	6.66	–	–	0.569	0.272	0.297	–	–	–	–	–	67.7	0.297	0.272
–	–	–	8.73	0.529	0.458	7.05	–	–	0.661	0.29	0.345	–	–	–	–	–	72	0.345	0.29
–	–	–	8.94	0.14	1.03	7.5	–	–	0.988	0.606	0.382	–	–	–	–	–	94.2	0.382	0.606
–	–	–	8.94	0.14	1.03	7.5	–	–	0.988	0.606	0.382	–	–	–	–	–	94.2	0.382	0.606
–	–	–	9.34	–	0.833	8.05	–	–	0.729	0.378	0.351	–	–	–	–	–	123	0.351	0.378
–	–	–	9.34	–	0.833	8.05	–	–	0.729	0.378	0.351	–	–	–	–	–	123	0.351	0.378
–	–	–	9.34	–	0.833	8.05	–	–	0.729	0.378	0.351	–	–	–	–	–	123	0.351	0.378
–	–	–	8.95	0.178	1.03	7.51	–	–	0.99	0.607	0.383	–	–	–	–	–	94.4	0.383	0.607
–	–	–	7.05	0.203	0.677	5.9	–	–	0.622	0.342	0.28	–	–	–	–	–	84.8	0.28	0.342
–	–	–	7.39	0.12	0.878	6.16	–	–	0.769	0.494	0.275	–	–	–	–	–	68.9	0.275	0.494
–	–	–	7.39	0.12	0.878	6.16	–	–	0.769	0.494	0.275	–	–	–	–	–	68.9	0.275	0.494

Table 13.4 (Continued)

Item	H_2O	Ash	Cals.	Prot.	Carb.	Fat-T	Fat-S	4:0	6:0	8:0	10:0	12:0	14:0	15:0	16:0	17:0	18:0
Queso Anejo (Aged Mexican cheese)	30.9	6.72	387	31.8	3.63	26.9	17.1	–	–	–	–	–	–	–	–	–	–
Queso Asadero cheese	48.4	2.91	318	21.6	2.47	24.6	15.6	–	–	–	–	–	–	–	–	–	–
Queso Chihuahua cheese	36.8	3.93	403	24.9	1.28	33.1	21.1	–	–	–	–	–	–	–	–	–	–
Romano cheese, cubic inch	30.9	6.73	387	31.8	3.64	26.9	17.1	1.36	0.508	0.271	0.671	0.911	3.04	0.352	7.27	0.318	2.4
Romano cheese, grated	30.9	6.73	387	31.8	3.64	26.9	17.1	1.36	0.508	0.271	0.671	0.911	3.04	0.352	7.27	0.318	2.4
Roquefort cheese	39.4	6.45	369	21.5	2.01	30.6	19.3	0.877	0.66	0.674	2.16	1.3	3.25	0.346	6.57	0.323	3.14
Roquefort cheese, crumbled	39.4	6.45	369	21.5	2.01	30.6	19.3	0.877	0.66	0.674	2.16	1.3	3.25	0.346	6.57	0.323	3.14
Roquefort cheese, cubic inch square	39.4	6.45	369	21.5	2.01	30.6	19.3	0.877	0.66	0.674	2.16	1.3	3.25	0.346	6.57	0.323	3.14
Samsor cheese, diced	37.2	3.54	376	28.4	3.39	27.5	17.8	1.1	0.488	0.289	0.625	0.521	3.06	0.346	7.79	0.331	3.25
Semi, soft goat cheese	45.5	2.95	364	21.6	2.55	29.8	20.6	1.5	0.656	0.806	2.88	1.32	3.03	–	7.81	–	2.65
Soft cream cheese (Philadelphia)	52.6	0.99	357	3.57	7.14	35.7	17.9	0.812	0.239	0.273	0.543	0.378	2.93	0.463	8.54	0.425	3.3
String cheese stick	53.8	3.28	254	24.3	2.78	15.9	10.1	0.516	0.1	0.117	0.251	0.164	1.6	0.282	4.86	0.27	1.94
Sweitzer cheese, diced	37.2	3.54	376	28.4	3.39	27.5	17.8	1.1	0.488	0.289	0.625	0.521	3.06	0.346	7.79	0.331	3.25
Swiss cheese	37.2	3.54	376	28.4	3.39	27.5	17.8	1.1	0.488	0.289	0.625	0.521	3.06	0.346	7.79	0.331	3.25
Swiss cheese, diced	37.2	3.54	376	28.4	3.39	27.5	17.8	1.1	0.488	0.289	0.625	0.521	3.06	0.346	7.79	0.331	3.25
Swiss cheese, melted	37.2	3.54	376	28.4	3.39	27.5	17.8	1.1	0.488	0.289	0.625	0.521	3.06	0.346	7.79	0.331	3.25
Swiss cheese, shredded	37.2	3.54	376	28.4	3.39	27.5	17.8	1.1	0.488	0.289	0.625	0.521	3.06	0.346	7.79	0.331	3.25
Swiss processed cheese food, slice	43.7	5.78	323	21.9	4.51	24.1	15.5	0.787	0.472	0.265	0.552	0.546	2.7	0.368	6.94	0.36	2.51
Swiss processed cheese, diced	42.3	5.86	334	24.7	2.11	25	16	0.815	0.489	0.275	0.572	0.566	2.79	0.355	7.19	0.348	2.6
Tilsit cheese, whole milk	42.9	4.88	340	24.4	1.89	26	16.8	0.878	0.536	0.381	0.748	1.07	2.61	0.369	7.68	0.338	2.19
Tilsit cheese, whole milk, shredded	42.9	4.88	340	24.4	1.89	26	16.8	0.878	0.536	0.381	0.748	1.07	2.61	0.369	7.68	0.338	2.19
White cheddar cheese	36.8	3.94	403	24.9	1.29	33.1	21.1	1.05	0.529	0.279	0.6	0.541	3.33	0.45	9.8	0.414	4.01
White cheddar cheese, shredded	36.8	3.94	403	24.9	1.29	33.1	21.1	1.05	0.529	0.279	0.6	0.541	3.33	0.45	9.8	0.414	4.01
Whole milk tilsit cheese, diced	42.9	4.88	340	24.4	1.89	26	16.8	0.878	0.536	0.381	0.748	1.07	2.61	0.369	7.68	0.338	2.19
Wilstermarsch cheese, diced	41.1	3.19	371	23.2	2.8	29.7	18.8	0.914	0.373	0.299	0.585	0.482	3.23	0.416	8.66	0.381	3.46
Zweiteitige cheese, diced	41.1	3.19	371	23.2	2.8	29.7	18.8	0.914	0.373	0.299	0.585	0.482	3.23	0.416	8.66	0.381	3.46

20:0	22:0	24:0	Fat-M	14:1	16:1	18:1	20:1	22:1	Fat-P	18:2	18:3	18:4	20:4	20:5	22:5	22:6	Chol., mg	Omeg3	Omeg6
–	–	–	7.84	–	–	–	–	–	0.593	–	–	–	–	–	–	–	104	–	–
–	–	–	7.03	–	–	–	–	–	0.778	–	–	–	–	–	–	–	89.4	–	–
–	–	–	9.39	–	–	–	–	–	0.942	–	–	–	–	–	–	–	105	–	–
–	–	–	7.84	0.099	0.411	6.94	–	–	0.593	0.284	0.31	–	–	–	–	–	104	0.31	0.284
–	–	–	7.84	0.099	0.411	6.94	–	–	0.593	0.284	0.31	–	–	–	–	–	104	0.31	0.284
–	–	–	8.47	0.124	0.73	7.46	–	–	1.32	0.616	0.704	–	–	–	–	–	90	0.704	0.616
–	–	–	8.47	0.124	0.73	7.46	–	–	1.32	0.616	0.704	–	–	–	–	–	90	0.704	0.616
–	–	–	8.47	0.124	0.73	7.46	–	–	1.32	0.616	0.704	–	–	–	–	–	90	0.704	0.616
–	–	–	7.27	0.158	0.879	6.02	–	–	0.972	0.62	0.352	–	–	–	–	–	91.7	0.352	0.62
–	–	–	6.81	–	0.709	6.1	–	–	0.709	0.709	–	–	–	–	–	–	79	–	0.709
–	–	–	10.1	0.215	1.01	8.57	–	–	1.3	0.792	0.502	–	–	–	–	–	107	0.502	0.792
–	–	–	4.51	0.102	0.44	3.87	–	–	0.472	0.335	0.137	–	–	–	–	–	57.8	0.137	0.335
–	–	–	7.27	0.158	0.879	6.02	–	–	0.972	0.62	0.352	–	–	–	–	–	91.7	0.352	0.62
–	–	–	7.27	0.158	0.879	6.02	–	–	0.972	0.62	0.352	–	–	–	–	–	91.7	0.352	0.62
–	–	–	7.27	0.158	0.879	6.02	–	–	0.972	0.62	0.352	–	–	–	–	–	91.7	0.352	0.62
–	–	–	7.27	0.158	0.879	6.02	–	–	0.972	0.62	0.352	–	–	–	–	–	91.7	0.352	0.62
–	–	–	7.27	0.158	0.879	6.02	–	–	0.972	0.62	0.352	–	–	–	–	–	91.7	0.352	0.62
–	–	–	6.8	0.196	0.653	5.69	–	–	0.6	0.33	0.27	–	–	–	–	–	81.8	0.27	0.33
–	–	–	7.05	0.203	0.677	5.9	–	–	0.622	0.342	0.28	–	–	–	–	–	84.8	0.28	0.342
–	–	–	7.14	0.179	0.668	6.07	–	–	0.721	0.398	0.323	–	–	–	–	–	102	0.323	0.398
–	–	–	7.14	0.179	0.668	6.07	–	–	0.721	0.398	0.323	–	–	–	–	–	102	0.323	0.398
–	–	–	9.39	0.209	1	7.91	–	–	0.942	0.577	0.365	–	–	–	–	–	105	0.365	0.577
–	–	–	9.39	0.209	1	7.91	–	–	0.942	0.577	0.365	–	–	–	–	–	105	0.365	0.577
–	–	–	7.14	0.179	0.668	6.07	–	–	0.721	0.398	0.323	–	–	–	–	–	102	0.323	0.398
–	–	–	8.6	0.171	0.817	7.4	–	–	0.784	0.491	0.293	–	–	–	–	–	94.4	0.293	0.491
–	–	–	8.6	0.171	0.817	7.4	–	–	0.784	0.491	0.293	–	–	–	–	–	94.4	0.293	0.491

Table 13.5 Lipid composition of fish and related products, contents/100 g

Item	H_2O	Ash	Cals.	Prot.	Carb.	Fat-T	Fat-S	4:0	6:0	8:0	10:0	12:0	14:0	15:0	16:0	17:0	18:0
Cod liver oil (fish oil)	–	–	902	–	–	100	22.6	–	–	–	–	–	3.57	–	10.6	–	2.8
Herring oil (fish oil)	–	–	902	–	–	100	21.3	–	–	–	–	0.157	7.19	–	11.7	–	0.818
Menhaden fish oil	–	–	902	–	–	100	30.4	–	–	–	–	–	7.96	1.2	15.1	1.8	3.78
Salmon oil (fish oil)	–	–	902	–	–	100	19.9	–	–	–	–	–	3.28	1	9.84	1.5	4.25
Sardine oil (fish oil)	–	–	902	–	–	100	29.9	–	–	–	–	0.103	6.53	0.8	16.6	0.8	3.89
Mehaden fish oil, fully hydrogenated	–	–	902	–	–	100	95.6	–	–	–	–	–	9.06	–	35.2	–	21.1
Shark oil	0.5	–	898	–	–	99.5	38.1	–	–	–	–	–	6.12	–	27.8	–	4.2
Anchovies in oil, canned, drained	50.3	2.05	210	28.9	–	9.72	2.2	–	–	–	–	0.02	0.428	–	1.33	–	0.42
Anchovies, cooked	50.3	2.05	210	28.9	–	9.72	2.2	–	–	–	–	0.02	0.428	–	1.33	–	0.42
Atlantic cod, baked/broiled	75.9	1.5	105	22.8	–	0.86	0.168	–	–	–	–	–	0.012	–	0.117	–	0.038
Atlantic cod, raw	81.2	1.17	82	17.8	–	0.67	0.131	–	–	–	–	–	0.009	–	0.091	–	0.03
Atlantic cod, dried, salted, 5.5 × 1.5 × .5 inch	16.1	4.11	290	62.8	–	2.38	0.462	–	–	–	–	–	0.032	–	0.321	–	0.106
Atlantic cod, not drained, can	75.6	1.5	105	22.8	–	0.86	0.167	–	–	–	–	–	0.012	–	0.116	–	0.038
Atlantic herring, baked/broiled	64.2	1.88	203	23	–	11.6	2.62	–	–	–	–	0.015	0.71	–	1.74	–	0.14
Atlantic herring, kippered, smoked	59.7	1.95	217	24.6	–	12.4	2.79	–	–	–	–	0.016	0.758	–	1.85	–	1.49
Atlantic herring, raw	72.1	1.47	158	18	–	9.05	2.04	–	–	–	0.005	0.012	0.554	–	1.35	–	0.109
Atlantic mackerel, baked/broiled	53.3	1.54	262	23.9	–	17.8	4.18	–	–	–	–	–	0.592	–	1.38	–	0.256
Atlantic mackerel, raw	63.6	1.36	205	18.6	–	13.9	3.26	–	–	–	–	0.017	0.674	–	2.13	–	0.423
Atlantic ocean perch, baked/broiled	72.7	1.55	121	23.9	–	2.1	0.313	–	–	–	–	0.004	0.081	–	0.178	–	0.044
Atlantic ocean perch/redfish, raw	78.7	1.21	94	18.6	–	1.64	0.244	–	–	–	–	0.003	0.063	–	0.139	–	0.34
Atlantic pollock, raw	78.2	1.42	92	19.4	–	0.98	0.135	–	–	–	–	–	0.005	–	0.084	–	0.043
Atlantic pollock, baked/broiled	72	1.82	118	24.9	–	1.27	0.17	–	–	–	–	–	0.006	–	0.108	–	0.055
Atlantic salmon, baked/broiled	59.6	3.27	182	25.4	–	8.14	1.26	–	–	–	–	–	0.176	–	0.81	–	0.272
Atlantic salmon, raw	68.5	2.55	142	19.8	–	6.35	0.981	–	–	–	–	–	0.137	–	0.632	–	0.212
Atlantic sardines in oil, canned, drained	59.6	3.39	208	24.6	–	11.5	1.53	–	–	–	–	–	0.192	–	0.993	–	0.343
Atlantic sardines, canned with liquid	60	0.9	245	24.6	–	14.5	1.95	–	–	–	–	–	–	–	–	–	–
Atlantic/Pacific halibut, baked/broiled	71.7	1.75	140	26.7	–	2.95	0.417	–	–	–	–	–	0.072	–	0.272	–	0.063
Atlantic/Pacific halibut, raw	77.9	1.37	110	20.8	–	2.3	0.325	–	–	–	–	–	0.56	–	0.212	–	0.049
Black/red caviar, granular	47.5	6.51	252	24.6	4.01	17.9	4.06	–	–	–	–	–	–	–	–	–	–
Bluefin tuna, baked/broiled	59.1	1.52	184	29.9	–	6.29	1.61	–	–	–	–	–	0.178	–	1.04	–	0.394
Bluefin tuna, cubic inch pieces, raw	68.1	1.19	144	23.3	–	4.91	1.26	–	–	–	–	–	0.139	–	0.81	–	0.307
Bluefish, baked/broiled	62.6	1.34	159	25.7	–	5.45	1.17	–	–	–	–	–	0.229	–	0.738	–	0.205
Bluefish, raw	70.9	1.05	124	20	–	4.25	0.915	–	–	–	–	–	0.179	–	0.576	–	0.16
Bluefish, fried in crumbs	60.8	2	205	22.7	4.7	9.8	2.14	–	–	–	–	–	–	–	–	–	–
Bluefish, steamed	68	2.6	148	24.4	–	5	1.07	–	–	–	–	–	–	–	–	–	–
Carp, baked/broiled	69.6	1.88	162	22.9	–	7.18	1.39	–	–	–	–	–	0.146	–	0.842	–	0.237
Carp, raw	76.3	1.47	127	17.8	–	5.61	1.08	–	–	–	–	–	0.114	–	0.657	–	0.185
Carp, breaded/flour fried	50	1.69	277	20.7	11.9	15.7	3.68	–	–	–	–	–	–	–	–	–	–
Catfish, battered/breaded, baked	48	2.81	284	20.7	12	16.6	3.54	–	–	–	–	–	–	–	–	–	–
Catfish, medium size, breaded/flour fried	49.7	2.61	269	21.1	12.1	14.6	3.62	–	–	–	–	–	–	–	–	–	–
Catfish, steamed/poached	70.5	1.47	145	22.7	–	5.32	1.23	–	–	–	–	–	–	–	–	–	–
Channel catfish, breaded, fried	58.8	1.27	229	18.1	8.05	13.3	3.29	–	–	–	–	–	0.091	–	2.01	–	1.17
Channel catfish, raw	80.4	0.96	95	16.4	–	2.83	0.722	–	–	–	–	–	0.064	–	0.442	–	0.15
Chinook salmon, raw	73.2	1.38	180	20.1	–	10.4	2.51	–	–	–	–	–	0.355	–	1.6	–	0.557

20:0	22:0	24:0	Fat-M	14:1	16:1	18:1	20:1	22:1	Fat-P	18:2	18:3	18:4	20:4	20:5	22:5	22:6	Chol., mg	Omeg3	Omeg6
–	4.7	–	46.7	–	8.31	20.7	10.4	7.33	22.5	0.935	0.935	0.935	0.935	6.9	0.935	11	570	18.8	1.87
–	–	–	56.6	–	9.64	12	13.6	20.6	15.6	1.15	0.763	2.31	0.289	6.27	0.619	4.21	766	11.2	1.44
0.2	0.1	0.2	26.7	–	10.5	14.5	1.33	0.352	34.2	2.15	1.49	2.74	1.17	13.2	4.92	8.56	521	23.3	3.32
–	–	–	29	–	4.82	17	3.86	3.38	40.3	1.54	1.06	2.8	0.675	13	2.99	18.2	485	32.3	2.22
0.4	0.2	0.1	33.8	–	7.51	14.8	5.99	5.59	31.9	2.01	1.33	3.03	1.76	10.1	1.97	10.7	710	22.1	3.77
–	–	–	–	–	–	–	–	–	–	–	–	–	–	–	–	–	500	–	–
–	–	–	25.4	0.44	5.86	16.2	2.89	–	25.2	0.27	–	5.08	–	3.5	–	16.4	–	19.9	0.27
–	–	–	3.77	–	0.59	2.94	0.009	0.163	2.56	0.362	0.017	0.078	0.01	0.763	0.041	1.29	85	2.07	0.372
–	–	–	3.77	–	0.59	2.94	0.009	0.163	2.56	0.362	0.017	0.078	0.01	0.763	0.041	1.29	85	2.07	0.372
–	–	–	0.124	–	0.021	0.078	0.019	0.004	0.292	0.006	0.001	0.001	0.028	0.004	0.013	0.154	55	0.159	0.34
–	–	–	0.094	–	0.016	0.061	0.015	0.003	0.231	0.005	0.001	0.001	0.022	0.064	0.01	0.12	43	0.185	0.27
–	–	–	0.342	–	0.056	0.215	0.053	0.011	0.804	0.018	0.004	0.004	0.078	0.011	0.035	0.423	152	0.438	0.096
–	–	–	0.124	–	0.02	0.078	0.019	0.004	0.291	0.006	0.001	0.001	0.028	0.004	0.013	0.153	55	0.158	0.034
–	–	–	4.79	–	0.797	1.94	0.924	1.08	2.74	0.167	0.132	0.274	0.077	0.909	0.071	1.11	77	2.15	0.244
–	–	–	5.11	–	0.851	2.07	0.986	1.15	2.92	0.178	0.141	0.293	0.082	0.97	0.075	1.18	82	2.29	0.26
–	–	–	3.74	–	0.622	1.52	0.721	0.84	2.13	0.13	0.103	0.214	0.06	0.709	0.055	0.862	60	1.67	0.19
–	–	–	7.01	–	0.534	1.2	1.6	2.5	4.3	0.147	0.113	–	0.051	0.504	0.106	0.699	75	1.32	0.198
–	–	–	4.06	–	0.727	2.28	1.04	1.41	4.76	0.219	0.159	0.278	0.183	0.898	0.212	1.4	70	2.46	0.402
–	–	–	0.801	–	0.11	0.265	0.128	0.294	0.547	0.036	0.073	0.031	0.005	0.103	0.029	0.271	54	0.447	0.041
–	–	–	0.625	–	0.086	0.207	0.1	0.229	0.427	0.028	0.057	0.024	0.004	0.08	0.023	0.211	42	0.348	0.032
–	–	–	0.112	–	0.012	0.067	0.022	0.01	0.483	0.009	–	0.005	0.026	0.071	0.022	0.35	71	0.421	0.035
–	–	–	0.143	–	0.015	0.086	0.028	0.013	0.622	0.012	–	0.006	0.033	0.091	0.028	0.451	91	0.542	0.045
–	–	–	2.7	–	0.322	1.73	0.286	0.358	3.26	0.22	0.378	0.106	0.342	0.411	0.368	1.43	71	2.22	0.562
–	–	–	2.1	–	0.251	1.35	0.223	0.279	2.54	0.172	0.295	0.083	0.267	0.321	0.287	1.12	55	1.74	0.439
–	–	–	3.87	–	0.22	2.15	0.423	1.08	5.15	3.54	0.498	0.125	–	0.473	–	0.509	142	1.48	3.54
–	–	–	4.92	–	–	–	–	–	6.55	–	–	–	–	–	–	–	142	–	–
–	–	–	0.967	–	0.209	0.463	0.159	0.131	0.94	0.038	0.083	0.05	0.178	0.091	0.121	0.374	41	0.548	0.216
–	–	–	0.652	–	0.163	0.361	0.124	0.102	0.835	0.03	0.065	0.039	0.139	0.071	0.094	0.292	32	0.428	0.169
–	–	–	4.64	–	–	–	–	–	7.41	–	–	–	–	–	–	–	588	–	–
–	–	–	2.05	–	0.208	1.19	0.355	0.304	1.84	0.068	–	0.05	0.055	0.363	0.16	1.14	49	1.5	0.123
–	–	–	1.36	–	0.162	0.924	0.277	0.237	1.68	0.053	–	0.039	0.043	0.283	0.125	0.89	38	1.17	0.096
–	–	–	2.3	–	0.355	0.876	0.436	0.63	1.36	0.077	–	0.214	–	0.323	0.079	0.665	76	0.988	0.077
–	–	–	1.79	–	0.277	0.684	0.34	0.492	1.06	0.06	–	0.167	–	0.252	0.062	0.519	59	0.771	0.06
–	–	–	4.34	–	–	–	–	–	2.46	–	–	–	–	–	–	–	60	–	–
–	–	–	2.23	–	–	–	–	–	1.23	–	–	–	–	–	–	–	63	–	–
–	–	–	2.99	–	0.84	1.47	0.091	0.515	1.84	0.663	0.346	0.074	0.195	0.305	0.105	0.146	84	0.797	0.858
–	–	–	2.33	–	0.655	1.15	0.071	0.402	1.43	0.517	0.27	0.058	0.152	0.238	0.082	0.114	66	0.622	0.669
–	–	–	6.69	–	–	–	–	–	3.99	–	–	–	–	–	–	–	99.6	–	–
–	–	–	6.96	–	–	–	–	–	4.74	–	–	–	–	–	–	–	90	–	–
–	–	–	6.05	–	–	–	–	–	3.61	–	–	–	–	–	–	–	92.1	–	–
–	–	–	2.02	–	–	–	–	–	1.25	–	–	–	–	–	–	–	72.5	–	–
–	–	–	5.61	–	0.171	5.35	0.049	0.002	3.33	2.62	0.178	0.03	0.098	0.119	0.057	0.222	81	0.519	2.71
–	–	–	0.844	–	0.176	0.594	0.021	0.008	0.865	0.101	0.071	0.013	0.149	0.13	0.1	0.234	58	0.435	0.25
–	–	–	4.48	–	0.884	2.8	0.452	0.346	2.08	0.106	0.086	0.144	0.154	0.788	0.231	0.567	66	1.44	0.26

Table 13.5 (Continued)

Item	H_2O	Ash	Cals.	Prot.	Carb.	Fat-T	Fat-S	4:0	6:0	8:0	10:0	12:0	14:0	15:0	16:0	17:0	18:0
Chinook salmon, baked/broiled	65.6	1.77	231	25.7	–	13.4	3.21	–	–	–	–	–	0.455	–	2.05	–	0.714
Chum salmon, baked/broiled	68.4	1.52	154	25.8	–	4.84	1.08	–	–	–	–	–	0.192	–	0.751	–	0.133
Chum salmon, canned, drained, #1 can	70.8	2.48	141	21.4	–	5.51	1.49	–	–	–	–	–	0.482	–	0.816	–	0.188
Chum salmon, raw	75.4	1.19	120	20.1	–	3.78	0.84	–	–	–	–	–	0.15	–	0.586	–	0.104
Cod, cubic inch pieces, batter fried	67.1	0.65	173	17.3	6.92	8.03	1.92	–	–	–	–	–	–	–	–	–	–
Cod, steamed/poached	76.5	0.343	103	22.3	–	0.837	0.164	–	–	–	–	–	–	–	–	–	–
Coho salmon, steamed/poached	65.4	1.54	184	27.4	–	7.51	1.6	–	–	–	–	–	0.334	–	0.95	–	0.262
Dolfinfish, baked/broiled	71.2	2.7	109	23.7	–	0.9	0.241	–	–	–	–	0.001	0.018	–	0.154	–	0.059
Dolfinfish/mahi mahi, raw	77.6	2.11	85	18.5	–	0.7	0.188	–	–	–	–	0.001	0.014	–	0.12	–	0.046
Dried mackerel	43	13.4	305	18.5	–	25.1	6.5	–	–	–	–	–	–	–	–	–	–
Eel, baked/broiled	59.3	1.81	236	23.7	–	15	3.02	–	–	–	–	–	0.74	–	2.03	–	0.245
Eel, raw	68.3	1.42	184	18.4	–	11.7	2.36	–	–	–	–	–	0.58	–	1.59	–	0.191
Smoked eel	47.1	2.68	281	21.1	14	15.2	4.15	–	–	–	–	–	–	–	–	–	–
Eel, steamed/poached	60.3	2.05	230	23	–	14.6	2.95	–	–	–	–	–	–	–	–	–	–
Farmed atlantic salmon, baked/broiled	64.8	1.16	206	22.1	–	12.4	2.5	–	–	–	–	–	0.571	–	1.49	–	0.315
Farmed atlantic salmon, raw	68.9	1.06	183	19.9	–	10.9	2.18	–	–	–	–	–	0.492	–	1.3	–	0.277
Farmed channel catfish, baked/broiled	71.6	1.18	152	18.7	–	8.03	1.79	–	–	–	–	–	0.112	–	1.31	–	0.319
Farmed channel catfish, raw	75.4	1.01	135	15.6	–	7.6	1.77	–	–	–	–	–	0.099	–	1.27	–	0.354
Farmed coho salmon, baked/broiled	67	1.41	178	24.3	–	8.24	1.94	–	–	–	–	–	0.308	–	1.18	–	0.33
Farmed coho salmon, raw	70.5	1.31	160	21.3	–	7.68	1.82	–	–	–	–	–	0.287	–	1.09	–	0.312
Farmed rainbow trout, baked/broiled	67.5	1.61	169	24.3	–	7.21	2.11	–	–	–	–	–	0.26	–	1.32	–	0.377
Farmed rainbow trout, raw	72.7	1.44	138	20.9	–	5.41	1.55	–	–	–	–	–	0.189	–	0.975	–	0.276
Fresh skipjack tuna, baked/broiled	62.3	1.68	132	28.2	–	1.3	0.42	–	–	–	–	–	0.051	–	0.298	–	0.07
Fresh skipjack tuna, raw	70.6	1.31	103	22	–	1.02	0.328	–	–	–	–	–	0.04	–	0.233	–	0.055
Fresh tuna, breaded/flour fried	51.2	5.42	234	24.4	7.88	11.1	2.81	–	–	–	–	–	–	–	–	–	–
Fresh yellowfin tuna, baked/broiled	62.8	1.73	139	30	–	1.23	0.301	–	–	–	–	0.012	0.014	–	0.2	–	0.065
Fresh yellowfin tuna, raw	71	1.35	108	23.4	–	0.95	0.235	–	–	–	–	0.009	0.011	–	0.156	–	0.051
Freshwater bass, baked/broiled	68.8	1.96	146	24.2	–	4.74	1	–	–	–	–	–	0.112	–	0.761	–	0.128
Freshwater bass, raw	75.7	1.53	114	18.9	–	3.7	0.78	–	–	–	–	–	0.087	–	0.593	–	0.1
Greenland halibut, baked/broiled	61.9	1.29	239	18.4	–	17.7	3.1	–	–	–	–	0.01	0.967	–	1.85	–	0.247
Greenland halibut, raw	70.3	1.01	186	14.4	–	13.8	2.42	–	–	–	–	0.008	0.754	–	1.44	–	0.193
Grouper, baked/broiled	73.4	1.42	118	24.8	–	1.31	0.299	–	–	–	–	–	0.025	–	0.14	–	0.059
Grouper, raw	79.2	1.18	92	19.4	–	1.03	0.233	–	–	–	–	–	0.012	–	0.169	–	0.049
Haddock, baked/broiled	74.3	1.56	112	24.2	–	0.93	0.167	–	–	–	–	–	0.01	–	0.113	–	0.042
Haddock, breaded, fried	55.1	1.7	234	19.4	12.4	11.4	2.87	–	–	–	–	–	–	–	–	–	–
Haddock, raw	79.9	1.22	87	18.9	–	0.72	0.13	–	–	–	–	–	0.008	–	0.088	–	0.033
Haddock, steamed/poached	74.9	0.6	109	23.6	–	0.9	0.162	–	–	–	–	–	–	–	–	–	–
Jack mackerel, canned, drained	69.2	1.74	156	23.2	–	6.31	1.86	–	–	–	–	0.007	0.289	–	1.19	–	0.376
King mackerel, baked/broiled	69	1.65	134	26	–	2.57	0.465	–	–	–	–	–	0.105	–	0.31	–	0.05
King mackerel, raw	75.9	1.29	105	20.3	–	2.01	0.363	–	–	–	–	–	0.082	–	0.242	–	0.039
Light tuna in oil, can, drained	59.8	2.77	198	29.1	–	8.22	1.53	–	–	–	–	–	0.027	–	1.42	–	0.092
Light tuna in water, can, drained	74.5	1.49	116	25.5	–	0.82	0.234	–	–	–	–	–	0.018	–	0.156	–	0.06
Ling fish, baked/broiled	73.9	1.8	111	24.4	–	0.82	–	–	–	–	–	–	–	–	–	–	–
Ling fish, raw	79.6	1.41	87	19	–	0.64	0.12	–	–	–	–	–	–	–	–	–	–

20:0	22:0	24:0	Fat-M	14:1	16:1	18:1	20:1	22:1	Fat-P	18:2	18:3	18:4	20:4	20:5	22:5	22:6	Chol., mg	Omeg3	Omeg6
–	–	–	5.74	–	1.13	3.59	0.58	0.444	2.66	0.136	0.11	0.185	0.197	1.01	0.296	0.727	85	1.85	0.333
–	–	–	1.98	–	0.283	1.05	0.236	0.41	1.15	0.077	0.044	0.087	0.038	0.299	0.101	0.505	95	0.848	0.115
–	–	–	1.92	–	0.237	1.16	0.278	0.245	1.52	0.057	0.049	0.073	0.073	0.473	0.09	0.702	39	1.22	0.13
–	–	–	1.54	–	0.221	0.816	0.184	0.32	0.898	0.06	0.034	0.068	0.03	0.233	0.079	0.394	74	0.661	0.09
–	–	–	3.19	–	–	–	–	–	2.34	–	–	–	–	–	–	–	56.8	–	–
–	–	–	0.117	–	–	–	–	–	0.289	–	–	–	–	–	–	–	53.8	–	–
–	–	–	2.7	–	0.64	1.52	0.316	0.185	2.52	0.26	0.199	0.15	0.169	0.543	0.294	0.831	57	1.57	0.429
–	–	–	0.155	–	0.029	0.11	0.004	0.012	0.211	0.046	0.006	0.006	0.003	0.026	0.012	0.113	94	0.145	0.049
–	–	–	0.121	–	0.023	0.86	0.003	0.009	0.165	0.036	0.005	0.005	0.002	0.02	0.009	0.088	73	0.113	0.038
–	–	–	9.8	–	–	–	–	–	6.7	–	–	–	–	–	–	–	95	–	–
–	–	–	9.22	–	1.61	3.55	3.85	–	1.21	0.251	0.554	–	0.122	0.108	0.095	0.081	161	0.743	0.373
–	–	–	7.19	–	1.26	2.77	3	–	0.947	0.196	0.432	–	0.095	0.084	0.074	0.063	1.26	0.579	0.291
–	–	–	6.33	–	–	–	–	–	3.51	–	–	–	–	–	–	–	100	–	–
–	–	–	8.99	–	–	–	–	–	1.18	–	–	–	–	–	–	–	158	–	–
–	–	–	4.43	–	0.767	2.05	1.37	–	4.43	0.666	0.113	0.184	1.27	0.69	–	1.46	63	2.26	1.94
–	–	–	3.87	–	0.67	1.79	1.19	–	3.93	0.586	0.094	0.157	1.15	0.618	–	1.29	59	2	1.74
–	–	–	4.16	–	0.292	3.74	0.106	–	1.39	1.03	0.082	0.016	0.041	0.049	–	0.128	64	0.259	1.07
–	–	–	3.59	–	0.283	3.17	0.074	0.049	1.57	0.876	0.096	0.037	0.085	0.067	0.09	0.207	47	0.37	0.961
–	–	–	3.62	–	0.482	1.84	0.551	0.665	1.96	0.373	0.076	0.118	0.094	0.408	–	0.871	63	1.36	0.467
–	–	–	3.33	–	0.446	1.72	0.507	0.585	1.86	0.349	0.75	0.112	0.095	0.385	–	0.821	51	1.28	0.444
–	–	–	2.1	–	0.319	1.43	0.326	–	2.33	0.949	0.82	0.068	0.036	0.334	–	0.82	68	1.24	0.985
–	–	–	1.54	–	0.227	1.06	0.24	–	1.81	0.71	0.058	0.05	0.025	0.26	–	0.668	59	0.986	0.735
–	–	–	0.243	–	0.046	0.168	0.022	0.008	0.403	0.02	–	0.005	0.033	0.091	0.017	0.237	60	0.328	0.053
–	–	–	0.19	–	0.036	0.131	0.017	0.006	0.315	0.016	–	0.004	0.026	0.071	0.013	0.185	47	0.256	0.042
–	–	–	4.58	–	–	–	–	–	2.87	–	–	–	–	–	–	–	65.8	–	–
–	–	–	0.197	–	0.032	0.138	0.009	0.018	0.364	0.01	0.015	0.006	0.036	0.047	0.017	0.232	58	0.294	0.046
–	–	–	0.154	–	0.025	0.108	0.007	0.0147	0.284	0.008	0.012	0.005	0.028	0.037	0.013	0.181	45	0.23	0.036
–	–	–	1.84	–	0.49	1.27	0.073	–	1.36	0.112	0.142	0.05	0.185	0.305	0.108	0.458	87	0.905	0.297
–	–	–	1.43	–	0.382	0.991	0.057	–	1.06	0.087	0.111	0.039	0.144	0.238	0.084	0.357	68	0.706	0.231
–	–	–	10.7	–	2.26	3.07	2.87	2.52	1.75	0.158	0.055	0.169	0.078	0.674	0.114	0.504	59	1.23	0.236
–	–	–	8.38	–	1.76	2.39	2.24	1.97	1.37	0.123	0.043	0.132	0.061	0.526	0.089	0.393	46	0.962	0.184
–	–	–	0.268	–	0.026	0.086	–	–	0.403	0.018	–	–	0.057	0.035	0.017	0.213	47	0.248	0.075
–	–	–	0.202	–	0.027	0.145	0.029	0.007	0.321	0.012	0.01	–	0.033	0.027	0.01	0.22	37	0.257	0.045
–	–	–	0.151	–	0.022	0.086	0.013	0.029	0.309	0.012	0.003	0.004	0.029	0.076	0.024	0.162	74	0.241	0.041
–	–	–	4.75	–	–	–	–	–	2.94	–	–	–	–	–	–	–	84.6	–	–
–	–	–	0.118	–	0.017	0.067	0.01	0.023	0.241	0.009	0.002	0.003	0.023	0.059	0.019	0.126	57	0.187	0.032
–	–	–	0.147	–	–	–	–	–	0.301	–	–	–	–	–	–	–	71.3	–	–
–	–	–	2.23	–	0.4	1.13	0.278	0.414	1.65	0.099	0.043	0.106	0.069	0.434	0.104	0.796	79	1.27	0.168
–	–	–	0.979	–	0.151	0.398	0.156	0.273	0.589	0.051	–	0.115	–	0.174	0.022	0.227	68	0.401	0.051
–	–	–	0.764	–	0.118	0.311	0.122	0.213	0.46	0.04	–	0.09	–	0.136	0.017	0.177	53	0.313	0.04
–	–	–	2.95	–	0.083	2.84	0.027	–	2.89	2.68	0.074	–	–	0.027	–	0.101	18	0.202	2.68
–	–	–	0.159	–	0.026	0.093	0.011	0.029	0.337	0.009	0.002	0.013	0.034	0.047	0.009	0.223	30	0.111	0.006
–	–	–	–	–	–	–	–	–	–	–	–	–	–	–	–	–	51	–	–
–	–	–	0.09	–	–	–	–	–	0.22	–	–	–	–	–	–	–	40	–	–

Table 13.5 (Continued)

Item	H_2O	Ash	Cals.	Prot.	Carb.	Fat-T	Fat-S	4:0	6:0	8:0	10:0	12:0	14:0	15:0	16:0	17:0	18:0
Lingcod, baked/broiled	75.7	1.56	109	22.6	–	1.37	0.255	–	–	–	–	–	0.047	–	0.151	–	0.045
Lingcod, raw	81	1.22	85	17.7	–	1.07	0.197	–	–	–	–	–	–	–	–	–	–
Mackerel, baked/broiled	61.7	2.09	201	25.7	–	10.1	2.88	–	–	–	–	0.012	0.436	–	1.78	–	0.565
Northern pike, baked/ broiled	73	1.55	113	24.7	–	0.88	0.151	–	–	–	–	0.001	0.023	–	0.1	–	0.023
Northern pike, raw	78.9	1.21	88	19.3	–	0.69	0.118	–	–	–	–	0.001	0.018	–	0.078	–	0.018
Ocean perch, breaded, fried	58.6	2.02	220	19.8	7.88	11.7	2.81	–	–	–	–	–	–	–	–	–	–
Ocean pout, baked/broiled	76.1	1.46	102	21.3	–	1.18	0.41	–	–	0.073	–	–	0.04	–	0.226	–	0.072
Orange roughy fish, raw	75.9	0.9	69	14.7	–	0.7	0.018	–	–	–	–	–	0.006	–	0.01	–	0.003
Orange roughy, baked/ broiled	69.1	1.16	89	18.9	–	0.9	0.023	–	–	–	–	–	0.007	–	0.013	–	0.004
Pacific cod, baked/broiled	76	1.55	105	23	–	0.81	0.104	–	–	–	–	–	0.004	–	0.085	–	0.015
Pacific cod, raw	81.3	1.21	82	17.9	–	0.63	0.081	–	–	–	–	–	0.003	–	0.066	–	0.012
Pacific halibut, steamed	75	–	131	22.1	–	2.94	0.417	–	–	–	–	–	–	–	–	–	–
Pacific herring, baked/ broiled	63.5	3.05	250	21	–	17.8	4.17	–	–	–	–	–	1.17	–	2.59	–	0.329
Pacific herring, raw	71.5	2.38	195	16.4	–	13.9	3.26	–	–	–	–	–	0.913	–	2.02	–	0.257
Pickled atlantic herring	55.2	2.96	262	14.2	9.65	18	2.38	–	–	–	–	–	0.721	–	1.56	–	0.105
Pink salmon, raw	76.4	1.23	116	19.9	–	3.46	0.558	–	–	–	–	–	0.105	–	0.316	–	0.137
Pink salmon with bone, #1 can, not drained	68.8	2.61	139	19.8	–	6.06	1.54	–	–	–	–	–	0.048	–	1.35	–	0.135
Pink salmon, baked/ broiled	69.7	1.57	149	25.6	–	4.43	0.715	–	–	–	–	–	0.135	–	0.405	–	0.176
Rainbow smelt, baked/ broiled	72.8	1.8	124	22.6	–	3.11	0.579	–	–	–	–	0.003	0.1	–	0.421	–	0.053
Rainbow smelt, raw	78.8	1.41	97	17.6	–	2.43	0.452	–	–	–	–	0.002	0.078	–	0.328	–	0.041
Rainbow trout, baked/ broiled	70.5	1.61	150	22.9	–	5.83	1.62	–	–	–	–	0.013	0.204	–	0.918	–	0.265
Raw shark, mixed species	73.6	1.4	130	21	–	4.52	0.925	–	–	–	–	–	0.08	–	0.726	–	0.117
Sea bass, baked/broiled	72.1	1.41	124	23.6	–	2.57	0.655	–	–	–	–	–	0.063	–	0.413	–	0.177
Striped bass, baked/ broiled	73.4	1.34	124	22.7	–	3	0.65	–	–	–	–	–	0.054	–	0.486	–	0.11
Sea bass, raw	78.3	1.1	97	18.4	–	2.01	0.511	–	–	–	–	–	0.049	–	0.322	–	0.138
Seatrout/steelhead, baked/broiled	71.9	1.63	133	21.5	–	4.64	1.29	–	–	–	–	–	0.095	–	0.82	–	0.214
Seatrout/steelhead, raw	78.1	1.27	104	16.7	–	3.62	1.01	–	–	–	–	–	0.074	–	0.64	–	0.167
Shark, baked/broiled with margarine + lemon juice + salt	65	1.87	179	24.5	0.37	8.21	1.66	–	–	–	–	–	–	–	–	–	–
Shark, batter fried	60.1	1.26	228	18.6	6.4	13.8	3.21	–	–	–	–	–	0.102	–	2.01	–	1.09
Skinless sardines, water pack	59.7	3.35	217	24.6	–	12.4	2.79	–	–	–	–	–	–	–	–	–	–
Smelt, breaded/flour fried	52	2.62	250	20.5	12.1	12.7	3.1	–	–	–	–	–	–	–	–	–	–
Smoked chinook salmon/ lox	72	2.63	117	18.3	–	4.33	0.929	–	–	–	–	–	0.183	–	0.56	–	0.186
Smoked cod, cubic inch pieces	71.5	1.63	116	25.2	–	0.96	0.173	–	–	–	–	–	0.011	–	0.117	–	0.044
Smoked haddock	71.5	1.63	116	25.2	–	0.96	0.173	–	–	–	–	–	0.011	–	0.117	–	0.044
Smoked sturgeon	62.5	1.91	173	31.2	–	4.41	1.04	–	–	–	–	–	0.177	–	0.761	–	0.099
Smoked tuna, cubic inch pieces	59.4	3.8	219	23.8	–	13	3.4	–	–	–	–	–	–	–	–	–	–
Snapper, baked/broiled	70.4	1.42	128	26.3	–	1.73	0.365	–	–	–	–	–	0.053	–	0.19	–	0.07
Snapper, raw	76.9	1.32	100	20.5	–	1.35	0.285	–	–	–	–	–	0.036	–	0.162	–	0.081
Sockeye salmon, baked/ broiled	61.8	1.38	216	27.3	–	11	1.92	–	–	–	–	–	0.258	–	1.01	–	0.158
Sockeye salmon, raw	70.2	1.19	168	21.3	–	8.57	1.5	–	–	–	–	–	0.29	–	1.01	–	0.155
Sockeye salmon, canned, drained, #1 can	68.7	2.71	153	20.5	–	7.32	1.64	–	–	–	–	–	0.359	–	1.08	–	0.205
Sole/flounder, baked/ broiled	73.2	1.51	117	24.2	–	1.54	0.363	–	–	–	–	–	0.057	–	0.226	–	0.055
Sole/flounder, breaded, fried	59	1.72	218	20.1	7.88	11.3	2.85	–	–	–	–	–	–	–	–	–	–
Sole/flounder, raw	79.1	1.21	91	18.8	–	1.2	0.283	–	–	–	–	–	0.048	–	0.158	–	0.071

20:0	22:0	24:0	Fat-M	14:1	16:1	18:1	20:1	22:1	Fat-P	18:2	18:3	18:4	20:4	20:5	22:5	22:6	Chol., mg	Omeg3	Omeg6
–	–	–	0.447	–	0.097	0.244	0.093	–	0.383	0.038	–	–	0.064	0.133	0.018	0.13	67	0.263	0.102
–	–	–	0.35	–	–	–	–	–	0.3	–	–	–	–	–	–	–	52	–	–
–	–	–	3.37	–	0.603	1.7	0.417	0.622	2.49	0.149	0.064	0.16	0.104	0.653	0.158	1.2	60	1.92	0.253
–	–	–	0.201	–	0.072	0.101	0.017	0.006	0.259	0.041	0.027	–	0.036	0.042	0.018	0.095	50	0.164	0.077
–	–	–	0.157	–	0.056	0.079	0.013	0.005	0.202	0.032	0.021	–	0.028	0.033	0.014	0.074	39	0.128	0.06
–	–	–	5.03	–	–	–	–	–	3.01	–	–	–	–	–	–	–	62.9	–	–
–	–	–	0.422	–	0.105	0.317	–	–	0.041	0.041	–	–	–	–	–	–	67	–	0.041
–	–	–	0.479	–	0.058	0.28	0.093	0.039	0.013	0.007	0.001	0.001	0.001	0.001	–	–	20	0.002	0.008
–	–	–	0.615	–	0.074	0.359	0.119	0.05	0.016	0.009	0.002	0.002	0.002	0.002	–	–	26	0.004	0.011
–	–	–	0.105	–	0.04	0.065	–	–	0.313	0.008	0.003	–	0.022	0.103	0.005	0.173	47	0.279	0.03
–	–	–	0.082	–	0.031	0.051	–	–	0.244	0.006	0.002	–	0.017	0.08	0.004	0.135	37	0.217	0.023
–	–	–	0.967	–	–	–	–	–	0.94	–	–	–	–	–	–	–	41	–	–
–	–	–	8.81	–	1.36	3.73	1.74	1.91	3.11	0.246	0.073	0.318	0.123	1.24	0.22	0.883	99	2.2	0.369
–	–	–	6.87	–	1.06	2.91	1.36	1.49	2.42	0.192	0.057	0.248	0.096	0.969	0.172	0.689	77	1.72	0.288
–	–	–	11.9	–	1.38	1.75	3.65	5.17	1.68	0.211	–	–	–	0.843	0.079	0.546	13	1.39	0.211
–	–	–	0.934	–	0.155	0.546	0.124	0.109	1.35	0.05	0.034	0.09	0.078	0.419	0.096	0.586	52	1.04	0.128
–	–	–	1.81	–	0.466	1.07	0.272	0.018	2.05	0.058	0.058	0.135	0.077	0.845	0.048	0.806	55	1.71	0.135
–	–	–	1.2	–	0.199	0.699	0.159	0.14	1.73	0.064	0.044	0.115	0.1	0.537	0.123	0.751	67	1.33	0.164
–	–	–	0.822	–	0.255	0.522	0.036	0.001	1.14	0.058	0.063	0.032	0.071	0.353	0.023	0.536	90	0.952	0.129
–	–	–	0.641	–	0.199	0.407	0.028	0.001	0.885	0.045	0.049	0.025	0.055	0.275	0.018	0.418	70	0.742	0.1
–	–	–	1.75	–	0.675	0.918	0.051	0.168	1.83	0.288	0.187	0.23	0.12	0.468	0.129	0.52	69	1.18	0.408
–	–	–	1.51	–	0.253	0.979	0.266	0.31	1.51	0.076	0.028	0.031	0.108	0.316	0.109	0.527	51	0.871	0.184
–	–	–	0.544	–	0.167	0.377	–	–	0.953	0.031	–	0.059	–	0.206	0.097	0.556	53	0.762	0.031
–	–	–	0.846	–	0.194	0.574	0.073	–	1.01	0.019	0.019	–	–	0.217	–	0.75	103	0.986	0.019
–	–	–	0.424	–	0.13	0.294	–	–	0.743	0.024	–	0.046	–	0.161	0.076	0.434	41	0.595	0.24
–	–	–	1.13	–	0.464	0.602	0.064	–	0.929	0.088	0.005	0.01	0.247	0.211	0.097	0.265	106	0.481	0.335
–	–	–	0.884	–	0.362	0.47	0.05	–	0.725	0.069	0.004	0.008	0.193	0.165	0.076	0.207	83	0.376	0.262
–	–	–	3.07	–	–	–	–	–	2.69	–	–	–	–	–	–	–	59.5	–	–
–	–	–	5.94	–	0.223	5.22	0.217	0.253	3.7	2.6	0.19	0.025	0.092	0.258	0.89	0.431	59	0.879	2.69
–	–	–	5.11	–	–	–	–	–	2.92	–	–	–	–	–	–	–	82	–	–
–	–	–	5.1	–	–	–	–	–	3.5	–	–	–	–	–	–	–	104	–	–
–	–	–	2.02	–	0.305	0.949	0.468	0.301	0.995	0.472	–	–	–	0.183	0.073	0.267	23	0.45	0.472
–	–	–	0.157	–	0.023	0.089	0.013	0.031	0.321	0.012	0.003	0.004	0.031	0.079	0.025	0.168	77	0.25	0.043
–	–	–	0.157	–	0.023	0.089	0.013	0.031	0.321	0.012	0.003	0.004	0.031	0.079	0.025	0.168	77	0.25	0.043
–	–	–	2.36	–	0.44	1.86	0.033	–	0.436	0.062	0.115	–	0.053	0.082	0.029	0.095	80	0.292	0.115
–	–	–	5.1	–	–	–	–	–	3.5	–	–	–	–	–	–	–	95	–	–
–	–	–	0.322	–	0.048	0.123	–	–	0.588	0.025	–	–	0.044	0.048	0.022	0.273	47	0.321	0.069
–	–	–	0.251	–	0.06	0.165	0.021	–	0.459	0.019	0.004	0.009	0.049	0.051	0.065	0.26	37	0.315	0.068
–	–	–	5.29	–	0.321	1.34	0.922	0.664	2.41	0.113	0.062	–	0.03	0.53	0.132	0.7	87	1.29	0.143
–	–	–	4.13	–	0.523	1.38	1.33	0.848	1.88	0.38	0.092	0.102	0.094	0.519	0.039	0.653	62	1.26	0.474
–	–	–	2.78	–	0.406	1.45	0.918	0.387	2.28	0.085	0.09	0.106	0.374	0.492	0.077	0.664	44	1.25	0.459
–	–	–	0.309	–	0.067	0.154	0.035	0.02	0.412	0.014	0.016	–	0.048	0.243	0.046	0.258	68	0.517	0.062
–	–	–	4.65	–	–	–	–	–	2.92	–	–	–	–	–	–	–	68.6	–	–
–	–	–	0.23	–	0.08	0.12	0.028	0.008	0.329	0.008	0.008	0.016	0.38	0.93	0.046	0.106	48	0.207	0.046

Table 13.5 (Continued)

Item	H_2O	Ash	Cals.	Prot.	Carb.	Fat-T	Fat-S	4:0	6:0	8:0	10:0	12:0	14:0	15:0	16:0	17:0	18:0
Sole/flounder, batterfried, cubic inch piece	64.8	–	189	20.3	7.07	8.19	1.52	–	–	–	–	–	–	–	–	–	–
Sole/flounder, steamed	73.8	1.11	114	23.6	–	1.49	0.354	–	–	–	–	–	–	–	–	–	–
Spanish mackerel, baked/broiled	68.5	1.54	158	23.6	–	6.33	1.8	–	–	–	–	–	0.204	–	1.16	–	0.442
Spanish mackerel, raw	71.7	1.28	139	19.3	–	6.31	1.83	–	–	–	–	–	0.173	–	1.24	–	0.416
Striped bass, raw	79.2	1.05	97	17.7	–	2.34	0.507	–	–	–	–	–	0.042	–	0.379	–	0.086
Sturgeon, baked/broiled	69.9	1.42	135	20.7	–	5.19	1.17	–	–	–	–	–	0.197	–	0.878	–	0.097
Sturgeon, breaded/flour fried	56.5	4.07	231	17.5	7.88	14	3.45	–	–	–	–	–	–	–	–	–	–
Sturgeon, raw	76.6	1.11	105	16.1	–	4.05	0.915	–	–	–	–	–	0.154	–	0.685	–	0.076
Swordfish, baked/broiled	68.8	1.9	155	25.4	–	5.15	1.41	–	–	–	–	–	0.138	–	0.965	–	0.271
Swordfish, raw	75.6	1.49	121	19.8	–	4.02	1.1	–	–	–	–	–	0.108	–	0.753	–	0.211
Swordfish, steamed/poached	69.1	0.734	153	25.1	–	5.08	1.39	–	–	–	–	–	–	–	–	–	–
Trout, steamed/poached	63.9	5.85	149	26	–	4.25	0.822	–	–	–	–	–	–	–	–	–	–
Trout, breaded/flour fried	50.5	2.33	267	23	9.47	14.7	3.22	–	–	–	–	–	–	–	–	–	–
Trout, mixed species, raw	71.4	1.18	148	20.8	–	6.62	1.15	–	–	–	–	–	0.185	–	0.815	–	0.148
Tuna, dried	59.5	1.37	210	30	–	9.13	2.5	–	–	–	–	–	–	–	–	–	–
Walleye pike, baked/broiled	73.5	1.55	119	24.5	–	1.57	0.319	–	–	–	–	–	0.023	–	0.252	–	0.044
Walleye pike, raw	79.3	1.21	93	19.1	–	1.23	0.249	–	–	–	–	–	0.018	–	0.197	–	0.034
Walleye pollock, baked/broiled	74.1	1.49	113	23.5	–	1.13	0.231	–	–	–	–	–	0.014	–	0.173	–	0.038
Walleye pollock, raw	81.6	1.22	81	17.2	–	0.8	0.164	–	–	–	–	–	0.008	–	0.123	–	0.028
Whale meat, raw	70.9	1	156	20.6	–	7.5	1.3	–	–	–	–	–	–	–	–	–	–
White meat tuna in water, drained, can	69.5	1.59	136	26.7	–	2.47	0.654	–	–	–	–	–	0.067	–	0.485	–	0.097
White meat tuna in oil, drained, can	64	2.19	186	26.5	–	8.09	1.65	–	–	–	–	–	–	–	–	–	–
Whitefish, baked/broiled	65.1	1.45	1.72	24.5	–	7.52	1.16	–	–	–	–	–	0.14	–	0.768	–	0.254
Whitefish, raw	72.8	1.13	134	19.1	–	5.87	0.906	–	–	–	–	–	0.109	–	0.599	–	0.198
Wild coho salmon, baked/broiled	71.5	1.31	139	23.5	–	4.31	1.05	–	–	–	–	–	0.15	–	0.683	–	0.158
Whild coho salmon, raw	72.7	1.22	146	21.6	–	5.94	1.26	–	–	–	–	–	0.264	–	0.751	–	0.207
Wild rainbow trout, raw	71.9	1.32	119	20.5	–	3.47	0.722	–	–	–	–	0.011	0.079	–	0.421	–	0.139
Yellowtail fish, baked/broiled	67.3	1.41	187	29.7	–	6.73	–	–	–	–	–	–	–	–	–	–	–
Yellowtail fish, raw	74.5	1.1	146	23.1	–	5.25	1.28	–	–	–	–	–	–	–	–	–	–

20:0	22:0	24:0	Fat-M	14:1	16:1	18:1	20:1	22:1	Fat-P	18:2	18:3	18:4	20:4	20:5	22:5	22:6	Chol., mg	Omeg3	Omeg6
–	–	–	3.1	–	–	–	–	–	2.8	–	–	–	–	–	–	–	47.8	–	–
–	–	–	0.291	–	–	–	–	–	0.411	–	–	–	–	–	–	–	60	–	–
–	–	–	2.14	–	0.309	1.1	0.123	0.07	1.81	0.108	0.116	–	0.159	0.294	0.095	0.952	73	1.36	0.267
–	–	–	1.53	–	0.291	1.12	0.072	0.05	1.74	0.097	0.034	–	0.166	0.329	0.101	1.01	76	1.37	0.263
–	–	–	0.66	–	0.151	0.448	0.057	–	0.784	0.015	0.015	–	–	0.169	–	0.585	80	0.769	0.015
–	–	–	2.49	–	0.523	1.84	0.069	0.029	0.885	0.087	0.128	0.164	0.079	0.249	0.058	0.119	77	0.496	0.166
–	–	–	6.28	–	–	–	–	–	3.26	–	–	–	–	–	–	–	80.1	–	–
–	–	–	1.94	–	0.408	1.43	0.054	0.023	0.69	0.068	0.1	0.128	0.062	0.194	0.045	0.093	60	0.387	0.13
–	–	–	1.98	–	0.322	1.39	0.156	0.11	1.18	0.037	0.238	–	0.087	0.138	–	0.681	50	1.06	0.124
–	–	–	1.55	–	0.251	1.09	0.122	0.86	0.922	0.029	0.186	–	0.068	0.108	–	0.531	39	0.825	0.097
–	–	–	1.96	–	–	–	–	–	1.17	–	–	–	–	–	–	–	49.4	–	–
–	–	–	1.31	–	–	–	–	–	1.52	–	–	–	–	–	–	–	72.2	–	–
–	–	–	5.91	–	–	–	–	–	4.37	–	–	–	–	–	–	–	84.7	–	–
–	–	–	3.25	–	0.701	1.44	0.28	0.83	1.5	0.175	0.155	0.064	0.189	0.202	0.183	0.528	58	0.885	0.364
–	–	–	2.38	–	–	–	–	–	3.5	–	–	–	–	–	–	–	68.8	–	–
–	–	–	0.377	–	0.126	0.251	–	–	0.573	0.033	0.018	–	0.074	0.11	0.049	0.288	110	0.416	0.107
–	–	–	0.294	–	0.098	0.196	–	–	0.447	0.026	0.014	–	0.058	0.086	0.038	0.225	86	0.325	0.084
–	–	–	0.174	–	0.019	0.124	0.024	0.008	0.524	0.012	0.008	–	0.016	0.185	0.016	0.283	96	0.476	0.028
–	–	–	0.125	–	0.011	0.092	0.018	0.004	0.412	0.009	0.004	–	0.011	0.15	0.011	0.222	71	0.376	0.02
–	–	–	4.1	–	–	1.2	–	–	1.6	0.2	–	–	–	–	–	–	50	–	0.2
–	–	–	0.648	–	0.118	0.424	0.064	0.036	0.918	0.045	0.058	0.042	0.042	0.191	0.015	0.515	42	0.764	0.087
–	–	–	2.48	–	–	–	–	–	3.38	–	–	–	–	–	–	–	31	–	–
–	–	–	2.56	–	0.667	1.73	0.133	0.032	2.76	0.349	0.235	0.064	0.286	0.406	0.209	1.21	77	1.85	0.634
–	–	–	2	–	0.52	1.35	0.104	0.025	2.15	0.272	0.183	0.05	0.223	0.317	0.163	0.941	60	1.44	0.495
–	–	–	1.58	–	0.272	0.86	0.275	0.153	1.27	0.056	0.055	0.52	0.022	0.401	–	0.658	55	1.11	0.078
–	–	–	2.13	–	0.506	1.2	0.25	1.46	1.99	0.206	0.157	0.119	0.133	0.429	0.232	0.656	45	1.24	0.339
–	–	–	1.13	–	0.203	0.614	0.112	0.138	1.24	0.239	0.119	0.066	0.109	0.167	0.106	0.42	59	0.706	0.348
–	–	–	–	–	–	–	–	–	–	–	–	–	–	–	–	–	71	–	–
–	–	–	1.99	–	–	–	–	–	1.42	–	–	–	–	–	–	–	55	–	–

Table 13.6 Lipid composition of shellfish and related products, contents/100 g

Item	H_2O	Ash	Cals.	Prot.	Carb.	Fat-T	Fat-S	4:0	6:0	8:0	10:0	12:0	14:0	15:0	16:0	17:0	18:0
Abalone, canned	80.2	–	80	16	2.3	5.67	0.27	–	–	–	–	–	–	–	–	–	–
Abalone, flour fried	60.1	1.78	189	19.6	11.1	6.79	1.65	–	–	–	–	–	0.049	–	0.953	–	0.648
Abalone, raw	74.6	1.58	105	17.1	6.02	0.76	0.149	–	–	–	–	–	0.023	–	0.103	–	0.021
Abalone, steamed/poached	49.1	3.14	210	34.2	12	1.52	0.298	–	–	–	–	–	–	–	–	–	–
Alaskan king crab leg, raw	79.6	1.81	84	18.3	–	0.6	0.09	–	–	–	–	–	–	–	–	–	–
Alaskan king crab leg, steamed/boiled	77.6	1.86	97	19.4	–	1.55	0.133	–	–	–	0.003	0.004	0.015	–	0.087	–	0.019
Blue crab, canned	76.2	2.06	99	20.5	–	1.24	0.252	–	–	–	–	–	–	–	0.16	–	0.092
Blue crab, raw	79	1.82	87	18.1	0.04	1.09	0.222	–	–	–	–	–	–	–	0.141	–	0.081
Blue crab, steamed/boiled	77.4	2.01	102	20.2	–	1.78	0.228	–	–	–	–	–	0.018	–	0.144	–	0.061
Blue mussels, raw	80.6	1.6	86	11.9	3.7	2.25	0.425	–	–	–	–	–	0.059	–	0.294	–	0.068
Blue mussels, steamed/boiled	61.2	3.19	172	23.8	7.4	4.49	0.85	–	–	–	–	–	0.118	–	0.588	–	0.136
Clam nectar, canned	97.7	1.81	2	0.4	0.1	0.02	0.002	–	–	–	–	–	–	–	–	–	–
Clams, baked/broiled, large	71.8	3.06	139	15	3.07	7.03	1.27	–	–	–	–	–	–	–	–	–	–
Clams, baked/broiled, small	71.8	3.06	139	15	3.07	7.03	1.27	–	–	–	–	–	–	–	–	–	–
Clams, breaded/flour fried, small	61.6	2.17	202	14.2	10.3	11.2	2.68	–	–	–	–	–	0.058	–	1.55	–	1.07
Clams, canned, drained	63.6	3.75	148	25.6	5.14	1.96	0.188	–	–	–	–	–	0.026	–	0.12	–	0.036
Crab, baked/broiled	73.4	1.26	138	19	0.053	6.37	1.14	–	–	–	–	–	–	–	–	–	–
Crayfish/crawdads, raw	82.2	1.35	77	16	–	0.95	0.159	–	–	–	–	–	0.002	–	0.102	–	0.042
Crayfish/crawdads, steamed/boiled	79.4	1.21	88	16.8	–	1.21	0.181	–	–	–	–	–	0.005	–	0.111	–	0.049
Cuttlefish, mixed species, raw	80.6	1.69	79	16.2	0.82	0.7	0.118	–	–	–	–	–	0.012	–	0.061	–	0.044
Cuttlefish, steamed/boiled	61.1	3.37	158	32.5	1.65	1.41	0.236	–	–	–	–	–	0.024	–	0.122	–	0.088
Dried calamari	20	6.29	339	57.3	11.3	5.08	1.32	–	–	–	–	–	–	–	–	–	–
Dried octopus	20	6.29	339	57.3	11.3	5.08	1.32	–	–	–	–	–	–	–	–	–	–
Dried octopus, boiled	56.7	3.41	183	31	6.13	2.74	0.712	–	–	–	–	–	0.078	–	0.464	–	0.157
Dried shrimp	30.5	3.47	304	58.4	2.61	4.96	0.944	–	–	–	–	–	–	–	–	–	–
Dungeness crab, raw	79.2	1.71	86	17.4	0.74	0.97	0.132	–	–	–	–	–	0.001	–	0.089	–	0.041
Dungeness crab, steamed/boiled	73.3	2.19	110	22.3	0.95	1.25	0.168	–	–	–	–	–	0.001	–	0.114	–	0.053
Eastern oysters, breaded, fried	64.7	1.73	197	8.78	11.6	12.6	3.2	–	–	–	–	–	0.15	–	1.91	–	1.11
Eastern oysters, canned, not drained	85.1	1.43	69	7.07	3.92	2.48	0.631	–	–	–	–	–	0.109	–	0.435	–	0.06
Eastern oysters, raw	85.2	1.43	68	7.06	3.92	2.47	0.772	–	–	–	–	–	0.125	–	0.535	–	0.073
Eastern oysters, steamed/boiled	70.3	2.85	137	14.1	7.83	4.92	1.54	–	–	–	–	–	0.249	–	1.07	–	0.146
Escargot (snails) steamed/poached	58.4	2.6	180	32.2	4	2.8	0.8	–	–	–	–	–	–	–	–	–	–
Farmed crayfish, mixed species, raw	84.1	1.01	72	14.9	–	0.97	0.163	–	–	–	–	–	0.002	–	0.103	–	0.044
Farmed crayfish, mixed species, steam/boil	80.8	1.08	87	17.5	–	1.31	0.216	–	–	–	–	–	0.006	–	0.137	–	0.05
Farmed eastern oysters, medium, baked/broiled	82	1.66	79	7.01	7.29	2.13	0.683	–	–	–	–	–	0.053	–	0.489	–	0.07
Farmed eastern oysters, medium, raw	86.2	1.51	59	5.23	5.54	1.56	0.443	–	–	–	–	–	0.035	–	0.314	–	0.045
King crab, leg, baked/broiled	73.4	1.26	138	19	0.053	6.37	1.14	–	–	–	–	–	–	–	–	–	–
Large clams, raw	81.8	1.88	74	12.8	2.58	0.97	0.94	–	–	–	–	–	0.013	–	0.06	–	0.018
Large clams, steamed/boiled	63.6	3.75	148	25.6	5.14	1.96	0.188	–	–	–	–	–	0.026	–	0.12	–	0.036
Large scallops, breaded, fried	58.4	1.84	215	18.1	10.1	10.9	2.67	–	–	–	–	–	0.056	–	1.55	–	1.06
Large shrimp, raw	75.9	1.21	106	20.3	0.91	1.74	0.328	–	–	–	0.009	0.005	0.021	–	1.84	–	0.103
Lobster tail, baked/broiled	73.8	2.2	116	19.8	1.24	3.01	1.63	–	–	–	–	–	–	–	–	–	–
Lobster tail, batter fried	60.3	1.71	211	19.1	7.78	11.1	2.68	–	–	–	–	–	–	–	–	–	–
Lobster, baked/broiled, diced	73.8	2.2	116	19.8	1.24	3.01	1.63	–	–	–	–	–	–	–	–	–	–
Lobster, batter fried	60.3	1.71	211	19.1	7.78	11.1	2.68	–	–	–	–	–	–	–	–	–	–

20:0	22:0	24:0	Fat-M	14:1	16:1	18:1	20:1	22:1	Fat-P	18:2	18:3	18:4	20:4	20:5	22:5	22:6	Chol., mg	Omeg3	Omeg6
–	–	–	2.25	–	–	–	–	–	3.15	–	–	–	–	–	–	–	80	–	–
–	–	–	2.74	–	0.022	2.72	–	–	1.68	1.47	0.095	0.007	–	0.054	0.046	–	94	0.149	1.47
–	–	–	0.107	–	0.02	0.087	–	–	0.104	0.007	–	0.006	–	0.049	0.041	–	85	0.049	0.007
–	–	–	0.214	–	–	–	–	–	0.208	–	–	–	–	–	–	–	170	–	–
–	–	–	0.08	–	–	–	–	–	0.13	–	–	–	–	–	–	–	42	–	–
–	–	–	0.185	–	0.041	0.089	0.014	0.038	0.536	0.02	0.014	0.015	0.043	0.295	0.031	0.118	53	0.427	0.063
–	–	–	0.218	–	0.067	0.118	0.033	–	0.44	0.014	–	–	0.063	0.193	–	0.17	89	0.363	0.077
–	–	–	0.192	–	0.059	0.104	0.029	–	0.387	0.012	–	–	0.055	0.17	–	0.15	78	0.32	0.067
–	–	–	0.28	–	0.089	0.151	0.038	–	0.68	0.028	0.021	0.019	0.084	0.243	0.054	0.231	100	0.495	0.112
–	–	–	0.507	–	0.138	0.205	0.138	0.026	0.606	0.018	0.02	0.035	0.07	0.188	0.022	0.253	28	0.461	0.088
–	–	–	1.01	–	0.276	0.41	0.276	0.052	1.21	0.036	0.04	0.07	0.14	0.276	0.044	0.506	56	0.822	0.176
–	–	–	0.002	–	–	–	–	–	0.005	–	–	–	–	–	–	–	3	–	–
–	–	–	2.71	–	–	–	–	–	2.19	–	–	–	–	–	–	–	39.8	–	–
–	–	–	2.71	–	–	–	–	–	2.19	–	–	–	–	–	–	–	39.8	–	–
–	–	–	4.55	–	0.042	4.47	0.017	0.006	2.87	2.46	0.16	0.015	0.044	0.066	0.05	0.07	61	0.296	2.5
–	–	–	0.172	–	0.044	0.068	0.036	0.012	0.552	0.032	0.008	0.032	0.082	0.138	0.104	0.146	67	0.292	0.114
–	–	–	2.36	–	–	–	–	–	2.12	–	–	–	–	–	–	–	93.6	–	–
–	–	–	0.174	–	0.028	0.113	0.01	0.018	0.293	0.052	0.032	0.005	0.045	0.104	0.01	0.038	114	0.174	0.097
–	–	–	0.244	–	0.027	0.147	0.012	0.05	0.367	0.09	0.028	0.001	0.056	0.119	0.023	0.047	133	0.194	0.146
–	–	–	0.081	–	0.012	0.031	0.036	0.001	0.134	0.002	0.001	0.002	0.013	0.039	0.006	0.066	112	0.106	0.015
–	–	–	0.162	–	0.024	0.062	0.072	0.002	0.268	0.004	0.002	0.004	0.026	0.078	0.012	0.132	224	0.212	0.03
–	–	–	0.394	–	–	–	–	–	1.93	–	–	–	–	–	–	–	857	–	–
–	–	–	0.39	–	–	–	–	–	1.93	–	–	–	–	–	–	–	857	–	–
–	–	–	0.213	–	0.087	0.081	0.032	0.013	1.04	0.039	–	0.122	0.17	0.331	0.026	0.353	463	0.684	0.209
–	–	–	0.742	–	–	–	–	–	1.91	–	–	–	–	–	–	–	438	–	–
–	–	–	0.167	–	0.043	0.093	0.011	0.02	0.317	–	–	–	–	0.219	0.01	0.088	59	0.307	–
–	–	–	0.214	–	0.055	0.119	0.014	0.026	0.407	–	–	–	–	0.281	0.013	0.113	76	0.394	–
–	–	–	4.7	–	0.092	4.44	0.057	0.009	3.31	2.44	0.156	0.092	0.07	0.202	0.048	0.218	81	0.576	2.51
–	–	–	0.25	–	0.074	0.097	0.06	0.009	0.739	0.049	0.037	0.096	0.068	0.211	0.05	0.228	55	0.476	0.117
–	–	–	0.314	–	0.09	0.122	0.075	0.011	0.968	0.058	0.05	0.117	0.08	0.268	0.062	0.292	53	0.61	0.138
–	–	–	0.628	–	0.181	0.244	0.15	0.022	1.94	0.117	0.1	0.233	0.16	0.536	0.125	0.584	105	1.22	0.277
–	–	–	0.8	–	–	–	–	–	0.6	–	–	–	–	–	–	–	100	–	–
–	–	–	0.178	–	0.031	0.134	0.005	0.002	0.313	0.076	0.025	0.005	0.05	0.117	0.005	0.027	107	0.169	0.126
–	–	–	0.251	–	0.037	0.194	0.007	–	0.413	0.156	0.022	0.002	0.062	0.124	–	0.038	137	0.184	0.218
–	–	–	0.23	–	0.059	0.111	0.049	–	0.713	0.043	0.063	0.118	0.044	0.229	–	0.211	38	0.503	0.087
–	–	–	0.152	–	0.038	0.074	0.033	–	0.591	0.028	0.044	0.092	0.033	0.188	–	0.203	25	0.435	0.061
–	–	–	2.36	–	–	–	–	–	2.12	–	–	–	–	–	–	–	93.6	–	–
–	–	–	0.08	–	0.022	0.034	0.018	0.006	0.282	0.016	0.004	0.016	0.041	0.069	0.052	0.073	34	0.146	0.057
–	–	–	0.172	–	0.044	0.068	0.036	0.012	0.552	0.032	0.008	0.032	0.082	0.138	0.104	0.146	67	0.292	0.114
–	–	–	4.5	–	0.031	4.46	0.006	–	2.86	2.44	0.156	0.011	0.027	0.086	0.016	0.103	61	0.345	2.47
–	–	–	0.253	–	0.083	0.147	0.017	0.005	0.669	0.028	0.014	0.006	0.087	0.258	0.046	0.222	152	0.494	0.115
–	–	–	0.86	–	–	–	–	–	0.179	–	–	–	–	–	–	–	76	–	–
–	–	–	4.65	–	–	–	–	–	3.03	–	–	–	–	–	–	–	79.7	–	–
–	–	–	0.86	–	–	–	–	–	0.179	–	–	–	–	–	–	–	76	–	–
–	–	–	4.65	–	–	–	–	–	3.03	–	–	–	–	–	–	–	79.7	–	–

Table 13.6 (Continued)

Item	H_2O	Ash	Cals.	Prot.	Carb.	Fat-T	Fat-S	4:0	6:0	8:0	10:0	12:0	14:0	15:0	16:0	17:0	18:0
Lobster, canned	76	1.61	98	20.5	1.29	0.59	0.107	–	–	–	–	–	0.006	–	0.081	–	0.02
Northern lobster, raw	76.8	2.21	90	18.8	0.5	0.9	0.18	–	–	–	–	–	–	–	–	–	–
Northern lobster, steamed/ boiled	76	1.61	98	20.5	1.29	0.59	0.107	–	–	–	–	–	0.006	–	0.081	–	0.02
Octopus, raw	80.3	1.61	82	14.9	2.21	1.05	0.227	–	–	–	–	–	0.025	–	0.148	–	0.05
Octopus, steamed/boiled	60.5	3.21	164	29.8	4.41	2.09	0.453	–	–	–	–	–	0.05	–	0.296	–	0.1
Oysters, baked/broiled	83.3	1.76	72	8.26	4.81	1.91	0.55	–	–	–	–	–	0.047	–	0.394	–	0.049
Pacific oysters, raw	82.1	1.24	81	9.46	4.96	2.31	0.51	–	–	–	–	–	0.082	–	0.357	–	0.071
Pacific oysters, steamed/ boiled	64.1	2.47	163	18.9	9.91	4.61	1.02	–	–	–	–	–	0.164	–	0.714	–	0.142
Pickled squid	74.6	4.85	91.5	15	4.23	1.33	0.344	–	–	–	–	–	–	–	–	–	–
Prawns/large shrimp, breaded, fried	52.9	2	242	21.4	11.5	12.3	2.09	–	–	–	–	–	0.035	–	1.46	–	0.541
Prawns/large, steamed/ boiled	77.3	1.58	99	20.9	–	1.09	0.289	–	–	–	–	–	0.017	–	0.143	–	0.096
Queen crab, raw	80.6	2.01	90	18.5	–	1.19	0.143	–	–	–	–	–	0.004	–	0.119	–	0.02
Queen crab, steamed/ boiled	75.1	2.57	115	23.7	–	1.52	0.183	–	–	–	–	–	0.005	–	0.152	–	0.026
Scallops, baked/broiled	70.3	2.65	133	20.3	2.88	3.95	0.692	–	–	–	–	–	0.096	–	0.473	–	0.096
Scallops, battered, fried	55.4	2.34	230	17.6	13.3	11.4	2.81	–	–	–	–	–	–	–	–	–	–
Scallops, breaded, fried	58.4	1.84	215	18.1	10.1	10.9	2.67	–	–	–	–	–	0.056	–	1.55	–	1.06
Scallops, steamed/boiled	76.2	2.12	107	16.2	2.3	3.16	0.55	–	–	–	–	–	0.077	–	00.376	–	0.077
Shrimp, batter fried, medium size	53.1	2.05	246	20.6	12.1	12.2	3.02	–	–	–	–	–	–	–	–	–	–
Shrimp, canned with liquid	73.5	4.34	102	19.6	0.891	1.67	0.317	–	–	–	–	–	–	–	–	–	–
Shrimp, canned, drained, each measure	72.6	1.37	120	23.1	1.04	1.97	0.373	–	–	–	–	0.006	0.024	–	0.209	–	0.117
Shrimp, medium size, baked/broiled	67	2.24	155	24.5	1.13	5.12	0.991	–	–	–	0.027	0.015	0.063	–	0.556	–	0.311
Shrimp, popcorn type, baked/broiled	67	2.24	155	24.5	1.13	5.12	0.991	–	–	–	0.027	0.015	0.063	–	0.556	–	0.311
Smoked clams in oil, canned, small	68.6	2.75	175	14.2	2.86	11.6	2.73	–	–	–	–	–	–	–	–	–	–
Smoked octopus	66.3	2.73	140	25.4	3.75	1.77	0.387	–	–	–	–	–	0.043	–	0.252	–	0.085
Snow crab, leg, baked/ broiled	73.4	1.26	138	19	0.053	6.37	1.14	–	–	–	–	–	–	–	–	–	–
Spiny lobster, raw	74.1	1.4	112	20.6	2.44	1.52	0.237	–	–	–	–	–	0.006	–	0.149	–	0.08
Spiny lobster, steamed/ boiled	66.8	1.79	143	26.4	3.13	1.95	0.303	–	–	–	–	–	0.008	–	0.192	–	0.103
Squid, canned	74.7	2.32	106	17.9	3.54	1.59	0.411	–	–	–	–	–	–	–	–	–	–
Squid, flour fried	64.5	1.6	175	17.9	7.8	7.49	1.88	–	–	–	–	–	0.064	–	1.13	–	0.689
Squid, raw	78.6	1.42	92	15.6	3.09	1.39	0.358	–	–	–	–	–	0.036	–	0.263	–	0.058
Squid/calamari, baked	70.2	2.5	138	18.8	3.74	4.7	1.03	–	–	–	–	–	–	–	–	–	–

20:0	22:0	24:0	Fat-M	14:1	16:1	18:1	20:1	22:1	Fat-P	18:2	18:3	18:4	20:4	20:5	22:5	22:6	Chol., mg	Omeg3	Omeg6
–	–	–	0.16	–	0.03	0.095	0.017	0.018	0.091	0.005	–	–	–	0.053	0.002	0.031	72	0.084	0.005
–	–	–	0.26	–	–	–	–	–	0.15	–	–	–	–	–	–	–	95	–	–
–	–	–	0.16	–	0.03	0.095	0.017	0.018	0.091	0.005	–	–	–	0.053	0.002	0.031	72	0.084	0.005
–	–	–	0.162	–	0.066	0.062	0.024	0.01	0.239	0.009	–	0.028	0.039	0.076	0.006	0.081	48	0.157	0.048
–	–	–	0.324	–	0.132	0.124	0.048	0.02	0.477	0.018	–	0.056	0.078	0.152	0.012	0.162	96	0.314	0.096
–	–	–	0.23	–	0.061	0.089	0.063	0.341	0.809	0.032	0.05	0.087	0.053	0.26	0.049	0.291	49	0.601	0.085
–	–	–	0.358	–	0.114	0.191	–	0.053	0.894	0.032	0.032	0.084	0.038	0.438	0.02	0.25	50	0.72	0.07
–	–	–	0.716	–	0.228	0.382	–	0.106	1.79	0.064	0.064	0.168	0.076	0.876	0.04	0.5	100	1.44	0.14
–	–	–	0.103	–	–	–	–	–	0.504	–	–	–	–	–	–	–	224	–	–
–	–	–	3.81	–	0.076	3.75	0.105	0.022	5.09	4.51	0.267	–	0.059	0.109	0.035	0.124	177	0.5	4.57
–	–	–	0.197	–	0.06	0.114	0.011	0.007	0.44	0.021	0.012	–	0.071	0.171	0.02	0.144	195	0.327	0.092
–	–	–	0.256	–	0.05	0.184	0.022	–	0.422	0.006	0.002	0.001	0.033	0.259	0.008	0.113	55	0.374	0.039
–	–	–	0.328	–	0.064	0.236	0.028	–	0.54	0.008	0.003	0.001	0.042	0.332	0.01	0.145	71	0.479	0.05
–	–	–	1.4	–	0.445	0.641	0.226	–	1.27	0.02	–	0.058	0.112	0.439	0.083	0.527	39.8	0.966	0.132
–	–	–	4.66	–	–	–	–	–	2.9	–	–	–	–	–	–	–	81.7	–	–
–	–	–	4.5	–	0.031	4.46	0.006	–	2.86	2.44	0.156	0.011	0.027	0.086	0.016	0.103	61	0.345	2.47
–	–	–	1.12	–	0.333	0.515	0.182	–	1.02	0.016	–	0.047	0.09	0.352	0.066	0.422	31.8	0.774	0.106
–	–	–	4.85	–	–	–	–	–	3.25	–	–	–	–	–	–	–	184	–	–
–	–	–	0.249	–	–	–	–	–	0.642	–	–	–	–	–	–	–	147	–	–
–	–	–	0.293	–	0.094	0.167	0.019	0.006	0.755	0.032	0.016	0.007	0.099	0.293	0.052	0.252	173	0.561	0.131
–	–	–	1.65	–	0.543	0.962	0.111	0.033	1.76	0.074	0.037	0.016	0.229	0.68	0.121	0.585	183	1.3	0.303
–	–	–	1.65	–	0.543	0.962	0.111	0.033	1.76	0.074	0.037	0.016	0.229	0.68	0.121	0.585	183	1.3	0.303
–	–	–	4.77	–	–	–	–	–	3.06	–	–	–	–	–	–	–	37.8	–	–
–	–	–	0.276	–	0.112	0.106	0.041	0.017	0.407	0.015	–	0.048	0.066	0.129	0.01	0.138	81.8	0.267	0.081
–	–	–	2.36	–	–	–	–	–	2.12	–	–	–	–	–	–	–	93.6	–	–
–	–	–	0.275	–	0.039	0.208	0.027	–	0.59	0.013	0.008	0.005	0.154	0.265	0.034	0.108	70	0.381	0.167
–	–	–	0.353	–	0.05	0.268	0.035	–	0.756	0.017	0.01	0.006	0.198	0.341	0.044	0.139	90	0.491	0.215
–	–	–	0.123	–	–	–	–	–	0.602	–	–	–	–	–	–	–	268	–	–
–	–	–	2.75	–	0.009	2.68	0.049	0.008	2.14	1.47	0.1	0.007	0.01	0.162	0.004	0.38	260	0.642	1.48
–	–	–	0.107	–	0.008	0.046	0.044	0.007	0.524	0.002	0.004	0.006	0.009	0.146	0.004	0.342	233	0.492	0.011
–	–	–	1.48	–	–	–	–	–	1.59	–	–	–	–	–	–	–	281	–	–

Table 13.7 Lipid composition of miscellaneous food products with a high fat content, contents/100 g

Item	H_2O	Ash	Cals.	Prot.	Carb.	Fat-T	Fat-S	4:0	6:0	8:0	10:0	12:0	14:0	15:0	16:0	17:0	18:0
Almond Joy candy bar	7.7	1.1	464	4.7	58.4	27.7	16.6	–	0.02	0.48	0.33	5.98	3.09	–	3.5	–	3.25
Bar None candy bar	4	1.3	521	8.2	52.2	33.9	–	–	–	–	–	–	–	–	–	–	–
Bittersweet chocolate, square	1.8	3.8	477	7.9	46.8	39.7	22.2	–	–	–	–	–	0.064	–	10.3	–	11.9
Butterscotch baking chips	1	0.9	520	2.2	66.8	29.1	24	–	0.06	0.97	1.1	13.8	4.83	–	2.41	–	0.83
Caramello candy bar	0.9	1.5	489	6.1	66	25.3	–	–	–	–	–	–	–	–	–	–	–
Carob candy bar	1.1	3.9	533	12.9	49.2	33	8.5	0.01	–	–	0.01	0.01	0.19	–	4.33	–	3.96
Chocolate candy kisses	0.9	2.2	520	7.7	56.9	32.3	19.2	–	–	–	–	–	–	–	–	–	–
Chocolate coated almonds	2	2.4	568	12.3	39.6	44.4	7.39	–	–	–	–	–	–	–	–	–	–
Chocolate malted milk balls (Whoppers)	1.7	1.6	504	6.5	62	28.2	16.8	–	–	–	–	–	–	–	–	–	–
Chunky candy bar, small	2.87	1.7	511	9.48	57.2	29.2	15.1	–	–	–	–	–	–	–	–	–	–
Demet's Turtles Candy	6.1	1.4	485	6.4	58	27.8	10.8	0.19	0.13	0.27	0.24	0.8	0.19	–	5.03	–	3.92
Dietetic peanut butter cup candy	4	0.8	549	12.6	43.2	39.4	32.1	–	–	–	–	–	–	–	–	–	–
English toffee candy bar (Skor)	4.3	1.5	527	4.5	55.1	34.4	22.1	–	–	–	–	–	–	–	–	–	–
Golden almond solitaires candy, pkg	1.4	2.6	535	11.6	47.1	37	–	–	–	–	–	–	–	–	–	–	–
Golden III chocolate candy bar	2.8	1.8	518	6.5	55.8	33	–	–	–	–	–	–	–	–	–	–	–
Goobers chocolate covered peanuts	1.85	1.79	520	15.2	48.1	33.2	12.6	–	0.033	0.066	0.066	0.133	0.332	–	5.91	–	4.95
Kit Kat candy bar	1.5	1.2	510	6.7	61.9	28.5	16.7	–	–	0.05	0.05	0.62	1.38	–	7.32	–	7.27
Krackel candy bar	2.2	1.5	502	6.2	61.9	27.8	11.8	–	–	–	0.24	4.32	1.62	–	2.96	–	2.71
M&M's Peanut chocolate candy, pieces	1.5	1.7	495	10.7	59	26.9	10.6	–	–	–	–	–	–	–	–	–	–
M&M's Peanut chocolate candy, pkg	1.5	1.7	495	10.7	59	26.9	10.6	–	–	–	–	–	–	–	–	–	–
Milk chocolate candy bar	1.3	1.5	513	6.9	59.3	30.6	18.4	–	–	–	–	–	0.82	–	8.93	–	8.55
Milk chocolate candy bar with rice cereal	1.9	1.6	496	6.3	63.4	26.5	15.9	0.15	0.04	0.05	0.11	0.11	0.52	–	7.02	–	7.92
Milk chocolate chips	1.3	1.5	513	6.9	59.3	30.6	18.4	–	–	–	–	–	0.82	–	8.93	–	8.55
Milk chocolate coated peanuts	1.9	1.8	519	13.1	49.4	33.5	14.6	0.13	0.03	0.04	0.09	0.09	0.44	–	7.01	–	6.8
Mr. Goodbar chocolate candy bar	1.8	1.7	514	12.6	51.3	32.3	18.1	–	–	–	–	0.43	0.43	–	7.85	–	9.37
Nestle's Crunch candy bar	1.19	1.63	510	6.63	64.1	26.4	15.3	0.106	0.106	0.079	0.159	0.211	0.608	–	5.37	–	8.43
Nestle's Crunch candy bar, snack size	1.19	1.63	510	6.63	64.1	26.4	15.3	0.106	0.106	0.079	0.159	0.211	0.608	–	5.37	–	8.43
Peanut butter cups candy (Reese's)	7.7	2.1	485	11	47.8	31.1	23.1	–	–	–	2.11	10.5	5.24	–	3.14	–	2.11
Peanut butter Twix cookie bar, pkg	1	1.9	506	10.9	57	29	–	–	–	–	–	–	–	–	–	–	–
Peanut candy bar, small	1.6	1.9	522	15.5	47.4	33.7	4.32	–	–	–	–	–	0.02	–	3.55	–	0.76
Planter's peanut candy bar	1.7	1.9	526	16.5	46.4	33.5	5.8	–	–	–	–	–	–	–	–	–	–
Semisweet chocolate chips	0.6	1.2	477	4.2	63.4	29.7	17.5	–	–	–	–	–	0.05	–	8.04	–	9.24
Sesame crunch candy	2.2	2.6	517	11.6	50.3	33.3	4.47	–	–	–	–	–	0.08	–	2.99	–	1.41
Small milk chocolate candy bar with almonds	1.5	1.7	526	9	53.2	34.4	17	0.16	0.04	0.05	0.11	0.11	0.58	–	7.67	–	8.28
Small milk chocolate candy bar with peanuts	1.44	2.81	553	16.1	38.5	41.1	12.1	–	–	–	–	–	0.383	–	6.87	–	4.46
Special dark sweet chocolate candy bar	1.2	1.5	476	4.7	61.6	30.2	20.8	–	–	–	–	–	–	–	–	–	–
Sweet chocolate	0.5	1	505	3.9	59.6	34.2	20.1	–	–	–	–	–	0.03	–	8.7	–	11.4
Sweet chocolate candy bar	0.5	1	505	3.9	59.6	34.2	20.1	–	–	–	–	–	0.03	–	8.7	–	11.4
Sweet dark chocolate	1.2	1.5	476	4.7	61.6	30.2	20.8	–	–	–	–	–	–	–	–	–	–
Symphony milk chocolate candy bar	1.3	1.5	522	7.8	56.8	32.4	–	–	–	–	–	–	–	–	–	–	–
Toffee candy, homemade	1.1	0.7	542	1.1	64.3	32.8	20.4	1.07	0.63	0.37	0.82	0.92	3.31	–	8.64	–	3.98
Truffles candy, homemade	13.8	1.3	488	5.7	45	34.3	21.5	0.31	0.18	0.2	0.33	1.02	1.88	–	9.36	–	7.96
Unsweetened baking chocolate, grated	1.3	3	522	10.3	28.3	55.3	32.6	–	–	–	–	–	0.1	–	14.9	–	17.2

20:0	22:0	24:0	Fat-M	14:1	16:1	18:1	20:1	22:1	Fat-P	18:2	18:3	18:4	20:4	20:5	22:5	22:6	Chol., mg	Omeg3	Omeg6
–	–	–	5.27	–	–	5.27	–	–	2.38	2.23	0.14	–	–	–	–	–	2	0.14	2.22
–	–	–	–	–	–	–	–	–	–	–	–	–	–	–	–	–	16	–	–
–	–	–	14.7	–	–	14.7	–	–	0.79	0.79	–	–	–	–	–	–	–	–	0.79
–	–	–	3.38	–	–	3.38	–	–	0.47	0.47	–	–	–	–	–	–	1	–	0.47
–	–	–	–	–	–	–	–	–	–	–	–	–	–	–	–	–	24	–	–
–	–	–	18.9	–	0.01	18.9	–	–	4.04	3.84	0.2	–	–	–	–	–	5	0.2	3.83
–	–	–	10.6	–	–	–	–	–	0.96	–	–	–	–	–	–	–	22	–	–
–	–	–	29.1	–	–	29.1	–	–	7.88	5.5	–	–	–	–	–	–	1	–	5.5
–	–	–	9.3	–	–	–	–	–	0.8	–	–	–	–	–	–	–	20	–	–
–	–	–	11.3	–	–	–	–	–	2.86	–	–	–	–	–	–	–	10.1	–	–
–	–	–	11.1	–	0.19	10.8	0.03	–	4.66	4.45	0.21	–	–	–	–	–	22	0.21	4.44
–	–	–	4.5	–	–	–	–	–	0.6	–	–	–	–	–	–	–	4	–	–
–	–	–	7.7	–	–	–	–	–	0.95	–	–	–	–	–	–	–	61	–	–
–	–	–	–	–	–	–	–	–	–	–	–	–	–	–	–	–	12	–	–
–	–	–	–	–	–	–	–	–	–	–	–	–	–	–	–	–	19	–	–
0.232	0.531	–	15.3	0.033	0.066	15	0.199	–	5.28	5.21	0.066	–	–	–	–	–	9	0.066	5.21
–	–	–	7.79	–	0.05	7.74	–	–	0.47	0.47	–	–	–	–	–	–	25	–	0.47
–	–	–	7.03	–	–	7.03	–	–	5.54	5.54	–	–	–	–	–	–	19	–	5.57
–	–	–	10.8	–	–	–	–	–	4.23	–	–	–	–	–	–	–	13	–	–
–	–	–	10.8	–	–	–	–	–	4.23	–	–	–	–	–	–	–	13	–	–
–	–	–	9.98	–	–	9.98	–	–	0.91	0.91	–	–	–	–	–	–	22	–	0.91
–	–	–	8.65	–	0.26	8.39	–	–	0.78	0.72	0.06	–	–	–	–	–	19	0.06	0.72
–	–	–	9.98	–	–	9.98	–	–	0.91	0.91	–	–	–	–	–	–	22	–	0.91
–	–	–	12.9	–	0.21	12.6	0.16	–	4.34	4.29	0.05	–	–	–	–	–	9	0.05	4.28
–	–	–	11.3	–	–	11.1	–	–	1.52	1.31	0.22	–	–	–	–	–	20	0.22	1.3
0.106	–	–	10.1	–	0.106	10	–	–	1.06	0.925	0.132	–	–	–	–	–	13.1	0.12	0.87
0.106	–	–	10.1	–	0.106	10	–	–	1.06	0.925	0.132	–	–	–	–	–	13.1	0.12	0.87
–	–	–	2.11	–	–	2.11	–	–	2.11	2.11	–	–	–	–	–	–	15	–	2.1
–	–	–	–	–	–	–	–	–	–	–	–	–	–	–	–	–	12	–	–
–	–	–	16.8	–	0.01	16.3	0.45	–	10.7	10.7	–	–	–	–	–	–	7	–	10.7
–	–	–	17.3	–	–	–	–	–	8.6	–	–	–	–	–	–	–	–	–	–
–	–	–	9.94	–	–	9.94	–	–	0.95	0.95	–	–	–	–	–	–	–	–	0.95
–	–	–	12.6	–	0.1	12.4	0.05	–	14.6	14.3	0.25	–	–	–	–	–	–	0.25	14.3
–	–	–	13.5	–	0.31	13.2	0.01	–	2.29	2.18	0.11	–	–	–	–	–	19	0.11	2.17
–	–	–	18.1	–	0.005	17.7	0.367	–	9.04	9.04	0.002	–	–	–	–	–	9.91	0.002	9.04
–	–	–	11.6	–	–	–	–	–	1.06	–	–	–	–	–	–	–	–	–	–
–	–	–	11.2	–	0.07	11.2	–	–	0.99	0.96	0.03	–	–	–	–	–	–	0.03	0.96
–	–	–	11.2	–	0.07	11.2	–	–	0.99	0.96	0.03	–	–	–	–	–	–	0.03	0.96
–	–	–	11.6	–	–	–	–	–	1.06	–	–	–	–	–	–	–	–	–	–
–	–	–	–	–	–	–	–	–	–	–	–	–	–	–	–	–	28	–	–
–	–	–	9.48	–	0.73	8.26	–	–	1.23	0.74	0.48	–	–	–	–	–	105	0.48	0.74
–	–	–	10.2	–	0.21	9.86	–	–	1.06	0.91	0.14	–	–	–	–	–	52	0.14	0.91
–	–	–	18.5	–	–	18.5	–	–	1.77	1.77	–	–	–	–	–	–	–	–	1.76

Table 13.7 (*Continued*)

Item	H_2O	Ash	Cals.	Prot.	Carb.	Fat-T	Fat-S	4:0	6:0	8:0	10:0	12:0	14:0	15:0	16:0	17:0	18:0
Unsweetened baking chocolate, square	1.3	3	522	10.3	28.3	55.3	32.6	–	–	–	–	–	0.1	–	14.9	–	17.2
Watchamacallit candy bar	2	1.5	503	9.3	58.9	25.9	–	–	–	–	–	–	–	–	–	–	–
White chocolate almond candybar (Alpine)	1.07	1.63	552	9.81	50.5	37	19.9	0.074	0.111	0.111	0.185	0.222	0.666	–	10.7	–	7.66
White chocolate chips	0.9	1.3	533	6.1	61.4	30.4	17.8	0.2	0.06	0.06	0.14	0.14	0.67	–	7.84	–	8.66
Yogurt covered peanuts	3.6	19.2	455	8.8	37.7	30.7	5.92	–	–	–	–	–	–	–	–	–	–
Almond granola bar, hard	3.1	1.6	495	7.7	62	25.5	12.5	–	0.08	0.94	0.75	5.61	2.13	–	2.44	–	0.53
Chocolate chip cookie, higher fat, unenriched, commercial	4.1	1.1	481	5.4	66.8	22.6	7.8	–	–	–	–	–	0.088	–	3.71	–	4.01
Chocolate chip cookie, refrigerated dough	12.7	1.1	443	4.4	61.4	20.4	6.98	–	–	–	–	–	0.078	–	3.39	–	3.51
Chocolate chip cookie, refrigerated dough, baked	3	1.2	492	4.9	68.2	22.6	7.76	–	–	–	–	–	0.086	–	3.77	–	3.9
Chocolate chip soft granola bar, chocolate coated	3.6	1.5	466	5.8	63.8	24.9	14.2	0.1	0.04	0.13	0.15	0.63	0.57	–	6.05	–	6.58
Cinnamon coffee cake with crumb topping	21.9	1.4	418	6.8	46.7	23.3	5.76	0.001	–	–	0.001	0.001	0.117	–	3.01	–	2.63
Cinnamon coffee cake with topping, recipe	21	1.9	400	6.5	50.3	20.2	3.63	0.005	0.003	0.002	0.004	0.006	0.088	–	2.34	–	1.17
Coconut frosting, can, no added phosphorus	21	0.8	412	1.5	52.8	24	7.04	–	0.02	–	0.3	1.89	0.74	–	2.52	–	1.59
Coconut nut frosting, can	21	0.8	412	1.5	52.8	24	7.04	–	0.02	–	0.3	1.89	0.74	–	2.52	–	1.59
Coffee cake dough, caramel roll + nuts PLB	2.81	22	399	5.05	49.2	20.9	4.18	–	–	–	–	–	–	–	–	–	–
Coffeecake, cinnamon + crumb top, unenriched, commercial	21.9	1.4	418	6.8	46.7	23.3	5.76	0.001	–	–	0.001	0.001	0.117	–	3.01	–	2.63
Creme filled chocolate sandwich cookie	2.3	1.9	472	4.7	70.3	20.6	4.15	0.001	–	–	0.001	0.001	0.003	–	2.2	–	1.83
Creme filled sugar cookie wafers	1	0.5	511	4.1	70.1	24.3	4.4	–	–	–	–	–	–	–	2.4	–	1.86
Kudos nutty fudge snack bar	–	–	542	8.13	54.2	32.5	–	–	–	–	–	–	–	–	–	–	–
Low sodium chocolate sandwich cookie with fructose	4.2	1.2	461	4.5	67.7	22.1	10.4	–	–	–	–	0.181	0.184	–	6.19	–	3.82
Low sodium peanut butter sandwich cookie with fructose	3.7	1.4	535	10	50.8	34	11.8	–	–	–	–	0.216	0.222	–	7.67	–	3.7
No sodium sugar wafer cookie + creme with fructose	4.4	0.9	502	3.1	66	25.7	11.7	–	–	–	–	0.243	0.243	–	7.32	–	3.88
Nut and Raisin granola bar, soft	6.1	1.9	454	8	63.6	20.4	9.55	–	0.06	0.67	0.53	4	1.55	–	2.1	–	0.66
Old fashion pound cake with butter	20.4	1	433	6.3	47.6	24.7	14.4	0.704	0.417	0.243	0.545	0.61	2.2	–	6.35	–	2.84
Old fashion pound cake with margarine	20.3	1	434	6.4	47.8	24.5	5.11	–	–	0.001	0.001	0.001	0.067	–	3.23	–	1.79
Peanut butter cookie, soft	11.5	1.1	457	5.3	57.7	24.4	5.37	–	–	–	–	0.002	0.089	–	3.01	–	2.27
Peanut butter soft granola bar, chocolate coated	3.2	1.9	509	10.2	53.4	31.1	17	0.07	0.04	0.42	0.49	5.51	2.16	–	4.43	–	3.93
Peanut butter + chocolate chip granola bar, soft	5.9	2	432	9.8	62.2	20	5.6	–	0.01	0.1	0.08	0.59	0.24	–	2.74	–	1.86
Soft chocolate chip cookie	11.6	1.3	458	3.5	59.1	24.3	7.48	–	–	–	–	–	0.105	–	3.64	–	3.73
Sugar cookie refrigerated dough	14.5	1.5	436	4.2	59	20.7	5.32	–	–	–	–	–	0.101	–	2.77	–	2.44
Sugar cookie refrigerated dough, baked	5	1.7	484	4.7	65.6	23.1	5.91	–	–	–	–	–	0.112	–	3.08	–	2.71
Whole wheat fruit and nut cookie	13.1	1.93	4.29	7.45	55.1	22.4	3.55	–	–	–	–	–	–	–	–	–	–
Apple fritter	37.2	0.603	364	5.9	32.4	23.8	6.44	–	–	–	–	–	–	–	–	–	–

20:0	22:0	24:0	Fat-M	14:1	16:1	18:1	20:1	22:1	Fat-P	18:2	18:3	18:4	20:4	20:5	22:5	22:6	Chol., mg	Omeg3	Omeg6
–	–	–	18.5	–	–	18.5	–	–	1.77	1.77	–	–	–	–	–	–	–	–	1.76
–	–	–	–	–	–	–	–	–	–	–	–	–	–	–	–	–	21	–	–
0.185	–	–	14.4	–	0.07	14.3	–	–	2.63	2.44	0.185	–	–	–	–	–	12.2	0.18	2.33
–	–	–	9.58	–	0.32	9.26	–	–	0.85	0.78	0.07	–	–	–	–	–	22	0.07	0.78
–	–	–	15	–	–	–	–	–	8.19	–	–	–	–	–	–	–	3	–	–
–	–	–	7.75	–	0.05	7.7	–	–	3.77	3.67	0.1	–	–	–	–	–	–	0.1	3.66
–	–	–	11.5	–	0.013	11.5	–	–	2.24	2.13	0.11	–	–	–	–	–	–	0.11	2.13
–	–	–	10.1	–	0.038	10.1	0.002	–	2.09	1.97	0.1	–	0.013	–	–	0.003	24	0.103	1.98
–	–	–	11.3	–	0.043	11.2	0.003	–	2.32	2.19	0.111	–	0.014	–	–	0.004	27	0.115	2.2
–	–	–	7.78	–	0.19	7.59	–	–	1.83	1.72	0.11	–	–	–	–	–	5	0.11	1.71
–	–	–	13.2	–	0.041	13.1	.004	–	2.86	2.71	0.137	–	0.013	–	–	0.003	32	0.14	2.72
–	–	–	7.61	–	0.061	7.5	0.048	–	7.75	6.95	0.774	–	0.022	0.001	–	0.005	60	0.78	6.97
–	–	–	12.2	–	0.04	12.1	0.06	–	3.42	3.26	0.16	–	–	–	–	–	–	0.16	3.26
–	–	–	12.2	–	0.04	12.1	0.06	–	3.42	3.26	0.16	–	–	–	–	–	–	0.16	3.25
–	–	–	14.3	–	–	–	–	–	2	–	–	–	–	–	–	–	–	–	–
–	–	–	13.2	–	0.041	13.1	0.004	–	2.86	2.71	0.137	–	0.013	–	–	0.003	32	0.14	2.72
–	–	–	11.8	–	0.022	11.8	–	–	2.69	2.64	0.046	–	–	–	–	–	–	0.046	2.64
–	–	–	14.3	–	0.024	14.3	–	–	3.28	3.23	0.54	–	–	–	–	–	–	0.054	3.23
–	–	–	–	–	–	–	–	–	–	–	–	–	–	–	–	–	–	–	–
–	–	–	9.09	–	0.007	9.09	–	–	1.49	1.4	0.084	–	–	–	–	–	–	0.084	1.4
–	–	–	15.3	–	0.002	15.1	0.151	–	5.11	5.02	0.091	–	–	–	–	–	–	0.091	5.02
–	–	–	10.9	–	0.001	10.9	–	–	1.82	1.71	0.105	–	–	–	–	–	–	0.105	1.71
–	–	–	4.23	–	0.04	4.15	0.03	–	5.53	5.37	0.17	–	–	–	–	–	1	0.17	5.36
–	–	–	7.32	–	0.566	6.42	0.008	0.001	1.28	0.903	0.331	–	0.038	0.001	–	0.01	173	0.342	0.941
–	–	–	10.6	–	0.08	10.5	0.008	0.001	7.28	6.92	0.309	–	0.038	0.001	–	0.01	114	0.32	6.96
–	–	–	13.3	–	0.001	13.2	0.098	–	4.37	4.24	0.113	–	0.018	–	–	–	–	0.113	4.26
–	–	–	6.55	–	0.12	6.4	0.03	–	1.91	1.84	0.07	–	0.01	–	–	–	12	0.07	1.84
–	–	–	8.37	–	0.02	8.18	0.18	–	4.62	4.55	0.03	–	0.03	–	–	–	1	0.03	4.57
–	–	–	13	–	0.009	13	–	–	2.65	2.52	0.132	–	–	–	–	–	–	0.132	2.52
–	–	–	11.7	–	0.026	11.7	0.002	–	2.6	2.46	0.128	–	0.012	–	–	0.003	29	0.131	2.47
–	–	–	13	–	0.028	13	0.003	–	2.89	2.73	0.143	–	0.013	–	–	0.003	32	0.146	2.74
–	–	–	7.71	–	–	–	–	–	9.83	–	–	–	–	–	–	–	52.8	–	–
–	–	–	10.2	–	–	–	–	–	5.79	–	–	–	–	–	–	–	87.2	–	–

Item	H_2O	Ash	Cals.	Prot.	Carb.	Fat-T	Fat-S	4:0	6:0	8:0	10:0	12:0	14:0	15:0	16:0	17:0	18:0
Banana fritter 2 inch long	38.8	0.865	342	5.45	33.4	21.5	5.72	–	–	–	–	–	–	–	–	–	–
Berry fritter	43.7	0.576	325	5.29	28.9	21.5	5.65	–	–	–	–	–	–	–	–	–	–
Cream puff shell, recipe 3-1/2′ diameter	40.5	1.8	362	9	22.8	25.9	5.6	–	–	0.001	0.001	0.001	0.068	–	3.6	–	1.9
Creme filled yeast doughnut	38.2	0.8	361	6.4	30	24.5	6.75	0.046	0.027	0.016	0.036	0.04	0.255	–	3.44	–	2.9
Large potato pancakes	47.3	2.7	272	6.17	28.6	15.2	3.04	–	–	0.001	0.002	0.005	0.047	–	2.32	–	0.656
Patty tart shells, frozen	8.5	1	551	7.3	45.1	38.1	5.43	–	–	–	–	–	0.038	–	3.97	–	1.42
Patty tart shells, frozen, baked	7.4	1	558	7.4	45.7	38.5	5.5	–	–	–	–	–	0.039	–	4.02	–	1.44
Pie crust, frozen, baked	11.2	2	514	4.4	49.6	32.8	10.6	–	–	–	0.023	0.045	0.295	–	6.35	–	3.81
Pie crust, frozen, ready to bake	21	1.8	457	3.9	44.1	29.2	9.42	–	–	–	0.02	0.04	0.263	–	5.65	–	3.39
Potato pancakes, pieces	47.3	2.7	272	6.17	28.6	15.2	3.04	–	–	0.001	0.002	0.005	0.047	–	2.32	–	0.656
Puff pastry, frozen	8.5	1	551	7.3	45.1	38.1	5.43	–	–	–	–	–	0.038	–	3.97	–	1.42
Puff pastry, frozen, baked	7.4	1	558	7.4	45.7	38.5	5.5	–	–	–	–	–	0.039	–	4.02	–	1.44
Wheat flour fritter, no syrup	35.4	0.917	448	7.24	17.3	39.1	9.05	–	–	–	–	–	–	–	–	–	–
Cheese filled crackers	3.9	4	477	9.3	61.7	21.1	5.65	0.031	0.017	0.009	0.02	0.02	0.196	–	2.91	–	2.45
Cheese filled rye crackers	3.8	3.9	481	9.2	60.8	22.3	5.76	0.022	0.011	0.006	0.013	0.013	0.175	–	2.97	–	2.55
Cheese filled wheat crackers	3.2	3.8	497	9.8	58.2	25	6.63	0.032	0.017	0.009	0.02	0.021	0.219	–	3.42	–	2.88
Cheese puffs (Cheetos)	1.5	2.7	554	7.6	53.8	34.4	6.6	0.03	–	–	–	0.06	0.24	–	4.26	–	2.01
Cheese puffs (Cheetos), grab bag	1.5	2.7	554	7.6	53.8	34.4	6.6	0.03	–	–	–	0.06	0.24	–	4.26	–	2.01
Corn chips (Fritos)	1	2.2	539	6.6	56.9	33.4	4.56	–	–	–	–	–	–	–	3.58	–	0.97
Fritos corn chips, grab bag	1	2.2	539	6.6	56.9	33.4	4.56	–	–	–	–	–	–	–	3.58	–	0.97
Light potato chips (lower fat)	1	4.1	471	7.1	66.9	20.8	4.17	–	–	–	–	–	0.12	–	3.53	–	0.51
Peanut butter filled snack crackers	3.1	3.2	488	11.1	58.7	23.9	5.1	–	–	–	–	0.004	0.079	–	2.92	–	2.1
Peanut butter filled wheat crackers	3.4	2.6	495	13.5	53.8	26.7	5.7	–	–	–	–	0.005	0.089	–	3.27	–	2.33
Potato chips, no salt added	2.54	3.82	523	6.43	51.9	35.4	9.07	–	–	–	0.001	0.004	0.28	–	7.95	–	0.805
Potato chips, sour cream and onion	1.8	4.7	531	8.1	51.5	33.9	8.9	0.01	0.01	–	0.01	0.01	0.3	–	7.74	–	0.82
Rectangle butter crackers (Club/Waverly)	3.5	2.8	502	7.4	61	25.3	4.85	–	–	–	0.04	0.008	0.077	–	3.12	–	1.64
Banana split with whipped cream	52.8	1.08	250	3.69	27.1	15.3	9.53	–	–	–	–	–	–	–	–	–	–
Chocolate coated ice cream bar with nuts	43.1	1.4	317	4.4	30.9	20.2	11.8	–	–	–	–	–	–	–	–	–	–
Chocolate dip vanilla ice cream cone with nuts	47.1	1.24	279	5.76	29	16.9	7.26	–	–	–	–	–	–	–	–	–	–
Chocolate ice cream bar with chocolate coating (Dove)	37.4	1.07	344	3.39	35	23.1	14.1	–	–	–	–	–	–	–	–	–	–
Chocolate mousse, homemade	62.1	0.8	221	4.3	16.4	16.3	9.19	0.32	0.19	0.11	0.25	0.28	1.03	–	4.18	–	2.66
Cocoa riche 800 foaming nondairy creamer	3	1	528	3.5	61	30	–	–	–	–	–	–	–	–	–	–	–
Dessert topping mix, dry (Dream Whip)	1.48	1.18	577	4.91	52.5	39.9	36.7	–	–	0.669	0.964	14.5	5.75	–	5.43	–	9.41
Drumstick ice cream bar	48.4	1.21	266	5.79	29.6	15	7.41	–	–	–	–	–	–	–	–	–	–
English toffee ice cream bar + chocolate coating (Heath)	48.1	0.893	307	3.04	25.7	22.3	17.1	–	–	–	–	–	–	–	–	–	–
Frozen dessert topping (Cool Whip)	50.2	0.18	318	1.26	23.1	25.3	21.8	–	–	0.638	0.905	8.84	3.76	–	3.09	–	4.58
Ice cream parfait with fruit and nuts	52.8	1.08	250	3.69	27.1	15.3	9.53	–	–	–	–	–	–	–	–	–	–
Ice cream pie, cookie crust, fudge topping	40.3	1.28	300	4.06	38.2	16.2	8.35	–	–	–	–	–	–	–	–	–	–
Imitation sour cream (IMO)	71.2	0.3	208	2.41	6.64	19.5	17.8	–	–	1.03	1.01	8.14	3.36	–	1.98	–	2.24

20:0	22:0	24:0	Fat-M	14:1	16:1	18:1	20:1	22:1	Fat-P	18:2	18:3	18:4	20:4	20:5	22:5	22:6	Chol., mg	Omeg3	Omeg6
–	–	–	9.17	–	–	–	–	–	5.33	–	–	–	–	–	–	–	73.8	–	–
–	–	–	9.18	–	–	–	–	–	5.37	–	–	–	–	–	–	–	74	–	–
–	–	–	11.1	–	0.137	11	0.013	0.001	7.38	6.99	0.309	–	0.065	0.002	–	0.017	196	0.328	7.06
–	–	–	13.5	–	0.057	13.4	0.002	–	2.97	2.8	0.164	–	0.009	–	–	0.002	24	0.166	2.81
0.008	0.005	–	4.64	–	0.094	4.54	0.006	0.001	6.54	6.12	0.381	–	0.032	0.001	–	0.008	96	0.39	6.15
–	–	–	8.71	–	0.077	8.56	0.074	–	21.9	19.4	2.56	–	–	–	–	–	–	2.56	19.4
–	–	–	8.83	–	0.078	8.67	0.075	–	22.2	19.6	2.59	–	–	–	–	–	–	2.59	19.6
–	–	–	15.7	–	0.622	15.1	–	–	4.03	3.79	0.245	–	–	–	–	–	–	0.245	3.79
–	–	–	14	–	0.554	13.4	–	–	3.59	3.37	0.218	–	–	–	–	–	–	0.218	3.37
0.008	0.005	–	4.64	–	0.094	4.54	0.006	0.001	6.54	6.12	0.381	–	0.032	0.001	–	0.008	96	0.39	6.15
–	–	–	8.71	–	0.077	8.56	0.074	–	21.9	19.4	2.56	–	–	–	–	–	–	2.56	19.4
–	–	–	8.83	–	0.078	8.67	0.075	–	22.2	19.6	2.59	–	–	–	–	–	–	2.59	19.6
–	–	–	17.1	–	–	–	–	–	10.7	–	–	–	–	–	–	–	162	–	–
–	–	–	11.5	–	0.029	11.4	–	–	2.74	2.59	0.15	–	–	–	–	–	2	0.15	2.59
–	–	–	12.1	–	0.023	12	–	–	2.87	2.72	0.157	–	–	–	–	–	9	0.157	2.72
–	–	–	13.6	–	0.032	13.5	–	–	3.31	3.13	0.181	–	–	–	–	–	7	0.181	3.13
–	–	–	20.3	–	0.13	20.2	–	–	4.77	4.58	0.19	–	–	–	–	–	4	0.19	4.57
–	–	–	20.3	–	0.13	20.2	–	–	4.77	4.58	0.19	–	–	–	–	–	4	0.19	4.57
–	–	–	9.67	–	0.09	9.58	–	–	16.5	15.2	1.27	–	–	–	–	–	–	1.26	15.2
–	–	–	9.67	–	0.09	9.58	–	–	16.5	15.2	1.27	–	–	–	–	–	–	1.26	15.2
–	–	–	4.81	–	0.5	4.31	–	–	10.9	10.8	0.19	–	–	–	–	–	–	0.19	10.8
–	–	–	12.6	–	0.001	12.4	0.115	–	4.57	4.42	0.11	–	0.035	–	–	–	–	0.11	4.46
–	–	–	14	–	0.003	13.8	0.128	–	5.16	5	0.126	–	0.039	–	–	–	–	0.126	5.04
0.001	0.001	–	6.23	–	0.28	5.95	–	–	18.2	18	0.07	–	0.035	–	–	–	–	0.07	18
0.001	0.001	–	6.13	–	0.28	5.85	–	–	17.4	17.3	0.08	–	–	–	–	–	7	0.08	17.3
–	–	–	10.8	–	0.19	10.6	–	–	8.28	7.71	0.575	–	–	–	–	–	–	0.575	7.71
–	–	–	4.55	–	–	–	–	–	0.483	–	–	–	–	–	–	–	49.6	–	–
–	–	–	4.79	–	–	–	–	–	1.4	–	–	–	–	–	–	–	1	–	–
–	–	–	7.05	–	–	–	–	–	1.68	–	–	–	–	–	–	–	34	–	–
–	–	–	7.26	–	–	–	–	–	0.764	–	–	–	–	–	–	–	37.5	–	–
–	–	–	5.1	–	0.31	4.66	0.01	–	0.84	0.63	0.16	–	0.04	–	–	0.01	148	0.17	0.67
–	–	–	–	–	–	–	–	–	–	–	–	–	–	–	–	–	–	–	–
–	–	–	0.6	–	–	0.6	–	–	0.447	0.447	–	–	–	–	–	–	–	–	0.447
–	–	–	5.17	–	–	–	–	–	1.63	–	–	–	–	–	–	–	35	–	–
–	–	–	3.33	–	–	–	–	–	0.567	–	–	–	–	–	–	–	35.3	–	–
–	–	–	1.62	–	0.241	1.38	–	–	0.523	0.305	0.218	–	–	–	–	–	–	0.218	0.305
–	–	–	4.55	–	–	–	–	–	0.483	–	–	–	–	–	–	–	49.6	–	–
–	–	–	5.26	–	–	–	–	–	1.73	–	–	–	–	–	–	–	27.8	–	–
–	–	–	0.588	–	–	0.589	–	–	0.056	0.056	–	–	–	–	–	–	–	–	0.056

Table 13.7 (Continued)

Item	H_2O	Ash	Cals.	Prot.	Carb.	Fat-T	Fat-S	4:0	6:0	8:0	10:0	12:0	14:0	15:0	16:0	17:0	18:0
Nondairy topping, pressurized	60.4	0.28	264	0.98	16.1	22.3	18.9	–	–	0.449	0.642	8.01	3.25	–	2.41	–	4.17
Nutty Buddy ice cream cone	47.1	1.24	279	5.76	29	16.9	7.26	–	–	–	–	–	–	–	–	–	–
Powdered coffee whitener/creamer	2.22	2.65	546	4.8	54.9	35.5	32.5	–	–	1.33	1.46	13.6	5.99	–	3.75	–	6.34
Rich vanilla ice cream	57.2	0.8	241	3.5	22.4	16.2	9.98	0.52	0.31	0.18	0.41	0.45	1.67	–	4.5	–	1.97
Rich vanilla ice cream, medium scoop/slice	57.2	0.8	241	3.5	22.4	16.2	9.98	0.52	0.31	0.18	0.41	0.45	1.67	–	4.5	–	1.97
Acorn flour, full fat	6.01	1.7	501	7.5	54.7	30.2	3.92	–	–	–	–	–	–	–	3.61	–	0.319
Ethiopian flaxseed cake	4.1	3.4	524	18.6	36.9	37	–	–	–	–	–	–	–	–	–	–	–
Full fat soy flour, stirred, raw	5.17	4.47	436	34.5	35.2	20.7	3.01	–	–	–	–	0.013	0.057	–	2.19	–	0.737
Full fat soy flour, stirred, roasted	3.82	5.87	441	34.8	33.7	21.9	3.19	–	–	–	–	0.014	0.061	–	2.32	–	0.78
High fat sesame flour	0.9	4.61	526	30.8	26.6	37.1	5.2	–	–	–	–	–	0.092	–	3.32	–	1.56
Roasted soy flour, full fat Nx6.25, stirred	3.81	5.86	439	38.1	30.4	21.9	3.19	–	–	–	–	0.014	0.061	–	2.32	–	0.78
Whole soy flour	5.16	4.44	433	37.8	31.9	20.7	3.01	–	–	–	–	0.013	0.057	–	2.19	–	0.737
Cocoa powder with alkali, high fat	3	8.2	295	16.8	45.4	23.7	13	–	–	–	–	–	0.033	–	5.75	–	7.03
Cocoa powder, high fat, dry	3	5	299	16.8	48.3	23.7	13	–	–	–	–	–	0.033	–	5.75	–	7.03
French coffee, instant, dry	2.4	4.8	499	4.5	57.5	29.6	25.6	–	0.17	2.22	1.79	13.2	4.97	–	2.42	–	0.823
Hollandaise sauce with butterfat, dry mix, pack	2.13	8.76	554	11	32.1	46	27.1	1.34	0.79	0.46	1.03	1.15	4.14	–	11.9	–	5.36
Banana Chips	4.3	1.4	519	2.3	58.4	33.6	29	–	0.2	2.52	2.02	14.9	4.63	–	2.8	–	0.94
Cauliflower, flowerets, batter fried	68.7	1.28	193	4.64	9.85	15.5	3.66	–	–	–	–	–	–	–	–	–	–
French fries, fried (animal + vegetable oil)	38	1.87	315	4.04	39.6	16.6	6.81	–	–	–	–	–	0.392	–	3.92	–	2.49
French fries, frozen, fried, cup measure	38	1.87	315	4.04	39.6	16.6	5.01	–	–	–	0.001	0.004	0.119	–	4.37	–	0.516
French fries, frozen, oven heated	35.4	2.67	326	3.56	37.8	18.7	7.57	–	–	–	0.012	0.024	0.169	–	4.85	–	2.1
French fries, frozen, restaurant fried	38	1.87	315	4.04	39.6	16.6	5.01	–	–	–	0.001	0.004	0.119	–	4.37	–	0.516
French fries, vegetable oil, regular order	39.5	1.88	309	4	38.6	16.1	5.01	–	–	–	0.018	0.399	0.102	–	3.39	–	1.1
Mushrooms, batter fried	66.3	1.36	211	3.24	11.8	17.4	2.99	–	–	–	–	–	–	–	–	–	–
Onion rings, frozen, heated	28.5	1.31	407	5.35	38.2	26.7	8.59	–	–	–	–	–	0.214	–	5.2	–	3.17
Potato, french fries, frozen, extruded	49.9	2.14	260	2.84	30.2	15	4.55	–	–	–	0.001	0.003	0.057	–	3.52	–	0.972
Cream of chicken soup, dry mix, packet	3.76	13	436	7.26	54.3	21.7	13.8	–	–	–	0.82	5.55	2.26	–	3.09	–	2.06
Cream of vegetable soup, dry, mix, packet	2.91	12.9	446	8.01	52.1	24.1	6.03	–	–	–	–	–	0.1	–	3.4	–	2.56
English muffin with cheese and sausage	37.7	2.51	342	13.3	25.4	21.1	8.57	–	–	0.063	0.15	0.171	0.727	–	4.94	–	2.32
Fried pork dumpling	39.5	1.44	350	12.7	24	22.3	6.11	–	–	–	–	–	–	–	–	–	–
Meat filled wontons, fried	43	1.68	328	12.8	21.5	21	4.1	–	–	–	–	–	–	–	–	–	–
Stuffed grape leaves (beef and rice roll)	61.5	1.16	251	8.04	8.58	20.7	4.84	–	–	–	–	–	–	–	–	–	–
Stuffed grape leaves (lamb and rice roll)	61.3	1.08	258	7.16	8.58	21.9	5.87	–	–	–	–	–	–	–	–	–	–
Twice baked potato, mix, prepared	45.7	3.27	278	7.02	28.4	15.5	7.7	–	–	–	–	–	–	–	–	–	–
Waldorf salad	59.3	0.9	290	2.28	8.96	28.6	3.81	–	–	–	–	0.027	0.05	–	2.78	–	0.95
Weiner wraps (franks in dough)	45.7	3.76	327	8.7	17.1	24.7	7.97	–	–	–	–	–	–	–	–	–	–
Vegetarian breakfast links	50.4	3.08	256	18.5	9.86	18.2	2.93	–	–	–	–	0.217	0.067	–	1.86	–	0.786
Vegetarian breakfast patties	50.4	3.08	256	18.5	9.86	18.2	2.93	–	–	–	–	0.217	0.067	–	1.86	–	0.786

20:0	22:0	24:0	Fat-M	14:1	16:1	18:1	20:1	22:1	Fat-P	18:2	18:3	18:4	20:4	20:5	22:5	22:6	Chol., mg	Omeg3	Omeg6
–	–	–	1.93	–	–	1.93	–	–	0.241	0.241	–	–	–	–	–	–	–	–	0.241
–	–	–	7.05	–	–	–	–	–	1.68	–	–	–	–	–	–	–	34	–	–
–	–	–	0.968	–	–	0.969	–	–	0.014	0.003	0.01	–	–	–	–	–	–	0.01	0.003
–	–	–	4.67	–	0.36	4.31	–	–	0.6	0.37	0.24	–	–	–	–	–	61	0.24	0.37
–	–	–	4.67	–	0.36	4.31	–	–	0.6	0.37	0.24	–	–	–	–	–	61	0.24	0.37
–	–	–	19.1	–	–	19.1	–	–	5.81	5.81	–	–	–	–	–	–	–	–	5.81
–	–	–	–	–	–	–	–	–	–	–	–	–	–	–	–	–	–	–	–
0.013	–	–	4.56	–	0.057	4.5	–	–	11.7	10.3	1.38	–	–	–	–	–	–	1.38	10.3
0.014	–	–	4.83	–	0.061	4.77	–	–	12.3	10.9	1.46	–	–	–	–	–	–	1.46	10.9
0.228	–	–	14	–	0.111	13.8	0.052	–	16.3	16	0.281	–	–	–	–	–	–	0.281	16
0.014	–	–	4.83	–	0.061	4.77	–	–	12.3	10.9	1.46	–	–	–	–	–	–	1.46	10.9
0.013	–	–	4.56	–	0.057	4.5	–	–	11.7	10.3	1.38	–	–	–	–	–	–	1.38	10.3
0.169	0.02	–	9.09	–	0.086	9	–	–	–	–	–	–	–	–	–	–	–	–	–
0.169	0.02	–	9.09	–	0.086	9	–	–	–	–	–	–	–	–	–	–	–	–	–
–	–	–	1.72	–	–	1.72	–	–	0.536	0.536	–	–	–	–	–	–	–	–	0.536
–	–	–	13.9	–	1.09	12.2	–	–	2.17	1.48	0.61	–	0.04	–	–	–	118	0.61	1.52
0.02	0.02	–	1.96	–	–	1.96	–	–	0.63	0.62	0.01	–	–	–	–	–	–	0.01	0.62
–	–	–	4.03	–	–	–	–	–	6.98	–	–	–	–	–	–	–	22.8	–	–
0.001	0.001	–	7.99	–	0.414	7.42	–	–	1.04	0.925	0.118	–	–	–	–	–	13	0.118	0.925
0.001	0.001	–	3.29	–	0.206	3.08	–	–	7.55	7.55	–	–	–	–	–	–	–	–	7.55
0.001	0.001	–	8.36	–	0.5	7.63	–	–	1.78	1.78	–	–	–	–	–	–	–	–	1.78
0.001	0.001	–	3.29	–	0.206	3.08	–	–	7.55	7.55	–	–	–	–	–	–	–	–	7.55
0.001	0.001	–	7.92	–	0.16	7.7	–	–	2.47	2.33	0.144	–	–	–	–	–	–	0.144	2.33
–	–	–	4.3	–	–	–	–	–	9.15	–	–	–	–	–	–	–	19.8	–	–
–	–	–	10.9	–	0.356	10.5	–	–	5.11	4.83	0.285	–	–	–	–	–	–	0.285	4.83
0.001	0.001	–	8.66	–	0.029	8.63	–	–	1.07	1.07	–	–	–	–	–	–	–	–	1.07
–	–	–	4.73	–	0.41	4.32	–	–	1.65	1.65	–	–	–	–	–	–	12	–	1.65
–	–	–	10.7	–	–	10.7	–	–	6.29	5.91	0.39	–	–	–	–	–	2	0.39	5.91
–	–	–	8.77	–	0.644	7.94	0.126	–	2.34	2.02	0.133	–	0.188	–	–	–	51	0.133	2.21
–	–	–	9.73	–	–	–	–	–	5.18	–	–	–	–	–	–	–	31.9	–	–
–	–	–	8.99	–	–	–	–	–	6.59	–	–	–	–	–	–	–	70.3	–	–
–	–	–	12.6	–	–	–	–	–	1.74	–	–	–	–	–	–	–	27.5	–	–
–	–	–	12.8	–	–	–	–	–	1.94	–	–	–	–	–	–	–	27.5	–	–
–	–	–	5.54	–	–	–	–	–	1.15	–	–	–	–	–	–	–	133	–	–
–	–	–	7.7	–	0.046	7.54	0.089	–	15.6	13.6	1.93	–	0.003	–	–	–	15.6	1.93	13.6
–	–	–	11.3	–	–	–	–	–	3.94	–	–	–	–	–	–	–	26.3	–	–
–	–	–	4.5	–	–	4.5	–	–	9.28	8.21	1.07	–	–	–	–	–	–	1.07	8.21
–	–	–	4.5	–	–	4.5	–	–	9.28	8.21	1.07	–	–	–	–	–	–	1.07	8.21

Table 13.7 (Continued)

Item	H_2O	Ash	Cals.	Prot.	Carb.	Fat-T	Fat-S	4:0	6:0	8:0	10:0	12:0	14:0	15:0	16:0	17:0	18:0
Vegetarian fillets	45	5	290	23	9	18	2.82	–	–	–	–	–	–	–	–	–	–
Vegetarian fish sticks	45	5	290	23	9	18	2.82	–	–	–	–	–	–	–	–	–	–
Vegetarian luncheon meat, 1/2 inch slice	46	4	280	25	9	16	2.5	–	–	–	–	–	–	–	–	–	–
Vegetarian scallops, breaded and fried	42.7	5.22	302	24	9.38	18.8	2.94	–	–	–	–	–	–	–	–	–	–

20:0	22:0	24:0	Fat-M	14:1	16:1	18:1	20:1	22:1	Fat-P	18:2	18:3	18:4	20:4	20:5	22:5	22:6	Chol., mg	Omeg3	Omeg6
–	–	–	4.32	–	–	–	–	–	9.41	–	–	–	–	–	–	–	–	–	–
–	–	–	4.32	–	–	–	–	–	9.41	–	–	–	–	–	–	–	–	–	–
–	–	–	3.84	–	–	–	–	–	8.36	–	–	–	–	–	–	–	–	–	–
–	–	–	4.5	–	–	–	–	–	9.8	–	–	–	–	–	–	–	–	–	–

Index

Note: For the fat content of individual foods, see the tables in Chapter 13, pp. 444–505.